Encyclopedia of
Plant Physiology

New Series Volume 3

Editors

A. Pirson, Göttingen

M. H. Zimmermann, Harvard

Transport in Plants III

Intracellular Interactions and Transport Processes

Edited by

C. R. Stocking and U. Heber

Contributors

J. Brachet   R.H. Falk   H. Fock   R. J. Graesser
M.D. Hatch   H.W. Heldt   Ph. Matile   R. E. McCarty
H.H. Mollenhauer   D.J. Morré   S. Murakami
C.B. Osmond   C. Schnarrenberger   C.R. Stocking
H. Strotmann   D.A. Walker   A. Wiemken
R.H. Wilson   D. Woermann

Springer-Verlag Berlin Heidelberg New York 1976

With 123 Figures

ISBN-13: 978-3-642-66419-9 e-ISBN-13: 978-3-642-66417-5
DOI: 10.1007/978-3-642-66417-5

Softcover reprint of the hardcover 1st edition 1976

Preface

The problems associated with the movement of water and solutes throughout the plant body have intrigued students of plants since Malpighi's conclusions in 1675 and 1679 that nutrient sap flows upward and downward in stems through vessels in both wood and bark. Steven Hale's ingenious experiments on the movement of water in plants in 1726 and Hartig's observations of sieve-tube exudation in the mid-19th century set the stage for continued intensive studies on long-range transport in plants.

In spite of this interest for more than 200 years in the movement of solutes and water in plants, it has only been within the last 20 to 30 years that extensive research effort has been directed toward a critical evaluation of the interactions among the various cellular organelles. The important roles played by the exchange of metabolites in the control and regulation of cellular processes is now widely recognized, but in most instances poorly understood.

Our knowledge of the structure and functions of individual organelles such as mitochondria, chloroplasts, nuclei, microbodies, etc., has increased, and in some instances become extensive. However, advances in our understanding of exchange processes among organelles in an individual cell have been slow and have awaited the development of techniques by which accurate measurements of actual pool sizes of metabolites and the quantitative determination of the kinetics of intracellular movement of solutes could be made. The loss of water-soluble solutes and the adsorption of others on membranes continue to be difficulties associated with the isolation of cellular organelles.

Structural differences and physical properties of membranes bounding individual organelles and the specific mechanisms involving transfer of solutes across membranes have all become fruitful areas of research. A picture is emerging, though still dim and hazy, of the intimate interactions and coordinated interdependency of the cellular organelles.

In planning for this volume, the problem arose as to whether even today enough factual information is available to warrant publication of a book on intracellular transport. The question was justifiably asked: "In such a rapidly advancing field with so many still unanswered questions, would the volume be a significant contribution?" As we considered this question, it seemed to us that although several excellent review articles covering specific areas of intracellular transport of solutes have appeared scattered through the literature, no attempt has been made to bring this material together. Thus, the publication in one volume of critical evaluations of current experimental results and theories in this field should be of service both to students and to research workers.

Our efforts have been to emphasize the importance of interactions among the various cellular compartments. We have therefore decided not to treat protoplasmic streaming in this volume. This appears justified as, within the dimensions of most cells, diffusion is a highly effective means of solute transfer within homogenous compartments and the relationship between protoplasmic streaming and interactions among membrane-bounded compartments is obscure. Diffusion has been considered previously in these series, so that in the first section of this volume some treatment of biological membranes appeared to be sufficient as an introduction to dealing with intracellular interactions across biomembranes. Since there have recently been conferences entirely devoted to the consideration of membrane structure and its biological applications, and several books published on the biochemistry and biology of cell membranes, the introductory chapter was limited to a brief consideration of membrane isolation, membrane composition and membrane models.

In the second section of this volume, each major cellular organelle has been considered in relation to its specific cellular environment and in so far as is known, the exchange processes that take place between it and other cellular compartments have been analyzed. In this connection, our knowledge has progressed much further for some organelles than for others. Thus, chloroplast transport is dealt with in several chapters while, for lack of knowledge on relevant plant systems, information obtained with animal systems is also discussed especially in the chapters on nuclei and mitochondria. This appears justified and is, in fact, necessary in view of the similar functional role of nuclei and mitochondria in different cells. In addition, some treatment of animal systems is hoped to stimulate research in those areas of the plant sciences where relevant information is still missing.

The third section of the volume deals with the mechanisms involved in the conservation of energy and its exchange between organelles (chloroplasts and mitochondria) and the rest of the cell. Again it is apparent that pictures emerge in differing detail. If one prefers to take a partisan's view, it is possible today to present an impressive unified picture of energy conservation in chloroplasts. For plant mitochondria, too many pieces of the puzzle are still missing or not yet shaped in sufficient detail for precise addition to a great structure. Though there is no doubt that energy is efficiently transferred in the cell from source to sink, the mechanisms of transfer across membrane barriers are known only in some instances, which are discussed in the final chapter of the third section.

It will become apparent to the reader that, although a wealth of piece-meal information on intracellular transport and cooperation is available, most of it is qualitative in nature. It remains a challenge for the future to proceed to a quantitative treatment. In the final section of this volume, the basis for this is given in a comprehensive mathematical consideration of transport processes which also gives insight into the thermodynamic analysis of transport processes.

Terminology, where not properly defined, may lead to ambiguity. Wherever the text does not clearly point to its broader definition, cytoplasm is used for cytosol, a term not yet universally accepted for the liquid matrix surrounding cell organelles.

The editors have been responsible for the selection of the topics to be included in this volume and for inviting the contributors to submit manuscripts. In a treatment of different parts and aspects of one broad theme by several specialists, some overlapping is natural. The question arose whether editorial interference should try to avoid it. For good reasons, this interference was kept to a minimum. While scientific discussion of a subject is based on factual observation, the weight of available evidence is open to interpretation. In a rapidly developing field, interpretation by different workers is bound to differ also. Overlapping in the few cases where it occurs will help the reader to form his own opinion on how firm the ground is in a particular field.

We wish to express our thanks to our friends and fellow scientists who readily accepted our invitation to contribute to this volume and whose efforts have made this volume possible. We are also deeply indebted to the staff of Springer-Verlag for their indispensible help with this publication.

Davis and Düsseldorf,
Oktober 1976

C.R. STOCKING
U. HEBER

Contents

I. Membrane Structure

Plant Membranes
R.H. Falk and C.R. Stocking (With 25 Figures)

1. Introduction . . . 3
2. Isolation of Membranes . . . 3
 2.1 Choice of Tissue . . . 4
 2.2 Identification of Specific Membranes . . . 4
  2.2.1 Microscopy . . . 4
  2.2.2 Histochemical Staining . . . 5
  2.2.3 Marker Enzymes . . . 5
  2.2.4 Antibodies . . . 8
 2.3 Tissue Disruption . . . 8
 2.4 Membrane Isolation . . . 9
3. Membrane Composition . . . 10
 3.1 Membrane Carbohydrates . . . 10
 3.2 Membrane Lipids . . . 11
 3.3 Membrane Proteins . . . 11
4. Specific Membranes . . . 13
 4.1 The Plasmalemma (Plasma Membrane) . . . 13
  4.1.1 Morphology . . . 13
  4.1.2 Chemical Composition . . . 14
 4.2 The Nuclear Envelope . . . 16
  4.2.1 Morphology . . . 16
  4.2.2 Chemical Composition . . . 17
 4.3 Plastid Membranes . . . 19
  4.3.1 Morphology . . . 19
   4.3.1.1 General . . . 19
   4.3.1.2 Proplastids . . . 20
   4.3.1.3 Etioplasts . . . 20
   4.3.1.4 Amyloplasts . . . 21
   4.3.1.5 Chromoplasts . . . 21
   4.3.1.6 The Higher Plant Chloroplasts . . . 21
  4.3.2 Chemical Composition of the Chloroplast Envelope . . . 22
 4.4 Mitochondrial Membranes . . . 26
  4.4.1 Morphology . . . 26
  4.4.2 Chemical Composition . . . 27
 4.5 Other Membranes in Plant Cells . . . 29
  4.5.1 The Microbody Membrane . . . 29

4.5.2 The Endoplasmic Reticulum Membranes 29
4.5.3 The Golgi Apparatus Membranes 29
4.5.4 The Vacuolar Membrane 31
4.5.5 The Spherosome Membrane 31
5. Membrane Models 31
5.1 The Lipid Bilayer Model 32
5.2 The Unit Membrane Model 32
5.3 Subunit Models 33
5.4 The Lipid-Globular Protein Mosaic Model 35
6. Membrane Structure and Solute Transport 37
6.1 Membranes with Hydrophilic Pores 37
6.2 Carriers and Ionophores 38
6.3 Bridging Proteins Acting as Channels through Membranes 38
References 40

II. Intracellular Interactions

1. Interactions between Nucleus and Cytoplasm
J. Brachet (With 7 Figures)

1. Old Theories and Present Ideas about the Biochemical Role of the Cell Nucleus 53
2. Experimental Approaches to the Study of Nucleocytoplasmic Interactions 54
2.1 Work on Intact Cells 54
2.1.1 Autoradiography 54
2.1.2 Studies on Chromatin and DNA in Intact Cells 55
2.2 Work on Anucleate Fragments of Unicellular Organisms, Eggs, and Cells 56
2.3 Biochemical Work on Isolated Organelles 58
3. Discussion of Some Recent Experimental Results 58
3.1 Work on Intact Cells 58
3.1.1 Autoradiography 58
3.1.2 Role of the Nuclear Membrane in Nucleocytoplasmic Interactions 60
3.1.3 Choice by the Cytoplasm between DNA Replication and Transcription 62
3.1.4 Effects of the Plasma Membrane on Nuclear Activity 63
3.2 Studies on Anucleate Fragments of Unicellular Organisms, Eggs, and Cells 64
3.2.1 Acetabularia 64
3.2.1.1 Introduction 64
3.2.1.2 Production and Distribution of Morphogenetic Substances 65
3.2.1.3 Energy Production, Circadian Rhythms 66
3.2.1.4 Chloroplastic and Cytoplasmic Protein Synthesis . . . 67
3.2.1.5 Nucleic Acid Synthesis 69
3.2.1.6 Cytoplasmic Effects on the Nucleus 70
3.2.2 Protozoa 71
3.2.3 Eggs 73

3.2.3.1 Sea Urchin Eggs 73
3.2.3.2 Other Eggs 74
3.2.4 Mammalian Cells 75
3.2.4.1 Reticulocytes 75
3.2.4.2 Cells Enucleated with Cytochalasin B 76
4. Conclusions 77
References 79

2. Plastids and Intracellular Transport
D.A. Walker (With 24 Figures)

1. Introduction 85
2. The Development of the Experimental Study of Metabolite Translocation in Chloroplasts 85
3. Methods 87
3.1 Distribution in vivo 87
3.1.1 Nonaqueous Techniques 87
3.1.2 Aqueous Separation 88
3.2 Distribution in vitro 88
3.2.1 Centrifugal Filtration 89
3.2.2 Chromatographic Analysis 89
3.3 Indirect Methods 89
3.3.1 Shortening of Induction and Reversal of Orthophosphate Inhibition 89
3.3.2 Catalysis by Intact and Ruptured Chloroplasts 90
3.3.3 Osmotic Volume Changes 92
3.4 Other Aspects of Work with Functional Chloroplasts 93
4. The Site of Sucrose Synthesis 93
5. Starch Synthesis 95
6. The Stimulation of Starch Synthesis in the Light by Exogenous Sugars 96
7. Starch Prints 99
8. The Transport of Metabolites and Coenzymes 101
9. The Transport of Cycle Intermediates 102
9.1 3-Phosphoglycerate and the Triose Phosphates 102
9.2 Pentose Monophosphates 104
9.3 Hexose and Heptose Monophosphates 105
9.4 Sugar Bisphosphates 105
9.4.1 Ribulose-1,5-bisphosphate (RBP) 105
9.4.2 Fructose-1,6-bisphosphate (FBP) and Sedoheptulose-1,7-bisphosphate (SBP) 105
10. Free Sugars 106
11. CO_2/Bicarbonate 106
12. Carboxylic Acids 109
12.1 Glycollate and Glyoxylate 109
12.2 Malate and Oxaloacetate 109
13. Amino Acids 109
14. Orthophosphate 111
15. Inorganic Pyrophosphate 113

16. ADP and ATP . . . 116
16.1 Direct Transfer . . . 116
16.2 The Evidence in Favour of Rapid Direct Transfer . . . 117
16.3 The Evidence against Rapid Direct Transfer . . . 118
17. NADP . . . 121
18. Shuttles . . . 122
18.1 The PGA/DHAP Shuttle . . . 122
18.2 The Malate/Oxaloacetate Shuttle . . . 123
19. Specific Transport . . . 124
19.1 ATP Transport . . . 124
19.2 Dicarboxylate Transport . . . 124
19.3 Phosphate Transport . . . 124
20. The Movement of Metabolites in C_4 Photosynthesis . . . 125
21. The Movement of Protons and Magnesium . . . 126
22. Concluding Remarks . . . 126
References . . . 127

3. Metabolite Carriers of Chloroplasts
H.W. HELDT (With 4 Figures)

1. Specific Transport into the Chloroplast . . . 137
2. The Phosphate Translocator . . . 137
3. The Dicarboxylate Translocator . . . 141
4. ATP Transport in Chloroplasts . . . 142
References . . . 142

4. Compartmentation and Transport in C_4 Photosynthesis
M.D. HATCH and C.B. OSMOND (With 7 Figures)

1. Introduction . . . 144
2. Radiotracer Kinetics and Compartments . . . 144
3. Inter- and Intracellular Compartmentation of Reactions . . . 148
3.1 Methods . . . 148
3.2 Activities of Isolated Mesophyll and Bundle Sheath Cells . . . 149
3.3 C_4 Pathway Enzymes and Their Intercellular Distribution . . . 151
3.4 Intracellular Location of Enzymes . . . 157
3.5 Detailed Schemes for C_4 Photosynthesis . . . 158
4. Intercellular Transport in C_4 Photosynthesis . . . 160
4.1 Structural Features of the Mesophyll-Bundle Sheath Cell Interface 161
4.2 Intercellular Transport of Metabolite Solutes . . . 161
4.3 Intercellular Transport of $CO_2+HCO_3^-$: the CO_2 Concentrating Mechanism . . . 164
4.4 Amino Group and Charge Balance during Intercellular Transport . . 166
5. Intracellular Transport in C_4 Photosynthesis . . . 167
5.1 Intracellular Transport in Mesophyll Cells . . . 167
5.2 Intracellular Transport in Bundle Sheath Cells . . . 168
5.3 Organelle Ultrastructure in Relation to Intracellular Transport . . . 170
6. Metabolite Transport in Relation to Chloroplast Photochemical Activities 170

7. Physiological Function of Compartmentation and Transport in C_4 Photosynthesis . . . 174
References . . . 177

5. Interactions among Organelles Involved in Photorespiration
C. SCHNARRENBERGER and H. FOCK (With 12 Figures)

1. Introduction . . . 185
2. Methodology . . . 186
2.1 Measurement of Photorespiration . . . 186
2.1.1 Definitions . . . 186
2.1.2 CO_2 Gas Exchange . . . 187
2.1.3 Estimates of Photorespiration by Leaf Models . . . 188
2.1.4 Estimates of Carbon Flux through the Glycolate Pathway . . . 190
2.2 Isolation of Leaf Peroxisomes and Other Cell Organelles . . . 191
3. Glycolate Metabolism and Its Intracellular Compartmentation . . . 193
3.1 Biosynthesis of Glycolate in Chloroplasts . . . 193
3.2 The Glycolate Pathway and Peroxisomes . . . 197
3.3 The Interconversion of Glycine and Serine in the Mitochondria . . . 200
3.4 Glycerate and Its Relationship to Phosphoglycerate . . . 201
4. Glycolate Metabolism in C_4-plants, CAM Plants, and Algae . . . 202
4.1 Glycolate Metabolism in C_4-plants . . . 202
4.2 Glycolate Metabolism in CAM Plants . . . 203
4.3 Glycolate Metabolism in Algae . . . 203
4.4 Glycolate Metabolism in Blue-Green Algae . . . 206
5. The Transport of Metabolites between Cell Organelles during Glycolate Metabolism . . . 206
5.1 Transport of Glycolate Pathway Intermediates between Peroxisomes, Chloroplasts, and Mitochondria . . . 206
5.2 Sites of CO_2 Evolution and O_2 Uptake during Photorespiration . . . 207
5.3 Insignificance of the Glycolate-Glyoxylate Shuttle . . . 208
5.4 The Malate-Aspartate Shuttle . . . 209
6. The Magnitude of Photorespiration with Respect to the Glycolate Pathway Metabolism and Metabolite Transfer . . . 211
6.1 Gas Exchange . . . 211
6.1.1 Effects of Oxygen and Carbon Dioxide . . . 211
6.1.2 Effects of Light and Temperature . . . 213
6.1.3 Effects of External Factors on the Gas Exchange of C_4-plants 214
6.1.4 The Specific Activity of the $^{14}CO_2$ Evolved in the Light and in the Dark . . . 214
6.2 Carbon Flux through the Glycolate Pathway . . . 216
6.2.1 C_3-plants in Air . . . 217
6.2.2 C_3-plants in a Medium with Low Oxygen . . . 220
6.2.3 C_4-plants in Normal and Low Oxygen . . . 220
6.2.4 Carbon Fluxes and Enzyme Activities . . . 220
6.3 The Function of Light Respiration . . . 222
7. Concluding Remarks . . . 223
References . . . 223

6. Transport of Metabolites between Cytoplasm and the Mitochondrial Matrix
H.W. HELDT (With 7 Figures)

1. Introduction . . . 235
2. Structure of Mitochondria . . . 235
3. Functional Organization of Mitochondrial Metabolism . . . 236
4. Transport of Adenine Nucleotides . . . 238
5. Transport of Inorganic Phosphate . . . 242
6. Transport of the Intermediates of the Tricarboxylic Acid Cycle . . . 242
7. Transport of Glutamate and Aspartate . . . 245
8. Transport of Pyruvate . . . 246
9. Transport of Phosphate and of Carboxylates in Plant Mitochondria . . 246
10. Uptake of Fatty Acids . . . 246
11. Oxidation of Extramitochondrial NADH by Mitochondria . . . 248
12. On a Possible Role of Mitochondrial Transport in Lipogenesis . . . 250
13. Conclusion . . . 250
References . . . 250

7. Interactions between Cytoplasm and Vacuole
PH. MATILE and A. WIEMKEN (With 7 Figures)

1. Introduction . . . 255
2. Vacuoles as Repositories of Substances . . . 257
2.1 The Internal Environment of the Cytoplasm . . . 257
2.2 Methods Used for Evaluating Vacuolar Compartmentation of Substances . . . 257
2.2.1 Isolation of Vacuoles . . . 258
2.2.2 Stepwise Extraction of Cells . . . 258
2.2.3 Indirect Evidence from Isotope Kinetics . . . 258
2.2.4 Electron Microscopy, Autoradiography, Electron Microprobe 259
2.2.5 Direct Microscopic Observations . . . 259
2.3 Substances Present in Vacuoles . . . 260
2.3.1 Metabolic Intermediates . . . 260
2.3.2 Reserve Substances . . . 261
2.3.3 Secondary Product of Metabolism . . . 261
2.3.4 Inorganic Substances . . . 263
2.4 Mechanisms of Transport in Vacuoles . . . 264
2.4.1 Transport of Ions . . . 264
2.4.2 Accumulation of Products of Secondary Metabolism . . . 266
3. Vacuoles as Digestive Compartments . . . 268
3.1 Hydrolases Localized in Vacuoles . . . 268
3.1.1 Remarks on Methods Employed . . . 268
3.1.2 Vacuoles as Lysosomes . . . 269
3.2 Autophagy . . . 272
3.2.1 Turnover . . . 272
3.2.2 Mechanisms of Autophagy . . . 273
3.2.3 Heterophagy . . . 277
3.2.4 Autolysis . . . 278
3.2.5 Digestive Processes in Plant Development . . . 279

4. Concluding Remarks 280
References 281

8. Interactions among Cytoplasm, Endomembranes, and the Cell Surface
D.J. MORRÉ and H.H. MOLLENHAUER (With 15 Figures)

1. Introduction 288
2. Cell Surfaces and Expansion Growth 288
3. Role of the Soluble Cytoplasm (Cytosol) 289
3.1 Biosynthetic Role of the Cytosol 290
3.2 Guide Elements of the Cytosol 291
3.3 Milieu for Biosynthesis and Transformation (Zones of Exclusion) . . 292
4. Role of Internal Membranes 295
4.1 Role in the Biosynthesis, Assembly, Transformation, and Transport of Membranes 295
4.1.1 The Endomembrane System 297
4.1.2 Membrane Differentiation 303
4.1.3 Membrane Flow 304
4.2 Role in the Biosynthesis, Assembly, Transformation, and Transport of Products for Secretion to the Cell's Exterior 306
4.2.1 Cell Walls 307
4.2.2 Slimes and Mucilages 314
4.2.3 Essential Oils 314
4.2.4 Encrusting Materials 315
4.2.5 Dilute Secretions Containing Lipids, Sugars, Mucilages, and Salts 315
4.2.6 Digestive Enzymes 315
4.2.7 Water Expulsion 316
4.3 Role in the Biosynthesis, Assembly, Transformation, and Transport of Products among Internal Compartments 316
4.3.1 Vacuoles 317
4.3.2 Vacuole Membranes 317
4.3.3 Vacuole Contents 319
5. Role of the Cell Surface 319
5.1 Role of the Plasma Membrane in Cell Wall Formation 319
5.2 Plasmodesmata 322
5.3 Role of Cell Walls in Cell Wall Formation 322
6. Specializations in the Formation of Plant Cell Surfaces 323
7. Summary 325
References 326

III. Intracellular Transport in Relation to Energy Conservation

1. Ion Transport and Energy Conservation in Chloroplasts
R.E. MCCARTY (With 5 Figures)

1. Introduction 347
2. Mechanisms of H^+ Uptake in Chloroplasts 348
2.1 H^+ Uptake in the Absence of Permeating Electron Donors or Acceptors 349
2.2 H^+ Uptake in the Presence of Permeating Electron Donors or Acceptors 354

3. Counter Ion Fluxes 357
4. The Magnitude of the Electrochemical Gradient 358
4.1 The H^+ Concentration Gradient 358
4.2 The Membrane Potential 364
5. Relation of H^+ Movements to Phosphorylation 365
6. Relation of the Rate of Electron Flow to H^+ Movements 368
7. Conclusion 370
References 370

2. Ion Transport in Plant Mitochondria
R.H. WILSON and R.J. GRAESSER (With 2 Figures)

1. Introduction 377
2. Monovalent Salt Transport 378
2.1 Monovalent Cation and Anion Transport in Plant Mitochondria . . 378
2.2 Chemiosmotic Theory of Ion Transport 379
2.3 Mechanism of Salt Flux 380
2.4 Monovalent Cation Transport with Ionophorous Antibiotics 381
2.5 Substrate Transport into Mitochondria 383
3. Divalent Cation Transport 384
3.1 Divalent Cation Transport in Plant Mitochondria 384
3.2 Theories of Ca^{2+} Transport in Animal Mitochondria 387
3.3 Regulation of Ca^{2+} Flux 389
3.4 Role for Mitochondria in the Transport and Regulation of Ca^{2+} 391
References 394

3. Energy Transfer between Cell Compartments
H. STROTMANN and S. MURAKAMI (With 3 Figures)

1. Introduction 398
2. Exchange of Energy between Mitochondria and Cytoplasm 400
3. Exchange of Energy between Chloroplasts and Cytoplasm 401
3.1 Evidence for Energy Exchange between Chloroplasts and Cytoplasm 401
3.1.1 Active Uptake of Sugars and Ions 401
3.1.2 Transient Changes of Intracellular Adenine Nucleotide Levels 402
3.2 Direct Translocation of Adenine Nucleotides across the Chloroplast Envelope via Adenine Nucleotide Translocator 403
3.3 Indirect Transport of Energy across the Chloroplast Envelope via Shuttle Systems 405
4. Conclusion 411
References 412

IV. Theory of Membrane Transport

Mass Transport across Membranes
D. WOERMANN (With 5 Figures)

1. Nonequilibrium Thermodynamics of Membrane Transport 419
1.1 Introduction 419

1.2 Transport of Nonelectrolytes 419
1.2.1 Description of the Membrane System 419
1.2.2 Entropy Balance 420
1.2.3 Energy Balance 421
1.2.4 Mass Balance 422
1.2.5 Entropy Production 422
1.2.6 Linear Laws 423
1.2.7 Linear Laws for Dilute Solutions 425
1.2.8 Influence of Nonpermeable Components 428
1.3 Transport of Nonelectrolytes with Chemical Reaction 429
1.3.1 Description of the Membrane System 429
1.3.2 Mass Balance 429
1.3.3 Entropy Production 430
1.3.4 Linear Laws 430
1.3.5 Active Transport 431
1.4 Transport of Electrolytes 432
1.4.1 Description of the Membrane System 432
1.4.2 Entropy Balance 433
1.4.3 Energy Balance 433
1.4.4 Mass Balance 434
1.4.5 Entropy Production 434
1.4.6 Generalization 436
1.4.7 Linear Laws for Dilute Solutions; Special Case 437
1.5 Summary 440
2. Transport Processes across Membranes with Narrow Pores 440
2.1 Introduction 440
2.2 Transport Equations 442
2.3 Equilibrium at the Phase Boundary Membrane/Solution 444
2.4 Integration of the Transport Equations 446
2.4.1 Bulk Phases of Equal Composition 447
2.4.1.1 Electroosmotic Volume Flow 448
2.4.1.2 Electric Conductivity 449
2.4.1.3 Electric Transference 449
2.4.1.4 Streaming Potential 450
2.4.2 Bulk Phases of Different Composition 450
2.4.2.1 Approximate Integration of the Transport Equations (Constant Field Assumption) 450
2.4.2.2 Unidirectional Flows and Flow Ratio 453
2.4.2.3 Membrane Potential 455
2.4.2.4 Osmotic Properties 456
2.5 Summary 459
List of Symbols 459
References 462

Author Index 465

Subject Index 507

List of Contributors

J. Brachet
Laboratoire de Cytologie et
Embryologie Moléculaires
Département de Biologie Moléculaire
Université libre de Bruxelles
67, rue des Chevaux
1640 Rhode St. Genèse/Belgique

R.H. Falk
Department of Botany
University of California
Davis, CA 95616/USA

H. Fock
Fachbereich Biologie
Universität Kaiserslautern
Postfach 3049
6750 Kaiserslautern
Federal Republic of Germany

R.J. Graesser
Delta Research Farm
CIBA-Geigy Corporation
Rt. 1 Box 540A
Greenville, MS 38701/USA

M.D. Hatch
Division of Plant Industry C.S.I.R.O.
P.O. Box 1600, Canberra City
A.C.T. 2601/Australia

H.W. Heldt
Institut für Physiologische Chemie
Physikalische Biochemie und Zell-
biologie der Universität München
Goethestraße 33
8000 München
Federal Republic of Germany

Ph. Matile
Institut für Allgemeine Botanik
Labor für Pflanzenphysiologie ETHZ
Sonneggstraße 5
8006 Zürich/Switzerland

R.E. McCarty
Section of Biochemistry,
Molecular and Cell Biology
Wing Hall, Cornell University,
Ithaca, N.Y. 14853/USA

H.H. Mollenhauer
VTERL, ARS, USDA
P.O. Drawer GE
College Station,
TX 77840/USA

D.J. Morré
Departments of Biological Sciences
and Medicinal Chemistry, Purdue University
West Lafayette, IN 47907/USA

S. Murakami
Department of Biology, University of Tokyo
Komaba, Meguro-ku
Tokyo/Japan

C.B. Osmond
Department of Environmental Biology
Research School of Biological Sciences
Australian National University
Box 475, Canberra City,
2601/Australia

C. Schnarrenberger
Fachbereich Biologie
Universität Kaiserslautern
Postfach 3049
6750 Kaiserslautern
Federal Republic of Germany

C.R. Stocking
Department of Botany
University of California
Davis, CA 95616/USA

H. Strotmann
Botanisches Institut
Abt. für Biochemie der Pflanzen
der Tierärztlichen Hochschule
Bünteweg 17
3000 Hannover
Federal Republic of Germany

D.A. Walker
Department of Botany
The University of Sheffield
Sheffield S10 2TN/Great Britain

A. WIEMKEN
Institut für Allgemeine Botanik
Labor für Pflanzenphysiologie ETHZ
Sonneggstraße 5
8006 Zürich/Switzerland

R.H. WILSON
Monsanto Agricultural Co.
800 N. Lindbergh Blvd.
St. Louis, MO 63166/USA

D. WOERMANN
Institut für Physikalische Chemie
Universität Köln
Luxemburger Straße 116
5000 Köln 41
Federal Republic of Germany

List of Abbreviations*

AAT	aspartate amino transferase
AMP, ADP, ATP	adenosine mono-, di-, triphosphate
cAMP	cyclic adenosine monophosphate
ADPG	adenosine diphosphoglucose
APS	apparent photosynthesis, see also definition on p. 187
Asp	aspartate
ATPase	adenosine triphosphatase
BrdUr	bromodeoxyuridine
BV	benzylviologen
C_3 plants	plants with 3-phosphoglycerate as the primary product of photosynthetic CO_2 assimilation
C_4 plants	plants with oxaloacetate or a related C_4 acid as the first identifiable product of photosynthetic CO_2 assimilation
CAM	crassulacean acid metabolism
CCCP	m-chlorocarbonylcyanide phenylhydrazone
CMP, CDP, CTP	cytidine mono-, di-, triphosphate
CDP-choline	cytidine diphosphocholine
CF_1	coupling factor 1 of thylakoid membranes
CoA	coenzyme A
Con A	concanavalin A
D	dictyosome
DAB	3,3′-diaminobenzidine
DBMIB	2,5-dibromo-3-methyl-6-isopropyl-p-benzoquinone
DCMU	dichlorophenyl-1,1-dimethylurea
DHAP	dihydroxyacetone phosphate
DNA	deoxyribonucleic acid
cDNA	complementary deoxyribonucleic acid
2,4-DNP	2,4-dinitrophenol
DPGA	1,3-diphosphoglycerate
EDTA	ethylenediamine tetraacetic acid
ER	endoplasmic reticulum
rER	rough endoplasmic reticulum
sER	smooth endoplasmic reticulum
FBP	fructose bisphosphate
FBPase	fructose bisphosphatase
FCCP	carbonylcyanide p-trifluoromethoxyphenylhydrazone
Fd	ferredoxin
FMN	flavine mononucleotide
$FMNH_2$	reduced flavine mononucleotide
F6P	fructose-6-phosphate
GAP	glyceraldehyde phosphate
GMP, GDP, GTP	guanosine mono-, di-, triphosphate
cGMP	cyclic guanosine monophosphate
Glu	glutamate
G1P	glucose 1-phosphate
G3P	glyceraldehyde 3-phosphate
G6P	glucose 6-phosphate
HP	hexose phosphate
IAA	indole 3-acetic acid
IMP, IDP, ITP	inosine mono-, di-, triphosphate
IPS	inhibition of photosynthesis, see also definition on p. 187
αKG	α-ketoglutarate
MDH	malic dehydrogenase
NAD, NADH	oxidized, and reduced nicotinamide adenine dinucleotide
NAD-ME	NAD-dependent malic enzyme
NADP, NADPH	oxidized, and reduced nicotinamide adenine dinucleotide phosphate
NADP-ME	NADP-dependent malic enzyme
NEM	N-ethylmaleimide
NPA	N-1-naphtylphthalamic acid
OAA	oxaloacetate
P_i	inorganic phosphate
p_{O_2}	partial pressure of O_2
p_{CO_2}	partial pressure of CO_2

* See also p. 459.

PCK	phosphoenolpyruvate carboxykinase
PCR cycle	photosynthetic carbon reduction cycle
PEP	phosphoenolpyruvate
PGA	3-phosphoglycerate
PHA	phytohemagglutinins
PM	perinuclear mastigonemes
PMF	proton motive force
PMS	N-methyl phenazonium methosulfate
PN	pyridine nucleotide
PP_i	inorganic pyrophosphate
PPiase	pyrophosphatase
PQ	plastoquinone
PR	photorespiration
PTA	phosphotungstic acid
R	ribosome
RA	reassimilation, see also definition on p. 187
RNA	ribonucleic acid
cRNA	complementary ribonucleic acid
hnRNA	heterogenous nuclear ribonucleic acid
mRNA	messenger ribonucleic acid
rRNA	ribosomal ribonucleic acid
tRNA	transfer ribonucleic acid
RNase	ribonuclease
RNP	ribonucleoprotein
R5P	ribose 5-phosphate
RSA	relative specific activity
Ru5P	ribulose 5-phosphate
RuBP	ribulose 1,5-bisphosphate
SBP	sedoheptulose 1,7-bis-phosphate
S7P	sedoheptulose 7-phosphate
SV	secretory vesicle
TCA-cycle	tricarboxylic acid cycle
TCPIP	trichlorophenolindophenol
THFA	tetrahydrofolic acid
TMPD	N,N,N′,N′-tetramethyl-phenylene diamine
TP	triose phosphate
TPP	thiamine pyrophosphate
TPS	true photosynthesis, see also definition on p. 187
TRIS	tris(hydroxymethyl) aminomethane
UMP, UDP, UTP	uridine mono-, di-, triphosphate
UDPG	uridine diphosphoglucose

I. Membrane Structure

Plant Membranes

R.H. FALK and C.R. STOCKING

1. Introduction

Within the past few years studies of the morphology and function of membranes have increased almost exponentially. Books and reviews discussing the biochemistry of membranes, the biochemical and biophysical properties of cell surfaces, the molecular organization of membranes, the organization of energy transducing membranes, membrane molecular biology, etc. have appeared with an increasing frequency that almost overwhelms the serious student of membrane structure and function. Although our understanding of the intricate details of the similarities and differences in structure and function among the various cellular membranes has not advanced as rapidly as the wealth of publications would lead one to believe, still a tremendous activity in the field is indicated, and rapid advances in our understanding of membranes are being made.

A primary function of membranes is the compartmentation of the cell. These barriers to the free diffusion of water-soluble solutes also have unique structural and chemical properties which facilitate the transport of specific polar solutes. Thus, it is logical that a discussion of membranes is included in a volume on the intracellular movement of solutes. The roles of membranes as energy transducers (NAKAO and PACKER, 1973; GREEN, 1972) and the organization of internal membranes in chloroplasts and mitochondria (RACKER, 1972; PARK and SANE, 1971; IKUMA, 1972; TREBST, 1974) will not be considered in detail. The goal of this chapter is to provide introductory background stressing similarities and differences in membranes that bear most specifically on intracellular solute transport.

2. Isolation of Membranes

Advances in our knowledge of membrane biochemistry and function depend, in part, on our ability to isolate and purify membranes from many diverse sources. Standard procedures that are currently employed and are reasonably successful in the isolation of plant cell organelles and membranes have been published (e.g., FLEISCHER and PACKER, 1974; see also specific articles in this volume) and need not be reviewed here in detail. However, it is well known that plant cells with relatively tough cell walls, and large vacuoles that frequently contain high concentrations of tannins and phenols (LOOMIS, 1974) and acids and hydrolases (MATILE, this volume) present particularly difficult problems in organelle and membrane isolation (PRICE, 1974). Contamination of membrane preparations commonly occur through occlusion, aggregation, denaturation, and adsorption. The tendency for fragile organelles and membranes to be broken and modified and to lose polar components during cell rupture and isolation are continuing and frequently unresolved problems. These problems are of particular importance and should be taken into consideration in the interpretation of reports on the physical properties and biochemical composition of isolated cell membranes.

DE PIERRE and KARNOVSKY (1973) have reviewed the methods for isolating and characterizing plasma membranes from mammalian cells. Their discussion and conclusions are apropos to the isolation and characterization of any membrane.

2.1 Choice of Tissue

Plant tissues from which membranes are to be isolated are often chosen on the basis of habit or availability. One must be very aware that starting materials, such as leaves, are composed of several tissue types and may therefore comprise quite heterogeneous membrane populations, and heterogeneity might persist in spite of efforts to obtain a homogeneous cell population. For example the portion of the plasma membrane situated in the vertical plane of a sieve tube may be quite different from that situated in the horizontal plane. Final isolates can easily be enriched in material from a very minor cell type. This possibility is usually only briefly mentioned or ignored completely by many investigators.

2.2 Identification of Specific Membranes

During isolation, one must be able to identify the desired organelle or organelle fragment containing the membrane being studied. If the organelle or organelle fragment has a distinct morphology, it may be possible to use microscopy to follow it during isolation. Or, if the desired component has some unique compound (e.g., chlorophyll) or specific enzyme associated with it, this feature may be exploited to follow the isolation. (See also Vol. 2, Part A: Chap. 10.4.)

2.2.1 Microscopy

Microscopy of sectioned pellets is not necessarily a good way to evaluate tissue fractionation. PEASE (1964) has calculated that an electron microscopist taking pictures at only $\times 1{,}500$ magnification on a two-inch plate records only 0.001 mm^2. If one takes 50 pictures a day (which is unlikely), it would take seven and one half years to photograph 1 cm^2. PEASE goes on to point out that the situation worsens if one considers the third dimension. Useful sections are no thicker than $^1/_{20}$ μ. It requires about 100 sections to get through a nucleus; 200 sections to get through a small cell. Sampling error is simply staggering when one tries to use electron microscopy as a means of evaluating homogeneity of pellets of subcellular material.

SJÖSTRAND (1957) suggested a technique which may partly alleviate these problems. He proposed that a representative sample of the isolated subcellular fraction that is to be monitored could be deposited, by centrifugation, as a very thin film on a flat surface. If this material were then fixed, embedded, and sectioned perpendicularly to the plane of the surface, the sampling difficulties involved in electron microscopical monitoring of the subcellular fraction either could be circumvented or at least significantly minimized. BAUDHUIN modified this technique to utilize thin samples that had been collected on the surface of Millipore filters (BAUDHUIN et al., 1967; BAUDHUIN, 1974). After fixation and dehydration of the sample, the filter was dissolved in propylene oxide and the fixed sample remained as a thin sheet. The embedded sample was then sectioned perpendicularly to the plane of the sheet. Since any stratification of the sample would have occurred along the depth of the sample, and since the sections observed in the electron microscope covered the entire depth of the sheet, this technique should be a useful method of monitoring the morpholgy of a mixed population of isolated subcellular fractions. The electron micrographs of the sample may be examined under a transparent overlay bearing lines 1 cm apart. If identifiable, the different membrane types that occur under the intersections of the lines may be counted and the composition of the sample estimated (LOUD, 1962; LEMBI et al., 1971; WILLIAMSON et al., 1975).

As if problems of obtaining representative samples were not enough, one quickly finds that morphological criteria for the identification of some types of organelles are not sufficiently rigorous. The membranes of dictyosomes, smooth endoplasmic reticulum and lysosomes can all look quite similar. Fragmentation of some organelles causes the membrane fragments to form smooth vesicles virtually indistinguishable from other naturally occurring smooth vesicles. Certainly, no isolation should ever rely solely on morphological markers as adequate criteria for isolate homogeneity.

2.2.2 Histochemical Staining

Histochemical staining is widely used by investigators to follow an organelle or organelle fragments during their isolation. For example, phosphatase activity, which is associated with specific membranes, is an often used marker. In the Gomori technique, the isolate is reacted with a substrate from which phosphatase releases inorganic phosphate. The released inorganic phosphate, in the presence of a heavy metal such as lead, precipitates and its intracellular position is subsequently identified by electron microscopy. SCHNITKA and SELIGMAN (1971) and others have noted a number of difficulties with this technique. The difficulties are common to any histochemical method, not just phosphatase histochemistry, and are as follows: (1) glutaraldehyde, a common and excellent fixative is also an excellent inhibitor of many enzymes; (2) the metal capture ion may also inhibit the enzyme being studied; (3) diffusion of substrate and/or capture ion into the sample may be impaired or nonexistent; (4) the enzyme-substrate product may diffuse some distance before finally being captured and precipitated and thereby preclude a sharp localization of the marker site.

2.2.3 Marker Enzymes

In addition to the use of morphological features and histochemical staining reactions as markers, if possible, fractions isolated from cells should be monitored by determining the activities of enzymes characteristically associated with specific

Table 1. Convenient markers and isopycnic densities for plant cell organelles and membranes after isolation by isopycnic centrifugation in sucrose density gradients

Organelles Markers	Isopycnic (bouyant) density $g \times 10^{-3}$ l	References
Plasmalemma	1.165	LEONARD and VAN DER WOUDE (1976)
5′-nucleotidases [a]		EMMELOT et al. (1964)
UDP-glucose and GDP-glucose transferases [b]		VAN DER WOUDE (1973)
K^+ stimulated ATP-ase (pH 6.5) (ouabain insensitive)	1.17	HODGES et al. (1972)
Phosphotungstic acid-chromic acid stain		ROLAND et al. (1972)
		VILLEMEZ et al. (1968)
Endoplasmic reticulum (rough and smooth microsomes)	1.10 1.14–1.19 1.19 (rough)	LEONARD and VAN DER WOUDE (1976) DONALDSON et al. (1972) WILLIAMSON et al. (1975)
glucose-6-phosphatase [a]	1.12–1.13	HUANG (1975)
NADH: cytochrome-c reductase [c]		HODGES et al. (1972) HUANG (1975)
Peroxisomes	1.17–1.25	TOLBERT (1970)
Catalase		NEWCOMB and BECKER (1974) TOLBERT (1971b)
Glycolate oxidase		TOLBERT (1971b)
Glyoxysomes	1.24–1.25	BREIDENBACH et al. (1968)
Isocitrate lyase		BREIDENBACH et al. (1968) BIEGLMAYER and RUIS (1974)
Lysosomes (animal)		
Acid phosphatase		DE PIERRE and KARNOVSKY (1973)
Mitochondria	1.17–1.22	
Cytochrome oxidase		APPLEMANS et al. (1955)
Succinate-cytochrome-c [d] reductase		TOLBERT et al. (1968)
Mitochondria inner membrane	1.21	PARSONS et al. (1966)
Mitochondria outer membrane	1.12–1.14	PARSONS et al. (1966) DOUCE et al. (1973)

[a] Used as a marker in animal cell preparations but activity is usually too low for use as a marker of plant membranes (HODGES et al., 1972) and its presence in the nuclear envelope in animal cells appears to be established (KASPER, 1974).

[b] Also thought to be associated with the Golgi apparatus (RAY et al., 1969).

[c] In animals this enzyme is reported to be associated with endoplasmic reticulum or outer mitochondrial membrane (LARDY and FERGUSON, 1969) but it is also present in the nuclear envelope (KASPER, 1974).

[d] Used to determine percent intact mitochondria. Intact mitochondria should not show succinate-cytochrome-c reductase activity but broken mitochondria should.

Table 1 (continued)

Organelles Markers	Isopycnic (bouyant) density $g \times 10^{-3}$ l	References
Chloroplasts—intact	1.20–1.25	
Chlorophyll peak at high density, absorbance at 663 and 645 nm		ARNON (1949)
NADPH: glyceraldehyde phosphate dehydrogenase		HEBER et al. (1963)
NADPH: glyoxylate reductase		TOLBERT (1971 b)
Triosephosphate isomerase		HUANG (1975)
Chloroplasts—broken (Thylaloid membranes)	1.14–1.18	
Chlorophyll peak at low density, absorbance at 663 and 645 nm		ARNON (1949)
Ca^{2+} dependent ATPase		DOUCE et al. (1973)
Chloroplast envelope		
Mg^{2+}-ATPase		DOUCE et al. (1973)
galactolipid synthesis		DOUCE et al. (1973)
Dictyosomes	1.12–1.15	RAY et al. (1969) LEONARD and VAN DER WOUDE (1976)
Inosine diphosphatase		DAUWALDER et al. (1969) GOLFISCHER et al. (1964) RAY et al. (1969) MORRÉ et al. (1971 b)
Nuclei	1.31–1.33	
Envelope	1.18–1.22	STECK (1972a)
DNA	1.69–1.71	
RNA	1.99–2.02	
Nucleoli		
RNA polymerase I (Nucleoli lack RNA polymerase II)		CHEN et al. (1975)
Vacuoles		
See pages 255–287 this volume		MATILE and WIEMKEN
Ribosomes and polysomes	1.55–1.59	
Absorbance at 254 nm		PRICE (1974)
Starch grains	1.61–1.63	PRICE (1974)

organelles. Frequently these enzymes are the most useful markers because their assays are quantitively precise and sensitive.

Table 1 is a brief list of commonly used marker enzymes for various plant cell organelles and membrane fractions. Also included in the table are approximate densities of some of the cell fractions. However, it should be realized that the exact equilibrium density of a given cell organelle may vary not only with the species of plant and the type of cell from which the organelle has been derived but also with the physiological state of the organelle. Even the method of cell fractionation and organelle isolation may modify the bouyant densities of some organelles (e.g. microbodies, TOLBERT, 1971a).

An invalid assumption often made regarding marker enzymes is that their distribution in leaf mesophyll cells is the same as that found in animal cells. For example, WALLACH and ULLREY (1962) found that 5′-nucleotidase, a commonly used animal plasma membrane marker enzyme, was localized exclusively in the nuclei of Ehrlich ascites carcinoma cells. Rat fat cells do not have *any* apparent 5′-nucleotidase activity (MCKEEL and JARETT, 1970), and the activity of this enzyme in preparations from plant cells is usually too low to be useful in monitoring (HODGES et al., 1972).

2.2.4 Antibodies

Antibodies specific to proteins associated with a particular organelle or organelle fragment can also be used as markers. TREBST (1974) reviewing the evidence for a sidedness of photosynthetic electron flow in chloroplast membranes, cited several studies where antibodies to various plant proteins have been used as markers. Several potential problems must be considered when using antibodies as markers for a membrane fraction: the antibody preparation must be specific for a protein component of the desired membrane and *no others*. Antibodies themselves are rather large proteins and many assays involve coupling a second large molecule to the antibody to allow visualization of the site of the antigen–antibody complex. These supermolecules diffuse very sluggishly. In spite of these problems, antibodies can provide the careful investigator with an elegant, sensitive marker system.

2.3 Tissue Disruption

The degree of cell disruption and organelle fragmentation are important considerations in evaluating any membrane isolation scheme. Disruption of the plasmalemma and endoplasmic reticulum is generally unavoidable, and STECK (1972a) points out that the endoplasmic reticulum is the most sensitive organelle with regard to homogenization, a commonly used method of cell disruption. Nuclear envelope damage releases polybasic proteins that absorb to polyanionic membrane components. This may promote aggregation. Lytic enzymes released from disrupted lysosomes of animal cells or from vacuoles of plant cells may alter the membrane fraction of interest. Since marker enzymes may be released and absorbed on membranes where they normally do not occur, the reliability of marker enzymes may become suspect if substantial organelle disruption takes place.

2.4 Membrane Isolation

Except for the plasmalemma and endoplasmic reticulum, which are unavoidably broken during cell rupture, generally the first step in membrane purification is the isolation and purification of the specific organelle containing the membrane being studied. Once the organelle has been isolated in as uncontaminated and unaltered a form as possible, various techniques, such as the use of detergents, sonication, and osmotic shock, are used to separate the envelope membrane(s) from the rest of the organelle. DE PIERRE and KARNOVSKY (1973) point out that organelles and membrane fragments have relatively few properties (e.g. mass, volume, density, and electrophoretic mobility) that one can exploit for separation purposes. Consequently the isolation of fractions relatively pure with respect to a single membrane is extremely difficult.

Many fractionation schemes call for an initial differential rate centrifugation followed by a differential density gradient centrifugation and/or an equilibrium (isopycnic) density gradient centrifugation. Differential rate centrifugation is most useful as an enrichment step but is rather limited in its resolving power and always produces an impure pellet enriched in the most rapidly sedimenting species present. These impure fractions are often referred to as the "chloroplast", "nuclear", or "mitochondrial" pellets.

Differential density gradient centrifugation, where a partially purified fraction is loaded in a thin layer on top of a shallow density gradient and then centrifuged for an empirically determined period of time, has proved to be a useful technique which, in some instances, can produce relatively pure "bands" of membrane fragments.

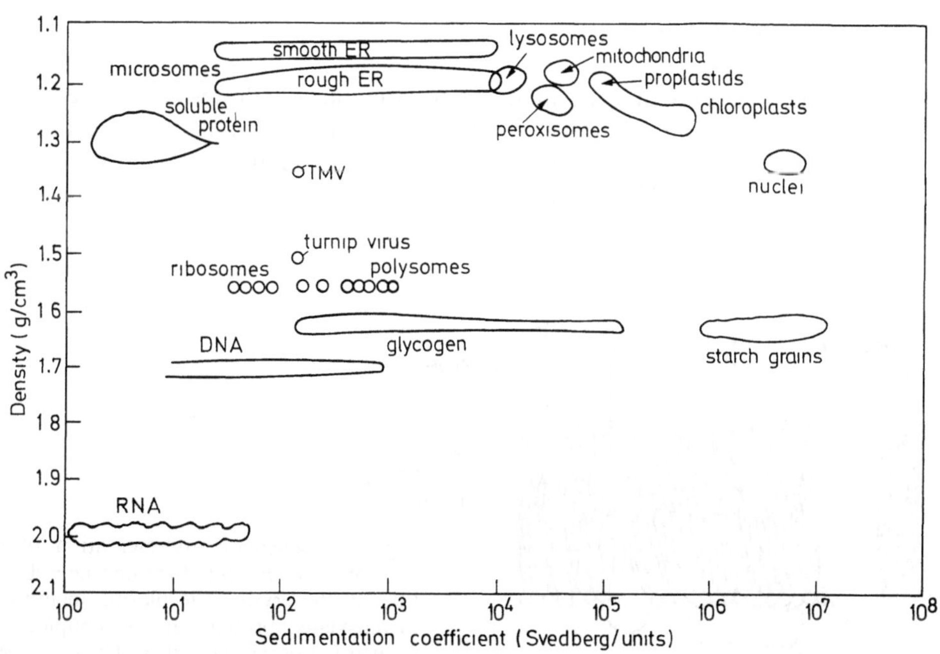

Fig. 1. A diagram showing the S, ρ relationships of several substances. (Modified from PRICE, 1974)

Equilibrium density gradient (isopycnic density gradient) centrifugation, a technique, where a fraction is centrifuged on a density gradient until its various components migrate to positions on the gradient equal to their densities, has also proven to be a powerful technique for membrane fraction isolation in many instances.

Generally, a combination of differential and isopycnic centrifugation methods is required for adequate separations. Nevertheless, because of the similarities in sedimentation velocities and bouyant densities of many of the organelles, purification of individual membranes is extremely difficult and evaluation of the purity of each fraction is essential. A useful figure (Fig. 1) which indicates the similarities and differences in the sedimentation coefficient (S) and bouyant density (ρ) values for many organelles and membrane fractions has been published (ANDERSON et al., 1966; PRICE, 1974).

3. Membrane Composition

Except for chloroplast and microbody membranes, most data available on membrane composition have been derived from studies of animal tissues (MORRÉ, 1975). The main constituents of membranes are protein (approximately 60%) and lipids (between 30 to 50% on a dry weight basis), but it is calculated that over half of the volume of most membranes is lipoidal. Some membranes, e.g., mammalian erythrocyte ghosts, may contain up to about 10% carbohydrate (BRETSCHER, 1973).

3.1 Membrane Carbohydrates

Several generalizations may be made about the carbohydrates sometimes found associated with membranes (STECK and FOX, 1972): (1) they consist of short heterosaccharides; (2) the preponderant monosaccharides found are galactose, glucose, N-acetylgalactosamine, N-acetylglucosamine, mannose, fucose, and sialic acid; (3)

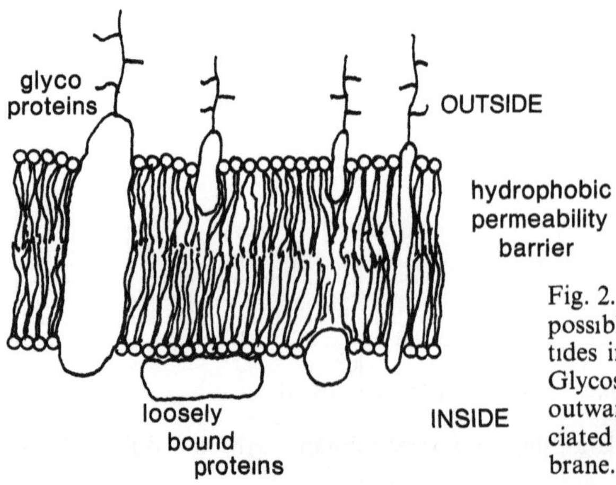

Fig. 2. A schematic representation of the possible arrangement of membrane polypeptides in the erythrocyte plasma membrane. Glycosylated proteins and glycolipids face outward. Loosely bound proteins are associated with the cytoplasmic face of the membrane. (Modified from STECK, 1972)

association of carbohydrates with membrane protein is either through O-glycoside linkages to the hydroxyls of serine or threonine or through N-glycoside linkages to asparagine; (4) the carbohydrate profile may vary characteristically among glycoproteins and organelles; (5) many membranes have no detectable associated carbohydrate. A study of the erythrocyte plasma membrane has shown that, in general, the carbohydrates are located on the outer surface of the membrane and are not found on the inner surface (STECK, 1972b; see also Fig. 2).

3.2 Membrane Lipids

Membrane lipids are characteristically bimodal, rodlike, amphipathic molecules, i.e. molecules with both hydrophilic (polar) and hydrophobic (nonpolar) regions. They are thus ideally suited to form interfaces between a polar solvent and a hydrophobic region, and they tend to align themselves in bilayers with their hydrophobic tails associated in such a way as to minimize their contact with the polar environment. This bilayer structure is thermodynamically the most stable arrangement for amphipathic lipids in a polar environment, and various lines of evidence indicate that the lipid bilayer is the predominant state of lipids in cell membranes (VANDERKOOI, 1972). However, a number of studies have also revealed that the lipid continuum is fluid-like, at least in the plasma membranes of animal and bacterial cells (LENARD and SINGER, 1966; WALLACH and ZAHLER, 1966; FRYE and EDIDIN, 1970; HUBBELL and McCONNELL, 1971; TAYLOR et al., 1971; EDIDIN, 1972; SINGER, 1972; SINGER and NICOLSON, 1972).

Lipids in cell membranes, although primarily amphipathic, are quite complex, consisting of a variety of classes such as phospholipids and glycolipids in prokaryotic cells, and phospholipids, sphingolipids, glycolipids, and sterols in eukaryotic cells (LAW and SNYDER, 1972). In contrast to animal cells, plant cell membranes lack both spingomyelin and cholesterol. About 40% by weight of the lipid found in leaves is made up of the glycolipids, mono- and diglactosyl diglyceride (BENSON, 1964; KIRK and TILNEY-BASSETT, 1967) with most or all of the galactosyl diglycerides being located in chloroplasts (WINTERMANS, 1960). A further difference between the lipids found in plant cell membranes and those found in animal cell membranes is that the predominant fatty acids found in the lipids of animal membranes are usually saturated, while those found in most plant tissues are linoleic and linolenic acids (GALLIARD, 1971). An exception to this generalization is that in animals the inner mitochondrial membrane (rat liver) is characterized by a high degree of unsaturation. Twenty percent of the total lipid in this membrane is reported to be cardiolipin with 90% unsaturated fatty acids, the main one being linoleic acid (see Sect. 4.4.2).

3.3 Membrane Proteins

Our knowledge of the organization of proteins in membranes has been greatly enhanced by the recent development of a number of techniques by which membranes can be probed and the location and physical state of some of the membrane components can be studied (Table 2).

Table 2. A list of some of the physical and chemical techniques used in studying biomembranes. Of the numerous studies that have been made, only one general article is referred to in each instance

Technique	Use	References
X-ray diffraction	Molecular motion in membranes	CAIN et al. (1972)
	Membrane lipid structure	KIETH and MEHLHORN (1972)
Antibodies	Rearrangement in plasma membrane	EDIDIN (1972)
	Topological structure of membrane	NICOLSON (1972)
Spin labeling	Lateral diffusion and phase separation in membranes	MCCONNELL et al. (1972)
	Interaction of molecular species	MEHLHORN and KIETH (1972)
Circular dichroism	Conformational changes in proteins in membranes	URRY (1972) HOLZWARTH (1972)
Optical rotatory dispersion	Conformational changes in proteins in membranes	URRY (1972) HOLZWARTH (1972)
Nuclear magnetic resonance spectroscopy	Phospholipids in membranes	HORWITZ (1972)
Freeze-etch electron microscopy	Physical structure of membranes	PARK (1972)
Ionic probes	Membrane conductance and membrane structure	SZABO et al. (1972)
Selective probing of resealed erythrocyte ghosts and inside-out vesicles	Organization of proteins in membranes	STECK (1972b)

A given membrane is apparently composed of unique proteins. Some membranes contain many protein species; others contain only a few identifiable species (STECK and FOX, 1972). It has been suggested that proteins associated with biomembranes may be divided into two classes: (1) peripheral proteins and (2) integral proteins (SINGER and NICOLSON, 1972; SINGER, 1974) or extrinsic and intrinsic proteins (VANDERKOOI, 1972). Peripheral proteins are characterized as being loosely bound, easily dissociated from the membrane by mild treatment (high ionic strength, metal ion chelating agents), and are soluble and molecularly dispersed in neutral aqueous buffers. They do not appear to interact with the membrane lipids. In contrast, integral proteins, which constitute about 70 to 80% of membrane proteins, are strongly anchored in the membrane by hydrophobic bonds. They are usually associated with lipids, are insoluble or aggregated in aqueous buffers, and are removed from the membrane by such hydrophobic bond breaking agents as organic solvents and detergents (SINGER, 1974).

Several studies indicate that each side of a membrane has characteristic, specific groups of peripheral proteins associated with it. In some cases (e.g., the periplasmic binding proteins of the plasma membrane of gram negative bacteria), these proteins are involved in the transport of ligands through the membrane (Boos, 1974; see Sect. 6). Very few integral proteins have been isolated, purified, and characterized, but it is believed that being amphipathic and globular, they possess hydrophobic regions which interact with the lipophilic region in the membrane. This non-polar interaction would then anchor the proteins in the membrane. Proteins so anchored may contain a hydrophilic end extending into the aqueous environment, or in some instances, they appear to contain two polar regions, one on either end. Such molecules could span the membrane and the polar ends would then be exposed to the aqueous environment on either side of the membrane (LENARD and SINGER, 1966; WALLACH and ZAHLER, 1966).

Glycophorin, the glycoprotein of the human erythrocyte membrane, is an example of an integral protein. It is oriented in the membrane so that the N-terminal half of the polypetide chain, bearing all of the covalently bound carbohydrate, extends into the external medium. The C-terminal half of the molecule has a unique segment of non-polar amino acids which are well adapted to interact with the hydrophobic area of the lipid in the membrane (MARCHESI et al., 1972).

The studies on the organization of proteins in human erythrocyte membranes carried out by STECK (1972b) on resealed ghosts and inside-out vesicles is an illustration of one of a number of observations that, taken together, have led to the general concept of membranes being lipid bilayers associated in various ways with proteins. STECK utilized non-penetrating reagents to probe selectively the two surfaces of the erythrocyte plasmalemma. The asymmetrical character of the membrane was readily demonstrated: some membrane components were found only on the outer surface, e.g. sialic acid and acetylcholine esterase; others, diaphorase and glyceraldehyde 3-phosphate dehydrogenase, were identified on the inner surface only; two glycoproteins appeared to span the membrane. Figure 2 diagrams a possible arrangement of the major membrane polypeptides as revealed by this study.

Such studies are of major significance to our understanding of transport across membranes since globular proteins that span the hydrophobic barrier could serve as channels for the conduction of polar solutes (see Sect. 6).

4. Specific Membranes

4.1 The Plasmalemma (Plasma Membrane)

4.1.1 Morphology

The plasmalemma (plasma membrane, ectoplast) appears in thin section as a three-layered structure about 60–100 Å across (Fig. 3). The outer electron dense layers are about 20 Å across; the clear, middle region is about 35 Å across. FINEAN (1961) pointed out that in cells fixed only in osmium tetroxide, the inner side (cytoplasmic) of the bilayer is much more readily seen. After post staining both sides of the bilayer are easily seen. Freeze-etch studies on yeast plasma membranes reveal 150 Å particles associated with the membrane matrix. These are frequently found arranged in a hexagonal array (MATILE, 1970; MOOR and MÜHLETHALER, 1963). BRANTON

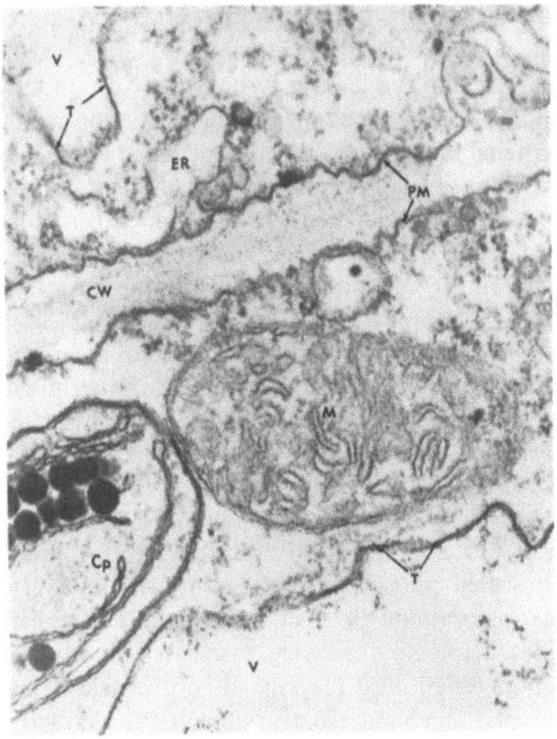

Fig. 3. Electron micrograph of two adjacent parenchyma cells from a wheat flower filament. *T* tonoplast, *V* vacuole, *ER* endoplasmic reticulum, *M* mitochondria, *PM* plasma membrane, *CW* cell wall, *Cp* chromoplast. Ca. ×46,000. (From LEDBETTER and PORTER, 1970)

and MOOR (1964) and NORTHCOTE and LEWIS (1968) have reported a mostly random arrangement of numerous small particles associated with the membrane matrix in freeze-fracture preparations of pea and onion root tip cells. Figure 4 is an example of the sort of image obtained by the freeze-etch technique.

4.1.2 Chemical Composition

Because of their heterogenous nature and the difficulty of isolating cell membranes free from contamination, analyses of the gross chemical composition of isolated membrane preparations reveals only a limited insight into their actual nature. In addition, the method of isolation often affects the results. For example, MATILE (1970) summarized the results obtained by several authors who used different methods of isolating the plasmalemma from the yeast *Saccharomyces cervisiae*. Although all of the analyses indicated the lipoprotein nature of the plasmalemma, analyses of membranes obtained from yeast sphaeroplasts (BOULTON, 1965; LONGLEY et al., 1968) yielded high values for protein (46 to 49.3% of the dry weight of the membranes), low values for carbohydrates (3.2 to 6.0%) and showed the presence of nucleic acids. In contrast, the plasmalemma isolated from yeast cells (MATILE, 1970), rather than yeast sphaeroplasts, had significantly lower protein levels (26.6% of the membrane dry weight), high carbohydrate levels (30.8%) and no nucleic acids. There was less variation in the lipid content (ranging from 37.8 to 45.6%) of the membranes isolated by the two different methods.

One of the most extensively studied and best-characterized membranes is the mammalian erythrocyte envelope which is usually isolated by osmotic lysis of

Table 3. Overall chemical composition of the plasmalemma of yeast, *S. cerevisiae* (isolated by two different methods) and human erythrocyte membranes. Values in percent dry weight of the membrane

Component	*S. cerevisiae*		
	Isolated from whole cells [a]	Isolated from sphaerosomes [b]	Human erythrocyte membranes [c]
Protein	26.6	46–49.3	49.2
Lipid	45.5	37.8–45.6	43.6
Ergosterol	8.1	5.6–6.0	–
Cholestrol	–	–	11.1
Carbohydrate	30.8	3.2–6.0	7.2

[a] MATILE (1970). [b] BOULTON (1965), LONGLEY et al. (1968). [c] ROSENBERG and GUIDOTTI (1969).

the red blood cells (STECK and WALLACH, 1970; FAIRBANKS et al., 1971; STECK et al., 1971; STECK, 1972b; WALLACH, 1972). A comparison between the composition of the plasmalemma of erythrocytes and of yeast cells is given in Table 3. The most striking qualitative difference between these two membranes is the presence of cholesterol in the erythrocyte membrane and ergosterol in the yeast membrane.

An example of differences in the properties of the plasma membranes in different plants is illustrated by the fact that the delicate plasmalemma of higher plant cells is usually damaged by plasmolysis with Na_2CO_3 solutions. In contrast, in desmids found in bogs, the plasmalemma is often resistant to damage and may withstand normal plasmolysis with hypertonic Na_2CO_3, $MnSO_4$, and even aluminum salts (STADELMANN, 1969).

The asymmetric nature of the erythrocyte plasmalemma has already been noted. In this membrane, glycoproteins are the only major polypeptides that have been identified at the outer surface. They appear to be firmly anchored in the membrane with their covalently bound sugar groups extending out of the membrane (WINZLER, 1969; MARCHESI et al., 1972). All of the membrane associated sugar appears to be located at the external surface. In some instances, e.g., glycophorin, the glycoproteins may extend from the external environment of the cell through the lipid barrier of the membrane and into the cytoplasm of the cell. That portion of the molecule which spans the lipid bilayer of the membrane is composed chiefly of nonpolar hydrophobic amino acids. Non-glycosylated proteins, less firmly bound, were found only at the inner surface of the erythrocyte plasmalemma (STECK, 1972b).

Although the morphological features of the plasmalemma of plant cells have been studied in detail by transmission and freeze-etch electron microscopy (FINEAN, 1961; BRANTON, 1969), and biochemical studies have been made on isolated and purified fragments of yeast plasmalemma (MATILE, 1970; FUHRMANN et al., 1974; SCHIBERI et al., 1973), it is only within the last five or six years that serious attempts have been made to isolate, purify, and characterize the plasmalemma from higher plant cells (LEMBI et al., 1971; HODGES et al., 1972; VAN DER WOUDE et al., 1972; HODGES and LEONARD, 1974; LEONARD and VAN DER WOUDE, 1976). To date, plant plasmalemma preparations have been severely contaminated (30%

Table 4. Some characteristics of the plant cell plasmalemma

Characteristic	References
High sterol: phospholipid ratio	HODGES et al. (1972)
K^+-stimulated ATPase with pH optimum 6.5	HODGES et al. (1972) LEONARD and VAN DER WOUDE (1976)
High binding capacity for N-1-naphthyl-pthalamic acid	LEMBI et al. (1971)
Periodic-chromi-phosphotungstic acid staining	ROLAND et al. (1972)
UDP-glucose and GPD-glucose transferase	VILLEMEZ et al. (1968) VAN DER WOUDE (1973) HODGES et al. (1972) RAY et al. (1969)

or more contamination) (LEONARD and VAN DER WOUDE, 1976), and little definite information is available on the exact chemical composition of the plasmalemma of plant cells. HODGES et al. (1972) reported that the plasmalemma-enriched fraction obtained from maize root cells had a sterol/phospholipid molar ratio of 1.1/1.2. Similarly in animal cells (erythrocyte and liver cells), the plasmalemma is unique compared to other cytoplasmic and organelle membranes in that it has a high sterol (in this case cholesterol) to phospholipid mol/mol ratio of 0.3/1.2 (COLEMAN and FINEAN, 1966; RAY et al., 1969; ROSENBERG and GUIDOTTI, 1969; WALLACH, 1972). The ratios reported for the endoplasmic reticulum and mitochondrial membranes are much lower ranging from 0.03 to 0.09 (WALLACH, 1972). Recently RUESINK (1971) has presented evidence that the plasmalemma of *Avena coleoptile* protoplasts contains little sterol and, in this respect, resembles the sterol-less bacterial plasma membrane rather than the erythrocyte plasma membrane. Thus sterols seem to be absent from the plasmalemmas of some, but not all plant tissues.

The presence of an ion stimulated ATPase is a characteristic, but not unique, feature of the plasmalemma. This is to be expected since ATPase has been implicated in transmembrane solute transport. Although several different ATPases are found associated with various membranes in plant cells (HODGES et al., 1972), the major portion of the potassium-stimulated ATPase with a pH optimum at 6.5 in maize roots has been shown to be associated with the plasmalemma, (HODGES et al., 1972; LEONARD and VAN DER WOUDE, 1976). Similarly, cation-activated ATPase activity has been observed in the plasmalemma fractions isolated from a wide variety of organisms (e.g. in rat, EMMELOT et al., 1964; in yeast, FUHRMANN et al., 1974; in mushroom, HOLTZ et al., 1972; in yeast, MATILE, 1970; in maize roots, LEIGH et al., 1975; in *Ochromonas*, PATNI et al., 1974).

Table 4 lists some biochemical properties of the plasmalemma of plant cells.

4.2 The Nuclear Envelope

4.2.1 Morphology

The nuclear membrane is more properly referred to as an envelope since it consists of two distinct membranes, each about 100 Å thick, that are separated by a space 100 to 150 Å wide, known as the perinuclear space or cisterna (Fig. 5).

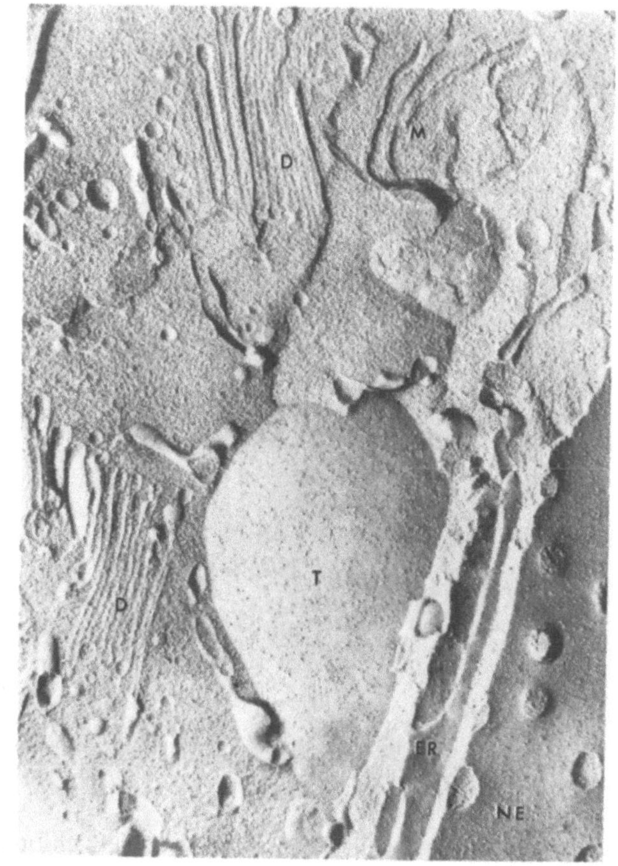

Fig. 4. Freeze etch image of onion root tip cell. *T* outer tonoplast surface, *D* dictyosome, *M* mitochondria, *ER* endoplasmic reticulum, *NE* nuclear envelope with associated pores. Ca. ×43,000. (From LEDBETTER and PORTER, 1970)

The nuclear envelope is interrupted by pore complexes, structures consisting of an apparent pore and poorly understood annular material. The pore is about 600 Å in diameter (see Fig. 4). The annular material which extends through the pore as a cylinder, effectively reduces its diameter to from 300 to 500 Å. There is some evidence that the annular material is proteinaceous (DUPRAW, 1965) and that it carries a positive charge (FELDHERR, 1964). The outer nuclear membrane is often found to be continuous with the endoplasmic reticulum and is frequently found to have ribosomes associated with it (FRANKE and SCHEER, 1974). Several studies have shown that the outer nuclear membrane has a role in protein synthesis (LEDUC et al., 1968, 1969; AVRAMEAS, 1970). BOUCK (1969, 1972) has hypothesized that glycoproteins synthesized in association with the outer nuclear membrane are secreted exocytotically by a membrane flow process to the cell exterior in some algae where they become associated with flagella. WELLS (1972) has suggested that ascospore-delimiting membranes are derived from a nuclear membrane to plasma membrane transfer. FELDHERR (1972) and FRANKE and SCHEER (1974) have provided recent reviews of the structure and function of the nuclear envelope.

4.2.2 Chemical Composition

Although the nucleus is the organelle most readily isolated and purified from animal tissue (STECK, 1972a), this is not true for the nuclei of mature green plant

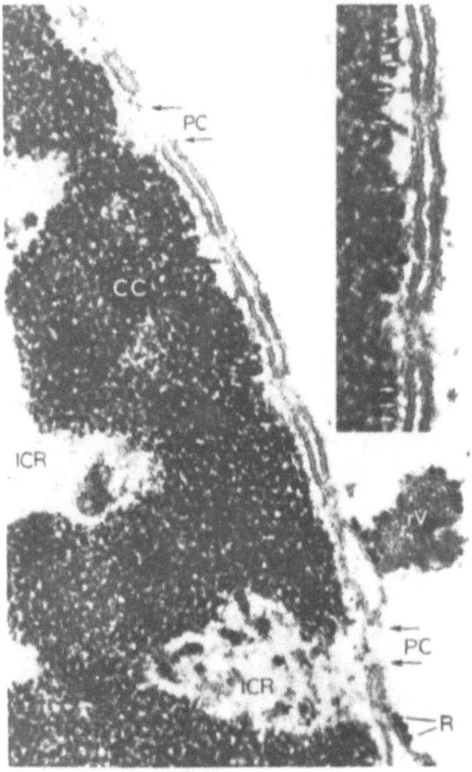

Fig. 5. Periphery of an isolated nucleus from an onion root tip cell. *PC* pore complex, *CC* condensed chromatin, *ICR* interchromatic regions, *rV* rough-surfaced vesicle, *R* ribosomes. Ca. ×80,000. (From PHILIPP et al., 1976)

cells where tough cell walls, large vacuoles, chloroplasts, and starch grains may make nuclear isolation and purification particularly difficult (STERN, 1968; MASCARENHAS et al., 1974). However, the isolation of nuclei from immature plant cells may be accomplished with reasonable ease (TAUTVYDAS, 1971), and methods have been described in which cell wall digesting enzymes are used to soften the walls of mature tissues such as leaf tissue. Tissue so treated may be broken up by very gentle grinding which has less damaging effects on the fragile nuclei than the vigorous grinding needed to break untreated tissue (D'ALESSIO and TRIM, 1968; MASCARENHAS et al., 1974).

It has been only fairly recently that the nuclear envelopes even from animal cells have been separated from the nuclear contents, thus permitting a chemical characterization of the nuclear envelope itself (STECK, 1972a). Little information is available on the actual chemical composition or precise relationship of the constitutive compounds within the two membranes of the nuclear envelope of animal cells, although recent studies have attacked this problem (KASPER, 1974). No studies of this nature appear to have been done yet on plant cell nuclear membranes.

Detergents, osmotic shock, ultrasound, high salt solutions, and deoxyribonuclease have all been used to break animal cell nuclei (BLOBEL and POTTER, 1966; BEREZNEY et al., 1970; FRANKE et al., 1970; KASHNING and KASPER, 1969; STECK, 1972a). These disruptive treatments are followed by isopycnic density gradient centrifugation and nuclear fragments are obtained at densities of about 1.18 to 1.22. Effective techniques of separating the inner and outer membranes of the

Table 5. Composition of nuclear envelope preparations obtained from various animal sources

Envelope component % (w/w)	Source of nuclei				
	Rat liver			Pig liver	Beef liver
Protein	25–47 [a]	73 [b]	64 [c]	74.8 [d]	70.4 [e]
Phospholipid	27–50	23	23	18.2	22.6
RNA	1–3	3	5	2.8	5.8
DNA	1–0.2	0.16	8	1.2	1.1

[a] ZBARSKY et al. (1969). [b] MONNERON et al. (1972). [c] AGUTTER (1972). [d] FRANKE et al. (1970). [e] BEREZNEY et al. (1972).

nuclear envelope are still lacking. Table 5 summarizes some of the information on the chemical properties of isolated nuclear envelope preparations and illustrate the wide range of the results for membrane preparations obtained from similar sources.

One point that is significant with reference to the structure and permeability of the nuclear envelope concerns whether the envelope pores represent a direct communication between the cytoplasm and the nucleoplasm. Since, at least in *Drosophilia* gland cell nuclei, the pores occupy between 20 and 25% of the total envelope area (LOEWENSTEIN et al., 1965; KLEINIG, 1970) (cf. with tobacco nuclear membranes, Fig. 5), if they represent areas for free diffusion between nucleoplasm and cytoplasm, the electrical resistance of the membrane would be shunted through the pores. Measurements made with microelectrodes of the nuclear membrane resistance of *Drosophilia* and *Chironomus* salivary gland cells yielded values at least three orders of magnitude higher than the resistance calculated ($10^{-3}\,\Omega\,cm^2$) on the assumption that the nuclear membrane pores were filled with a material with a resistivity similar to that of the nucleoplasm and cytoplasm (LOEWENSTEIN and KANNO, 1963). Several enzymes have been identified as being present in the nuclear envelopes of animal cells and frequently similar enzymes are identified with microsomal preparations as well. Among the enzymes found to be present in both membrane systems are NADPH and NADP-cytochrome reductases, aryl hydrolase, and cytochrome b_5 (KASPER, 1974). The biochemical nature of the pore complex is still unknown, consequently, the mechanism involved in the transport of macromolecules between the nucleus and the cytoplasm remains to be determined.

4.3 Plastid Membranes

4.3.1 Morphology

4.3.1.1 General

Substantial morphological diversity is present in the several plastid types found in higher plants. Furthermore, considerable differences in morphology are found when one contrasts higher and lower plant chloroplasts. Plastids are bounded by a double membrane, the plastid envelope. The inner membrane is concentric with the outer membrane, and both are about 60 Å wide. DNA, ribosomes, starch grains, so-called osmiophilic droplets and thylakoids may be found within the stroma or matrix material bounded by the plastid envelope.

4.3.1.2 Proplastids

Proplastids (Fig. 6) are small (about 1 μ in diameter), colorless, undifferentiated plastids that occur in meristematic tissues of higher plants. Because of their small size relatively little is known about their structure. They are bounded by a plastid envelope and have a uniform matrix. Often one may find invaginations of the inner membrane of the envelope protruding into the matrix. Proplastids are difficult to distinguish from other small bodies found in the cell (BOGORAD, 1967).

4.3.1.3 Etioplasts

If a plant is deprived of light, the plastids that develop in the etiolated leaves are called etioplasts (Fig. 7). Etioplasts are quite different from proplastids (cf. Figs. 5 and 6) in that the former have prolamellar bodies. These structures have been extensively studied (WETTSTEIN, 1958; WEHRMEYER, 1965a, b, c; GUNNING, 1965; WEIER and BROWN, 1970). Their exact manner of development is disputed.

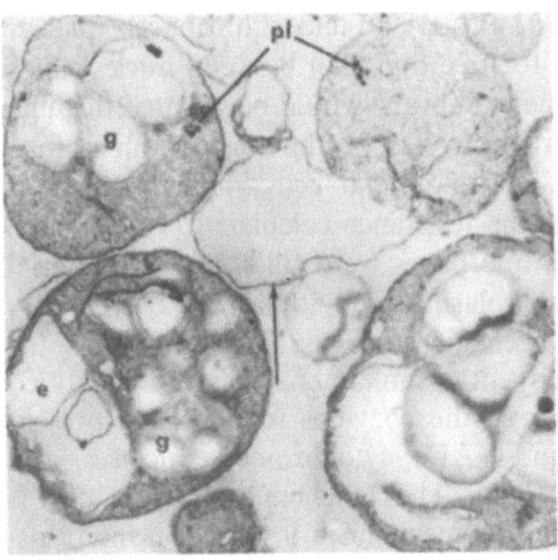

Fig. 6. Electron micrograph of isolated, pelleted proplastids. *g* starch grains, *pl* plastiglobuli, *e* membrane-bounded encalves. Ca. ×13,500. (From THOMSON et al., 1972)

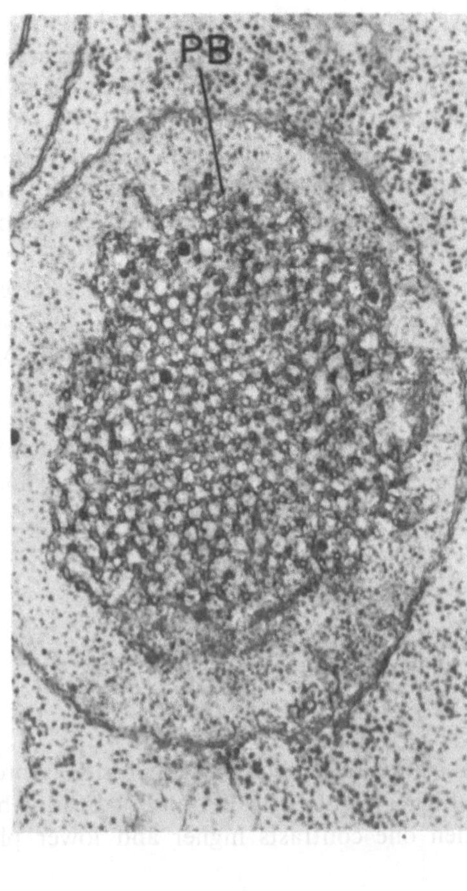

Fig. 7. Electron micrograph of an etioplast ▷ from an etiolated bean seedling. *PB* prolamellar body. Ca. ×41,000. (From NEWCOMB, 1967)

WEIER and BROWN (1970) regard the prolamellar body as an accumulation of membrane products due to the blockage of light reactions which would normally occur during the process of greening.

4.3.1.4 Amyloplasts

Plastids in which most of the organelle volume is taken up by starch are called amyloplasts. These are commonly found in storage tissues (cotyledons, endosperm) and in roots. They are bounded by a double membrane envelope. Few internal membrane structures remain. The properties of the outer envelope have not been characterized.

4.3.1.5 Chromoplasts

Chromoplasts (see Fig. 3) are plastids that contain pigments other than chlorophyll. They are commonly found in fruits, leaves and flower petals and exhibit a diverse morphology (SPURR and HARRIS, 1968). In some plants a highly crystalline structure is found; in others globular osmiophilic structures are observed (HARRIS and SPURR, 1969).

4.3.1.6 The Higher Plant Chloroplasts

The outer membrane of higher plant chloroplasts has a smooth profile; the inner membrane is concentric with the outer and may occasionally show finger-like invaginations. Both membranes are about 60 Å wide. In some plants that have a high photosynthetic capacity and a low CO_2 compensation point (C_4 plants), an interesting variation on inner membrane structure has been reported (SHUMWAY and WEIER, 1967; ROSADO-ALBERIO et al., 1968; LAETSCH, 1968; 1969a, b). The inner membrane has many invaginations which appear as a system of anastomosing tubules which apparently connect with the internal membrane system. The tubules have been named the peripheral reticulum (ROSADO-ALBERIO et al., 1968).

The organization of the internal membranes is complex. Using the terminology of WEIER (1961) Figure 8 shows the structures normally observed in a thin section. This view is deceptively simple. PAOLILLO and co-workers (PAOLILLO and FALK, 1966; PAOLILLO and REIGHARD, 1967; PAOLILLO et al., 1969; PAOLILLO, 1970) have shown that the frets are more like flexible channels that are arranged in a helical manner around grana (Fig. 9).

Chloroplast DNA and ribosomes are found in the stroma region of the chloroplast.

Electron microscopic studies of freeze-etched chloroplast membranes of higher plants have shown that there is only one fracture plane associated with membranes. The fracture produces two matching faces and is through the hydrophobic region of the membrane (DEAMER and BRANTON, 1967; PARK and PFEIFHOFER, 1969). PARK and PFEIFHOFER'S (1969) model of a thylakoid is shown in Figure 10. The external surface A′, is coated with substantial protein but presents little surface relief. B, the adjacent fracture plane has large particles associated with it (up to 175 Å) which show little surface relief and may be the quantasome of PARK

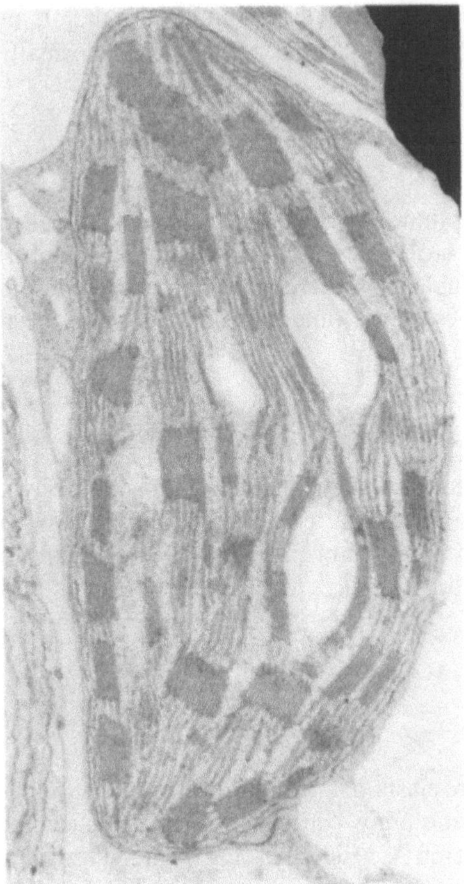

Fig. 8. Electron micrograph of a mesophyll chloroplast from maize. Ca. ×17,000

and BIGGINS (1964). The adjacent fracture plane, C, has small particles associated with it (about 120 Å in diameter). SANE et al. (1970) and others have pointed out that B face particle is found only in partition regions of thylakoids. The C face particles are found only on fret lamellae and the margins of grana stacks.

4.3.2 Chemical Composition of the Chloroplast Envelope

Chloroplasts are probably the most easily isolated and purified of all of the plant cell organelles, yet until recently surprisingly few attempts were made to isolate and characterize chloroplast envelopes. In 1970, MACKENDER and LEECH successfully prepared membrane vesicles from envelopes of isolated *Vicia faba* chloroplasts. These preparations contained both single and double membrane-bound vesicles, but chlorophyll determinations suggested that there was between 8 and 16% contamination by thylakoid membranes. Subsequently, MACKENDER and LEECH (1972, 1974) modified their technique to obtain preparations of higher purity and reported that the chloroplast envelope is characterized by a significantly different lipid composition than that found in the thylakoid lamellae. Although as much as 61% of the total envelope lipid is galactolipid, this is substantially less than the proportion (about 92%) present in the lamellae.

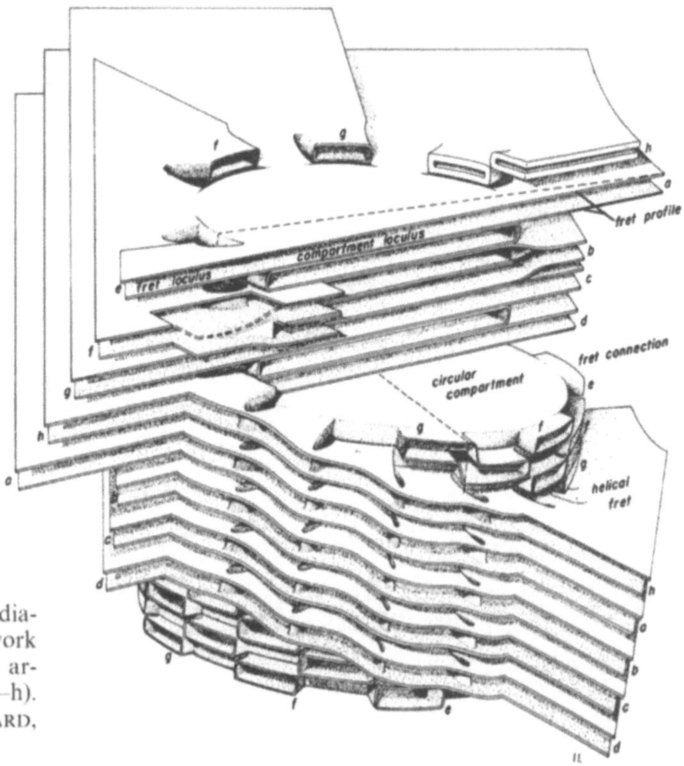

Fig. 9. Three-dimensional diagram of a granum with fretwork connections. Note the helical arrangement of the frets (a–h). (From PAOLILLO and REIGHARD, 1967)

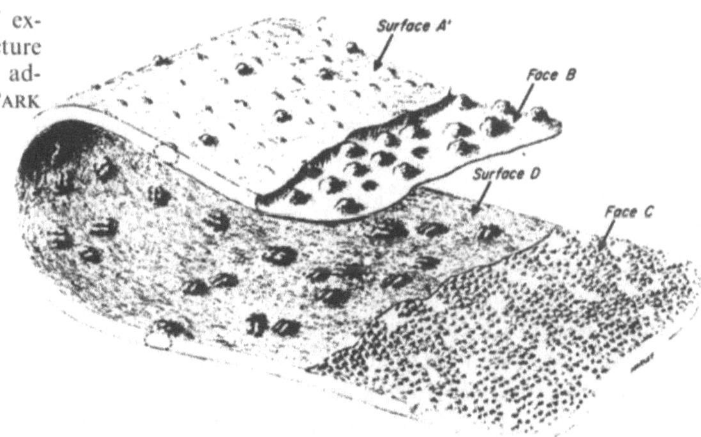

Fig. 10. Thylakoid model. *A'* external surface, *B* adjacent fracture plane, *D* interior surface, *C* adjacent fracture plane. (From PARK and PFEIFHOFER, 1969)

Chloroplast envelope fragments have now been prepared from spinach chloroplasts in three other laboratories (POINCELOT, 1973; POINCELOT and DAY, 1974; RACUSEN and POINCELOT, 1976; DOUCE et al., 1973; HASHIMOTO and MURAKAMI, 1975). In all instances, the envelopes showed lower levels of galactosyl diglycerides and higher levels of phosphatidylcholine than the corresponding thylakoids. In contrast to the situation in mitochondria and microsomal membranes in which phosphatidylethanolamine may represent as much as 33% and 20% of their respective total lipid content (Table 6) (MACKENDER and LEECH, 1974), the chloroplast

Table 6. Phosphilipid composition of mitochondrial membranes and microsomal fractions from different sources (after MOREAU et al., 1974)

Material	PC	PE	DPG	PG	PI
Inner membrane					
Cauliflower [a]	41	37	14	3	5
White potato [b]	37	29	19.5	–	24.5
White potato [c]	33	33	19	5	7
Rat liver [d]	39.2	39.9	15.4	–	3.6
Guinea pig liver [e]	44.5	27.2	21.5	2.2	4.2
Outer membrane					
Cauliflower [a]	42	24	3	10	21
White potato [b]	52.6	25	12.1	–	10.3
White potato [c]	36	64	–	–	–
Rat liver [d]	49.4	34.9	4.2	–	9.2
Guinea pig liver [e]	55.2	25.3	3.2	2.5	13.5
Microsomes					
Cauliflower [a]	50	35	1	8	6
White potato [b]	44.3	17.6	19.0	–	19.1
Rat liver [d]	58.7	25.5	1.6	–	8.2
Guinea pig liver [e]	62.8	18.3	0.5	1.1	13.4

PC phosphatidylcholine, PE phosphatidylethanolamine, DPG diphosphatidylglycerol, PG phosphatidylglycerol, PI phosphatidylinositol.
[a] MOREAU et al. (1974). [b] MEUNIER and MAZLIAK (1972). [c] MCCARTY et al. (1973). [d] COLBEAU et al. (1971). [e] PARSONS et al. (1967).

envelope appears to lack or to have only very small amounts of phosphatidylethanolamine (MACKENDER and LEECH, 1974; DOUCE et al., 1973; HASHIMOTO and MURAKAMI, 1975; POINCELOT, 1973). Chlorophyll was absent, and the carotenoid composition of the envelope differed drastically from that of thylakoid membranes (DOUCE and BENSON, 1974).

Between 60 and 90% of the total leaf hexoseamine has been reported to be associated with chloroplast envelope preparations from maize, sunflower, and spinach leaves. However, the possible errors involved in arriving at these values are quite significant, and estimates of as much as 18% of the total chloroplast protein being located in these envelope preparations are surprisingly high (RACUSEN and POINCELOT, 1976). In a subsequent paper (POINCELOT and DAY, 1976) chloroplast envelopes isolated from spinach, sunflower and maize were reported to contain between 2 and 4% of the total chloroplast protein.

A further characteristic of the chloroplast envelope that is undoubtedly related to solute transport is the presence of nonlatent, N, N′, dicyclohexylcarbodiimide insensitive, Mg^{2+}-dependent ATPase (SABNIS et al., 1970; DOUCE et al., 1973; POINCELOT, 1973). Histochemical tests indicate that this enzyme is located between the membranes of the envelope of pea tendril chloroplasts (SABNIS et al., 1970). It is probably more firmly attached to the inner membrane of the envelope, although a significant portion of the enzyme is lost if the envelope membranes are separated during isolation (POINCELOT and DAY, 1974).

Differences in the structure and or chemical composition of the two membranes are indicated by the fact that the outer but not the inner membrane of the chloroplast envelope is nonspecifically permeable to sucrose (HELDT and RAPLEY, 1970a). Our understanding of the permeability properties of the chloroplast envelope will be significantly increased when each of the envelope membranes is separately isolated and chemically characterized. This has not yet been accomplished.

Transport studies indicate that the inner membrane is the site of specific translocators for phosphate (HELDT and RAPLEY, 1970b; WERDAN and HELDT, 1972), dicarboxylate (HELDT and RAPLEY, 1970b; HELDT et al., 1972), amino acids (NOBEL and WANG, 1970; NOBEL and CHEUNG, 1972; but see GIMMLER et al., 1974), and adenine nucleotides (HELDT, 1969; STROTMANN and BERGER, 1969). The Mg^{2+}-stimulated ATPase may be involved in the transport of ions and other solutes across the inner membrane, although no membrane components that specifically bind solutes and thus might act as permeases or translocators have been isolated from or identified as being part of the chloroplast envelope. The recent successful isolation of an envelope fraction enriched in double membrane-bound vesicles should be useful in clarifying some of the perplexing problems of solute transport across the chloroplast envelope (POINCELOT, 1975; POINCELOT and DAY, 1976).

Possible changes in chemical composition and permeability properties of the chloroplast envelope during plastid development have not yet been investigated with isolated envelopes, but COCKBURN and WELLBURN (1974) present evidence that suggest that selective changes in the permeability of plastid envelopes may occur during plastid development.

A further complicating feature of the chloroplast envelope in reference to transport is that, in contrast to the plasmalemma, endoplasmic reticulum, and microbodies, chloroplasts, like nuclei and mitochondria, are double membrane-bound structures. Although the outer membrane of the envelope is permeable to many solutes, it probably represents an important barrier to the movement of high molecular weight molecules such as peptides. There is abundant evidence, nevertheless, that some of the thylakoid proteins (THORNBER, 1975) as well as chloroplast enzymes or enzyme subunits (BOULTER et al., 1972) are synthesized on 80S cytoplasmic ribosomes and transported into the plastid. Essentially nothing is known about the mechanism of movement of peptides across the chloroplast envelope. It is tempting to speculate that during the synthesis of polypeptides destined to enter the chloroplast, 80S ribosomes are bound to the outer chloroplast membrane in a manner similar to the binding of ribosomes to the membrane of the rough endoplasmic reticulum (PALADE, 1975), to the outer membrane of yeast mitochondria (KELLEMS et al., 1975), to the inner surface of the plasmalemma during protein excretion by microorganisms (LAMPEN, 1974), or to the single membrane of protein bodies during protein synthesis (BURR and BURR, 1976). The polypeptide that is being synthesized might then be extruded vectorially through the envelope before becoming folded. In the case of the chloroplast, any synthesized polypeptide would have to pass through both membranes as well as the intermembrane area before it could be released into the chloroplast.

Transmission electron micrographs of higher plant chloroplasts usually do not show cytoplasmic ribosomes specifically associated with chloroplast envelopes. Developing chloroplasts (transformed etioplasts) which are observed at a time when protein synthesis is rapidly occurring, about six to 8 h after etioplasts have been exposed to light, occasionally appear to have cytoplasmic ribosomes associated with the outer surface of the chloroplast envelope (STOCKING and LIN, unpublished)

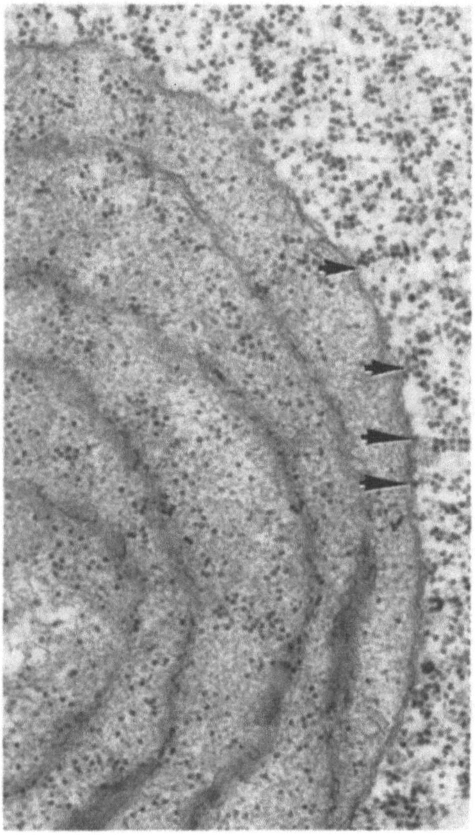

Fig. 11. Electron micrograph of a section of a developing bean leaf. A 7-day old etiolated bean plant was transferred to the light 5.5 h before the leaf section was fixed. Note (*black arrows*) the arrangement of several rows of cytoplasmic ribosomes in close association with and perpendicular to the plastid envelope

(Fig. 11). However, whether this relationship is related to peptide synthesis and transenvelope transport is not yet known.

The postulated extremely rapid shuttling of metabolites between mesophyll and bundle sheath chloroplasts in C_4 plants is another area in which the role of the chloroplast envelope is poorly understood. It has been suggested that the peripheral reticulum, which is frequently extensively developed in chloroplasts of C_4 plants may be involved in solute transport (ROSADO-ALBERIO et al., 1968) but no convincing data is yet available on this subject.

4.4 Mitochondrial Membranes

4.4.1 Morphology

The mitochondrion is also an envelope-bounded organelle. Its morphology was initially described by PALADE (1952; 1953) and SJÖSTRAND (1953). The outer membrane profile is usually quite smooth; the inner membrane, however, is thrown into a series of folds called cristae which project into a central matrix (see Fig. 3). Both membranes are about 60 Å in thickness. The compartment or space between the inner and outer membranes is referred to as the intermembrane space and is continuous with the intercristae space. The central matrix may be dense or

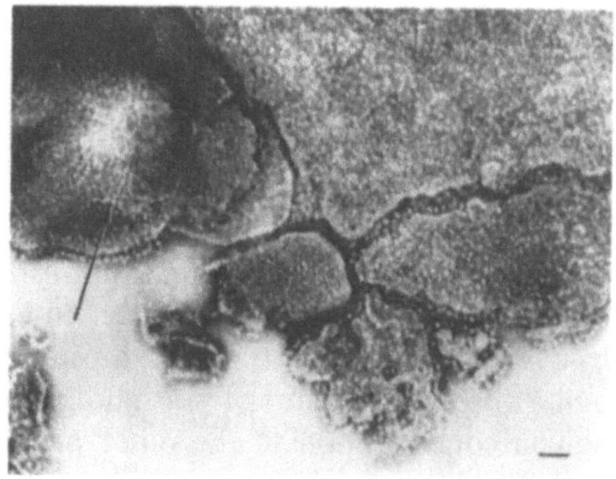

Fig. 12. Fragments of beef heart mitochondrial cristae showing elementary particle arrangement. The arrow points out a linear array of particles on the membrane surface. Marker 500 Å. (From STILES et al., 1968)

not, often contains dense bodies known as intramitochondrial dense granules, and houses mitochondrial DNA and ribosomes.

Isolated mitochondrial membranes are often examined as negatively stained preparations (Fig. 12). The inner membrane, when prepared and examined in this manner, reveals stalked, spherical particles about 90 Å in diameter (FERNÁNDEZ-MORÁN, 1962). These particles were initially considered to be the sites of the respiratory enzymes (BLAIR et al., 1963; FERNÁNDEZ-MORÁN, 1963) but this viewpoint was modified when it was pointed out that the particles were too small to house the requisite enzymes (FERNÁNDEZ-MORÁN et al., 1964). RACKER and his colleagues in a series of studies (RACKER et al., 1964; RACKER et al., 1965; KAGAWA and RACKER, 1966) eventually isolated the particles and showed them to be the coupling factor for phosphorylation.

4.4.2 Chemical Composition

Several methods have been used in the separation of the outer and inner membranes of the mitochondrial envelope. Probably the most effective and widely used technique, first described for rat liver and mung bean mitochondria by PARSONS and coworkers, involves swelling isolated and purified mitochondria in hypotonic phosphate solution (PARSONS et al., 1966; 1967). This results in a rupture of the outer envelope which can then be separated from the inner membrane by centrifugation through a sucrose density gradient. It should be remembered that the outer membrane, but not the inner membrane, is readily permeable to certain ions and small molecules such as sugars, nucleotides, and salts (WERKHEISER and BARTLEY, 1957; PFAFF et al., 1968). The relatively high permeability of the outer membrane may be related to its high phospholipid content. Cardiolipin, which is restricted to the inner membrane of rat liver mitochondria reduces the permeability of phospholipid vesicles to anions (CHAPPELL, 1968).

PARSON'S method of isolating the inner and outer membranes of mitochondrial envelopes has been modified by several workers (RACKER, 1972). Unfortunately, although rather extensive studies have been made on animal mitochondrial envelopes, until recently few attempts have been made to isolate and characterize the mitochondrial membranes of plant cells. However, CASSADY and WAGNER (1971), MEUNIER et al. (1971), BANDLOW (1972), MOREAU and LANCE (1972) and MOREAU et al. (1974) have separated the two membranes from plant mitochondria. Table 6 lists some of the differences in lipid composition between the inner and outer mitochondrial membranes from plant as well as animal cells. Plant mitochondria appear to be similar to mammalian mitochondria in their utilization of glycerolphosphatides as membrane components (BENSON and STRICKLAND, 1960; BENSON, 1964; MOREAU et al., 1974).

The two major phospholipids of most plant mitochondria are phosphatidylcholine and phosphatidylethanolamine. In cauliflower (*Brassica oleraceae*) mitochondria, the outer, in contrast to the inner, membrane is characterized by having higher absolute amounts of phospholipids, a higher phosphatidylinositol level (also high in concentration in the outer membranes of animal mitochondria), a high level of phosphatidyl glycerol, little diphosphatidyl glycerol and most of the fatty acids are saturated. In contrast, the inner membrane of cauliflower contains most of the envelope diphosphatidyl glycerol and the fatty acids are generally unsaturated. It has been suggested that the relative rigidity of the outer membrane and the plasticity of the inner membrane, which can undergo passive and active swelling and contraction, may be related to the degree of saturation of fatty acids in each membrane. The high degree of saturation of the fatty acids may lend rigidity to the outer membrane (MOREAU et al., 1974).

Analyses of the two mitochondrial membranes from animal cells (rat liver) show that, in general, they are distinctly different in structure, enzyme content, and lipid composition. Detailed information on the localization of over 50 enzymes on the outer membrane, intermembrane space, inner membrane, and the matrix have been published (ERNSTER and KUYLENSTIERNA, 1970). Earlier controversies in characterizing the two membranes of animal mitochondria on the basis of specific proteins that they contain (PARSONS et al., 1966; GREEN and PERDUE, 1966; SCHNAITMAN and GREENAWALT, 1968) largely have been resolved (ERNSTER and KUYLENSTIERNA, 1970).

The problem of the transport of peptides across the chloroplast envelope has already been discussed (Sect. 4.3.2). Similarly, a central problem with reference to mitochondrial structure and transport concerns the method by which peptides synthesized in the cytoplasm are transported across the two membranes (SCHATZ and MASON, 1974). There have been some early suggestions that polypeptides might move as phospholipid complexes (SIEGEL and SISLER, 1965; KADENBACH, 1969). However, if yeast mitochondria are isolated in the presence of Mg^{2+}, they retain bound cytoplasmic ribosomes that are removed with 2 μM EDTA. Thus, vectorial translocation during peptide synthesis is a possibility (KELLEMS and BUTOW, 1972; KELLEMS et al., 1975).

Experiments by PRESSMAN (1970) and CHAPPELL (1968) clearly indicate that there are specific channels in mitochondrial envelopes for the translocation of cations such as K^{+}, Mg^{2+}, Ca^{2+}, and anions such as adenine nucleotide phosphate and the Krebs intermediates (see Sect. 5.2). The way in which such specific channels relate to the two different membranes that make up the mitochondria envelope is still not resolved.

4.5 Other Membranes in Plant Cells

4.5.1 The Microbody Membrane

Microbodies (peroxisomes, glyoxisomes) are organelles showing circular profiles in section, ranging from 0.2–1.5 μ in diameter (Fig. 13). They are bounded by a single membrane rather than a double membrane envelope, have a granular matrix, and often contain a crystalline inclusion. See TOLBERT (1971b) for a recent review.

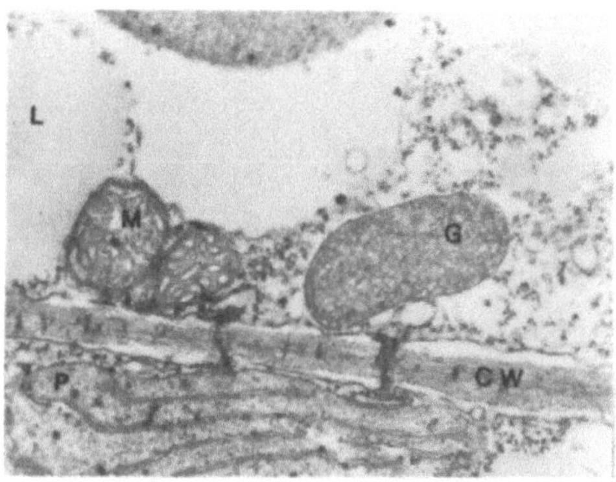

Fig. 13

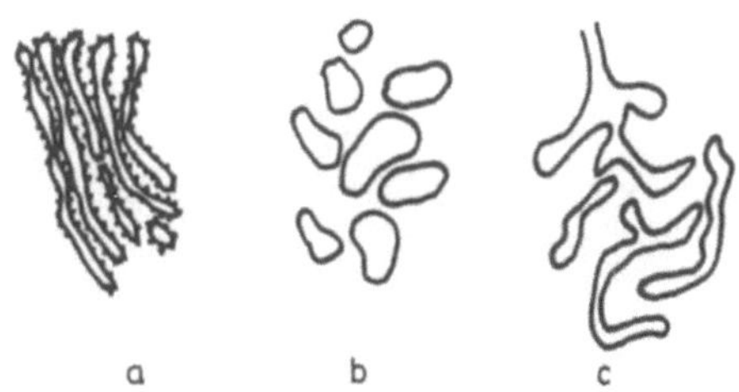

Fig. 14

Fig. 13. Electron micrograph of a microbody (glyoxysome) from a cucumber cotyledon. *G* glyoxysome, *CW* cell wall, *M* mitochondrion, *L* lipid body, *P* plastid. Ca. × 19,500. (From TRELEASE et al., 1974)

Fig. 14a–c. Types of endoplasmic reticulum. (a) cisternal (b) vesicular (c) tubular. (Adapted from COHN, 1969)

4.5.2 The Endoplasmic Reticulum Membranes

In its most familiar form, the endoplasmic reticulum (ER) consists of vast amounts of flattened or sheet-like sacs. The bounding membranes of the ER are unit membranes about 50–60 Å across (see Figs. 3, 4). The cavities between membranes are cisternae. Some refer to this type of ER as cisternal ER (FAWCETT, 1961). Tubular and vesicular ER formation may also be seen in some tissues (HELENDER, 1964; CHRISTENSEN and FAWCETT, 1961). Figures 14a, b, c, illustrate these ER types. If ribosomes are associated with ER, it is known as rough or granular ER. Ribosome-free ER is known as smooth or agranular ER. Microsomes, a biochemist's product, are fragments of rough ER.

4.5.3 The Golgi Apparatus Membranes

The classical electron micrographic view of a dictyosome is shown in Figure 15 (see Fig. 4 for the freeze-etch image). According to MORRÉ et al. (1971a), three levels of organization are apparent: the cisterna, the dictysome, and the Golgi apparatus. Groups of cisternae make up a dictyosome; groups of dictyosomes

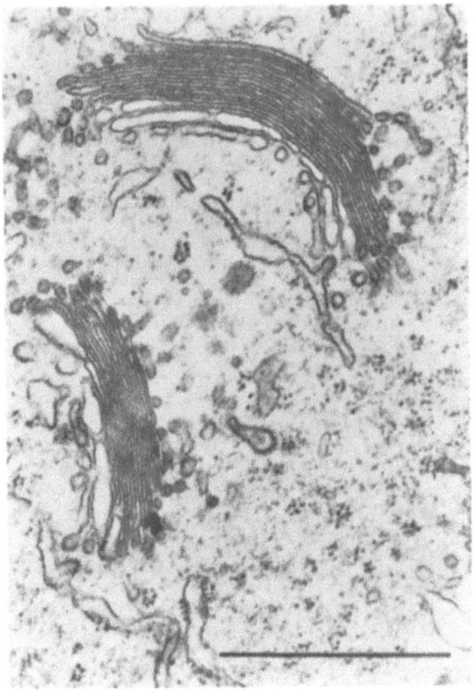

Fig. 15. Two dictyosomes from *Helix aspersa*. Marker 1 μ. Ca. ×30,000. (From WHALEY et al., 1971)

acting in functional concert are called a Golgi apparatus. Many, however, refer to individual dictyosomes as Golgi or a Golgi apparatus.

Dictyosomal cisternae are flattened vesicles with membranes about 60 Å wide. They lack ribosomes and often have a system of peripheral tubules 300–500 Å in diameter which lends a fenestrated appearance to the isolated organelle (Fig. 16). Secretory vesicles are attached to the tubules at the periphery. The number of cisternae comprising a dictyosome varies but is usually in the range of 5–8 (MORRÉ et al., 1971 b). The space between cisternae is about 100–150 Å. A layer of parallel structures called intercisternal elements 70–80 Å in diameter have been reported to occur in the intercisternal regions of dictyosomes (MOLLENHAUER, 1965). Examination of the cytoplasm near a dictyosome shows it to be differentiated – to be a zone of exclusion where ribosomes, mitochondria, and plastids are not allowed.

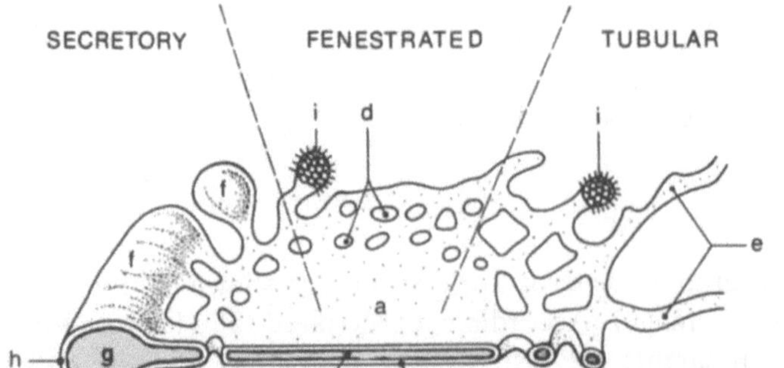

Fig. 16. Diagram of dictyosome. *a* central plate region, *b* cisternal membrane, *c* cisternal lumen, *d* fenestrae, *e* peripheral tubules, *f* secretory vesicles, *g* vesicle lumen, *h* vesicle membrane, *i* coated vesicle. (From MORRÉ et al., 1971 a)

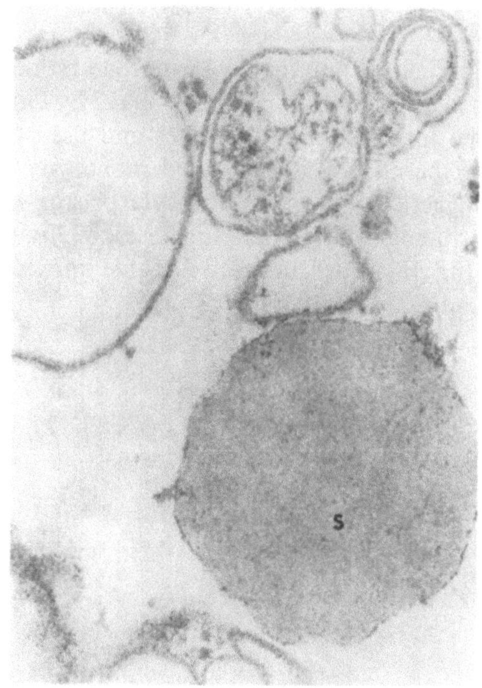

Fig. 17. Electron micrograph of a mature spherosome. Note the "half membrane" as compared to the usual unit membranes seen elsewhere in the picture. *S* spherosome. Ca. ×68,000. (From SCHWARZENBACH, 1971)

4.5.4 The Vacuolar Membrane

Fully developed plant cells usually have a large central vacuole that is bounded by a single membrane, the vacuolar membrane (tonoplast). As is the case with plasmalemma (Sect. 4.1.) a sideness for the vacuolar membrane is seen in tissues prepared in osmium tetroxide, i.e., the cytoplasmic side appears more dense (see Fig. 3).

4.5.5 The Spherosome Membrane

Organelles that are regarded as the oil storage depots of plant cells, spherosomes originate from the ER (FREY-WYSSLING et al., 1963) and show a distinct developmental pattern (SCHWARZENBACH, 1971). Prospherosomes, single unit membrane bounded organelles, begin to accumulate material between the dense layers of their bounding membranes. Continued accumulation of material (lipid?) results in obliteration of the inner portion of the membrane leaving a "half membrane" bounded organelle at maturity (Fig. 17).

5. Membrane Models

Over the years, several models for membrane structure at the molecular level have evolved. These models have been the subject of recent exhaustive reviews (KORN, 1969; STOECKENIUS and ENGELMAN, 1969; HENDLER, 1971; ROTHFIELD, 1971; GREEN, 1972; WALLACH, 1972).

5.1 The Lipid Bilayer Model

The most enduring of membrane models is the lipid bilayer model. This model has its origin in experiments made by GORTER and GRENDEL (1925). They extracted the lipids from erythrocytes and used a Langmuir trough to obtain an estimate of the ratio of area of lipid monolayer to erythrocyte surface area. Their results suggested that the erythrocyte plasma membrane was two lipid molecules thick; a bimolecular leaflet. In the early 30s, a protein layer was added to this model (Fig. 18) which formally became known as the Danielli-Davson model (DANIELLI and DAVSON, 1935).

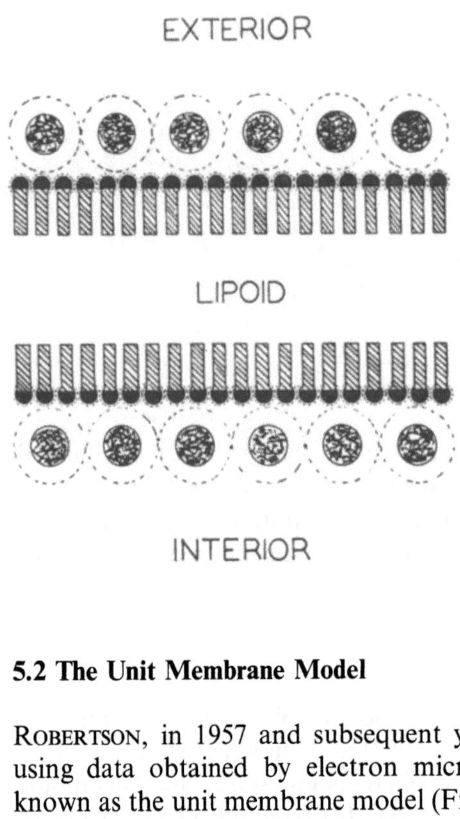

Fig. 18. The Danielli-Davson membrane model. A lipid bilayer interior with associated protein. (From DANIELLI and DAVSON, 1935)

5.2 The Unit Membrane Model

ROBERTSON, in 1957 and subsequent years, modified the Danielli-Davson model using data obtained by electron microscopy. The modified model came to be known as the unit membrane model (Fig. 19). This model enjoyed almost universal acceptance until the early 1960s.

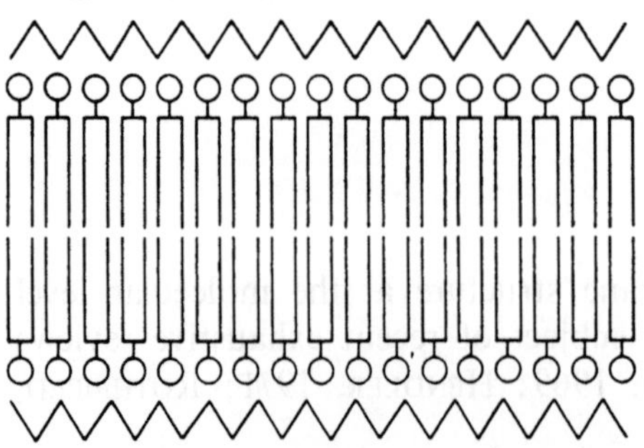

Fig. 19. The Robertson unit membrane model. A lipid bilayer interior (*tuning forks*) covered by non-lipid material (*zig-zag lines*). (From ROBERTSON, 1959)

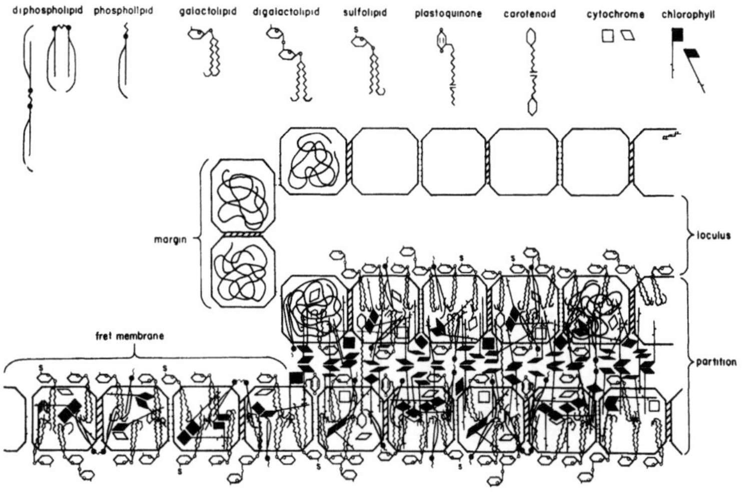

Fig. 20. The Weier-Benson chloroplast membrane model. Membranes are considered as sheets of lipoprotein subunits. Lipids are external and associated with coiled proteins as shown. Partitions are regions where two thylakoids are appressed. A loculus is a space within a granal membrane. A margin is the edge of a granal membrane. Fret membranes are intergranal membranes. (From WEIER and BENSON, 1967)

5.3 Subunit Models

In 1953, FREY-WYSSLING and STEINMANN reported a subunit structure for thylakoid membranes based on electron microscopic observations of metal shadowed preparations. PARK and PON (1961) also reported observing a subunit structure for thylakoid membranes, but their measurements indicated that the subunits were larger than FREY-WYSSLING and STEINMANN had observed. PARK (1962) gave these subunit particles the name "quantasomes". High resolution electron microscopy done later by PARK and BIGGINS (1964) revealed that each quantasome consisted of four 90 Å subunits.

Several subunit models have been proposed. SJÖSTRAND (1963) working with mouse kidney tissue fixed in potassium permanganate, showed that mitochondrial membranes exhibited subunit structure.

In 1967, WEIER and BENSON published micrographs of leaf tissue fixed in permanganate-glutaraldehyde that showed a subunit structure for thylakoid membranes. WEIER and BENSON postulated that the subunits were identical with PARK's quantasome subunits and advanced a complex model (Fig. 20) which exemplified a lipid coated protein subunit model where the hydrophobic regions of the lipid are deeply embedded in the protein particles. In those regions where the thylakoid membranes are appressed, WEIER and BENSON placed chlorophyll. The phytol groups are embedded in the protein; the porphyrin rings remain associated with the more exterior hydrophilic regions of lipids.

KREUTZ and MENKE (1962) and KREUTZ (1963a, 1963b, 1964) advanced a model for thylakoid membranes (Fig. 21) that can be thought of as an extreme variation on the Danielli-Davson model. Their model, derived from x-ray diffraction studies of chloroplast membranes, has a lipid layer only one molecule thick,

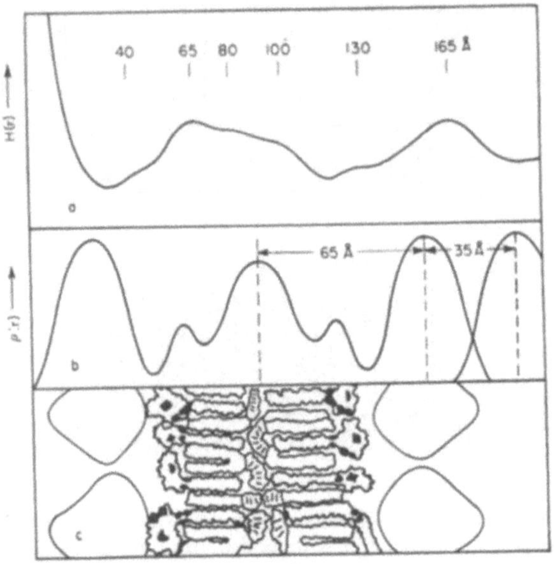

Fig. 21a–c. The Kreutz-Menke chloroplast membrane model. (a) Characteristic function $H(r)$ versus distance in Å along a plane normal to a thylakoid. (b) Electron density distribution $\rho(r)$ versus distance along a plane normal to a thylakoid (two membranes). (c) Membrane model derived from (b). *From left to right*: protein–porphyrin ring monolayer–lipid layer–water–lipid layer porphyrin ring monolayer–protein. (From KREUTZ, 1966). Electron density distribution functions are obtained by Fourier inversions and require that the square roots of measured x-ray scattering intensities be known. These values may be positive or negative and in any case are not easily determined, KREUTZ and co-workers empirically calculated these scattering functions using assumed models until agreement with their experimental scattering data was obtained. The amplitude signs found in this manner were then used to calculate an electron density distribution as is seen in (b). To validate their work further, KREUTZ and co-workers calculated the characterisitc function of their experimental system (a) which in this case is a multiplicity function of the distance distribution in a one dimensional point arrangement. From the electron density distribution shown in (b), multiplicity of occurring periods can be predicted and should agree with those shown by the characteristic function (a)

a particulate protein layer, and has the prophyrin rings of chlorophyll in a layer between the lipid and protein. The Kreutz-Menke model, while compatible with freeze-etch studies and metal shadowed preparations, differs considerably from models proposed by workers using these latter techniques as a basis for data for their models. Recently RADUNZ (1971) using an antibody to chlorophyll a has provided some support for the Kreutz-Menke model.

MÜHLETHALER (1966), BRANTON and PARK (1967), and ARNTZEN et al. (1969) have proposed rather similar membrane models based on data obtained by the electron microscopic examination of freeze-etch replicas of chloroplast membranes. Figure 22 is a diagram of Mühlethaler's model and shows the membrane as a sheet of lipid with proteinaceous particles inserted partially or entirely into the lipid matrix. BRANTON (1969) has suggested that the lipid matrix of his model might be the Danielli-Davson type greatly modified by the presence of protein particles.

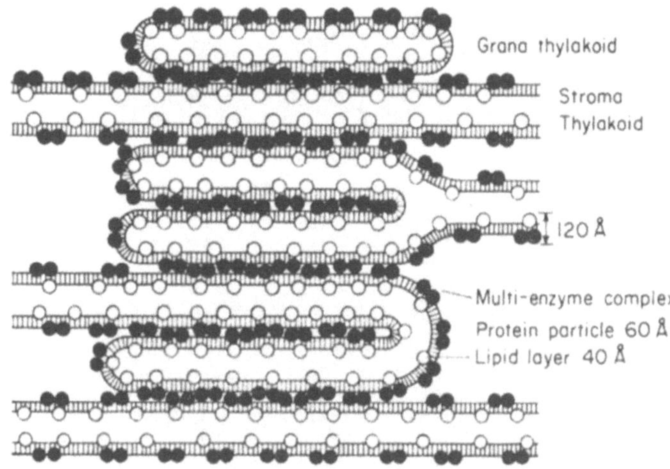

Fig. 22. The Mühlethaler chloroplast membrane model. A granum showing thylakoid structure. The dark particles are attached to the outer membrane surface; the white particles to the inner membrane surface. (From MÜHLETHALER, 1966)

5.4 The Lipid-Globular Protein Mosaic Model

WALLACH and ZAHLER (1966) and LENARD and SINGER (1966) have independently arrived at a very similar model for membrane structure based in part on thermodynamic considerations and in part on experimental evidence obtained by use of biophysical probes such as optical rotatory dispersion and circular dichroism (Table 2). SINGER has been a vocal champion of his model, calling it the lipid-globular protein mosaic (fluid-mosaic) model. Figure 23 is a diagram of the model and reveals its similarity to the models proposed by MÜHLETHALER (1966), BRANTON and PARK (1967), and ARNTZEN et al. (1969).

Several arguments have been advanced for dismissing the Danielli-Davson-Robertson model. If one assumes that the steady-state structure of a membrane is the structure with the lowest free energy, then the Danielli-Davson-Robertson model is not acceptable because it is not an arrangement that would lead to the lowest free energy state for a membrane in an aqueous environment. SINGER (1971) points out that the two most serious thermodynamic problems with the Danielli-Davson-Robertson model are (1) that the nonpolar residues of the membrane proteins are exposed to rather than protected from water and (2) the ionic groups of the lipids and proteins are largely in contact with each other rather than water. The burying of these ionic groups in such a manner is expensive in terms of free energy.

The Danielli-Davson-Robertson model postulates an identical structure for all membranes. One cannot readily explain the functional diversity exhibited by membranes by means of this model. However, it has been judged as acceptable in part because membranes fixed with osmium tetroxide often show a trilaminar image when sectioned and viewed with an electron microscope. The dark lines were thought to represent osmium bound to the polar ends of phospholipid. KORN (1966a, b, 1967) has shown that osmium does not bind to phospholipid and further,

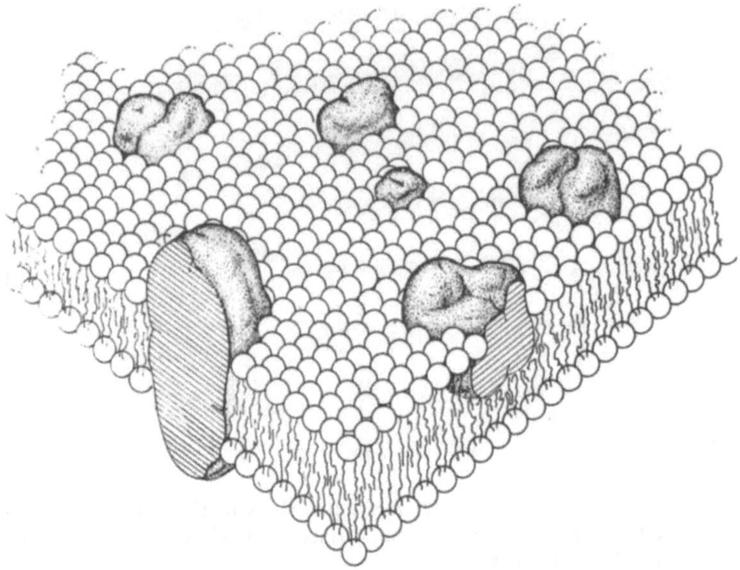

Fig. 23. The Singer lipid-globular protein mosaic membrane model. A lipid bilayer with various proteins embedded in or projecting through the bilayer. (From LENARD and SINGER, 1966)

that if one removes the lipid from membranes and then fixes with osmium tetroxide, one still obtains a trilaminar image in sectioned material. BRETSCHER (1973) mentions that osmium tetroxide may oxidize amino groups which could account for the trilaminar image seen in lipid-free membranes.

The subunit model as exemplified by the Benson model is thermodynamically more likely than the Danielli-Davson-Robertson model but has too high free energy to comprise the general organizational pattern for membranes (SINGER, 1971).

STOECKENIUS and ENGELMAN (1969) objected to a subunit model for the following reasons: (1) membranes are not uniform in their composition, (2) self assembly from "building blocks" has not been demonstrated, and (3) difficulties in explaining growth and differentiation with the development of new functions. ROBERTSON (1966) has shown that if a membrane is tilted with respect to the electron beam, surface features of the membrane begin to contribute to the image and may present a pseudo-subunit structure.

Thermodynamically, the lipid-globular protein mosaic model appears to be the most plausible model. There is also a large body of freeze-etch literature on membrane structure that can be used to support the lipid-globular protein mosaic model.

Three principal polypeptide conformations are recognized for protein molecules: the righthanded α-helix, the antiparallel pleated sheet or β-conformation and the aperiodic or random coil form. The optical rotatory dispersion and circular dichroism spectra of model polypeptides possessing these conformations are markedly different. The Danielli-Davson-Robertson model leads to expectation of a membrane protein mostly in the β-conformation. A major argument of SINGER (1971) against this model is found in his interpretation of circular dichroism spectra: he fails to find evidence for significant β-conformation in membrane protein.

CRAIG (1968) has shown that optical rotatory dispersion analysis of cyclic peptide can show helical curves. CLOULES and BJÖRKLUND (1970) and WALLACH et al. (1969) have shown that considerable β strucure could be present in a membrane and not be detected by either optical rotatory dispersion or circular dichroism analyses.

Experiments have confirmed several aspects of the "fluid-mosaic" model (OSEROFF et al., 1973). It appears that in most membranes the majority of the lipid is in the form of a fluid bilayer matrix and that there are at least two general classes of membrane associated proteins: (1) proteins that are loosely bound, the periplasmic proteins of gram negative bacteria (OXENDER, 1971; OXENDER and QUAY, 1975; HEPPEL, 1971; PARDEE, 1968) or other peripheral proteins (SINGER, 1974) and (2) integral proteins (Sect. 5.4) that are deeply or transversely imbedded in the lipid matrix. In addition, there is little doubt that at least some areas in some membranes are fluid in nature and lateral movement of lipids and proteins in the plane of the membrane may occur (SINGER, 1974). However, EDIDIN (1972) pointed out that tissue architecture and cytochemistry suggest that mechanisms must exist that act to restrict the mobility of membrane molecules. For example, saccharidases appear to be present at the luminal ends and sodium pump enzymes are localized at the basal ends of intestinal epithelial cells. This indicates that these proteins are restricted in their lateral movement in the plane of the membrane.

6. Membrane Structure and Solute Transport

Probably the most universal feature of biological membranes is the presence of the lipid bilayer structure (Sect. 5). The evidence is convincing that active transport across most biological membranes is mediated by proteins (PARDEE, 1968; OSEROFF et al., 1973). This is logical since the transport of highly hydrophilic solutes such as ions, sugars, and amino acids must be intimately associated with some system that is capable of bridging this lipophilic barrier. Many different mechanisms have been proposed and tested. In addition to the process of pinocytosis which has been well documented, particularly for some animal cells, most proposals for transport mechanisms can be considered under one of three different categories: (1) the existence of hydrophilic pores in the membranes, (2) the presence of carriers which are capable of rotating or moving through the membranes, (3) the presence of bridging proteins that can act as channels through which hydrophilic solutes may be moved. (See also Vol. 2, Part A: Chap. 1.)

6.1 Membranes with Hydrophilic Pores

Early modifications of the Danielli-Davson model of membrane structure (Sect. 5.1) envisioned a lipid-protein mosaic structure in which the lipid bilayer was bridged by hydrophilic gaps or "pores" protected by proteins whose non-polar amino acids were associated with the polar heads of the lipids (DANIELLI, 1975). This structure, which is not considered thermodynamically likely by SINGER (1974), would provide a channel through which polar solutes could move.

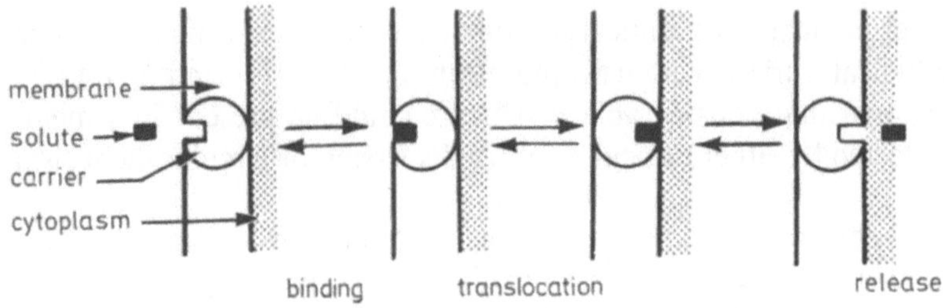

Fig. 24. A schematic representation of the carrier hypothesis of solute transport across a membrane

6.2 Carriers and Ionophores

As early as 1933, OSTERHOUT suggested that small molecular weight ion carriers were involved in solute transport across membranes. Since then, there have been many suggestions concerning the possible mechanism of such carrier action. The discovery of low molecular weight (ca. 500 to 1,500) compounds, ionophores, such as valinomycin, nigericin, cyclohexyl ester, etc. capable of forming lipid-soluble complexes with ions and thus being vehicles for cation transport across lipid barriers (PRESSMAN, 1965; PRESSMAN and DE GUZMAN, 1975), provided a tool by which some of the mechanisms of transport in biological membranes could be studied (SHAMOO, 1975). There are increasingly more numerous reports of the isolation and characterization of components of biological membranes that have demonstrable ionophoretic properties, e.g., cardiolipin, lecithin, free fatty acids (BLONDIN, 1975; GREEN, 1975; SHAMOO and RYAN, 1975; URRY et al., 1975). (See also Vol. 2, Part B: Chaps. 3.2, p. 70, and 7.7.5, p. 326.)

One of the most common models of the action of a carrier is diagrammed in Figure 24. In this model the carrier, which may extend through the membrane, is represented as chelating with the solute at one surface of the membrane. In the second stage, a conformational change is postulated in which the carrier rotates in the membrane bringing the solute-carrier site to the opposite surface of the membrane. Finally, the solute is released and the carrier is freed to return to its original receptive configuration.

This model has been questioned on thermodynamic grounds (SINGER, 1974). It is pointed out that protein, the postulated carrier, would have its hydrophilic groups extending into the aqueous environment in contact with the membrane. SINGER contends that an amphipathic molecule, with polar regions extending to the surface of the membrane and with the non-polar regions anchored in the lipophilic region of the lipid bilayer, would not logically rotate in the membrane. The energy necessary for such a rotation of a carrier protein would be too great (SINGER and NICOLSON, 1972; SINGER, 1975).

6.3 Bridging Proteins Acting as Channels through Membranes

An alternate model proposed by SINGER (1974, 1975) is that those integral proteins that bridge the entire membrane (Sect. 4.1) may act as channels through which the ligand is translocated. In this model, an extension of the "fluid-mosaic" model, the integral proteins that are associated with transport are visualized as specific aggregates of subunits which span the membrane and associate in such a way that a continuous hydrophilic channel traverses the membrane. An example of such a central hydrophilic pore is the pore in the hemoglobin molecule (SINGER, 1972). This hypothesis further postulates that there is a specific binding site on the surface of one end of the pore. Once a solute is bound to this site, rearrangement in the protein subunits could result in an exposure of a binding site in the pore near the other surface of the membrane. Thus solute transport would involve three phases: (1) solute recognition (solute binding), (2) transmembrane transport (subunit rearrangement), and (3) energy coupling (associated with active transport and accumulation) (SIMONI, 1972; BOOS, 1974).

SINGER extends this model to include the extrinsic proteins that may be associated with membrane transport. These highly soluble, loosely membrane associated proteins (e.g., the periplasmic proteins of gram negative bacteria) bind specific ligands, thus they would be involved in the solute recognition step. Subsequently, it is postulated that they could attach to the hydrophilic ends of specific integral proteins which are at the surface of the membrane. In this model, the binding protein does not itself cross the membrane but does account for the specificity of transport (Fig. 25).

It is unfortunate that the evidence for this type of transport across biomembranes is derived almost entirely from mammalian erythrocyte, mitochondria, and bacterial membranes. The existence of specific binding sites on plant cell membranes is well documented by kinetic studies, particularly with reference to salt absorption by roots (EPSTEIN, 1971), and mitochondria and chloroplasts (HELDT et al., 1972). However specific binding proteins have not yet been isolated and characterized from higher plant membranes. See however, the report by GREEN (1975) of the possible isolation of nonphospholipid ionophores from chloroplast membranes.

Since both mitochondria and chloroplasts (as well as nuclei) are organelles with double membrane envelopes, and since the outer envelopes of mitochondria and chloroplasts are permeable to small molecules, specific binding sites (or extrinsic binding proteins) involved in the inward movement of solutes across these envelopes should be located on the outer face of the inner membrane or in the intermembrane space.

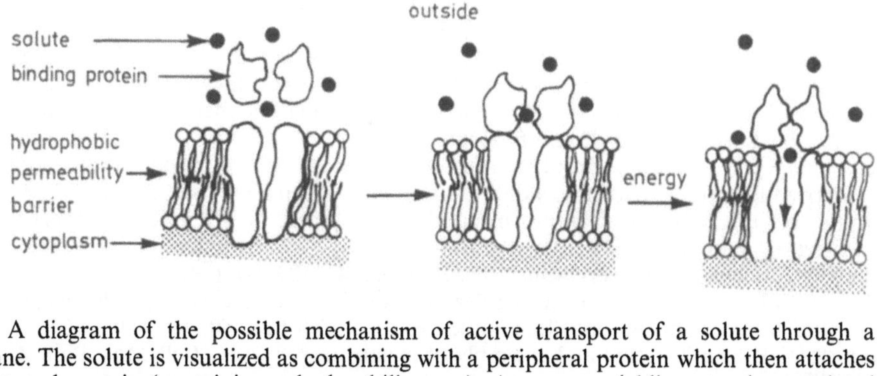

Fig. 25. A diagram of the possible mechanism of active transport of a solute through a membrane. The solute is visualized as combining with a peripheral protein which then attaches to an integral protein (containing a hydrophilic pore). An energy-yielding step is postulated during which a conformational change in the protein takes place. This causes a release of the solute across the membrane. (Modified from SINGER, 1974)

Plausable as the "fluid-mosaic" hypothesis appears, WALLACH and associates point out that proteins in some membranes are quite restricted in their mobility and that fluid-lipid regions may be quite small in some membranes. Consequently, the "fluid-mosaic" hypothesis should be viewed with some reserve and alternate modes of lipid-protein interactions in biomembranes should continue to be considered (WALLACH et al., 1975).

References

AGUTTER, P.S.: The isolation of envelopes of rat liver nuclei. Biochim. Biophys. Acta **255**, 397–401 (1972)

ANDERSON, N.G., HARRIS, W.W., BARBER, A.A., RANKIN, C.T., JR., CANDLER, E.L.: Separation of subcellular components and viruses by combined rate- and isopycnic zonal centrifugation. Natl. Cancer Inst. Monograph **21**, 253–269 (1966)

APPELMANS, F., WATTIAUX, R., DE DUVE, C : Tissue fraction studies. 5. The association of acid phosphatase with a special class of cytoplasmic granules in rat liver. Biochem. J. **59**, 438–445 (1955)

ARNON, D.I.: Copper enzymes in isolated chloroplasts. Polyphenoloxidase in *Beta vulgaris*. Plant Physiol. **24**, 1–15 (1949)

ARNTZEN, C.J., DILLEY, R.A., CRANE, F.L.: A comparison of chloroplast membrane surfaces visualized by freeze-etch and negative staining techniques; and ultrastructural characterization of membrane fractions from digitonin-treated spinach chloroplasts. J. Cell Biol. **43**, 16–31 (1969)

AVRAMEAS, S.: Immunoenzyme techniques: Enzymes as markers for the localization of antigens and antibodies. Intern. Rev. Cytol. **27**, 349–385 (1970)

BANDLOW, W.: Membrane separation and biogenesis of the outer membrane of yeast mitochondria. Biochim. Biophys. Acta **282**, 105–122 (1972)

BAUDHUIN, P.: Morphometry of subcellular fractions. Methods Enzymol. **32**, 3–20 (1974)

BAUDHUIN, P., EVARARD, P., BERTHET, J.: Electron microscopic examination of subcellular fractions I. The preparation of representative samples from suspensions of particles. J. Cell Biol. **32**, 181–191 (1967)

BENSON, A.A.: Plant membrane lipids. Ann. Rev. Plant Physiol. **15**, 1–16 (1964)

BENSON, A.A., STRICKLAND, E.H.: Plant phospholipids III. Identification of diphosphatidyl glycerol. Biochim. Biophys. Acta **41**, 328–333 (1960)

BEREZNEY, R., FUNK, L.K., CRANE, F.L.: The isolation of nuclear membrane from a large-scale preparation of bovine liver nuclei. Biochim. Biophys. Acta **203**, 531–546 (1970)

BEREZNEY, R., MACAULAY, L.K., CRANE, F.L.: The purification and biochemical characterization of bovine liver nuclear membranes. J. Biol. Chem. **247**, 5549–5561 (1972)

BIEGLMAYER, C., RUIS, H.: Cytochemical localization of catalase activity in glyoxysomes from castor bean endosperm. Plant Physiol. **53**, 276–278 (1974)

BLAIR, P.V., ODA, T., GREEN, D.E., FERNÁNDEZ-MORÁN, H.: Studies on the electron transfer system. LIV. Isolation of the unit of electron transfer. Biochemistry **2**, 756–764 (1963)

BLOBEL, G., POTTER, V.R.: Nuclei from rat liver: Isolation method that combines purity with high yield. Science **154**, 1662–1665 (1966)

BLONDIN, G.A.: Isolation, properties, and structural features of divalent cation ionophores derived from beef heart mitochondria. Ann. N.Y. Acad. Sci. **264**, 98–111 (1975)

BOGORAD, L.: Control mechanisms in plastid development. I. The role of cytoplasmic units. Develop. Biol., Suppl. **1**, 1–31 (1967)

BOOS, W.: Bacterial transport. Ann. Rev. Biochem. **43**, 123–146 (1974)

BOUCK, G.B.: Extracellular microtubules. The origin, structure and attachment of flagellar hairs in Fucus and Ascophyllum antherozoids. J. Cell Biol. **40**, 446–460 (1969)

BOUCK, G.B.: Architecture and assembly of mastigonemes. Adv. Cell Mol. Biol. **2**, 237–271 (1972)

BOULTER, D., ELLIS, R.J., YARWOOD, A.: Biochemistry of protein synthesis in plants. Biol. Rev. **47**, 113–175 (1972)

BOULTON, A.A.: Some observations on the chemistry and morphology of the membranes released from yeast protoplasts by osmotic shock. Exptl. Cell Res. **37**, 343–359 (1965)

BRANTON, D.: Membrane structure. Ann. Rev. Plant Physiol. **20**, 209–238 (1969)

BRANTON, D., MOOR, D.: Fine structure in freeze-etched *Allium cepa* L. root tips. J. Ultrastruct. Res. **11**, 401–411 (1964)

BRANTON, D., PARK, R.B.: Subunits in chloroplast lamellae. J. Ultrastruct. Res. **19**, 283–303 (1967)

BREIDENBACH, R.W., KAHN, A., BEEVERS, H.: Characterization of glyoxysomes from castor bean endosperm. Plant Physiol. **43**, 705–713 (1968)

BRETSCHER, M.S.: Membrane structure: Some general principles. Science **181**, 622–629 (1973)

BURR, B., BURR, F.A.: Zein synthesis in maize endosperm by polyribosomes attached to protein bodies. Proc. Natl. Acad. Sci. U.S. **73**, 515–519 (1976)

CAIN, J., SANTILLAN, G., BLASIE, J.K.: Molecular motion in membranes as indicated by X-ray diffraction. In: Membrane Research. Fox, C.F. (ed.), 1st ICN-UCLA Symp. Mol. Biol., 1972, pp. 3–14

CASSADY, W.E., WAGNER, R.P.: Separation of mitochondrial membranes of *Neurospora crassa*. I. Localization of l-hynurenine-3-hydroxylase. J. Cell Biol. **49**, 536–541 (1971)

CHAPPELL, J.B.: Systems used for the transport of substrates into mitochondria. Brit. Med. Bull. **24**, 150–157 (1968)

CHEN, Y., LIN, C., CHANG, H., GUILFOYLE, T.H., KEY, J.L.: Isolation and properties of nuclei from control and auxin-treated soybean hypocotyl. Plant Physiol. **56**, 78–82 (1975)

CHRISTENSEN, A.K., FAWCETT, D.W.: The normal fine structure of opossum testicular interstitial cell. J. Biophys. Biochem. Cytol. **9**, 653–670 (1961)

CLOULES, G.L., BJÖRKLUND, R.F.: Evidence of beta structure in mycoplasma membranes: circular dichroism, optical rotatory dispersion and infrared studies. Biophys. Soc. Abstr. (1970), p. 47a

COCKBURN, B.J., WELLBURN, A.R.: Changes in the envelope permeability of developing chloroplasts. J. Exptl. Botany **25**, 36–49 (1974)

COHN, N.S.: Elements of cytology, 2nd ed. New York-Chicago-San Francisco-Atlanta: Harcourt, Brace and World, Inc. 1969

COLBEAU, A., NACHBAUR, J., VIGNAIS, P.M.: Enzymic characterization and lipid composition of rat liver subcellular membranes. Biochim. Biophys. Acta **249**, 462–492 (1971)

COLEMAN, R., FINEAN, J.B.: Preparation and properties of isolated plasma membranes from guinea-pig tissues. Biochim. Biophys. Acta **125**, 197–206 (1966)

CRAIG, L.C.: Conformation studies with polypeptides by rotatory dispersion and thin-film analysis. Proc. Natl. Acad. Sci. U.S. **61**, 152–159 (1968)

D'ALESSIO, G., TRIM, A.R.: A method for the isolation of nuclei from leaves. J. Exptl. Botany **19**, 831–839 (1968)

DANIELLI, J.F.: The bilayer hypothesis of membrane structure. In: Cell Membranes. Biochemistry, Cell Biology, and Pathology. WEISSMAN, C., CLAIBORNE, R. (eds.), pp. 3–11. New York: H.P. Publishing Co. 1975

DANIELLI, J.F., DAVSON, H.: A contribution to the theory of permeability of thin films. J. Cell. Physiol. **5**, 495–508 (1935)

DAUWALDER, M., WHALEY, W.G., KEPHART, J.E.: Phosphatases and differentiation of the Golgi apparatus. J. Cell Sci. **4**, 455–497 (1969)

DEAMER, D.W., BRANTON, D.: Fracture planes in an ice-bilayer model membrane system. Science **158**, 655–657 (1967)

DONALDSON, R.P., TOLBERT, N.E., SCHNARRENBERGER, C.: A comparison of microbody membranes with microsomes and mitochondria from plant and animal tissue. Arch. Biochem. Biophys. **152**, 199–215 (1972)

DOUCE, R., BENSON, A.A.: Isolation, purification and properties of the envelope of spinach chloroplasts. Separata de Portugaliae Acta Biol. A **14**, 45–64 (1974)

DOUCE, R., HOLTZ, R.B., BENSON, A.A.: Isolation and properties of the envelope of spinach chloroplasts. J. Biol. Chem. **248**, 7215–7222 (1973)

DUPRAW, E.J.: The organization of nuclei and chromosomes in honeybee embryonic cells. Proc. Natl. Acad. Sci. U.S. **53**, 161–168 (1965)

EDIDIN, M.: Aspects of plasma membrane fluidity. In: 1st ICN-UDLA Symp. Membrane Biol. Membrane Res. Fox, C.F. (ed.), pp. 15–25 (1972)

EMMELOT, P., BOS, C.J., BENEDETTI, E.L., RÜMKE, P.H.: Studies on plasma membranes. I. Chemical composition and enzyme content of plasma membranes isolated from rat liver. Biochim. Biophys. Acta **90**, 126–145 (1964)

EPSTEIN, E.: Mineral Nutrition of Plants: Principles and Perspectives, p. 412. New York: John Wiley and Sons 1971

ERNSTER, L., KUYLENSTIERNA, B.: Structure, composition and function of mitochondrial membranes. FEBS Symp. **17**, 5–31 (1969)

ERNSTER, L., KUYLENSTIERNA, B.: Outer membrane of mitochondria. In: Membranes of Mitochondria and Chloroplasts. RACKER, E. (ed.), pp. 172–212. New York: Van Nostrand-Reinhold, 1970

FAIRBANKS, G., STECK, T.L., WALLACH, D.F.H.: Electrophoretic analysis of the major polypeptides of the human erythrocyte membrane. Biochemistry **10**, 2606–2617 (1971)

FAWCETT, D.W.: The membranes of the cytoplasm. Lab. Invest. **10**, 1162–1188 (1961)

FELDHERR, C.M.: Binding within the nuclear annuli and its possible effect on nucleocytoplasmic exchanges. J. Cell Biol. **20**, 188–192 (1964)

FELDHERR, C.M.: Structure and function of the nuclear envelope. In: Advances in Cell Molecular Biology. DUPRAW, E.J. (ed.), Vol. 2, pp. 274–307. New York: Academic Press 1972

FERNÁNDEZ-MORÁN, H.: Cell-membrane ultrastructure. Low-temperature electron microscopy and X-ray diffraction studies of lipoprotein components in lamellar systems. Circulation **26**, 1039–1065 (1962)

FERNÁNDEZ-MORÁN, H.: Subunit organization of mitochondrial membranes. Science **140**, 381 (1963)

FERNÁNDEZ-MORÁN, H., ODA, T., BLAIR, P.V., GREEN, D.E.: A macromolecular repeating unit of mitochondrial structure and function. J. Cell Biol. **22**, 63–100 (1964)

FINEAN, J.B.: Chemical ultrastructure in living tissues. Springfield, Illinois: Ch. Thomas (1961)

FLEISCHER, S., PACKER, L.: Biomembranes. Methods Enzymol. **31A**, (1974)

FRANKE, W.W., DEUMLING, B., ERMEN, B., JARASCH, E.-D., KLEINIG, H.J.: Nuclear membranes from mammalian liver I. Isolation procedure and general characterization. J. Cell Biol. **46**, 379–395 (1970)

FRANKE, W.W., SCHEER, V.: Structure and functions of the nuclear envelope. In: The Cell Nucleus. BUSCH, H. (ed.), Vol. 1, pp. 220–348. New York-London: Academic Press 1974

FREY-WYSSLING, A., GRIESHABER, E., MÜHLETHALER, K.: Origin of spherosomes in plant cells. J Ultrastruct. Res. **8**, 506–516 (1963)

FREY-WYSSLING, A., STEINMANN, E.: Fine structure analysis of chloroplasts. Naturf. Ges. Zurich, Vjhschr. **98**, 20–29 (1953) as translated by D. BRANTON, In: Papers on Biological Membrane Structure. BRANTON, D., PARK, R. (eds.), p. 311. Boston: Little, Brown and Co. 1968

FRYE, L.D., EDIDIN, M.: The rapid intermixing of cell surface antigens after formation of mouse-human heterokaryons. J. Cell Sci. **7**, 319–335 (1970)

FUHRMANN, G.F., WEHRLI, E., BOEHM, C.: Preparation and identification of yeast plasma membrane vesicles. Biochim. Biophys. Acta **363**, 295–310 (1974)

GALLIARD, T.: Enzymic deacylation of lipids in plants. The effects of free fatty acids on the hydrolysis of phospholipids by lipolytic acyl hydrolase of potato tubers. Eur. J. Biochem. **21**, 90–98 (1971)

GIMMLER, H., SCHÄFER, G., KRAMINER, A., HEBER, U.: Amino acid permeability of the chloroplast envelope as measured by light scattering, volumetry and amino acid uptake. Planta **120**, 47–61 (1974)

GOLFISCHER, S., ESSNER, E., NOVIKOFF, A.B.: The localization of phosphatase activities at the level of ultrastructure. J. Histochem. Cytochem. **12**, 72–95 (1964)

GORTER, E., GRENDEL, F.: On biomolecular layers of lipoids on the chromocytes of the blood. J. Exptl. Med. **41**, 433–439 (1925)

GREEN, D.E.: Membrane Structure and Its Biological Applications. Ann. N.Y. Acad. Sci. (1972) 195 p.

GREEN, D.E., PERDUE, J.F.: Correlation of mitochondrial structure and function. Ann. N.Y. Acad. Sci. **137**, 667–684 (1966)

GREEN, D.G.: Role of ionophores in energy coupling. Ann. N.Y. Acad. Sci. **264**, 61–82 (1975)

GUNNING, B.E.S.: The greening process in plastids. I. The structure of the prolamellar body. Protoplasma **60**, 111–130 (1965)

HARRIS, W.M., SPURR, A.R.: Chromoplasts of tomato fruits. II. The red' tomato. Am. J. Botany **56**, 380–389 (1969)

HASHIMOTO, H., MURAKAMI, S.: Dual character of lipid composition of the envelope membrane of spinach chloroplasts. Plant Cell Physiol. **16**, 895–902 (1975)

HEBER, U., PON, N.G., HEBER, M.: Localization of carboxydismutase and triosephosphate dehydrogenases in chloroplasts. Plant Physiol. **38**, 355–360 (1963)

HELDT, H.W.: Adenine nucleotide translocation in spinach chloroplasts. FEBS Lett. **5**, 11–14 (1969)

HELDT, H.W., RAPLEY, L.: Unspecific permeation and specific uptake of substances in spinach chloroplasts. FEBS Lett. **7**, 139–172 (1970a)

HELDT, H.W., RAPLEY, L.: Specific transport of inorganic phosphate, 3-phosphoglycerate and dihydroxyacetonephosphate, and of dicarboxylates across the inner membrane of spinach chloroplasts. FEBS Lett. **10**, 143–148 (1970b)

HELDT, H.W., SAUER, F., RAPLEY, L.: Differentiation of the permeability properties of the two membranes of the chloroplast envelope. In: Proc. 2nd. Intern. Congr. Photosyn. Res. FORTI, G., AVRON, M., MELANDRI, A. p. 1345–1355. The Hague: Dr. W. Junk 1972

HELENDER, H.F.: Ultrastructure of secreting cells in the pyloric gland area of the mouse gastric mucosa. J. Ultrastruct. Res. **10**, 145–159 (1964)

HENDLER, R.W.: Biological membrane structure. Physiol. Rev. **51**, 66–97 (1971)

HEPPEL, L.A.: The concept of periplasmic enzymes. Structure and Function of Biological Membranes. ROTHFIELD, L.I. (ed.), pp. 234–247. New York: Academic Press 1971

HODGES, T.K., LEONARD, R.T.: Purification of plasma membrane-bound adenosine triphosphate from plant roots. Methods Enzymol. **32**, 392–406 (1974)

HODGES, T.K., LEONARD, R.T., BRACKER, C.E., KEENAN, T.W.: Purification of an ion-stimulated adenosine triphosphatase from plant roots: associated with plasma membranes. Proc. Natl. Acad. Sci. US **69**, 3307–3311 (1972)

HOLTZ, R.B., STEWART, P.S., PATTON, S., SCHISLER, L.C.: Isolation and characterization of membranes from the cultivated mushroom. Plant Physiol. **50**, 541–546 (1972)

HOLZWARTH, G.: Ultraviolet spectroscopy of biological membranes. In: Membrane Molecular Biology. FOX, C.F., KEITH, A.C. (eds.), pp. 228–286. Stamford, Conn: Sinauer Assoc. Inc. 1972

HORWITZ, A.F.: Nuclear magnetic resonance studies on phospholipids and membranes. In: Membrane Molecular Biology. FOX, C.F., KEITH, A.C. (eds.), pp. 164–191. Stamford, Conn.: Sinauer Assoc. Inc. 1972

HUANG, A.H.C.: Comparative studies of glyoxysomes from various fatty seedlings. Plant Physiol. **55**, 870–874 (1975)

HUBBELL, W.L., MCCONNELL, H.M.: Molecular motion in spin-labeled phospholipids and membranes. J. Am. Chem. Soc. **93**, 314 (1971)

IKUMA, H.: Electron transport in plant respiration. Ann. Rev. Plant Physiol. **23**, 419–436 (1972)

KADENBACH, B.: Biosynthesis of mitochondrial enzymes. FEBS Symp. **17**, 179–188 (1969)

KAGAWA, Y., RACKER, E.: Partial resolution of the enzymes catalyzing oxidative phosphorylation. X. Correlation of morphology and function in submitochondrial particles. J. Biol. Chem. **241**, 2475–2482 (1966)

KASHNING, D.M., KASPER, C.B.: Isolation, morphology, and composition of the nuclear membrane from rat liver. J. Biol. Chem. **244**, 3786–3792 (1969)

KASPER, C.B.: Chemical and biochemical properties of the nuclear envelope. In: The Cell Nucleus. BUSCH, H. (ed.), Vol. 1, pp. 349–384. New York-London: Academic Press 1974

KELLEMS, R.E., ALLISON, V.F., BUTOW, R.A.: Cytoplasmic type 80S ribosomes associated with yeast mitochondria IV. Attachment of ribosomes to the outer membrane of isolated mitochondria. J. Cell Biol. **65**, 1–14 (1975)

KELLEMS, R.E., BUTOW, R.A.: Cytoplasmic-type 80S ribosomes associated with yeast mitochondria I. Evidence for ribosome binding sites on yeast mitochondria. J. Biol. Chem. **247**, 8043–8050 (1972)

KIETH, A.D., MEHLHORN, R.J.: Membrane lipid structure as revealed by X-ray diffraction. In: Membrane Molecular Biology. FOX, C.F., KEITH, A.C. (eds.), pp. 117–122. Stamford, Conn: Sinauer Assoc. Inc. 1972

KIRK, J.T.O., TILNEY-BASSETT, R.A.E.: The Plastids. 608 p. San Francisco: W.H. Freeman and Co. 1967

KLEINIG, H.: Nuclear membranes from mammalian liver II. Lipid composition. J. Cell Biol. **46**, 396–402 (1970)

KORN, E.D.: II. Synthesis of bis(methyl 9, 10-dihydroxystearate) osmate from methyl oleate and osmium tetroxide under conditions used for fixation of biological material. Biochim. Biophys. Acta **116**, 317–324 (1966a)

KORN, E.D.: III. Modification of oleic acid during fixation of amoebae by osmium tetroxide. Biochim. Biophys. Acta **116**, 325–335 (1966b)

KORN, E.D.: A chromatographic and spectrophotometric study of the products of the reaction of osmium tetroxide with unsaturated lipids. J. Cell Biol. **34**, 627–638 (1967)

KORN, E.D.: Current concepts of membrane structure and function. Fed. Proc. **28**, 6–11 (1969)

KREUTZ, W.: Strukturuntersuchungen an Plastiden. IV. Über das zweidimensionale Gitter in der Proteinlamelle. Z. Naturforsch. **18b**, 567–571 (1963a)

KREUTZ, W.: Strukturuntersuchungen an Plastiden. V. Bestimmung der Elektronendichte-Verteilung längs der Flächennormalen im Thylalkoid der Chloroplasten. Z. Naturforsch. **18b**, 1098–1104 (1963b)

KREUTZ, W.: Strukturuntersuchungen an Plastiden. VI. Über die Struktur der Lipoprotein-lamellen in Chloroplasten lebender Zellen. Z. Naturforsch. **19b**, 441–446 (1964)

KREUTZ, W.: The structure of the lamellar system of chloroplasts. In: Biochemistry of Chloroplasts. GOODWIN, T.W. (ed.), Vol. 1, pp. 83–88. New York: Academic Press 1966

KREUTZ, W., MENKE, W.: Strukturuntersuchungen an Plastiden. III. Röntgenographische Untersuchungen an isolierten Chloroplasten und Chloroplasten lebender Zellen. Z. Naturforsch. **17b**, 675–683 (1962)

LAETSCH, W.M.: Chloroplast specialization in dicotyledons possessing the C_4-dicarboxylic acid pathway of photosynthetic CO_2 fixation. Am. J. Botany **55**, 875–883 (1968)

LAETSCH, W.M.: Relationship between chloroplast structure and photosynthetic carbon fixation pathways. Sci. Progr. Oxf. **57**, 323–351 (1969a)

LAETSCH, W.M.: Specialized chloroplast structure of plants exhibiting the dicarboxylic acid pathway of photosynthetic CO_2 fixation. In: Progress in Photosynthetic Research. METZNER, H. (ed.), Vol. 1, pp. 36–46. Tübingen: Intern. Union Biol. Sci. 1969b

LAMPEN, J.O.: Movement of extracellular enzymes across cell membranes. Symp. Soc. Exptl. Biol. **27**, 351–374 (1974)

LARDY, H.A., FERGUSON, S.M.: Oxidative phosphorylation in mitochondria. Ann. Rev. Biochem. **38**, 991–1034 (1969)

LAW, J.H., SNYDER, W.R.: Membrane lipids. In: Membrane Molecular Biology. FOX, C.F., KEITH, A.D. (eds.), pp. 3–26. Stamford, Conn: Sinauer Assoc. Inc. 1972

LEDBETTER, M.C., PORTER, K.R.: Introduction to the fine structure of plant cells. New York-Heidelberg-Berlin: Springer 1970

LEDUC, E.N., AVRAMEAS, S., BOUTEILLE, M.: Ultrastructural localization of antibody in differentiating plasma cells. J. Exptl. Med. **127**, 109–118 (1968)

LEDUC, E.H., SCOTT, G.B., AVRAMEAS, S.: Ultrastructural localization of intracellular immune globulins in plasma cells and lymphoblasts by enzymelabeled antibodies. J. Histochem. Cytochem. **17**, 211–224 (1969)

LEIGH, R.A., WILLIAMSON, F.A., WYN JONES, R.G.: Presence of two different membrane-bound, KCl-stimulated adenosine triphosphatase activities in maize roots. Plant Physiol. **55**, 678–685 (1975)

LEMBI, C.A., MORRÉ, D.J., ST.THOMSON, K., HERTEL, H.: N-l-Naphylphthalomic acid-binding activity of a plasma membrane-rich fraction from maize coleoptiles. Planta **99**, 37–45 (1971)

LENARD, J., SINGER, S.J.: Protein conformation in cell membrane preparations as studied by optical rotatory dispersion and circular dichroism. Proc. Natl. Acad. Sci. US **56**, 1828–1835 (1966)

LEONARD, R.T., VAN DER WOUDE, W.J.: Isolation of plasma membranes form corn roots by sucrose density gradient centrifugation. An anomalous effect of ficoll. Plant Physiol. **57**, 105–114 (1976)

LOEWENSTEIN, W.R., KANNO, Y.: Some electrical properties of a nuclear membrane examined with a microelectrode. J. Gen. Physiol. **46**, 1123–1140 (1963)

LOEWENSTEIN, W.R., KANNO, Y., ITO, S.: Permeability of nuclear membranes. Ann. N.Y. Acad. Sci. **137**, 708–716 (1965)

LONGLEY, R.P., ROSE, A.H., KNIGHTS, B.A.: Composition of the protoplast membrane of *Saccharomyces cerevisiae*. Biochem. J. **108**, 401–412 (1968)

LOOMIS, W.D.: Overcoming problems of phenolics and quinones in the isolation of plant enzymes and organelles. Methods Enzymol. **31A**, 528–544 (1974)

LOUD, A.V.: A method for the quantitative estimation of cytoplasmic structures. J. Cell Biol. **15**, 481–487 (1962)

MACKENDER, R.O., LEECH, R.M.: The isolation of chloroplast envelope membranes. Nature **228**, 1347–1349 (1970)

MACKENDER, R.O., LEECH, R.M.: The isolation and characterization of plastid envelope membranes. In: Proc. 2nd Intern. Congr. Photosyn. Res. FORTI, G., AVRON, M., MELANDRI, A., Vol. 2, pp. 1431–1440. The Hague: Dr. W. Junk 1972

MACKENDER, R.O., LEECH, R.M.: The galactolipid, phospholipid, and fatty acid composition of the chloroplast envelope membranes of *Vicia faba* L. Plant Physiol. **53**, 496–502 (1974)

MARCHESI, V.T., SEGREST, J.P., KAHANE, I.: Molecular features of human erythrocyte glycophorin. In: Membrane Research. FOX, C.F. (ed.), pp. 41–51. 1st ICN-UCLA Symp. Mol. Biol. New York: Academic Press 1972

MASCARENHAS, J.P., BERMAN-KURTZ, M., KULIKOWSKI, R.R.: Isolation of plant nuclei. Methods Enzymol. **31**, 558–565 (1974)

MATILE, PH.: Properties of the purified cytoplasmic membrane of yeast. In: Membrane Structure and Function. VILLANUEVA, J.R., PONZ, F. (eds.). FEBS Symp. **20**, 39–49 (1970)

MCCARTY, R.E., DOUCE, R., BENSON, A.A.: The acyl lipids of highly purified plant mitochondria. Biochim. Biophys. Acta **316**, 266–270 (1973)

MCCONNELL, H.M., DEVAUX, P., SCANDELLA, C.: Lateral diffusion and phase separation in biological membranes. In: Membrane Research. 1st ICN-UCLA Symp. Mol. Biol. Fox, C.F. (ed.), pp. 27–37. New York: Academic Press 1972

MCKEEL, D.W., JARETT, L.: Preparation and characterization of a plasma membrane fraction from isolated fat cells. J. Cell Biol. **44**, 417–432 (1970)

MEHLHORN, R.J., KIETH, A.D.: Spin labeling of biological membranes. In: Membrane Molecular Biology. FOX, C.F., KEITH, A.C. (eds.), pp. 192–227. Stamford, Conn: Sinauer Assoc. Inc. 1972

MEUNIER, D., MAZLIAK, P.: Compt. Rend. Ser. D **275**, 213–216 (1972)

MEUNIER, D., PIANETA, C., COULOMB, P.: Obtention de membranes externes et de membranes internes à partir de mitochondries issues du tubercule de pomme de terre (*Solanum tuberosum*) Compt. Rend. **272**, Ser. D 1376–1379 (1971)

MOLLENHAUER, H.H.: An intercisternal structure in the Golgi apparatus. J. Cell Biol. **24**, 504–511 (1965)

MONNERON, A., BLOBEL, G., PALADE, G.E.: Fraction of the nucleus by divalent cations. J. Cell Biol. **55**, 104–125 (1972)

MOOR, H., MÜHLETHALER, K.: Fine structure in frozen-etched yeast cells. J. Cell Biol. **17**, 609–628 (1963)

MOREAU, F., DUPONT, J., LANCE, C.: Phospholipid and fatty acid composition of outer and inner membranes of plant mitochondria. Biochim. Biophys. Acta **345**, 294–304 (1974)

MOREAU, F., LANCE, C.: Isolement et propriétés des membranes externes et internes de mitochondries végétales. Biochimie **54**, 1335–1348 (1972)

MORRÉ, D.J.: Membrane biogenesis. Ann. Rev. Plant Physiol. **26**, 441–481 (1975)

MORRÉ, D.J., KEENAN, T.W., MOLLENHAUER, H.H.: Golgi apparatus function in membrane tranformations and product compartmentalization; studies with cell fractions isolated from rat liver. Adv. Cytopharm. **1**, 159–182 (1971 b)

MORRÉ, D.J., MOLLENHAUER, H.H., BRACKER, C.E.: Origin and continuity of Golgi apparatus. In: Origin and Continuity of Cell Organelles. REINERT, J., URSPRUNG, H. (eds.), pp. 82–126. Berlin-Heidelberg-New York: Springer 1971 a

MÜHLETHALER, K.: The ultrastructure of the plastid lamellae. In: Biochemistry of Chloroplasts. GOODWIN, T.W. (ed.), Vol. 1, pp. 49–64. New York: Academic Press 1966

NAKAO, M., PACKER, L.: Organization of Energy-Transducing Membranes. Tokyo: Univ. Park Press 1973

NEWCOMB, E.H.: Fine structure of protein-storing plastids in bean root tips. J. Cell Biol. **33**, 143–163 (1967)

NEWCOMB, E.H., BECKER, W.M.: The diversity of plant organelles with special reference to the electron cytochemical localization of catalse in plant microbodies. Methods Enzymol. **31** (part A), 489–500 (1974)

NICOLSON, G.L.: Topological studies on the structure of cell membranes. In: Membrane Research. 1st ICN-UCLA Symp. Mol. Biol. Fox, C.F. (ed.), 1972, pp. 53–70

NOBEL, P.S., CHEUNG, Y.-N.S.: Two amino acid carriers in pea chloroplasts. Nature New Biol. **237**, 207–208 (1972)

NOBEL, P.S., WANG, C.-T.: Amino acid permeability of pea chloroplasts as measured by osmotically determined reflection coefficients. Biochim. Biophys. Acta **211**, 79–87 (1970)

NORTHCOTE, D.H., LEWIS, D.R.: Freeze-etched surfaces of membranes and organelles in the cells of pea root tips. J. Cell Sci. **3**, 199–206 (1968)

OSEROFF, A.R., ROBBINS, P.W., BURGER, M.M.: The cell surface membrane: Biochemical aspects and biophysical probes. Ann. Rev. Biochem. **42**, 646–682 (1973)

OSTERHOUT, W.J.V.: Ergeb. Physiol. **35**, 967 (1933)

OXENDER, D.L.: Membrane transport. Ann. Rev. Biochem. **41**, 777–814 (1971)

OXENDER, D.L., QUAY, S.: Binding proteins and membrane transport. Ann. N.Y. Acad. Sci. **264**, 358–372 (1975)

PALADE, G.E.: The fine structure of mitochondria. Anat. Rec. **114**, 427–452 (1952)

PALADE, G.E.: An electron microscope study of the mitochondrial structure. J. Histochem. Cytochem. **1**, 188–211 (1953)

PALADE, G.E.: Intracellular aspects of the process of protein synthesis. Science **189**, 347–358 (1975)

PAOLILLO, D.J., JR.: The three-dimesional arrangement of intergranal lamellae in chloroplasts. J. Cell Sci. **6**, 243–255 (1970)

PAOLILLO, D.J., JR., FALK, R.H.: The ultrastructure of grana in mesophyll plastids of Zea mays. Am. J. Botany **53**, 173–180 (1966)

PAOLILLO, D.J., JR., MACKAY, N.C., REIGHARD, J.A.: The structure of grana in flowering plants. Am. J. Botany **56**, 344–347 (1969)

PAOLILLO, D.J., JR., REIGHARD, J.A.: On the relationship between mature structure and ontogeny in the grana of chloroplasts. Can. J. Botany **45**, 773–782 (1967)

PARDEE, A.B.: Membrane transport proteins. Science **162**, 632–637 (1968)

PARK, R.B.: Advances in photosynthesis. J. Chem. Educ. **39**, 424–429 (1962)

PARK, R.B.: Freeze etching: A classical view. Ann. N.Y. Acad. Sci. **195**, 262–290 (1972)

PARK, R.B., BIGGINS, J.: Quantasome: Size and composition. Science **144**, 1009–1011 (1964)

PARK, R.B., PFEIFHOFER, A.O.: Ultrastructural observations on deep-etched thylakoids. J. Cell Sci. **5**, 299–311 (1969)

PARK, R.B., PON, N.G.: Correlation of structure with function in *Spinacea oleracea* chloroplasts. J. Mol. Biol. **3**, 1–10 (1961)

PARK, R.B., SANE, P.V.: Distribution of function and structure in chloroplast lamellae. Ann. Rev. Plant Physiol. **22**, 395–430 (1971)

PARSONS, D.F., WILLIAMS, G.R., CHANCE, B.: Characteristics of isolated and purified preparations of the outer and inner membranes of mitochondria. Ann. N.Y. Acad. Sci. **135**, 2, 643–666 (1966)

PARSONS, D F., WILLIAMS, G.R., THOMPSON, W., WILSON, D., CHANCE, B.: Improvement in the procedure for purification of mitochondrial outer and inner membrane. Comparison of the outer membrane with the endoplasmic reticulum. In: Mitochondrial Structure and Compartmentation. QUAGLIARIELLO, E., PAPPA, S., SLATER, E.C., TAGER, J.T. (eds.), pp. 29–70. Bari: Adriatica Editrice 1967

PATNI, N.J., BILLMIRE, J.E., AARONSON, S.: Isolation of the *Ochromonas danica* plasma membrane and identification of several membrane enzymes. Biochim. Biophys. Acta **373**, 347–355 (1974)

PEASE, D.C.: Histological Techniques for Electron Microscopy. 2nd ed. 381 p. New York: Academic Press 1964

PFAFF, E., KLINGENBERG, M., RITT, E., VOGELL, W.: Korrelation des unspezifisch permeablen mitochondrialen Raumes mit dem Intermembranraum. Eur. J. Biochem. **5**, 222–232 (1968)

PHILIPP, E.-I., FRANKE, W.W., KEENAN, T.W., STADLER, J., JARASCH, E.-D.: Characterization of nuclear membranes and endoplasmic reticulum isolated from plant tissue. J. Cell Biol. **68**, 11–29 (1976)

PIERRE, J.W. DE, KARNOVSKY, M.L.: Plasma membranes of mammalian cells. J. Cell Biol. **56**, 275–305 (1973)

POINCELOT, R.P.: Isolation and lipid composition of spinach chloroplast envelope membranes. Arch. Biochem. Biophys. **159**, 134–142 (1973)

POINCELOT, R.P.: Transport of metabolites across isolated envelope membranes of spinach chloroplasts. Plant Physiol. **55**, 849–852 (1975)

POINCELOT, R.P., DAY, P.R.: An improved method for the isolation of spinach chloroplast envelope membranes. Plant Physiol. **54**, 780–783 (1974)

POINCELOT, R.P., DAY, P.R.: Isolation and bicarbonate transport of chloroplast envelope membranes from species of differing net photosynthetic efficiencies. Plant Physiol. **57**, 334–338 (1976)

PRESSMAN, B.C.: Induced transport of ions in mitochondria. Proc. Natl. Acad. Sci. US **53**, 1076–1083 (1965)

PRESSMAN, B.C.: Energy-linked transport in mitochondria. In: Membranes of Mitochondria and Chloroplasts. RACKER, E. (ed.), pp. 213–250. New York: Van Nostrand-Reinhold 1970

PRESSMAN, B.C., DE GUZMAN, N.T.: Biological applications of ionophores: theory and practice. Ann. N.Y. Acad. Sci. **264**, 373–385 (1975)

PRICE, C.A.: Plant cell fractionation. Methods Enzymol. **31A**, 501–519 (1974)

RACKER, E.: Membranes of Mitochondria and Chloroplasts ACS Monograph. New York: Van Nostrand-Reinhold Co. 1972

RACKER, E., CHANCE, B., PARSONS, D.F.: Correlation of structure and function of submitochondrial units in oxidative phosphorylation. Fed. Proc. **23**, 431 (1964)

RACKER, E., TYLER, D.D., ESTABROOK, R.W., CONNOVER, T.E., PARSONS, D.F., CHANCE, B.: In: Oxidases and Related Redox Systems. KING, T.E., MASON, H.S., MORRISON, M., Vol. 2, pp. 1077–1101. New York: John Wiley and Sons 1965

RACUSEN, D., POINCELOT, R.P.: Distribution of protein-bound hexoseamine in chloroplasts. Plant Physiol. **57**, 53–54 (1976)

RADUNZ, A.: Antiserum to chlorophyll a and its reactions with chloroplasts. In: Proc. 2nd Intern. Congr. Photosyn. Res. FORTI, G., AVRON, A., MELANDRI, A. (eds.), Vol. 2, pp. 1613–1618. The Hague: Dr. W. Junk 1971

RAY, P.M., SHININGER, T.L., RAY, M.M.: Isolation of β-glucan synthetase particles from plant cells and identification with Golgi-membranes. Proc. Natl. Acad. Sci. US **64**, 605–612 (1969)

ROBERTSON, J.D.: New observations on the ultrastructure of the membranes of frog peripheral nerve fibers. J. Biophys. Biochem. Cytol. **3**, 1043–1047 (1957)

ROBERTSON, J.D.: The ultrastructure of cell membrane and their derivatives. Biochem. Soc. Symp. **16**, 3–43 (1959)

ROBERTSON, J.D.: Granulo-fibrillar and globular substructure in unit membranes. Ann. N.Y. Acad. Sci. **137**, 421–440 (1966)

ROLAND, J.-C., LEMBI, C.A., MORRÉ, D.J.: Phosphotungstic acid-chromic acid as a selective electron-dense stain for plasma membranes of plant cells. Stain Technol. **47**, 195–200 (1972)

ROSADO-ALBERIO, J., WEIER, T.E., STOCKING, C.R.: Continuity of the chloroplast membrane systems in *Zea mays* L. Plant Physiol. **43**, 1325–1331 (1968)

ROSENBERG, S.A., GUIDOTTI, G.: The proteins of the erythrocyte membrane: Structure and arrangement in the membrane. In: Red Cell Membrane: Structure and Function. JAMIESON, G.A., GREENWALT, T.J. (eds.), pp. 93–109. Philadelphia: L.B. Lippincott Co. 1969

ROTHFIELD, L.I.: Structure and Function of Biological Membranes 468 p. New York: Academic Press 1971

RUESINK, A.W.: The plasma membranes of *Avena* coleoptile protoplasts. Plant Physiol. **47**, 192–195 (1971)

SABNIS, D.D., GORDON, M., GALSTON, A.W.: Localization of adenosine triphosphatase activity on the chloroplast envelope in tendrils of *Pisum sativum*. Plant Physiol. **45**, 25–32 (1970)

SANE, P.V., GOODCHILD, D.J., PARK, R.B.: Characterization of chloroplast photosystems 1 and 2 separated by a non-detergent method. Biochim. Biophys. Acta **216**, 162–168 (1970)

SCHATZ, G., MASON, T.L.: The biosynthesis of mitochondrial proteins. Ann. Rev. Biochem. **43**, 51–87 (1974)

SCHIBERI, A., RATTRAY, J.B.M., KIDBY, D.K.: Isolation and identification of yeast plasma membrane. Biochim. Biophys. Acta **311**, 15–25 (1973)

SCHNAITMAN, C., GREENAWALT, J.W.: Enzymatic properties of the inner and outer membranes of rat liver mitochondria. J. Cell Biol. **38**, 158–175 (1968)

SCHNITKA, T.K., SELIGMAN, A.M.: Ultrastructural localization of enzymes. Ann. Rev. Biochem. **40**, 375–396 (1971)

SCHWARZENBACH, A.M.: Observations on spherosomal membranes. Cytobiologie **4**, 145–147 (1971)

SHAMOO, A.F. (ed.): Carriers and channels in biological systems. Ann. N.Y. Acad. Sci. Vol. **264** (1975)

SHAMOO, A.E., RYAN, T.E.: Isolation of ionophores from ion-transport systems. Ann. N.Y. Acad. Sci. **264**, 83–96 (1975)

SHUMWAY, L.K., WEIER, T.E.: The chloroplast structure of iojap maize. Am. J. Botany **54**, 773–780 (1967)

SIEGEL, M.R., SISLER, H.D.: Site of action of cycloheximide in cells of *Saccharomyces pastorianus* III. Further studies on the mechanism of action and the mechanism of resistance in Saccharomyces species. Biochim. Biophys. Acta **103**, 558–567 (1965)

SIMONI, R.D.: Macromolecular characterization of bacterial transport systems. In: Membrane Molecular Biology. FOX, C.F., KEITH, A. (eds.), pp. 289–322. Stamford, Conn: Sinauer Associates Inc. 1972

SINGER, S.J.: The molecular organization of biological membranes. In: Structure and Function of Biological Membranes. ROTHFIELD, L.I. (ed.), pp. 145–222. New York: Academic Press 1971

SINGER, S.J.: A fluid lipid-globular protein mosaic model of membrane structure. Ann. N.Y. Acad. Sci. **195**, 16–23 (1972)

SINGER, S.J.: The molecular organization of membranes. Ann. Rev. Biochem. **43**, 805–833 (1974)

SINGER, S.J.: Architecture and topography of biology membranes. In: Cell Membranes Biochemistry, Cell Biology, and Pathology. WEISSMANN, G., CLAIBORNE, R. (eds.), pp. 3–11. New York: HP Publishing Co. 1975

SINGER, S.J., NICOLSON, G.: The fluid mosaic model of the structure of cell membranes. Science **175**, 720–731 (1972)

SJÖSTRAND, F.S.: Electron microscopy of mitochondria and cytoplasmic double membranes. Ultrastructure of rod-shaped mitochondria. Nature **171**, 30–31 (1953)

SJÖSTRAND, F.S.: Electron microscopy of cellular constituents. Methods Enzymol. **4**, 391–422 (1957)

SJÖSTRAND, F.S.: A new repeat structural element of mitochondrial and certain cytoplasmic membranes. Nature **199**, 1262–1264 (1963)

SPURR, A.R., HARRIS, W.M.: Ultrastructure of chloroplasts and chromoplasts in *Capsicum annuum*. I. Thylakoid membrane changes during fruit ripening. Am. J. Botany **55**, 1210–1224 (1968)

STADELMANN, E.J.: Permeability of the plant cell. Ann. Rev. Plant Physiol. **20**, 585–606 (1969)

STECK, T.L.: Membrane isolation. In: Membrane Molecular Biology. FOX, C.F., KEITH, A.D. (eds.), pp. 76–114. Stamford, Conn.: Sinauer Associates Inc. 1972a

STECK, T.L.: The organization of proteins in human erythrocyte membranes. In: Membrane Research. 1st ICN-UCLA Symp. Mol. Biol. FOX, C.F. (ed.), pp. 71–93. New York: Academic Press 1972b

STECK, T.L., FAIRBANKS, G., WALLACH, D.F.H.: Disposition of the major proteins in the isolated erythrocyte membrane. Proteolytic dissection. Biochemistry **10**, 2617–2624 (1971)

STECK, T.L., FOX, C.F.: Membrane proteins. In: Membrane Molecular Biology. FOX, C.F., KEITH, A.D. (eds.), pp. 27–75. Stamford, Conn.: Sinauer Associates Inc. 1972

STECK, T.L., WALLACH, D.F.H.: The isolation of plasma membranes. Methods Cancer Res. **5**, 93–153 (1970)

STERN, H.: Isolation and purification of plant nucleic acids from whole tissues and from isolated nuclei. Methods Enzymol. **12B**, 100–112 (1968)

STILES, J.W., WILSON, J.T., CRANE, F.L.: Membranefibrils in cristae and grana. Biochim. Biophys. Acta **162**, 631–634 (1968)

STOCKING, C.R., LIN, C.H.: Unpublished (1976)

STOECKENIUS, W., ENGELMAN, D.M.: Current models for the structure of biological membranes. J.Cell Biol. **42**, 613–646 (1969)

STROTMANN, H., BERGER, S.: Adenine nucleotide translocation across the membrane of siolated *Acetabularia* chloroplasts. Biochem. Biophys. Res. Commun. **35**, 20–26 (1969)

SZABO, G., EISENMAN, G., MCLAUGHLIN, S.G.A., KRASNE, S.: Ionic probes of membrane structure. Ann. N.Y. Acad. Sci. **195**, 273–290 (1972)

TAUTVYDAS, K.J.: Mass isolation of pea nuclei. Plant Physiol. **47**, 499–503 (1971)

TAYLOR, R.B., DUFFUS, W.P.H., RAFF, M.C., DEPETRIS, S.: Redistribution and pinocytosis of lymphocyte surface immunoglobulin molecules induced by anti-immunoglobulin antibody. Nature (New Biol.) **233**, 225–229 (1971)

THOMSON, W.W., FOSTER, P., LEECH, R.M.: The isolation of proplastids from roots of *Vicia faba*. Plant Physiol. **49**, 270–272 (1972)

THORNBER, J.P.: Chlorophyll-proteins: light-harvesting and reaction center components of plants. Ann. Rev. Plant Physiol. **26**, 127–158 (1975)

TOLBERT, N.E.: Isolation of leaf peroxisomes. In: Particle Separation from Plant Materials. (Compiled by C.A. Price.) Oak Ridge National Laboratory Microsymposium. Conf. 700119 (1970)

TOLBERT, N.E.: Isolation of leaf peroxisomes. Methods Enzymol. **23**, 665–682 (1971a)

TOLBERT, N.E.: Microbodies, peroxisomes, and glyoxysomes. Ann. Rev. Plant Physiol. **22**, 45–74 (1971b)

TOLBERT, N.E., OESER, A., KISAKI, T., HAGEMAN, R.H., YAMAZAKI, R.K.: Peroxisomes from spinach leaves containing enzymes related to glycolate metabolism. J. Biol. Chem. **243**, 5179–5184 (1968)

TREBST, A.: Energy conservation in photosynthetic electron transport of chloroplasts. Ann. Rev. Plant Physiol. **25**, 423–458 (1974)

TRELEASE, R.N., BECKER, W.M., BURKE, J.J.: Cytochemical localization of malate synthase in glyoxysomes. J. Cell Biol. **60**, 483–495 (1974)

URRY, D.W.: Conformation of protein in biological membranes and a model transmembrane channel. Ann. N.Y. Acad. Sci. **195**, 108–125 (1972)

URRY, D.W., LONG, M.M., JACOBS, M., HARRIS, R.D.: Conformation and molecular mechanisms of carriers and channels. Ann. N.Y. Acad. Sci. **264**, 203–219 (1975)

VANDERKOOI, G.: Molecular architecture of biological membranes. Ann. N.Y. Acad. Sci. **195**, 6–15 (1972)

VAN DER WOUDE, W.J.: Significance of the specific staining of plant plasma membranes by treatment with chromic acid-phosphotungstic acid. Plant Physiol. **51**, S82 (1973)

VAN DER WOUDE, W.J., LEMBI, C.A., MORRÉ, D.J.: Auxin (2,4-D) stimulation (in vivo and in vitro) of polysaccharide synthesis in plasma membrane fragments isolated from onion stems. Biochem. Biophys. Res. Commun. **46**, 245–253 (1972)

VILLEMEZ, C.L., MCNAB, J.M., ALBERSHEIM, P.: Formation of plant cell wall polysaccharides. Nature **218**, 878–880 (1968)

WALLACH, D.F.H.: The Plasma Membrane. 186 p. Berlin-Heildelberg-New York: Springer 1972

WALLACH, D.F.H., BIERI, V., VERMA, S.P., SCHMIDT-ULLRICH, R.: Modes of lipid-protein interactions in biomembranes. Ann. N.Y. Acad. Sci. **264**, 142–160 (1975)

WALLACH, D.F.H., GRAHAM, J.M., FERNBACH, B.R.: Beta-conformation in mitochondrial membranes. Arch. Biochem. Biophys. **131**, 322–324 (1969)

WALLACH, D.F.H., ULLREY, D.: The hydrolysis of ATP and related nucleotides by Ehrlich ascites carcinoma cells. Cancer Res. **22**, 228–234 (1962)

WALLACH, D.F.H. ZAHLER, P.H.: Protein conformations in cellular membranes. Proc. Natl. Acad. Sci. US **56**, 1557–1559 (1966)

WEHRMEYER, W.: Zur Kristallgitterstruktur der sogenannten Prolamellarkörpes in Proplastiden etiolierter Bohnen. I. Pentagondodekaeder als Mittelpunkt konzentrischer Prolamellarkörper. Z. Naturforsch. **20b**, 1270–1278 (1965a)

WEHRMEYER, W.: Zur Kristallgitterstruktur der sogenannten Prolamellarkörper in Proplastiden etiotierter Bohnen. II. Zinkblendegitter als Muster tubulärer Anordnungen in Prolamellarkörper. Z. Naturforsch. **20b**, 1278–1288 (1965b)

Wehrmeyer, W.: Zur Kristallgitterstruktur der sogenannten Prolamellarkörper in Proplastiden etiolierter Bohnen. III. Wurtzitgitter als Muster tubulärer Anordnungen in Prolamellarkörper. Z. Naturforsch **20b**, 1288–1296 (1965c)

Weier, T.E.: The ultramicro structure of starch-free chloroplasts of fully expanded leaves of *Nicotiana rustica*. Am. J. Botany **48**, 615–630 (1961)

Weier, T.E., Benson, A.A.: The molecular organization of chloroplast membranes. Am. J. Botany **54**, 389–402 (1967)

Weier, T.E., Brown, D.L.: Formation of the prolamellar body in 8-day, dark-grown seedlings. Am. J. Botany **57**, 267–275 (1970)

Wells, K.: Light and electron microscopic studies of *Ascobolus stercorarius*. Univ. Calif. Publ. Bot. **62**, 1–93 (1972)

Werdan, K., Heldt, H.W.: The phosphate translocator of spinach chloroplasts. In: Proc 2nd. Intern. Congr. Photosyn. Res. Forti, G., Avron, M., Melandri, A., (eds.), Vol. 2, pp. 1337–1344. The Hague: Dr. W. Junk 1972

Werkheiser, W.C., Bartley, W.: The study of steady-state concentrations of internal solutes of mitochondria by rapid centrifugal transfer to a fixation medium. Biochem. J. **66**, 79–91 (1957)

Wettstein, D. von: The formation of plastid structures. In: The photochemical apparatus: Its structure and function. Brookhaven Symp. Biol. **2**, 138–159 (1958)

Whaley, W.G., Dauwalder, M., Kephart, J.E.: Assembly, continuity, and exchanges in certain cytoplasmic membrane systems. In: Origin and Continuity of Cell Organelles. Reinert, J., Ursprung, H. (eds.), pp. 1–45. New York-Heidelberg-Berlin: Springer 1971

Williamson, F.A., Morré, D.J., Jaffe, M.J.: Association of phytochrome with rough-surfaced endoplasmic reticulum fractions from soybean hypocotyls. Plant Physiol. **56**, 738–743 (1975)

Wintermans, J.F.G.M.: Concentrations of phosphatides and glycolipids in leaves and chloroplasts.. Biochim. Biophys. Acta **44**, 49–54 (1960)

Winzler, R.J.: A glycoprotein in human erythrocyte membranes. In: Red Cell Membrane: Structure and Function. Jameson, G.A., Greenwalt, T.J. (eds.), pp. 157–171. Philadelphia, Pa.: J.B. Lippincott 1969

Zbarsky, I.B., Perevoshchikova, K.A., Delektorskaya, L.N., Delektorsky, V.V.: Isolation and biochemical characteristics of the nuclear envelope. Nature **221**, 257–259 (1969)

II. Intracellular Interactions

1. Interactions between Nucleus and Cytoplasm

J. BRACHET

1. Old Theories and Present Ideas about the Biochemical Role of the Cell Nucleus

When experimenters, at the beginning of this century, cut large Protozoa like *Amoeba proteus* or *Stentor caeruleus,* or sea urchin or starfish eggs into two, they were surprised to discover that anucleate fragments always survive for some time (see WILSON, 1925, for an excellent review of the earlier literature): removal of the nucleus is never followed by immediate death. However, the biological activities characteristic of the cell (ciliary movement in Ciliates, locomotion in Amebae) sooner or later fade out, and the anucleate fragments always die before their nucleate counterparts. Anucleate cytoplasm from a mammalian cell survives for a few hours only; the anucleate half of an ameba remains alive for more than one week; in the unicellular *Acetabularia,* anucleate fragments can survive as long as three months. Longevity, in anucleate cytoplasm, can thus vary considerably, and this fact requires explanation. Anucleate cytoplasm has of course lost the major property of all living organisms, genetic continuity; the nucleate fragments, given proper conditions for feeding, will, on the contrary, grow and divide.

Many ideas have been proposed in the past to explain the inferiority of anucleate cytoplasm as compared to nucleate fragments of cells. For instance, LOEB (1899) thought that the cytoplasm dies in the absence of the nucleus because the latter is the main center of energy production. This idea was based on inadequate cytochemical evidence about the intracellular localization of respiratory enzymes; we know now that the mitochondria (and the chloroplasts in green cells) are centers of energy production. The question of whether isolated nuclei are also capable of energy production remains a debated one. Conflicting results have been obtained with nuclei isolated from different cell types. In any event, energy production by the nucleus is, as we shall see, negligible in eggs and unicellular organisms.

A different idea suggested by WILSON (1925) was that the main function of the nucleus is the synthesis of enzymes or coenzymes. This idea was generalized for all proteins by CASPERSSON (1950), who proposed that the cell nucleus is the main center of protein synthesis.

Few biochemists, if any, would still accept these conclusions. We know that most, if not all the proteins (which include many enzymes) present in the nucleus are synthesized in the cytoplasm by the polyribosomes. Good evidence, which will be discussed later, has been presented to demonstrate that cytoplasmic proteins can migrate to and even accumulate in the nucleus. However, as in the case of energy production, published results about in vitro protein synthesis in isolated nuclei are conflicting. While nuclei isolated from cells which possess little cytoplasm (lymphocytes, thymocytes) seem to be capable of limited energy production and protein synthesis, these processes are almost absent in nuclei isolated from larger cells, hepatocytes for instance. In any event, it is now clear that, as we shall see, neither energy production nor protein synthesis is the major activity of the cell nucleus.

In reality, the nucleus is specialized in nucleic acid synthesis. It contains the major part of the cell DNA, except in huge cells (eggs, *Acetabularia*) which possess many mitochondria or chloroplasts.

We know that DNA is the genetic material which contains all the information needed to build up an organism of a given species. DNA can replicate and this replication, in diploid organisms, is usually followed by cell division. In order to allow the expression of the genes, DNA must be transcribed in RNA molecules. This transcription is catalyzed by the RNA polymerases, which often, but not always, accumulate in the nuclei. The result of this transcription is the synthesis of all kinds of RNAs: 28S, 18S and 5S ribosomal RNAs (rRNAs), messenger RNAs (mRNAs) and transfer RNAs (tRNAs). Since all these RNAs are involved in the assembly of the polyribosomes in the cytoplasm, it follows that the RNA molecules synthesized in the nucleus must be transferred to the cytoplasm.

2. Experimental Approaches to the Study of Nucleocytoplasmic Interactions

2.1 Work on Intact Cells

2.1.1 Autoradiography

Autoradiography is a very valuable method for the study of macromolecule synthesis, and has the great advantage that it can be used with intact cells. The principle and the methodology are simple (FICQ, 1959): radioactive precursors, usually tritium (^3H)-labeled, are added to living cells and their incorporation into macromolecules is observed by a photographic procedure. The classical precursors are ^3H-amino acids for the study of protein synthesis, ^3H-uridine for RNA synthesis and ^3H-thymidine for DNA synthesis. Autoradiography shows with great precision, especially when it is used at the electron microscope level, the localization of newly synthesized proteins and nucleic acids. While autoradiography can be used quantitatively to some extent, it has its limitations. For instance, it does not allow an easy measurement of the uptake of precursors, because the latter are too soluble in cytological fixatives. This measurement is important when one wishes to compare results obtained with cells placed in different experimental conditions. Uptake of the precursor and its dilution by preexisting pools are often limiting factors for incorporation into the corresponding macromolecule. For instance, if permeability to amino acids was strongly decreased in anucleate cytoplasm, one might wrongly conclude, if autoradiography shows decreased incorporation, that enucleation diminishes protein synthesis. Another drawback of autoradiography is that it does not, in general, allow the measurement of the specific activity of the newly synthesized macromolecule. For instance, it would be nonsense to study by autoradiography the incorporation of tryptophane into histones, since biochemistry teaches us that most histones do not contain tryptophane. For the interpretation of the results, autoradiography thus often requires the help of biochemical analyses.

2.1.2 Studies on Chromatin and DNA in Intact Cells

Besides the classical and still widely used Feulgen reaction (FEULGEN and ROSSENBECK, 1924), new methods have been proposed for the cytochemical analysis of chromatin: binding of fluorescent dyes or of ^{3}H-actinomycin D (RINGERTZ and BOLUND, 1969; RINGERTZ et al., 1969; HULIN and BRACHET, 1969). Acridine orange gives a bright green fluorescence with DNA and a red fluorescence with RNA. The work of RINGERTZ and his colleagues has shown that there is a close correlation between the amount of dye bound by chromatin and its genetic activity. The same is true of the binding of ^{3}H-actinomycin D, which can be detected by autoradiography. Actinomycin D binds preferentially to the guanylic acid residues of DNA. If chromatin is in a repressed, inactive state, DNA is covered with basic and acidic proteins and there will be little or no actinomycin D binding. If, for some reason, the genome is activated, binding of actinomycin D or acridine orange increases. This has been clearly demonstrated by RINGERTZ and his team (1969) in studies on phytohemagglutinine-treated lymphocytes and on the "reactivation" of chick erythrocyte nuclei. If an avian red blood cell is fused with a normal cell, the condensed erythrocyte nucleus swells as a result of an intake of cytoplasmic proteins, and it soon begins to synthesize RNA. This reactivation is accompanied by a marked increase in actinomycin D and acridine orange binding. There is, on the other hand, a decrease in actinomycin D binding when embryonic cells differentiate and become specialized in the synthesis of a major specific protein-like hemoglobin (HULIN and BRACHET, 1969). Actinomycin D binding can also be studied at the ultrastructural level (STEINERT and VAN GANSEN, 1971).

Another, more delicate technique must be mentioned: in vitro hybridization, which allows the detection in chromosomes and interphase nuclei of highly repetitive DNA sequences (review by HENNIG, 1973). These DNA sequences must first be isolated by biochemical and biophysical means; they are then "copied" into the corresponding radioactive RNA (complementary RNA:cRNA) with RNA polymerase and radioactive precursors. Fixed preparations of nuclei or chromosomes are treated, after denaturation of their DNA by heating or alkali treatment, with the radioactive cRNA; the latter can be detected by autoradiography. Using this technique PARDUE and GALL (1969), JOHN et al. (1969), and others have elegantly demonstrated the localization, in giant chromosomes of *Drosophila,* of the genes coding for the various kinds of rRNAs, for the tRNAs and the histones. This powerful technique has also demonstrated the centromeric localization of the adenine-thymine (A-T)-rich satellite DNA of the mouse and that of the ribosomal genes in the amplified nucleolar organizers of *Xenopus* oocytes (BIRNSTIEL et al., 1968).

2.2 Work on Anucleate Fragments of Unicellular Organisms, Eggs, and Cells

Comparison between nucleate and anucleate halves of living cells is of course one of the best experimental approaches to the understanding of nucleocytoplasmic interactions. Removal of the nucleus by micro-surgery or others means is a more elegant and efficient way of suppressing all transcription than treatments with inhibitors of RNA synthesis such as actinomycin D or α-amanitin.

Acetabularia (Fig. 1) is unique among unicellular organisms because of its exceptional capacities for growth and regeneration (formation of the species-specific "cap" or umbrella) in the absence of the nucleus (HÄMMERLING, 1934). The subject has been extensively reviewed by HÄMMERLING (1953), BRACHET and LANG (1965) and BRACHET (1968) among others. Recently, WERZ (1974) has reviewed in detail our present knowledge about the ultrastructure of this giant alga. Methods for culture sectioning into two halves, etc. have been described by KECK (1964). The great advantages of *Acetabularia* are its huge size and its enormous regeneration potentiality, which facilitate biochemical and biological work. The presence of chloroplasts simplifies the culture problems and gives an exceptional opportunity for analyzing the interactions between nuclear and chloroplastic genomes.

Among Protozoa, *Stentor* and *Amoeba* remain the two favorite objects for study of nucleocytoplasmic interactions. Their relatively large size makes surgical operations fairly easy, but the fragments obtained after section are too small for classic biochemical analysis. One has to use either ultramicromethods of analysis (which are not very popular now) or autoradiography, in order to observe the biochemical changes which follow removal of the nucleus.

Fig. 1. Large caps formed by anucleate fragments of *Acetabularia mediterranea* 3 weeks after section

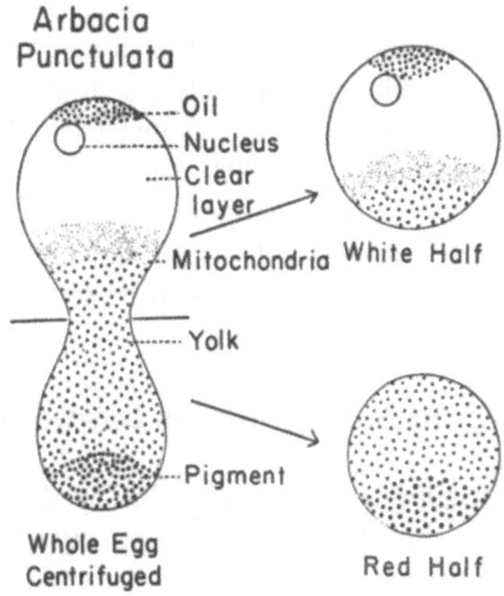

Fig. 2. Separation in nucleate (white) and anucleate (red) halves of unfertilized eggs of the sea urchin *Arbacia*. (HARVEY, 1933)

Sea urchin eggs and, to a lesser extent, those of the mollusks *Ilyanassa* and *Mytilus*, are more favorable than Protozoa for biochemical studies. Unfertilized eggs of *Arbacia* (HARVEY, 1933) and *Sphaerechinus* (AIMI, 1974) can easily be cut into two halves (nucleate and anucleate) by centrifugation at high speed in a density gradient (Fig. 2); large amounts of fragments can be collected in this way. In the mollusk eggs, the first cleavage is unequal, due to temporary formation of an anucleate polar lobe, which is soon taken up by the large CD dorsal blastomere (Fig. 3); it is easy to separate this polar lobe at the so-called "trefoil" stage.

Until a few years ago, the only mammalian cells which could be used for our present purpose were those in which enucleation occurs spontaneously: the red blood cells. The biochemical activity of the reticulocytes (immature red blood cells which have just lost their nucleus) has often been compared to that of mature erythrocytes. But, in 1970, LADDA and ESTENSEN, then POSTE and REEVE (1971) reported that treatment with cytochalasin B produces the expulsion of the nucleus, together with a small amount of cytoplasm, in a large variety of cells. Cytochalasin B is a drug which affects the permeability of the plasma membrane and disrupts

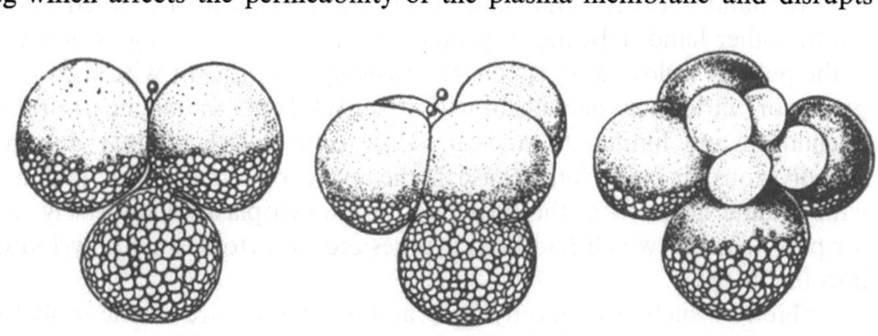

Fig. 3a–c. Cleavage of the egg of the Mollusk Ilyanassa. (a) 2 blastomere stage showing the first polar lobe (trefoil stage). (b) 4 blastomere stage with the second polar lobe. (c) 8 blastomere stage

the bundles of contractile microfilaments which are present in most cells (review by WESSELLS et al., 1971). In many kinds of cells, cytochalasin B induces a constriction which pinches off the nucleus; the latter is finally expulsed: a caryoplast and a much larger cytoplast can be separated from each other. They represent about 20% and 80% of the cell respectively (EGE et al., 1974a). In this way it is possible to separate caryoplasts from cytoplasts by gradient centrifugation with a good yield. More and more work is being done presently on cytochalasin B-enucleated cells; it will be reviewed at the end of this article.

2.3 Biochemical Work on Isolated Organelles

Isolation of cell organelles by differential centrifugation, which was initiated by CLAUDE (1940), is the most important technique for cellular biochemistry: nuclei, mitochondria, lysosomes, peroxisomes, microsomes, ribosomes, chloroplasts, Golgi, plasma membranes, etc. can be isolated, in purified form and in sufficient amounts for biochemical analysis.

The greatest part of our knowledge about the biochemical constitution and the biological role of the cell organelles comes from work done on such purified cellular fractions obtained by differential centrifugation of homogenates. It is beyond the scope of the present article to discuss the enormous wealth of results obtained with that approach.

3. Discussion of Some Recent Experimental Results

3.1 Work on Intact Cells

3.1.1 Autoradiography

Effects of Inhibitors. Incorporation of labeled amino acids, uridine and thymidine has been observed by means of autoradiography in many kinds of cells. If the cells are submitted to short "pulses" (1 h or less) with these precursors, it is usually found that thymidine and uridine are quickly incorporated into the nuclei. On the other hand, labeling of proteins with amino acids begins in the cytoplasm. If the pulse is followed by a "chase" (usually a treatment with the nonradioactive precursor) various changes occur with time. While thymidine incorporation remains unchanged and limited to nuclear DNA (unless it is diluted out, by repeated cell divisions, in very long chases), the nuclear RNAs which had been labeled with uridine move from the nucleus into the cytoplasm. Conversely, the proteins (or part of them) which had been synthesized on cytoplasmic polyribosomes move into the nuclei.

Inhibitors such as actinomycin D and α-amanitin are classic tools for blocking DNA transcription. Small doses of actinomycin D (0.25 μg/ml) selectively arrest rRNA synthesis in most cells (PERRY, 1962), probably because it binds preferentially to the DNA segments rich in guanine and cytosine (G+C). The nucleolar organizers, i.e. the chromosomal regions where the genes coding for 28S and 18S rRNA synthesis are clustered, are precisely made of G+C-rich DNA in most (but not all) animal and plant species. Higher concentrations of actinomycin D arrest the synthesis of all RNA species. α-amanitin (FIUME and WIELAND, 1970) is a specific inhibitor of one of the RNA polymerases (the very important enzymes responsible for DNA transcription). Usually, three different kinds of RNA polymerases are found in the nucleus (I, II, and III, also called A, B, and C).

RNA polymerase I is localized in the nucleolus and specialized in rRNA synthesis. It is α-amanitin-insensitive. RNA polymerases II and III are present in the nuclear sap and differ from each other by their α-amanitin sensitivity: RNA polymerase II, which is inhibited by α-amanitin, is thought to be responsible for mRNA synthesis.

Other inhibitors of transcription allow a more delicate analysis of mRNA synthesis. For instance, camptothecin inhibits the "processing" of mRNA, i.e. the transformation of the giant precursor molecules present in the nucleus (heterogeneous nuclear RNA:HnRNA) into the finished mRNA molecules, which can be isolated from the cytoplasmic polyribosomes (ABELSON and PENMAN, 1972; BUSHNELL et al., 1974). Cordycepin (3′-d-adenosine) selectively blocks the addition to mRNAs of the polyadenylate (poly A) sequences that most of them possess at their 3′ end. The role of these 3′ poly A sequences is probably to stabilize the mRNA molecules, as recently shown by HUEZ et al. (1974). It was known from the work of GURDON et al. (1971) that injection of mouse globin mRNA into oocytes of the toad *Xenopus* is followed by synthesis of mouse hemoglobin (whether the oocyte's nucleus is present or not). If, as was done by HUEZ et al. (1974), one injects mouse globin mRNA which previously has been deprived of its poly A sequence by specific enzymatic digestion, perfectly normal mouse hemoglobin is synthesized by the toad oocyte. The fidelity of translation is thus not affected by the presence or absence of the poly A sequence, but translation of the injected mRNA stops much earlier if it has been deprived of its poly A sequence prior to injection. This is due to the fact that poly A-less mRNA is degraded by the oocytes within 1–2 days while normal mRNA is still intact one week after its injection into *Xenopus* oocytes (MARBAIX et al., 1975).

Discussion of the intranuclear processing of the various RNA species is beyond the scope of this review, but it is worth mentioning here that a mutation (nu-o, in *Xenopus*) has played a major role in our understanding of the control of rRNA synthesis. This mutation is a deletion of the nucleolar organizer region. In the homozygous state, nu-o-fertilized eggs neither form nucleoli, nor synthesize 28S and 18S rRNAs (GURDON and BROWN, 1965). This was the first demonstration that the macromolecular precursors of these rRNAs are synthesized by the nucleolar organizers. It is also worth mentioning that nu-o-homozygotes survive until they reach the feeding stage. This is possible because the egg cytoplasm has accumulated, during oogenesis, a huge stock of ribosomes, sufficient to ensure development until this fairly advanced larval stage.

Little will be said here about DNA synthesis. It is mainly a nuclear event correlated with cell division, and takes place during the so-called S phase of the cell cycle. Two interesting questions, in our present context, may be asked. Is there coordination between the synthesis of nuclear DNA and that of nuclear proteins, histones for instance? Is there a temporal correlation between nuclear DNA synthesis and that of cytoplasmic (mitochondrial, chloroplastic) DNAs?

It is impossible to give a general answer to these two questions. However, in a number of cases such correlations have been found. In general, both histone and mitochondrial DNA syntheses are continuous during the whole interphase (in contrast to that of chromosomal DNA), but their rates increase during the S phase of nuclear DNA replication. Some cytoplasmic change apparently stimulates a coordinate synthesis of nuclear DNA, mitochondrial DNA and histones during this phase of the cell cycle. In the exceptional case of Trypanosomes, nuclear and mitochondrial DNA syntheses are even temporarily linked (STEINERT and STEINERT, 1962). In *Acetabularia,* chloroplast DNA synthesis is also independent of nuclear DNA synthesis since, as we shall see, it can take place in the absence of the nucleus. During cleavage in *Xenopus,* both DNA and histones are synthesized, but their synthesis is not strictly coordinated.

Protein synthesis is essentially a cytoplasmic event which takes place in the polyribosomes. It can be arrested by addition to the cells of inhibitors such as cycloheximide or puromycin. Pulse experiments with labeled amino acids, followed by chases in the presence of inhibitors of protein synthesis, leave little doubt that many, if not all nuclear proteins are of cytoplasmic origin. Another proof of this conclusion is provided by experiments in which labeled proteins have been injected into amphibian oocytes (review by GURDON and WOODLAND, 1970): it was found that small basic proteins (histones, lysozyme, etc.) penetrate quickly into the nucleus, where they accumulate. Larger globular proteins (albumins, hemoglobin) with MW around 60,000 daltons also penetrate into the nucleus, but their concentration does not become higher than in the cytoplasm. Still larger molecules, like ferritine, do not penetrate into the nucleus and are arrested by the nuclear membrane barrier (FELDHERR, 1974; BONNER, 1975).

Finally, it should be added that we have at our disposal valuable inhibitors of cytoplasmic DNA, RNA and protein synthesis: they are ethidium bromide for mitochondrial and chloroplast DNA synthesis; rifampicin for the transcription of these two kinds of DNAs; and chloramphenicol for the synthesis of mitochondrial and chloroplast proteins. Although one should be careful about the interpretation of the results obtained when these inhibitors are added to living cells (BRACHET et al., 1964), they remain very useful tools for the analysis of nucleocytoplasmic interactions.

3.1.2 Role of the Nuclear Membrane in Nucleocytoplasmic Interactions

The structure of the nuclear membrane and its "pore complexes" is well known (Fig. 4). The upper limit for the transport from the cytoplasm to the nucleus of globular proteins is around 60,000 to 70,000 daltons MW.

This raises an interesting question: how can the large rRNA molecules with MW of 1.2×10^{6} and 0.7×10^{-6} daltons, which are synthesized in the nucleus, cross the nuclear membrane and reach the cytoplasm? An ingenious answer to this puzzling question has been proposed by FRANKE et al. (1971), who have studied in detail the ultrastructure and the biochemical composition of nuclear envelopes isolated from various kinds of cells. As shown in Figure 5, RNA molecules are thought to be coiled and associated with proteins in the nuclear sap. When they come into contact with the inner layer of the nuclear membrane, they uncoil and pass as a thread through the nuclear pore complex. In fact,

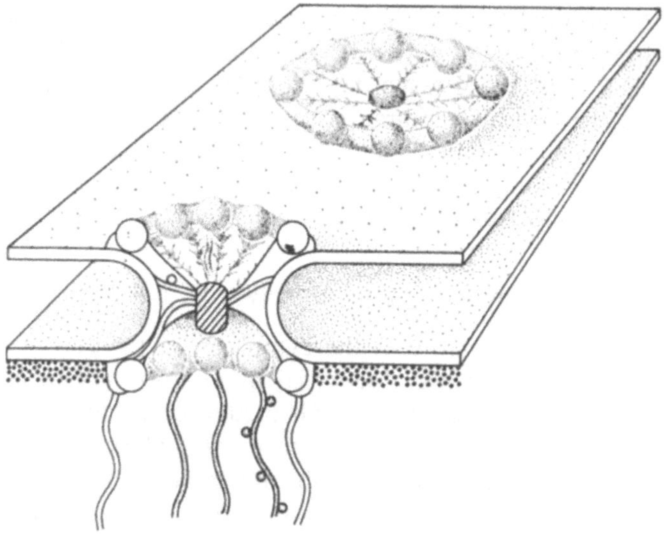

Fig. 4. Diagram illustrating the structural components of the nuclear pore complex of the amphibian oocyte. Annuli lie upon the cytoplasmic and nucleoplasmic margin of the nuclear pore which consists of eight symmetrically distributed granules and some amorphous material. Amorphous material extends also into the pore interior in which the central granule and the internal fibrils appear as particulate structures. Similar fibrils, often studded with small granules, are attached to the nucleoplasmic annulus. Peripheral chromatin underlies the inner nuclear membrane. (FRANKE and SCHEER, 1970)

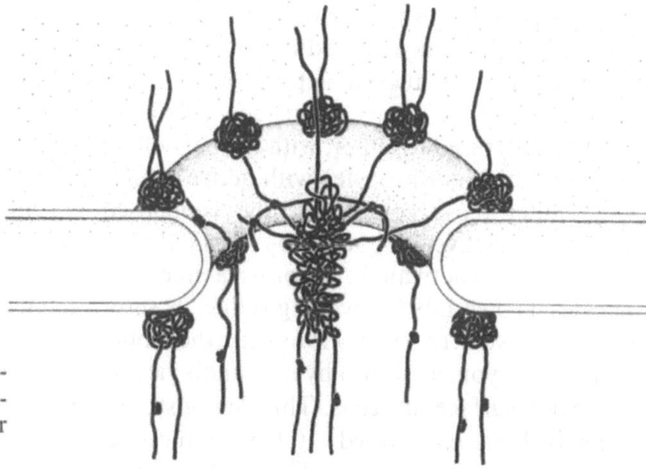

Fig. 5. Passage of fibrillar molecules (RNAs) from the nucleus to the cytoplasm. (After FRANKE, 1970)

the so-called central granule of this complex would correspond to RNA molecules in the process of migrating from the nucleus toward the cytoplasm. Outside the nuclear membrane, the rRNA molecules coil again and combine with ribosomal proteins. The same scheme remains valid for the passage of mRNA molecules. It is known that these molecules are bound to proteins, which are different in the nuclear sap and in the cytoplasm. These mRNA-protein complexes are the so-called *informofers* in the nucleus (GEORGIEV, 1972) and *informosomes* (SPIRIN, 1966) in the cytoplasm.

Since it is known that the nucleus contains many breakdown products of the large precursors of both mRNAs (HnRNA) and rRNAs (45S RNA in mammalian cells), which never reach the cytoplasm, it seems difficult to escape the conclusion that the nuclear membrane is more than a passive permeability border. It might operate a selection among the molecules which come into contact with it and exert control (like policemen at the border between two states) over the movement of these molecules in both directions (AARONSON and WILT, 1969).

3.1.3 Choice by the Cytoplasm between DNA Replication and Transcription

The determining role of the cytoplasm in the choice between DNA replication and transcription was discovered by GURDON (1970) in experiments in which he injected nuclei taken from adult tissues (brain, for instance) into *Xenopus* oocytes before and after maturation (meiosis). Labeled precursors (thymidine, uridine) were injected together with the nuclei and their incorporation into DNA and RNA was observed by autoradiography. In situ, brain nuclei synthesize actively all kinds of RNAs, but no DNA. This situation remains unchanged when they are injected into oocytes, but if they are injected into unfertilized eggs (which have eliminated their first polar body) they take up cytoplasmic proteins, swell and soon begin to replicate their DNA. Simultaneously, the nucleoli disappear and RNA synthesis stops. In other words, the injected nuclei behave, in unfertilized eggs just like spermatozoa, which undergo DNA synthesis soon after their penetration into the eggs. Like the nuclei of cleaving eggs, the injected brain nuclei do not synthesize RNA and do not form nucleoli until the end of cleavage. At that time, rRNA synthesis and formation of nucleoli begin simultaneously in both blastula nuclei and injected adult nuclei.

This control of nuclear activity by the surrounding cytoplasm is not specific to eggs. Exactly the same conclusions have been drawn from experiments in which HARRIS (1970) fused together, with the help of inactivated Sendai virus, biochemically inactive chick erythrocytes with active human cells (fibroblasts or HeLa cancer cells). Under these conditions, the condensed erythrocyte nucleus, which was completely inactive in both DNA and RNA syntheses, swells and, as already mentioned, is reactivated. It forms a nucleolus, begins to synthesize RNA and replicates its own DNA. Swelling is due, as shown in elegant immunological experiments by RINGERTZ et al. (1969), to the uptake of human cytoplasmic proteins. The heterokaryon, as shown by immunological methods, becomes capable of synthesizing chick surface antigens. This demonstrates that the chick erythrocyte nucleus has really been reactivated: it has synthesized its own mRNAs, which have in turn been correctly translated. Curiously enough, the synthesis of such chick surface antigens is suppressed when the chick nucleoli are inactivated, in the heterokaryons, by localized UV radiation (DEAK and DEFENDI, 1975). This suggests a hitherto unknown role of the nucleolus in the synthesis, processing or transport of mRNAs. To return to DNA synthesis in heterokaryons, it has been shown that the cytoplasm controls its initiation, but not the length of the S phase of DNA replication (MARSHALL-GRAVES, 1972; RAO et al., 1975).

There are also important effects of cell fusion on cell differentiation. As shown by EPHRUSSI and his colleagues (reviews in DAVIDSON, 1971; DAVIS and ADELBERG, 1973), fusion of a differentiated cell (a pigmented cell, for instance) with an undifferentiated cell almost invariably produces the "extinction" of the differentiated phenotype. In example the hybrid no longer synthesizes pigments and the various enzymes involved in melanin formation. When the two nuclei have fused together, forming a synkaryon, cell division occurs and a hybrid cell line can be established. During the repeated cell divisions, a number of chromosomes are lost (there is a preferential loss of human chromosomes in man-mouse somatic hybrids) and the differentiated phenotype often reappears. These experiments, as well as those in which the differentiated phenotype is lost when differentiated cells are allowed to incorporate bromodeoxyuridine (BrdUr) into their DNA (HOLTZER et al., 1972), provide promising means for a better understanding of the genetic control of cell differentiation. How BrdUr arrests specifically the synthesis of the "luxury" proteins characteristic of the differentiated state (melanin, hemoglobin, etc.) remains unknown.

3.1.4 Effects of the Plasma Membrane on Nuclear Activity

Certain cells, the lymphocytes in particular, react strongly to contact with certain foreign proteins. Besides antigens, phytohemagglutinins (PHA) and concanavalin A (Con A) induce a stimulation of RNA and DNA synthesis, followed by cell division. These proteins of plant origin bind to specific sites of the plasma membrane and are therefore called *lectins* or *mitogens*. Con A, in particular, binds to carbohydrate residues of the cell surface.

Limited digestion of the proteins present in the cell membrane by trypsin or other proteolytic enzymes also induces proliferation in tissue culture cells (BURGER et al., 1972). Inhibitors of trypsin, such as proteases, on the other hand, arrest cell proliferation (SCHNAEBLI and BURGER, 1973) and inhibit reactivation of chick erythrocyte nuclei after fusion with fibroblasts (DARZYNKIEWICZ et al., 1974). Since proteases and lectins do not penetrate into cells (except by pinocytosis), it is now generally accepted that signals received at the level of the cell surface affect the activity of chromatin: its derepression can lead to both DNA transcription and replication.

A similar mechanism probably takes place in embryonic inductions. For instance, induction of the exocrine and endocrine parts of pancreas (which is accompanied by synthesis of trypsin, chymotrypsin, insulin, glucagon, etc.) is mediated by a protein produced by the surrounding mesenchyme cells (LEVINE et al., 1973). This protein is still active when it is bound to sepharose beadles; this strongly suggests that it acts on the plasma membrane. A comparable situation exists in plant cells. There is good evidence that auxins bind to specific receptors present in the plasma membrane (THIMANN, 1969).

Other substances which also act on the cell surface and produce very important effects are the cyclic nucleotides (cAMP and cGMP). They have opposite effects: in general, cAMP slows down mitotic activity and favors cell differentiation, whereas cGMP, like the lectins, stimulates cell proliferation. The enzymes involved in the synthesis (cyclases) and the breakdown (phosphodiesterases) of the c-nucleotides are, to a large extent, localized on the cell surface. Cell physiology is thus largely controlled by the cAMP/cGMP ratio, which would be responsible for what HERSHKO et al. (1971) have called the pleiotypic control of cell activities. Such parameters as the uptake of amino acids and nucleosides, synthesis and degradation of RNAs and proteins would all be regulated by a single effector, the ratio between the two c-nucleotides (MOENS et al., 1975).

How can the signals received at the cell membrane level be transmitted to the cell nucleus? There is good evidence, based on the use of inhibitors such as colchicin and cytochalasin B, that microtubules and microfilaments are involved in this process. One of the main functions of the c-nucleotides is to modify the activity of the proteinkinases, which phosphorylate proteins. Such phosphorylation probably takes place during the assembly of tubulin molecules for the formation of microtubules.

From the viewpoint of cell differentiation, the appearance of complex structures such as centrioles, microtubules, cilia and flagellae has long been a baffling problem; present evidence suggests that these are probably less complicated than they appeared, because self-assembly from preexisting subunits dispersed in the cytoplasm is a common phenomenon. For instance, centrioles might arise, in eggs, from preexisting procentrioles (VAN ASSEL and BRACHET, 1968; KATO and SUGIYAMA, 1971). Particularly striking are the experiments of REBHUN et al. (1974) where maturation spindles isolated from eggs of the worm *Chaetopterus* have been found to grow in vitro when tubulin molecules prepared from mammalian brain are added to them.

The long-distance interactions between cell membrane and genetic material form an exciting subject; great progress can be expected, in the near future, in this new and important field.

3.2 Studies on Anucleate Fragments of Unicellular Organisms, Eggs, and Cells

3.2.1 Acetabularia

3.2.1.1 Introduction

Acetabularia (Fig. 1) has been a fascinating subject for biologists ever since the discovery by HÄMMERLING (1934, 1953) that anucleate fragments of this giant alga not only survive for months, but can even regenerate species-specific caps. Regeneration is due to the production by the nucleus of species-specific morphogenetic substances. It is now generally accepted (but not yet definitely proved) that these morphogenetic substances are families of mRNAs transcribed on the nuclear genes and their translation products. Such mRNAs should be very stable and distributed along a decreasing apico-basal gradient.

Acetabularia has been the object of many review articles, of several Symposia and even of a whole book by PUISEUX-DAO (1970). The synthesis of macromolecules in relationship with morphogenesis was discussed in detail by BRACHET in 1968. More recent information can be found in Symposia edited by BRACHET and BONOTTO (1970), and BONOTTO et al. (1971). Two recent and very complete reviews concerning ultrastructure and cytochemistry of the alga are available (WERZ, 1974; SPRING et al., 1974). In view of the existing wealth of information, the present review is limited to the most significant recent findings.

3.2.1.2 Production and Distribution of Morphogenetic Substances

Morphogenetic substances must be translocated from the nucleus to the apex of the stalk, since the concentration of cap-forming substances gradually decreases from the apical to the basal part of the stalk. Recent experiments by SANDAKHIEV et al. (1972) have shown that centrifugation of anucleate fragments does not modify the polarity of the alga. Centrifugation displaces chloroplasts and other cell organelles. According to LÜSCHER and MATILE (1974), 90% of total aldolase is sedimentable in vivo. The time when the morphogenetic substances become unsedimentable coincides with genetic expression, i.e. the beginning of cap formation. These experiments substantiate a hypothesis by BONOTTO et al. (1971), who proposed that the plasma membrane might contain specific receptors for the morphogenetic substances; the concentration of these receptors would decrease along an apico-basal gradient. It is indeed known, from the cytochemical work of WERZ (reviewed in WERZ, 1974), that the tip of the alga contains specific glycoproteins. This tip, in contrast with the rest of the cell wall, gives a strong metachromatic staining characteristic of acid polysaccharides, when algae are vitally stained with toluidine blue (personal observations).

Recently, CHIRKOVA et al. (1970) have shown that it is possible to separate by centrifugation and dissection the main constituents of the alga and to "reconstitute" a living alga by grafting the previously isolated nucleus and chloroplasts into the cell wall. Such reconstituted algae are capable of cap and even cyst formation. This kind of work is now possible because media in which isolated nuclei remain biologically active for at least 10 min have been devised (BRÄNDLE and ZETSCHE, 1973).

Biochemical studies on such isolated *Acetabularia* nuclei are much needed, in view of the generally accepted idea that the morphogenetic substances are stable mRNAs of nuclear origin. All we know is that isolated *Acetabularia* nuclei are capable of RNA synthesis and that the latter, as one would expect, is sensitive to actinomycin D and α-amanitin, but not to rifampicin (BRÄNDLE and ZETSCHE, 1971, 1973). One would like to know more about the nature of the RNA species synthesized by isolated *Acetabularia* nuclei. Are they mainly ribosomal? Could newly synthesized mRNAs be isolated thanks to the presence of poly A sequences? Would these mRNAs be associated with proteins? Could they be translated in an in vitro or a *Xenopus* oocyte system?

One should not forget, however, that the ultrastructure of the perinuclear space is particularly complex in *Acetabularia;* this suggests the existence in vivo of very complex mechanisms for the control of nucleocytoplasmic exchanges. Readers interested in this perinuclear "labyrinthum" are referred to the papers by BOLOUKHÈRE (1965), WERZ (1974) and particularly FRANKE et al. (1974).

Recently, an attempt was made by ALEXEEV et al. (1974) to prove experimentally that the morphogenetic substances are indeed ribonucleoprotein (RNP) particles. The authors separated, by ultracentrifugation of homogenates, RNP particles of various sizes (20–25 S, 40–45 S and 80–100 S). The 40 S particles were injected into basal fragments of anucleate stalks and it was found that, in 10% of the cases, some morphogenesis took place (formation of sterile whorls and rudimentary caps). Morphogenesis, in the experiments, was very limited, but one can hope that continuation of these interesting and bold experiments will lead to more convincing results.

In view of the suggestion by BONOTTO et al. (1971) that the cell membrane might contain specific receptors for such RNP particles, more information is needed about the biochemistry of the cell wall. All we know is that the pattern of the newly synthesized proteins changes during development of the alga and that there are differences in this respect between nucleate and anucleate halves (CERON and JOHNSON, 1971). Very recently, we found (BRACHET, 1975) that organomercurials, which react with sulfhydril groups without penetrating easily into cells, block growth and cap formation in both nucleate and anucleate fragments. These experiments suggest that proteins containing -SH groups and present in (or near) the cell membrane are involved in morphogenesis.

3.2.1.3 Energy Production, Circadian Rhythms

It is well known (review in BRACHET, 1968) that enucleation has little or no effect on oxygen consumption and photosynthesis. More recently, CRAIG and GIBOR (1970) have confirmed that photosynthesis remains normal in both nucleate and anucleate halves for at least two weeks. Parallel work by SHEPHARD and BIDWELL (1973) has demonstrated that isolated chloroplasts synthesize in vitro the pigments involved in photosynthesis.

Virtually nothing is known about the regulation of energy production in nucleate and anucleate halves. Of potential interest is the finding by LEGROS et al. (1974) that addition of insulin increases oxygen consumption in both nucleate and anucleate fragments; aldosterone decreases respiration in nucleate halves and has no effect on anucleate ones. These observations suggest that the cell membrane possesses, in *Acetabularia,* receptors for hormones. While the insulin receptors are retained in anucleate halves, aldosterone receptors would be lost in the absence of the nucleus.

It has been known for more than ten years that anucleate fragments of *Acetabularia* retain a rhythm in photosynthetic capacity for several weeks. However, it is the nucleus which "sets the clock", as shown by experiments where nucleate and anucleate halves of fragments which were at opposite phases of the cycle were combined (SCHWEIGER et al., 1964). The same conclusion has been drawn from experiments where nucleate fragments of algae which had lost the rhythm were combined with anucleate ones which had retained it and vice versa. The presence or absence of the rhythm is thus nucleus-dependent (VANDEN DRIESSCHE, 1967).

Experiments with inhibitors, actinomycin D in particular, indicate that RNA synthesis is involved in the rhythm of photosynthetic capacity. Actinomycin D abolishes the rhythm in nucleate halves, but not in anucleate ones (VANDEN DRIESSCHE, 1967). Similar experiments by BRACHET et al. (1964) dealing with morphogenesis had shown that actinomycin D inhibits regeneration in nucleate halves and not in anucleate ones. This has been taken as support for the hypothesis that the morphogenetic substances are mRNAs or proteins coded for by these mRNAs.

More recent work (reviewed by VANDEN DRIESSCHE, 1973) has shown that many circadian rhythms superimpose themselves upon the rhythm in photosynthetic capacity. They deal with the size and ultrastructure of the chloroplasts, their content in polysaccharides and ATP, the Hill reaction, etc. In all the cases which have been analyzed from that particular viewpoint, it seems that anucleate fragments retain these rhythms for a long time, but their phase is modulated by factors of nuclear origin, probably mRNAs.

3.2.1.4 Chloroplastic and Cytoplasmic Protein Synthesis

As discussed in detail in a previous review (BRACHET, 1968), many proteins, including a number of enzymes, are synthesized in the absence of the nucleus. The fact that specific enzymes are synthesized in anucleate fragments was at first taken as evidence that their production is controlled by stable mRNAs formed in the nucleus, but the discovery that chloroplasts contain DNA and are able to synthesize RNA and proteins obviously complicates this interpretation.

However, the fact that *Acetabularia* contains a single nucleus, which can easily be taken away, makes this alga an exceptionally favorable material for the study of interactions between the nucleus and the chloroplasts. Full advantage of this favorable situation has been taken by SCHWEIGER and his colleagues in Wilhelmshaven.

One should first be reminded of the fact that the activity of many enzymes markedly increases in both anucleate and nucleate halves at the time of cap formation. Enzyme synthesis (as shown by experiments with inhibitors of protein synthesis) is thus under the same control whether the nucleus is present or not. This would obviously speak for a control at the translational level, if autonomous protein synthesis in isolated chloroplasts did not weaken this conclusion (GOFFEAU and BRACHET, 1965). Experiments on the regulation of phosphatase activity, described in a review article (BRACHET, 1968), show that the idea of a controlling role of the nucleus should not be hastily rejected. Enzyme activity undergoes different changes in nucleate and anucleate halves when they are transferred from normal to phosphate-deficient medium. Of interest, in this respect, is a recent paper by BANNWARTH and SCHWEIGER (1975) dealing with thymidine phosphorylation, an event which is closely linked to DNA synthesis. Phosphorylation considerably increases when the cap reaches its full size and the stalk becomes yellowish. This is the stage at which, in nucleate algae, the large vegetative nucleus breaks down and gives rise to daughter nuclei, which are the site of repeated DNA replication. Nevertheless, the same increase in thymidine phosphorylation occurs in anucleate halves, where there is of course no multiplication of secondary nuclei. It must be concluded that the phosphorylation process is regulated at the posttranscriptional level and that, contrary to expectations, it is not coupled with DNA synthesis.

Elegant interspecific grafting experiments by SCHWEIGER and his team have clearly demonstrated that most, if not all chloroplast proteins are under nuclear control. Chloroplast enzymes such as lactic and malic dehydrogenases can be distinguished, in various *Acetabularia* species, by the electrophoretic pattern of their isozymes. When a nucleate half of a given species is grafted on an anucleate stalk of another species, the isozyme pattern becomes that of the nuclear type after a few days (SCHWEIGER et al., 1967). Similar results were obtained for insoluble chloroplast proteins by APEL and SCHWEIGER (1972, 1973), except that replacement of the initial species-specific proteins by those which are synthesized under nuclear control takes a longer time (6 weeks). This delay might be due to the fact that isolated chloroplasts mainly synthesize insoluble (membrane) proteins. Chloroplast protein synthesis is thus probably under double (nuclear and chloroplastic) genetic control. In fact, analysis of protein synthesis inhibition in isolated chloroplasts by chloramphenicol (an inhibitor of chloroplast protein synthesis) and cycloheximide (which arrests cytoplasmic protein synthesis) has led APEL and SCHWEIGER (1973) to the following conclusions: two out of the three main proteins forming the chloroplast membranes are synthesized by 70S and 80S chloroplast ribosomes; the third one is synthesized by 80S cytoplasmic ribosomes. But even the synthesis of chloroplast ribosomal proteins (at least those of their 44S subunits) is under nuclear control, as shown by KLOPPSTECH and SCHWEIGER (1973a) in similar experiments.

It is worth reminding the reader, at this point, that morphogenesis is inhibited by cycloheximide and puromycin, but not by chloramphenicol, in both nucleate and anucleate fragments of *Acetabularia*. There is no doubt that protein synthesis carried out by polyribosomes, made of cytoplasmic 80S ribosomes and stable mRNAs of nuclear origin, is much more important for morphogenesis than chloroplast protein synthesis. This view is shared by BRÄNDLE and ZETSCHE (1973), who conclude that chloroplasts are dependent on the nucleus, but less than mitochondria, and that the arrest of growth in anucleate halves is due to the arrest of cell wall and protein formation: this is probably the stage at which cytoplasmic polyribosomes can no longer function and are degraded, because some essential factor of nuclear origin (mRNAs?) is exhausted.

Cytoplasmic ribosomes have long remained elusive, because they are quantitatively much less abundant than the chloroplast ribosomes. KLOPPSTECH and SCHWEIGER (1973b, 1975) have succeeded in isolating cytoplamic ribosomes from *Acetabularia*. Their sedimentation constant is 80S and they contain 26S and 18S rRNAs. Uridine incorporation into those two rRNAs is entirely dependent on the nucleus; it is not inhibited by rifampicin, which arrests chloroplast RNA synthesis. Ribosome synthesis is rapid in young algae. The newly made cytoplasmic ribosomes accumulate first in the neighborhood of the nucleus. A recent paper by KLOPPSTECH and SCHWEIGER (1975) confirms that the 80S cytoplasmic ribosomes first appear in the proximity of the nucleus. It takes them 5–7 days to attach to membranes. They are then translocated to the apex, where after 2–3 weeks they are more concentrated than elsewhere. The half-life of their rRNAs is long (80 days). Calculations show that, in order to account for the number of ribosomes present in the alga, the redundancy of the ribosomal genes (to which we shall return in the next Section) must be greater than 10,000 copies. KLOPPSTECH and SCHWEIGER (1974) have also studied the effects of amputation on the synthesis of 80S ribosomes by the nucleus. Removal of part of the stalk speeds up their synthesis after 1 day; the rate of ribosomal synthesis becomes maximal after 2–4 days and is not affected by illumination. In such nucleate halves, the synthesis of chloroplast ribosomes lags behind that of cytoplasmic ribosomes for 2 days and does not occur in the absence of light. These experiments clearly show that chloroplast and cytoplasmic ribosome syntheses are two distinct events in *Acetabularia*.

3.2.1.5 *Nucleic Acid Synthesis*

Little will be said here about chloroplast DNA and RNA synthesis, for two reasons: there has not been major progress since the 1968 review (BRACHET), and there is more and more evidence that these nucleic acids are not directly involved in morphogenesis. Their main function is probably to contribute to the synthesis of chloroplast proteins, as discussed in the preceding section.

Chloroplast DNA, in *Acetabularia,* is made of long fibrous molecules; their renaturation, after denaturation, is slow, suggesting a high degree of kinetic complexity (GREEN, 1973). Chloroplast DNA can replicate autonomously, as has been known for a long time; its synthesis is inhibited by hydroxyurea and, more specifically, by ethidium bromide (HEILPORN and LIMBOSCH, 1971 a, b). But the experiments with these inhibitors lead to the conclusion that chloroplast DNA replication is not required for initiation of cap formation; at best, it may play a role in the growth of already formed caps. These negative conclusions reinforce the view that the morphogenetic substances are synthesized as the result of nuclear gene transcription.

Fortunately, our knowledge of transcription of nuclear genes (especially the ribosomal cistrons) has recently made rapid and important progress, thanks to refined electron microscope work by TRENDELENBURG et al. (1974), BERGER and SCHWEIGER (1975b) and SPRING et al. (1974). They have spread and positively stained preparations from isolated nuclei and nucleoli of *Acetabularia* and found that the nucleoli are formed from an increasing number of subunits. This number reaches about 100 in a fully grown vegetative nucleus. Each of these subunits contains about 130 ribosomal cistrons. The number of the ribosomal genes, which increases through an amplification process during the growth of the alga, thus finally reaches a value of the order of 13,000. The arrangement of the cistrons for rRNA precursor is basically similar in *Acetabularia* and in animal cells such as amphibian oocytes. Fibril-covered matrix units alternate with untranscribed "spacers", giving the cistrons the well-known "Christmas tree" appearance (Fig. 6). However, in *Acetabularia,* the length of the repeated units is highly variable (from 2 to 6 μm) and this is due to variation in the length of both transcribed matrix units and spacers. Many of the shorter matrix units have a length corresponding to a molecular weight of approximately 2×10^6 daltons; this corresponds to the sum of the molecular weights of the two cytoplasmic rRNAs (1.3 and 0.7×10^6 daltons). It therefore follows that there is little or no loss of material during the processing of the rRNA precursors, in contrast to what happens in animal cells (in mammalian cells, the molecular weight of the rRNA precursor is about 4.5×10^6 daltons, i.e. more than twice that found for the *Acetabularia* precursor).

Much less, unfortunately, is known about chromosomal DNA in *Acetabularia,* but definite progress has been made recently by SPRING et al. (1975). They found that, during the growth of the vegetative nucleus, lampbrush chromosomes (similar to those of the amphibian oocytes) can often be detected. These lampbrush chromosomes are made of chromomeres and loops; the former measure 1–2 μm, the latter 20 μm. The similarity of the lampbrush chromosomes in oocytes and in *Acetabularia* suggests that the vegetative nucleus of this alga might be in meiotic prophase.

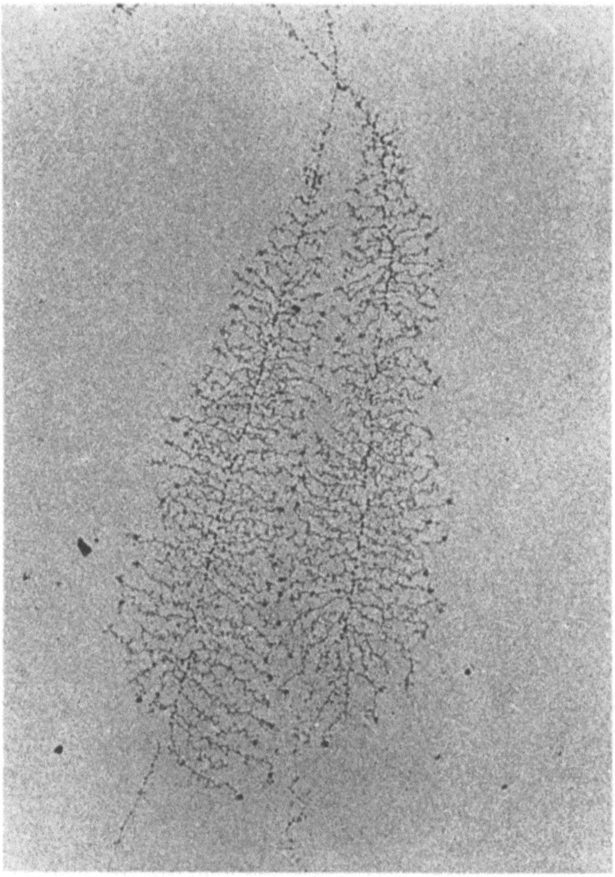

Fig. 6. Nucleolar organizers isolated from nucleoli of *Xenopus* oocytes. "Christmas tree" aspect (also found for ribosomal genes in *Acetabularia*) due to growth of precursor rRNA chains missing at lower and upper ends of DNA fibers, these regions correspond to untranscribed spacer. (Courtesy of Dr. O.L. MILLER)

3.2.1.6 Cytoplasmic Effects on the Nucleus

It is known from the work of STICH (1951) that a decrease in energy production in the cytoplasm deeply affects the structure and chemical composition of the nucleus and nucleolus in *Acetabularia*. The nucleus shrinks, the nucleolus becomes spherical and loses much of its basophily when the algae are cultivated in the dark. These changes are reversible when the algae are brought back to the light. Similar results are obtained when algae are cultivated in the presence of uncouplers of oxidative phosphorylations (BRACHET, 1968). Another interesting effect of the cytoplasm on the nucleus was found by HÄMMERLING (1953). Cutting the stalk each time a cap forms completely prevents the breakdown of the vegetative nucleus and its entry into mitosis. As in amphibian oocytes and somatic hybrids (see Sect. 3.1.3), the choice between DNA transcription and replication is decided by the quality of the surrounding cytoplasm in *Acetabularia*.

Recently, BERGER and SCHWEIGER (1975a) have studied the ultrastructure of the nucleus and perinuclear cytoplasm in young and old algae and found marked differences. These morphological differences are controlled by the surrounding cytoplasm. If an old nucleus is transplanted into young cytoplasm, it "rejuvenates" and acquires, after 10 days the morphology of a young nucleus. The converse experiment leads to ageing of the grafted young nucleus. The cytoplasmic control of ageing of the nucleus is faster than that of its rejuvenation.

3.2.2 Protozoa

Most of the work has been done on amebae (especially *A. proteus*) and, to a lesser extent, on the Ciliate *Stentor ceruleus*. This work is much less extensive than that done on *Acetabularia*, and its development is hampered by the fact that most, if not all *A. proteus* strains harbor endosymbionts which retain their own synthetic capacities in anucleate cytoplasm.

Section of an ameba into two halves leads to rapid changes in the anucleate fragment: it becomes spherical, no longer attaches to the substratum and becomes incapable of locomotion and phagocytosis within a few minutes. Grafting a nucleus into the anucleate half is followed by a fast and spectacular reappearance of motility; this occurs even when a heterologous nucleus (i.e. belonging to another species of amebae is transplanted into the anucleate fragment (FLICKINGER, 1973). But in the case of such heterologous transplantations the anucleate cytoplasm is not saved from death, which usually occurs after 10–12 days. The molecular bases of these facts remain completely unknown: it is known that amebae contain contractile proteins (actin and myosin), but why they should contract as soon as the nucleus is removed and relax when a nucleus (even from a different species) is reintroduced remains a mystery.

It is possible to go further and, as has been done by JEON et al. (1970), to isolate by microdissection the nucleus, the endoplasm and the contractile cortex of an ameba. These constituents can then be reassembled and a normal ameba will be obtained, capable not only of locomotion, but even of division if it is properly fed.

Enucleation alters the ultrastructure of the cytoplasm. The most conspicuous changes are the disappearance of the Golgi bodies, a swelling of the mitochondria and a decrease in the number of ribosomes. Re-introduction of a nucleus is followed by quick reappearance of the Golgi bodies (FLICKINGER, 1973).

The biochemical changes induced by removal of the nucleus have been studied mainly by the author of this article and have been described in a previous review (BRACHET, 1974). Briefly, these changes are the following: the oxygen consumption and the ATP content remain unaffected for several days after removal of the nucleus. However, the ATP content falls much faster in anucleate than in nucleate halves under anaerobic conditions. Anaerobic energy-producing reactions (glycolysis probably) are thus less efficient in the absence of the nucleus. It has indeed been found that anucleate amebae hardly use their glycogen and fat supplies. This metabolic deficiency of the anucleate halves might be due to a decreased content in NAD, since NAD synthesis is often a function of the nucleus. The protein content of the anucleate halves slowly goes down. Not all proteins are under a single general nuclear control, since the activity of certain enzymes remains constant for 10 days, while that of others quickly and markedly decreases after removal of the nucleus. It is interesting to note that the enzyme which removes sialic acid from cell surface glycoproteins (cytidine 5′-monophosphosialic synthetase) is accumulated in the nucleate halves (KEAN and BRUNER, 1971). This suggests a role of the nucleus in the composition of the cell surface. Finally, the total RNA content quickly and strongly decreases in anucleate halves of *A. proteus*.

The experiments on the incorporation of precursors for the synthesis of macromolecules have little significance in view of the already mentioned presence of endosymbionts in amebae. Much more interesting and important are the nuclear transplantation experiments by GOLDSTEIN (review in 1974). The nucleus of a radioactive ameba, labeled with either a RNA or a protein precursor, is injected into a normal "cold" ameba. The movement of the labeled material from the "hot" nucleus to the "cold" cytoplasm and eventually back to the nucleus is observed by autoradiography or biochemical techniques.

These experiments have demonstrated the existence of cytonucleoproteins, i.e. of a class of proteins which migrate back and forth from the nucleus to the cytoplasm (PRESCOTT and GOLDSTEIN, 1969). These proteins vary in molecular size and in turnover speed. The proteins which move very quickly from the nucleus to the cytoplasm and vice versa have a low molecular weight (about 2,300 daltons only). They represent as much as 17% of all soluble proteins and 3% of total proteins (JELINEK and GOLDSTEIN, 1973). Their physiological function remains unknown, but GOLDSTEIN'S hypothesis that they play a role in gene regulation by binding to chromatin segments is certainly attractive and plausible.

Recent experiments by GOLDSTEIN et al. (1973) leave little doubt that certain RNA molecules also shuttle back and forth between the nucleus and the cytoplasm. According to GOLDSTEIN and TRESCOTT (1970), RNA species with sedimentation constants of 4–6S, 19S and even 30S easily move from the cytoplasm into the nucleus. This migration of a 30S RNA into the nucleus is surprising, particularly since the same authors found that certain 4S RNAs never leave the nucleus. These experimental results, if correct, speak strongly in favor of some selectivity or control exerted by the nuclear membrane on nucleocytoplasmic interchanges (cf. Sect. 3.1.2). It should be added that, according to GOLDSTEIN (1974) and YUDIN and NEYFAKH (1973), nuclear RNA leaves the nucleus at the beginning of mitosis and goes back into the nuclei of the daughter cells after cell division. The function of these RNAs which migrate back and forth from nucleus to cytoplasm remains totally unknown.

Protozoa have also proved useful for studying the control of mitosis: nuclear transplantation experiments, analyzed by autoradiography after ^{3}H-thymidine incorporation, have shown that, in amebae, neither the continuation nor the termination of DNA synthesis is controlled by the cytoplasm (ORD, 1971). However, similar transplantation experiments, but with heterologous nuclei, have led RAO and CHATTERJEE (1974) to the conclusion that chromosome replication, in amebae, is under cytoplasmic control. The question obviously deserves more experimental work, but there is little doubt that, in many cells, signals received at the cell membrane or underlying cortex levels are of fundamental importance for the initiation of cell division. This has been clearly shown in *Stentor* by DE TERRA (1974) who pointed out that the same conclusions are valid for eggs. The question of cytoplasmic control on DNA replication has already been discussed, in the case of tissue culture cells and somatic hybrids, in Sect. 3.1.2 and 3.1.3. The general rule seems to be that the cytoplasm, after receiving signals from the outer layers of the cell, controls the initiation, but not the length, of the DNA replication period (S phase of the cell cycle).

3.2.3 Eggs

3.2.3.1 Sea Urchin Eggs

A good deal of work has been done on unfertilized sea urchin eggs which, as already mentioned, can easily be separated into nucleate and anucleate halves by gradient centrifugation. Unfertilized sea urchin eggs are in a strongly repressed condition so far as oxygen consumption and protein, RNA and DNA synthesis are concerned. Both nucleate and anucleate fragments can be activated by treatment with parthenogenetic agents (hypertonic sea water, for instance) which induce cortical changes leading to the formation of a fertilization membrane. Development of such activated anucleate halves is very poor. They give, at best, irregularly cleaved morulae or blastulae.

The results of biochemical work on sea urchin egg fragments have been presented in detail in a previous review (BRACHET, 1974). They will be briefly summarized. *Oxygen* consumption, after parthenogenetic activation, is higher in anucleate than in nucleate halves. This definitely disposes of the old Loeb theory (Sect. 1). The difference is probably due to the fact that gradient centrifugation of intact eggs segregates, in the two halves, two different populations of mitochondria which can easily be recognized under the electron microscope (GEUSKENS, 1965). Treatment with hypertonic sea water strongly stimulates *protein synthesis* in both halves (BRACHET et al., 1963; DENNY and TYLER, 1964; BALTUS et al., 1965). Since this stimulation is stronger in anucleate halves than in nucleate ones, one can rule out the possibility that the increase in protein synthesis at fertilization is due to the production of new mRNA molecules by the nucleus. Unfertilized sea urchin eggs contain very few polyribosomes. Their number greatly increases when the anucleate halves are treated with hypertonic sea water. It can be concluded from these experiments that unfertilized sea urchin eggs must contain a store of "masked", inactive mRNA molecules which have been synthesized during oogenesis. These maternal mRNAs, after fertilization or parthenogenetic activation, are able to bind to the preexisting ribosomes and to form fully active polyribosomes.

More recent work has confirmed these conclusions. Unfertilized sea urchin eggs indeed contain a large variety of mRNAs bound to proteins in the form of RNP particles smaller than the ribosomes (GROSS et al., 1973). Among these mRNAs, those coding for histones and tubulin have been definitely identified (GROSS et al., 1973; SKOULTCHI and GROSS, 1973; RAFF et al., 1971, 1972). For instance, it has been shown that anucleate halves of unfertilized sea urchin eggs already contain the tubulin messenger, but it is not translated unless the anucleate fragments have undergone parthenogenetic activation. The result of this translation is the formation, in such activated anucleate fragments, of asters which are responsible for their irregular cleavage. Cytasters also appear when anucleate halves are treated with heavy water (D_2O). Electron microscopy has shown that these cytasters are centered around a small centriole which has apparently formed de novo (KATO and SUGIYAMA, 1971).

Fertilization of sea urchin eggs is followed by a doubling of poly A sequences in the preexisting maternal mRNAs (Slater and Slater, 1974). This polyadenylation also occurs in anucleate halves (Wilt, 1973) and is thus a posttranscriptional cytoplasmic event. It also occurs when anucleate fragments are treated with ammonia (Wilt and Mazia, 1974). The biological significance of polyadenylation, as already mentioned, remains to be discussed; but recent work by Huez et al. (1974) provides very strong evidence for the view that the main function of polyadenylation is to stabilize the mRNAs and to protect them against degradation in the cytoplasm.

Another peculiarity of sea urchin oocytes is that some of their polyribosomes contain, according to Giudice et al. (1974), giant mRNA molecules (MW up to 8×10^6 daltons). They are present in the cytoplasm of oocytes which, after having been fully labeled with uridine, have been cut into two halves by microsurgery. This rules out the possibility that one is dealing with giant nuclear pre-mRNAs (HnRNA).

Anucleate fragments of sea urchin eggs synthesize, after activation, small amounts of RNAs. Their sedimentation constants are 11 S, 13 S and 15 S (Chamberlain and Metz, 1972). They are entirely located in the mitochondria. The 11 S and 13 S species are certainly mitochondrial rRNAs, while the 15 S mitochondrial RNA might either be a precursor for the mitochondrial rRNAs or a mitochondrial mRNA. In fact, Selvig et al. (1972) found that mitochondria synthesize, in anucleate fragments of sea urchin eggs, a RNA which can diffuse out into the cytoplasm, but is unable to bind to cytoplasmic ribosomes. In whole eggs, mitochondrial RNA synthesis begins a few minutes after fertilization (Cantatore et al., 1974).

On the other hand, there is no evidence for replication of mitochondrial DNA either in whole eggs or in activated anucleate halves. This negative conclusion is based on the biochemical experiments of Bresch (1972) and the electron microscope analysis of mitochondrial DNA by Matsumoto et al. (1974).

3.2.3.2 Other Eggs

Similar experiments, but on a much smaller scale, have been nade on the eggs of the mollusk *Ilyanassa* and on amphibian eggs (*Rana, Xenopus*).

As already mentioned, a purely cytoplasmic body called the *polar lobe* (Fig. 3) forms during the first cleavage ("trefoil" stage) in *Ilyanassa*. It can easily be isolated by mechanical means. This polar lobe, like the anucleate fragments of sea urchin eggs, is the site of important protein synthesis (Clement and Tyler, 1967) and of limited RNA synthesis (Geuskens and de Jonghe d'Ardoye, 1971).

Amphibian oocytes react to progesterone by completing their maturation and by increasing the rate of protein synthesis (review by Smith and Ecker, 1970). This increase in protein synthesis, which affects many different proteins, occurs even when the nucleus has been taken out of the oocyte prior to the addition of progesterone. It is thus, as in activated anucleate fragments of sea urchin eggs, controlled at the level of translation. Anucleate frog eggs treated with parthenogenetic agents can, like their sea urchin counterparts, cleave several times and form partial blastulae. The pattern of protein synthesis in such anucleate frog eggs has been compared with that of normally fertilized eggs by Malacinski (1972). He found that, in both cases, the pattern of protein synthesis changes when the eggs reach the 64-cell stage. But when the blastula stage is reached and development stops in the anucleate eggs, the latter are unable to synthesize some of the proteins at the normal rate.

In conclusion, it can be said that anucleate fragments of eggs have a much lower capacity for development than *Acetabularia*. They are nevertheless capable of increasing the rate of protein synthesis under circumstances where an increase occurs in normal egg development. Unfertilized eggs are loaded with stable mRNAs and ribosomes. Protein synthesis during the early stages of development is regulated at the translational level. Development stops when transcription of new messages is needed (at the end of cleavage) for further development. RNA synthesis is probably purely mitochondrial in anucleate fragments.

3.2.4 Mammalian Cells

3.2.4.1 Reticulocytes

Nature itself performed an enucleation experiment during mammalian erythropoiesis, when nuclei and mitochondria were cast out of the cells. The resulting anucleate reticulocytes still contain many polyribosomes. Their main function is the synthesis of a single specific protein, of hemoglobin. As in anucleate *Acetabularia* and sea urchin eggs, synthesis of a functional protein, coded by a nuclear gene, continues for several days in the absence of the nucleus. Later on, the polyribosomes disintegrate into individual ribosomes. Finally, the latter also undergo complete degradation and one is left with the adult red blood cell (erythrocyte), which is little more than a bag surrounded by a membrane and filled with hemoglobin.

Innumerable papers have been devoted to reticulocytes. The most important point is that their specialization in hemoglobin synthesis allowed CHANTRENNE and his colleagues (1967) to isolate, for the first time, a mRNA from an eukaryotic source. It is globin mRNA, which is in fact a mixture of two mRNAs coding respectively for the α- and β-chains of globin. As already mentioned in this review, globin mRNA (which is associated with proteins in the polyribosomes) can be translated into globin in a very efficient way after injected into *Xenopus* oocytes (GURDON et al., 1971). The injected mRNA is very stable in these oocytes and hemoglobin synthesis can go on for several days. Removal of the poly A sequence present at the 3′ end of globin mRNA considerably shortens the length of time during which the messenger remains translatable (HUEZ et al., 1974; MARBAIX et al., 1975).

The isolation of hemoglobin mRNA in pure form has made it possible to answer important biological questions: is the number of copies of the hemoglobin genes the same in all cells? Is there a selective amplification of these genes in hematopoietic cells, the only ones in the organism which will produce hemoglobin? Molecular hybridization experiments, using labeled globin cDNA as a probe, have given a clear answer to these questions. The number of hemoglobin genes is small (estimates vary between two and five copies per haploid genome) and is the same in all cells, whether they will synthesize hemoglobin or will never do so. It follows that the probable reason why erythropoietic cells (BISHOP et al., 1972) synthesize hemoglobin to the exclusion of all other cells present in the organism is that these cells are the only ones in which the hemoglobin genes are derepressed.

Their chromatin must be organized in such a way that the hemoglobin genes are fully available for transcription by RNA polymerase, while these genes are protected by proteins against this enzyme in all other cells. Work on in vitro transcription of chromatin isolated from hematopoietic and other cells has provided good evidence that this is indeed the case (GILMOUR and PAUL, 1973).

3.2.4.2 Cells Enucleated with Cytochalasin B

Treatment with cytochalasin B (a drug which disrupts the organization of microfilaments and affects cell permeability cf. Sect. 2.2) induces, in many cells, the extrusion of the nucleus together with a small amount of cytoplasm. The cells are divided into small caryoplasts and large cytoplasts (the latter representing 80–90% of the initial cell volume). Caryoplasts and cytoplasts can be isolated, on a large scale, by taking advantage of their difference in size and density (centrifugation methods) or of the fact that cytoplasts stick much better to the substratum than caryoplasts (LADDA and ESTENSEN, 1970; POSTE and REEVE, 1971; WIGLER and WEINSTEIN, 1974). Anucleate cytoplasts can be fused with other cells in the presence of inactivated Sendai virus. Such hybrids between a whole cell and cytoplasm from another cell can survive and divide (LADDA and ESTENSEN, 1970; VEOMATT et al., 1974; EGE et al., 1974a, b).

The ultrastructure of caryoplasts and cytoplasts has been studied by several authors. WISE and PRESCOTT (1973) did not find any modification in the ultrastructure of cytoplasts 30 min after cytochalasin B treatment. However, 24 h later, vesiculization of the endoplasmic reticulum and the Golgi bodies was conspicuous. Caryoplasts are formed by the nucleus surrounded by a thin ribosome-containing layer. Studies by SHAY et al. (1974) show that caryoplasts lack microtubules and centrioles. These cell organelles are present in the cytoplasts where, as in amebae, the Golgi complex soon breaks down.

These findings explain some of the biological properties of caryo- and cytoplasts. In contrast to what is found in amebae (where nucleate and anucleate halves are of the same size), the small caryoplasts soon round up, while the locomotion of the cytoplasts remains normal. Caryoplasts survive longer (72 h) than cytoplasts (48 h). SHAY et al. (1974) conclude that maintenance of shape and motility are cytoplasmic functions.

This conclusion is supported by experiments showing that the reactions of cytoplasts to agents which induce modifications in whole cells are perfectly normal. Among such agents are dibutyryl-cAMP and prostaglandin E_1. The induced changes in shape are inhibited by colchicin and vinblastin. Cytoplasts are thus capable of responding to hormonal treatments in a normal fashion. Assembly of microtubules is a prerequisite for such responses (SCHRÖDER and HSIE, 1973).

Similar conclusions have been reached by GOLDMAN et al. (1973) for epithelial cells and by MILLER and RUDDLE (1974) for neuroblastoma cells. Neuroblastoma cytoplasts respond normally to addition of cAMP by neurite formation. Epithelial and fibroblast cytoplasts behave normally as regards locomotion, substratum attachment, pinocytosis and contact inhibition. It has also been shown that the cell surface of cytoplasts binds lectins such as Con A in a normal way, despite the loss of the Golgi bodies (WISE, 1974) and that viral DNA (isolated from avian sarcoma virus) can replicate in cytoplasts made from fibroblasts (VARMUS et al., 1974).

Finally, the biochemical properties of cytoplasts and caryoplasts have been studied by several authors. In cytoplasts, protein synthesis continues but it is lower than in whole cells and its rate decreases continuously. DNA and RNA synthesis stops very quickly (FOLLET, 1974; POSTE, 1972; SHAY et al., 1974). Caryoplasts synthesize RNA and proteins, but for a few hours only (EGE, 1974a, b; SHAY et al., 1974). They are unable to regenerate and to divide. The poor abilities of caryoplasts for macromolecular synthesis are probably a result of their small size and their very abnormal nucleocytoplasmic volume ratio. A similar condition prevails in the polar bodies expulsed during oocyte maturation and is probably responsible for their fast degeneration.

As mentioned above, production of the coenzyme NAD is generally thought to be a nuclear function, mainly because the enzyme responsible for its synthesis is usually associated with the nuclear fraction in centrifuged homogenates. Recent experiments by RECHSTEINER and CATANZARITE (1974) on cytoplasts and caryoplasts have clearly shown that the enzyme responsible for NAD synthesis (NAD pyrophosphorylase) is only present in the caryoplasts and is thus really localized in the nucleus. However, there is no fast destruction or turnover of NAD in cytoplasts.

Cytoplasts are definitely inferior to whole cells when a more refined activity, amino acid transport, is analyzed. HUME et al. (1975) found that cytoplasts are unable properly to control amino acid transport when the amino acid concentration of the medium is experimentally changed. This finding is reminiscent of observations (BRACHET and LIEVENS, in BRACHET, 1968; see Sect. 3.2.1.4) that phosphatase activity is not regulated in the same way, in nucleate and anucleate *Acetabularia* halves, when the phosphate content of the medium is modified.

4. Conclusions

Figure 7 summarizes, in a very simplified way, what has been said in the preceding pages. It shows that the nucleus is the main center of the synthesis of all kinds of nucleic acids. DNA is replicated during the S phase of the cell cycle and can be transcribed in precursors of mRNAs, tRNAs and rRNAs. The processing of these precursors is a function of the nucleus, and the degradation products probably never reach the cytoplasm. This loss of RNA by processing is particularly important for the mRNAs and the 28S and 18S rRNAs, when more than 50% of the precursor is often degraded. When processing is over, the RNAs which have been synthesized on chromatin and nucleolar DNAs cross the nuclear membrane; they are associated with proteins (informofers in the nucleus, informosomes in the cytoplasm, in the particular case of mRNAs) and finally assemble to form polyribosomes. Some of the proteins synthesized on the polyribosomes (histones, nonhistone proteins of chromatin) migrate back into the nucleus and combine with DNA to exert positive or negative controls on gene transcription. A similar regulatory role has been assigned to "chromatin RNA", but its very existence remains controversial.

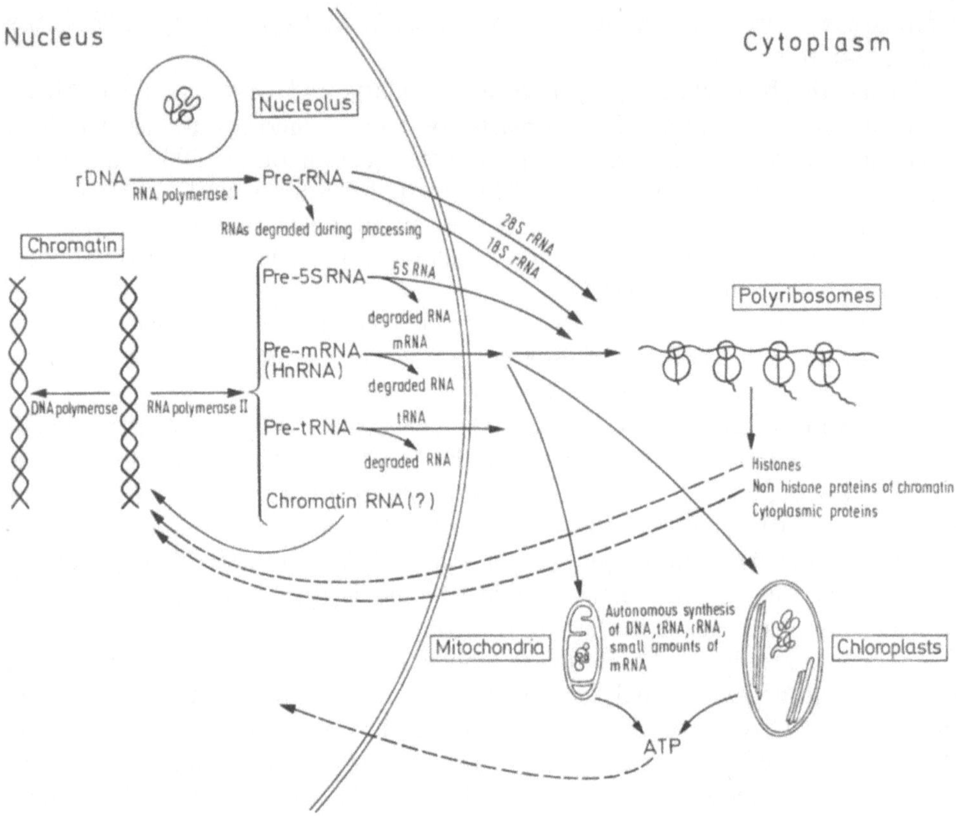

Fig. 7. Scheme of nucleocytoplasmic interactions. Explanation in text

Full activity for mitochondria and chloroplasts, despite the fact that they contain DNA, requires "help" from the nucleus. This help is probably provided in the form of mRNAs of nuclear origin. In return, chloroplasts and mitochondria provide the nucleus with the energy it requires for synthetic purposes. Another important cytoplasmic function is to control the choice between DNA replication and transcription. We do not know the molecular bases of this control, but cytonucleoproteins, which shuttle back and forth between cytoplasm and nucleus, are likely candidates for this important job.

We can now return to our initial question: why is anucleate cytoplasm inferior to its nucleate counterpart? As we have seen, survival of anucleate cytoplasm greatly varies according to the biological system chosen. *Acetabularia* anucleate fragments survive for months, while cytoplasts from mammalian cells die after one or two days, depending on the cell type. Anucleate halves of amebae and eggs remain alive for one or two weeks. In all cases, RNA synthesis stops (except in mitochondria and chloroplasts) as soon as the nucleus has been taken away, but protein synthesis continues at a slowly decreasing rate so long as anucleate cytoplasm remains alive. This means that the amount and stability of the RNA of nuclear origin present in the cytoplasm at the time of section were sufficient to allow continuation of protein synthesis. Death probably occurs when the machinery for protein synthesis becomes incapable of providing sufficient amounts of one or several key proteins which are essential for life. Differences in the stability of the RNAs (especially the mRNAs) in different kinds of cells would thus be responsible for the variable longevity of anucleate fragments from different living organisms. Investigation of how the stability of mRNAs is controlled (polyadenylation, combination with specific proteins, binding to organelles such as the cell membranes, etc.) is an important task for the future.

References

AARONSON, A.I., WILT, F.H.: Properties of nuclear RNA in sea urchin embryos. Proc. Natl. Acad. Sci. US **62**, 186–193 (1969)

ABELSON, H.T., PENMAN, S.: Selective interruption of high molecular weight RNA synthesis in HeLa cells by camptothecin. Nature New Biol. **237**, 144–146 (1972)

AIMI, J.: Isolement de fragments nucléés et anucléés d'oeufs de l'oursin Sphaerechinus granularis. Experientia **30**, 837–838 (1974)

ALEXEEV, A.B., BETINA, M.I., SWOLINSKY, S.L., YAZIKOV, A., ZUBAREV, T.N.: Evidence for a relationship between some ribonucleoprotein complexes and the morphogenetic substances of Acetabularia. Plant Sci. Lett. **3**, 297–302 (1974)

APEL, K., SCHWEIGER, H.G.: Nuclear dependency of chloroplast proteins in Acetabularia. Eur. J. Bioch. **25**, 229–238 (1972)

APEL, K., SCHWEIGER, H.G.: Sites of synthesis of chloroplast membrane protein. Eur. J. Bioch. **38**, 373–383 (1973)

ASSEL, S. VAN, BRACHET, J.: Metabolisme des acides nucléiques et formation de cytasters dans les oeufs d'Amphibiens sous l'action de l'eau lourde. J. Embryol. Exptl. Morphol. **19**, 261–272 (1968)

BALTUS, E., QUERTIER, J., FICQ, A., BRACHET, J.: Biochemical studies of nucleate and anucleate fragments isolated from sea urchin eggs. Biochim. Biophys. Acta **95**, 408–417 (1965)

BANNWARTH, H., SCHWEIGER, H.G.: Regulation of thymidine phosphorylation in nucleate and anucleate cells of Acetabularia. Proc. Roy. Soc. London **188**, 203–219 (1975)

BERGER, S., SCHWEIGER, H.G.: Cytoplasmic induction of changes in the ultrastructure of the Acetabularia nucleus and perinuclear cytoplasm. J. Cell Sci. **17**, 517–529 (1975a)

BERGER, S., SCHWEIGER, H.G.: 80S ribosomes in Acetabularia major. Redundancy of r-RNA cistrons. Protoplasma **83**, 41–50 (1975b)

BIRNSTIEL, M., SPEIRS, J., PURDOM, I., JONES, K., LOENING, U.E.: Properties and composition of the isolated ribosomal DNA satellite of Xenopus laevis. Nature **219**, 454–463 (1968)

BISHOP, J.O., PEMBERTON, R., BAGLIONI, C.: Reiteration frequency of haemoglobin genes in the duck. Nature New Biol. **235**, 231–234 (1972)

BOLOUKHERE, M.: Effet de l'actinomycine D sur l'ultrastructure des chloroplastes et du noyau d'Acetabularia. J. Microsc. **4**, 363–372 (1965)

BONNER, W.M.: Protein migration into nuclei. I. Frog oocyte nuclei in vivo accumulate injected histones, allow entry to small proteins and exclude large proteins. J. Cell Biol. **64**, 421–430 (1975)

BONOTTO, S., PUISEUX-DAO, S., KIRCHMANN, M., BRACHET, J.: Faits et hypothèses sur le contrôle de l'alternance morphogénétique: croissance végétative et différenciation de l'appareil reproducteur chez les Acetabularia mediterranea. Compt. Rend. **272**, 392–396 (1971)

BRACHET, J.: Synthesis of macromolecules and morphogenesis in Acetabularia. Current Topics Develop. Biol. **3**, 1–35 (1968)

BRACHET, J.: Relations nucléocytoplasmiques au cours de la différenciation cellulaire chez quelques organismes unicellulaires. Colloque CNRS, Paris (1974)

BRACHET, J.: The effect of some inhibitors of RNA synthesis and proteolysis on Morphogenesis in Acetabularia mediterranea. Biochem. Physiol. Pflanzen **168**, 493–510 (1975)

BRACHET, J., BONOTTO, S.: Biology of Acetabularia. New York: Academic Press 1970

BRACHET, J., DENIS, H., VITRY, F. DE: The effects of actinomycin D and puromycin on morphogenesis in amphibian eggs and Acetabularia mediterranea. Develop. Biol. **9**, 398–434 (1964)

BRACHET, J., FICQ, A., TENCER, R.: Amino acid incorporation into proteins of nucleate and anucleate fragments of sea-urchin eggs. Effect of parthenogenetic activation. Exptl. Cell Res. **32**, 168–170 (1963)

BRACHET, J., LANG, A.: The role of the nucleus and the nucleocytoplasmic interactions in morphogenesis. Handbuch der Pflanzenphysiol. **15**, 1 (1965)

BRÄNDLE, E., ZETSCHE, K.: Die Wirkung von Rifampicin auf die RNA und Proteinsynthesis sowie die Morphogenesis und den Chlorophyllgehalt knhaltiger und kernloser Acetabularia-Zellen. Planta **9**, 46–55 (1971)

BRANDLE, E., ZETSCHE, K.: Zur Localisation der α-Amanitin-sensitiven ARN-Polymerase in Zellkernen von Acetabularia. Planta **11**, 209–219 (1973)

BRESCH, H.: Biology and radiobiology of anucleate systems. I. Animal cells. BONOTTO, S., KIRCHMANN, R., GOUTIER, R., MAISIN, J.R. (eds.). New York, London: Academic Press 1972

BURGER, M.M., BOMBIK, B.M., BRECKENRIDGE, B.M.L., SHEPPARD, J.R.: Growth control and cyclic alterations of cAMP in the cell cycle. Nature New Biol. **239**, 161–163 (1972)

BUSHNELL, D.E., BECKER, J.E., POTTER, V.R.: The role of messenger RNA in tyrosine aminotransferase superinduction; effects of camptothecin on hepatoma cells in culture. Biochem. Biophys. Res. Commun. **56**, 815–821 (1974)

CANTATORE, P., NICOTRA, A., LORIA, P., SACCONE, C.: RNA synthesis in isolated mitochondria from sea urchin embryos. Cell Differentiation **3**, 45–53 (1974)

CASPERSSON, T.: Cell Growth and Cell Function. New York: Norton 1950

CERON, G., JOHNSON, E.M.: Control of protein synthesis during the development of Acetabularia J. Embryol. Exptl. Morphol. **26**, 323–338 (1971)

CHAMBERLAIN, J.T., METZ, C.B.: Mitochondrial RNA synthesis in sea urchin embryos. J. Mol. Biol. **64**, 593–607 (1972)

CHANTRENNE, H., BURNY, A., MARBAIX, G.: The search for the messenger RNA of haemoglobin. Progr. Nucleic Acid Res. Molec. Biol. **7**, 173–194 (1967)

CHIRKOVA, L.I., PIKALOV, A.V., KISSELEVA, E.V., BETINA, N.I., KHRISTOLUBOVA, N.B., NIKORO, Z.S., SANDAKHIEV, L.S.: Subcellular localization of the morphogenetic factors in Acetabularia mediterranea. Ontogenesis **1**, 42–55 (1970)

CLAUDE, A.: Particulate components of normal and tumor cells. Science **91**, 77–78 (1940)

CLEMENT, A.C., TYLER, A.: Protein-synthesizing activity of the anucleate polar lobe of the mud snail Ilyanassa obsoleta. Science **158**, 1457–1458 (1967)

CRAIG, I.W., GIBOR, A.: Biosynthesis of proteins involved with photosynthetic activity in anucleated Acetabularia. Biochim. Biophys. Acta **217**, 488–495 (1970)

DARZYNKIEWICZ, Z., CHEMILKA, E., ARNASON, B.: Chick erythrocyte nucleus reactivation in Heterokaryons: suppression by inhibitors of proteolytic enzymes. Proc. Natl. Acad. Sci. US **71**, 644–647 (1974)

DAVIDSON, R.L.: Regulation of differentiation in cell hybrids. Federation Proc. **30**, 926–929 (1971)

DAVIS, E.M., ADELBERG, E.A.: Use of the somatic cell hybrids for analysis of the differentiated state. Bacteriol. Rev. **37**, 197–214 (1973)

DEAK, I.I., DEFENDI, V.: Effects of discrete nuclear UV microbeam irradiation on herpes virus and SV40 infection. J. Cell Science **17**, 531–538 (1975)

DENNY, P., TYLER, A.: Activation of protein biosynthesis in non nucleate fragments of sea urchin eggs. Biochem. Biophys. Res. Commun. **14**, 245–249 (1964)

EGE, T., HAMBERG, H., KRONDAHL, U., ERICSSON, J., RINGERTZ, N.R.: Characterization of minicells (nuclei) obtained by cytochalasin enucleation. Exptl. Cell Res. **87**, 365–377 (1974a)

EGE, T., KRONDAHL, U., RINGERTZ, N.R.: Introduction of nuclei and of micronuclei into cells and enucleated cytoplasm by Sendai virus induced fusion. Exptl. Cell Res. **88**, 428–432 (1974b)

FELDHERR, C.M.: The uptake of cytoplasmic proteins by germinal vesicles in *Xenopus laevis*. J. Cell Biol. **63**, 98a (1974)

FEULGEN, R., ROSSENBECK, H.S.: Z. Physiol. Chem, p. 21 (1924)

FICQ, A.: Autoradiography. In: The Cell, Vol. 1, BRACHET, J., MIRSKY, A.E. (eds.). New York: Academic Press 1959

FIUME, L., WIELAND, T.: Amanitins, Chemistry. FEBS Letters **8**, 1–5 (1970)
FLICKINGER, C.J.: Maintenance and regeneration of cytoplasmic organelles in hybrid amoebae by nuclear transplantation. Exptl. Cell Res. **80**, 31–46 (1974)
FOLLETT, E.A.C.: A convenient method for enucleating cells in quantity. Exptl. Cell Res. **84**, 72–78 (1974)
FRANKE, W., BERGER, W., FALK, H., SPRING, H., SCHEER, U., HERTH, W., TRENDELENBURG, M., SCHWEIGER, H.: Morphology of the nucleocytoplasmic interactions during the development of Acetabularia cells. Protoplasma **82**, 249–282 (1974)
FRANKE, W., KARTENBECK, J., ZEUTGRAF, H., SCHEER, U., FALK, H.: Membrane-to-membrane cross-bridges. A means to orientation and interaction of membrane faces. J. Cell Biol. **51**, 881–888 (1971)
FRANKE, W.W.: On the universality of nuclear pore complex structure. Z. Zellforsch. Mikroskop. Anat. **105**, 405–429 (1970)
FRANKE, W.W., SCHEER, U.: The ultrastructure of the nuclear envelope of amphibian oocytes. A reinvestigation. J. Ultrastruct. Res. **30**, 288–327 (1970)
GEORGIEV, G.P.: The structure of transcriptional units in eukaryotic cells. Current Topics Develop. Biol. **7**, 1–53 (1972)
GEUSKENS, M.: A study of the ultrastructure of nucleate and anucleate fragments of unfertilized sea urchin eggs. Exptl. Cell Res. **39**, 413–417 (1965)
GEUSKENS, M., JONGHE D'ARDOYE, V. DE: Metabolic patterns in Ilyanassa polar lobes. Exptl. Cell Res. **67**, 61–72 (1971)
GILMOUR, R.S., PAUL, J.: Tissue-specific transcription of the globin gene in isolated chromatin. Proc. Natl. Acad. Sci. US **70**, 3440–3442 (1973)
GIUDICE, G., SCONZO, G., ALBANESE, I., ORTOLANI, G., CAMARATA, M.: Cytoplasmic giant RNA in sea urchin embryos. Proof that it is not derived from artifactual nuclear leakage. Cell Differentiation **3**, 287–295 (1975)
GOFFEAU, A., BRACHET, J.: Deoxyribonucleic acid-dependent incorporation of amino acids into proteins of chloroplasts isolated from anucleate *Acetabularia* fragments. Biochim. Biophys. Acta **95**, 302–313 (1965)
GOLDMAN, R.D., POLLACK, R., HOPKINS, N.H.: Preservation of normal behaviour by enucleated cells in culture. Proc. Natl. Acad. Sci. US **70**, 750–754 (1973)
GOLDSTEIN, L.: Stable nuclear RNA returns to post-division nuclei following release to the cytoplasm during mitosis. Exptl. Cell Res. **89**, 421–425 (1974)
GOLDSTEIN, L., TRESCOTT, D.H.: Characterization of RNA's that do and do not migrate between cytoplasm and nucleus. Proc. Natl. Acad. Sci. US **67**, 1367–1374 (1974)
GOLDSTEIN, L., WISE, G.E., BEESON, M.: Proof that certain RNA's shettle nonrandomly between cytoplasm and nucleus. Exptl. Cell Res. **76**, 281–288 (1973)
GREEN, B.R.: The genetic potential of Acetabularia chloroplasts. J. Cell Biol. **59**, 123a (1973)
GROSS, K.W., JACOBS-LORENA, M., BAGLIONI, C., GROSS, P.R.: Cell free translation of maternal m-RNA from sea-urchin eggs. Proc. Natl. Acad. Sci. US **70**, 2614–2618 (1973)
GURDON, J.B.: Nuclear transplantation and the control of gene activity in animal development. Proc. Roy. Soc. B **176**, 303–314 (1970)
GURDON, J.B., BROWN, D.D.: Cytoplasmic regulation of RNA synthesis and nucleolus formation in developing embryos of *Xenopus laevis*. J. Mol. Biol. **12**, 27–35 (1965)
GURDON, J.B., LANE, C.D., WOODLAND, H.R., MARBAIX, G.: Use of frog eggs and oocytes for the study of messenger RNA and its translation in living cells. Nature **233**, 177–182 (1971)
GURDON, J.B., WOODLAND, H.R.: On the long term control of nuclear activity during cell differentiation. Current Topics Develop. Biol. **5**, 39–70 (1970)
HÄMMERLING, J.: Arch. Entwicklungsmech. Organ. **131**, 1–81 (1934)
HÄMMERLING, J.: Nucleocytoplasmic relationships in the Development of Acetabularia. Intern. Rev. Cytol. **2**, 475–498 (1953)
HARRIS, H.: Cell Fusion. Oxford: Clarendon Press 1970
HARVEY, E.B.: Development of the parts of sea urchin eggs separated by centrifugal force. Biol. Bull. **64**, 125–148 (1933)
HEILPORN, V., LIMBOSCH, S.: Les effets du bromure d'éthidium sur *Acetabularia mediterranea.* Biochim. Biophys. Acta **240**, 94–108 (1971)
HEILPORN, V., LIMBOSCH, S.: Recherches sur les acides désoxyribonucléiques d'*Acetabularia mediterranea.* Eur. J. Biochem. **22**, 573–579 (1971)

HENNIG, W.: Molecular hybridization of DNA and RNA *in situ*. Intern. Rev. Cytol. **36**, 1–44 (1973)
HERSHKO, R., MAMONT, P., SHIELDS, R., TOMKINS, G.M.: Pleiotypic response. Nature New Biol. **232**, 206–211 (1971)
HOLTZER, H., WEINTRAUB, H., MAYNE, R., MOCHAN, B.: The cell cycle, cell lineages and cell differentiation. Current Topics Develop. Biol. **7**, 229–254 (1972)
HUEZ, G., MARBAIX, G., HUBERT, E., LECLERCQ, M., NUDEL, V., SOREQ, H., SALOMON, R., LEBLEU, B., REVEL, M., LITTAUER, U.J.: Role of the polyadenylate segment in the translation of globin messenger-RNA in Xenopus oocytes. Proc. Natl. Acad. Sci. US **71**, 3143–3146 (1974)
HULIN, N., BRACHET, J.: Actinomycin binding in differentiating and dividing cells. Exptl. Cell Res. **59**, 486–488 (1970)
HUME, S.P., LAMB, J.F., WEINGART, R.: Evidence for nuclear control of amino acid transport in cultured cells. Nature **255**, 73–74 (1975)
JELINEK, W., GOLDSTEIN, L.: Isolation and characterization of some of the proteins that shuttle between cytoplasm and nucleus in *Amoeba p.* J. Cell Physiol. **81**, 181–197 (1973)
JEON, K.W., LORCH, I.J., DANIELLI, J.F.: Reassembly of living cells from dissociated components. Science **167**, 1626–1629 (1970)
JOHN, A.A., BIRNSTIEL, M.L., JONES, K.W.: RNA-DNA hybrids at the cytological level. Nature **223**, 582–587 (1969)
KATO, K.H., SUGIYAMA, M.: On the *de novo* formation of the centriole in the activated sea urchin egg. Develop. Growth Differentiation **13**, 359–366 (1971)
KEAN, E.L., BRUNER, W.E.: Cytidine-5′-monophosialic acid synthetase. Activity and localization in the nucleate fragments of the unfertilized sea urchin eggs. Exptl. Cell Res. **69**, 384–392 (1971)
KECK, K. (ed.): Methods in Cell Physiology. New York: Prescott Academic Press 1964
KLOPPSTECH, K., SCHWEIGER, H.G.: Synthesis of chloroplast and cytosol ribosomes in regenerating Acetabularia cells. Differentiation **1**, 331–337 (1973)
KLOPPSTECH, K., SCHWEIGER, H.G.: 80S ribosomes from Acetabularia. Biochim. Biophys. Acta **324**, 365–374 (1973b)
KLOPPSTECH, K., SCHWEIGER, H.G.: The site of synthesis of chloroplast ribosomal proteins. Plant Sci. Letters **2**, 101–105 (1974)
KLOPPSTECH, K., SCHWEIGER, H.G.: 80S ribosomes in Acetabularia major: distribution and transportation within cell. Protoplasma **83**, 27–40 (1975)
LADDA, R., ESTENSEN, R.D.: Introduction of a heterologous nucleus into enucleated cytoplasms of cultured mouse L-cells. Proc. Natl. Acad. Sci. US **67**, 1528–1533 (1970)
LEGROS, F., SAINES, M., RENARD, M., CONARD, V.: Nuclear influence of oxygen uptake modifications induced by insulin and aldosterone in Acetabularia mediterranea. Plant Sci. Letters **2**, 339–345 (1974)
LEVINE, S., PICTET, R., RUTTER, W.: Control of cell proliferation and cytodifferentiation by a factor reacting with the cell surface. Nature New Biol. **246**, 49–52 (1973)
LOEB, J.: Warum ist die Regeneration kernloser Protoplasmastücke unmöglich oder erschwert? Arch. Entwicklungsmech. Organ. **8**, 689–693 (1899)
LÜSCHER, K., MATILE, P.: Distribution of acid ribonuclease and other enzymes in stratified Acetabularia. Planta **118**, 323–332 (1974)
MALACINSKI, G.M.: Deployment of maternal template during early Ambibian embryogenesis. J. Exptl. Zool. **181**, 409–420 (1972)
MARBAIX, G., HUEZ, G., BURNY, A., CLEUTER, Y., HUBERT, E., LECLERCQ, M., CHANTRENNE, H., SOREQ, H., NUDEL, U., LITTAUER, U.Z.: Absence of polyadenylate segment in globin messenger RNA accelerates its degradation in Xenopus oocytes. Proc. Natl. Acad. Sci. US **72**, 3065–3067 (1975)
MARSHALL-GRAVES, J.A.: DNA synthesis in heterokaryons formed by fusion of mammalian cells from different species. Exptl. Cell Res. **72**, 393–403 (1972)
MATSUMOTO, L., KASAMATSU, H., PIKO, L., VINOGRAD, J.: Mitochondrial DNA replication in the sea urchin oocytes. J. Cell Biol. **63**, 146–159 (1974)
MILLER, R.A., RUDDLE, F.H.: Enucleated neuroblastoma cells from neurites when treated with dibutyril cAMp. J. Cell Biol. **63**, 295–299 (1974)
MOENS, W., VOKAER, A., KRAM, R.: cAMP and cGMP concentrations in serum and density restricted fibroblast cultures. Proc. Natl. Acad. Sci. US **72**, 1063–1067 (1975)

ORD, M.J.: The initiation maintenance and termination of DNA synthesis: a study of nuclear DNA replication using Amoeba proteus as a cell model. J. Cell Sci. **9**, 1–21 (1971)

PARDUE, M.L., GALL, J.G.: Molecular hybridization of radioactive DNA to the DNA of cytological preparations. Proc. Natl. Acad. Sci. US **64**, 600–604 (1969)

PERRY, R.: The cellular sites of synthesis of ribosomal and 4S RNA. Proc. Natl. Acad. Sci. US **48**, 2179–2186 (1962)

POSTE, G.: Enucleation of mammalian cells by cytochalasin B. I. Characterization of anucleate cells. Exptl. Cell Res. **73**, 273–286 (1972)

POSTE, G., REEVE, P.: Formation of hybrid cells and heterokaryons by fusion of enucleated and nucleated cells. Nature **229**, 123–125 (1971)

PRESCOTT, D., GOLDSTEIN, L.: Proteins in nuclear-cytoplasmic interactions in Amoeba proteus. Ann. Embryol. Morphol., Suppl. 181–188 (1969)

PUISEUX-DAO, S.: Acetabularia and Cell Biology. London: Logos Press 1970

RAFF, R.A., COLOT, H.V., SELVIG, S.E., GROSS, P.R.: Oogenetic origin of messenger RNA for embryonic synthesis of microtubule proteins. Nature **235**, 211–214 (1972)

RAFF, R.A., GREENHOUSE, G., GROSS, K.W., GROSS, P.R.: Synthesis and storage of microtubule proteins by sea-urchin embryos. J. Cell Biol. **50**, 516–527 (1971)

RAO, M.V., CHATTERJEE, S.: Regulation of nuclear DNA synthesis in Amoeba interspecific hybrids. Exptl. Cell Res. **88**, 371–374 (1974)

RAO, P.N., HITTELMAN, W.N., WILSON, B.A.: Mammalian cell fusion. Exptl. Cell Res. **90**, 40–46 (1975)

REBHUN, L., ROSENBAUM, J., LEFEBRE, P., SMITH, G.: Reversible restoration of the birefringence of cold treated isolated mitotic apparatus of surf clam eggs with chick brain tubulin. Nature **249**, 113–115 (1974)

RECHSTEINER, M., CATANZARITE, V.: The biosynthesis and turnover of nicotinamide adenine dinucleotide in enucleated culture cells. J. Cell Physiol. **84**, 409–421 (1974)

RINGERTZ, N.R., BOLUND, L.: Activation of hen erythrocytes desoxyribonucleoproteins. Exptl. Cell Res. **55**, 205–214 (1969)

RINGERTZ, N.R., DARZYNKIEWICZ, Z., BOLUND, L.: Actinomycin binding properties of stimulated human lymphocytes. Exptl. Cell Res. **56**, 411–417 (1969)

SANDAKHIEV, L.S., PUCHKOVA, L.I., PIKALOV, A.V.: Biology and Radiobiology of anucleate system, Vol. II, p. 297, (eds., S. BONOTTO, R. KIRCHMANN, R. GOUTIER, T.R. MAISIN). New York and London: Acad. Press 1972

SCHNAEBLI, H.P., BURGER, M.M.: Selective inhibition of growth of transformed cells by protease inhibitors. Proc. Natl. Acad. Sci. US **69**, 3825–3827 (1972)

SCHRÖDER, H., HSIE, A.W.: Morphological transformation of enucleated chinese hamster cells by dibutyril cAMP and hormones. Nature **246**, 58–60 (1973)

SCHWEIGER, E., WALLRAFT, H.G., SCHWEIGER, H.G.: Endogenous circadian rhythm in cytoplasm of Acetabularia: influence of the nucleus. Science **146**, 658–659 (1964)

SCHWEIGER, H.G., MASTER, R.W.P., WERZ, G.: Nuclear control of a cytoplasmic enzyme in Acetabularia. Nature **216**, 554–557 (1967)

SELVIG, S.A., GREENHOUSE, G.E., GROSS, P.R.: Cytoplasmic synthesis of RNA in the sea urchin embryo. II. Mitochondrial transcription. Cell Differentiation **1**, 5–14 (1972)

SHAY, J.W., PORTER, K.R., PRESCOTT, D.M.: The surface morphology and fine structure of CHO (Chinese Hamster Ovary) cells following enucleation. Proc. Natl. Acad. Sci. US **71**, 3059–3063 (1974)

SHEPHARD, D.C., BIDWELL, R.G.S.: Photosynthesis and carbon metabolism in a chloroplast preparation from Acetabularia. Protoplasma **76**, 289–308 (1973)

SKOULTCHI, A., GROSS, P.R.: Maternal histone messenger RNA: detection by molecular hybridization. Proc. Natl. Acad. Sci. US **70**, 2840–2844 (1973)

SLATER, I., SLATER, D.W.: Polyadenylation and transcription following fertilization. Proc. Natl. Acad. Sci. US **71**, 1103–1107 (1974)

SMITH, L.D., ECKER, R.E.: Regulatory processes in the maturation and early cleavage of amphibian eggs. Current Topics Develop. Biol. **5**, 1–38 (1970)

SPIRIN, A.S.: On masked forms of messenger RNA in early embryogenesis and in other differentiating systems. Current Topics Develop. Biol. **1**, 1–36 (1966)

SPRING, H., SCHEER, U., FRANKE, W.W., TRENDELENBURG, M.F.: Lampbrush type chromosomes in the primary nucleus of the green alga Acetabularia mediterranea. Chromosoma **50**, 25–43 (1975)

SPRING, H., TRENDELENBURG, M.F., SCHEER, U., FRANKE, W.W., HERTH, W.: Structural and biochemical studies on the primary nucleus of two green algal species; Acetabularia mediterranea and Acetabularia major. Cytobiologie **60**, 1–45 (1974)

STEINERT, G., VANGANSEN, P.: Binding of ^{3}H-actinomycin to vitelline platelets of amphibian oocytes. A high resolution autoradiographic investigation. Exptl. Cell Res. **64**, 355–365 (1971)

STEINERT, M., STEINERT, G.: La synthèse de l'acide désoxyribonucléique au cours du cycle de division de Trypanosoma mega. J. Protozool. **9**, 203–211 (1962)

STICH, H.: Experimentelle karyologische und cytochemische Untersuchungen an Acetabularia mediterranea. Z. Naturforsch. **6b**, 319–326 (1951)

TERRA, N. DE: Cortical control of cell division. Science **184**, 530–537 (1974)

THIMANN, K.: The Auxins in the Physiology of Plant Growth and Development. (M.B. WILKINS, ed.), pp. 1–47. London: McGraw Hill 1969

TRENDELENBURG, M.F., SPRING, H., SCHEER, U., FRANKE, W.W.: Morphology of nucleolar cistrons in a plant cell (Acetabularia mediterranea). Proc. Natl. Acad. Sci. US **71**, 3626–3630 (1974)

VANDEN DRIESSCHE, T.: The nuclear control of the chloroplasts circadian rhythms. Science Progr. **55**, 293–303 (1967)

VANDEN DRIESSCHE, T.: The chloroplasts of Acetabularia. The control of their multiplication and activities. Subcellular Biochem. **2**, 33–67 (1973)

VARMUS, H.E., GUNTAKA, R.V., FAN, W.J.W., HEASLEY, S., BISHOP, J.A.: Synthesis of viral DNA in the cytoplasm of duck embryo fibroblasts and in enucleated cells after infection by avian sarcoma virus. Proc. Natl. Acad. Sci. US **71**, 3874–3878 (1974)

VEOMATT, G., PRESCOTT, D.M., SHAY, J., PORTER, K.R.: Reconstruction of mammalian cells from nuclei and cytoplasmic components separated by treatment with cytochalasin B. Proc. Natl. Acad. Sci. US **71**, 1999–2000 (1974)

WERZ, G.: Fine structural aspects of morphogenesis in Acetabularia. Intern. Rev. Cytol. **38**, 319–365 (1974)

WESSELLS, N.K., SPOONER, D.F., ASH, G.F., MACBRADLEY, M.A., TAYLOR, E.L., WRENG, J., YAMADA, K.M.: Microfilaments in cellular and developmental processes. Science **171**, 135–143 (1971)

WIGLER, M., WEINSTEIN, S.: A preparative method for obtaining enucleated mammalian cells. J. Cell Biol. **63**, 371a (1974)

WILSON, E.B.: The Cell in Development and Heredity, p. 1232. London: MacMillan 1925

WILT, F.H.: Polyadenylation of maternal RNA of sea urchin eggs after fertilization. Proc. Natl. Acad. Sci. US **70**, 2345–2349 (1973)

WILT, F.H., MAZIA, D.: The stimulation of cytoplasmic polyadenylation in sea urchin eggs by ammonia. Develop. Biol. **37**, 422–424 (1974)

WISE, G.E.: Maintenance of surface concanavalin A binding sites in enucleated mammalian cells. J. Cell Biol. **63**, 375a (1974)

WISE, G.E., PRESCOTT, D.M.: Ultrastructure of enucleated mammalian cells in culture. Exptl. Cell Res. **81**, 63–72 (1973)

YUDIN, A.L., NEYFAKH, A.A.: Migration of newly synthesized RNA during mitosis. Exptl. Cell Res. **82**, 210–214 (1973)

2. Plastids and Intracellular Transport

D.A. WALKER

1. Introduction

The chloroplast always seems to be faced with reconciling the irreconcilable. While evolving oxygen, it must simultaneously produce an intermediate more reducing than hydrogen. While reducing NADP it can simultaneously generate ATP, an achievement which, when first reported, seemed almost as remarkable as making water flow uphill. No less striking in their own way are two seemingly conflicting roles in carbon metabolism. On the one hand, the chloroplast must operate its carbon cycle as an autocatalytic breeder reaction, while on the other, it must export elaborated carbon and chemical energy to its cellular environment. In order to export, it must produce more than it uses, but it can only do this by returning newly synthesised intermediates to the cycle. Conversely, in order to satisfy the needs of the cell, it must release newly made products to the cytoplasm. Clearly, these processes could not be efficiently accomplished unless it were possible to strike a delicate balance between recycling, export and internal storage. Precisely how this is done is still largely a matter for speculation, but this Chapter will attempt to show how contemporary work has provided a factual basis for conjecture.

Like a previous article (WALKER, 1974b), this chapter draws heavily on two excellent and comprehensive papers by HEBER (1970, 1974).

2. The Development of the Experimental Study of Metabolite Translocation in Chloroplasts

The classic equation for photosynthesis was often written

$$6CO_2 + 6H_2O \rightarrow C_6H_{12}O_6 + 6O_2$$

although it was conceded that the empirical $C_6H_{12}O_6$ might be something other than free glucose. Originally starch seemed the more likely end-product, and certainly the historic observations of SACHS, PFEFFER and GODLEWSKI established an intimate relationship between photosynthesis and the process and location of starch accumulation in green leaves (RABINOWITCH, 1945). Even in Sachs' day, however, it was known that some plants seem incapable of accumulating starch within their leaves in any circumstances. Moreover, an almost insoluble polysaccharide could quite clearly not be moved about the plant unchanged. As sucrose emerged as the major compound to be transported within the plant (e.g. THOMAS et al.,

1973), it became increasingly accepted as the real end-product of photosynthesis (RABINOWITCH, 1956), whereas starch was relegated to the role of a temporary storage compound. Even when it became evident that sugar phosphates played a central role in photosynthetic carbon metabolism (BENSON et al., 1950, 1952; BENSON and CALVIN, 1950; BASSHAM and CALVIN, 1957), there was no suggestion that these compounds should be regarded as end-products rather than intermediates. The percentage of radioactive carbon in sucrose extrapolated to zero at zero time, and thereafter it increased in a progressive manner in precisely the way which might have been predicted for an end-product awaiting shipment to other parts of the plant. The first real hint that sucrose was not necessarily synthesised within the chloroplast prior to export came from HEBER and WILLENBRINK (1964). Using freeze-drying and nonaqueous fractionation, these workers produced good evidence that a large part of sucrose synthesis occurred not inside the chloroplast but within the cytoplasm.

Almost all workers in the field were also slow to realise the significance of the use of sugars in the isolation of chloroplasts capable of relatively fast rates of photosynthesis. Sucrose had been used as an osmoticum by HILL, as it had been by ENGELMANN et al. before him (see HILL, 1965), but the correlation between chloroplast intactness and the ability to assimilate carbon, while recognised by WHATLEY et al. (1956), was not related to envelope integrity until the mid-1960s (LEECH, 1964; JAMES and LEECH, 1964; WALKER, 1965a). Following the pioneering work of ARNON and his colleagues (see e.g. ARNON, 1967), the first real improvement in the rate of photosynthesis by isolated organelles had followed the substitution of sorbitol for (the then ubiquitous) Tris-NaCl. Salt solutions had been preferred by ARNON et al. because there was no possibility that they could be used as metabolites leading to the generation of ATP by oxidative processes (see KALBERER et al., 1967). Similarly sorbitol was selected because it was thought to be relatively inert and, in preliminary experiments, marginally superior to glucose (WALKER, 1964). Nevertheless, though sorbitol has continued to find preference as an osmoticum (see e.g. JENSEN and BASSHAM, 1966; WALKER, 1971), it is in no way superior in this particular respect to other sugar alcohols (KALBERER et al., 1967), related hexoses or free sucrose (WALKER, 1971). Conversely, it was shown that as the concentration of sucrose in the medium was decreased below 0.09 M, chloroplast envelopes first became distended and then burst. Envelope rupture was associated with an increased ability to phosphorylate exogenous ADP and a decreased ability to fix CO_2 (WHATLEY et al., 1956; WALKER, 1965a, b). In retrospect it is clear from this work (see e.g. WALKER, 1974; HEBER, 1974) that if sucrose is an effective osmoticum for isolated chloroplasts it cannot readily penetrate the chloroplast envelope. This has also been established independently by HELDT and RAPLEY (1970a) and HELDT and SAUER (1971), whose results show that it is the inner of the two envelopes which is impermeable.

By 1965 then, there were already good indications that neither sucrose nor reducing power in the form of NADP and ATP could move freely across both chloroplast envelopes. Conversely there was at least presumptive evidence that the chloroplast might be more readily permeated by some of the intermediates of the Benson-Calvin cycle. Since this time the picture which has emerged (Fig. 1) is that the principal imports are CO_2 and orthophosphate and that the principal export is triose phosphate (HEBER, 1974; WALKER, 1974b). This view is based

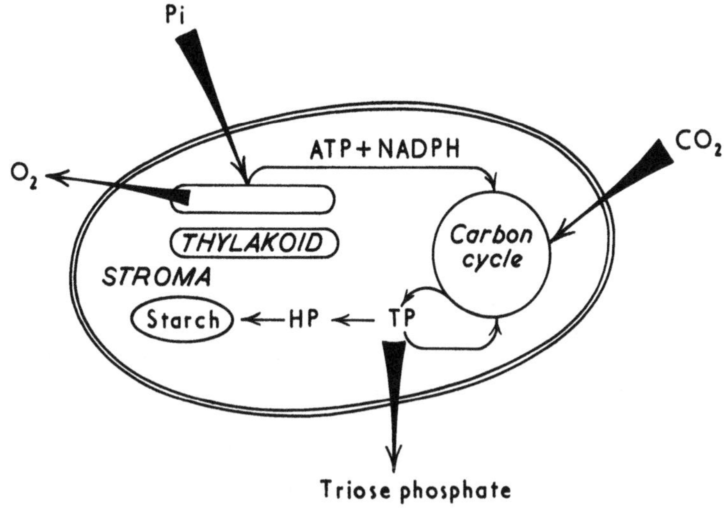

Fig. 1. Movement of principal metabolites across chloroplast envelopes in spinach. Carbon dioxide and orthophosphate enter and triose phosphate is exported. Oxygen escapes. Reduction of NADP and formation of ATP from ADP associated with thylakoids, which house pigments and other constituents of photochemical apparatus. Benson-Calvin or carbon cycle located in the stroma. Triose phosphate (TP) is made in the cycle and is available for export, recycling, and starch synthesis via hexose phosphate (HP)

largely on three types of evidence which are described in Sect. 3.1–3. Each, in itself, can be criticised on many grounds, and it is most unlikely that the present view will not be refined as work progresses. Nevertheless, when several independent groups, using entirely different procedures, point to the same conclusion, then their proposals must merit serious consideration until such time as there is irrefutable evidence to the contrary.

3. Methods

Heber's summary of methods (Heber, 1974) has been reproduced in outline in Table 1 and is followed here. As previously noted (Heber, 1974; Walker, 1974b) each method has certain advantages and certain disadvantages.

3.1 Distribution in vivo

Visual examination allied to iodine staining shows that starch is found in chloroplasts (Rabinowitch, 1945). No other photosynthetic product is quite so readily located, but information can be derived from fractionation of leaf tissues.

3.1.1 Nonaqueous Techniques

Nonaqueous techniques rest on the concept that if a leaf is quickly frozen, the distribution of compounds within its tissues at that moment will remain unchanged throughout subsequent

Table 1. Methods used in study of intracellular transport. (After HEBER, 1974)

A. Information on distribution and flow of compounds in vivo
- I. Nonaqueous cell fractionation
  - a) Kinetic measurements of substrate levels and substrate distribution
  - b) Kinetic measurements of tracer distribution
- II Aqueous cell fractionation
  - Measurements of chloroplast contents in intact chloroplasts after "fast" isolation

B. Information on metabolite distribution and flow in vitro
- I. Direct methods using aqueously isolated chloroplasts
  - a) Distribution of metabolites or tracer between chloroplasts and medium
  - b) Measurements of substrate concentration after centrifugation of chloroplasts through silicone oil
- II. Indirect methods using aqueously isolated chloroplasts
  - a) Response of chloroplast metabolism to the addition of metabolites
  - b) Measurement of the activity of "cryptic" enzymes
  - c) Osmotic response of chloroplasts to additives

nonaqueous fractionation (see e.g. STOCKING, 1959, 1971). This approach has been used in enzyme studies and it is here that its credibility is most suspect because of the very real probability that, e.g., denatured cytoplasmic protein can be precipitated on to chloroplasts and then separated with this fraction. Nevertheless, when applied with the scrupulous attention to detail which has been exercised in some laboratories, it is a useful technique expecially in relation to the distribution of metabolites.

3.1.2 Aqueous Separation

To avoid redistribution, aqueous separation often depends on rapidity of separation (see e.g. NOBEL, 1969) and on the assumption, as above, that fractionation can be achieved without cross-contamination.

Before the development of procedures better suited to the maintenance of envelope integrity, chloroplasts were found to be particularly leaky. It was concluded that major damage must occur during isolation in aqueous media (SMILLIE and FULLER, 1959). More recently the separation of really active chloroplasts (see e.g. WALKER, 1971; LILLEY et al., 1975) indicates that massive damage can be avoided. Even so there is no real doubt that chloroplasts isolated in aqueous media may be contaminated by enzymes from other organelles (see Sect. 4). Similarly any smaller molecules or ions which are free to cross the envelopes are likely to do so to some extent however rapidly separation is achieved. However, careful washing may remove contaminating enzymes and there seems every prospect that the development of increasingly sophisticated types of centrifugation (MIFLIN and BEEVERS, 1974; MORGENTHALER et al., 1974) could lead to considerable advances in this field. (This chapter is not concerned with enzyme distribution as such, but the location of enzymes, and their accessibility to exogenous substrates is often relevant to metabolite transport.)

3.2 Distribution in vitro

Direct analysis is used to determine the ways in which various compounds partition themselves between isolated chloroplasts (and chloroplast compartments) and the suspending medium under different conditions.

3.2.1 Centrifugal Filtration

This procedure, first devised by WERKHEISER and BARTLEY (1957), and used extensively with mitochondria (see e.g. KLINGENBERG and PFAFF, 1967), has been applied to chloroplasts with great success by HELDT and his colleagues (see e.g. HELDT et al., 1972). In principle it usually involves loading chloroplasts with a labelled metabolite and following the redistribution of label between the plastid and a bathing solution after centrifugal acceleration through a filtering layer of silicone oil. Its disadvantage, if it has any, is that it has not yet proved possible to apply the method to chloroplasts which are actively engaged in photosynthesis under relatively normal conditions.

3.2.2 Chromatographic Analysis

Mixtures containing intact isolated chloroplasts are analysed during and after photosynthesis under a variety of conditions. This is a natural extension of the work on *Chlorella* on which the Benson-Calvin cycle is based (BENSON and CALVIN, 1950; BASSHAM and CALVIN, 1957), and in the hands of experts like Martha Kirk is patently an extremely useful technique. Its principal disadvantage is that it is extremely laborious, which inevitably restricts the number of observations on which conclusions are based; also, like other procedures involving isolated chloroplasts, it cannot readily distinguish between events which take place inside the chloroplast and those which may occur in the surrounding medium.

3.3 Indirect Methods

These are often based on attempts to influence the course of photosynthesis in intact isolated chloroplasts by the addition of exogenous reagents. If, for example, an added metabolite can produce a rapid and profound effect, the simplest explanation is that it does so following penetration. The advantages of this approach are that many experiments can be carried out in a short time, so that a large body of information can be elicited, and that the experiments relate to chloroplasts engaged in active photosynthesis. The disadvantages derive from the fact that interpretation is largely presumptive. Clearly it does not necessarily follow that an additive must penetrate the chloroplast envelope in order to produce an effect, or that if it does penetrate, it does so unchanged.

3.3.1 Shortening of Induction and Reversal of Orthophosphate Inhibition

In whole plants and isolated chloroplasts, photosynthesis does not reach its full rate immediately upon illumination (unlike photosynthetic electron transport), but only after a lag or induction period which is believed to represent the time taken for Benson-Calvin cycle intermediates to build up to the steady state concentration dictated by the prevailing light intensity (RABINOWITCH, 1956; WALKER, 1973, 1975). This lag can be shortened by certain intermediates (see e.g. WALKER et al., 1967) and shortening of induction has been taken as an indication of penetration (WALKER and CROFTS, 1970; GIBBS, 1971; HEBER, 1974; WALKER, 1974b). Similarly the initial lag can be lengthened almost indefinitely by exogenous Pi, and reversal of Pi inhibition by metabolites (Fig. 2) has therefore also been interpreted as evidence of entry (COCKBURN et al., 1967a, b; COCKBURN et al., 1968; WALKER, 1969). Inhibition by Pi and its reversal by PGA and triose phosphates is believed to involve the Pi translocator (see e.g. HELDT and RAPLEY, 1970b; WALKER and CROFTS, 1970).

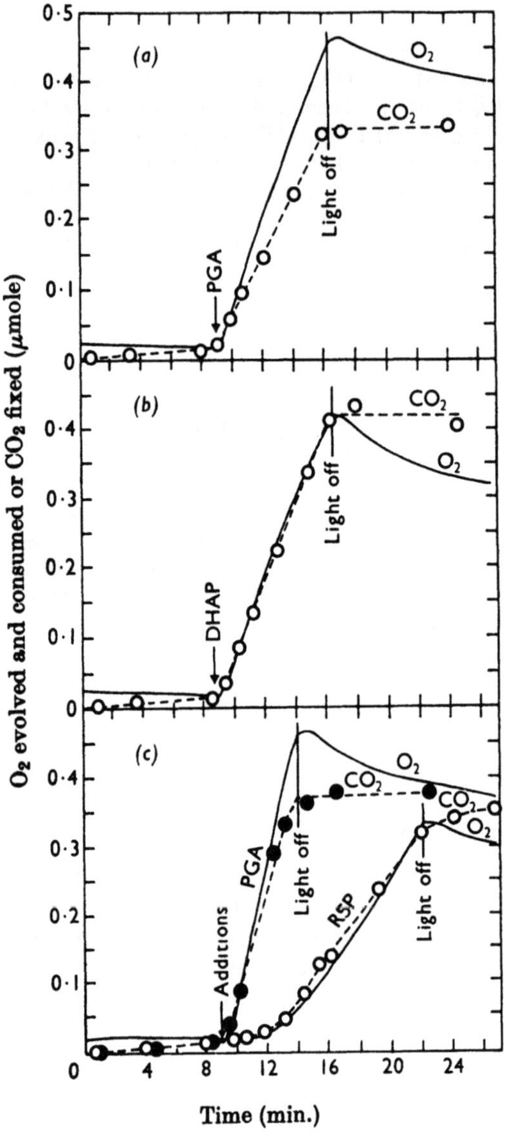

Fig. 2a–c. Reversal of orthophosphate inhibition. Simultaneous measurements of CO_2 fixation (*points*) and O_2 evolution (*continuous lines*) in intact illuminated chloroplasts in which photosynthesis has been inhibited by high concentrations of orthophosphate (see Fig. 10). Reversal, exhibiting characteristic kinetics, follows addition of PGA, DHAP or R5P as indicated. Orthophosphate inhibition believed to be caused by enforced export of cycle intermediates; reversal taken as evidence of penetration by effective intermediate (Sect. 14). (From COCKBURN et al., 1968)

3.3.2 Catalysis by Intact and Ruptured Chloroplasts

Techniques developed with the aim of separating intact chloroplasts from leaf tissues (WALKER, 1971) yield preparations containing, on average, some 70–80% of Class A chloroplasts (HALL, 1971; see Fig. 3). These may be readily stripped of their envelopes by brief osmotic shock and reaction mixtures can be prepared which differ only in the degree of envelope integrity displayed by the chloroplasts within them. Envelope rupture exposes cryptic catalytic sites to reagents which do not cross the envelope, and this has been used for some years as a criterion of permeability (see e.g. WALKER, 1965b; HEBER et al., 1967a; MATHIEU, 1967). Osmotic shock abolishes the ability of intact chloroplasts to assimilate carbon at rapid rates (see e.g. WALKER, 1965b), principally because of dilution of stromal coenzymes and cofactors. Conversely (Fig. 4) it accelerates the Hill reaction with nonpenetrating oxidants such as ferricyanide (HEBER and SANTARIUS, 1970; COCKBURN et al., 1967b) and dark CO_2-fixation with ribulose bisphosphate or ribose-5-phosphate+ATP as substrates.

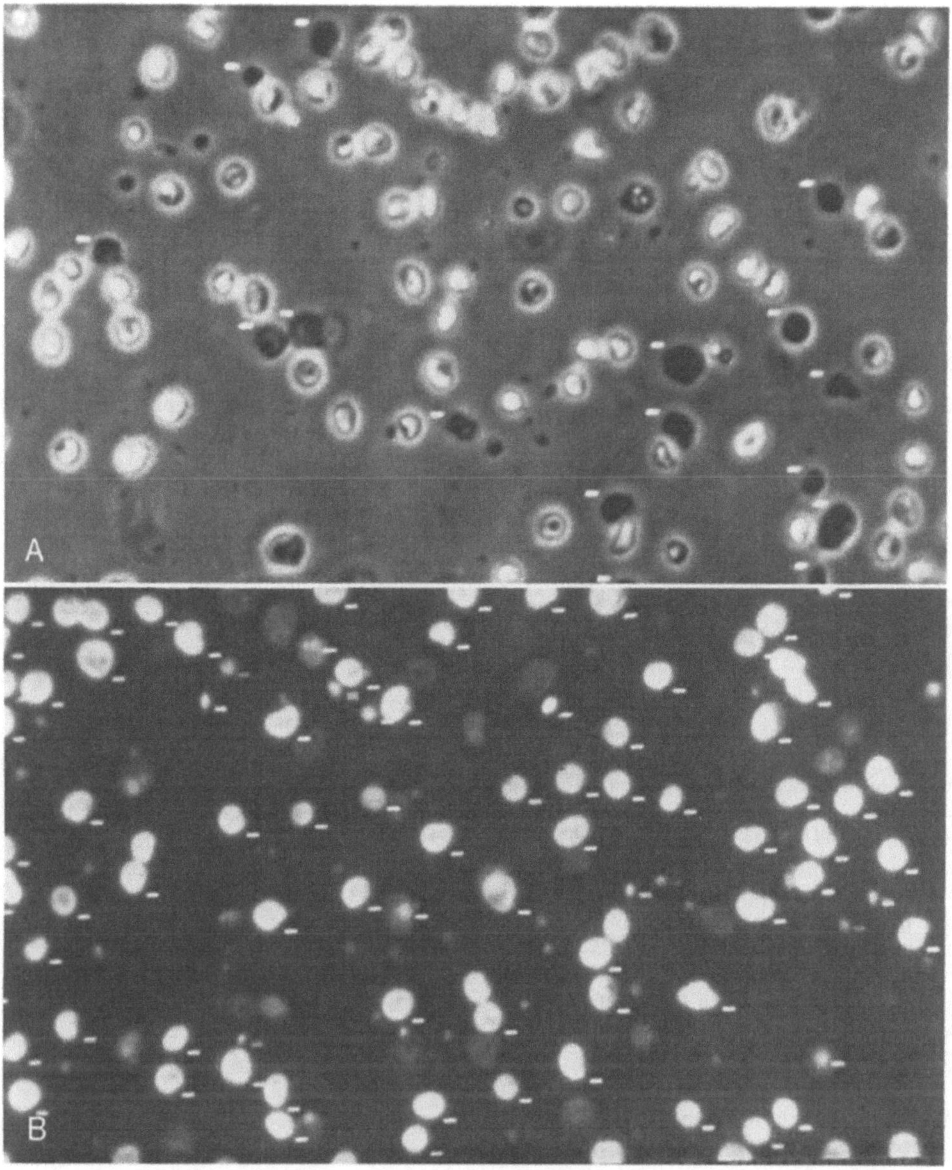

Fig. 3A and B. Chloroplasts as seen in phase contrast. Positive (A) and negative or anatropal phase contrast (B) used in conjunction with microdensiometer and image analyser computer (Quantimet 720). (A) Class C chloroplasts marked with white bar. (B) Class A chloroplasts marked with white bar. In both, intermediate or uncertain categories are discernible. (After LILLEY et al., 1975, by courtesy of P. FRASER, of Imanco, Melbourn, Royston, Herts., U.K.)

Hazards in interpretation derive largely from the assumption that osmotic shock only brings about envelope rupture and that the chloroplasts are uncontaminated (e.g. by enzymes released from damaged chloroplasts). For example, it is almost impossible to wash intact chloroplasts free of inorganic pyrophosphatase activity (see Sect. 15), but this can be elimated in the external medium by omission of Mg (SCHWENN et al., 1973; LILLEY et al., 1973).

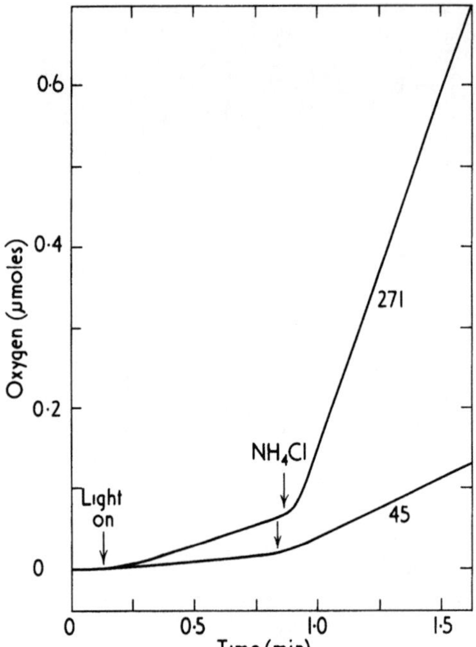

Fig. 4. Oxygen evolution from whole and osmotically shocked chloroplasts. Ferricyanide is unable to enter chloroplasts with intact envelopes (*lower trace*); all O_2 recorded is thought to evolve from the ruptured chloroplasts which are always present. Note that the osmotically shocked chloroplasts (*upper trace*) show a larger response to the uncoupler (NH_4Cl) than those which have lost their stroma during preparation (cf. Fig. 17). (From LILLEY et al., 1975)

3.3.3 Osmotic Volume Changes

Intact chloroplasts shrink in hypertonic media and swell in hypotonic media in which the osmoticum is a nonpenetrating compound such as sucrose (NOBEL and WANG, 1970; HELDT and SAUER, 1971; WANG and NOBEL, 1971). Thus sorbitol causes a fast and irreversible increase in apparent absorbance at 535 nm (Fig. 5), indicating chloroplast shrinkage brought about by exosmosis of water. Ribose induces a very similar initial response followed by a recovery as it enters the chloroplast. Alanine behaves in much the same way as sorbitol, suggesting that it is taken up slowly if at all. However, as GIMMLER et al. (1974) have pointed out, a rapid exchange process would not necessarily bring about an osmotic response. At present there are also unexplained discrepancies between some conclusions based on direct measurements and corresponding data for changes in absorbance at 535 nm.

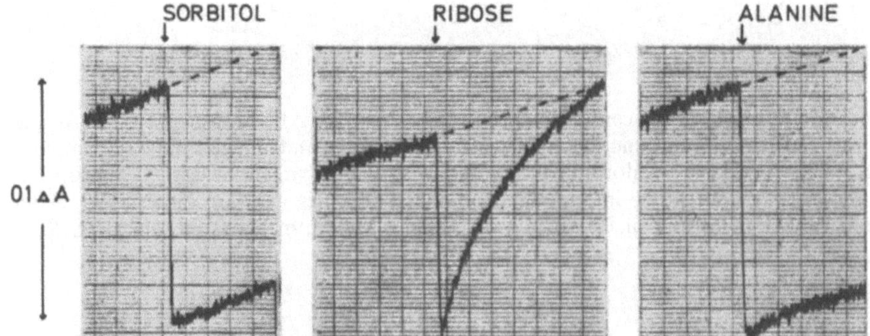

Fig. 5. Changes in apparent absorbance at 535 nm following the addition of sorbitol, ribose and alanine to intact chloroplasts. The initial rapid decrease is caused by exosmosis of water and consequent shrinkage. Recovery (swelling) implies penetration of the solute followed by endosmosis of water. (From GIMMLER et al., 1974)

3.4 Other Aspects of Work with Functional Chloroplasts

In the preceding Section, several methods were discussed which involve chloroplasts isolated in aqueous media, and some general comment on their use would seem appropriate. In principle there is much to be gained in terms of accessibility etc. by simply isolating the chloroplast from its cellular environment as a prelude to studies on metabolite transfer. The objections are as old as biochemistry itself and hinge on the question: can isolation be achieved without damage? The answer at present is an unequivocal: "no", but this is not to say that information cannot be usefully derived from a damaged or imperfect system provided that approach and interpretation are suitably circumspect. Certainly most of the work on photosynthetic electron transport and phosphorylation has involved the use of organelles which are much more damaged.

The question of rate and chloroplast intactness is not unimportant in this context. If an isolated chloroplast is capable of rapid electron transport and photophosphorylation but has lost the ability to assimilate CO_2 when illuminated in a suitable reaction mixture, then it has obviously been exposed to damage or irreversible inhibition during isolation. Because there is often a clear correlation between envelope integrity and function (see e.g. WALKER, 1969), results obtained with relatively inactive chloroplasts must be viewed with more suspicion than those obtained with chloroplasts capable of achieving the same rates of photosynthesis as the parent tissue. What the critic cannot do, with justification, is to invoke the possibility of damage to explain some facts but not others, particularly if all the results under consideration have been obtained with good Class A chloroplasts from the same species.

In the past, one of the greatest hazards of interpretation has sprung from inadequate recognition of the contribution made by the presence of ruptured chloroplasts (see e.g. Sect. 16.2) and enzymes released from ruptured chloroplasts and/or other organelles. Perhaps the clearest example of the latter is the fact that intact chloroplasts, as usually prepared, exhibit high catalase activity, although this is almost certainly an enzyme which is restricted to microbodies in vivo (NEWCOMB and FREDERICK, 1971). Similarly it is almost impossible to prepare suspensions free of external pyrophosphatase activity, because of the release of this enzyme from damaged chloroplasts (SCHWENN et al., 1973; LILLEY et al., 1973). Recently, it has been suggested (LILLEY et al., 1975) that preparations capable of rapid photosynthesis (see Fig. 3) will normally contain chloroplasts intermediate in form between fully intact (Class A) and fully stripped (Class C). These would appear to be impermeable to ferricyanide (cf. HEBER and SANTARIUS, 1970) and this, together with measurements of protein content, suggests that chloroplast envelopes may rupture and then reseal following the escape of a proportion of the stroma. Chloroplasts will also exist in a transient form in which envelope loss is not immediately followed by complete loss of stroma (RIDLEY and LEECH, 1968).

4. The Site of Sucrose Synthesis

The concept that sucrose is an end-product of photosynthesis (in the sense that it is synthesised within the chloroplast and then transported to other parts of the plant) was difficult to sustain in view of its effectiveness as an osmoticum (see Sect. 2), but equally difficult to abandon while it seemed likely that the enzymes concerned in sucrose synthesis were located in the chloroplast (BIRD et al., 1965). Moreover, sucrose synthesis in chloroplast preparations has been unequivocally demonstrated (EVERSON et al., 1967; GIBBS et al., 1967b) even though this was an apparently seasonal effect which could not be repeated at will and is, indeed, contrary to general experience. This, together with the direct evidence of HELDT and colleagues (HELDT and SAUER, 1971; HELDT et al., 1972) that sucrose is not translocated across the inner envelope, and the conclusions of HEBER and WILLENBRINK (1964) that the cytoplasm is the primary site of sucrose synthesis, led to the compromise conclusion that if some sucrose was synthesised within the chloroplast it would not then be exported from it (see e.g. WALKER, 1974b). More recently this view has been strengthened by BIRD et al. (1974), who conclude that if the enzyme distribution which they observed occurs in vivo, "sucrose synthesis in leaves cannot take place in the chloroplasts". In the light of this new evidence the most likely interpretation of the observations of GIBBS et al. (1967b) is that some seasonal variation in material favoured cytoplasmic contamination of their chloroplast fraction. Certainly the evidence in toto favours the concept that sucrose synthesis is a cytoplasmic event and that it occurs at the expense of triose phosphate exported from the chloroplast (Fig. 6).

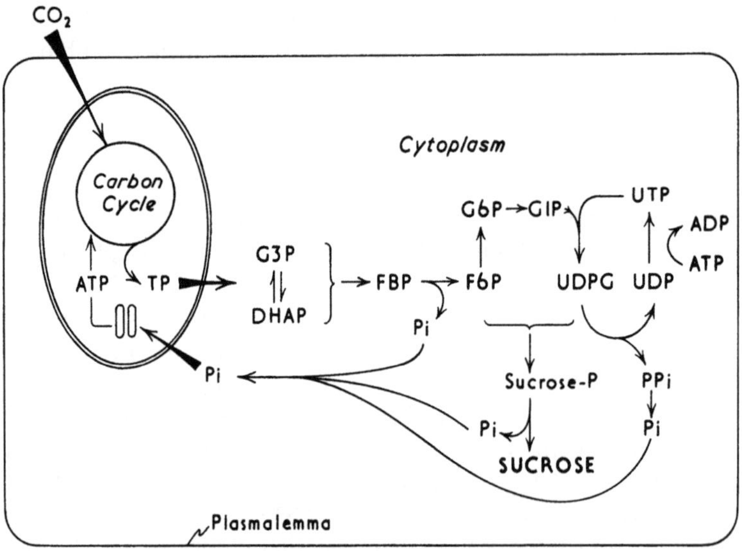

Fig. 6. Sucrose synthesis. Envisaged as cytoplasmic process in which substrate is triose phosphate (TP) exported from chloroplast. Triose phosphates dihydroxyacetone phosphate (DHAP) and glyceraldehyde-3-phosphate (G3P) are in equilibrium in presence of cytoplasmic TP isomerase and undergo aldol condensation to yield fructose-1,6-bisphosphate (FBP) This gives fructose-6-phosphate (F6P) glucose-6-phosphate G6P, glucose-1-phosphate (G1P) and finally uridine diphosphate glucose and inorganic pyrophosphate after reaction with uridine triphosphate (UTP). UTP is reformed from UDP at the expense of ATP. Uridine diphosphate glucose (UDPG) donates a glucose unit to F6P yielding sucrose phosphate and finally free sucrose. Orthophosphate released in 3 of these reactions passes into cytoplasmic pool, from which it can re-enter chloroplast. Orthophosphate release in cytoplasm will favour triose phosphate export. (See Sect. 4)

5. Starch Synthesis

The evidence that starch is synthesised within the chloroplast is, of course, visual and unequivocal (SACHS, 1862, 1887; RABINOWITCH, 1945). In the light, synthesis from a precursor such as triose phosphate would raise no problems with regard to transport, and starch formation could be favoured by the activation of ADP glucose pyrophosphorylase by a combination of relatively high concentrations of PGA and relatively low concentrations of Pi (PREISS et al., 1967b; PREISS and KOSUGE, 1970). Starch synthesis can however be promoted in the dark by exogenous sugars such as glucose (see e.g. BOEHM, 1883; BROWN and MORRIS, 1893; PARKIN, 1899; TOLLENAAR, 1925; PHILLIS and MASON, 1937; MACLACHLAN and PORTER, 1959; CHEN-SHE et al., 1975), and if it is accepted that the chloroplast is more or less impermeable to free hexose and ATP (Sect. 10, 16) it seems highly probable that import must be preceded by phosphorylation (cf. MACLACHLAN and PORTER, 1959) within the cytoplasm (Fig. 7). Entry as triose phosphate is favoured in Figure 7 because in spinach there is no doubt that triose phosphate can enter the

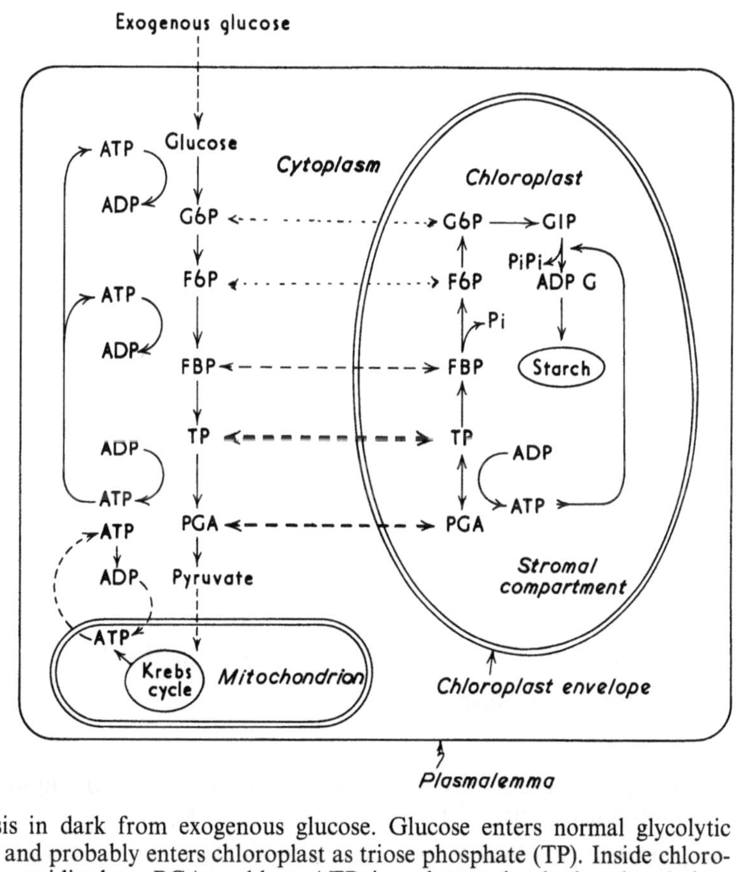

Fig. 7. Starch synthesis in dark from exogenous glucose. Glucose enters normal glycolytic sequence in cytoplasm and probably enters chloroplast as triose phosphate (TP). Inside chloroplast some TP will be oxidised to PGA yielding ATP in substrate-level phosphorylation. Remainder will condense to give FBP and, after a series of reactions (similar to those outlined in Fig. 6), G1P will combine with ATP to give ADPglucose (ADPG). ADPG will donate a glucose unit to an acceptor, lengthening an existing polysaccharide chain. (After WALKER, 1974b)

chloroplast freely, whereas the evidence relating to hexose phosphate is much less clear. However, starch synthesis in the dark from exogenous sugars is a relatively slow process and if direct penetration of hexose phosphate is as high as the 1–10 µmol/mg chlorophyll/h that some results suggest (WALKER, 1974b), then the proportion of starch derived from hexose phosphate could be appreciable and the emphasis laid on the entry of triose phosphate in Figure 6 would need to be changed. Indeed, in tobacco (MACLACHLAN and PORTER, 1959) and some other species the relatively small redistribution of label between C1 and C6, which occurred when specifically labelled sugars were fed, points more to direct utilisation than to prior conversion to triose phosphate. The extent of redistribution could well be variable (see e.g. SHIBKO and EDELMAN, 1957), depending on the respective rates of all of the partial reactions involved, including the cytoplasmic conversion of hexose to triose phosphate. Rapid glycolytic degradation of glucose would favour the present emphasis in Figure 7. Slow glycolysis could increase the possibility of direct utilisation.

Some additional ATP will be required within the chloroplast during the final stages of synthesis, but this could be provided by the shuttle mechanisms illustrated in Figures 22 and 23. If starch synthesis from triose phosphate in the dark involves the "sugar phosphate shuffle" of the Benson-Calvin cycle then there must be enough fructose bisphosphatase activity in the dark to allow this sequence to occur at the observed rates, even though there is evidence that this is normally a light-activated enzyme (PEDERSEN et al., 1966; BUCHANAN et al., 1967; PREISS et al., 1967a; PREISS and KOSUGE, 1970; BASSHAM, 1971; BUCHANAN et al., 1971).

If fructose bisphosphatase is totally inactive in the dark, then the redistribution of label between C1 and C6 referred to above is not readily explained, particularly since more extensive randomisation (involving carbons 2–5) does not apparently occur to any large extent (MACLACHLAN and PORTER, 1959). Similarly the most obvious route for starch degradation in the dark would involve fructose-6-phosphate kinase, an enzyme which has been reported to be absent from the chloroplast (GIBBS et al., 1967b; HEBER et al., 1967a). Recently, however, evidence of phosphofructokinase activity in chloroplast extracts has been published by KELLY and LATZKO (1975). If, as it is sometimes supposed, the chloroplast operates the oxidative pentosephosphate pathway in the dark (BASSHAM, 1971), triose phosphate could also be produced via this sequence.

6. The Stimulation of Starch Synthesis in the Light by Exogenous Sugars

Exogenous sugars can also stimulate starch synthesis in the light (see e.g. MACLACHLAN and PORTER, 1959; CHEN-SHE et al., 1975). It might easily be assumed that the processes involved would be a combination of those which occurred in light and dark (see Sect. 15), i.e. triose phosphate would be formed from sugar in the cytoplasm and would then enter the chloroplast and the Benson-Calvin cycle. To an extent, this may be true for glucose (cf. MACLACHLAN and PORTER, 1959) but there is now evidence that additional factors may be involved (CHEN-SHE

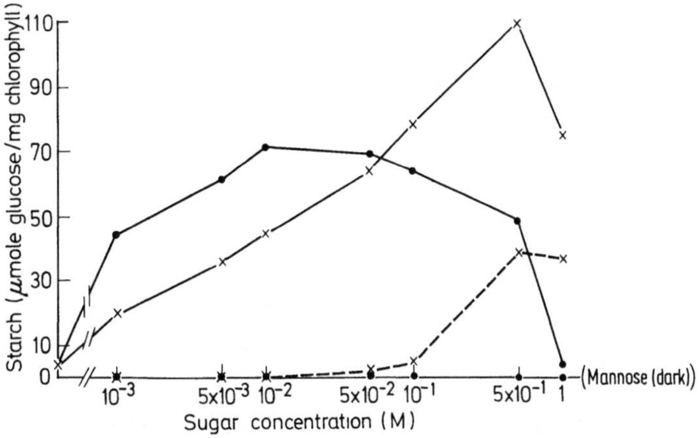

Fig. 8. Stimulation by mannose of light-dependent starch synthesis in spinach beet leaves. Unlike glucose (×) mannose (●) does not promote starch synthesis in dark but brings about massive stimulation in light with optimum at about 10^{-2}M. Carbon from mannose not incorporated into starch; stimulation believed to be brought about by sequestration of cytoplasmic orthophosphate (Fig. 9). (From CHEN-SHE et al., 1975)

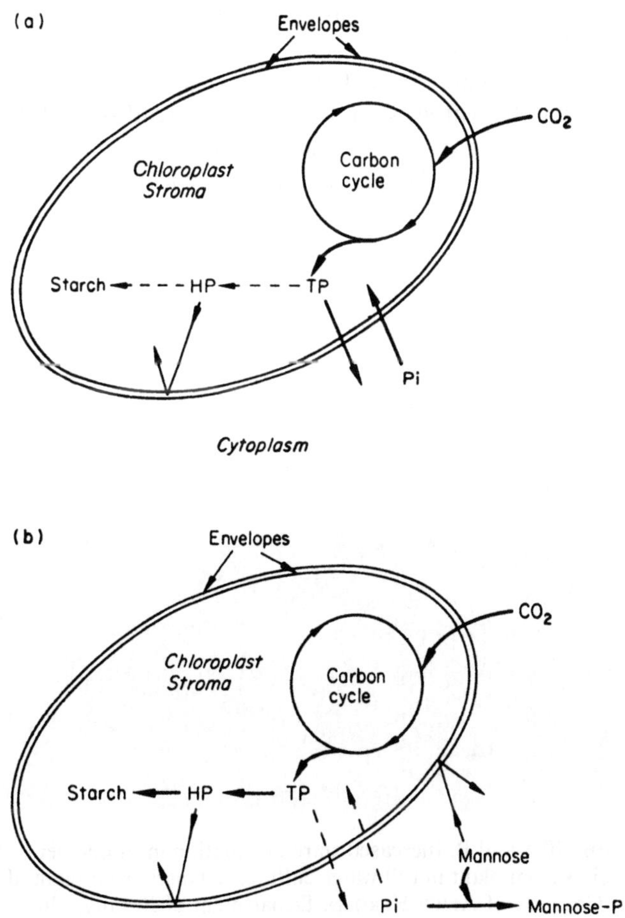

Fig. 9a and b. Stimulation of starch synthesis by orthophosphate sequestration. (a) Normal situation in which TP is divided between export, recycling and starch synthesis. (b) Abnormal situation in which starch synthesis is promoted by mannose (cf. Fig. 8). Mannose combines with cytoplasmic Pi which is then unable to exchange with TP via phosphate translocator (Fig. 21; Sect. 14, 19.3). Consequently, TP is retained within chloroplast and starch synthesis favoured. (Hypothesis proposed by CHEN-SHE et al., 1975). TP, triose phosphate; HP, hexose phosphate et al., 1975). The operation of these is seen most clearly if mannose is fed to a species such as spinach beet at a concentration of 10^{-2} M. At this concentration, with this species, there is no starch formation in the dark but as much as a ten-fold stimulation in the light (Fig. 8). Experiments with labelled mannose show that there is no incorporation of radioactivity into the starch fraction. Instead the major product appears to be mannose-6-phosphate. The stimulation is completely inhibited by DCMU at 10^{-6} M (CHEN-SHE, unpublished). These experiments were undertaken because it was believed that mannose might bring about a sequestration of Pi in the same way as it does when supplied to the roots of certain species (LOUGHMAN, 1956; GOLDSWORTHY and STREET, 1965). If this occurred it was supposed that starch stimulation might ensue as a consequence of suppression of the operation of the phosphate translocator (Sect. 19.3.) and inhibition of triose phosphate export (Fig. 9). The fact that the results bear out the prediction does not, of course, constitute proof that the mechanism is, in fact, that illustrated in Figure 9, but it is at least difficult to escape the conclusion that the level of cytoplasmic orthophosphate may have an important bearing on starch formation within the chloroplast (cf. DE FEKETE and VIEWEG, 1974). A deficiency of orthophosphate, if it extended from cytoplasm to chloroplast, would also be expected to facilitate starch synthesis by activation of ADP glucose pyrophosphorylase (PREISS et al., 1967b) and suppression of phosphorylytic degradation. Abnormally high levels of starch synthesis can be induced by growing spinach beet in Pi-deficient media (Fig. 10).

Preliminary work (CHEN-SHE et al., unpublished) suggests that the stimulation of starch synthesis in the light by readily metabolised sugars such as glucose may result partly from incorporation of carbon (presumably entering the chloroplast as triose phosphate) and partly as a consequence of Pi sequestration.

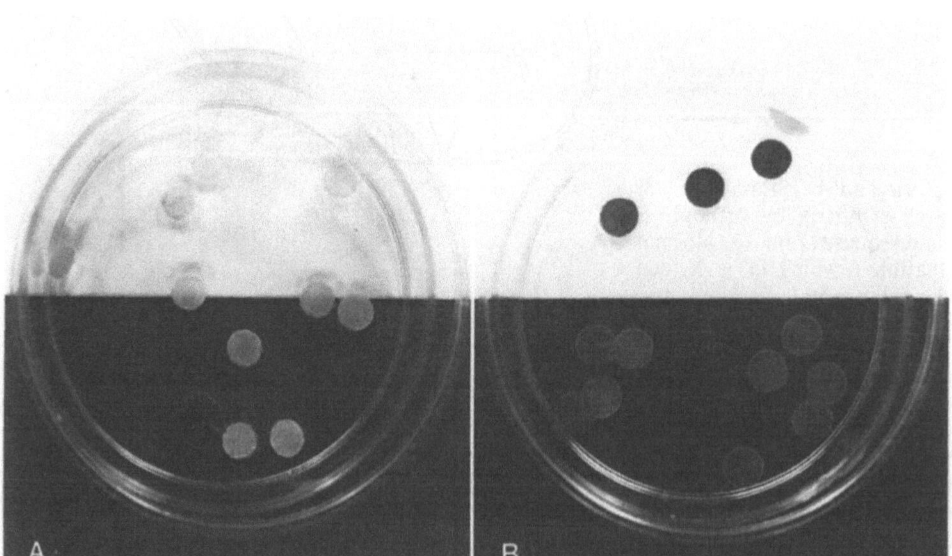

Fig. 10A and B. Increased starch formation in Pi-deficient spinach beet leaf. (A) iodine-stained discs from plant in full water-culture. (B) discs from plant after several days in Pi-free medium. (Courtesy of ALICE HEROLD, Department of Botany, University of Sheffield)

7. Starch Prints

In the 1850s SACHS (1862) demonstrated that starch formation in leaves was restricted to chloroplasts, and described (1887) how he used starch prints to illustrate his elementary teaching:

"I now employ this form of experiment in my lectures on vegetable physiology, in order to demonstrate the influence of light on the formation of starch; or, better, of darkness on the disappearance of starch. It suffices for instance to fasten a broad band of tinfoil or lead in summer on plants with conveniently large leaves and growing in pots e.g. Tobacco, Maize, *Canna,* etc., without depriving the plants of light. After a few days, the leaves so treated are cut off, and thrown for a few minutes into boiling water in order to kill them, and to cause the starch in the chlorophyll to swell. They are then placed for some hours in strong alcohol, which removes the chlorophyll colouring-matter, and the now colourless leaves are finally placed in a vessel containing a weak, pale brown, alcoholic solution of iodine. After a short time, the parts of the leaf which were not shaded from the light appear blue-black, owing to the formation of iodide of starch: the place shaded by the band of tinfoil, on the other hand, remains colourless, simply because the chlorophyll corpuscles there contain no more starch."

This technique was taken a stage further by MOLISCH (1922) who employed photographic negatives to produce starch pictures in leaves. More recent examples are illustrated in Figure 11.

What impressed SACHS about starch prints was the fact that the polysaccharide was formed in illuminated chloroplasts and, of course, in the last century this was an extremely important observation. What is just as important in retrospect is the fact that starch prints can be made with such startling definition. In other words there is no spread of starch formation beyond the illuminated areas even though leaf discs will readily form starch when floated on glucose in the dark. Obviously a chloroplast must export some end product or intermediate of photosynthesis to the cytoplasm and, as already mentioned, contemporary work indicates that this is probably triose phosphate. Why is it that adjacent chloroplasts cannot apparently synthesise starch from this source in the dark although they can from exogenous sugars? Although this question still cannot be answered it is a little less baffling than it once was. If triose phosphate from the chloroplast is rapidly converted into sucrose in the adjacent cytoplasm and, if the sucrose cannot cross the envelope of neighbouring chloroplasts, the problem is at least partly resolved. The operation of such a sequence would also do much to facilitate the transport of elaborated carbon to the vascular tissues. What remains unanswered is why exogenous sucrose can initiate starch synthesis in the dark. It is possible, of course, that it may be simply a matter of concentration (which could conceivably be higher during sucrose feeding). Exogenous sucrose may also be exposed to invertase and converted into more readily metabolised glucose, whereas the route to the phloem followed by internally produced sucrose (via the endoplasmic reticulum?) might not bring it into contact with this enzyme. Whatever the answer, the starch print continues to provide a remarkable demonstration of the relative sophistication of metabolite transport in leaves and of the existence of mechanisms for its control.

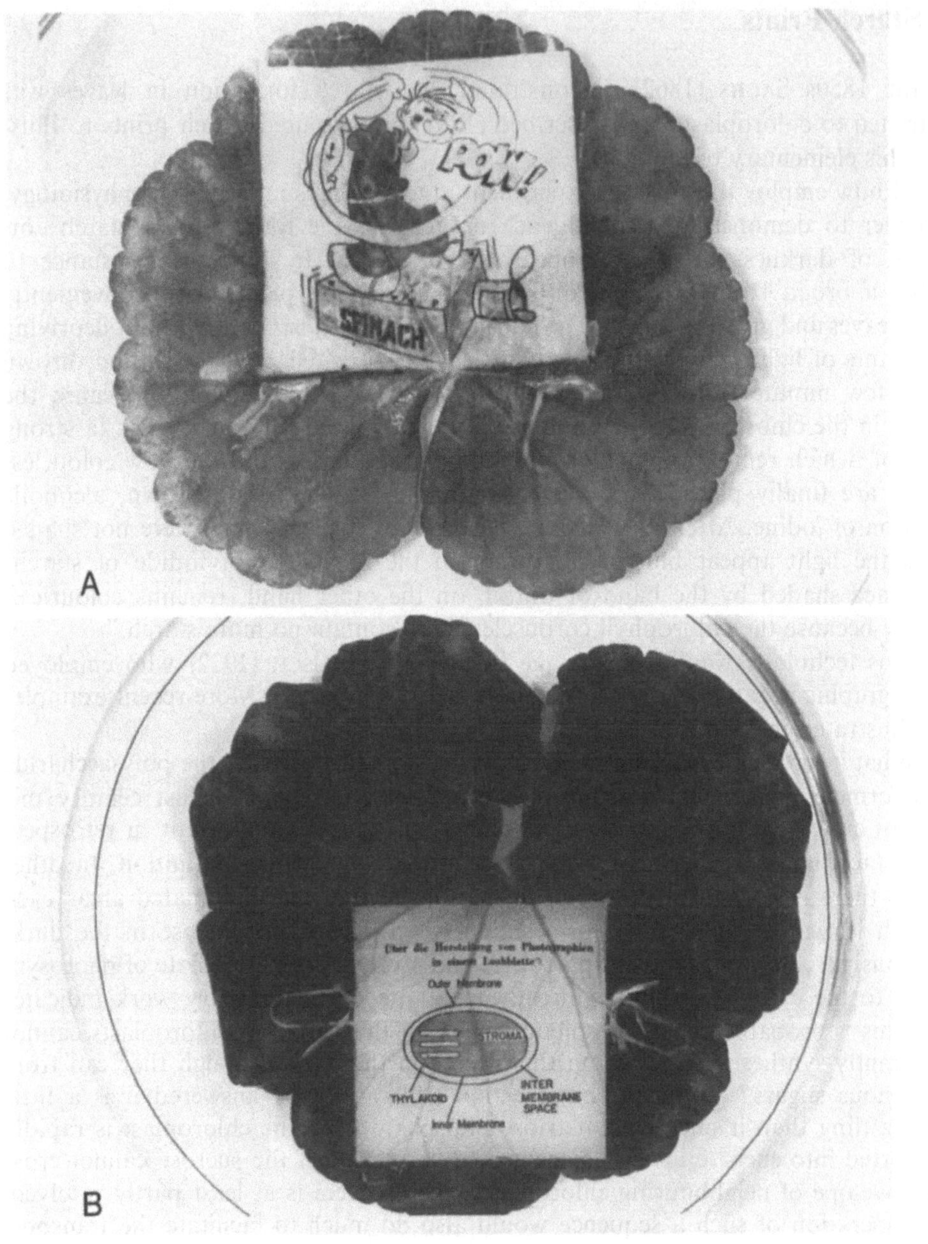

Fig. 11 A and B

Fig. 11 A–D. Examples of starch prints prepared by illuminating starch-free *Pelargonium* leaves through photographic negatives (see Text). Negative used in (B) (C) (D) taken from HELDT and incidentally illustrates the principal compartments of the chloroplast. Also incorporates title of article by MOLISCH (1922) describing this procedure. (D) detail from (C) in which starch which persists in guard cells in nonilluminated areas appears as small rings, providing convenient scale to measure definition of remainder Result: machine-made dots used by draughtsman in original figure reproduced in leaf tissue with remarkable clarity; no discernible spread of starch formation beyond cells actually illuminated. Conversely, discs from *Pelargonium* leaves will readily form starch from exogenous sucrose in dark

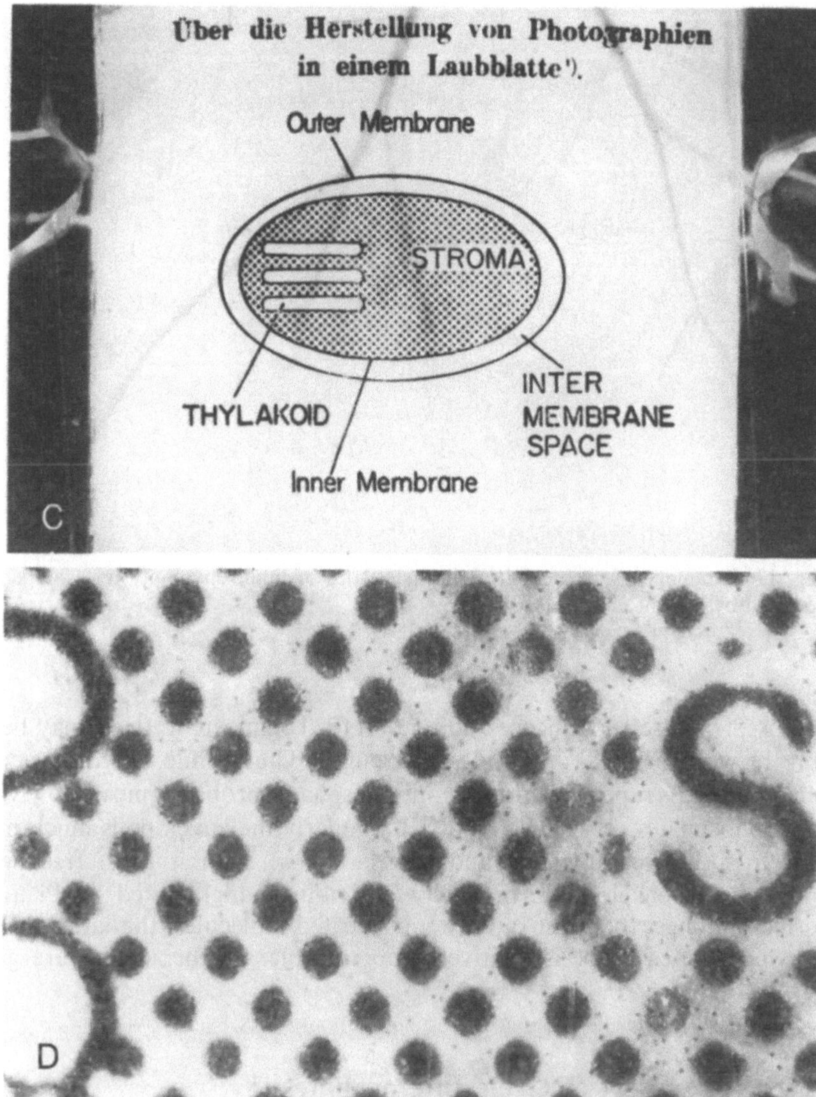

Fig. 11C and D

8. The Transport of Metabolites and Coenzymes

The position as it is presently understood is summarised in Figure 12. Carbon dioxide and Pi enter freely and dihydroxyacetone phosphate is the major export. The inner envelope is probably largely impermeable to bisphosphates and polyphosphates (including sugar bisphosphates, ADP and inorganic pyrophosphate), to

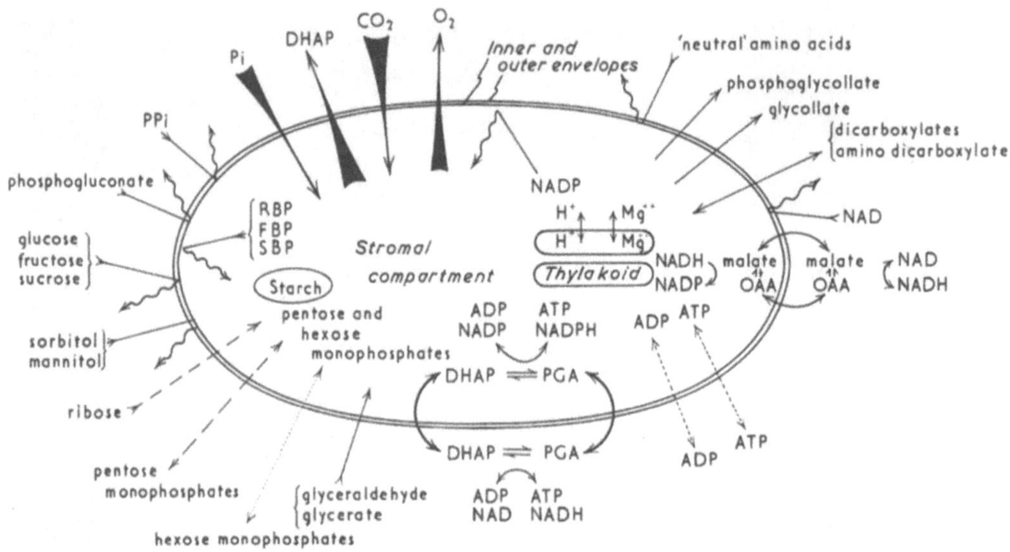

Fig. 12. Summary of permeability characteristics of inner envelope. Ease of penetration represented by magnitude of arrows

sucrose, fructose and glucose and to NADP. To an extent, there may be a relationship between permeability and chain length. Thus, while dihydroxyacetone phosphate can cross rapidly, pentose monophosphates probably move slowly and hexose monophosphates very slowly, if at all. Similarly the envelope is much more permeable to D,L-glyceraldehyde and to free ribose than it is to free hexoses. The specificity of envelope permeability is strikingly highlighted by Pi and PPi, the former amongst the most rapidly translocated molecules, the latter amongst those to which the envelope is relatively impermeable. Further details are given below.

9. The Transport of Cycle Intermediates

9.1 3-Phosphoglycerate and the Triose Phosphates

At the time when rapid rates of CO_2-dependent oxygen evolution by isolated chloroplasts were first demonstrated (WALKER and HILL, 1967), it was also shown that PGA would act as a Hill oxidant and subsequently (COCKBURN et al., 1967b) that this ability was lost following osmotic shock, implying that PGA must cross the envelope at least twice as fast as the rate of the observed O_2 evolution (i.e. at about 80 µmol/mg chlorophyll/h). This was in accord with a previous report of PGA reduction by intact chloroplasts (URBACH et al., 1965). Recently, rates of PGA reduction as high as 450 µmol/mg chlorophyll/h have been observed (HEBER, unpublished; cf. HEBER, 1974). In leaves, HEBER et al. found labelled PGA and DHAP in the cytoplasm within seconds of the onset of photosynthesis (HEBER and WILLENBRINK, 1964; HEBER et al., 1964; URBACH et al., 1965) and

Table 2. Release of orthophosphate or 3-phosphoglycerate from chloroplasts on addition of various anions. (After HELDT and RAPLEY, 1970b)

Anion added (0.5 mM)	Amount of radioactive substance released (nmol/mg chlor., 20 s 4°)	
	^{32}P-phosphate	^{14}C-3PGA
Inorganic phosphate (Pi)	47.0	11.1
3-Phosphoglycerate (PGA)	47.2	12.9
Dihydroxyacetonephosphate (DHAP)	57.3	13.1
Glyceraldehydephosphate (G3P)	48.1	8.9
α-Glycerophosphate	19.1	4.6
Phosphoenolpyruvate	11.2	0.4
Inorganic pyrophosphate (PPi)	6.3	0.2
2-Phosphoglycerate	5.1	1.9
2,3-Diphosphoglycerate	4.2	0.2
Fructose-1,6-bisphosphate (FBP)	1.7	1.3
Glucose-6-phosphate (G6P)	1.2	0.1
Fructose-6-phosphate (F6P)	1.1	0.6
Ribose-5-phosphate (R5P)	3.0	1.0
6-Phosphogluconate	0.9	0.2
ATP	3.0	0.8
Malate	0.2	0.6
Succinate	0.8	0.9
Acetate	1.0	0.6
Arsenate	36.3	8.6

similarly fast movements of PGA have been reported for *Chlorella*. BASSHAM et al. have also shown that PGA, DHAP and G3P pass freely from isolated spinach chloroplasts into the suspending medium (BASSHAM and JENSEN, 1967; BASSHAM et al., 1968a). HELDT and RAPLEY (1970b) have proposed a specific translocator for PGA, DHAP and Pi (cf. Table 2). DHAP, G3P, PGA and 2-phosphoglycerate are among those compounds which shorten induction (WALKER, 1964; BAMBERGER and GIBBS, 1965; BUCKE et al., 1966; BALDRY et al., 1966b; GIBBS et al., 1967a; SCHACTER, et al., 1968). The first three compounds have also been shown to reverse Pi inhibition.

HEBER and SANTARIUS (1970) have proposed that DHAP is exported, but not G3P. This conclusion was based on results obtained during an investigation of a shuttle mechanism for the export of ATP (STOCKING and LARSON, 1969) which involves external oxidation of G3P (see also KRAUSE, 1971). In the experiments by HEBER and SANTARIUS (1970) this system did not work in the absence of added triose phosphate isomerase, and the simplest explanation appeared to be that G3P was formed externally following export of DHAP. The reversal of orthophosphate inhibition (COCKBURN et al., 1967a; COCKBURN et al., 1968; WALKER, 1969) works about as well with G3P as it does with DHAP, but rapid external interconversion catalysed by isomerase released from damaged chloroplasts cannot be completely ruled out, even though this evidently did not occur in Heber's experiments. If both sets of results are accepted at their face value, it could be inferred that G3P moves in but not out, or that the factors controlling the movement of the triose phosphates are even more complex than presently supposed (STROT-

MANN and HELDT, 1969; HELDT and RAPLEY, 1970b; WERDAN and HELDT, 1972; HELDT et al., 1972). On the other hand DHAP is recognised to be a major product of short-term photosynthesis (BASSHAM and JENSEN, 1967; BASSHAM et al., 1968a) by isolated chloroplasts (its preponderance over G3P perhaps reflecting the equilibrium position of triose phosphate isomerase, which favours DHAP by a factor of 20), and the failure of the shuttle in the absence of the isomerase could merely result from limited export of G3P rather than from any innate inability of G3P to move across the envelope. Chromatography showed that virtually no G3P was retained in the chloroplasts, but that (after 3 min) the concentration of G3P in the medium was only about 4% of the DHAP concentration (BASSHAM et al., 1968a). Because of its higher concentration within the chloroplast more DHAP than G3P will be translocated in most circumstances.

9.2 Pentose Monophosphates

After a relatively brief lag, all of the pentose monophosphates of the Benson-Calvin cycle will reverse Pi inhibition (COCKBURN et al., 1967a; COCKBURN et al., 1968; WALKER, 1969), and R5P has also been shown to shorten induction (BUCKE et al., 1966; BALDRY et al., 1966b; WALKER et al., 1967; WALKER and CROFTS, 1970; GIBBS, 1971; WALKER, 1973) and to reverse inhibition by iodoacetate (SCHACTER et al., 1968). Chromatographic analysis also indicates release of pentose monophosphates to the medium (Table 3, BASSHAM and JENSEN, 1967; BASSHAM et al., 1968a). The possibility of external conversion of pentose phosphates to triose phosphates (or vice versa) by enzymes releases from ruptured chloroplast cannot be discounted at this stage. A slow response could indicate slow penetration or relatively slow metabolism within the chloroplast, but it would be equally consistent with slow external conversion and rapid translocation of an end-product.

Table 3. Distribution of labelled metabolites between chloroplast and supernatant following 3-min photosynthesis. (After BASSHAM et al., 1968a)

Compound	Chloroplast μmol ^{14}C (mg chlorophyll/h)	Supernatant μmol ^{14}C (mg chlorophyll/h)	Supernatant/chloroplast Ratio
Dihyroxyacetonephosphate (DHAP)	0.56	23.8	42.5
Fructose-1,6-bisphosphate (FBP)	0.19	6.66	34.0
Pentose monophosphates	0.21	1.79	8.5
Glycollate	0.71	3.94	5.5
3-Phosphoglycerate (PGA)	8.3	39.6	4.8
Sedoheptulose-1,7-bisphosphate (SBP)	0.06	0.12	2.0
Ribulose-1,5-bisphosphate (RuBP)	0.07	0.03	0.4
Fructose-6-phosphate (F6P)	0.38	0.17	0.4
Glucose-6-phosphate (G6P)	1.60	0.21	0.13
Sedoheptulose-7-phosphate (S7P)	4.30	0.27	0.06

9.3 Hexose and Heptose Monophosphates

Fructose-6-phosphate (F6P) and glucose-6-phosphate (G6P) appear fairly quickly in the cytoplasm of leaves illuminated in the presence of $^{14}CO_2$ (HEBER and WILLENBRINK, 1964; HEBER et al., 1967b), indicating export from the chloroplast, but (as with FBP) there is ambiguity because of the possibility of external synthesis from exported triose phosphate. Chromatography of isolated chloroplasts (BASSHAM and JENSEN, 1967; BASSHAM et al., 1968a) indicates a low permeability to F6P and G6P. Fructose-6-phosphate has been reported to bring about a moderately fast reversal of Pi inhibition under conditions in which G6P gave no detectable response (COCKBURN et al., 1968). Subsequently, smaller (sometimes negligible and even inhibitory) responses have also been observed (WALKER, unpublished) and the possibility that F6P actually moves, as such, at significant rates seems increasingly doubtful. There seems no real likelihood that F6P could be converted externally to FBP and triose phosphate but, in the presence of appreciable transketolase from ruptured chloroplasts, it might donate 2-carbon units to exported triose phosphate and enter as erythrose-4-phosphate and ribose-5-phosphate.

Sedoheptulose-7-phosphate (S7P) does not appear rapidly in the cytoplasm (HEBER and WILLENBRINK, 1964; HEBER et al., 1967b) or in the medium (BASSHAM and JENSEN, 1967; BASSHAM et al., 1968a), suggesting that it is slow to leave the chloroplast. Similarly, 6-phosphogluconate (which is known both to stimulate and inhibit RBP carboxylase, according to concentration) (TABITA and MCFADDEN, 1972; CHU and BASSHAM, 1972, 1973, 1974; BUCHANAN and SCHÜRMANN, 1973) apparently fails to cross the envelope because it does not inhibit CO_2 fixation by intact chloroplasts, nor does it reverse Pi inhibition or interfere with the subsequent reversal of Pi inhibition by triose phosphate (COCKBURN et al., 1968) or facilitate the release of PGA etc. (HELDT and RAPLEY, 1970b).

9.4 Sugar Bisphosphates

9.4.1 Ribulose-1,5-bisphosphate (RuBP)

There is general agreement that RBP does not move through the envelope. It does not appear in appreciable quantities in the cytoplasm (HEBER and WILLENBRINK, 1964; HEBER, 1967; HEBER et al., 1967b) or in the medium (BASSHAM and JENSEN, 1967; BASSHAM et al., 1968a). It neither shortens the lag (WALKER, 1969) nor reverses orthophosphate inhibition (COCKBURN et al., 1968) and only slightly affects inhibition by arsenate etc. (SCHACTER et al., 1968). Intact chloroplasts supplied with RuBP fix less CO_2 in the dark than similar chloroplasts which have been osmotically shocked in the reaction mixture (SMILLIE and FULLER, 1959; HEBER et al., 1967a; WALKER, 1969). Dark deactivation of RuBP carboxylation (JENSEN and BASSHAM, 1968) together with the failure of RuBP to move out of the chloroplast could ensure that enough of this metabolite is retained to restart photosynthetic carboxylation in the next light period (WALKER, 1974a; LILLEY et al., 1974; WALKER, 1974b).

9.4.2 Fructose-1,6-bisphosphate (FBP) and Sedoheptulose-1,7-bisphosphate (SBP)

Both substances are formed in relatively large quantities in illuminated chloroplasts in situ but SBP does not pass readily into the cytoplasm (HEBER and WILLENBRINK, 1964; HEBER et al., 1967b) nor from isolated chloroplasts into the medium (BASSHAM et al., 1968a). Apparent movement of FBP is considerable (HEBER, 1967; HEBER et al., 1967a) but is questioned in view of the possibility of external condensation of freely permeable triose phosphates (BASSHAM and JENSEN, 1967; BASSHAM et al., 1968a; COCKBURN et al., 1968). FBP-shortened induction (BALDRY et al., 1966b), but again external lysis followed by endodiffusion of DHAP, G3P, or both, cannot be discounted and (in the absence of more definitive evidence) it may be supposed that the envelope is as impermeable to FBP as it is to RuBP and SBP. Direct measurements (HELDT and RAPLEY, 1970b) support this view.

10. Free Sugars

There is clear evidence that sucrose does not cross the inner envelope of the chloroplast (Sect. 2 and 4). Intact chloroplasts have also been successfully isolated in solutions in which a variety of sugars and sugar alcohols have been used to maintain the osmotic pressure, and it may be concluded, therefore, that the inner envelope is also largely impermeable to glucose, fructose, sorbitol and mannitol. Conversely, ribose enters the chloroplast relatively freely (WANG and NOBEL, 1971; GIMMLER et al., 1974). In early experiments with intact chloroplasts both ribose and fructose caused slight diminution of the initial induction period and it is therefore possible that fructose also penetrates at a very slow rate.

The triose D,L-glyceraldehyde, which is a potent and specific inhibitor of the Benson-Calvin cycle (STOKES and WALKER, 1972), produces very rapid decreases in CO_2-dependent O_2 evolution when added to intact photosynthesising chloroplasts, and must therefore enter the stroma with corresponding rapidity. The related acid (D,L-glycerate) can support O_2 evolution by intact chloroplasts at rates of the order of 10 μmol/mg chlorophyll/h (HEBER et al., 1974).

11. CO_2/Bicarbonate

Dissolved CO_2 undergoes hydration in a reaction catalysed by carbonic anhydrase and by various inorganic ions such as orthophosphate (RABINOWITCH, 1945)

$$CO_2 + H_2O \rightleftharpoons H_2CO_3 \rightleftharpoons H^+ + HCO_3^- .$$

As the pH is raised, the distribution of carbon between the molecular species changes, and above pH 9 an appreciable and increasing proportion of the total CO_2 is present as the carbonate ion. At physiological pH values, however, the CO_2 is present in the forms indicated in the above equation with the proportion of bicarbonate ion (HCO_3^-) increasing from ca. 2% at pH 5 to 95% at pH 8 (RABINOWITCH, 1945).

Because of the ready interconversion of these species it is not easy to say whether green plants utilise free CO_2 or bicarbonate ion in photosynthesis, but a great many observations such as the fact that CO_2 saturation in *Chlorella* relates to dissolved CO_2 rather than bicarbonate ion (RABINOWITCH, 1945) imply that it is CO_2 rather than bicarbonate which readily penetrates plant membranes. Similarly, a weakly buffered bicarbonate medium containing aquatic plants (BLINKS and SKOW, 1938; NEUMANN and LEVINE, 1971) or isolated chloroplasts (HEBER and KRAUSE, 1971) becomes more alkaline on illumination and the kinetics and magnitude of this pH shift are consistent with the preferential uptake of CO_2 from the medium. (See Vol. 2, Part A: Chaps. 6.5.5.3, p. 148; 6.6.4.4, p. 163; 12.5.2.1, p. 333).

WERDAN and HELDT (1973a) also observed that the bicarbonate concentration within the chloroplast increased from 0 to 2 mM within 10 s of being placed on 0.5 mM bicarbonate. The authors (WERDAN and HELDT, 1973a; WERDAN et al., 1972; WERDAN and HELDT, 1973b; HELDT and WERDAN, 1973; HELDT et al., 1973) found that the accumulation of bicarbonate follows the pH gradient across the envelope, and that the internal bicarbonate concentration is in accord with the Henderson-Hasselbach equation if it is assumed that the external and internal concentration of CO_2 are unchanged. This, they concluded, would be consistent with free and rapid movement of CO_2 as follows:

$$HCO_3^- + H^+ \xrightleftharpoons{\text{Outside}} H_2O + CO_2 \overset{\text{Envelope}}{\dashrightarrow} CO_2 + H_2O \xrightleftharpoons{\text{Inside}} HCO_3^- + H^+.$$

POINCELOT (1974) has also measured bicarbonate accumulation within isolated envelopes at rates equivalent to penetration at 400 µmol/mg chlorophyll/h. However, when he compared $H^{14}CO_3^- + {}^{12}CO_2$ with $H^{12}CO_3^- + {}^{14}CO_2$ (a device applied by COOPER et al., 1968, 1969, and FILMER and COOPER, 1970, in establishing the fact that ribulose bisphosphate carboxylase uses CO_2 rather than bicarbonate), his results clearly indicated that $H^{14}CO_3^-$ uptake was more rapid than ${}^{14}CO_2$ uptake.

Irrespective of the penetrating species, there seems no doubt that CO_2/bicarbonate can traverse the envelopes at extremely rapid rates. A problem which remains is whether or not there is active transport of CO_2 and what, if any, is the role of carbonic anhydrase in this process. At a time when the activity of ribulose bisphosphate carboxylase seemed entirely inadequate for its role, it was difficult to escape the conclusion that, if the extracted enzyme retained its in vivo characteristics, there must exist some means of ensuring that the CO_2 concentration at the carboxylation site was greater than atmospheric.

The basic problem (see e.g. LILLEY and WALKER, 1975) may be stated as follows. The average plant, in favourable conditions in its natural environment, fixes CO_2 in the light at rates in the region of 100 µmol/mg chlorophyll/h. Assuming a chlorophyll content of 0.045 mg/cm^2 of leaf surface, which is a little lower than that in spinach as grown in Sheffield (cf. HEATH, 1969), this is equivalent to an uptake of 20 mg CO_2/100 cm^2 of leaf surface per hour or about 0.1 ml/cm^2/h. Clearly CO_2 must enter the leaf at these rates in order to be fixed. At an external concentration of CO_2 of 0.033%, (as in pure air) it is then possible to calculate the highest concentration of CO_2 which could be reasonably maintained at the carboxylation site by unassisted diffusion. Assuming a concentration of zero, or near zero at the carboxylation site, most workers would accept a rate of influx of about 0.15 ml/cm^2/h (e.g. HEATH, 1969). This rate would fall as the internal concentration was increased. It is probably rash to assume that the fall in the rate of influx would be strictly proportional to the difference in partial pressure between the internal and external CO_2 concentration, but it will suffice as a first approximation. On this basis the influx of 0.10 ml/cm^2/h needed to maintain average photosynthesis could be achieved with an internal concentration of CO_2 of about 0.01%. Like the average plant, spinach cannot match the high performance of species like the sunflower (e.g. HESKETH, 1963; EL-SHARKAWAY and HESKETH, 1965) but if the concentration of CO_2 within its chloroplast is no more than 0.01% (or the equivalent in the liquid phase) its carboxylase must be capable of fixing 100 μmol CO_2/mg chlorophyll/h at this concentration. As first measured by PETERKOFSKY and RACKER (1961), spinach ribulose bisphosphate carboxylase was inadequate by a factor of approximately 500. Because its maximum velocity was 150 μmol/mg chlorophyll/h its apparent inadequacy was usually attributed to its surprisingly high CO_2 requirement which, at that time, was thought to be about 6% for half maximal velocity. Over the intervening years this value has gradually fallen with a growing awareness of the activation which can be achieved by preincubation with Mg (see e.g. PON et al., 1963; SUGIYAMA et al., 1968; BASSHAM et al., 1968b, c; JENSEN, 1971; LIN and NOBEL, 1971; STOKES et al., 1971; WALKER, 1973; LILLEY et al., 1974). In addition the separation of intact chloroplasts and the development of methods of estimating percentage intactness have allowed a more realistic basis for relating ribulose bisphosphate carboxylase to chlorophyll content. If, as is now supposed (WALKER and LILLEY, 1975; LILLEY and WALKER, 1975), the enzyme in vivo has a *Km* (CO_2) equivalent to about 0.12% in the gas phase (cf. BAHR and JENSEN, 1974) and a V_{max} of 1,000 μmol/mg chlorophyll/h or more, then the rate which could be achieved in 0.01% CO_2 is close to that achieved by the parent tissue in air. It does not follow, of course, that there is no active transport or facilitated diffusion of CO_2 into the chloroplast, but the necessity to invoke such processes (see e.g. GIBBS et al., 1967b; WALKER and CROFTS, 1970) in order to equate in vitro and in vivo rates of CO_2 fixation (by C3 plants) has disappeared.

If CO_2/bicarbonate enters the chloroplast as CO_2 and is used by the carboxylase in this form, it is difficult to find a role for carbonic anhydrase, particularly if the enzyme is present in the stroma (POINCELOT, 1972) and is not directly concerned with passage across the envelope (POINCELOT, 1974). WERDAN et al. (1972) have suggested that because the carboxylation is acid-forming there could be a local decrease in pH at the enzyme surface. This would result in the release of CO_2 from bicarbonate accumulating in the main body of the stroma as the overall pH of this compartment rose in response to electron transport and its associated proton gradient. If, on the other hand, bicarbonate is the transported species, carbonic anhydrase could, as before, facilitate the release of CO_2 required for carboxylation (see also EVERSON, 1969; GRAHAM and REED, 1971).

In rejecting the necessity to invoke a CO_2 pump (while continuing to recognise that some sort of facilitated diffusion might still exist) it is implicit that the spinach chloroplast has a carboxylation potential in excess of its normal performance. If this were not the case it could not, in common with many other plants, exhibit higher rates of photosynthesis in augmented CO_2 at high light intensities. The fact that some C_3 species such as sunflower can exceed the performance of the average plant by a factor of 3 or 4 could mean that they have a greatly superior photosynthetic machinery. For many reasons, which can not be explored here, it now seems more likely that their superiority in this respect might be a consequence of decreased diffusive resistance to CO_2. There is a distinct possibility that the sunflower can out-perform spinach in normal air simply because it has a thinner leaf (cf. HESKETH, 1963).

12. Carboxylic Acids

12.1 Glycollate and Glyoxylate

At its most simple, photorespiration is a process involving light-stimulated CO_2 release and O_2 uptake (see e.g. GOLDSWORTHY, 1970; JACKSON and VOLK, 1970; GIBBS, 1971; ZELITCH, 1971, 1973). It is a characteristic of C_3 plants (such as tobacco) which evolve CO_2 when brightly illuminated in CO_2-free air and have CO_2 compensation points of about 50 ppm or greater. Factors which favour photorespiration (e.g. high light, high temperature and low CO_2) also favour glycollate production. Photorespiration involves the oxidation of glycollate, but glycollate oxidase is believed to be located in microbodies or peroxisomes (TOLBERT et al., 1969; KISAKI and TOLBERT, 1969; KISAKI et al., 1971), whereas glycollate itself is thought to be derived from the Benson-Calvin cycle. The precise source of glycollate still remains to be found. The 2-carbon glycoaldehyde moiety involved in the transketolase reactions of the Benson-Calvin cycle remains a possibility (BRADBEER and RACKER, 1961; PLAUT and GIBBS, 1970; SHAIN and GIBBS, 1971), although more recently much importance has been attached to the fact that ribulose bisphosphate carboxylase can function as an oxidase yielding one molecule of phosphoglycerate and another of phosphoglycollate (BOWES et al., 1971; OGREN and BOWES, 1971; BOWES and OGREN, 1972; ANDREWS et al., 1973; TOLBERT, 1973; LORIMER and ANDREWS, 1973). Comparative biochemistry would suggest that either glycollate or phosphoglycollate might penetrate the envelope freely, and there is a certain amount of direct evidence which favours this view (see e.g. KEARNEY and TOLBERT, 1962; BASSHAM et al., 1968a, but cf. HEBER and KRAUSE, 1972).

12.2 Malate and Oxaloacetate

The dicarboxylate translocator (HELDT and RAPLEY, 1970a; HELDT and SAUER, 1971; HELDT et al., 1972) which also facilitates transport of succinate, α-oxoglutarate, fumarate, aspartate and glutamate, allows malate and oxaloacetate to exchange at rates similar to those recorded for phosphoglycerate. These findings, based on direct measurement, are supported by experiments in which the observed rates of oxaloacetate-dependent oxygen evolution demanded rates of penetration by oxaloacetate of up to 300 μmol/mg chlorophyll/h (HEBER and KRAUSE, 1971). KIRK and LEECH (1972) found that the enzymes necessary for oxaloacetate synthesis are missing from intact chloroplasts and concluded that oxaloacetate transport from the cytoplasm is necessary for aspartate synthesis. They also demonstrated stimulation of amino acid synthesis in the presence of an external oxaloacetate generating system.

13. Amino Acids

Conversion of exogenous aspartate and α-oxoglutarate to glutamate and oxaloacetate by intact chloroplasts is accelerated from ca. 10 to 40 μmol/mg chlorophyll/h following osmotic shock, indicating that the intact envelope limits transport of the slowest of these reactants to the lower rate (HEBER et al., 1967a).

Other evidence (see AACH and HEBER, 1967; ONGUN and STOCKING, 1965; ROBERTS et al., 1970) based on the distribution of labelled amino acids between cytoplasm and chloroplasts suggests rapid transport (especially of glycine and serine), but again metabolic conversion on one side of the envelope and resynthesis on the other cannot be excluded (HEBER, 1970). Difficulties of interpretation are also reflected in the conclusions of ROBERTS et al. (1970) that while, in their experience, glycine and serine move freely from chloroplasts to cytoplasm, the same is not true of sugar phosphates. In view of considerable evidence that triose phosphates *can* move readily from the chloroplast, it could be equally well concluded that sugar phosphates move rapidly but do not readily *accumulate* in the cytoplasm. Indeed, the metabolic route from triose phosphate to sucrose is short and direct (Fig. 6), and there is no reason why accumulation of sucrose (formed from exported triose phosphate) should necessarily be accompanied by accumulation of an appreciable quantity of sugar phosphates. The practical difficulties associated with nonaqueous extraction of this nature are also exemplified by the finding of ONGUN and STOCKING (1965) that 15% of total starch was associated with the nonplastid fraction.

NOBEL et al. (NOBEL and WANG, 1970; WANG and NOBEL, 1971; NOBEL and CHEUNG, 1972) proposed two carriers (translocators) on the basis of light-dependent shrinkage in chloroplasts caused by osmotic extraction of water (cf. PACKER et al., 1965; PACKER and CROFTS, 1967). It was proposed (NOBEL and CHEUNG, 1972) that one carrier transports glycine, L-alanine, L-leucine, L-isoleucine and L-valine, and the other L-serine, L-threonine and L-methionine. Conversely, GIMMLER et al. (1974) found no evidence for the existence of specific carriers which might bring about the rapid uptake of neutral amino acids. The slow diffusion which

Table 4. Release of radioactive malate (a) or alanine (b) from intact chloroplasts (Integrity 95%) upon addition of unlabelled malate and amino acids

Preloaded with ^{14}C-labelled	Compound added	Radioactivity released (cpm/mg chlorophyll)
Malate (a)	Malate	18,300
	Aspartate	16,100
	Glutamate	8,900
	Asparagine	3,400
	Glutamine	2,500
Alanine (b)	Alanine	40
	Glycine	0
	Serine	160
	Proline	0
	Threonine	0

All compounds added at zero time to give final concentration of 0.5 mM; aliquots of supernatant measured after 3 min. Values corrected for unspecific leaking of labelled substances. (From GIMMLER et al., 1974)

they observed, however, was more than sufficient to permit the 0.1 μmol/mg chlorophyll/h uptake demanded by protein turnover as observed by HELLEBUST and BIDWELL (1964).

GIMMLER et al. (1974) were also unable to support the notion of ROBERTS et al. (1970) that glycine and serine might play an important role in carbon translocation, concluding that at physiological concentrations the maximum rate of movement of these compounds was probably well below 1 μmol/mg chlorophyll/h, whereas the rate for triose phosphates can be as high as 500. On the other hand aspartate could be transferred via the dicarboxylate translocator almost as rapidly as malate and glutamate, asparagine and glutamine at 50–15% of this rate (Table 4).

14. Orthophosphate

Isolated chloroplasts are essentially Pi-consuming organelles, and if illuminated in Pi-free media will readily cease to photosynthesise as the endogenous Pi is consumed (COCKBURN et al., 1967b). Under the conditions usually employed, the release of Pi associated with starch synthesis (which is known to occur in chloroplasts in vitro, see e.g. ARNON et al., 1954a; GIBBS and CYNKIN, 1958) is insufficient, even in the short term, to offset the Pi incorporated into sugar phosphates and lost to the medium in this form. In vivo it must be assumed that Pi which is lost to the cytoplasm in this way is then released in processes such as sucrose synthesis (Fig. 6) and constantly recycled. In vitro a constant supply of Pi may be maintained by the inclusion of PPi in the reaction mixture (JENSEN and BASSHAM, 1966; WALKER, 1971). Chloroplasts contain an active PPiase and, in the presence of exogenous Mg, there is normally enough of this enzyme released from damaged chloroplasts to supply Pi at a noninhibitory concentration (LILLEY et al., 1973; SCHWENN et al., 1973).

If photosynthesis by isolated chloroplasts has been allowed to cease for lack of Pi, it can then be restarted without appreciable delay by addition of Pi (COCKBURN et al., 1967c). In the short term there is then a rough stoichiometry of three molecules of O_2 evolved or three molecules of CO_2 for each molecule of Pi added. This is consistent with observations (BALDRY et al., 1966a) which show that P^{32} is initially incorporated into triose phosphate according to the overall equation

$$3CO_2 + Pi + 2H_2O \rightarrow 1 \text{ triose phosphate} + 3O_2.$$

What is perhaps surprising is that photosynthesis by isolated chloroplasts shuts down to the extent that it does in the absence of exogenous Pi, and that this will occur equally readily in the presence of sugar phosphate (COCKBURN et al., 1967b). If, for example, triose phosphates were able to enter freely in the absence of exogenous Pi, then the subsequent release of Pi as a consequence of bisphosphatase activity ought to be sufficient to make good any Pi deficiency. The fact that this apparently does not occur to any real extent implies that the stoichiometry of the phosphate translocator (Sect. 19.3) may be rigid. On the other hand, an influx of triose phosphate might also lead to the development of abnormal internal Pi-sinks. For instance, it has been demonstrated that R5P is actually inhibitory in the presence of low Pi and that this inhibition can be reversed by the addition of external Pi (COCKBURN et al., 1967b). At the time that this observation was made, the inhibition was attributed to the added Pi consumption which might follow if Ru5P kinase were supplied with an abnormal concentration of its substrate's precursor. At present it seems likely (LILLEY and WALKER, 1974) that a contributory and possibly crucial factor is a consequent increase in the ADP concentration under these conditions (cf. Fig. 21).

The reversal of SO_4^{2-} inhibition by Pi is believed to relate to the competition between SO_4^{2-} and Pi in photophosphorylation (BALDRY et al., 1968) and implies that the envelope is also permeable to inorganic SO_4^{2-}.

As the Pi concentration in the external medium is increased, an optimum is reached at about 10^{-4}M and thereafter increasing concentrations at first extend the induction period and then depress the maximal rate until photosynthesis is ultimately almost entirely suppressed (Fig. 13) within the first hour of illumination (COCKBURN et al., 1967a, b). The precise nature of the response to Pi depends on the activity of the individual chloroplast preparation and the pretreatment of the parent tissue. Thus very active chloroplasts from preilluminated leaves are difficult to inhibit, whereas at the other extreme, poor chloroplasts from dark-stored leaves are easy to inhibit. Inhibition is reversed, often completely, by the addition of certain cycle intermediates (COCKBURN et al., 1968). The response to PGA is very rapid (Fig. 2) and the two trioses also produce a very fast response. Ribose-5-phosphate (Fig. 2) and fructose-1,6-bisphosphate produce reversal after a lag. Ribulose-1,5-bisphosphate is ineffective. The lag extension by Pi and its reversal by cycle intermediates has been attributed to the operation of the phosphate transloca-

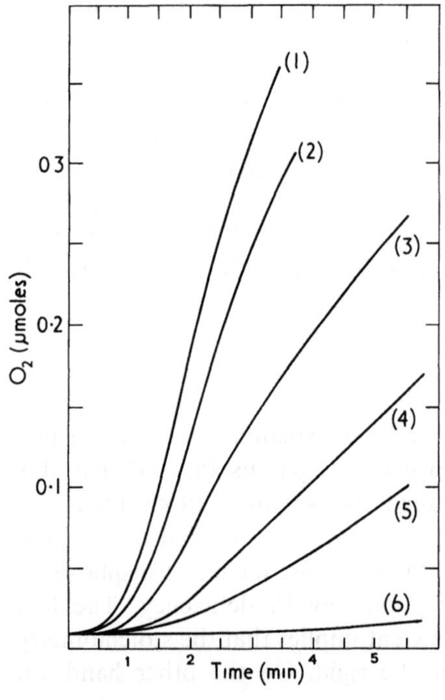

Fig. 13. Inhibition of CO_2-dependent O_2 evolution by orthophosphate. Intact chloroplasts illuminated in presence of increasing quantities of Pi (1) 0.1 (2) 0.25 (3) 0.5 (4) 1.0 (5) 2.5 (6) 5.0 mM. (Previously unpublished but essentially similar to COCKBURN et al., 1968)

tor (HELDT and RAPLEY, 1970b; WALKER and CROFTS, 1970). Thus Pi inhibition is seen as an exaggeration of the normal process in which newly synthesised triose phosphate is exchanged for external Pi (Fig. 24). If external Pi produces its inhibitory effect solely by withdrawing intermediates from the cycle, the pull which it exerts must presumably increase as the concentration of cycle intermediates builds up within the system. Otherwise, there would only be increasing lag extension but no secondary depression of the rate. If, on the other hand, the secondary depression is associated with inhibition of some other process within the chloroplast, it would appear that Pi must be able to enter independently of the translocator at high concentrations. Conversely, if Pi entry is strictly controlled by the translocator (i.e. if it can only enter by exchange with internal triose phosphate etc.), it is not entirely inconceivable that the chloroplast could suffer as much from Pi deficiency in the presence of excess exogenous Pi as it does in its absence. The role of Antimycin A in this respect (see, e.g., CHAMPIGNY and MIGINIAC-MASLOW, 1971) remains to be resolved, but it is of interest that at low concentrations of Antimycin A the kinetics of Pi inhibition approach much more closely those which would result from simple lag extension (WALKER, unpublished).

If high external Pi encourages export, low internal Pi should aid retention of product within the chloroplast. Preliminary experiments provide support for this proposition, suggesting that more $^{14}CO_2$ is incorporated into insoluble products in low external Pi than in high (HELDT et al., private communication; FITZGERALD et al., unpublished). Orthophosphate may also be sequestered within the parent tissue by feeding mannose with consequent increase in starch synthesis (Sect. 6).

15. Inorganic Pyrophosphate

In the original experiments in which it was demonstrated that Pi would restore photosynthesis in Pi-deficient chloroplasts with an approximate stoichiometry of 1 Pi to $3O_2$, it was also established that PPi was doubly effective (COCKBURN et al., 1967c). There was a slight delay (about 15 s) in the response, but then one PPi produced twice as much O_2 as Pi at equal concentration. This could have been caused by penetration of PPi followed by internal hydrolysis, but it left unexplained the fact that PPi did not inhibit at higher concentrations in the same way as Pi. Subsequently, with very carefully washed chloroplasts, longer delays were observed before a response to PPi could be detected, and on one or two occasions (Fig. 14) there was no response at all (LUDWIG and WALKER, unpublished). The reason for this anomalous and variable behaviour became apparent when it was demonstrated (Fig. 15) that PPi produced a response in the presence of external Mg but not in its absence (SCHWENN et al., 1973). In short, PPi complexed with Mg, was hydrolysed externally by PPiase released from ruptured chloroplasts, and then penetrated the envelope as Pi. The related conclusion that PPi itself must be unable to enter the envelope (because there is Mg and PPiase within the chloroplast but no response to external PPi in the absence of external Mg) was confirmed by the centrifugal filtration technique (HELDT and LILLEY, unpublished; see also SCHWENN et al., 1973).

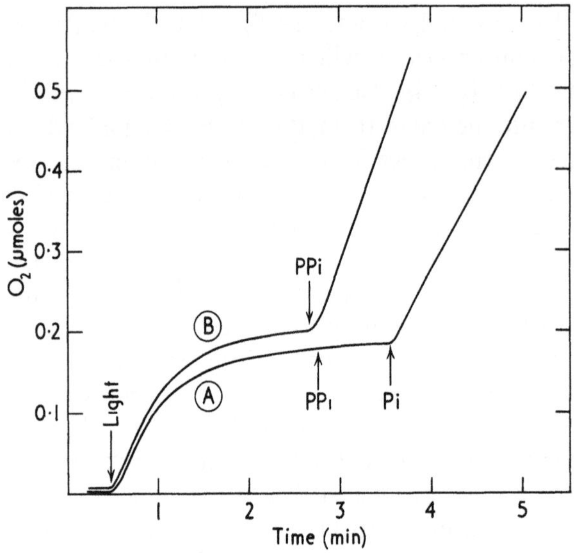

Fig. 14. Failure of intact washed chloroplasts to respond to inorganic pyrophosphate. *A* and *B*: pen-recordings of CO_2-dependent O_2 evolution by well-washed intact chloroplasts in Pi-free media. In both, O_2 evolution ceases as endogenous Pi is consumed. *A* shows that chloroplasts will then respond to addition of Pi but not PPi. Reaction mixture *B* also contained small proportion of chloroplasts deliberately exposed to osmotic shock in reaction vessel. These release PPiase so that intact chloroplasts now respond to PPi following external hydrolysis to Pi. (LUDWIG and WALKER, unpublished)

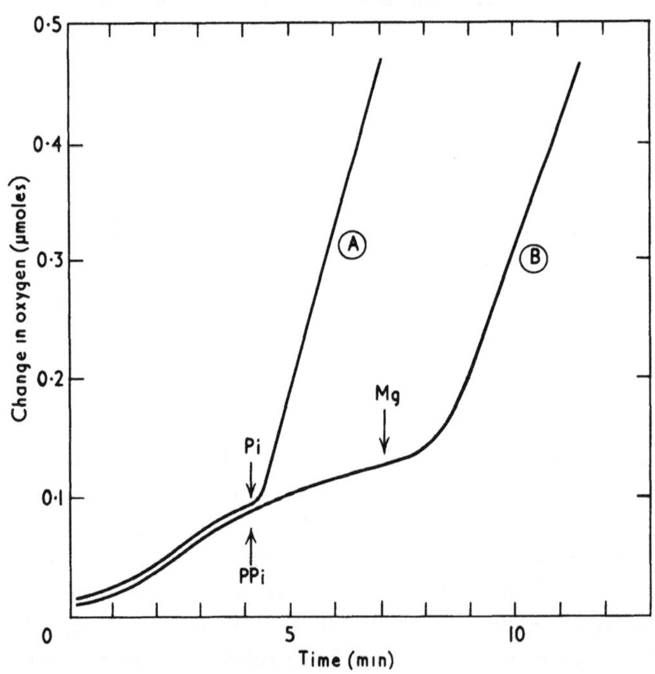

Fig. 15. Failure of unwashed intact chloroplasts to respond to inorganic pyrophosphate. For details cf. Fig. 14, except that chloroplasts are not washed free of PPiase released from damaged chloroplasts and are illuminated in medium containing neither Pi nor Mg. After photosynthesis has ceased for want of Pi it can be restored *A* by addition of Pi but not by PPi *B*. However, if Mg is then added, Pi is released from PPi by hydrolysis and O_2 evolution is restored. (After SCHWENN et al., 1973)

In one respect the role of PPi (in maintaining maximal photosynthesis by intact chloroplasts), now became clear. Evidently PPi at 5 mM, as used by JENSEN and BASSHAM (1966), served as a reservoir from which Pi would be slowly made available be hydrolysis. The PPiase required to catalyse this reaction would be released from ruptured chloroplasts and the rate of Pi formation would be held in check by the balance between the Mg-PPi complex in the external medium and a relative excess of anionic PPi which is inhibitory. What remains to be established is why PPi also ameliorates the Pi inhibition (see Fig. 16; LILLEY et al., 1973). The Pi inhibition is increased to an extent by Mg, but PPi will act in the absence of external Mg and it seems unlikely that Mg chelation is the answer. Competition between Pi and PPi for access to the phosphate translocator remains an attractive possibility. Recent evidence obtained by BAMBERGER et al. (1974) that there is less export of G3P in the presence of inorganic pyrophosphate would be entirely consistent with this proposal (cf. WALKER, 1974b) and the results illustrated in Figure 16.

It is perhaps apposite at this point to add some comment on the use of chelating agents in media used for chloroplast isolation and assay. The first reasonably active intact chloroplasts were prepared in Pi media containing Mg and Mn (WALKER, 1964). To avoid precipitation of magnesium and manganese phosphates, EDTA was added on the assumption that these cations would partition between the EDTA and binding sites within the chloroplast. Following the synthesis and employment of Goods buffers (GOOD et al., 1966) it had become customary to continue this practice and, in addition, PPi was added to the assay medium (JENSEN and

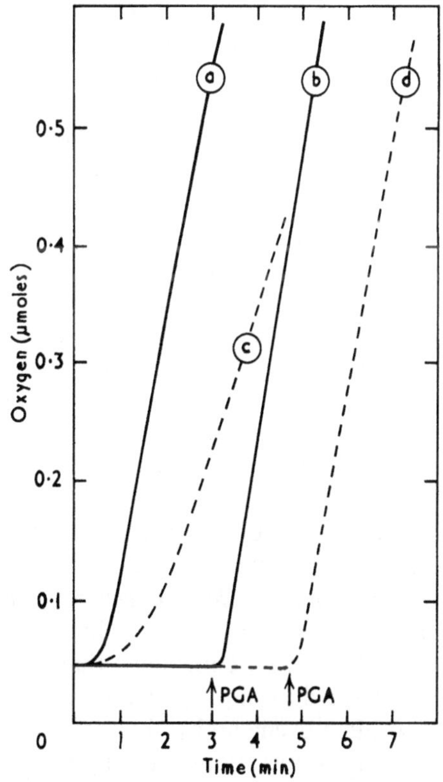

Fig. 16. Amelioration of orthophosphate inhibition by excess pyrophosphate. *a* Intact chloroplasts continue to evolve oxygen when illuminated in 40 mM PPi. *b* and *d* Inhibition is complete in 4 mM Pi until addition of PGA as indicated. *c* When added together there is considerable amelioration, by PPi, of Pi inhibition (cf. LILLEY et al., 1973)

BASSHAM, 1966) with the consequences which have now been established. If EDTA and PPi are omitted from the external medium (cf. AVRON and GIBBS, 1974) Mg may become inhibitory by combining with Pi (an effect not seen in phosphorylation mixtures because of chelation with ADP).

The interaction between Pi, PPi and PPiase has taken a long time to elucidate and the problems which it presents are still far from being completely resolved. For a time, it looked as though FBPase might also be involved (see e.g. BASSHAM et al., 1970) but there is now no real doubt that the regulatory role assigned to this enzyme in this particular context is related to the PPiase content of the preparations used (LILLEY et al., 1973; LEVINE and BASSHAM, 1974). It is not suggested, of course, that PPiase regulates photosynthesis in vivo in the way that it influences the course of photosynthesis by isolated chloroplasts. In this respect it is probably a complete artefact but nevertheless a most useful one because it has pointed to the way in which the exchange of triose phosphate and Pi between chlorplast and cytoplasm must influence photosynthesis at the cellular level.

16. ADP and ATP

16.1 Direct Transfer

Prior to the discovery of photosynthetic phosphorylation (see e.g. ARNON et al., 1954a; ARNON et al., 1954b; ALLEN et al., 1955; ARNON, 1961; ARNON, 1967) the plant biochemist was obliged to consider the possibility that ATP utilised in photosynthesis might be generated in the mitochondrion, albeit from a photosynthetically reduced coenzyme or newly formed product of photosynthesis (see e.g. VISHNIAC and OCHOA, 1952). When it became clear that isolated chloroplasts could support very rapid rates of cyclic photophosphorylation (JAGENDORF and AVRON, 1958a, b; ALLEN et al., 1958; HILL and WALKER, 1959) the converse line of thought was favoured and it was suggested that ATP formed within the chloroplast might be used in reactions other than CO_2 assimilation (ARNON, 1958; MACLACHLAN and PORTER, 1959). In the sense that all metabolites are ultimately derived from photosynthesis, all ATP is light-generated, but the important possibility of direct transfer of adenylates across the chloroplast envelopes called for investigation. In this respect only a reasonably fast exchange of ATP with ADP (or AMP) would be of major physiological significance. Rapid unilateral movement of ATP or ADP would deplete the concentration of these compounds within given cellular compartments with equal rapidity.

Despite some evidence to the contrary (Sect. 16.2) it seems quite clear that a direct rapid exchange of this sort does not occur. *Whatever its precise role, the adenylate translocator evidently does not play a major part in the transfer of energy from the chloroplast.* Nowhere is this stated more explicitly than by HELDT et al. (1972) who conclude "that direct transfer of ATP from the stroma to the cytoplasm does not appear to exist to any considerable extent". Their estimate of ADP movement by this route would appear to set the upper limit for the continuous operation of the translocator at about 0.2 μmol/mg chlorophyll/h.

16.2 The Evidence in Favour of Rapid Direct Transfer

Rapid penetration of ADP into intact chloroplasts was implicit in the work of WEST and WISKICH (1968), who reported photosynthetic control (i.e. acceleration of electron transport by ADP) in what they described as Class I chloroplast (i.e. chloroplasts with intact envelopes (cf. SPENCER and UNT, 1965; HALL, 1972). If, however, it is accepted that ferricyanide (which was used as the oxidant in these experiments) does not penetrate the intact envelope (HEBER and SANTARIUS, 1970) it must follow that the effects which they observed must have been produced by envelope-free chloroplasts in their preparations. Certainly there is no doubt that envelope-free chloroplasts can exhibit photosynthetic control (see e.g. WHITEHOUSE et al., 1971). The increase in ferricyanide-dependent O_2 evolution and its response to ADP which follows osmotic shock is illustrated in Figure 17.

JENSEN and BASSHAM (1968) detected a rapid accumulation of ^{32}P labelled ATP in the medium in which intact chloroplasts were illuminated (implying export of ATP) and an increase in the concentration of RuBP when ATP was supplied to chloroplasts in the dark (implying penetration of ATP). Again, however, alternative explanations could be suggested. For example, the export of ribose-5-phosphate

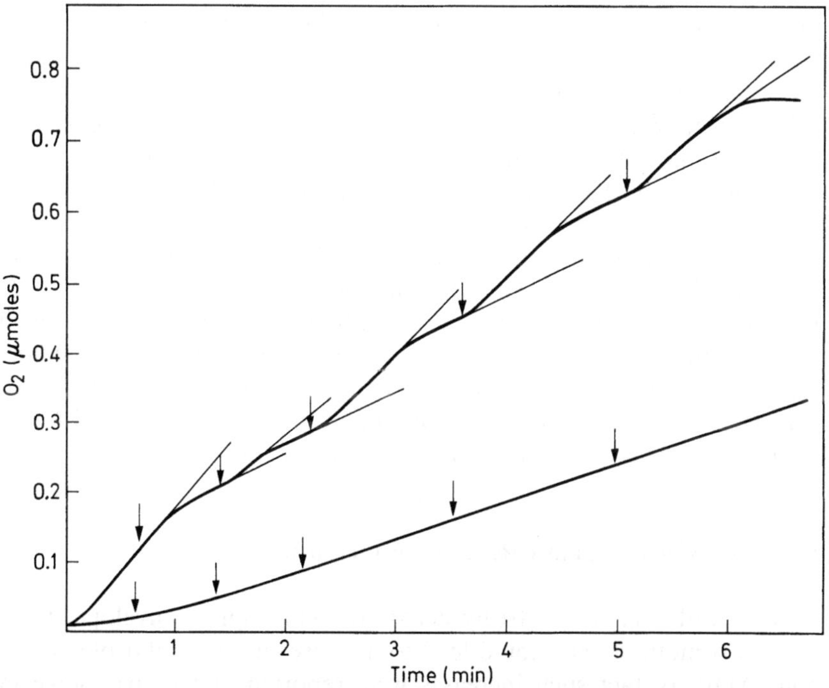

Fig. 17. Photosynthetic control (cf. WEST and WISKICH, 1968) in envelope-free chloroplasts. Traces represent O_2 evolution in Hill reaction with ferricyanide as oxidant. *Lower trace:* intact chloroplasts (i.e. approximately 80% intact and 20% ruptured). *Upper trace:* otherwise identical chloroplasts, osmotically shocked to give 100% envelope-free chloroplasts. Ferricyanide unable to enter intact chloroplast so all O_2 recorded is attributed to chloroplasts deliberately or inadvertently ruptured. Small quantities of ADP (0.1, 0.1 then 0.3 µmol successively) added as indicated (*arrows*). Response of small proportion of chloroplasts inadvertently ruptured during preparation (*lower trace*) negligible (cf. small response to uncoupler in Fig. 4), response of 100% deliberately ruptured (*upper trace*) large

from the chloroplast is not in question and external formation of ribulose bisphosphate would follow in the presence of added ATP if an appreciable concentration of ribulose-5-phosphate kinase escaped from damaged chloroplasts.

It is also extremely easy to show an effect of added ATP on O_2 evolution but, in our own experience, much more difficult to produce one which cannot be copied by free Pi. Thus ATP and ADP and Pi all restore O_2 evolution which has ceased in Pi-deficient reaction mixtures (COCKBURN et al., 1967b) but external hydrolysis followed by uptake of Pi seems a more credible explanation than rapid direct entry, in view of the other contrary evidence. Slow direct entry could conceivably account in part for the report by SCHÜRMANN et al. (1971, 1972) that ATP shortens induction because their rates (ca. 2–12 μmoles O_2/mg chlorophyll/h) are only a little faster at maximum than the fastest recorded entry of ATP based on direct measurement. Once more, however, it must be pointed out that chloroplasts suspended in media containing inorganic pyrophosphate (as were those of SCHÜRMANN et al. (1971, 1972), but deficient in Pi will sometimes show a positive response to the addition of *small* quantities of Pi if the external pyrophosphatase activity is inadequate to produce optimal Pi by hydrolysis. Because R5P failed to stimulate the rate of phosphate esterification by "intact" chloroplasts, AVRON and GIBBS (1974) concluded that the rate of entry of ADP was not rate-limiting in their experiments, but did not apparently entertain the possibility that all of the phosphorylation to which exogenous ADP contributed might have been catalysed by the envelope-free chloroplasts in their mixtures. In fact Figure 1 in their paper shows that the rate of esterification of Pi brought about by added R5P was appreciably faster after osmotic shock. Such an increase would be explained if the esterification of exogenous ADP was catalysed by the ruptured chloroplasts and most of the phosphorylation of R5P was accomplished by endogenous adenylate within the intact plastids. Moreover, AVRON and GIBBS (1974) also reported experiments in which "EDTA abolished phosphorylation without affecting CO_2 fixation". This they attribute to the absence of free Mg, but again it must be assumed that this is an inhibition of a process brought about by ruptured chloroplasts. If this had also occurred within the intact chloroplast, there seems to be no reason why it should not also have inhibited the Mg-dependent carboxylase and the endogenous photophosphorylation required for the operation of the cycle, whereas CO_2 fixation was unaffected.

16.3 The Evidence against Rapid Direct Transfer

If ADP and ATP move freely across the chloroplast envelope, it is difficult to see why osmotic shock should lead to increases in photophosphorylation of exogenous ADP. In fact such increases were reported at an early stage (WHATLEY et al., 1956) and became considerably more marked when improved techniques yielded preparations containing larger proportions of intact chloroplasts. In the best preparations (shown by light and electron microscopy to contain only approximately 10% chloroplasts without envelopes) the rate of cyclic photophosphorylation was only 102 μmol of Pi-esterified/mg chlorophyll/h but this could be increased to 864 by osmotic shock, so that it was possible to conclude that most, if not all, of the phosphorylation may have been mediated by the membrane (envelope)-free chloroplasts (WALKER, 1965b). This conclusion was subsequently supported by HEBER and SANTARIUS (1970), who measured pseudocyclic phosphorylation of exogenous ADP by spinach chloroplasts in the absence of cofactors such as pyrocyanine (which was used in the above work). In their experiments chloroplasts which had been osmotically shocked gave rates of ca. 20 μmol/mg chlorophyll/h, whereas with chloroplasts which were ca. 70–90% intact the rates ranged from 0 to 4 μmol/mg chlorophyll/h.

When supplied with PGA in the light, intact chloroplasts will evolve O_2 (WALKER and HILL, 1967). In a sense this is an ATP-requiring Hill reaction in which PGA first undergoes phosphorylation and is then reduced at the expense of light-generated ATP. This reaction can be inhibited by uncouplers such as ammonium chloride or nigericin (Fig. 18), but evolution is not restarted by the addition of ATP (STOKES and WALKER, 1970). In a reconstituted chloroplast system, containing envelope-free chloroplasts supplemented with stromal protein, the same reaction can be demonstrated and is again inhibited by the uncoupler. In contrast to the behaviour of the intact chloroplasts, however, the addition of ATP not only restores O_2 evolution (Fig. 19) but increases it to that of the uncoupled rate seen in the presence of substrate concentrations of NADP (STOKES and WALKER, 1970). The principal difference between the two systems is that in the former the reaction centres are separated from the remainder of the reaction mixture by intact envelopes, whereas in the latter these have been deliberately ruptured. Again, if ATP were able to penetrate the intact chloroplasts, it is difficult to understand why there is no discernible restitution of O_2 evolution. Figure 20 shows a similar comparison in which the addition of ADP is entirely without effect on the course of O_2 evolution by intact chloroplasts provided with R5P. In a reconstituted system

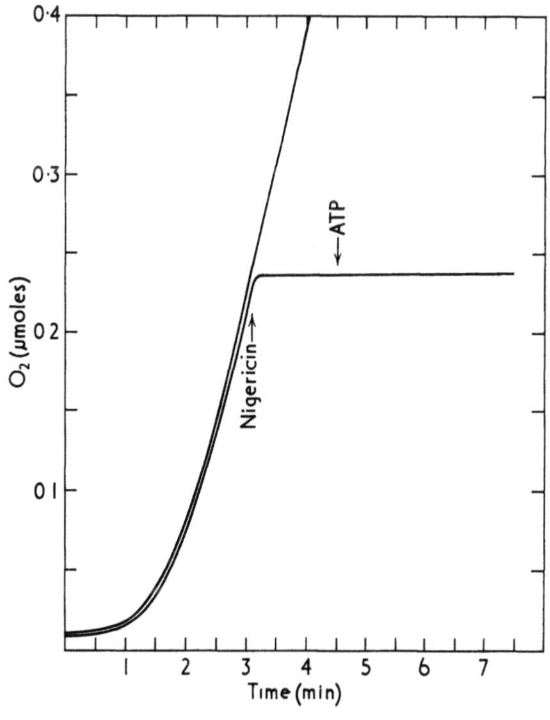

Fig. 18. Apparent failure of ATP to enter intact chloroplast. Addition of uncoupler nigericin (*lower trace*) to intact chloroplasts displaying CO_2-dependent O_2 evolution results in rapid inhibition. Addition of ATP does not then restore activity (cf. Fig. 16)

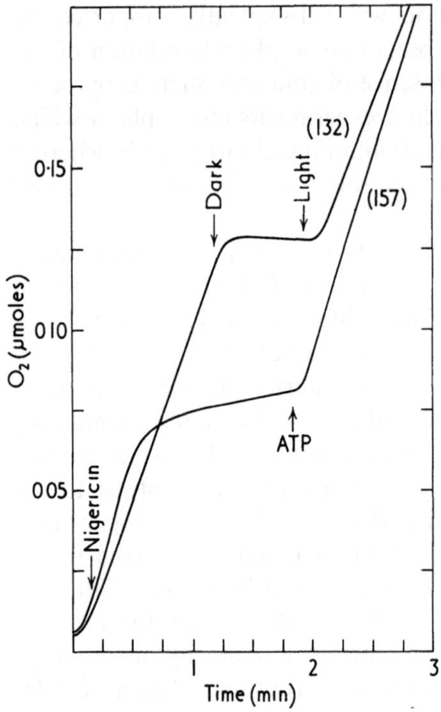

Fig. 19. Restoration of CO_2-dependent O_2 evolution in reconstituted chloroplast system with ribulose bisphosphate as substrate. System contains envelope-free chloroplasts supplemented with stromal protein, bicarbonate, additional Mg^{2+} and catalytic quantities of ADP and NADP, but believed to differ from system in Fig. 18 only in that all chloroplasts are without envelopes. Addition of nigericin to one mixture causes acceleration as it uncouples electron transport, followed by inhibition as reaction almost ceases for lack of ATP (cf. dark/light response of control). Addition of ATP brings about restoration of O_2 evolution at higher, uncoupled rate. This, together with Fig. 18, strongly suggests ATP is unable to penetrate intact envelope at rapid rates

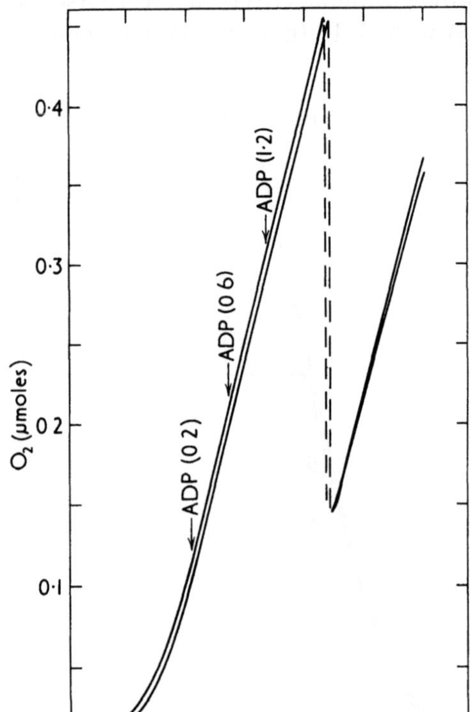

Fig. 20. Inability of exogenous ADP to influence course of CO_2-dependent O_2 evolution by intact chloroplasts. *Upper trace*. ADP added as indicated. *Lower trace*. control without additions (cf. Fig. 18)

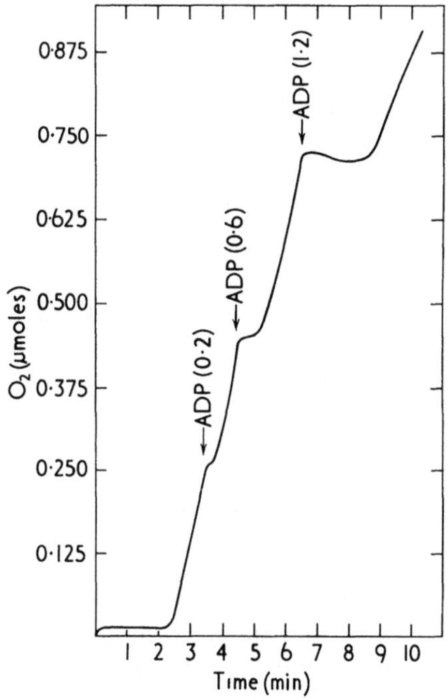

Fig. 21. Effect of ADP in reconstituted system. Time course of CO_2-dependent O_2 evolution from reconstituted chloroplast system (see Fig. 19) with ribose-5-phosphate as substrate. ADP causes inhibition persisting until added ADP is converted to ATP (cf. Fig. 20). Figs. 20 and 21 constitute evidence that ADP is unable to cross intact envelope at rapid rates

(cf. WALKER et al., 1971; LILLEY et al., 1974), similarly provided with R5P, the addition of ADP causes a break in O_2 evolution of a length which is directly proportional to the quantity of ADP and commensurate with the time which would elapse if photophosphorylation continued unchanged at the rate needed to maintain the previous CO_2-dependent reaction (Fig. 21, previously unpublished). The precise nature of this transient inhibition is at present unknown (although it almost certainly involves the known ADP inhibition of PGA reduction (LILLEY and WALKER, 1974). What is important in this context is that it provides another example of a situation in which the reconstituted (envelope-free) chloroplast will react to added adenylate (this time ADP rather than ATP), whereas the intact chloroplast will not.

17. NADP

NADP will act as a Hill oxidant in the presence of ruptured (Class C) chloroplasts supplemented with ferredoxin (see e.g. HALL, 1972), whereas with intact (Class A) chloroplasts this reaction is minimal or nonexistent prior to osmotic shock (HEBER and SANTARIUS, 1965; COCKBURN et al., 1967b; MATHIEU, 1967; HEBER et al., 1967a). This indicates that movement of reducing equivalents between the chloroplast and cytoplasm can only take place indirectly, e.g. via the PGA/DHAP shuttle (Sect. 18.1) or the dicarboxylate shuttle (Sect. 18.2).

18. Shuttles

If the permeability characteristics of the chloroplast envelope preclude rapid export of reducing power in the shape of ATP and NADPH (Sect. 16, 17), can this be achieved indirectly? More or less simultaneously and independently several authors (see e.g. HEBER, 1970; WALKER and CROFTS, 1970) have postulated the existence of "shuttles" which would serve this purpose. The first evidence in support of these exercises in paper biochemistry came from STOCKING and LARSON (1969) who observed external reduction of NAD by chloroplasts in the presence of PGA, PGA kinase and triose phosphate dehydrogenase (Sect. 18.1).

18.1 The PGA/DHAP Shuttle

As Figure 22 indicates, this involves export of triose phosphate, external oxidation to PGA, and finally reentry followed by internal reduction to triose phosphate. Indirect export of ATP and reducing equivalents by this procedure (which is believed to involve the phosphate translocator—see Sect. 19.3) is decreased by high external ratios of ATP/ADP and NADH/NAD but has been observed to reach 50 µmol/mg chlorophyll/h (HEBER and SANTARIUS, 1970; KRAUSE, 1971). Should it occur at this rate in vivo it would be more than adequate to account for observed light-dependent fluctuation in cytoplasmic adenylates (HEBER, 1974) and would need to be regarded as a major feature of cellular energy economy.

An analogous shuttle, involving the irreversible glyceraldehyde-3-phosphate dehydrogenase, has been proposed by KELLY and GIBBS (1973) and would bring about the indirect export of NADPH.

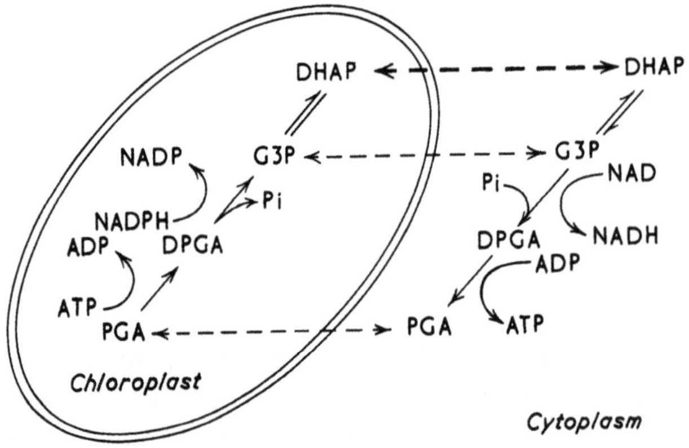

Fig. 22. PGA/DHAP shuttle. Inside chloroplast PGA is reduced to TP at expense of NADPH and ATP. Externally NADH and ATP are formed from TP released from chloroplast. PGA can re-enter chloroplast to continue shuttle. (After HEBER, 1974; WALKER, 1974b)

18.2 The Malate/Oxaloacetate Shuttle

The equilibrium position of the reaction catalysed by malic dehydrogenase overwhelmingly favours oxaloacetate reduction, and oxaloacetate will serve as a Hill oxidant if supplied to intact chloroplasts. In a system containing intact chloroplasts supplemented with malic dehydrogenase, NAD/NADH and malate/oxalacetate illumination can nevertheless bring about a 30% reduction of the external nicotinamide nucleotide (HEBER and KRAUSE, 1971, 1972; HEBER, 1974), indicating the export of reducing equivalents according to Figure 23. Conversely, if the steady state concentration of internal NADPH is decreased by the addition of PGA, external NADH is oxidised (HEBER, 1974). There are grounds for believing that the chloroplast contains both NAD and NADP-specific malic dehydrogenases (HEBER, 1960; JOHNSON and HATCH, 1970).

The operation of the PGA/DHAP shuttle would result in the indirect transfer of both ATP and reducing equivalents to the cytoplasm. If it operated in conjunction with the malate/oxaloacetate shuttle, then it could export ATP alone. Again, it must be assumed that this shuttle incorporates the dicarboxylate translocator, and related amino acids such as aspartate might participate in vivo (HEBER, 1974). Linked function of the two shuttles would permit ATP generation within the unilluminated chloroplast. In leaf discs, starch synthesis from exogenous sugars (Sect. 5) in the dark requires the expenditure, in the penultimate step, of one ATP per glucose incorporated, and rates of synthesis as high as 10 µmol glucose incorporated/mg chlorophyll/h have been observed. If the continuous operation of the ATP translocator (Sect. 16, 19.1) is governed by the rate of movement which ADP imposes (about 0.2 µmol/mg chlorophyll/h) then there is a strong implication that the required ATP must be synthesised internally by this mechanism. If the chloroplast operates the oxidative pentose phosphate cycle in the dark (BASSHAM, 1971), NADPH might also be reoxidised in this way.

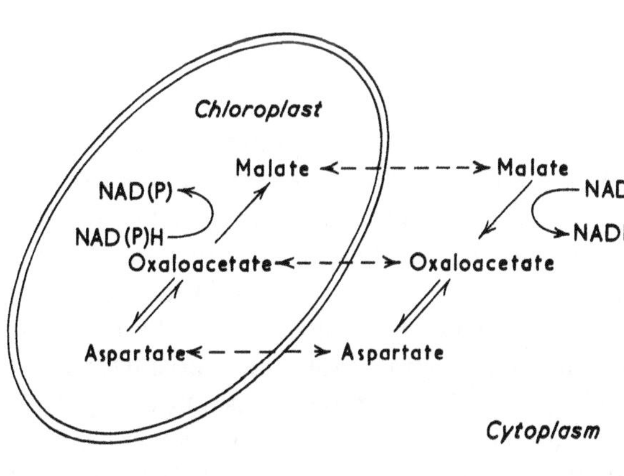

Fig. 23. Malate/oxaloacetate shuttle. Inside chloroplast, oxaloacetate is reduced to malate at expense of NADPH. Externally it is reoxidised by cytoplasmic malic dehydrogenase. Reducing equivalents thus exported. (After HEBER, 1974; WALKER, 1974b)

19. Specific Transport

The concept of translocators which separately and specifically facilitate the exchange of metabolites and coenzymes (see e.g. HELDT, 1969; HELDT and RAPLEY, 1970a, b; WERDAN and HELDT, 1972; HELDT et al., 1972) has proved to be an extremely useful development in this field. It is considered only briefly here, but in more detail by HELDT in Chap. II, 3.

Three translocators have been defined. They catalyse the transport of adenylates, dicarboxylates and phosphate or phosphate esters between chloroplasts and external medium.

19.1 ATP Transport

ATP in the external medium will initiate export of internal adenylates at a maximum rate of 5 µmol/mg chlorophyll/h at 20°. ADP induces a similar response at about 12% of this rate. The translocator responsible for transfer is not believed to play an important role in ATP export (see Sect. 16).

19.2 Dicarboxylate Transport

A special translocator facilitates the exchange of malate, oxaloacetate etc. and related amino acids at rapid rates and is believed to moderate the exchange of reducing equivalents between the chloroplast and cytoplasm.

19.3 Phosphate Transport

Another translocator facilitates a counter-exchange of Pi with PGA and DHAP (HELDT and RAPLEY, 1970b; WERDAN and HELDT, 1972) at rates of 100 µmol/mg chlorophyll/h or more (perhaps at considerably faster rates at 20°, judging from recently observed rates of PGA reduction by intact chloroplasts; see HEBER, 1974). Pentose and hexose monophosphates are not carried and neither is PPi. Measurements relate to the uptake (and release) of labelled compounds into (and from) the sucrose-impermeable space of chloroplasts (Table 2), and the translocator is believed to be located in the inner envelope (Fig. 24), where it may constitute of the order of 10% of the membrane protein (HEBER, 1974).

The phosphate translocator would appear to play a major role in metabolite transfer, exchanging inorganic phosphate for triose phosphate. Orthophosphate inhibition of photosynthesis (ARNON et al., 1954a; COCKBURN et al., 1967b; COCKBURN et al., 1968) is seen as an exaggeration of this process (HELDT and RAPLEY, 1970b; WALKER and CROFTS, 1970) in which the normal autocatalytic build-up of intermediates during the induction period (WALKER, 1973) is prevented by the abnormal export imposed by high external Pi concentrations. The induction period is, therefore, lengthened indefinitely at completely inhibitory concentrations. Sup-

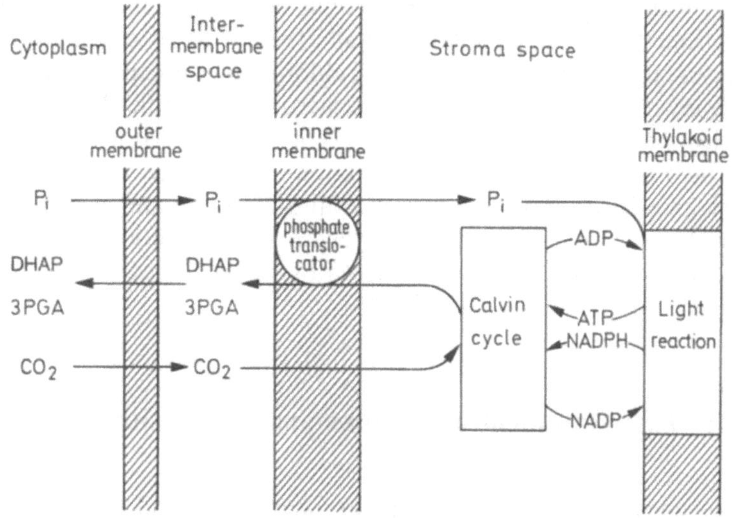

Fig. 24. Phosphate translocator. Exchange of external orthophosphate for internal DHAP and PGA. (From HELDT et al., 1972)

port for this interpretation comes from the immediate reversal of Pi inhibition by PGA and triose phosphate (COCKBURN et al., 1968), and the fact that Pi is much less inhibitory if added after the induction period is completed. Inhibition by Pi is also less marked at a given concentration if the chloroplasts are isolated from preilluminated tissue. The fact that the course of Pi inhibition is influenced by the external Mg concentration suggests some interaction possibly associated with charge separation (LILLEY et al., 1973). Amelioration of Pi inhibition by PPi (LILLEY et al., 1973) may be caused in part by chelation of Mg, but the fact that PPi is effective in Mg-free media suggests that it may act also as a straight competitive inhibitor. The fact that the translocator is essentially an anion exchanger also suggests that its operation may be influenced by the charge differences which could result from a high concentration of nonpermeable PPi in the external medium.

20. The Movement of Metabolites in C_4 Photosynthesis

This topic is considered in Chap. II, 4 and will not be pursued in detail here. At present much of the evidence for movement is based on enzyme localisation. For example, it now seems probable that oxaloacetate is formed from PEP in the mesophyll cytoplasm and transferred as malate or aspartate to the bundle sheath where decarboxylation produces CO_2 for RuBP carboxylation. PEP must then be regenerated and it seems likely that the pyruvate released by decarboxylation in the bundle sheath will be transferred back to the mesophyll. Until definitive evidence is forthcoming it seems sensible to assume, as a working hypothesis, that the permeability characteristics of C_4 chloroplasts are similar to those from

C_3 plants. On this basis there would be no difficulty in accommodating existing schemes (see e.g. GUTIERREZ et al., 1974; HATCH et al., 1975) within the known permeability characteristics of C_3 chloroplasts, except where it is thought necessary to invoke the transfer of pyruvate across the bundle sheath and mesophyll chloroplast envelope (HEBER, 1974). Recent work by HUBER and EDWARDS (1975), in which mesophyll carbon assimilation is shown to be stimulated in a synergistic manner by mixtures of pyruvate and oxaloacetate, has been interpreted by the authors as an initiation of noncyclic photophosphorylation yielding ATP for PEP formation in the reaction catalysed by pyruvate phosphate dikinase.

21. The Movement of Protons and Magnesium

Ion movements are dealt with in Chap. III, 1 and are largely outside the scope of this chapter. It may be noted however that there is now strong evidence that, as a consequence of light-induced electron transport, protons move into the thylakoid compartment(s) from the stroma and magnesium moves in the opposite direction as the principal counter ion (see e.g. HIND et al., 1974; BARBER et al., 1974b; BARBER et al., 1974a; and Fig. 12). This means that in the light the concentration of Mg in the stroma will increase and simultaneously conditions within this compartment will become more alkaline and more reducing. Activation of RuBP carboxylase by Mg (for a review see WALKER, 1973, and cf. LILLEY et al., 1974) has been touched on (Sect. 11), but it seems clear that these internal changes must affect several stromal enzymes (see e.g. PREISS and KOSUGE, 1970, and, for a recent consideration of pH effects, WERDAN et al., 1975) and that the consequences in relation to carbon metabolism and transport (cf. WERDAN and HELDT, 1972) will be far-reaching.

22. Concluding Remarks

In a sense the wheel has turned full circle. At first, it seemed that the act of separating the chloroplast from its cellular environment might rob it of its ability to assimilate carbon; that the chloroplast must depend on the cytoplasm or the mitochondrion for some essential component such as ATP. Then the discovery of photophosphorylation and the demonstration of CO_2 fixation by isolated chloroplasts pointed increasingly to self-sufficiency, and the last lingering doubts were swept away with the demonstration of rates of carbon assimilation (and commensurate CO_2-dependent O_2 evolution) comparable to those achieved by the parent tissue. What is clear in retrospect, however, is that the improvement in the performance of the isolated chloroplast depended only partly on better extraction procedures. These ensured the isolation of preparations containing large proportions of intact (or Class A) chloroplasts, but although they offered a ten-fold improvement over the average performance of chloroplasts separated in Tris-NaCl, the best rates were only modest compared with those which are commonplace today. Even so, chloroplasts prepared according to the 1964 procedures can match present performance if they are washed free of excess Pi or supplemented with cycle intermediates. Further improvement in the rate comes mostly from the way in which the chloroplasts are treated after separation. Undoubtedly some benefit derives from the avoidance of deleterious buffers, but the most important feature is the inclusion of PPi. This ingredient works its magic by regulating the supply of orthophosphate to the chloroplast in a way which must approach that employed in the cell. In short, the chloroplast is not so self-sufficient as had been supposed, and maximal in vitro performance has only been achieved in the presence of an entirely artificial substitute for one aspect of the cytoplasm's supporting role.

The study of transport between the chloroplast and the cytoplasm is still in its infancy, but the broad outlines have been established. It would seem that the vital coenzymes are an integral part of the machinery and do not drift freely in an uncontrolled manner between the organelle and its cellular environment. Certainly the chloroplast is able to export reducing power, but it does this indirectly and principally in the same way as it exports elaborated carbon. In all probability the classic equation for photosynthesis should be modified so that it reads

$$3CO_2 + Pi + 2H_2O \rightarrow \text{triose phosphate} + 3O_2.$$

It would then imply (in addition to the traditional stoichiometry) that the principal imports into the chloroplast are CO_2 and Pi, and that the immediate end product and principal export is triose phosphate. Starch synthesis would be relegated to an important but nevertheless secondary role, and there is no doubt that in some actively growing species such as spinach only a minor fraction of the carbon fixed is diverted into temporary storage products of this nature. Similarly sucrose is not an end-product of chloroplast photosynthesis but a transport molecule synthesised from triose phosphate in the cytoplasm. Finally the chloroplast in vivo, like the chloroplast in vitro is in many circumstances a phosphate-consuming organelle, dependent on the cytoplasm for a continuous supply of this compound at one third of the rate of imported carbon. As sucrose is synthesised in the cytoplasm, from exported triose phosphate, orthophosphate will be released for recycling. The level of cytoplasmic Pi, itself influenced by sinks in adjacent tissues, will, in turn, affect the release of elaborated carbon from the chloroplast and hence the activity of the photosynthetic cycle and the rate of starch synthesis.

Acknowledgement. This chapter owes its existence to U. HEBER, whose many important contributions to this field are matched only by his powers of persuasion.

References

AACH, H.G., HEBER, U.: Kompartimentierung von Aminosäuren in der Blattzelle. Z. Pflanzenphysiol **57**, 317–328 (1967)

ALLEN, M.B., ARNON, D.I., CAPINDALE, J.B., WHATLEY, F.R., DURHAM, L.J.: Photosynthesis by isolated chloroplasts. III. Evidence for complete photosynthesis. J. Am. Chem. Soc. **77**, 4149–4155 (1955)

ALLEN, M.B., WHATLEY, F.R., ARNON, D.I.: Photosynthesis by isolated chloroplasts. VI. Rates of conversion of light into chemical energy in photosynthetic phosphorylation. Biochim. Biophys. Acta **27**, 16–23 (1958)

ANDREWS, T.J., LORIMER, G.H., TOLBERT, N.E.: Ribulose diphosphate oxygenase. I. Synthesis of phosphoglycolate by fraction-1 protein of leaves. Biochemistry **12**, 11–18 (1973)

ARNON, D.I.: Chloroplasts and Photosynthesis. In: The Photochemical Apparatus—its Structure and Function. (Brookhaven Symp. Biol. **11**), (R.C. FULLER, J.A. BERGERON, L.G. AUGENSTINE, M.E. KOSHLAND, H.J. CURTIS, eds.), pp. 181–235. Upton N.Y.: Brookhaven Nat. Lab. 1958

ARNON, D.I.: Light and Life, Cell-free Photosynthesis and the Energy Conversion Process. Baltimore: Johns Hopkins 1961

ARNON, D.I.: Photosynthetic activity of isolated chloroplasts. Physiol. Rev. **47**, 317–358 (1967)

ARNON, D.I., ALLEN, M.B., WHATLEY, F.R.: Photosynthesis by isolated chloroplasts. Nature **174**, 394–396 (1954a)

ARNON, D.I., WHATLEY, F.R., ALLEN, M.B.: Photosynthesis by isolated chloroplasts. II. Photosynthetic phosphorylation, the conversion of light into phosphate bond energy. J. Am. Chem. Soc. **76**, 6324–6329 (1954b)

AVRON, M., GIBBS, M.: Carbon dioxide fixation in the light and in the dark by isolated spinach chloroplasts. Plant Physiol. **53**, 140–143 (1974)

BAHR, J.T., JENSEN, R.G.: Ribulose diphosphate carboxylase from freshly ruptured spinach chloroplasts having an in vivo *Km*[CO_2]. Plant Physiol. **53**, 39–44 (1974)

BALDRY, C.W., BUCKE, C., WALKER, D.A.: Incorporation of inorganic phosphate into sugar phosphates during carbon dioxide fixation by illuminated chloroplasts. Nature **210**, 793–796 (1966a)

BALDRY, C.W., COCKBURN, W., WALKER, D.A.: Inhibition by sulphate of the oxygen evolution associated with photosynthetic carbon assimilation. Biochim. Biophys. Acta **153**, 476–483 (1968)

BALDRY, C.W., WALKER, D.A., BUCKE, C.: Calvin-cycle intermediates in relation to induction phenomenon in photosynthetic carbon dioxide fixation by isolated chloroplasts. Biochem. J. **101**, 641–646 (1966b)

BAMBERGER, E.S., EHRLICH, B.A., GIBBS, M.: The D-glyceraldehyde-3-phosphate and 3-phosphoglycerate shuttle system and carbon dioxide assimilation in intact spinach chloroplasts. Proc. 3rd Intern. Cong. Photosynth. Res., (M. AVRON, ed.), pp. 1349–1362. Amsterdam: Elsevier 1974

BAMBERGER, E.S., GIBBS, M.: Effects of phosphorylated compounds and inhibitors on CO_2 fixation by intact spinach chloroplasts. Plant Physiol. **40**, 919–926 (1965)

BARBER, J., MILLS, J., NICOLSON, J.: Studies with cation specific ionophores show that within the intact chloroplast Mg^{++} acts as the main exchange cation for H^+ pumping. FEBS Letters **49**, 106–110 (1974a)

BARBER, J., TELFER, A., NICOLSON, J.: Evidence for divalent cation movement within isolated whole chloroplasts from studies with ionophore A23187. Biochim. Biophys. Acta **357**, 161–165 (1974b)

BASSHAM, J.A.: Control of photosynthetic carbon metabolism. Science **172**, 526–534 (1971)

BASSHAM, J.A., CALVIN, M.: The Path of Carbon in Photosynthesis. Englewood Cliffs, N.J.: Prentice-Hall Inc. 1957

BASSHAM, J.A., EL-BADRY, A.M., KIRK, M R., OTTENHEYM, H.C.J., SPRINGER-LEDERER, H.: Photosynthesis of isolated chloroplasts. V. Effect on fixation rate and metabolite transport from the chloroplasts caused by adding fructose-1,6-diphosphatase Biochim. Biophys. Acta **223**, 261–274 (1970)

BASSHAM, J.A., JENSEN, R.G.: Photosynthesis of carbon compounds. In: Harvesting the Sun (A SAN PIETRO, F.A. GREER, T.J. ARMY, eds.), pp. 79–110. New York: Academic Press 1967

BASSHAM, J.A., KIRK, M., JENSEN, R.G.: Photosynthesis by isolated chloroplasts. I. Diffusion of labelled photosynthetic intermediates between isolated chloroplasts and suspending medium. Biochim. Biophys. Acta **153**, 211–218 (1968a)

BASSHAM, J.A., SHARP, P., MORRIS, I.: The effect of Mg^{2+} concentration on the pH optimum and Michaelis constants of the spinach chloroplast ribulose diphosphate carboxylase. Biochim. Biophys. Acta **153**, 898–900 (1968b)

BASSHAM, J.A., SHARP, P., MORRIS, I.: The effect of Mg^{2+} concentration on the pH optimum and Michaelis constants of the spinach chloroplast ribulose diphosphate carboxylase (carboxydismutase). Biochim. Biophys. Acta **153**, 901–902 (1968c)

BENSON, A.A., BASSHAM, J.A., CALVIN, M., GOODALE, T.C., HAAS, V.A., STEPKA, W.: The path of carbon in photosynthesis. V. Paper chromatography and radioautography of the products. J. Am. Chem. Soc. **72**, 1710–1718 (1950)

BENSON, A.A., BASSHAM, J.A., CALVIN, M., HALL, A.G., HIRSCH, H.E., KAWAGUCHI, S., LYNCH, V., TOLBERT, N.E.: The path of carbon in photosynthesis. XV. Ribulose and sedoheptulose. J. Biol. Chem. **196**, 703–716 (1952)

BENSON, A.A., CALVIN, M.: Carbon dioxide fixation by green plants. Ann. Rev. Plant Physiol. **1**, 25–42 (1950)

BIRD, I.F., CORNELIUS, M.J., KEYS, A.J., WHITTINGHAM, C.P.: Intracellular site of sucrose synthesis in leaves. Phytochemistry **13**, 59–64 (1974)

BIRD, I.F., PORTER, H.K., STOCKING, C.R.: Intracellular localisation of enzymes associated with sucrose synthesis in leaves. Biochim. Biophys. Acta **100**, 366–375 (1965)

BLINKS, L.R., SKOW, R.K.: The time course of photosynthesis as shown by the glass electrode, with anomalies in the acidity changes. Proc. Natl. Acad. Sci. US **24**, 413–419 (1938)

BOEHM, J.: Ueber Stärkebildung aus Zucker. Bot. Z. **41**, 33–38 und 49–54 (1883)

BOWES, G., OGREN, W.L.: Oxygen inhibition and other properties of soybean ribulose-1,5-diphosphate carboxylase. J. Biol. Chem. **247**, 2171–2176 (1972)

BOWES, G., OGREN, W.L., HAGEMAN, R.H.: Phosphoglycolate production catalysed by ribulose diphosphate carboxylase. Biochem. Biophys. Res. Commun. **45**, 716–722 (1971)

BRADBEER, J.W., RACKER, E.: Glycolate formation from fructose 6-P by cell free preparations. Federation Proc. **20**, 80–88 (1961)

BROWN, H.T., MORRIS, G.H.: A contribution to the chemistry and physiology of foliage leaves. J. Chem. Soc. **63**, 604–677 (1893)

BUCHANAN, B.B., KALBERER, P.P., ARNON, D.I.: Ferredoxin-activated fructose diphosphatase in isolated chloroplasts. Biochem. Biophys. Res. Commun. **29**, 74–79 (1967)

BUCHANAN, B.B., SCHÜRMANN, P.: Regulation of ribulose-1,5-diphosphate carboxylase in the photosynthetic assimilation of carbon dioxide. J. Biol. Chem. **248**, 4956–4964 (1973)

BUCHANAN, B.B., SCHÜRMANN, P., KALBERER, P.P.: Ferredoxin-activated fructose diphosphatase of spinach chloroplasts. J. Biol. Chem. **246**, 5952–5959 (1971)

BUCKE, C., WALKER, D.A., BALDRY, C.W.: Some effects of sugars and sugar phosphates on carbon dioxide fixation by isolated chloroplasts. Biochem. J. **101**, 636–641 (1966)

CHAMPIGNY, M.L., MIGINIAC-MASLOW, M.: Relations entre l'assimilation photosynthétique de CO_2 et la photophosphorylation des chloroplastes isolés. 1. Stimulation de la fixation de CO_2 par l'antimycine A, antagoniste de son inhibition par le phosphate. Biochim. Biophys. Acta **234**, 335–343 (1971)

CHEN-SHE, S.H., LEWIS, D.H., WALKER, D.A.: Stimulation of photosynthetic starch formation by sequestration of cytoplasmic orthophosphate. New Phytologist **74**, 381–390 (1975)

CHU, D.K., BASSHAM, J.A.: Inhibition of ribulose-1,5-diphosphate carboxylase by 6-phosphogluconate. Plant Physiol. **50**, 224–227 (1972)

CHU, D.K., BASSHAM, J.A.: Activation and inhibition of ribulose-1,5-diphosphate carboxylase by 6-phosphogluconate. Plant Physiol. **52**, 373–379 (1973)

CHU, D.K., BASSHAM, J.A.: Activation of ribulose-1,5-diphosphate carboxylase by nicotinamide adenine dinucleotide phosphate and other chloroplast metabolites. Plant Physiol. **54**, 556–559 (1974)

COCKBURN, W., BALDRY, C.W., WALKER, D.A.: Photosynthetic induction phenomena in spinach chloroplasts in relation to the nature of the isolating medium. Biochim. Biophys. Acta **143**, 603–613 (1967a)

COCKBURN, W., BALDRY, C.W., WALKER, D.A.: Some effects of inorganic phosphate on O_2 evolution by isolated chloroplasts. Biochim. Biophys. Acta **143**, 614–624 (1967b)

COCKBURN, W., BALDRY, C.W., WALKER, D.A.: Oxygen evolution by isolated chloroplasts with carbon dioxide as the hydrogen acceptor. A requirement for orthophosphate or pyrophosphate. Biochim. Biophys. Acta **131**, 594–596 (1967c)

COCKBURN, W., WALKER, D.A., BALDRY, C.W.: Photosynthesis by isolated chloroplasts. Reversal of orthophosphate inhibition by Calvin cycle intermediates. Biochem. J. **107**, 89–95 (1968)

COOPER, T.G., FILMER, D., WISHNICK, M., LANE, M.D.: The active species of "CO_2" utilized by ribulose diphosphate carboxylase. J. Biol. Chem. **244**, 1081–1083 (1969)
COOPER, T.G., TCHEN, T.T., WOOD, H.G., BENEDICT, C.R.: The carboxylation of phosphoenolpyruvate. J. Biol. Chem. **243**, 3857–3863 (1968)
EL-SHARKAWAY, M., HESKETH, J.: Photosynthesis among species in relation to characteristics of leaf anatomy and CO_2 diffusion resistances. Crop Sci. **5**, 517–521 (1965)
EVERSON, R.G.: Bicarbonate equilibria and apparent *Km* (HCO_2) of isolated chloroplasts. Nature **222**, 876 (1969)
EVERSON, R.G., COCKBURN, W., GIBBS, M.: Sucrose as a product of photosynthesis in isolated spinach chloroplasts. Plant Physiol. **42**, 840–844 (1967)
FEKETE, M.A.R. DE, VIEWEG, G.H.: Starch metabolism: Synthesis versus degradation pathways. In: Plant Carbohydrate Biochemistry (J.B. PRIDHAM, ed.), pp. 127–144. (Proc. Phytochem. Soc. Symp. Edinburgh 1973.) London: Academic Press 1974
FILMER, D., COOPER, T.G.: Effect of varying temperature and pH upon the predicted rate of "CO_2" utilisation by carboxylases. J. Theoret. Biol. **29**, 131–145 (1970)
GIBBS., M. (ed.): Structure and Function of Chloroplasts, pp. 169–214. Berlin-Heidelberg-New York: Springer 1971
GIBBS, M., BAMBERGER, E.S., ELLYARD, P.W., EVERSON, R.G.: Assimilation of carbon dioxide by chloroplast preparations. In: Biochemistry of Chloroplasts (T.W. GOODWIN, ed.), (Proc. NATO Adv. Study Inst., Aberystwyth, 1965), pp. 3–38. London: Academic Press 1967a
GIBBS, M., CYNKIN, M.A.: Conversion of carbon-14 dioxide to starch and glucose during photosynthesis by spinach chloroplasts. Nature **182**, 1241–1242 (1958)
GIBBS, M., LATZKO, E., EVERSON, R.G., COCKBURN, W.: Carbon mobilization by the green plant. In: Harvesting the Sun (A. SAN PIETRO, F.A. GREEN, T.J. ARMY, eds.), pp. 111–130. New York: Academic Press 1967b
GIMMLER, H., SCHÄFER, G., KRAMINER, H., HEBER, U.: Amino acid permeability of the chloroplast envelope as measured by light scattering, volumetry and amino acid uptake. Planta **120**, 47–61 (1974)
GOLDSWORTHY, A.: Photorespiration. Botan. Rev. **36**, 321–340 (1970)
GOLDSWORTHY, A., STREET, H.E.: The carbohydrate nutrition of tomato roots. VIII. The mechanism of the inhibition by D-mannose of the respiration of excised roots. Ann. Botany (London), New Ser. **29**, 45–58 (1965)
GOOD, N.E., WINGET, G.D., WINTER, W., CONOLLY, T.N., IZAWA, S., SINGH, R.M.M.: Hydrogen ion buffers for biological research. Biochemistry **5**, 467–477 (1966)
GRAHAM, D., REED, M.L.: Carbonic anhydrase and the regulation of photosynthesis. Nature New Biol. **231**, 81–82 (1971)
GUTIERREZ, M., HUBER, S.C., KU, S.B., KANAI, R., EDWARDS, G.E.: Intracellular localisation of carbon metabolism in mesophyll cells of C4 plants. Proc. 3rd Intern. Congr. Photosynth. Res. Amsterdam: Elsevier 1974
HALL, D.O.: Nomenclature for isolated chloroplasts. Nature New Biol. **235**, 125–126 (1972)
HATCH, M.D., KAGAWA, T., CRAIG, S.: Subdivision of C4-pathway species based on differing C4 acid decarboxylating systems and ultrastructural features. Australian J. Plant Physiol. **2**, 111–128 (1975)
HEATH, O.V.S.: The Physiological Aspects of Photosynthesis. London: Heinemann, 1969
HEBER, U.: Vergleichende Untersuchungen an Chloroplasten, die durch Isolierungsoperationen in nicht-wässrigem und in wässrigem Milieu erhalten wurden. II. Kritik der Reinheit und Enzymlokalisationen in Chloroplasten. Z. Naturforsch. **15b**, 100–109 (1960)
HEBER, U.: Transport metabolites in photosynthesis. In: Biochemistry of Chloroplasts (T.W. GOODWIN, ed.), Vol. II, pp. 71–78. New York: Academic Press 1967
HEBER, U.: Flow of metabolites and compartmentation phenomena in chloroplasts. In: Transport and Distribution of Matter in Cells of Higher Plants K. MOTHES, E. MÜLLER, A. NELLES, D. NEUMANN, (eds.), pp. 152–184. Berlin: Akademie-Verlag 1970
HEBER, U.: Metabolite exchange between chloroplasts and cytoplasm. Ann. Rev. Plant Physiol. **25**, 393–421 (1974)
HEBER, U., HALLIER, U.W., HUDSON, M.A.: Lokalisation von Enzymen des reduktiven und oxydativen Pentosephosphat-Zyklus in den Chloroplasten und Permeabilität der Chloroplastenmembran gegenüber Metaboliten. Z. Naturforsch. **22b**, 1200–1215 (1967a)
HEBER, U, KIRK, M.R., GIMMLER, H., SCHÄFER, G.: Uptake and reduction of glycerate by isolated chloroplasts. Planta **120**, 31–46 (1974)

HEBER, U., KRAUSE, G.H.: Transfer of carbon, phosphate energy and reducing equivalents across the chloroplast envelope. In: Photosynthesis and Photorespiration M.D. HATCH, C.B. OSMOND, R.O. SLATYER (eds.), pp. 218–225. (Proc. Conf. Australian Nat. Univ. Canberra, 1970.) New York: Wiley Interscience 1971
HEBER, U., KRAUSE, G.H.: Hydrogen and proton transfer across the chloroplast envelope. In: Progress in Photosynthesis G. FORTI, M. AVRON, A. MELANDRI (eds.), Vol. II, pp. 1023–1033 (Proc. 2nd Intern. Congr. Photosyn. Res. 1971). The Hague: Junk N.V. 1972
HEBER, U., SANTARIUS, K.A.: Compartmentation and reduction of pyridine nucleotides in relation to photosynthesis. Biochim. Biophys. Acta **109**, 390–408 (1965)
HEBER, U., SANTARIUS, K.A.: Direct and indirect transport of ATP and ADP across the chloroplast envelope. Z. Naturforsch. **25b**, 718–778 (1970)
HEBER, U., SANTARIUS, K.A., HUDSON, M.A., HALLIER, U.W.: Intrazellulärer Transport von Zwischenprodukten der Photosynthese im Photosynthese-Gleichgewicht und im Dunkel-Licht-Wechsel. Z. Naturforsch. **22b**, 1189–1199 (1967b)
HEBER, U., SANTARIUS, K.A., URBACH, W., ULLRICH, W.: Photosynthese und Phosphathaushalt. Intrazellulärer Transport von ^{14}C- und ^{32}P-markierten Intermediärprodukten zwischen den Chloroplasten und dem Zytoplasma und seine Folgen für die Regulation des Stoffwechsels. Z. Naturforsch. **19b**, 576–587 (1964)
HEBER, U., WILLENBRINK, J.: Sites of synthesis and transport of photosynthetic products within the leaf cell. Biochim. Biophys. Acta **82**, 313–324 (1964)
HELDT, H.W.: Adenine nucleotide translocation in spinach chloroplasts. FEBS Letters **5**, 11–14 (1969)
HELDT, H.W., RAPLEY, L.: Unspecific permeation and specific uptake of substances in spinach chloroplasts. FEBS Letters **7**, 139–142 (1970a)
HELDT, H.W., RAPLEY, L.: Specific transport of inorganic phosphate, 3-phosphoglycerate and dihydroxyacetonephosphate, and of dicarboxylates across the inner membrane of spinach chloroplasts. FEBS Letters **10**, 143–148 (1970b)
HELDT, H.W., SAUER, F.: The inner membrane of the chloroplast envelope as the site of specific metabolite transport. Biochim. Biophys. Acta **234**, 83–91 (1971)
HELDT, H.W., SAUER, F., RAPLEY, L.: Differentiation of the permeability properties of the two membranes of the chloroplast envelope. In: Progress in Photosynthesis. G. FORTI, M. AVRON, A. MELANDRI (eds.), pp. 1345–1355. (Proc. 2nd Intern. Congr. Photosynth. Res.) The Hague: Junk N.V. 1972
HELDT, H.W., WERDAN, K.: Die Bedeutung der Lichtabhängigen pH-Änderungen im Stroma- und im Thylakoidraum intakter Chloroplasten. Z. Physiol. Chem. **354**, 224 (1973)
HELDT, H.W., WERDAN, K., MILOVANCEV, M., GELLER, G.: Alkalization of the chloroplast stroma caused by light-dependent proton flux into the thylakoid space. Biochim. Biophys. Acta **314**, 224–241 (1973)
HELLEBUST, J.A., BIDWELL, R.G.S.: Protein turnover in attached wheat and tobacco leaves. Can. J. Botany **42**, 1–12 (1964)
HESKETH, J.D.: Limitations to photosynthesis responsible for differences among species. Crop Sci. **3**, 493–496 (1963)
HILL, R.: The biochemists' green mansions: The photosynthetic electron-transport chain in plants. Essays Biochem. **1**, 121–151 (1965)
HILL, R., WALKER, D.A.: Pyocyanine and phosphorylation with chloroplasts. Plant Physiol. **34**, 240–245 (1959)
HIND, G., NAKATANI, H.Y., IZAWA, S.: Light-dependent redistribution of ions in suspensions of chloroplast thylakoid membranes. Proc. Natl. Acad. Sci. US **71**, 1484–1488 (1974)
HUBER, S.C., EDWARDS, G.E.: C4 photosynthesis: Light dependent CO_2 fixation by mesophyll cells, protoplasts and chloroplasts of *Digitaria sanquinalis*. Plant Physiol. (In press) **55**, 835–844 (1975)
JACKSON, W.A., VOLK, R.J.: Photorespiration. Ann. Rev. Plant Physiol. **21**, 385–432 (1970)
JAGENDORF, A.T., AVRON, M.: Pathways in photosynthetic phosphorylation by spinach chloroplasts. Federation Proc. **17**, 248 (1958a)
JAGENDORF, A.T., AVRON, M.: Cofactors and rates of photosynthetic phosphorylation by spinach chloroplasts. J. Biol. Chem. **231**, 277–290 (1958b)
JAMES, W.O., LEECH, R.M.: The cytochromes of isolated chloroplasts. Proc. Roy. Soc. B **160**, 13–24 (1964)

JENSEN, R.G.: Activation of CO_2-fixation in isolated spinach chloroplasts. Biochim. Biophys. Acta **234**, 360–370 (1971)
JENSEN, R.G., BASSHAM, J.A.: Photosynthesis by isolated chloroplasts. Proc. Natl. Acad. Sci. US **56**, 1095–1101 (1966)
JENSEN, R.G., BASSHAM, J.A.: Photosynthesis by isolated chloroplasts. III. Light activation of the carboxylation reaction. Biochim. Biophys. Acta **153**, 227–234 (1968)
JOHNSON, H.S., HATCH, M.D.: Properties and regulation of leaf nicotinamide-adenine dinucleotide phosphate-malate dehydrogenase and 'Malic' enzyme in plants with the C4-dicarboxylic acid pathway of photosynthesis. Biochem. J. **119**, 273–280 (1970)
KALBERER, P.P., BUCHANAN, B.B., ARNON, D.I.: Rates of photosynthesis by isolated chloroplasts. Proc. Natl. Acad. Sci. US **57**, 1542–1549 (1967)
KEARNEY, P.C., TOLBERT, N.E.: Appearance of glycolate and related products of photosynthesis outside of chloroplasts. Arch. Biochem. Biophys. **98**, 164–171 (1962)
KELLY, G.J., GIBBS, M.: Nonreversible D-glyceraldehyde 3-phosphate dehydrogenase in plant tissues. Plant Physiol. **52**, 111–118 (1973)
KELLY, G.J., LATZKO, E.: Evidence for phosphofructokinase in chloroplasts. Nature **256**, 429–430 (1975)
KIRK, P.R., LEECH, R.M.: Amino acid biosynthesis by isolated chloroplasts during photosynthesis. Plant Physiol. **50**, 228–234 (1972)
KISAKI, T., IMAI, A., TOLBERT, N.E.: Intracellular localisation of enzymes related to photorespiration in green leaves. Plant Cell Physiol. **12**, 267–273 (1971)
KISAKI, T., TOLBERT, N.E.: Glycolate and glyoxylate metabolism by isolated peroxisomes or chloroplasts. Plant Physiol. **44**, 242–250 (1969)
KLINGENBERG, M., PFAFF, E.: Means of terminating reactions. Methods Enzymol. **10**, 680–684 (1967)
KRAUSE, G.H.: Indirekter ATP-Transport zwischen Chloroplasten und Zytoplasma während der Photosynthese. Z. Pflanzenphysiol. **65**, 13–23 (1971)
LEECH, R.M.: The isolation of structurally intact chloroplasts. Biochim. Biophys. Acta **79**, 637–639 (1964)
LEVINE, G., BASSHAM, J.A.: Inhibition of photosynthesis in isolated spinach chloroplasts by inorganic phosphate or inorganic pyrophosphatase in the presence of pyrophosphate and magnesium ions. Biochim. Biophys. Acta **333**, 136–140 (1974)
LILLEY, R.MCC., FITZGERALD, M.P., RIENITS, K.G., WALKER, D.A.: Criteria of intactness and the photosynthetic activity of spinach chloroplast preparations. New Phytologist **75**, 1–10 (1975)
LILLEY, R.MCC., HOLBOROW, K., WALKER, D.A.: Magnesium activation of photosynthetic CO_2 fixation in a reconstituted chloroplast system. New Phytologist **73**, 659–664 (1974)
LILLEY, R.MCC., SCHWENN, J.D., WALKER, D.A.: Inorganic pyrophosphatase and photosynthesis by isolated chloroplasts. II. The controlling influence of orthophosphate. Biochim. Biophys. Acta **325**, 596–604 (1973)
LILLEY, R.MCC., WALKER, D.A.: The reduction of 3-phosphoglycerate by reconstituted chloroplasts and by chloroplast extracts. Biochim. Biophys. Acta **368**, 269–278 (1974)
LILLEY, R.MCC., WALKER, D.A.: Carbon dioxide assimilation by leaves, isolated chloroplasts and ribulose bisphosphate carboxylase from spinach. Plant Physiol. **55**, 1087–1092 (1975)
LIN, D.C., NOBEL, P.S.: Control of photosynthesis by Mg^{2+}. Arch. Biochem. Biophys. **145**, 622–632 (1971)
LORIMER, G.H., ANDREWS, T.J.: Plant photorespiration—an inevitable consequence of the existence of atmospheric oxygen. Nature **243**, 359 (1973)
LOUGHMAN, B.C.: The mechanism of absorption and utilisation of phosphate by barley plants in relation to subsequent transport to the shoot. New Phytologist **65**, 388–397 (1956)
MACLACHLAN, G.A., PORTER, H.K.: Replacement of oxidation by light as the energy source for glucose metabolism in tobacco leaf. Proc. Roy. Soc. B **150**, 460–473 (1959)
MATHIEU, Y.: Sur l'isolement, en milieu aqueux, de chloroplastes "intacts" à partir de feuilles de plantules d'Orge. Photosynthetica **1** (1–2), 57–63 (1967)
MIFLIN, B.J., BEEVERS, H.: Isolation of intact plastids from a range of plant tissues. Plant Physiol. **53**, 870–874 (1974)
MOLISCH, H.: Populäre biologische Vorträge. Jena: Fischer 1922

MORGENTHALER, J.J., PRICE, C.A., ROBINSON, M., GIBBS, M.: Photosynthetic activity of spinach chloroplasts after isopycnic centrifugation in gradients of silica. Plant Physiol. **54**, 532–534 (1974)

NEUMANN, J., LEVINE, R.P.: Reversible pH changes in cells of *Chlamydamonas reinhardi* resulting from CO_2 fixation in the light and its evolution in the dark. Plant Physiol. **47**, 700–704 (1971)

NEWCOMB, E.H., FREDERICK, S.E.: Distribution and structure of plant microbodies (peroxisomes). In: Photosynthesis and Photorespiration (M.D. HATCH, C.B. OSMOND, R.O. SLATYER, eds.), pp. 442–457. (Proc. Conf. Aust. Nat. Univ. Canberra, 1970.) New York: Wiley-Interscience 1971

NOBEL, P.S.: Light-induced changes in the ionic content of chloroplasts in *Pisum sativum*. Biochim. Biophys. Acta **172**, 134–143 (1969)

NOBEL, P.S., CHEUNG, Y.S.: Two amino acid carriers in pea chloroplasts. Nature **237**, 207–208 (1972)

NOBEL, P.S., WANG, C.T.: Amino acid permeability of pea chloroplasts as measured by osmotically determined reflection coefficients. Biochim. Biophys. Acta **211**, 79–87 (1970)

OGREN, W.L., BOWES, G.: Ribulose diphosphate carboxylase regulates soybean photorespiration. Nature New Biol. **230**, 159–160 (1971)

ONGUN, A., STOCKING, C.R.: Effect of light and dark on the intracellular fate of photosynthetic products. Plant Physiol. **40**, 825–831 (1965)

PACKER, L., CROFTS, A.R.: The energized movement of ions and water by chloroplasts. Current Topics Bioenerg. **2**, 23–64 (1967)

PACKER, L., SIEGENTHALER, P.A., NOBEL, P.S.: Light induced volume changes in spinach chloroplasts. J. Cell Biol. **26**, 593–599 (1965)

PARKIN, J.: Contributions to our knowledge of the formation, storage and depletion of carbohydrates in Monocotyledons. Roy. Soc. Lond. Phil. Trans. B **CXCI**, 35–74 (1899)

PEDERSEN, T.A., KIRK, M., BASSHAM, J.A.: Light-dark transients in levels of intermediate compounds during photosynthesis in air-adapted *Chlorella*. Physiol. Plantarium **19**, 219–231 (1966)

PETERKOFSKY, A., RACKER, E.: The reductive pentose phosphate cycle. III. Enzyme activities in cell-free extracts of photosynthetic organisms. Plant Physiol. **36**, 409–414 (1961)

PHILLIS, E., MASON, T.G.: On the effects of light and oxygen on the uptake of sugar by the foliage leaf. Ann. Botany (London), (N.S.) **1**, 231–237 (1937)

PLAUT, Z., GIBBS, M.: Glycolate formation in intact spinach chloroplasts. Plant Physiol. **45**, 470–474 (1970)

POINCELOT, R.P.: Intracellular distribution of carbonic anhydrase in spinach leaves. Biochim. Biophys. Acta **258**, 637–642 (1972)

POINCELOT, R.P.: Uptake of bicarbonate ion in darkness by isolated chloroplast envelope membranes and intact chloroplasts of spinach. Plant Physiol. **54**, 520–526 (1974)

PON, N.G., RABIN, B.R., CALVIN, M.: Mechanism of the carboxydismutase reaction. I. The effect of preliminary incubation of substrates, metal ion and enzyme on activity. Biochem. Z. **338**, 7–19 (1963)

PREISS, J., BIGGS, M.L., GREENBERG, E.: The effect of magnesium ion concentration on the pH optimum of the spinach leaf alkaline fructose diphosphatase. J. Biol. Chem. **242**, 2292–2294 (1967a)

PREISS, J., GHOSH, H.P., WITTKOP, J.: Regulation of the biosynthesis of starch in spinach leaf Chloroplasts. In: The Biochemistry of Chloroplasts (T.W. GOODWIN, ed.), Vol. II, pp. 131–153. New York: Academic Press 1967b

PREISS, J., KOSUGE, T.: Regulation of enzyme activity in photosynthetic systems. Ann. Rev. Plant Physiol. **21**, 433–466 (1970)

RABINOWITCH, E.I.. Photosynthesis and Related Processes, Vol. 1. New York: Wiley Interscience 1945

RABINOWITCH, E.I.: Photosynthesis and Related Processes, Vol. II, Part 2. New York: Wiley Interscience 1956

RIDLEY, S.M., LEECH, R.M.: The survival of chloroplasts in vitro. Particle volume distribution patterns as a criterion for assessing the degree of integrity of isolated chloroplasts. Planta **84**, 20–34 (1968)

ROBERTS, G.R., KEYS, A.J., WHITTINGHAM, C.P.: The transport of photosynthetic products from the chloroplasts of tobacco leaves. J. Exptl. Botany **21**, 683–692 (1970)

SACHS, J.: Über den Einfluß des Lichtes auf die Bildung des Amylums in den Chlorophyllkörnern. Botan. Z. **20**, 365–373 (1862)

SACHS, J.: Lectures on the Physiology of Plants. Translated by H.M. WARD, p. 309 seq. Oxford: Clarendon Press 1887

SCHACTER, B., ELEY, J.H., JR., GIBBS, M.: Effect of sugar phosphates and photosynthesis inhibitors on CO_2 fixation and O_2 evolution by chloroplasts. Plant Physiol. **43**, 8–30 (1968)

SCHÜRMANN, P., BUCHANAN, B.B., ARNON, D.I.: Role of cyclic photophosphorylation in photosynthetic carbon dioxide assimilation by isolated chloroplasts. Biochim. Biophys. Acta **267**, 111–124 (1971)

SCHÜRMANN, P., BUCHANAN, B.B., ARNON, D.I.: Role of cyclic photophosphorylation in photosynthetic carbon dioxide assimilation by isolated chloroplasts. In: Progress in Photosynthesis (G. FORTI, M. AVRON, A. MELANDRI, eds.), Vol. II, pp. 1283–1291. (Proc. 2nd Intern. Congr. Photosyn. Res.) The Hague: Junk, N.V. 1972

SCHWENN, J.D., LILLEY, R.McC., WALKER, D.A.: Inorganic pyrophosphatase and photosynthesis by isolated chloroplasts. I. Characterisation of chloroplast pyrophosphatase and its relation to the response to exogenous pyrophosphate. Biochim. Biophys. Acta **325**, 586–595 (1973)

SHAIN, Y., GIBBS, M.: Formation of glycolate by a reconstituted spinach chloroplast preparation. Plant Physiol. **48**, 325–330 (1971)

SHIBKO, S., EDELMAN, J.: Randomization of the carbon atoms in glucose and fructose during their metabolism in barley seedlings. Biochim. Biophys. Acta **25**, 642–644 (1957)

SMILLIE, R.M., FULLER, R.C.: Ribulose-1,5-diphosphate carboxylase activity in relation to photosynthesis by intact leaves and isolated chloroplasts. Plant Physiol. **34**, 651–656 (1959)

SPENCER, D.L., UNT, H.: Biochemical and structural correlations in isolated spinach chloroplasts under isotonic and hypotonic conditions. Australian J. Biol. Sci. **18**, 197–210 (1965)

STOCKING, C.R.: Chloroplast isolation in non-aqueous media. Plant Physiol. **34**, 56–61 (1959)

STOCKING, C.R.: Chloroplasts: Nonaqueous. Methods Enzymol. **23A**, 221–228 (1971)

STOCKING, C.R., LARSON, S.: A chloroplast cytoplasmic shuttle and the reduction of extraplastid NAD. Biochem. Biophys. Res. Commun. **37**, 278–282 (1969)

STOKES, D.M., WALKER, D.A.: Relative impermeability of the intact chloroplast envelope to ATP. In: Photosynthesis and Photorespiration (M.D. HATCH, C.B. OSMOND, R.O. SLATYER, eds.), pp. 226–231. Proc. Conf. Australian Natl. Univ. Canberra 190. New York: Wiley Interscience 1971

STOKES, D.M., WALKER, D.A.: Photosynthesis by isolated chloroplasts. Inhibition by DL-glyceraldehyde of carbon dioxide assimilation. Biochem. J. **128**, 1147–1157 (1972)

STOKES, D.M., WALKER, D.A., MCCORMICK, A.V.: Photosynthetic oxygen evolution in a reconstituted chloroplast system. In: Progress in Photosynthesis (G. FORTI, M. AVRON, A. MELANDRI, eds.), pp. 1779–1785. (Proc. 2nd Intern. Congr. Photosyn. Res.) The Hague: Junk N.V. 1972

STROTMANN, H., HELDT, H.W.: Phosphate containing metabolites participating in photosynthetic reactions of *Chlorella pyrtnoidosa*. In: Progress in Photosynthesis Research. (H. METZNER, ed.), Vol. III, pp. 1131–1140. (Proc. 1st Intern. Congr. Photosynth. Res.) Tübingen: Intern. Union Biol. Sci. 1969

SUGIYAMA, T., NAKAYAMA, N., AKAZAWA, T.: Structure and function of chloroplast proteins. V. Homotropic effect of bicarbonate in RuDP carboxylase reaction and the mechanism of activation by magnesium ions. Arch. Biochem. Biophys. **126**, 737–745 (1968)

TABITA, F.R., MCFADDEN, B.A.: Regulation of ribulose-1,5-diphosphate carboxylase by 6-phospho-D-gluconate. Biochem. Biophys. Res. Commun. **48**, 1153–1159 (1972)

THOMAS, M., RANSON, S.L., RICHARDSON, J.A.: Plant Physiology. London: Longman 1973

TOLBERT, N.E.: Glycolate biosynthesis. In: Current Topics in Cellular Regulation (B.L. HORECKER, E.R. STADTMAN, eds.), Vol. VII, pp. 21–49. New York: Academic Press 1973

TOLBERT, N.E., OESER, A., YAMAZAKI, R.K., HAGEMAN, R.H., KISAKI, T: A survey of plants for leaf peroxisomes. Plant Physiol. **44**, 135–147 (1969)

TOLLENAAR, D.: Omzettingen van Koolhydraten in het blad van *Nicotiana tabacum* L. Wageningen: H. Veenman and Zonen 1925

URBACH, W., HUDSON, M.A., ULLRICH, W., SANTARIUS, K.A., HEBER, U.: Verteilung und Wanderung von Phosphoglycerat zwischen den Chloroplasten und dem Zytoplasma während der Photosynthese. Z. Naturforsch. **20b**, 890–898 (1965)

VISHNIAC, W., OCHOA, S.: Phosphorylation coupled to photochemical reduction of pyridine nucleotides by chloroplast preparations. J. Biol. Chem. **198**, 501–506 (1952)

WALKER, D.A.: Improved rates of carbon dioxide fixation by illuminated chloroplasts. Biochem. J. **92**, 22c–23c (1964)

WALKER, D.A.: Photosynthetic activity of isolated pea chloroplasts. In: Biochemistry of the Chloroplast (T.W. GOODWIN, ed.), Vol. II, pp. 53–69. (Proc. NATO Adv. Study Inst., Aberystwyth 1965a.) New York: Academic Press 1967

WALKER, D.A.: Correlation between photosynthetic activity and membrane integrity in isolated pea chloroplasts. Plant Physiol. **40**, 1157–1161 (1965b)

WALKER, D.A.: Permeability of the chloroplast envelope. In: Progress in Photosynthesis (H. METZNER, ed.). Vol. I, pp. 250–257. (Proc. 1st Intern. Congr. Photosyn. Res., Freudenstadt 1968.) Tübingen: Intern. Union Biol. Sci. 1969

WALKER, D.A.: Chloroplasts (and Grana)—Aqueous (including high carbon fixation ability). In: Methods in Enzymology (A SAN PIETRO, ed.), Vol. XXIII, pp. 211–220. London-New York: Academic Press 1971

WALKER, D.A.: Photosynthetic induction phenomena and the light activation of ribulose diphosphate carboxylase. New Phytologist **72**, 209–235 (1973)

WALKER, D.A.: Some characteristics of a primary carboxylating mechanism. In: Plant Carbohydrate Biochemistry (J.B. PRIDHAM, ed.), pp. 7–26. (Ann. Proc. Phytochem. Soc., No. 10. Proc. Phytochem. Soc. Symp., Edinburgh 1973.) London-New York: Academic Press 1974a

WALKER, D.A.: Chloroplast and cell. Concerning the movement of certain key metabolites etc. across the chloroplast envelope. In: Med. Tech. Publ. Int. Rev. Sci. Biochem. Ser. I. (D.H. NORTHCOTE, ed.), Vol. XI, pp. 1–49. London: Butterworths 1974b

WALKER, D.A.: Photosynthetic induction and its relation to transport phenomena in chloroplasts. In: The Intact Chloroplast (J. BARBER, ed.). Amsterdam: ASP Biol. Med. Press. B.V. 1975 (in press)

WALKER, D.A., COCKBURN, W., BALDRY, C.W.: Photosynthetic oxygen evolution by isolated chloroplasts in the presence of carbon cycle intermediates. Nature **216**, 597–599 (1967)

WALKER, D.A., CROFTS, A.R.: Photosynthesis. Ann. Rev. Biochem. **39**, 389–428 (1970)

WALKER, D.A., HILL, R.: The relation of oxygen evolution to carbon assimilation with isolated chloroplasts. Biochim. Biophys. Acta **131**, 330–338 (1967)

WALKER, D.A., LILLEY, R.McC.: Ribulose Bisphosphate Carboxylase—an Enigma Resolved. (N. SUNDERLAND, ed.). (Proc. 50th Ann. Meeting Soc. Exp. Biol., Cambridge, 1974.) Oxford: Pergamon Press 1975

WALKER, D.A., MCCORMICK, A.V., STOKES, D.M.: CO_2-dependent oxygen evolution by envelope-free chloroplasts. Nature **233**, 346–347 (1971)

WANG, C.-T., NOBEL, P.S.: Permeability of pea chloroplasts to alcohols and aldoses as measured by reflection coefficients. Biochim. Biophys. Acta **241**, 200–212 (1971)

WERDAN, K., HELDT, H.W.: The phosphate translocator of spinach chloroplasts. In: Progress in Photosynthesis Research. (G. FORTI, M. AVRON, A. MELANDRI, eds.), pp. 1337–1344. (Proc. 2nd Intern. Conf. Photosyn. Res.) The Hague: Junk N.V. 1972

WERDAN, K., HELDT, H.W.: Bicarbonate uptake into the chloroplast stroma. In: Mechanisms in Bioenergetics. (G.F. AZZONE, L. ERNSTER, S. PAPA, E. QUAGLIARIELLO, N. SILIPRANDI, eds.), pp. 285–292. (Proc. Conf. Mechanisms Bioenerg. Pugnochiuso, Italy, 1972.) New York-London: Academic Press 1973a

WERDAN, K., HELDT, H.W.: Messung des pH-Wertes im Stroma- und im Thylakoidraum intakter Chloroplasten. Z. Physiol. Chem. **354**, 223–224 (1973b)

WERDAN, K., HELDT, H.W., GELLER, G.: Accumulation of bicarbonate in intact chloroplasts following a pH gradient. Biochim. Biophys. Acta **283**, 430–441 (1972)

WERDAN, K., HELDT, H.W., MILOVANCEV, M.: The role of pH in the stroma of chloroplasts in the regulation of carbon fixation studies on CO_2 fixation in the light and dark. Biochim. Biophys. Acta **396**, 276–292 (1975)

WERKHEISER, W.C., BARTLEY, W.: The study of steady-state concentrations of internal solutes of mitochondria by rapid centrifugal transfer to a fixation medium. Biochem. J. **66**, 79–91 (1957)

WEST, K.R., WISKICH, J.T.: Photosynthetic control by isolated pea chloroplasts. Biochem. J. **109**, 527–532 (1968)

WHATLEY, F.R., ALLEN, M.B., ROSENBERG, L.L., CAPINDALE, J.B., ARNON, D.I.: Photosynthesis by isolated chloroplasts. V. Phosphorylation and carbon dioxide fixation by broken chloroplasts. Biochim. Biophys. Acta **20**, 462–468 (1956)

WHITEHOUSE, D.G., LUDWIG, L.J., WALKER, D.A.: Participation of the Mehler reaction and catalase in the oxygen exchange of chloroplast preparations. J. Exptl. Botany **23**, No. 73, 772–791 (1971)

ZELITCH, I.: Photosynthesis, Photorespiration and Plant Productivity. New York-London: Academic Press 1971

ZELITCH, I.: The biochemistry of photorespiration. Current Advan. Plant Sci. **6**, 44–54 (1973)

3. Metabolite Carriers of Chloroplasts

H.W. HELDT

1. Specific Transport into the Chloroplast

The chloroplast stroma represents a separate metabolic compartment within the plant cell. The envelope which surrounds it consists of two membranes. The *outer* one has been shown to be freely permeable to small molecules, such as nucleotides and intermediates of the Calvin cycle (HELDT and SAUER, 1971). In this respect the outer membrane of the chloroplast is very similar to the outer mitochondrial membrane (see Chap. II, 6). The functional barrier between the chloroplast stroma and the cytoplasm is the *inner* membrane. The metabolic function of the chloroplast is to fix CO_2. The main products of CO_2 fixation delivered by the chloroplast to the plant cell are triosephosphates (see Chap. II, 2). Therefore, CO_2 fixation occurring in the chloroplast stroma involves the uptake of CO_2 and inorganic phosphate and the release of triosephosphates. Whereas the inner membrane of the envelope appears to be freely permeable to CO_2 (WERDAN et al., 1972), it is impermeable to ions. This points to the physiological necessity of a specific transport of inorganic phosphate from the cytoplasm into the chloroplast stroma and of triosephosphates in the other direction.

2. The Phosphate Translocator

There have been many indications that inorganic phosphate, triosephosphates and 3-phosphoglycerate may readily pass the chloroplast envelope (see Chap. II, 2). Direct studies of metabolite uptake using silicon layer filtering centrifugation showed that all substances mentioned were indeed very rapidly transported into the chloroplast stroma (HELDT and RAPLEY, 1970). Figure 1 shows the uptake of inorganic phosphate (Pi). Since it is very rapid, measurements are usually carried out at a low temperature in order to obtain a better resolution of the kinetics. The

Table 1. Transport of phosphates into the sorbitol impermeable space of spinach chloroplasts as measured by back-exchange. Temp. 4°. (FLIEGE and HELDT, unpubl.)

	Km (mM)	V_{max} (µmol/mg chlor./h)
Phosphate	0.23	48
3-Phosphoglycerate	0.11	32
Dihydroxyacetonephosphate	0.08	45

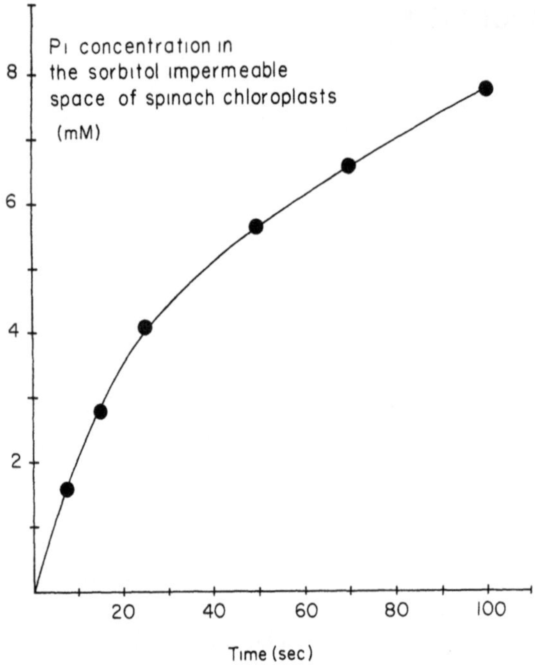

Fig. 1. Transport of inorganic phosphate (Pi) into the sorbitol impermeable space of spinach chloroplasts. Temp. 4°, Pi concentration in medium: 1 mM. (WERDAN and HELDT, unpubl.)

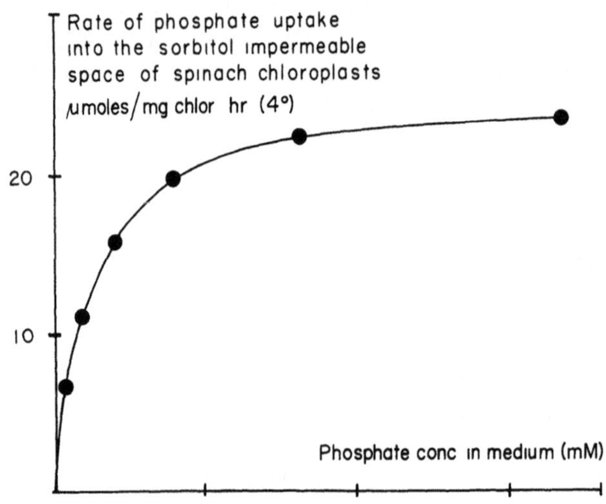

Fig. 2. Rate of transport of inorganic phosphate (Pi) into the sorbitol impermeable space of spinach chloroplasts depending on phosphate concentration in medium. Temp. 4°. (FLIEGE and HELDT, unpubl.)

rate of uptake is determined from the initial linear slope within a range of 10 s. It depends on the concentration in the medium (Fig. 2). The concentration dependence reveals hyperbolic saturation characteristics, indicating substrate saturation of uptake. A double reciprocal plot of the data yields a linear function, which enables the determination of *Km* (substrate concentration causing half maximal saturation of the transport) and of V_{max} (maximal velocity) as shown in Table 1. For glyceraldehyde phosphate exact kinetic data are not yet available. Preliminary data indicate that it is transported in a similar manner to dihydroxyacetone phosphate (DHAP).

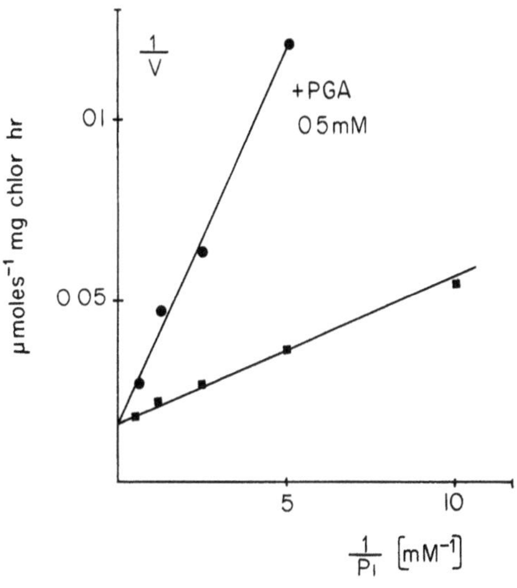

Fig. 3. Concentration dependence of the transport of inorganic phosphate (Pi) into the sorbitol impermeable space of spinach chloroplasts. Inhibition by 3-phosphoglycerate (PGA). Temp. 4°. (FLIEGE and HELDT, unpubl.)

The compounds taken up compete with each other for transportation (Fig. 3). The *Km* for transportation of a certain compound is equal to the inhibition constant *Ki* of this particular compound for competitive inhibition of transport (WERDAN and HELDT, 1972). Therefore the specificity of the carrier can be described from inhibition studies (Table 2). Transport by the phosphate translocator is highly specific. It requires either Pi or a phosphate esterified to the end of a three carbon molecule. Thus neither 2-phosphoglycerate nor inorganic pyrophosphate are transported. There appears to be some interaction of the carrier with erythrose-4-phosphate, whereas hexose-phosphates do not react with the carrier.

The concentration of phosphates taken up into the stroma is usually much higher than in the medium (Fig. 1), especially if the concentration in the medium is kept very low. The apparent accumulation is caused by a counter-exchange of anions. For each phosphate [Pi, DHAP or 3-phosphoglycerate (PGA)] taken up, another phosphate is released from the stroma. Coupling of inward and outward movement is very strict (HELDT et al., 1975). In this way the total amount of phosphate in the stroma is kept constant. Leakage of phosphates from intact chloroplasts is therefore very slow; it was found to be about three orders of magnitude slower than the rate of counter-exchange (FLIEGE and HELDT, unpubl.).

Table 2. Specificity of the phosphate translocator in spinach chloroplasts, as measured by inhibition of phosphate transport (*Km* 0.20 mM). Temp. 4°. (HELDT et al., 1975)

	Ki (mM)		*Ki* (mM)
Arsenate	0.35	Erythrose-4-phosphate	2.30
Pyrophosphate	1.8	Ribose-5-phosphate	10.0
2-Phosphoglycerate	6.5	Fructosediphosphate	8.5
1-Glycerophosphate	1.3	Fructose-6-phosphate	12.5
2-Glycerophosphate	7.7	Glucose-6-phosphate	40.0
Phosphoenolpyruvate	4.7	6-Phosphogluconate	20.0

Measurements of the counter-exchange proved to be a valuable tool for studying the kinetic properties of the phosphate translocator. Instead of measuring the uptake of radioactively labeled phosphates directly, one can obtain the same information indirectly by following the release of ^{32}P-labeled phosphates from the chloroplasts after the addition of unlabeled compounds. By such back-exchange one can determine even kinetic constants of those compounds which are not available radioactively labeled. The data of Table 1 were obtained in this way. Kinetic data obtained from direct transport and from back-exchange were identical.

Transport by the phosphate translocator is very strongly temperature-dependent, from 0–12° an activation energy of 16 Kcal was determined (FLIEGE and HELDT, unpublished results). From this one can extrapolate that the rate of phosphate transport at 20° may be about five times higher than that measured at 4°. Parachloromercuriphenylsulphonic acid, a reagent for SH groups, strongly inhibits phosphate transport (WERDAN and HELDT, 1972). The transport of dicarboxylates in chloroplasts is much less affected by this inhibitor.

The rate of PGA transport into the chloroplasts, but not of Pi or DHAP transport, is enhanced when the chloroplasts are illuminated. This stimulation is due to an increase of the V_{max} of PGA; the *Km* remains unchanged (WERDAN, 1975). The stimulation is diminished by the addition of the uncoupler m-chlorocarbonylcyanide phenylhydrazone (CCCP). It seems that the transport of PGA is influenced by the pH gradient across the inner membrane of the envelope. Illumination of intact chloroplasts leads to proton transport from the stroma into the thylakoid space. Consequently there is an alkalization of the stroma and an acidification of the thylakoid space. With pH 7.6 in the medium, the pH in the stroma changes from 7.0 in the dark to 8.0 in the light (WERDAN et al., 1975). In order to correlate the pH gradient between the stroma and the medium with a stimulation of PGA transport, the balance of charges during a counter-transport of PGA with DHAP or Pi has to be considered. At physiological pH the anion of PGA has one more negative charge than Pi or DHAP. Therefore, a counter-exchange of PGA with either Pi or DHAP is likely to involve a transfer of protons. This would explain, how PGA transport into the chloroplast might be stimulated by light induced alkalization of the stroma. It still has to be decided experimentally whether this is the only explanation for the observed effect.

The stimulation of PGA transport leads to an uneven distribution of PGA and DHAP between the stroma and the external space. In illuminated chloroplast suspensions, the ratio DHAP/PGA in the medium was found to be much higher than in the stroma (BASSHAM et al. 1968, HELDT et al., unpubl.). Similarly, in whole leaf cells the DHAP/PGA ratio in the cytoplasm was reported to be higher than in the chloroplasts (HEBER, 1975). In isolated chloroplasts this difference of ratios is diminished after the addition of CCCP, which is known also to diminish the light-dependent alkalization in the stroma. The trapping of PGA in chloroplasts in the light will cause a preferential export of DHAP, ensuring the reduction of PGA in the stroma. The asymmetric distribution of PGA and DHAP will also favour a DHAP/PGA shuttle across the inner membrane. In this way phosphorylating and/or reducing equivalents may be exported from the stroma (STOCKING and LARSON, 1969; HEBER and SANTARIUS, 1970; BAMBERGER et al., 1975). For further details about these shuttles see Chapter II, 2.

3. The Dicarboxylate Translocator

A number of dicarboxylates are rapidly taken up into the chloroplasts. This transport also shows substrate saturation (Fig. 4). Table 3 shows values for *Km* and V_{max} obtained at 4°. There is competition between these compounds for transportation (Fig. 4). This clearly indicates that all dicarboxylates mentioned are transported by the same carrier. This carrier is specific for C_4 and C_5 dicarboxylates. In contrast to mitochondrial dicarboxylate transport (see Chap. II, 6) malonate is not transported, and there is no connection between dicarboxylate transport and transport of phosphate or tricarboxylates. From the temperature dependence of dicarboxylate transport between 0° and 19° an activation energy of about 7 Kcal was calculated (LEHNER and HELDT, unpubl.). This is considerably lower than the activation energy of the phosphate translocator.

Table 3. Transport of dicarboxylates into the sorbitol impermeable space of spinach chloroplasts. Temp. 4°. (HELDT et al., 1975)

	Km (mM)	V_{max} (µmoles/mg chlor./h)
L-Malate	0.39	18.6
Succinate	0.26	14.0
Fumarate	0.21	18.6
L-Aspartate	0.72	31.1
L-Glutamate	1.17	15.8

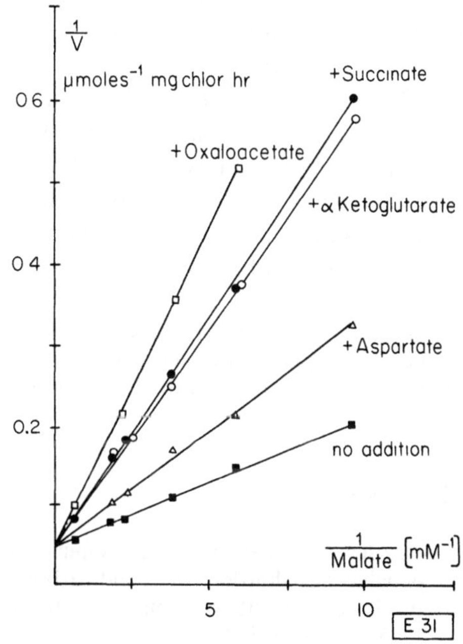

Fig. 4. Concentration dependence of the transport of malate into the sorbitol impermeable space of spinach chloroplasts. Inhibition by various dicarboxylates (1 mM). Temp. 4°. (LEHNER and HELDT, unpubl.)

Dicarboxylate transport proceeds also as a counter-exchange. Therefore, the dicarboxylate concentration in the chloroplasts may be much higher than in the medium. However, in contrast to the transport of phosphate, the counter-exchange of dicarboxylates is not strictly coupled (HELDT et al., 1975). Transport in one direction only, either additional uptake, or loss of dicarboxylates without accompanying influx, appears to be possible, but the rates of the uni-directional transport are two orders of magnitude lower than the rates of counter-transport (LEHNER and HELDT, unpubl.). A specific inhibitor of the dicarboxylate translocator hat not yet been reported.

The dicarboxylate translocator may play a role in the transfer of reducing equivalents from the stroma into the cytoplasm, either by malate/oxaloacetate or by malate/glutamate/aspartate/α-ketoglytarate shuttles (HEBER, 1975).

4. ATP Transport in Chloroplasts

There is also direct transport of adenine nucleotides into the stroma proceeding by counter-exchange (STROTMANN and HELDT, 1969; HELDT, 1969). It is highly specific for ATP. Transport of ADP is about one order of magnitude slower; AMP, GTP, CTP, UTP and ITP are transported very little, if at all. In contrast to mitochondrial adenine nucleotide transport, ATP transport in chloroplasts is not inhibited by atractyloside. The rate of ATP transport is very low, being only about 1/100 of the rate of phosphate transport. The metabolic function of the ATP translocator is not fully known. It may play a role in providing the chloroplast during the dark phase with ATP, as required for the mobilization of starch.

Acknowledgements. The author is grateful to Mrs. D. MARONDE for her technical help during the preparation of the manuscript. Our own experimental work shown here has been supported by a grant from the Deutsche Forschungsgemeinschaft.

References

BAMBERGER, E.S., EHRLICH, B.A., GIBBS, M.: The D-glyceraldehyde-3-phosphate and 3-phosphoglycerate shuttle system and carbon dioxide assimilation in intact spinach chloroplasts. In: Proc. 3rd Intern. Congr. Photosynth. Res. M. AVRON (ed.), Amsterdam: Elsevier, 1975 pp. 1349–1362

BASSHAM, J.A., KIRK, M., JENSEN, R.G.: Photosynthesis by isolated chloroplasts. I. Diffusion of labelled photosynthetic intermediates between isolated chloroplasts and suspending medium. Biochim. Biophys. Acta **153**, 211–218 (1968)

HEBER, U.: Energy transfer within leaf cells. In: Proc. 3rd Inter. Congr. Photosynth. Res. M. AVRON (ed.), Amsterdam: Elsevier, 1975, pp. 1335–1348

HEBER, U., SANTARIUS, K.A.: Direct and indirect transfer of ATP and ADP across the chloroplast envelope. Z. Naturforsch. **25**b, 718–728 (1970)

HELDT, H.W.: Adenine nucleotide translocation in spinach chloroplasts. FEBS Lett. **5**, 11–14 (1969)

HELDT, H.W., FLIEGE, R., LEHNER, K., MILOVANCEV, M., WERDAN, K.: Metabolite movement and CO_2 fixation in spinach chloroplasts. In: Proc. 3rd Intern. Congr. Photosynth. Res. M. AVRON (ed.), Amsterdam: Elsevier, 1975, pp. 1369–1379

HELDT, H.W., RAPLEY, L.: Specific transport of inorganic phosphate, 3-phosphoglycerate and dihydroxyacetonephosphate across the inner membrane of spinach chloroplasts. FEBS Lett. **10**, 143–148 (1970)

HELDT, H.W., SAUER, F.: The inner membrane of the chloroplast envelope as the site of specific metabolite transport. Biochim. Biophys. Acta **234**, 83–91 (1971)

STOCKING, C.R., LARSON, S.: A chloroplast cytoplasmic shuttle and the reduction of extraplastid NAD. Biochem. Biophys. Res. Commun. **37**, 278–282 (1969)

STROTMANN, H., HELDT, H.W.: Phosphate containing metabolites participating in photosynthetic reactions of *Chlorella pyrenoidosa*. In: Progress in Photosynthesis Research. H. METZNER (ed.), Tübingen: Metzner, 1969, pp. 1131–1140

WERDAN, K.: Dissertation, Medizin. Fakultät, Universität München. Regulation des Stoffwechsels durch Änderungen des pH. Untersuchungen an Chloroplasten. 1975

WERDAN, K., HELDT, H.W.: The phosphate translocator of spinach chloroplasts. In: Proc. 2nd Intern. Congr. Photosynth. Res. G. FORTI, M. AVRON, A. MELANDRI (eds.). The Hague: A.W. Junk, 1972, pp. 1337–1344

WERDAN, K., HELDT, H.W., GELLER, G.: Accumulation of bicarbonate in intact chloroplasts following a pH gradient. Biochim. Biophys. Acta **283**, 430–441 (1972)

WERDAN, K., HELDT, H.W., MILOVANCEV, M.: The role of pH in the regulation of carbon fixation in the chloroplast stroma. Biochim. Biophys. Acta **396**, 276–292 (1975)

4. Compartmentation and Transport in C_4 Photosynthesis

M.D. HATCH and C.B. OSMOND

1. Introduction

The C_4 dicarboxylic acid pathway of photosynthetic carbon assimilation (the C_4 pathway) is a complex biochemical and physiological elaboration of the common photosynthetic carbon reduction cycle (PCR cycle, C_3 pathway; BASSHAM and CALVIN, 1962). We define the C_4 pathway as the complete reaction sequence in which CO_2 is transferred via the C-4 carboxyl of C_4 acids to the reactions of the PCR cycle and there reduced to the level of carbohydrate (HATCH and SLACK, 1970; HATCH et al., 1971; BLACK, 1973; HATCH, 1976a). The distinctive biochemical features of this process are the carboxylation and associated reactions leading to the synthesis of C_4 acids, and those concerned with the subsequent decarboxylation of these C_4 acids to supply CO_2 for the PCR cycle. Unlike the PCR cycle, in which carboxylation and carbon reduction is restricted to the chloroplast, the C_4 pathway involves the operation of reactions in cytoplasm, mitochondria, and chloroplasts, and the transport of intermediates between intracellular compartments. In this sense it may be compared with another well-established elaboration of the PCR cycle, the glycolate pathway of photorespiration (see Chap. II,5). However, an additional and distinctive feature of the C_4 pathway is the mandatory exchange of photosynthetic intermediates between adjacent cells. These exchanges constitute one of the most rapid and complex forms of symplastic transport known.

KARPILOV (1970) introduced the term 'cooperative photosynthesis' to describe the integrated metabolism of adjacent cells during photosynthetic CO_2 fixation in leaves of C_4 plants. Cooperative photosynthesis is a useful term, for, although the component partial processes can fix CO_2 in isolation, only when integrated do they comprise the C_4 pathway of photosynthesis. This review will therefore briefly examine the kinetic definition of the C_4 dicarboxylic acid pathway of photosynthesis, establish the location of component processes, and specify the inter- and intracellular transport phenomena implied by these reactions. These transport events will then be examined in relation to the limited data available, and discussed in the context of similar transport processes in other tissues. Finally, speculations on the physiological significance of these cooperative phenomena will be presented.

2. Radiotracer Kinetics and Compartments

The kinetics of radiotracer experiments provide the most unequivocal evidence for the operation of the C_4 pathway (KORTSCHAK et al., 1965; HATCH and SLACK,

Abbreviations: PEP=phosphoenolpyruvate; 3-PGA=3-phosphoglycerate; DHAP=dihydroxyacetone phosphate; RuBP=ribulose-1,5-bisphosphate.

1966; also see HATCH, 1976a), and also provide an indispensible framework for understanding the component processes. During steady state photosynthesis in air labeled with $^{14}CO_2$, the C_4 acids oxaloacetate, malate and aspartate appear both as initial products and as primary intermediates of the pathway. These characteristics are implied from the observation that in continuous feeding experiments the C_4 acids show a concave negative slope which extrapolates to approximately 100% of incorporated ^{14}C at zero time (Fig. 1a). Conversely, the curve for 3-PGA

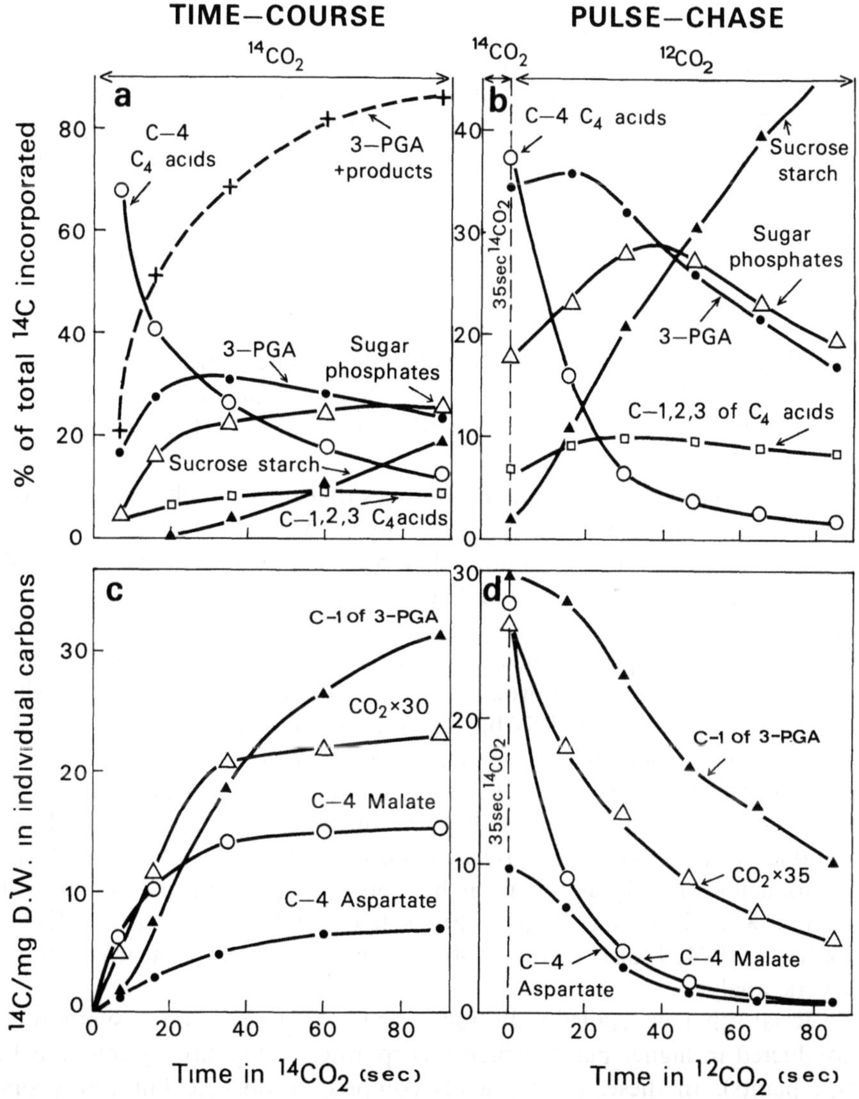

Fig. 1a–d. Radiotracer kinetic analysis of C_4 pathway metabolism (*Z. mays,* from HATCH, 1971a). (a) Time-course of $^{14}CO_2$ incorporation and (b) chase in $^{12}CO_2$ following 35 sec in $^{14}CO_2$ plotted as % of total ^{14}C fixed. (c) Labeling kinetics for specific carbon atoms during $^{14}CO_2$ incorporation and (d) a chase in $^{12}CO_2$ for the same experiments described in (a) and (b) respectively. Experiments were conducted under steady-state conditions at a light intensity of approx. 3×10^5 erg/cm^2/s. Addition of $^{14}CO_2$ increased the total CO_2 concentration from 330 ppm to 390 ppm

plus its products shows a convex-upwards slope extrapolating to approximately zero percent of incorporated ^{14}C at zero time. Incorporation of radioacitvity into 3-PGA and its products shows a substantial lag, whereas labeling of C_4 acids is linear from zero time. These kinetic characteristics are discussed in more detail elsewhere (HATCH, 1976b). Similar kinetic relationships are seen in pulse-chase experiments (Fig. 1b), but the extrapolated percentage of label in C_4 acids and 3-PGA depends on the duration of the exposure to $^{14}CO_2$.

When the labeling of specific carbon atoms of the C_4 acids and 3-PGA is considered, it is clear that the C-4 carboxyl of the C_4 acids attains constant specific activity more rapidly than the C-1 carboxyl of 3-PGA (Fig. 1c). In a chase the C-4 of C_4 acids initially loses label more quickly than C-1 of 3-PGA which shows a pronounced initial lag (Fig. 1d). Figures 1c and 1d show a free pool of $^{14}CO_2$ with labeling kinetics typical of an intermediate between the C-4 of C_4 acids and C-1 of 3-PGA. These data are consistent with the operation of the following metabolic sequence during C_4 photosynthesis:

$$\text{External } CO_2 \rightarrow \text{C-4 of } C_4 \text{ acids} \rightarrow \text{intermediate } CO_2 \text{ pool} \rightarrow \text{C-1 of 3-PGA} \rightarrow \text{products.}$$

Direct measurements of the intermediate CO_2 pool by radiotracer kinetic analysis (see Figs. 1c and 1d) have shown that it is much larger than expected by simple diffusion of CO_2 from air and that it only develops in the light (HATCH, 1971a).

The above sequence of reactions implies that the C_4 pathway is a series formulation of carboxylation reactions and defines the process in kinetic terms. Direct evidence from radiotracer studies that parallel fixation (direct access of external CO_2 to the PCR cycle) is negligible (JOHNSON and HATCH, 1969; GALMICHE, 1973; also see HATCH, 1976a) is further supported by time course $^{14}CO_2$ fixation data for several other species, showing that curves for C_4 acids and for 3-PGA plus products extrapolate to approximately 100% and 0%, respectively, when plotted as in Figure 1a (HATCH, 1976b). Suggestions that up to 15% of the carbon fixed by *Digitaria sanguinalis* may be incorporated directly by the PCR cycle (EDWARDS and BLACK, 1971) have not been substantiated by kinetic analyses. At 10,000 ppm CO_2, however, evidence of some direct carboxylation of RuDP has been obtained in *Pennisetum purpureum* (COOMBS et al., 1973a). Species alleged to show components of CO_2 assimilation by both the C_4 pathway and the PCR cycle (KHANNA and SINHA, 1973; SHOMER-ILAN and WAISEL, 1973; KENNEDY and LAETSCH, 1974; LABER et al., 1974) were not subjected to steady-state radiotracer kinetic analysis.

Relatively few detailed radiotracer kinetic analyses of photosynthesis have been conducted in higher plants which incorporate $^{14}CO_2$ directly into the PCR cycle (C_3 plants). In these, the C_4 acids comprise a variable but low proportion of the initial products and show kinetic responses very different from that described in Figure 1 (OSMOND et al., 1969; GALMICHE, 1973; OSMOND and ALLAWAY, 1974).

Changes in pool sizes of intermediates during CO_2 and light-dark transients also support the metabolic sequence illustrated above. In leaves of *Zea mays* substantial lags occur before 3-PGA and RuBP pools respond to transfer to CO_2 free air following steady-state photosynthesis (FARINEAU, 1971). These experiments imply a substantial internal source of CO_2 for the PCR cycle and demonstrate the product-precursor relationship between aspartate and PEP. When leaves are maintained in CO_2 free air and then exposed to CO_2, changes in the 3-PGA and RuBP pools are not substantially different from those of C_3 plants (FARINEAU, 1971; LATZKO et al., 1971; LABER et al., 1974). However, these experiments cannot distinguish the source of CO_2 fixed by the PCR cycle or the behavior of the C_4 acid pools involved in photosynthesis.

One of several features invariably correlated with C_4 pathway labeling kinetics is an unusual leaf anatomy. The leaves of C_4 plants show a high degree of structural differentiation and organization of chloroplast-containing cells, in marked contrast to the more-or-less uniform mesophyll cells of C_3 plants. HABERLANDT (1884) described the 'sheath cells' surrounding vascular elements and the adjacent 'girdle cells' as characteristic features of leaves now known to be capable of C_4 photosynthesis. This arrangement, termed Kranz anatomy, may take many forms (LAETSCH, 1971, 1974) and is difficult to define in precise anatomical terms. In the Gramineae the most common arrangement is that of two adjacent concentric hollow cylinders of cells, now described as inner bundle sheath and outer mesophyll. In the Cyperaceae the mesophyll cells are frequently separated from the bundle sheath by a layer of cells without chloroplasts (LAETSCH, 1971). Dicotyledons may have more complex arrangements, but in all of these mesophyll and bundle sheath cells are arranged adjacent to each other (BJÖRKMAN et al., 1973).

The Kranz complex appears to embody an essential structural component of the C_4 pathway. HABERLANDT (1884) speculated on the possibility of 'division of labor' between the mesophyll and bundle sheath cells in the Kranz complex. These speculations were reiterated in the studies of RHOADES and CARVALHO (1944), particularly in relation to sites of starch synthesis, and in correlations of features associated with the C_4 pathway (DOWNTON and TREGUNNA, 1968; LAETSCH, 1969). Early experimental support for the intercellular compartmentation of C_4 pathway reactions came from the differential grinding experiments of BJÖRKMAN and GAUHL (1969) and the nonaqueous separation of chloroplasts from mesophyll and bundle sheath cells (SLACK et al., 1969). A wealth of evidence now supports the view that the mesophyll cells contain enzymes of C_4 acid synthesis and the bundle sheath cells the enzymes of C_4 acid decarboxylation and of the PCR cycle (Sect. 3). An intriguing qualitative confirmation of this separation of the two carboxylases is found in experiments in which $^{14}CO_2$ was supplied via the stomata and $H^{14}CO_3^-$ solution via the vascular system of *Sorghum* and *Amaranthus* species. C_4 acids were the predominant products of 3 sec $^{14}CO_2$, fixation, while 3-PGA and phosphorylated compounds were the principal products after 3 sec feeding of $H^{14}CO_3^-$ to the vascular system (FARINEAU, 1972).

Wide surveys have been made correlating the carbon isotope discrimination ratios characteristic of C_4 plants with leaf anatomy (SMITH and BROWN, 1973). These surveys, based on many hundreds of species, indicate that all plants capable of C_4 photosynthesis have the Kranz complex. The possibility that the undifferentiated callus of *Froelichia gracillus* is an exception (LAETSCH, 1974) must await publication of radiotracer kinetics. On the other hand, some hybrids of the cross between *Atriplex patula* ssp. *hastata* (C_3) × *Atriplex rosea* (C_4) appear to have a normal Kranz complex and incorporate substantial amounts of $^{14}CO_2$ into C_4 acids during photosynthesis. However, these hybrids do not show tracer kinetics of the form shown in Figure 1 and do not display other characteristics of the C_4 pathway (BJÖRKMAN et al., 1971). These studies serve as a reminder that the relationship between the kinetics and compartments of wild-type C_4 plants is an exceedingly complex affair. The division of labor envisaged by HABERLANDT (1884) now appears to involve cooperativity (KARPILOV, 1970) at practically all levels of cellular activity.

3. Inter- and Intracellular Compartmentation of Reactions

Further definition of the C_4 pathway beyond that provided by radiotracer studies, requires information about the enzymes involved and the distribution of component reactions between and within mesophyll and bundle sheath cells. Such data, together with comments on methods, are considered in this section.

3.1 Methods

The key technical problem in these studies is to distinguish between the enzymes or organelles derived from the two types of photosynthetic cells. One early approach to this problem involved density fractionation of leaf extracts in a nonaqueous medium (SLACK, 1969; SLACK et al., 1969). For maize in particular, this method provided almost complete separation of mesophyll and bundle sheath chloroplasts and yielded information about enzymes associated with each chloroplast type, as well as their cellular origin. However, with some other species the chloroplasts were not as definitively separated as in maize. Another approach, initiated by BJÖRKMAN and GAUHL (1969), depended upon the fact that mesophyll cells are generally relatively fragile and loosely attached to bundle sheath cells, while the latter cells are highly resistant to breakage and firmly attached to vascular strands. These authors achieved a large degree of fractionation of mesophyll and bundle sheath cell contents by differential grinding of leaf tissue. Unfortunately, it also turned out that this simple graded mechanical extraction of leaves provides only a partial separation of mesophyll and bundle sheath cell contents with many C_4 plants, a fact resulting in considerable confusion and conflicting data in the literature (see Sect. 3.3). Results with this procedure also vary with leaf age and seasonal factors (probably depending on whether the epidermis is readily removed), so that the method is a highly empirical one. With those species in which mesophyll cells are readily broken by mild blending, this procedure has proved particularly useful for preparing bundle sheath cell strands with well-preserved metabolic activities (KAGAWA and HATCH, 1974b and unpublished results).

A basically similar graded mechanical treatment of leaves has provided intact mesophyll cells and bundle sheath cell strands from *Digitaria* species (EDWARDS and BLACK, 1971; KANAI and BLACK, 1972). While providing much valuable information for *Digitaria* species, poor yields of intact mesophyll cells are obtained from other species. Recently, a more versatile procedure for cell separation, depending on pretreatment of leaf tissue slices with cellulase and pectinase (CHEN et al., 1973; KANAI and EDWARDS, 1973a), has been described. This procedure yields either mesophyll protoplasts after lengthy enzyme treatment (KANAI and EDWARDS, 1973a, b), or mesophyll cells after brief incubation followed by mild mechanical blending (CHEN et al., 1973). In each case bundle sheath cells remain as strands attached to vascular tissue. Very brief treatment of tissue with these enzymes, in some instances with a pH of about 6.8 instead of 5.0–5.5, allows the successful application of graded mechanical blending for separating cell contents (HATCH et al., 1975).

3.2 Activities of Isolated Mesophyll and Bundle Sheath Cells

As will be further discussed in the following Section (3.3), analysis of the enzyme content of separated mesophyll and bundle sheath cells has made a vital contribution towards understanding the intercellular partitioning of photosynthetic processes in C_4 plants. However, the demonstration of component processes in isolated cells, at rates commensurate with those of intact leaves, has been slow in forthcoming. The metabolic capacities of isolated cells reported to date are summarized in Table 1.

In earlier studies, low rates of assimilation of ^{14}C were observed with both mesophyll and bundle sheath cells when $H^{14}CO_3^-$ was added alone. With bundle sheath cells of several species $^{14}CO_2$ fixation was increased many-fold by adding ribose-5-P, and further still by adding ADP or ATP (EDWARDS and BLACK, 1971; CHOLLET and OGREN, 1973; HUBER et al., 1973; CHEN et al., 1974). Under these conditions 3-PGA was the major product but activity was only partially light-dependent, particularly when ADP or ATP were also added. In view of the evidence for the low permeability of isolated chloroplasts to sugar phosphates, and particularly to adenylates (HEBER, 1974), these requirements raise some doubts about the integrity of these cells. However, rapid light-dependent incorporation of $^{14}CO_2$ into PCR cycle intermediates, as well as sucrose and starch, have now been reported for *Atriplex spongiosa* bundle sheath cells provided with $H^{14}CO_3^-$ alone (KAGAWA and HATCH, 1974b), and they also evolve O_2 rapidly in response to HCO_3^-. Similar responses to adding only HCO_3^- have now been observed with *Panicum miliaceum* and *Panicum maximum* (HATCH and KAGAWA, unpublished), and cells from several other species have been shown to assimilate CO_2 at reasonable rates without addition of other metabolites (GUTIERREZ et al., 1974b). These results are consistent with the PCR cycle being located in bundle sheath cells.

With mesophyll cells, CO_2 fixation is not stimulated by adding ribose-5-P, with or without ADP, supporting the view that they lack the PCR cycle (Table 1). However, these cells rapidly fix CO_2 into C_4 acids when PEP is provided in the light or the dark (EDWARDS and BLACK, 1971; GUTIERREZ et al., 1974b; HUBER and EDWARDS, 1975). Mesophyll cells from *D. sanguinalis* show light-dependent fixation of CO_2 when pyruvate is added (HUBER and EDWARDS, 1975), and light-dependent evolution of O_2 when either oxaloacetate or PEP plus HCO_3^- is added (SALIN et al., 1973). These results are consistent with the primary assimilation of CO_2 into C_4 acids occurring in mesophyll cells.

D. sanguinalis bundle sheath cells were shown to decarboxylate malate in the dark, but only when NADP and Mg^{2+} were also provided (HUBER et al., 1973).

Table 1. Photosynthetic carbon metabolism of isolated mesophyll and bundle sheath cells

Cell type and activity	Additions	Species	Comments	Rate (μmol/min/mg chlorophyll) Light	Dark	References[a]
Mesophyll cells						
$^{14}CO_2$ fixation	$H^{14}CO_3^-$ ± ribose-P, ADP	Several spp.	Negligible activity	—	—	1, 6, 7, 9
	Pyruvate + oxalo-acetate	*D. sanguinalis*	Light-dependent	0.9	0	7
	PEP	Several spp.	High rate, light or dark	7–17	7–17	1, 6, 7, 9
O_2 evolution	Oxalacetate or PEP + HCO_3^-	*Digitaria* sp.	Light-dependent	2.3–3	0	4
Bundle sheath cells						
$^{14}CO_2$ fixation	$H^{14}CO_3^-$ only	*A. spongiosa*	To PCR cycle products	2.3	0	10
	$H^{14}CO_3^-$ only	Several species	Products not determined	0.5–1.5	0	2, 9
	Ribose-P, ADP	*D. sanguinalis*	{Endogenous activity low, 3-PGA major product, only partially light-dependent with ADP, variety of PCR cycle products with *C. gayana*}	{1.1 / 2–2.8}	{0.4 / —}	1 / 5
	Ribose-P, ADP	*Cyperus, rotundus*		1.4	1.0	6
	Ribose-P	*Z. mays*		0.5–1.1	<0.1	3, 11
	Ribose-P	*C. gayana*		1.16	0	11
	PEP	Several spp.	Low versus mesophyll	—	—	1, 6, 9
O_2 evolution	HCO_3^- or C_4 acids	*A. spongiosa*	Dependent on additions	1.9–2.0	0	10
C_4 acid decarb-oxylation	Malate	*A. spongiosa*	Aspartate, 2-oxoglutarate also required	2.3	1.4	10
	Malate, NADP, Mg^{2+}	*D. sanguinalis*	NADP and Mg^{2+} essential	5	5	5
C-4 of C_4 acids to PCR cycle	Malate	*A. spongiosa*	{Fixation following C_4 acid decarboxylation}	1.6	0	10
		Z. mays		0.9	0	11

[a] References: (1) EDWARDS and BLACK (1971); (2) EDWARDS and GUTIERREZ (1972); (3) CHOLLET and OGREN (1973); (4) SALIN et al. (1973); (5) HUBER et al. (1973); (6) CHEN et al. (1974); (7) HUBER and EDWARDS (1975); (8) GUTIERREZ et al. (1975); (9) GUTIERREZ et al. (1974b); (10) KAGAWA and HATCH (1974b); (11) HATCH and KAGAWA, Arch. Biochem. Biophys., in press.

However, *Z. mays* cells catalyze a rapid light-dependent decarboxylation of malate requiring only HCO_3^- or 3-PGA (HATCH and KAGAWA, unpublished results). Rapid decarboxylation of supplied malate or aspartate occurs with bundle sheath cells from *A. spongiosa,* and in the light the CO_2 released is incorporated into PCR cycle intermediates and products (KAGAWA and HATCH, 1974b). These results support the view, originally developed from studies on enzyme localization (see Sect. 3.3), that C_4 acids derived from mesophyll cells are decarboxylated in bundle sheath cells.

3.3 C_4 Pathway Enzymes and Their Intercellular Distribution

Initially at least, progress in the elucidation of the C_4 pathway depended upon interpretations of radiotracer data combined with studies on the identification of enzymes capable of catalysing the reactions predicted from the labeling data. Involvement of enzymes in the C_4 pathway was assessed on the basis of several criteria. High or unique activity of appropriate enzymes in C_4 plants relative to C_3 plants was usually the first feature noted. Other criteria for photosynthetic activity of some of these enzymes included light-dark regulation of activity, specific intercellular location, and association with chloroplasts (see references in Table 2; HATCH, 1976a). In addition, like other enzymes involved in photosynthesis, the level of these enzymes is low in leaves of dark-grown plants but increases many-fold following illumination and greening of leaves (HATCH et al., 1969; GRAHAM et al., 1970; JOHNSON and HATCH, 1970; HATCH and MAU, 1973; HATCH and KAGAWA, 1974b).

Table 2 summarizes data on the activity and location of enzymes implicated in photosynthesis in C_4 plants. This, and other information on properties and regulation of these enzymes, is contained in the references cited in Table 2 and has been discussed in detail elsewhere (HATCH, 1976a). It will be useful for the following discussion to introduce here a simplified scheme showing the reactions of C_4 photosynthesis (Fig. 2), based on radiotracer kinetic data and evidence relating to the intercellular partitioning of reactions summarized in Tables 1 and 2. More detailed schemes will be introduced in Sect. 3.5. The scheme in Figure 2 proposes that CO_2 is initially fixed in mesophyll cells with the initial product, oxaloacetate, being converted into malate and aspartate. Either malate or aspartate is then transferred to bundle sheath cells, where the C-4 carboxyl is released as CO_2 via one of three alternative decarboxylating systems (see below) and refixed into the PCR cycle through RuBP carboxylase. The C_3 compound remaining after C-4 decarboxylation is returned to the mesophyll cells to provide the precursor for PEP formation. Details of these latter reactions, and the basis for separating C_4 plants according to the mechanisms utilized for C_4 acid decarboxylation in bundle sheath cells, will be considered later (see Table 2 and Fig. 5).

The assumption of a discrete compartmentation of PEP carboxylase and hence C_4 acid formation in mesophyll cells, and of RuBP carboxylase and associated PCR cycle enzymes in bundle sheath cells (Fig. 2), has been the subject of considerable controversy. Early data supporting such a conclusion was provided by nonaqueous fractionation of chloroplasts (SLACK et al., 1969), graded mechanical extraction of leaves (BJÖRKMAN and GAUHL, 1969; BERRY et al., 1970; ANDREWS et al., 1971; OSMOND and HARRIS, 1971) and analysis of separated mesophyll and bundle sheath cells (EDWARDS and BLACK, 1971; EDWARDS et al., 1971; KANAI

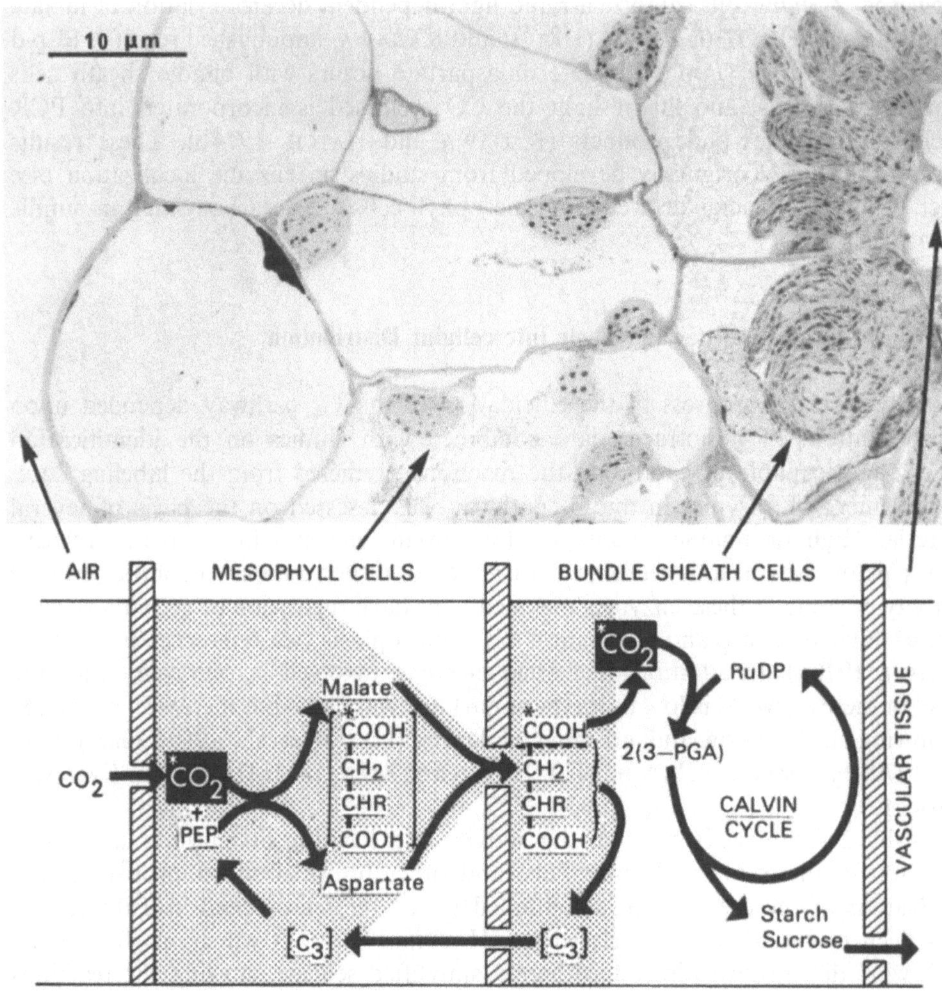

Fig. 2. A simplified scheme for C_4 pathway photosynthesis showing the basic reactions and their intercellular location. The shaded reactions are those unique to C_4 photosynthesis. The accompanying electron micrograph shows mesophyll and bundle sheath cells and their relationship to the epidermis and vascular tissue (*P. miliaceum*, kindly provided by S. CRAIG)

and BLACK, 1972). It was generally recognized that these procedures did not provide absolute separation of cells or their contents. However, it could be reasonably deduced from these data that at least 90–95% of the total leaf PEP carboxylase activity was located in mesophyll cells and that a similar proportion of RuBP carboxylase was located in bundle sheath cells.

Meanwhile, other studies were interpreted to indicate that PEP carboxylase was located in either buliform or mesophyll cells, that RuBP carboxylase and other PCR cycle enzymes were also located in mesophyll cells, and that bundle sheath cells operated largely or solely for starch synthesis (BALDRY et al., 1971; BUCKE, and LONG, 1971; COOMBS and BALDRY, 1971; COOMBS et al., 1973a). Another interpretation was that RuBP carboxylase was about equally distributed between the two cell types (POINCELOT, 1972). Significantly, all these studies were based solely on data provided by the graded mechanical extraction of leaves (see Sect. 3.1). These interpretations have been discussed elsewhere in several articles (BLACK, 1973; HATCH, 1976a) and papers (CHEN et al., 1973; HATCH and KAGAWA, 1973; KU et al., 1974a; also others quoted below), in which they have been specifically and completely refuted. We would simply reiterate (BLACK, 1973; HATCH and MAU, 1973; HATCH and KAGAWA, 1973) that graded mechanical extraction is an empirical procedure which must be interpreted with an appreciation of its limitations, and that its effectiveness for providing separation of cell contents varies widely with different species.

Data now available for the photosynthetic capabilities and enzyme content of mesophyll and bundle sheath cells separated from a large number of species (KANAI and EDWARDS, 1973a, b; CHEN et al., 1974; KU et al., 1974a; GUTIERREZ et al., 1975) confirm the essentially absolute partitioning of PEP carboxylase in mesophyll cells and of RuBP carboxylase, together with other PCR cycle enzymes, in bundle sheath cells (Sect. 3.2). There is also clear evidence that mesophyll chloroplasts, active in several light-dependent transformations of carbon compounds, totally lack RuBP carboxylase and PCR cycle activity (HATCH and KAGAWA, 1973; KAGAWA and HATCH, 1974a).

Notable exceptions to this specific compartmentation of PCR cycle enzymes are those enzymes responsible for converting 3-PGA to triose phosphates (3-PGA-kinase, NADP glyceraldehyde-3-P dehydrogenase, and triose-P isomerase). These enzymes are about equally distributed between mesophyll and bundle sheath chloroplasts isolated in nonaqueous medium (SLACK et al., 1969), and there is evidence for their occurrence or operation in both mesophyll and bundle sheath cells (EDWARDS and BLACK, 1971; HATCH and KAGAWA, 1973; CHEN et al., 1974; KAGAWA and HATCH, 1974a, b; KU et al., 1974a). The possible significance of this distribution will be considered later (Sect. 3.5 and 4).

Accepting the compartmentation of carboxylases proposed in Figure 2, the role and location of the enzymes responsible for the other reactions in this scheme can be considered. Radiotracer studies (see Sect. 2) indicate that oxaloacetate, formed via PEP carboxylase in mesophyll cells, is rapidly converted into large pools of malate and aspartate. These transformations are attributed to NADP malate dehydrogenase and aspartate aminotransferase, respectively, both being freely reversible reactions. In this interconvertible pool of C_4 acids malate and aspartate predominate, but vary between species in terms of their relative labeling after short periods in $^{14}CO_2$ (DOWNTON, 1970), and also with respect to which acid appears from radiotracer kinetics to be directly involved in transfer of C-4 label to other products (CHEN et al., 1971; HATCH, 1971a). In fact, it is now clear that the C_4 acid moving to bundle sheath cells, and the metabolic steps leading to its decarboxylation, differ among C_4 plants. These differences provide the basis for the grouping of C_4 plants shown in Table 2. The nomenclature used to define these subgroups, 'NADP-ME-type', 'PCK-type' and 'NAD-ME-type', is that suggested by HATCH et al. (1975) and is similar to that used by GUTIERREZ et al. (1974a).

For one group, the NADP-ME-type, malate derived from mesophyll cells is decarboxylated in bundle sheath cells via a chloroplast-located NADP malic enzyme. Besides containing very high NADP malic enzyme activity, this group also contains much more NADP malate dehydrogenase than other groups of C_4 or C_3 plants (Table 2). Other C_4 plants lack high activities of these enzymes but are noted for their exceptionally high aspartate and alanine aminotransferase activities (ANDREWS et al., 1971; HATCH and MAU, 1973). These activities are distributed about equally between mesophyll and bundle sheath cells and occur as quite distinct and separate isoenzymes in each cell type (HATCH and MAU, 1973; HATCH, 1973).

Table 2. Activity and location of C_4 pathway enzymes in sub-groups of C_4 species[a]. Activity unit = 1 µmol/min. Abbreviations: M, mesophyll cell; BS, bundle sheath cell; cyto, cytoplasm; chloro, chloroplast; mito, mitochondria

Enzymes with similar activities in all C_4 species

Enzymes	Activity (units/mg chl.)	Location		Activity in C_3 species (units/mg chl.)	References to identification activity, properties and location[b]
		Cell type	Within cell		
PEP carboxylase	15–40	M	cyto	0.4–1.5	1, 2, 3, 4, 5, 6, 7
Pyruvate, Pi dikinase	3–10	M	chloro	0	8, 9, 10, 7, 11, 12, 28
Adenylate kinase	17–45	>M	chloro in M	0.5–1.0	13, 10, 7
Pyrophosphatase	20–60	>M	chloro in M	2–4	13, 10, 7
3-PGA to triose-*P* enzyme	similar to C_3	both	chloro	—	10, 14, 22, 7, 11
Other PCR cycle enzymes	similar to C_3	BS	chloro	—	10, 2, 14, 22, 7, 4, 6

Enzymes with varying activity in sub-groups of C_4 species[c]

Enzymes	'NADP-ME-type' species			'PCK-type' species			'NAD-ME-type' species			Activity in C_3 species (units/ mg chl.)	References to identification activity, properties and localisation[b]
	Activity (units/ mg chl.)	Location		Activity (units/ mg chl.)	Location		Activity (units/ mg chl.)	Location			
		Cell type	Within cell		Cell type	Within cell		Cell type	Within cell		
NADP malate dehydrogenase	10–17	M	chloro	1–3	—	—	1–2	both	chloro	0.5–1.2	15, 10, 16, 7, 11, 19
NADP malic enzyme	9–14	BS	chloro	0.3–0.4	—	—	0.2–0.8	—	—	0.1–0.8	1, 10, 17, 16, 18, 4, 19
PEP carboxykinase	<0.2	—	—	10–14	BS	?	<0.2	—	—	<0.2	21, 22, 6, 19, 23, 12
NAD malic enzyme	0.2–0.4	—	—	0.2–0.5	—	—	5–9	BS	mito	0.05–0.1	24, 25, 26, 12, 23, 27
Aspartate aminotransferase	5–7	>M	chloro	45–60	both	cyto?	25–45	M BS	cyto mito	1–2.4	10, 18, 20, 12, 19
Alanine aminotransferase	2–4	—	—	38–45	both	cyto?	30–60	both	cyto	2–3	18, 20, 12, 19

For footnotes see opposite page.

[a] For the enzymes common to all C_4 species the ranges of activity are for a large selection of species. Ranges of activity are also shown for 'NADP-ME-type' C_4 species (*Z. mays, Sorghum sudanense, Pennisetum typhoides, D. sanguinalis* and *Saccharum officinarum*), 'PCK-type' C_4 species (*P. maximum, Sporobolus fimbriatus, Chloris gayana*), 'NAD-ME-type' C_4 species (*A. spongiosa, Portulaca oleracea, A. edulis, P. miliaceum, Eragrostis curvula*), and C_3 plants (wheat, spinach, pea, soybean and *Atriplex patula*). In some instances low values reported in the literature have been excluded, where higher activities have been reported elsewhere for the same species. For comparison, maximum photosynthesis rates for C_4 species range between 3 and 5 µmol/min/mg chlorophyll. The prefix (>) indicates the predominant location, 'Both' indicates approximately equal distribution between two cell types (see text).

[b] References in Table are: (1) SLACK and HATCH (1967); (2) BJÖRKMAN and GAUHL (1969); (3) EDWARDS and BLACK (1971); (4) CHEN et al. (1973); (5) TING and OSMOND (1973); (6) GUTIERREZ et al. (1974b); (7) HATCH and KAGAWA (1973); (8) HATCH and SLACK (1968); (9) ANDREWS and HATCH (1969); (10) SLACK et al. (1969); (11) KAGAWA and HATCH (1974a); (12) HATCH et al. (1975); (13) HATCH et al. (1969); (14) see HATCH and SLACK (1970); (15) HATCH and SLACK (1969b); (16) JOHNSON and HATCH (1970); (17) BERRY et al. (1970); (18) ANDREWS et al. (1971); (19) GUTIERREZ et al. (1975); (20) HATCH and MAU (1973); HATCH (1973); (21) EDWARDS et al. (1971); (22) KANAI and BLACK (1972); (23) GUTIERREZ et al. (1974a); (24) HATCH and KAGAWA (1974a); (25) HATCH and KAGAWA (1974b); (26) HATCH et al. (1974); (27) KAGAWA and HATCH (1975); (28) SUGIYAMA (1973).

[c] See HATCH et al. (1975) for the sub-grouping terminology used.

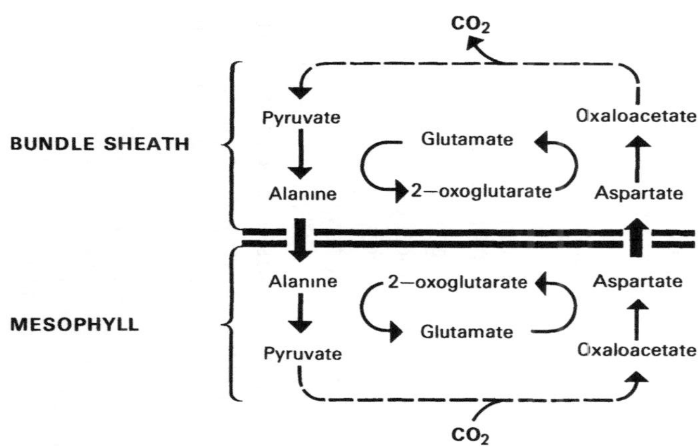

FIG. 3. Aminotransferase-catalyzed transformations in species that transfer aspartate to bundle sheath cells and return alanine to mesophyll cells. Note the interdependency of the reactions in each cell type through the cycling of glutamate and 2-oxoglutarate

The proposal that these enzymes operate to support an aspartate-alanine interchange between mesophyll and bundle sheath cells by the mechanism shown in Figure 3 is supported by further evidence discussed below.

For some of the species in this second group, the C_4 acid decarboxylating function is accounted for by PEP carboxykinase (Table 2, PCK-type). Both the activity of this enzyme and its location are appropriate for supporting the C_4 acid decarboxylation process shown in Figure 2 (EDWARDS et al., 1971; GUTIERREZ et al., 1974b; HATCH et al., 1975). For these species it is proposed that aspartate derived from mesophyll cells is converted to oxaloacetate in bundle sheath cells (see Fig. 3) and then decarboxylated via PEP carboxykinase to PEP plus CO_2. As will be discussed later, there are sound reasons for proposing that PEP is converted to alanine via pyruvate, and that alanine is returned to mesophyll cells (shown in Fig. 3). However, to date, an appropriate enzyme to account for the conversion of PEP to pyruvate in bundle sheath cells has not been found.

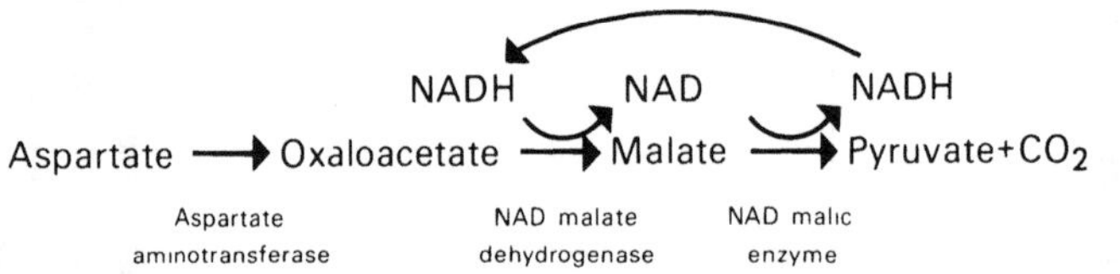

Fig. 4. Reaction sequence for aspartate decarboxylation in the bundle sheath cells of NAD-ME-type species. These reactions all occur in mitochondria

There remains a third group of C_4 plants containing high aminotransferase activities but lacking significant activities of the C_4 acid decarboxylases already discussed. This group (designated NAD-ME-type in Table 2) has now been found to contain uniquely high NAD-malic enzyme activity also confined to bundle sheath cells (HATCH and KAGAWA, 1974a, b). It is proposed that aspartate derived from mesophyll cells is metabolized via the sequence of reactions shown in Figure 4, all of which are located in the mitochondria of bundle sheath cells (see Sect. 3.4). Pyruvate is then converted via alanine aminotransferase to alanine which is then returned to the mesophyll cells (see Fig. 3).

To sustain the cycle shown in Figure 2 the C_3 compound formed by C_4 acid decarboxylation must be returned to mesophyll cells to regenerate PEP. The reaction leading to PEP formation is catalysed by pyruvate,P_i dikinase, an enzyme common to all groups of C_4 species but absent from C_3 species. This enzyme is confined to mesophyll chloroplasts (SLACK et al., 1969; HATCH and KAGAWA, 1973) and catalyses the reaction of pyruvate with ATP and P_i to give PEP plus AMP and pyrophosphate (HATCH and SLACK, 1968; ANDREWS and HATCH, 1969). The enzyme is difficult to isolate in its fully active form (HATCH and SLACK, 1969a) but occurs with adequate activity in species from all the different sub-groups of C_4 plants (HATCH et al., 1975). The mesophyll chloroplasts contain high activities of adenylate kinase and pyrophosphatase which would serve as part of a sequence of reactions to convert AMP and pyrophosphate, respectively, back to ATP. For NADP-ME-type species pyruvate,P_i dikinase is provided directly with pyruvate returned from bundle sheath cells. For the other two groups the conversion of alanine to pyruvate via alanine aminotransferase (see Fig. 3) would be a prerequisite for PEP formation.

The intercellular distribution of enzymes concerned with synthesis of sucrose and starch may also be significant in terms of metabolite transport between mesophyll and bundle sheath cells. The fact that the PCR cycle is confined to bundle sheath cells would suggest that these cells are also the most likely sites for sucrose and starch synthesis during photosynthesis. Certainly, the preferential synthesis of starch in bundle sheath cells is a widely observed phenomenon in C_4 plants, with starch appearing in mesophyll chloroplasts only after heavy deposition in bundle sheath chloroplasts (DOWNTON, 1971a; LAETSCH, 1974). Starch and sucrose synthesizing enzymes apparently occur in both cells types. However, starch synthase is located predominantly in bundle sheath cells, whereas the enzymes involved in sucrose synthesis are either equally distributed between cells or are more active in mesophyll cells (DOWNTON and HAWKER, 1973; DE FEKETE and VIEWEG, 1973; VIEWEG and DE FEKETE, 1973; CHEN et al., 1974; BUCKE and OLIVER, 1975). Such a distribution could imply transport of precursors between the two types of photosynthetic cells at least under some conditions.

3.4 Intracellular Location of Enzymes

Information on the intracellular location of C_4-pathway enzymes is incomplete (see Table 2). Of those enzymes located in meophyll cells, pyruvate, P_i dikinase, most of the mesophyll component of the ancillary enzymes adenylate kinase and pyrophosphatase, and the NADP malate dehydrogenase of NADP-ME-type species, are associated with chloroplasts isolated in non-aqueous media (SLACK et al., 1969) and aqueous osmotic media (HATCH and KAGAWA, 1973; KAGAWA and HATCH, 1974a; GUTIERREZ et al., 1975). Most of the aspartate aminotransferase of NADP-ME-type species is located in mesophyll cells and this is largely located in chloroplasts (SLACK et al., 1969; HATCH and KAGAWA, 1973; GUTIERREZ et al., 1975). However, for other groups (PCK-type and NAD-ME-type), typified by their high aminotransferase activities, the major aspartate and alanine aminotransferase iso-enzymes of mesophyll cells are not associated with chloroplasts or mitochondria (HATCH and MAU, 1973).

The photosynthetic enzymes concerned with transforming 3-PGA to triose phosphates (3-PGA kinase, NADP glyceraldehyde-3-P dehydrogenase, and triose-P isomerase) are present in both mesophyll and bundle sheath chloroplasts of C_4 plants (SLACK et al., 1969; HATCH and KAGAWA, 1973; GUTIERREZ et al., 1975). However, mesophyll chloroplasts lack RuBP carboxylase and other PCR cycle enzymes and will not fix CO_2 or evolve O_2 when HCO_3^- is added (SLACK et al., 1969; HATCH and KAGAWA, 1973; KAGAWA and HATCH, 1974a); they will catalyse the light-dependent conversion of pyruvate to PEP, oxaloacetate to malate, and 3-PGA to DHAP (KAGAWA and HATCH, 1974a). PEP carboxylase is not associated with intact and functionally active mesophyll chloroplasts (HATCH and KAGAWA, 1973; GUTIERREZ et al., 1975) or preparations of intact chloroplasts from whole leaves (O'NEAL et al., 1972), and is apparently a cytoplasmic enzyme.

Bundle sheath chloroplasts of *Z. mays* (NADP-ME-type species) separated in non-aqueous medium contain RuBP carboxylase, other PCR cycle enzyme, and NADP malic enzyme (SLACK et al., 1969). To date, there have been no reports of the isolation of intact chloroplasts from bundle sheath cells, possibly due to the vigorous cell-breakage procedures required to release them. Consequently, direct confirmation of their predicted functions has not been possible. However, chloroplast preparations capable of PCR cycle photosynthesis have been obtained from immature primary leaves of *Z. mays* (O'NEAL et al., 1972), and these probably included a substantial proportion of plastids from bundle sheath cells which may be more readily broken in this particular tissue.

For PCK-type C_4 plants the intracellular location of PEP carboxykinase and the bundle sheath cell components of aspartate and alanine aminotransferases is uncertain. However, these enzymes are apparently not associated with mitochondria (HATCH et al., 1975). This is significant, because the enzymes concerned with transforming and decarboxylating C_4 acids in the bundle sheath cells of NAD-ME-type species (see Fig. 4) are located in mitochondria (HATCH and KAGAWA, 1974a, b; KAGAWA and HATCH, 1975), and isolated mitochondria from these cells rapidly decarboxylate added C_4 acids (KAGAWA and HATCH, 1975). Since the alanine aminotransferase is not located in mitochondria, the pyruvate formed in mitochondria is apparently transported to the cytoplasm for conversion to alanine.

3.5 Detailed Schemes for C_4 Photosynthesis

On the basis of the evidence discussed in the preceding Sections the general scheme for C_4 photosynthesis in Figure 2 can be replaced by more detailed formulations (Fig. 5). These define both the inter- and intracellular location of particular reactions and hence the metabolite transport processes demanded by this partitioning of processes. Enzymes responsible for particular reactions are listed in the legend. Reactions operative in mesophyll cells are largely common to all C_4 plants with the exceptions noted below. The basic role of mesophyll cells is to assimilate CO_2 into C_4 acids for transport to bundle sheath cells. While CO_2 is assimilated in the cytoplasm via PEP carboxylase, PEP is provided from pyruvate via reactions occurring in chloroplasts.

The major route of net utilization of oxaloacetate varies in different groups of C_4 plants. In NADP-ME-type C_4 plants oxaloacetate is reduced to malate in chloroplasts via NADP malate dehydrogenase, and the malate then moves to bundle sheath cells and is decarboxylated leaving pyruvate to be returned (Fig. 5). In other groups, aspartate replaces malate as the C_4 acid moving to bundle sheath cells, and alanine is apparently the C_3 compound returned to mesophyll cells. It appears that aspartate transfer occurs when the provision of oxaloacetate is a prerequisite for bundle sheath decarboxylating systems. During this process, the operation of aminotransferases catalysing aspartate-oxaloacetate and pyruvate-alanine interconversions is linked in each cell type by cycling of glutamate and 2-oxoglutarate (see also Fig. 3). In

Fig. 5. Schemes outlining the reactions of C_4 pathway photosynthesis in mesophyll (*lower*) ▷ and bundle sheath (*upper*) cells and the intracellular location of these reactions. Separate schemes for bundle sheath cells show the different C_4 acid decarboxylation mechanisms for NADP-ME-type, PCK-type and NAD-ME-type species (*see text*). The malate-pyruvate shuttle from mesophyll cells applies to NADP-type species and the aspartate-alanine shuttle to PCK-type and NAD-ME-type species. The enzymes involved are: *1* PEP carboxylase; *2* NADP malate dehydrogenase; *3* aspartate aminotransferase; *4* alanine aminotransferase; *5* pyruvate, Pi dikinase; *6* adenylate kinase; *7* pyrophosphatase; *8* 3-PGA kinase, NADP glyceraldehyde-3-P dehydrogenase and triose-P isomerase; *9* NADP malic enzyme; *10* PEP carboxykinase; *11* NAD malate dehydrogenase; *12* NAD malic enzyme

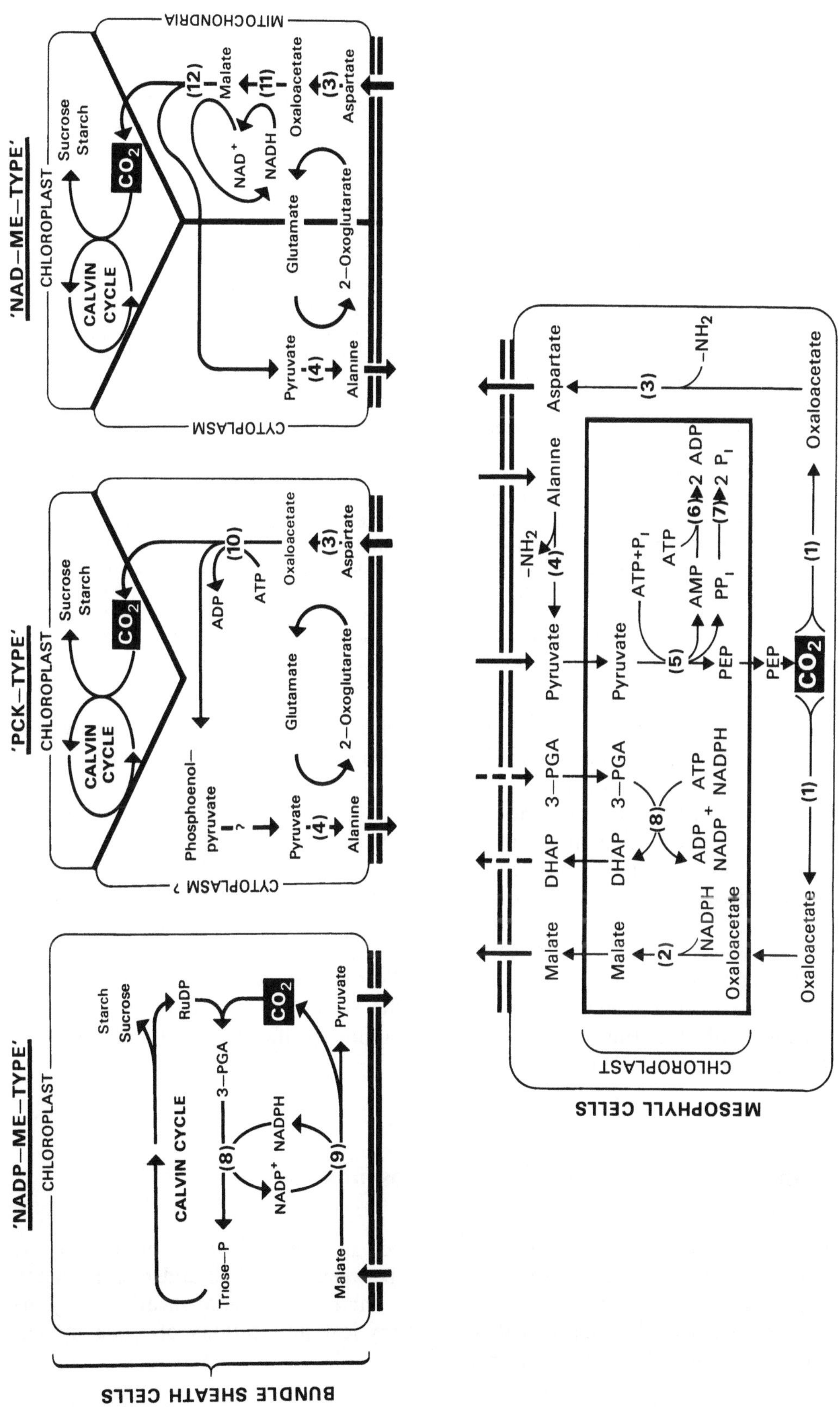
'NADP–ME–TYPE'
CHLOROPLAST
Starch
Sucrose
RuDP
3–PGA
CO_2
CALVIN CYCLE
(8)
NADP⁺
NADPH
(9)
Triose–P
Malate
Pyruvate
BUNDLE SHEATH CELLS
'PCK–TYPE'
CHLOROPLAST
CALVIN CYCLE
Sucrose
Starch
CO_2
ADP
ATP
(10)
Oxaloacetate
(3)
Aspartate
Phosphoenol-pyruvate
?
Pyruvate
(4)
Alanine
Glutamate
2–Oxoglutarate
CYTOPLASM ?
'NAD–ME–TYPE'
CHLOROPLAST
MITOCHONDRIA
CALVIN CYCLE
Sucrose
Starch
CO_2
(12)
Malate
(11)
NAD⁺
NADH
Oxaloacetate
(3)
Aspartate
Glutamate
2–Oxoglutarate
Pyruvate
(4)
Alanine
CYTOPLASM
MESOPHYLL CELLS
CHLOROPLAST
Malate
DHAP
3–PGA
Pyruvate
Alanine
Aspartate
(2)
NADPH
Oxaloacetate
(8)
ADP
NADP⁺
ATP
NADPH
(5)
ATP+P$_i$
AMP
PP$_i$
PEP
(6)
(7)
2 ADP
2 P$_i$
–NH$_2$
(4)
(3)
(1)
Oxaloacetate
CO_2

terms of intercellular organization it may be mandatory to return alanine to mesophyll cells when aspartate is transferred to bundle sheath cells, so that an overall amino-group balance is maintained (see Sect. 4.4). The basis for suggesting that part of the 3-PGA formed in bundle sheath cells may be cycled to mesophyll cells for reduction to DHAP has already been outlined (Sect. 3.3 and 3.4), and has been discussed in more detail elsewhere (HATCH and KAGAWA, 1973). Possible reasons for this particular intercellular cycling will be considered below (Sect. 6).

The basic photosynthetic functions of bundle sheath cells are C_4 acid decarboxylation and re-assimilation of the CO_2 so formed via the PCR cycle. In NADP-ME-type C_4 plants all these reactions occur within chloroplasts, and the continuing operation of NADP malic enzyme is almost certainly coupled via an NADPH-NADP$^+$ cycle to the reductive steps of the PCR cycle (Fig. 5). In PCK-type species oxaloacetate, derived from aspartate via aspartate aminotransferase, is decarboxylated via the ATP-requiring PEP carboxykinase reaction. The exact location of these reactions is uncertain, as is the nature of the proposed reaction leading from PEP to pyruvate. If pyruvate kinase catalysed this transformation it would also serve the purpose of regenerating ATP for PEP carboxykinase. Possible reasons why these species utilize an apparently complicated system of aspartate-alanine exchange instead of a more simple one involving movements of oxaloacetate and pyruvate will be considered in Section 4.4.

In the remaining group of C_4 plants (NAD-ME-type, Fig. 5) aspartate is decarboxylated via a series of reactions located in mitochondria. Significantly, this group has a notably high frequency of mitochondria in bundle sheath cells (DOWNTON, 1971b; HATCH et al., 1975). A feature of these reactions is the interdependence of the conversion of oxaloacetate to malate (malate dehydrogenase) and malate decarboxylation (NAD malic enzyme) through an NAD$^+$-NADH couple. Likewise, the aminotransferase-mediated interconversions are coupled by a glutamate-2-oxoglutarate cycle. Arguments against the alternative possibility that malate could be transferred directly from mesophyll cells and decarboxylated, with pyruvate being returned, are considered in Section 4.4.

The current evidence outlined in this section has necessitated the subdivision of C_4 species into three metabolically distinguished sub-groups (Table 2 and Fig. 5). This subdivision renders the task of generalizing about the C_4 pathway, or the species in which it operates, more difficult. Even the generalizations made here may have to be further qualified or modified. In terms of enzyme content and other features, the great majority of species examined to date fall clearly into one of these three sub-groups (GUTIERREZ et al., 1974a; HATCH and KAGAWA, 1974b; HATCH et al., 1975). However, *Gomphrena celosioides*, for instance, could prove to be a sub-variant of NADP-ME-type species in that it may transport both malate and aspartate to bundle sheath cells (HATCH et al., 1975). Other data suggest that, in a few instances, two of the separate C_4 acid decarboxylating systems defined in Figure 5 may be contributing significantly in a single species (GUTIERREZ et al., 1974a).

4. Intercellular Transport in C_4 Photosynthesis

The formulations of the C_4 pathway outlined in Figures 2 and 5 necessitate the rapid transport of a variety of metabolites between mesophyll and bundle sheath cells. They imply that the C_4 acids are CO_2 carriers to bundle sheath cells. This section considers the anatomical and biophysical implications of these models for C_4 acid and $CO_2 + HCO_3^-$ transport.

4.1 Structural Features of the Mesophyll-Bundle Sheath Cell Interface

Bundle sheath cells of C_4 plant leaves are thick walled and arranged with few intercellular spaces, whereas mesophyll cells are thin walled with intercellular spaces similar to those of spongy mesophyll cells of other leaves. The mesophyll and bundle sheath cells are connected by numerous plasmodesmata of well-defined structure. In the monocotyledons, the wall between mesophyll and bundle sheath cells contains a clearly identified suberin-lamella (O'BRIEN and CARR, 1970; LAETSCH, 1971, 1974) which is particularly thick in the region of the plasmodesmata. Although the properties of this suberin lamella have not been determined directly, it is generally accepted that they are similar to cutin in the leaf cuticle. The cuticular resistance of leaves to CO_2 and H_2O exchange may be about 10^2 times greater than the resistance of open stomata. Wound suberin of potato tissue is responsible for a 10-fold increase in resistance to water loss in this tissue (KOLATTUKUDY and DEAN, 1974). One can infer that the suberin lamella between bundle sheath and mesophyll cells in the Kranz complex is a substantial barrier to the diffusion of CO_2 to or from the bundle sheath cells. In the dicotyledons the suberin lamella is not evident in ultrastructural studies and the possible implication of this in relation to diffusion pathways will be discussed below.

The plasmodesmata are evidently quite permeable to low molecular weight solutes but there is no evidence for the movement of macromolecules between cells (ROBARDS, 1975). Movement of viruses between cells only occurs after substantial modification of the pore itself. These observations are consistent with the fact that cytosol enzymes of the C_4 pathway, such as PEP carboxylase, PEP carboxykinase, and aminotransferases, are strictly compartmented in the cytoplasm of either mesophyll or bundle sheath cells. It is generally agreed that a tubular structure (endoplasmic reticulum or desmotubule) passes through the plasmodesmata (O'BRIEN and CARR, 1970; ROBARDS, 1975) and cytosol enzymes may well be excluded from this compartment. Even in the absence of a suberin lamella, it is generally agreed that the symplast pathway through the plasmodesmata is a low resistance pathway for the movement of ions and metabolites (SPANSWICK, 1974; ROBARDS, 1975). The permeability of plasmodesmata for ions may be 10^3 that of the permeability of two cell membranes and a cell wall (TYREE, 1970). The suberin lamella may further increase the relative permeability of the plasmodesmata to a factor of 10^4 or 10^5 that of the cell membrane pathway at the mesophyll bundle sheath interface. (See also Vol. 2, Part B: Chaps. 1 and 2.)

4.2 Intercellular Transport of Metabolite Solutes

The relatively simple geometry of the Kranz complex permits estimates of the biophysical properties of the intercellular transport systems required to sustain the schemes of C_4 photosynthesis outlined in Figures 2 and 5. Two parameters, the metabolite flux between mesophyll and bundle sheath cells, and the apparent diffusion coefficient for the metabolites involved, are particularly useful. The flux calculation permits comparison of C_4 acid transport with that of $CO_2+HCO_3^-$ flux in the same system. The estimate of apparent diffusion coefficient permits evaluation of the probable mechanism of C_4 acid transport (i.e. simple diffusion vs some form of 'facilitated transport'). The treatment given below, and the conclusions reached, are essentially similar to those published earlier (OSMOND, 1971a), but are based on more accurate estimates of C_4 acid and $CO_2+HCO_3^-$ pool sizes (HATCH, 1971a, and unpublished data).

The flux of C_4 acids or C_3 metabolites may be estimated in two ways. Minimum fluxes of these compounds across the mesophyll bundle sheath cell interface, required to support photosynthesis, may be calculated from the photosynthesis rate and the surface area of the bundle sheath cylinder as follows,

$$J_M = \frac{P}{2\pi r l} \ \mu\text{mol/s/cm}^2 \qquad (1)$$

where J_M is the minimum flux, P=net photosynthesis, r and l=radius and length of the bundle sheath cylinder. If the flux is restricted to the plasmodesmata, as indicated in Sect. 4.1, then J_M must be corrected by a factor α, corresponding to the area of the interface occupied by plasmodesmata. In Table 3, the corrected flux J_{MP} was estimated using a value of $\alpha=3\times10^{-2}$, a median value from the range of estimates reported by TYREE (1970). Estimates of α for maize (O'BRIEN, personal communication) are comparable with those of the 'intermediate' vascular bundles of wheat for which $\alpha=2.8\times10^{-2}$ (KUO et al., 1974).

An independent estimate of the potential C_4 acid flux (which does not include assumptions as to the value of α) may be made from the maximum concentration gradients of C_4 acids likely to exist in the Kranz complex during C_4 photosynthesis. The photosynthetic pool sizes of C_4 acids can be accurately estimated in radiotracer experiments (HATCH, 1971a), and a maximum possible concentration estimated if these acids are confined to the cytoplasm of mesophyll cells (BERRY et al., 1970). Flux may be estimated from the resultant concentration gradients as follows (PITMAN, 1965):

$$J_C = \frac{D(C_m - C_b)}{r_m\left(l_n \frac{r_m}{r_b}\right)} \qquad (2)$$

where J_C=flux, D=the diffusion coefficient of compounds similar in molecular weight and charge to the C_4 acids and C_3 metabolites (taken as 8×10^{-6}/s/cm^2; WEAST, 1963), C_m and C_b=concentration of C_4 acids in mesophyll and bundle sheath cells and r_m and r_b are the average radii of the chloroplast layers in mesophyll and bundle sheath cells relative to the central vascular strand. The values of J_C in Table 3 are much larger than the values of J_M obtained above, but only 2 to 3 times larger than values of J_{MP} calculated after allowing for a plasmodesmata area of 3% of the cell wall interface.

This difference between J_C and J_{MP} indicates that the extreme assumption of zero C_4 acid concentration in bundle cells during steady-state photosynthesis is not necessary. The concentration gradient actually required to maintain a C_4 acid flux necessary to support photosynthesis, can be calculated by rearranging equation 2 and substituting J_{MP} for J_C. As shown in Table 3, the concentration gradient required in *Z. mays* and *Amaranthus edulis* would be only about 25% and 50%, respectively, of the potential concentration gradient, so that a significant proportion of the total C_4 acid pool could reside in bundle sheath cells during

Table 3. Properties of intercellular transport of C_4 acids and CO_2 during C_4 photosynthesis. (After OSMOND, 1971a)

Parameter	Equation symbol and units		*Z. mays*	*A. edulis*
Net photosynthesis per leaf volume	P	µmol/s/cm³	0.185	0.180
Surface of bundle sheath cells	$2\pi rl$	cm²/cm³	125	220
Required C_4 acid flux, total surface [Eq. (1)]	J_M	µmol/s/cm²	1.48×10^{-3}	0.82×10^{-3}
Required C_4 acid flux, *via* plasmodesmata [a]	J_{MP}	µmol/s/cm²	4.9×10^{-2}	2.7×10^{-2}
Pool size, C_4 acids involved in photosynthesis		µmol/cm³	1.85	1.82
Estimated maximum C_4 acid concentration, mesophyll [b]	C_m	µmol/cm³	38	60
Radius, chloroplast layer, mesophyll cells	r_m	cm	60×10^{-4}	80×10^{-4}
Radius, chloroplast layer, bundle sheath cells	r_b	cm	45×10^{-4}	25×10^{-4}
Calculated C_4 acid flux [c] [Eq. (2)]	J_C	µmol/s/cm²	17.8×10^{-2}	5.2×10^{-2}
Required C_4 acid gradient [Eq. (2)] [d]	C_m–C_b	µmol/cm³	10.5	31.3
Estimated $CO_2+HCO_3^-$ concentration [e]		µmol/cm³	0.6	2.0
Calculated back-flux $CO_2+HCO_3^-$ [Eq. (2)] [f]		µmol/s/cm²	3.5×10^{-3}	2.2×10^{-3}

[a] Assuming plasmodesmata 3×10^{-2} of total surface (TYREE, 1970); equation 2.
[b] Assuming C_4 acids restricted to cytoplasm (10%) of mesophyll cells.
[c] Assuming $D=8\times10^{-6}$/s/cm² (WEAST, 1963), bundle sheath cell C_4 acid concentration zero.
[d] Assuming plasmodesmata flux and $D=8\times10^{-6}$/s/cm².
[e] Assuming $CO_2+HCO_3^-$ restricted to cytoplasm and chloroplasts of bundle sheath cells.
[f] Assuming $D=10^{-5}$/s/cm² (NOBEL, 1974) and mesophyll cell $CO_2+HCO_3^-$ concentration zero.

C_4 photosynthesis. In fact, autoradiographic data for *A. spongiosa* shows substantial ^{14}C in bundle sheath cells after 2 s in $^{14}CO_2$, at a time when about 90% of the label is in C_4 acids (OSMOND, 1971a).

If diffusion is restricted to the central desmotubule of the plasmodesmata, then smaller values of α must be applied (OLESEN, 1975; OSMOND and SMITH, 1976). However, if this adjustment is made, and Eq. (2) is replaced by an equation which accounts for the fact that the restricted area for diffusion is confined to only a very small proportion of the total diffusion path, the values given in Table 3 are scarcely affected. OSMOND and SMITH (1976) have considered a variety of additional refinements without substantially influencing the principal conclusions arising from Table 3. Three points are important in relation to the minimum fluxes of C_4 acids from mesophyll to bundle sheath cells during C_4 photosynthesis:

1. These fluxes may be sustained by simple diffusion as deduced earlier (OSMOND, 1971a). The apparent diffusion coefficients for C_4 acids estimated previously (2 to 16×10^{-6}/s/cm) are close to the published values used in the above calculations.

2. The fluxes evidently require concentration gradients less than the potential gradients created if C_4 acids are confined to mesophyll cells.

3. The fluxes apparently involve movement largely through the plasmodesmata, not the whole cell interface. This is supported by the fact that the potential fluxes, J_C (calculated with no assumptions about plasmodesmata), are about 100 times the estimated fluxes required to support photosynthesis, assuming movement over the total cell surface (J_M).

A. edulis and other dicotyledons have no identifiable suberin lamella in the wall between mesophyll and bundle sheath cells, so that the plasma membrane-cell wall pathway may make some contribution to total metabolite flux. Another feature of these plants is the greater separation of mesophyll and bundle sheath chloroplasts (the latter are centripetally located, DOWNTON, 1971b), and metabolites would probably have to move over greater distances as a consequence. The higher diffusion resistance imposed by the length of the diffusion path may be compensated for partly by lower mesophyll-bundle sheath cell wall resistance, and partly by the larger concentration gradients of the metabolites (see Table 3).

A similar analysis of the C_3 metabolite fluxes proposed to operate between bundle sheath to mesophyll cells (Fig. 5) is more difficult because radiotracer studies do not provide direct estimates of pool sizes. However, in those species transporting alanine (PCK-type and NAD-ME-type), both the quantity and kinetics of alanine labeling closely parallel that of the C-1, C-2 and C-3 of the C_4 acids (CHEN et al., 1971; HATCH, 1971a). This suggests that the photosynthetic pools of alanine and C_4 acids are comparable in size and hence that concentration gradients and fluxes of these intermediates could also be comparable. The models require that the fluxes of C_4 acids and C_3 metabolites should be equal but in the opposite direction. Diffusion of different metabolites in opposite directions presents no difficulties, but some interaction between diffusion and water flow in this system could be involved. The C_4 acid flux is opposed to, and the C_3 metabolite flux is in the same direction as, the mainstream water flux from the vascular tissue during transpiration.

4.3 Intercellular Transport of $CO_2+HCO_3^-$: the CO_2 Concentrating Mechanism

Radiotracer kinetic studies suggest that the direct diffusion of external $^{14}CO_2$ to bundle sheath cells provides an insignificant proportion of the ^{14}C incorporated by RuBP carboxylase during C_4 photosynthesis (Sect. 2). Structural features of the bundle sheath cylinder such as thickened walls, absence of intercellular spaces, and the suberin lamella, probably ensure that the symplasm is the only path for CO_2 or HCO_3^- exchange between the bundle sheath cells and the external environment. The high activities of PEP carboxylase and carbonic anhydrase in mesophyll cell cytoplasm constitute a strong sink for CO_2, whether it is derived from the external pool or from decarboxylation reactions in the bundle sheath. Together, these features have been presumed to account for a CO_2 tight space in the Kranz complex of C_4 plants. The physiological manifestations of this phenomenon are many, and include the observations that C_4 plants normally release little CO_2 into CO_2 free air (TROUGHTON, 1971), and that they show CO_2 compensation points of near zero (DOWNTON and TREGUNNA, 1968). Furthermore, the carbon isotope discrimination ratio of C_4 plants ($\delta^{13}C$ value), reflects that of PEP carboxylase in vitro (WHELAN et al., 1973). If the Kranz complex were not CO_2 tight, further isotope discrimination by exchange with external CO_2 would occur during CO_2 fixation by RuBP carboxylase, and C_4 plants would display $\delta^{13}C$ values characteristic of this enzyme. This is clearly not the case and the $\delta^{13}C$ value has emerged as one of the most reliable criterior for detecting the C_4 pathway (LERMAN, 1975).

Direct measurements of the intermediate pool of $CO_2+HCO_3^-$ from radiotracer studies (Fig. 1c, d) permits estimates of the concentration of total CO_2 in bundle sheath cells. The values given in Table 3, 0.6 mM for *Z. mays* and 2.0 mM for *A. edulis* (HATCH, 1971a, and unpublished data), correspond to minimum equilibrium concentrations of free CO_2 equal to 13 μM and 44 μM at pH 8.0, respectively. These are absolute minimum estimates for the free CO_2 concentration in bundle sheath cell cytoplasm plus chloroplasts, since the CO_2 released by the decarboxylase systems is unlikely to reach equilibrium with HCO_3^- in these cells which apparently contain little carbonic anhydrase (EVERSON and SLACK, 1968; GRAHAM et al., 1971; POINCELOT, 1972). Nevertheless, these CO_2 concentrations are about two to five times higher than the free CO_2 concentration in air-saturated water (approximately 8 μM at 25° C) and very much higher than the substomatal CO_2 concentrations prevailing during steady photosynthesis in C_4 species (see Sect. 7).

It has been proposed that these higher CO_2 concentrations in bundle sheath cells would saturate RuBP carboxylase, thereby permitting the high, light-saturated rates of CO_2 fixation in C_4 plants (BJÖRKMAN, 1971; HATCH, 1971b). These CO_2 concentrations are also about sufficient to abolish the inhibition of RuBP carboxylase due to air levels of O_2 (250 μM) in vitro, and presumably account for the lack of an O_2 inhibition of photosynthesis in C_4 plants in vivo (Sect. 7). Thus, the C_4 acid carriers of CO_2 to the bundle sheath cells and the decarboxylase systems in these cells are thought to act as a CO_2 concentrating mechanism (BJÖRKMAN, 1971, 1973; HATCH, 1971b, 1976a; HOCHACHKA and SOMERO, 1973) which effectively increases the capacity of the PCR cycle in the bundle sheath cells. However, as pointed out by SMITH (1971), the symplast pathway for C_4 acid transport must be equally accessible to CO_2 and HCO_3^- movement and the apparent diffusion coefficients of $CO_2+HCO_3^-$ (0.7 to 1.0×10^{-5}/s/cm^2; NOBEL, 1974) are a little faster than that of the C_4 pathway metabolites. Clearly it is important that the physical basis of the CO_2 concentrating mechanism be examined as explicitly as possible.

In considering the operation of a CO_2 concentrating mechanism it is significant to note that a $CO_2+HCO_3^-$ concentration of only about 1–2 mM (pH 8.0) is required to saturate RuBP carboxylase in bundle sheath cells, a value close to the concentrations estimated to occur in bundle sheath cells (see above). Even if the concentration of $CO_2+HCO_3^-$ is near zero in mesophyll cells, the prevailing concentration gradient would result in a back flux of $CO_2+HCO_3^-$ to mesophyll cells [from Eq. (2)] which is less than 10% of the fluxes of C_4 acid and C_3 metabolites (Table 3). This implies that little of the CO_2 released in bundle sheath cells would be lost by outward diffusion and that a relatively strict stoichiometry between the C_4 acid carboxylase-decarboxylase system and the PCR cycle may be preserved. If the C_4 acid cycle was required to turn over much more rapidly than the PCR cycle simply to maintain high CO_2 concentrations in the bundle sheath cells, the additional ATP consumption and reduced pyridine nucleotide requirements would impose serious energetic problems.

Hence the essential components of the CO_2 concentrating mechanism in the Kranz complex are:

1. The structural organization, i.e. the separation of C_4 acid carboxylation and decarboxylation sites by a mean distance of about 20–60 μ, interposed by a thick cell wall (often containing suberin) which largely confines CO_2 and metabolite exchange to the plasmodesmata.

2. The kinetic properties and activity of enzymes which function to maintain a concentration gradient of C_4 acids from mesophyll to the bundle sheath of about 10–30 mM and an outward gradient of about 1–2 mM $CO_2+HCO_3^-$. This combination of characteristics would represent a reasonable compromise in which the carboxylation-decarboxylation cycle would have to operate only slightly faster than the PCR cycle to refix the small proportion (less than 10%) of the CO_2 that diffuses back to mesophyll cells.

4.4 Amino Group and Charge Balance during Intercellular Transport

The intercellular transport of C_4 and C_3 metabolites proposed to operate in the C_4 pathway (Fig. 5) poses no inconsistencies in carbon or amino group balance. However, an examination of the schemes for C_4 photosynthesis in Figure 5 shows that apparently simpler processes of intercellular C_4 and C_3 metabolite exchange are possible in PCK-type and NAD-ME type species. For PCK-type species oxaloacetate formed in mesophyll cells could be transferred directly to bundle sheath cells and decarboxylated via PEP carboxykinase, with PEP being returned to mesophyll cells. Likewise, in NAD-ME-type species malate derived from mesophyll cells could be directly decarboxylated by the NAD-malic enzyme in bundle sheath cells with the resulting pyruvate being returned. Both these alternatives replace the aspartate-alanine flux systems and hence circumvent the need for the complex sequence of amino acid-keto acid interconversions (see Figs. 3 and 5).

The operation of these alternatives is unlikely for the following reasons. There is strong evidence from radiotracer studies with intact leaves and cells, as well as enzyme studies, implicating these amino acids as intermediates of photosynthesis. Radiotracer studies indicate that the oxaloacetate pool in a wide variety of C_4 species is rarely more than 5% and often less than 1% of the malate and aspartate pools. Under these circumstances this compound would be unlikely to provide the concentration gradients to support appropriate rates of flux to bundle sheath cells. For NAD-ME-type species, if malate derived directly from mesophyll cells were decarboxylated by NAD-malic enzyme, then an alternative system for oxidising NADH to NAD^+, other than oxaloacetate reduction to malate, would be required (see Fig. 5). The nature of such a system, operating in mitochondria at rates commensurate with photosynthesis, is not readily apparent. For these reasons, aspartate synthesis and transport provides a more rational explanation of the data. When aspartate is transferred to bundle sheath cells the return of alanine, rather than pyruvate, would serve to maintain intercellular amino group balance as well as coupling for the transamination reactions within cells (see Fig. 3).

The present formulations of the C_4 pathway (see Fig. 5) propose large intercellular fluxes of dicarboxylic and monocarboxylic acids, and possibly other charged metabolites such as 3-PGA and DHAP. Comparable fluxes of metabolites to and from chloroplasts, and also mitochondria in NAD-ME-type species, are also required. Clearly, during the overall operation of the C_4 pathway there must be a balance of charges between cells and intracellular compartments. At the simplest level there is an apparent imbalance of negative charges across the mesophyll-bundle sheath cell junction with coupled fluxes of C_4 acid^{2-} and C_3 acid^{1-}, and also of 3-PGA^{3-} and DHAP^{2-} in opposite directions. However, in terms of negative charges on metabolites only, there is a balance between cells if the complete metabolic sequences are considered. Thus, for the C_4-C_3 acid shuttle, a balance is maintained when it is taken into account that one of the carboxyl charges on the C_4 acid is derived initially from CO_2 in mesophyll cells and ultimately appears as CO_2 in bundle sheath cells. Likewise, a balance of negative charges is satisfied if the 3-PGA-DHAP shuttle is considered in terms of their complete metabolism via the PCR cycle. However, we should emphasize that a complete charge balance analysis of these fluxes would almost certainly require consideration of H^+, or other counter-ion movements, as well as component phosphorylation reactions and pyridine nucleotide mediated oxidation-reductions.

5. Intracellular Transport in C_4 Photosynthesis

Within both mesophyll and bundle sheath cells a complex of transport processes involving metabolite movement between cytoplasm, chloroplast, and mitochondrion are apparently essential to the carbon flux of C_4 photosynthesis (Fig. 5). As pointed out in Sect. 3, the reactions in mesophyll cells leading to C_4 acid formation are basically similar in all C_4 species and evidently involve similar transport events. However, in bundle sheath cells there are substantial differences in the site and mechanism of C_4 acid decarboxylation, requiring different transport processes in different groups of species.

5.1 Intracellular Transport in Mesophyll Cells

The unique metabolic capabilities of the mesophyll cell chloroplast have been detailed in Sect. 3. Transport of pyruvate into the mesophyll cell chloroplast, and its conversion to PEP dependent on concurrent photophosphorylation, have been observed in preparations from *Z. mays, A. spongiosa* and *D. sanguinalis* (KAGAWA and HATCH, 1974a; HUBER and EDWARDS, 1975). The maximum rates attained, which also provide minimum estimates of transport, were comparable to the net carbon flux during photosynthesis by leaves of these plants (Table 4). HUBER and EDWARDS (1975) showed that the mesophyll chloroplasts evidently have a high affinity for pyruvate (*Km* 0.1 mM), comparable with that of pyruvate,P_i dikinase in vitro (HATCH and SLACK, 1968). Rapid transport of pyruvate is apparently another distinctive feature of mesophyll chloroplasts of C_4 species since, although spinach chloroplasts transport pyruvate, rates are about 10^2 slower (HEBER and KRAUSE, 1971; H.W. HELDT, unpublished). Pyruvate metabolism by the pyruvate P_i dikinase system in intact mesophyll chloroplast also requires an uptake of orthophosphate. It is conceivable that an orthophosphate—PEP exchange mediated by a phosphate translocator of the spinach chloroplast type (HELDT et al., 1975) could be involved, but the mechanism of pyruvate uptake remains an important and interesting problem.

Table 4. Minimum rate of metabolite transport in intact mesophyll chloroplasts and bundle sheath mitochondria of C_4 plants as indicated by reaction rates with exogenously supplied substrates. (After KAGAWA and HATCH, 1974a, 1975) (μmol/min/mg chlorophyll). *Z. mays*, NADP-ME-type; *A. spongiosa*, NAD-ME-type

Organelle and process	Reaction rate	
	Z. mays	*A. spongiosa*
Mesophyll chloroplasts		
Pyruvate influx/PEP efflux	1.5	3.8
oxaloacetate influx	2.5	3.0
malate efflux	3.1	1.2
3-PGA influx	1.6	4.5
DHAP efflux	—	4.4
Bundle sheath mitochondria		
C_4 acid influx	0.03	5–8

The exchange uptake of OAA and excretion of malate in the mesophyll chloroplast presumably involves the high affinity, high rate dicarboxylate translocator similar to that in spinach chloroplasts (HEBER, 1974; HELDT et al., 1975). Minimum rates of OAA uptake and malate efflux measured in chloroplasts from *Z. mays* and *A. spongiosa* (KAGAWA and HATCH, 1974a) are comparable to the net photosynthesis rates in leaves of these plants (Table 4). Mesophyll chloroplasts from these species also have a substantial capacity for 3-PGA-dependent oxygen evolution and this was accompanied by the formation and efflux of DHAP (Table 4; KAGAWA and HATCH, 1974a). Presumably, this transformation also involves a 3-PGA-DHAP shuttle similar to that in spinach chloroplasts which is in turn mediated by the phosphate translocator.

5.2 Intracellular Transport in Bundle Sheath Cells

In NADP-ME-type C_4 plants the decarboxylation of malate takes place in the bundle sheath chloroplast (Sect. 3). To date, there have been no reports of the isolation of intact chloroplasts of specified bundle sheath origin, so that direct studies of permeability characteristics have not been possible. However, a mixed population of plastids prepared from immature primary leaves of maize were capable of CO_2 assimilation via the PCR cycle (O'NEAL et al., 1972). From the scheme outlined in Figure 5, one would predict that the bundle sheath chloroplasts from NADP-ME-type species should be capable of substantial malate transport in exchange for pyruvate. Although the malate transporting system of these chloroplasts may be similar to the dicarboxylate translocator of C_4 mesophyll chloroplasts and spinach chloroplasts, it is difficult to conceive a scheme in which any of the known complimentary exchange species (HEBER, 1974; HELDT et al., 1975) are involved. Like mesophyll chloroplasts, the movement of pyruvate from these bundle sheath chloroplasts should also differ from the slow exchange system for this compound in spinach chloroplasts. However, the complimentary malate efflux-pyruvate influx of mesophyll chloroplasts and malate influx-pyruvate efflux of bundle sheath chloroplasts in C_4 plants may represent a unique transport system which cannot be evaluated in terms of our current understanding of chloroplast metabolite movement in C_3 plants.

In NAD-ME-type C_4 plants, enzyme distribution indicates a more complex sequence of metabolite exchanges involving mitochondria as the principal site of decarboxylation (Fig. 5). Leaf mitochondria from these plants decarboxylate C_4 acids at rates in excess of the rate of carbon flux in C_4 photosynthesis (Table 4). Decarboxylation in vitro requires the simultaneous addition of aspartate and 2-oxoglutarate, malate and orthophosphate (KAGAWA and HATCH, 1975) and involves the net consumption of aspartate and production of pyruvate. As shown in Figure 5, a 2-oxoglutarate-glutamate shuttle is required for the aminotransferase reactions responsible for utilising aspartate and forming alanine. With isolated mitochondria, added malate would be required to establish the NAD^+-NADH cycle linking NAD malic enzyme and NAD malate dehydrogenase. Orthophosphate could be required in a carrier system for malate similar to those operative in other mitochondria (CHAPPELL and HAAROFF, 1967; WISKICH, 1974), since none of the enzymes involved in decarboxylation requires orthophosphate. Alternatively, malate and orthophosphate may facilitate the transport of aspartate and 2-oxoglutarate into the mitochondria as described in other systems (PALMER and HALL, 1972).

In C_4 plants these decarboxylation activities are at least 10 times faster than oxidative respiratory processes, which in these mitochondria are similar to the activities recorded for mitochondria from other green tissues (CHAPMAN and OSMOND, 1974; KAGAWA and HATCH, 1975). The involvement of mitochondria in decarboxylation reactions associated with photosynthesis has an important parallel in C_3 plants. In these, leaf mitochondria evidently contain the glycine decarboxylase system responsible for CO_2 release during photorespiration (TOLBERT, 1971).

In vivo experiments with NADP-ME and NAD-ME type C_4 plants may be interpreted to show a great deal of mobility of C_4 acids between intracellular compartments. For example, in *Sorghum bicolor* and *A. spongiosa* the C_4 acids labeled by metabolism of succinate-^{14}C in the dark (mitochondrial pools, or mitochondria + cytoplasm) are rapidly converted to photosynthetic products in the light (CHAPMAN and OSMOND, 1974). These events appear to involve the C_4 acid decarboxylase systems of the mitochondria and chloroplasts, rather than tricarboxylic acid cycle metabolism. In C_3 plants the same dark to light experiments show that very little label is transferred from the succinate-derived C_4 acids to photosynthetic products.

Enzyme distribution studies have so far failed to localise PEP carboxykinase, the decarboxylase reaction of PCK-type C_4 plants, in any organelle of the bundle sheath cells (Table 2). The decarboxylation probably takes place in the cytoplasm of these cells, and CO_2 fixation by bundle sheath chloroplasts is presumably analogous to that in C_3 plants. However, the PCK-type C_4-plants are the least well understood at this time and substantial problems, such as the source of ATP for PEP carboxykinase and the mechanism of pyruvate formation from PEP, remain to be resolved.

5.3 Organelle Ultrastructure in Relation to Intracellular Transport

The inner membrane of chloroplasts and mitochondria is the limiting membrane for metabolite transport (Chaps. II,3, II,5). In leaves of C_4 plants these membranes often show unusually complex development which would provide a substantial increase in surface area. Proliferation of cell membranes in glands (LÜTTGE, 1971) and transfer cells (PATE and GUNNING, 1972) appears to be correlated with increased transport capacity, and it is reasonable to apply similar arguments to the limiting membranes of organelles participating in C_4 photosynthesis.

The mesophyll cell chloroplasts of most C_4 plants investigated, and bundle sheath chloroplasts of some species, show an invagination of the inner membrane which has been described as the peripheral reticulum (LAETSCH, 1971, 1974). It has been proposed that the increased surface area of the chloroplast envelope may be related to the low mesophyll resistance to CO_2 of C_4 plants (OSMOND et al., 1969) or to the transport of metabolites (LAETSCH, 1974). For example, if the peripheral reticulum was composed of a series of uniform spherical invaginations on the surface of a sphere it would result in a 4-fold increase in surface area. However, the peripheral reticulum appears to take the form of anastomising tubules (LAETSCH, 1974) and to be rather irregular in form, so that realistic estimates of increase in metabolite exchange surface are impossible. There is no substantial evidence for a connection between the peripheral reticulum and any tubular system which may pass through plasmodesmata as speculated earlier (SLACK et al., 1969).

The large and numerous mitochondria of bundle sheath cells in NAD-ME-type C_4 plants also show elaborate development of the inner membrane (OSMOND et al., 1969; DOWNTON, 1971b; LAETSCH, 1971, 1974; NEWCOMB and FREDERICK, 1971; HATCH et al., 1975). It is likely that the large number of mitochondria and the proliferation of the inner, transport-limiting membrane could also be related to demands of metabolite transport.

6. Metabolite Transport in Relation to Chloroplast Photochemical Activities

From the earliest conceptions of cooperative processes in C_4 photosynthesis it has been recognised that the intercellular transport of malate in NADP-ME-type species contributes both CO_2 and reducing potential to bundle sheath cells (KARPILOV, 1969; SLACK et al., 1969; BERRY et al., 1970). The possibility that NADPH formed via NADP malic enzyme might contribute directly to the requirements of the carbon reduction cycle for reductant has been correlated with the structural features of bundle sheath chloroplasts in these species. Dimorphic chloroplasts were described in early studies of *Z. mays* (HODGE et al., 1955) and the presence of nonappressed thylakoids in bundle sheath chloroplasts (described for convenience as agranal) has since attracted a good deal of attention (LAETSCH, 1968, 1971, 1974). Studies of tobacco mutants suggest that agranal chloroplasts are deficient in photosystem II capacity (HOMAN and SCHMID, 1967) and histo-chemical studies by DOWNTON et al. (1970) indicated that agranal bundle sheath chloroplasts were also deficient in photosystem II activity in vivo. This deficiency in photosystem II activity is evident in Table 5, which is drawn from the results of assays of photochemical reactions in mesophyll and bundle sheath chloroplasts of three well researched species in several laboratories (also see ARNTZEN and BRIANTAIS, 1974).

Recent more extensive surveys of photochemical properties of chloroplasts isolated from C_4 plants confirm the photosystem II deficiency in the bundle sheath cells of NADP-ME-type species (KU et al., 1974b; MAYNE et al., 1974). However, plants of the NAD-ME-type and PCK-type show normal granal development in both mesophyll and bundle sheath cells (GUTIERREZ et al., 1974a; HATCH et al., 1975). The data in Table 5 indicate that, for *A. spongiosa* (NAD-ME-type), chloroplasts of both cell layers are fully competent in noncyclic electron transport and photophosphorylation. Similar data have been obtained for other species in this group, and for PCK-type C_4 plants (KU et al., 1974b; MAYNE et al., 1974).

A great deal of controversy, much of it semantic, has developed in the literature in relation to the statement that plants of the NADP-ME-type, such as *Z. mays* and *S. bicolor*, show substantial deficiencies in noncyclic electron transport and photosystem II mediated photophosphorylation (Table 5). For example, deficiency in photosystem II activity has been expressed as a relative enrichment of photosystem I activity (MAYNE et al., 1971; BAZZAZ and GOVINDJEE, 1973). Some early assays of photosystem II activity were obviously made under sub-optimal conditions (WOO et al., 1970; cf. ANDERSON et al., 1972; BISHOP et al., 1971, 1972a, b), and the ratio of NADP photoreduction from water by mesophyll versus bundle sheath chloroplasts declines with leaf maturity in *Z. mays* (ANDERSEN et al., 1972; KARPILOV, 1974). Further-

Table 5. Comparisons of photochemical activities of isolated mesophyll (M) and bundle sheath cell (B) chloroplasts from leaves of C_4 plants (µmol/min/mg chl.)

Component process	Assay system	*A. spongiosa*		*Z. mays*		*S. bicolor*	
		M	B	M	B	M	B
Electron transport							
Photosystem I and II	($H_2O \rightarrow NADP$)	2.6	2.3 [a]	3.2	0.8 [b]	2.7	0 [c]
						1.3	0.9 [d]
Photosystem II	$K_3Fe(CN)_6$			10.3	2.7 [c]	13.6	0.9 [c]
				21.0	6.0 [b]		
	TCPIP, DCPIP			7.0	2.3 [c]	8.7	0.6 [c]
						4.5	0.7 [e]
Photosystem I	DCPIP Ascorbate → NADP			7.2	14.7 [b]	3.0	7.1 [e]
	DCPIP Ascorbate → Methylviologen	5.5	7.8 [f]			4.5	7.0 [f]
Photophosphorylation							
Photosystem II	$K_3Fe(CN)_6$	6.3	3.8 [g]	1.6	0.3 [h]	4.1	0.3 [g]
				3.6	0.3 [g]	4.0	0.4 [e]
Photosystem I	PMS	15.7	8.7 [g]	10.1	5.2 [g]	10.6	4.8 [g]
				8.7	7.3 [h]	10.3	9.5 [e]

[a] WOO et al. (1970). [b] BISHOP et al. (1972b). [c] ANDERSON et al. (1972). [d] SMILLIE et al. (1972a). [e] ARNTZEN et al. (1971). [f] OSMOND (1974). [g] POLYA and OSMOND (1972). [h] ANDERSON et al. (1971).

more, there is no doubt that NADP-ME-type C_4 plants differ in the extent of the deficiency of photosystem II activity in bundle sheath cells with respect to mesophyll cells. In *Z. mays* and *D. sanguinalis* the ratio of photosystem II activity, bundle sheath to mesophyll, may be about 0.3, whereas in *S. bicolor* this ratio is about 0.1 or less. The most important conclusion from these data is that in *Zea* and in *Sorghum* the maximum potential for photochemical generation of $NADPH_2$ in agranal bundle sheath chloroplasts is considerably less than 50% of that required to support the measured rates of light-saturated photosynthetic carbon reduction.

The limited photosystem II activity of agranal chloroplasts in vitro is not saturated with light at intensities approaching full sunlight (ANDERSON et al., 1972; SMILLIE et al., 1972b), but even then the rates of the Hill reaction remain substantially below those in granal mesophyll chloroplasts. It appears unreasonable to suppose (SMILLIE et al., 1972b) that chloroplasts in bundle sheath cells of these plants, which are shielded by mesophyll cells containing 60% or more of total leaf chlorophyll, would be exposed to light intensities anywhere near saturating for photosystem II activity. The high light requirement for photosystem II activity in these chloroplasts appears to be unrelated to the fact that high light intensities are required for saturation of photosynthesis in C_4 plants, since factors other than photochemical capacity have been implicated (GIFFORD, 1971).

The significance, in vivo, of the deficiency in photosystem II activity, and the hypothesis that NADP malic enzyme may contribute $NADPH_2$ for carbon reduction in photosystem II deficient bundle sheath cells, were assessed during in vivo pulse-chase experiments under conditions of narrow band illumination (OSMOND, 1974). It was predicted that in *A. spongiosa,* photosystem II activity would be obligatory for the transfer of ^{14}C from C_4 acids to carbohydrate, but that in *S. bicolor* the carbon reduction might occur under far-red illumination in which only photosystem I was activated. The selective activation of photosystems using illumination of different wavelengths was confirmed with the partial reactions of isolated chloroplasts. As shown in Figure 6, label in C_4 acids did not move into reduced products in leaves of *Atriplex* illuminated with far-red light, but in *Sorghum,* substantial carbon reduction was observed (OSMOND, 1974). These data suggest that in far-red light, which supports only low levels of photosystem II activity in vitro, the agranal bundle sheath chloroplasts of *Sorghum* are capable of photophosphorylation and carbon reduction utilising NADPH generated by malic enzyme.

Further support of these conclusions is provided by studies on the light-dependent assimilation of $H^{14}CO_3^-$ by *Z. mays* bundle sheath cells (HATCH and KAGAWA, unpublished). With only $H^{14}CO_3^-$ and ribose-5-P added, most of the label fixed (rate 1.05 μatoms ^{14}C/min/mg chlorophyll) appeared in 3-PGA and the flow of label to reduced products was only 0.13 μatoms ^{14}C/min/mg chlorophyll. When malate was also provided the proportion of label appearing in reduced products increased several fold, giving a reduction rate of 0.8 μatom ^{14}C/min/mg chlorophyll. As anticipated, there was negligible O_2 evolution from these cells while they were fixing CO_2.

If CO_2 assimilation by the PCR cycle is stoichiometrically linked with malate decarboxylation via NADP malic enzyme, then the latter process can only provide half the required NADPH. In fact, the rate of carbon reduction in *Sorghum* leaves during the far-red illumination chase (Fig. 6) is about 50% of the rate in red light. In NADP-ME-type species, part of the 3-PGA generated in bundle sheath chloroplasts may be returned to mesophyll chloroplasts for reduction to DHAP (HATCH, 1971b), and this may also be true for other C_4 species (HATCH and KAGAWA, 1973). This suggestion is supported by the fact that mesophyll chloro-

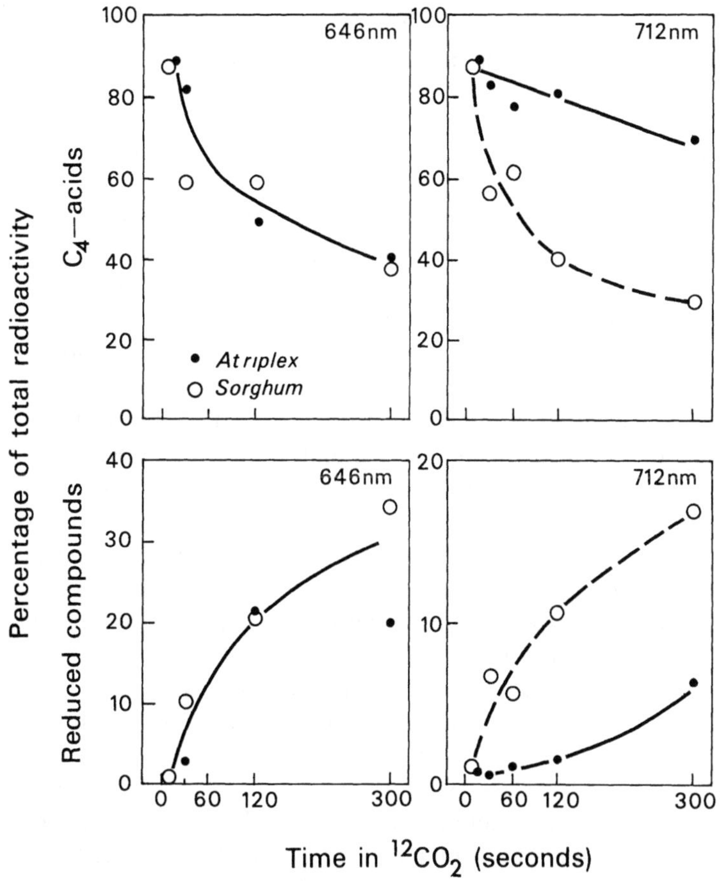

Fig. 6. Pulse chase experiments in *A. spongiosa* and *S. bicolor* under different illumination conditions (after OSMOND, 1974). Intact leaves were exposed to $^{14}CO_2$ (approx. 370 ppm) in red light (646 nm) for 7 s (*Sorghum*) and 15 s (*Atriplex*), then chased in $^{12}CO_2$ with either red (646 nm) or far-red (712 nm) light. The loss of label from C_4 acids in the chase in 646 nm light, and the transfer of label to reduced compounds was similar in *Atriplex* and *Sorghum*. In 712 nm light transfer of label from C_4 acids to reduced compounds is much faster in *Sorghum*

plasts contain about half of the total leaf content of photosynthetic enzymes involved in 3-PGA reduction (Table 2). In this way, the responsibility for providing ATP and reducing power for CO_2 reduction may be shared between the mesophyll and bundle sheath chloroplasts of all C_4 plants.

The cooperation evident between C_4 acid transport and photochemical activity in the NADP-ME-type C_4 plants apparently represents the highest level of development of the C_4 pathway. Mesophyll and bundle sheath cells of these plants display extreme modifications of the familiar, photochemically competent chloroplast containing the PCR cycle enzymes which may be prepared from spinach leaves. In *Sorghum*, photochemically competent chloroplasts lacking most enzymes of the PCR cycle are found adjacent to chloroplasts with a normal complement of PCR cycle enzymes, but deficient in the photochemical activities associated with photosystem II. An understanding of regulated development of this complex system of cooperative photosynthesis is a major challenge for future studies of the C_4 pathway.

7. Physiological Function of Compartmentation and Transport in C_4 Photosynthesis

Although few enzymes of the C_4 pathway are unique to this process, the levels and specific compartmentation of these enzymes combine to create a novel metabolic complex. Basically, this metabolic complex serves as a mechanism for transporting CO_2 to the bundle sheath cells, and for concentrating this CO_2 in the vicinity of RuBP carboxylase. Arguments that this concentrating of CO_2 facilitates the operation of the RuBP carboxylase at rates commensurate with C_4 photosynthesis, as well as reducing the inhibitory effect of O_2 on photosynthesis, are discussed below.

The original rationale for proposing that concentrating of CO_2 in bundle sheath cell of C_4 species may be mandatory for the operation of RuBP carboxylase at adequate rates (BJÖRKMAN, 1971; HATCH, 1971b) was apparently negated by recent reports that the physiological form of RuBP carboxylase has a high affinity for CO_2 (BADGER and ANDREWS, 1974; BAHR and JENSEN, 1974a, b). However, the information summarised in Figure 7 shows that concentrating of CO_2 in bundle sheath cells is still likely to be essential for the operation of the PCR cycle at appropriate rates in C_4 species.

Figure 7 shows the relationship of carboxylase enzyme activity to CO_2 concentration and the maximum photosynthetic rate of C_4 and C_3 plants in air. Although HCO_3^- rather than CO_2 is the substrate for PEP carboxylase (COOPER and WOOD, 1971; COOMBS et al., 1975), the CO_2 equivalent of the average $Km[HCO_3^-]$ values at pH 7.8 are used here to simplify the following discussion. The CO_2 concentration in solution in equilibrium with air is about 8 μM. At this concentration projected RuBP carboxylase activities would be adequate to account for carbon assimilation in C_3 photosynthesis but somewhat deficient for C_4 photosynthesis. However, during photosynthesis CO_2 gradients exist between air and the mesophyll cell surface such that for C_3 plants the internal CO_2 concentration would be about 5–6 μM. At this concentration, estimated RuBP carboxylase activities are still about sufficient to account for carbon assimilation in these plants (Fig. 7). However, for C_4 plants the combination of greater stomatal resistance and higher photosynthetic rate result in larger CO_2 gradients into leaves, and values of about 1 μM are estimated for the CO_2 concentration at the mesophyll cell surface. Significantly, the potential activity of PEP carboxylase at concentrations of about 1 μM remains sufficient to account for photosynthesis in C_4 species, but the potential activity of RuBP carboxylase at about 1 μM CO_2 is well below the rates of C_4 photosynthesis (Fig. 7).

This problem would be resolved by reactions of the C_4 pathway operating to concentrate CO_2 in bundle sheath cells. Figure 7 shows that minimal concentrations of about 15–20 μM CO_2 would be required to give RuBP carboxylase activities that match photosynthesis rates in C_4 plants. Evidence reviewed above (Sect. 4.3) shows that free CO_2 concentrations in bundle sheath cells are at least of this magnitude. Thus, the operation of the C_4 pathway, on the one hand, permits the rapid initial assimilation of CO_2 at the low concentrations of CO_2 prevailing in the region of mesophyll cells while on the other, also facilitates the adequate operation of RuBP carboxylase by concentrating CO_2 in bundle sheath cells.

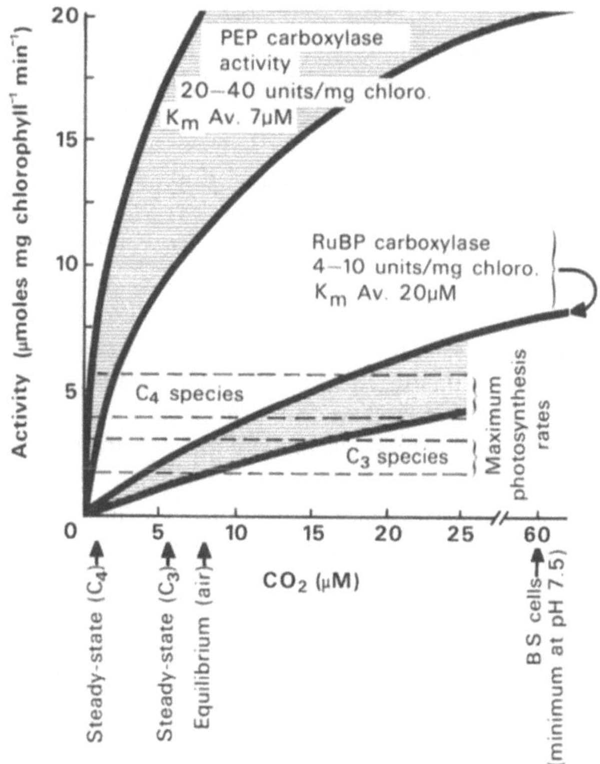

Fig. 7. Range of potential PEP carboxylase (C_4 plants) and RuBP carboxylase (C_3 and C_4 plants) activities as a function of carbon dioxide concentration compared with the range of maximum photosynthesis rates for C_3 and C_4 species. Literature sources for activities and substrate affinities were: PEP carboxylase, maximum activities adjusted in some instances for glucose-6-P activation (HATCH and SLACK, 1970; TING and OSMOND, 1973; GUTIERREZ et al., 1975) and *Km* for HCO_3^- or CO_2 (WALKER and BROWN, 1957; MARAYAMA et al., 1966; WAYGOOD et al., 1969; COOMBS et al., 1973b; TING and OSMOND, 1973); RuBP carboxylase, maximum velocities including some values for the 'low *Km*' enzyme (BJÖRKMAN and GAUHL, 1969; BADGER and ANDREWS, 1974 and unpublished results; BADGER et al., 1975; GUTIERREZ et al., 1975) and *Km* (CO_2) values for the 'low *Km*' form (BADGER and ANDREWS, 1974; BAHR and JENSEN, 1974a, b; BADGER et al., 1975); the range of maximum photosynthesis rates (HATCH and SLACK, 1970; HATCH, 1971a and unpublished data). The carbon dioxide concentration in bundle sheath (*BS*) cells was estimated from measurements of total carbon dioxide plus bicarbonate pools developed during steady photosynthesis in C_4 species (HATCH, 1971a). Other arrows indicate the aqueous carbon dioxide concentration in equilibrium with air (UMBREIT et al., 1964) and average carbon dioxide concentrations in the region of mesophyll cells during steady-state photosynthesis of C_3 and C_4 plants computed from data on stomatal and boundary layer resistances to water vapor diffusion (DOWNES, 1969, 1970; LUDLOW, 1970; SLATYER, 1970)

BOWES et al. (1971) suggested that an additional function of concentrating CO_2 in bundle sheath cells of C_4 plants may be to reduce photorespiration (see JACKSON and VOLK, 1970). The synthesis of glycolate, the substrate for photorespiration, is at least largely due to an oxygenation of RuBP yielding 3-PGA and phosphoglycolate. Recent studies have shown that RuBP carboxylase is a multifunctional enzyme, catalysing both carboxylation and oxygenation of RuBP (BOWES and

Ogren, 1972; Andrews et al., 1973). Oxygen and CO_2 are reciprocally competitive substrates and inhibitors for this enzyme from C_3 plants (Badger and Andrews, 1974; Bahr and Jensen, 1974b; Laing et al., 1974) and the enzyme from C_4 plants shows similar properties (Badger et al., 1975). The oxygenation of RuBP in air saturated solutions in vitro is substantially reduced by increasing the free CO_2 concentration to 20–40 μM. Likewise, the inhibition of photosynthesis in C_3 plants in 21% O_2 (approx. 250 μM) may be abolished by increasing intracellular CO_2 concentrations to 30 μM (approx. 900 ppm CO_2) (Jackson and Volk, 1970), and the rate of C_3 photosynthesis under these conditions is comparable to that of C_4 plants (Osmond and Björkman, 1972). These CO_2 concentrations are similar to the estimates of minimum CO_2 concentration in the vicinity of RuBP carboxylase-oxygenase in bundle sheath cells of C_4 plants (Sect. 4.3). Significantly, CO_2 fixation in C_4 plants is insensitive to O_2 concentration between 2 and 21% (approx. 25 μM to the estimates of minimum CO_2 concentration in the vicinity of RuBP carboxylase- of O_2 inhibition in C_4 plants, as a result of concentrating CO_2 in bundle sheath cells, would be a major contributor to the higher photosynthesis rates in these plants (Bowes and Ogren, 1972; Osmond and Björkman, 1972; Chollet and Ogren, 1973).

Associated with this consequence of C_4 pathway compartmentation and transport is the question of the extent of glycolate production and metabolism in bundle sheath cells of C_4 plants (Osmond, 1971b). The labeling of the glycolate pathway intermediates glycine and serine in C_4 plants is slow compared with their labeling in C_3 plants (Osmond and Harris, 1971; Osmond, 1972; Osmond and Björkman, 1972; Mahon et al., 1974). These observations, and the low capacity of the glycolate pathway enzymes, which are largely restricted to bundle sheath cells (Osmond and Harris, 1971; Huang and Beevers, 1972), suggest that carbon flux through this pathway is much lower in C_4 plants than in C_3 plants. Such CO_2 as is released during glycolate metabolism in C_4 plants presumably mixes with the CO_2 derived from the C_4 acid decarboxylation systems in bundle sheath cells and some may be refixed in mesophyll cells (Osmond and Harris, 1971; Liu and Black, 1972). However, Osmond and Björkman (1972) were unable to demonstrate the additional energy requirements expected of light-limited C_4 photosynthesis in 21% O_2 if recycling of photorespiratory CO_2 was a significant process. Claims for a different form of photorespiration in C_4 plants (Zelitch, 1973; Mahon et al., 1974) must await definitive evidence.

In terms of plant performance, the ability to fix CO_2 by PEP carboxylase at low intercellular concentrations enables C_4 plants to sustain higher photosynthetic rates, associated with high stomatal resistance and low rates of water loss. In fact, these plants expend less than half as much water as C_3 plants per unit of dry weight gain (Black et al., 1969; Downes, 1969, 1970). The ability to concentrate CO_2 in the vicinity of RuBP carboxylase in bundle sheath cells optimizes the activity of this enzyme and overcomes the inhibition due to atmospheric O_2, which appears to be an inevitable property of this carboxylation reaction (Lorimer and Andrews, 1973). The intricate complex of cooperative processes, based on compartmentation and transport which constitute the C_4 pathway, apparently represents a major evolutionary adaptation of higher plants to the water status and prevailing O_2 levels of terrestrial environments.

References

ANDERSEN, K.S., BAIN, J.M., BISHOP, D.G., SMILLIE, R.M.: Photosystem II activity in agranal bundle sheath chloroplasts from *Zea mays*. Plant Physiol. **49**, 461–466 (1972)

ANDERSON, J.M., BOARDMAN, N.K., SPENCER, D.: Photophosphorylation by intact bundle sheath chloroplasts from maize. Biochim. Biophys. Acta **245**, 253–258 (1971)

ANDERSON, J.M., WOO, K.C., BOARDMAN, N.K.: Deficiency of photosystem II in agranal bundle sheath chloroplasts of *Sorghum bicolor* and *Zea mays*. In: Photosynthesis, Two Centuries After its Discovery by Joseph Priestley (FORTI, G., AVRON, M., MELANDRI, A., eds.), pp. 611–619. The Hague: W. Junk 1972

ANDREWS, T.J., HATCH, M.D.: Properties and mechanism of action of pyruvate, phosphate dikinase from leaves. Biochem. J. **114**, 117–125 (1969)

ANDREWS, T.J., JOHNSON, H.S., SLACK, C.R., HATCH, M.D.: Malic enzyme and aminotransferases in relation to 3-phosphoglycerate formation in plants with the C_4-dicarboxylic acid pathway of photosynthesis. Phytochemistry **10**, 2005–2013 (1971)

ANDREWS, T.J., LORIMER, G.H., TOLBERT, N.E.: Ribulose diphosphate oxygenase. I. Synthesis of phosphoglycolate by Fraction-1 protein of leaves. Biochemistry **12**, 11–18 (1973)

ARNTZEN, C.J., BRIANTAIS, J.-M.: Chloroplast structure and function. In: Bioenergetics of Photosynthesis (GOVINDJEE, ed.). New York: Academic Press 1974

ARNTZEN, C.J., DILLEY, R.A., NEUMANN, J.: Localization of photophosphorylation and proton transport activities in various regions of the chloroplast lamellae. Biochim. Biophys. Acta **245**, 409–424 (1971)

BADGER, M.R., ANDREWS, T.J.: Effects of CO_2, O_2 and temperature on the high-affinity form of ribulose diphosphate carboxylase-oxygenase from spinach. Biochem. Biophys. Res. Commun. **60**, 204–210 (1974)

BADGER, M.R., ANDREWS, T.J., OSMOND, C.B.: Detection in C_3, C_4 and CAM plant leaves of a low *km* (CO_2) form of RuDP carboxylase, having high RuDP oxygenase at physiological pH. In: Proc. 3rd Intern. Congress on Photosynthesis. (AVRON, M., ed.), pp. 1421–1429. Amsterdam: Elsevier 1975

BAHR, J.T., JENSEN, R.G.: Ribulose diphosphate carboxylase from freshly ruptured spinach chloroplasts having an in vivo *Km* (CO_2). Plant Physiol. **53**, 39–44 (1974a)

BAHR, J.T., JENSEN, R.G.: On the activity of ribulose diphosphate carboxylase with CO_2 and O_2 from leaf extracts of *Zea mays*. Biochem. Biophys. Res. Commun. **57**, 1180–1185 (1974b)

BALDRY, C.W., BUCKE, C., COOMBS, J.: Progressive release of carboxylating enzymes during mechanical grinding of sugarcane leaves. Planta **97**, 310–319 (1971)

BASSHAM, J.A., CALVIN, M.: The Photosynthesis of Carbon Compounds. New York: W.A. Benjamin Inc. 1962

BAZZAZ, M.B., GOVINDJEE, I.: Photochemical properties of mesophyll and bundle sheath chloroplasts of maize. Plant Physiol. **52**, 257–262 (1973)

BERRY, J.A., DOWNTON, W.J.S., TREGUNNA, E.B.: The photosynthetic carbon metabolism of *Zea mays* and *Gomphrena globosa*: The location of the CO_2 fixation and carboxyl transfer reactions. Can. J. Botany **48**, 777–786 (1970)

BISHOP, D.G., ANDERSEN, K.S., SMILLIE, R.M.: Incomplete membrane-bound photosynthetic electron transfer pathway in agranal chloroplasts. Biochem. Biophys. Res. Commun. **42**, 74–81 (1971)

BISHOP, D.G., ANDERSEN, K.S., SMILLIE, R.M.: Photoreduction and oxidation of cytochrome f in bundle sheath cells of maize. Plant Physiol. **49**, 467–470 (1972a)

BISHOP, D.G., ANDERSEN, K.S., SMILLIE, R.M.: pH dependence and cofactor requirements of photochemical reactions in maize chloroplasts. Plant Physiol. **50**, 774–777 (1972b)

BJÖRKMAN, O.: Comparative photosynthetic CO_2 exchange in higher plants. In: Photosynthesis and Photorespiration (HATCH, M.D., OSMOND, C.B., SLATYER, R.O., eds.), pp. 18–32. New York: Wiley-Interscience 1971

BJÖRKMAN, O.: Comparative studies on photosynthesis in higher plants. In: Current Topics in Photobiology, Photochemistry and Photophysiology (GIESE, A., ed.). Vol. VIII, pp. 1–63. New York: Academic Press 1973

BJÖRKMAN, O., GAUHL, E.: Carboxydismutase activity in plants with and without β-carboxylation photosynthesis. Planta **88**, 197–203 (1969)

BJÖRKMAN, O., NOBS, M., PEARCY, R., BOYNTON, J., BERRY, J.: Characteristics of hybrids between C_3 and C_4 species of *Atriplex*. In: Photosynthesis and Photorespiration (HATCH, M.D., OSMOND, C.B., SLATYER, R.O., eds.), pp. 105–119. New York: Wiley Interscience 1971

BJÖRKMAN, O., TROUGHTON, J.H., NOBS, M.: Photosynthesis in relation to leaf structure. In: Basic Mechanisms in Plant Morphogenesis: Brookhaven Symposia in Biology No. 25, pp. 206–226. Upton: Brookhaven National Laboratory 1973

BLACK, C.C.: Photosynthetic carbon fixation in relation to net CO_2 uptake. Ann. Rev. Plant Physiol. **24**, 253–286 (1973)

BLACK, C.C., CHEN, T.M., BROWN, R.H.: Biochemical basis for plant competition. Weed Science **17**, 338–344 (1969)

BOWES, G., OGREN, W.L.: Oxygen inhibition and other properties of ribulose 1,5-diphosphate carboxylase. J. Biol. Chem. **247**, 2171–2176 (1972)

BOWES, G., OGREN, W.L., HAGEMAN, R.H.: Phosphoglycolate production catalysed by ribulose diphosphate carboxylase. Biochem. Biophys. Res. Commun. **45**, 716–722 (1971)

BUCKE, C., LONG, S.P.: Release of carboxylating enzymes from maize and sugarcane leaf tissue during progressive grinding. Planta **99**, 199–210 (1971)

BUCKE, C., OLIVER, I.R.: Location of enzymes metabolizing sucrose and starch in the grasses *Pennisetum purpureum* and *Muhlenbergia montana*. Planta **122**, 45–52 (1975)

CHAPMAN, E.A., OSMOND, C.B.: The effect of light on the tricarboxylic acid cycle in green leaves III. A comparison between some C_3 and C_4 plants. Plant Physiol. **53**, 893–898 (1974)

CHAPPELL, J.B., HAAROFF, K.N.: The penetration of the mitochondrial membrane by anions and cations. In: Biochemistry of Mitochondria. (SLATER, E.C., KANIUGA, Z., WOJTCSAK, L., eds.), pp. 75–91. New York: Academic Press 1967

CHEN, T.M., BROWN, R.H., BLACK, C.C.: Photosynthetic $^{14}CO_2$ fixation products and activities of enzymes related to photosynthesis in bermuda grass and other plants. Plant Physiol. **47**, 199–203 (1971)

CHEN, T.M., CAMPBELL, W.H., DITTRICH, P., BLACK, C.C.: Distribution of carboxylation and decarboxylation enzymes in isolated mesophyll cells and bundle sheath strands of C_4 plants. Biochem. Biophys. Res. Commun. **51**, 461–467 (1973)

CHEN, T.M., DITTRICH, P., CAMPBELL, W.H., BLACK, C.C.: Metabolism of epidermal tissues, mesophyll cells and bundle sheath strands resolved from mature nutsedge leaves. Arch. Biochem. Biophys. **163**, 246–262 (1974)

CHOLLET, R., OGREN, W.L.: Photosynthetic carbon metabolism in isolated maize bundle sheath strands. Plant Physiol. **51**, 787–792 (1973)

COOMBS, J., BALDRY, C.W.: C_4 pathway in *Pennisetum purpureum*. Nature **238**, 268–270 (1972)

COOMBS, J., BALDRY, C.W., BROWN, J.E.: The C_4 pathway in *Pennisetum purpureum* III. Structure and photosynthesis. Planta **110**, 121–129 (1973a)

COOMBS, J., BALDRY, C.W., BUCKE, C.: The C_4 pathway in *Pennisetum purpureum*. I. The allosteric nature of PEP carboxylase. Planta **110**, 95–107 (1973b)

COOMBS, J., MAW, S.L., BALDRY, C.W.: Metabolic regulation in C_4 photosynthesis: the inorganic carbon substrate for PEP carboxylase. Plant Sci. Letters **4**, 97–102 (1975)

COOPER, T.G., WOOD, H.G.: The carboxylation of phosphoenolpyruvate and pyruvate. II. The active species of 'CO_2' utilized by phosphoenolpyruvate carboxylase and pyruvate carboxylase. J. Biol. Chem. **246**, 5488–5490 (1971)

DOWNES, R.W.: Differences in transpiration rates between tropical and temperate grasses under controlled conditions. Planta **88**, 261–273 (1969)

DOWNES, R.W.: Effect of light intensity and leaf temperature on photosynthesis and transpiration in wheat and Sorghum. Australian J. Biol. Sci. **23**, 775–782 (1970)

DOWNTON, W.J.S.: Preferential C_4-dicarboxylic acid synthesis, the post-illumination CO_2 burst, carboxyl transfer step, and grana configurations in plants with C_4-photosynthesis. Can. J. Botany **48**, 1795–1800 (1970)

DOWNTON, W.J.S.: Adaptive and evolutionary aspects of C_4 photosynthesis. In: Photosynthesis and Photorespiration (HATCH, M.D., OSMOND, C.B., SLATYER, R.O., eds.), pp. 3–17. New York: Wiley-Interscience 1971a

DOWNTON, W.J.S.: Chloroplasts and mitochondria of bundle sheath cells in relation to C_4-photosynthesis. In: Photosynthesis and Photorespiration (HATCH, M.D., OSMOND, C.B., SLATYER, R.O., eds.), pp. 419–425. New York: Wiley-Interscience 1971b

DOWNTON, W.J.S., BERRY, J.A., TREGUNNA, B.: C_4 photosynthesis: Non-cyclic electron flow and grana development in bundle sheath chloroplasts. Z. Pflanzenphysiol. **63**, 194–199 (1970)

DOWNTON, W.J.S., HAWKER, J.S.: Enzymes of starch and sucrose metabolism in *Zea mays* leaves. Phytochemistry **12**, 1551–1556 (1973)

DOWNTON, W.J.S., TREGUNNA, E.B.: Carbon dioxide compensation—its relation to photosynthetic carboxylation reactions, systematics of the Gramineae and leaf anatomy. Can. J. Botany **46**, 207–215 (1968)

EDWARDS, G.E., BLACK, C.C.: Photosynthesis in mesophyll cells and bundle sheath cells isolated from *Digitaria sanguinalis* leaves. In: Photosynthesis and Photorespiration (HATCH, M.D., OSMOND, C.B., SLATYER, R.O., eds.), pp. 153–168. New York: Wiley-Interscience 1971

EDWARDS, G.E., GUTIERREZ, M.: Metabolic activities in extracts of mesophyll and bundle sheath cells of *Panicum miliaceum* in relation to the C_4 decarboxylic acid pathway of photosynthesis. Plant Physiol. **50**, 728–732 (1972)

EDWARDS, G.E., KANAI, R., BLACK, C.C.: Phosphoenolpyruvate carboxykinase in leaves of certain plants which fix CO_2 by the C_4-dicarboxylic acid cycle of photosynthesis. Biochem. Biophys. Res. Commun. **45**, 278–285 (1971)

EVERSON, R.G., SLACK, C.R.: Distribution of carbonic anhydrase in relation to the C_4 pathway of photosynthesis. Phytochemistry **7**, 581–584 (1968)

FARINEAU, J.: A comparative study of the activities of photosynthetic carboxylation in a C_4 and a Calvin type plant (the sites of CO_2 fixation in C_4 plants). In: Photosynthesis and Photorespiration (HATCH, M.D., OSMOND, C.B., SLATYER, R.O., eds.), pp. 202–210. New York: Wiley-Interscience 1971

FARINEAU, J.: Studies of light and dark fixation of CO_2 in C_4 plants. In: Photosynthesis, Two Centuries After its Discovery by Joseph Priestly (FORTI, G., AVRON, M., MELANDRI, A., eds.), pp. 1971–1979. The Hague: W. Junk 1972

FEKETE, M.A.R. DE, VIEWEG, G.H.: Synthesis of sucrose in *Zea mays* leaves. Ber. Deut. Botan. Ges. **86**, 227–231 (1973)

GALMICHE, J.M.: Studies on the mechanism of glycerate 3-phosphate synthesis in tomato and maize leaves. Plant Physiol. **51**, 512–519 (1973)

GIFFORD, R.M.: The light response of CO_2 exchange: on the source of differences between C_3 and C_4 species. In: Photosynthesis and Photorespiration (HATCH, M.D., OSMOND, C.B., SLATYER, R.O., eds.), pp. 51–56. New York: Wiley-Interscience 1971

GRAHAM, D., ATKINS, C.A., REED, M.L., PATTERSON, B.D., SMILLIE, R.M.: Carbonic anhydrase, photosynthesis and light-induced pH changes. In: Photosynthesis and Photorespiration (HATCH, M.D., OSMOND, C.B., SLATYER, R.O., eds.), pp. 267–274. New York: Wiley-Interscience 1971

GRAHAM, D., HATCH, M.D., SLACK, C.R., SMILLIE, R.M.: Light-induced formation of enzymes of the C_4-dicarboxylic acid pathway of photosynthesis in detached leaves. Phytochemistry **9**, 521–532 (1970)

GUTIERREZ, M., GRACEN, V.E., EDWARDS, G.E.: Biochemical and cytological relationships in C_4 plants. Planta **119**, 279–300 (1974a)

GUTIERREZ, M., HUBER, S.C., KU, S.B., KANAI, R., EDWARDS, G.E.: Intracellular localization of carbon metabolism in mesophyll cells of C_4 plants. In: Proc. 3rd Intern. Congress on Photosynthesis (AVRON, M., ed.), pp. 1219–1230. Amsterdam: Elsevier 1975

GUTIERREZ, M., KANAI, R., HUBER, S.C., KU, S.B., EDWARDS, G.E.: Photosynthesis in mesophyll protoplasts and bundle sheath cells of various types of C_4 plants. I. Carboxylases and CO_2 fixation studies. Z. Pflanzenphysiol. **72**, 305–319 (1974b)

HABERLANDT, G.: Physiological Plant Anatomy. (Transl. M. DRUMMOND. London: Macmillan) 1884

HATCH, M.D.: The C_4-pathway of photosynthesis. Evidence for an intermediate pool of carbon dioxide and the identity of the donor C_4-dicarboxylic acid. Biochem. J. **125**, 425–432 (1971a)

HATCH, M.D.: Mechanism and function of the C_4-pathway of photosynthesis. In: Photosynthesis and Photorespiration (HATCH, M.D., OSMOND, C.B., SLATYER, R.O., eds.), pp. 139–152. New York: Wiley-Interscience 1971b

HATCH, M.D.: Separation and properties of leaf aspartate aminotransferase and alanine aminotransferase isoenzymes operative in the C_4 pathway of photosynthesis. Arch. Biochem. Biophys. **156**, 207–214 (1973)

HATCH, M.D.: Photosynthesis: the path of carbon. In: Plant Biochemistry (BONNER, J., VARNER, J., eds.). In press. New York: Academic Press 1976a

HATCH, M.D.: The C_4 pathway of photosynthesis: mechanism and function. In: CO_2 Metabolism and Plant Productivity (BURRIS, R.H., BLACK, C.C., eds.), pp. 59–81. Baltimore: University Park Press 1976b

HATCH, M.D., KAGAWA, T.: Enzymes and functional capacities of mesophyll chloroplasts from plants with C_4-pathway photosynthesis. Arch. Biochem. Biophys. **159**, 842–853 (1973)

HATCH, M.D., KAGAWA, T.: NAD malic enzyme in leaves with C_4-pathway photosynthesis and its role in C_4 acid decarboxylation. Arch. Biochem. Biophys. **160**, 346–349 (1974a)

HATCH, M.D., KAGAWA, T.: Activity, location and role of NAD malic enzyme in leaves with C_4-pathway photosynthesis. Australian J. Plant Physiol. **1**, 357–369 (1974b)

HATCH, M.D., KAGAWA, T., CRAIG, S.: Subdivision of C_4-pathway species based on differing C_4 acid decarboxylating systems and ultrastructural features. Australian J. Plant Physiol. **2**, 111–128 (1975)

HATCH, M.D., MAU, S.: Activity location and role of aspartate aminotransferase and alanine aminotransferase isoenzymes in leaves with C_4 pathway photosynthesis. Arch. Biochem. Biophys. **156**, 195–206 (1973)

HATCH, M.D., MAU, S., KAGAWA, T.: Properties of leaf NAD malic enzyme from plants with C_4-pathway photosynthesis. Arch. Biochem. Biophys. **165**, 188–200 (1974)

HATCH, M.D., OSMOND, C.B., SLATYER, R.O.: Photosynthesis and Photorespiration. New York: Wiley-Interscience 1971

HATCH, M.D., SLACK, C.R.: Photosynthesis by sugarcane leaves. A new carboxylation reaction and the pathway of sugar formation. Biochem. J. **101**, 103–111 (1966)

HATCH, M.D., SLACK, C.R.: A new enzyme for the interconversion of pyruvate and phosphoenolpyruvate and its role in the C_4 dicarboxylic acid pathway of photosynthesis. Biochem. J. **106**, 141–146 (1968)

HATCH, M.D., SLACK, C.R.: Studies on the mechanism of activation and inactivation of pyruvate, Pi dikinase: A possible regulatory role for the enzyme in the C_4 dicarboxylic acid pathway of photosynthesis. Biochem. J. **112**, 549–558 (1969a)

HATCH, M.D., SLACK, C.R.: NADP-Specific malate dehydrogenase and glycerate kinase in leaves and evidence for their location in chloroplasts. Biochem. Biophys. Res. Commun. **34**, 589–593 (1969b)

HATCH, M.D., SLACK, C.R.: The C_4-carboxylic acid pathway of photosynthesis. In: Progress in Phytochemistry (REINHOLD, L., LIWSCHITZ, Y., eds.), pp. 35–106. London: Wiley-Interscience 1970

HATCH, M.D., SLACK, C.R., BULL, T.A.: Light-induced changes in the content of some enzymes of the C_4 decarboxylic acid pathway of photosynthesis and its effect on other characteristics of photosynthesis. Phytochemistry **8**, 697–706 (1969)

HEBER, U.: Metabolite exchange between chloroplasts and cytoplasm. Ann. Rev. Plant Physiol. **25**, 393–421 (1974)

HEBER, U., KRAUSE, G.H.: Transfer of carbon, phosphate energy, and reducing equivalents across the chloroplast envelope. In: Photosynthesis and Photorespiration (HATCH, M.D., OSMOND, C.B., SLATYER, R.O., eds.), pp. 218–225. New York: Wiley-Interscience 1971

HELDT, H.W., FLIEGE, R., LEHNER, K., MILOVANCEV, M., WERDEN, K.: Metabolite movement and CO_2 fixation in spinach chloroplasts. In: Proc. 3rd Intern. Congress on Photosynthesis Research (AVRON, M., ed.), pp. 1369–1379. Amsterdam: Elsevier 1975

HOCHACHKA, P.W., SOMERO, G.N.: Strategies of Biochemical Adaptation. Philadelphia: W.B. Saunders, 1973

HODGE, A.J., MCLEAN, J.D., MERCER, F.V.: Ultrastructure of the lamellae and grana in the chloroplasts of *Zea mays* L. J. Biophys. Biochem. Cytol. **1**, 605–617 (1955)

HOMAN, P.H., SCHMID, G.H.: Photosynthetic reactions of chloroplasts with unusual structures. Plant Physiol. **42**, 1619–1632 (1967)

HUANG, A.H.C., BEEVERS, H.: Microbody enzymes and carboxylases in sequential extracts from C_4 and C_3 leaves. Plant Physiol. **50**, 242–248 (1972)

HUBER, S.C., EDWARDS, G.E.: C_4 photosynthesis: light-dependent CO_2 fixation by mesophyll cells, protoplasts and chloroplasts of *Digitaria sanguinalis*. Plant Physiol. **55**, 835–844 (1975)

HUBER, S.C., KANAI, R., EDWARDS, G.E.: Decarboxylation of malate by isolated bundle sheath cells of certain plants having the C_4-dicarboxylic acid cycle of photosynthesis. Planta **113**, 53–66 (1973)

JACKSON, W.A., VOLK, R.J.: Photorespiration. Ann. Rev. Plant Physiol. **21**, 385–432 (1970)

JOHNSON, H.S., HATCH, M.D.: The C_4-dicarboxylic acid pathway of photosynthesis: Identification of intermediates and products and quantitative evidence for the route of carbon flow. Biochem. J. **114**, 127–134 (1969)

JOHNSON, H.S., HATCH, M.D.: Properties and regulation of leaf NADP malate dehydrogenase and malic enzyme in plants with the C_4 dicarboxylic acid pathway of photosynthesis. Biochem. J. **119**, 273–280 (1970)

KAGAWA, T., HATCH, M.D.: Light-dependent metabolism of carbon compounds by mesophyll chloroplasts from plants with the C_4 pathway of photosynthesis. Australian J. Plant Physiol. **1**, 51–64 (1974a)

KAGAWA, T., HATCH, M.D.: C_4-acids as the source of carbon dioxide for Calvin cycle photosynthesis by bundle sheath cells of the C_4-pathway species *Atriplex spongiosa*. Biochem. Biophys. Res. Commun. **59**, 1326–1332 (1974b)

KAGAWA, T., HATCH, M.D.: Mitochondria as a site of C_4 acid decarboxylation in C_4-pathway photosynthesis. Arch. Biochem. Biophys. **167**, 687–696 (1975)

KANAI, R., BLACK, C.C.: Biochemical basis for net CO_2 assimilation in C_4 plants. In: Net Carbon Dioxide Assimilation in Higher Plants (BLACK, C.C., ed.), pp. 75–93. Symp. South. Sect. Am. Soc. Plant Physiol. 1972

KANAI, R., EDWARDS, G.E.: Enzymatic separation of mesophyll protoplasts and bundle sheath cells from leaves of C_4 plants. Naturwissenschaften **60**, 157–158 (1973a)

KANAI, R., EDWARDS, G.E.: Separation of mesophyll protoplasts and bundle sheath cells of maize leaves for photosynthetic studies. Plant Physiol. **51**, 1133–1137 (1973b)

KARPILOV, Y.S.: Pecularities of the functions and structure of the photosynthetic apparatus in some species of plants of tropical origin. Proc. Mold. Inst. Irrigation Vegetable Res. **11**, 1–34 (1969)

KARPILOV, Y.S.: Cooperation photosynthesis in xerophytes. Proc. Mold. Inst. Irrigation Vegetable Res. **11**(3), 3–66 (1970)

KARPILOV, Y.S. (ed.): Photosynthesis in maize. Structural and functional properties of photosynthetic apparatus. Puschino-on-oka: Academy of Sciences (USSR) 1974

KENNEDY, R.A., LAETSCH, W.M.: Plant species intermediate for C_3, C_4 photosynthesis. Science **184**, 1087–1089 (1974)

KHANNA, R., SINHA, S.K.: Change in the predominance from C_4 to C_3-pathway following anthesis in *Sorghum*. Biochem. Biophys. Res. Commun. **52**, 121–124 (1973)

KOLATTUKUDY, P.E., DEAN, B.B.: Structure, gas chromatographic measurement and function of suberin synthesised by potato tuber tissue slices. Plant Physiol. **54**, 116–121 (1974)

KORTSCHAK, H.P., HART, C.E., BURR, G.O.: Carbon dioxide fixation in sugarcane leaves. Plant Physiol. **40**, 209–213 (1965)

KU, S.B., GUTIERREZ, M., EDWARDS, G.E.: Localization of the C_4 and C_3 pathways of photosynthesis in the leaves of *Pennisetum purpureum* and other C_4 species. Planta **119**, 267–278 (1974a)

KU, S.B., GUTIERREZ, M., KANAI, R., EDWARDS, G.E.: Photosynthesis in mesophyll protoplasts and bundle sheath cells of various types of C_4 plants. II. Chlorophyll and Hill reaction studies. Z. Pflanzenphysiol. **72**, 320–337 (1974b)

KUO, J., O'BRIEN, T.P., CANNY, M.J.: Pit-field distribution, plasmodesmatal frequency, and assimilate flux in the mestome sheath cells of wheat leaves. Planta **121**, 97–118 (1974)

LABER, L.J., LATZKO, E., GIBBS, M.: Photosynthetic path of carbon dioxide in spinach and corn leaves. J. Biol. Chem. **249**, 3436–3441 (1974)

LAETSCH, W.M.: Chloroplast specialization in dicotyledons possessing the C_4-dicarboxylic acid pathway of photosynthetic CO_2 fixation. Am. J. Botany **55**, 875–883 (1968)

LAETSCH, W.M.: Relationship between chloroplast structure and photosynthetic carbon fixation pathways. Sci. Prog. Oxf. **57**, 323–351 (1969)

LAETSCH, W.M.: Chloroplasts structural relationships in leaves of C_4 plants. In: Photosynthesis and Photorespiration (HATCH, M.D., OSMOND, C.B., SLATYER, R.O., eds.), pp. 323–349. New York: Wiley-Interscience 1971

LAETSCH, W.M.: The C_4 syndrome: a structural analysis. Ann. Rev. Plant Physiol. **25**, 27–52 (1974)

LAING, W.A., OGREN, W.L., HAGEMAN, R.H.: Regulation of soybean net photosynthetic CO_2 fixation by the interaction of CO_2, O_2 and ribulose 1,5-diphosphate carboxylase. Plant Physiol. **54**, 678–685 (1974)

LATZKO, E., LABER, L., GIBBS, M.: Transient changes in levels of some compounds in spinach and maize leaves. In: Photosynthesis and Photorespiration (HATCH, M.D., OSMOND, C.B., SLATYER, R.O., eds.), pp. 196–201. New York: Wiley-Interscience 1971

LERMAN, J.C.: How to interpret variations in the carbon isotope ratio of plants. In: Environmental and Biological Control of Photosynthesis (MARCELLE, J., ed.), pp. 323–335. The Hague: W. Junk 1975

LIU, A.Y., BLACK, C.C.: Glycolate metabolism in mesophyll cells and bundle sheath cells isolated from Crab grass, *Digitaria sanguinalis* (L) Scop leaves. Arch. Biochem. Biophys. **149**, 269–280 (1972)

LORIMER, G.H., ANDREWS, T.J.: Plant photorespiration—an inevitable consequence of the existence of an oxygen atmosphere. Nature **243**, 359–360 (1973)

LUDLOW, W.M.: Effect of oxygen concentration on leaf photosynthesis and resistance to carbon dioxide diffusion. Planta **91**, 285–290 (1970)

LÜTTGE, U.: Structure and function of plant glands. Ann. Rev. Plant Physiol. **22**, 23–44 (1971)

MAHON, J.D., FOCK, H., HÖHLER, T., CANVIN, D.T.: Changes in specific radioactivities of corn-leaf metabolites during photosynthesis in $^{14}CO_2$ and $^{12}CO_2$ at normal and low oxygen. Planta **120**, 113–123 (1974)

MARAYAMA, H., EASTERDAY, R.L., CHANG, H.C., LANE, M.D.: The enzymic carboxylation of phosphoenolpyruvate. I. Purification and properties of phosphoenolpyruvate carboxylase. J. Biol. Chem. **241**, 2405–2412 (1966)

MAYNE, B.C., DEE, A.M., EDWARDS, G.E.: Photosynthesis in mesophyll protoplasts and bundle sheath cells of various types of C_4 plants. III. Fluorescence emission spectra, delayed light emission, and P_{700} content. Z. Pflanzenphysiol. **74**, 275–291 (1974)

MAYNE, B.C., EDWARDS, G.E., BLACK, C.C.: Spectral physical and electron transport activities in the photosynthetic apparatus of mesophyll cells and bundle sheath cells of *Digitaria sanguinalis* (L.) Scop. Plant Physiol. **47**, 600–605 (1971)

NEWCOMB, E.H., FREDERICK, S.E.: Distribution and structure of plant microbodies (peroxisomes). In: Photosynthesis and Photorespiration (HATCH, M.D., OSMOND, C.B., SLATYER, R.O., eds.), pp. 442–457. New York: Wiley-Interscience 1971

NOBEL, P.S.: Introduction to Biophysical Plant Physiology. San Francisco: W.H. Freeman 1974

O'BRIEN, T.P., CARR, D.J.: A suberized layer in the cell walls of the bundle sheath of grasses, Australian J. Biol. Sci. **23**, 275–287 (1970)

OLESEN, P.: Plasmodesmata between mesophyll and bundle sheath cells in relation to the exchange of C_4-acids. Planta **123**, 199–202 (1975)

O'NEAL, D., HEW, C.-S., LATZKO, E., GIBBS, M.: Photosynthetic carbon metabolism of isolated corn chloroplasts. Plant Physiol. **49**, 607–614 (1972)

OSMOND, C.B.: Metabolite transport in C_4 photosynthesis. Australian J. Biol. Sci. **24**, 159–163 (1971a)

OSMOND, C.B.: The absence of photorespiration in C_4 plants: real or apparent? In: Photosynthesis and Photorespiration (HATCH, M.D., OSMOND, C.B., SLATYER, R.O., eds.), pp. 472–482. New York: Wiley-Interscience 1971b

OSMOND, C.B.: Glycolate metabolism in C_4 plants. In: Photosynthesis, Two Centuries after its Discovery by Joseph Priestley (FORTI, G., AVRON, M., MELANDRI, A., eds.), pp. 2233–2239. The Hague: W. Junk 1972

OSMOND, C.B.: Carbon reduction and photosystem II deficiency in leaves of C_4 plants. Australian J. Plant Physiol. **1**, 41–50 (1974)

OSMOND, C.B., ALLAWAY, W.G.: Pathways of CO_2 fixation in the CAM plant *Kalanchoe daigremontiana*. I. Patterns of $^{14}CO_2$ fixation in the light. Australian J. Plant Physiol. **1**, 503–511 (1974)

OSMOND, C.B., BJÖRKMAN, O.: Simultaneous measurements of oxygen effects on net photosynthesis and glycolate metabolism in C_3 and C_4 species of *Atriplex* Carnegie Inst. Wash. Yearbook **71**, 141–148 (1972)

OSMOND, C.B., HARRIS, B.: Photorespiration during C_4-photosynthesis. Biochim. Biophys. Acta **234**, 270–282 (1971)

OSMOND, C.B., SMITH, F.A : Symplastic transport of metabolites during C_4 photosynthesis. In: Intercellular Communication in Plants: Studies on Plasmodesmata (GUNNING, B.E.S., ROBARDS, A.W., eds.), pp. 229–241. Berlin-Heidelberg-New York: Springer 1976

OSMOND, C.B., TROUGHTON, J.H., GOODCHILD, D.J.: Physiological, biochemical and structural studies of photosynthesis and photorespiration in two species of *Atriplex*. Z. Pflanzenphysiol. **61**, 218–237 (1969)

PALMER, J.M., HALL, D.O.: The mitochondrial membrane system. In: Progress in Biophysics and Molecular Biology (BUTLER, J.A.V., NOBLE, D., eds.), Vol. 24, pp. 125–176. New York: Pergamon Press 1972

PATE, J.S., GUNNING, B.E.S.: Transfer cells. Ann. Rev. Plant Physiol. **23**, 173–196 (1972)

PITMAN, M.G.: Sodium and potassium uptake by seedlings of *Hordeum vulgare*. Australian J. Biol. Sci. **18**, 10–24 (1965)

POINCELOT, R.P.: The distribution of carbonic anhydrase and ribulose diphosphate carboxylase in maize leaves. Plant Physiol. **50**, 336–340 (1972)

POLYA, G.M., OSMOND, C.B.: Photophosphorylation by mesophyll and bundle sheath chloroplasts of C_4 plants. Plant Physiol. **49**, 267–269 (1972)

RHOADES, M.M., CARVALHO, A.: The function and structure of the parenchyma sheath plastids of the maize leaf. Bull. Torrey Botan. Club **71**, 335–346 (1944)

ROBARDS, A.W.: Plasmodesmata. Ann. Rev. Plant Physiol. **26**, 13–29 (1975)

SALIN, M.L., CAMPBELL, W.H., BLACK, C.C.: Oxaloacetate as the Hill oxidant in mesophyll cells of plants possessing the C_4 dicarboxylic acid cycle of leaf photosynthesis. Proc. Natl. Acad. Sci. US **70**, 3730–3734 (1973)

SHOMER-ILAN, A., WAISEL, Y.: The effect of sodium chloride on the balance between the C_3- and C_4-carbon fixation pathways. Physiol. Plantarum **29**, 190–193 (1973)

SLACK, C.R.: Localization of photosynthetic enzymes in mesophyll and parenchyma sheath chloroplasts of maize and *Amaranthus palmeri*. Phytochemistry **8**, 1387–1391 (1969)

SLACK, C.R., HATCH, M.D.: Comparative studies on the activity of carboxylases and other enzymes in relation to the new pathway of photosynthetic carbon dioxide fixation in tropical grasses. Biochem. J. **103**, 660–665 (1967)

SLACK, C.R., HATCH, M.D., GOODCHILD, D.J.: Distribution of enzymes in mesophyll and parenchyma sheath chloroplasts of maize in relation to the C_4 dicarboxylic acid pathway of photosynthesis. Biochem. J. **114**, 489–498 (1969)

SLATYER, R.O.: Comparative photosynthesis, growth and transpiration of two species of *Atriplex*. Planta **93**, 175–189 (1970)

SMILLIE, R.M., ANDERSEN, K.S., TOBIN, N.F., ENTSCH, B., BISHOP, D.G.: Nicotinamide adenine dinucleotide phosphate reduction from water by agranal chloroplasts isolated from bundle sheath cells of maize. Plant Physiol. **49**, 471–475 (1972a)

SMILLIE, R.M., BISHOP, D.G., ANDERSEN, K.S.: The photosynthetic electron transfer system in agranal chloroplasts. In: Photosynthesis, Two Centuries After its Discovery by Joseph Priestly (FORTI, G., AVRON, M., MELANDRI, A., eds.), pp. 779–788. The Hague: W. Junk 1972b

SMITH, B.N., BROWN, W.V.: The Kranz syndrome in the Gramineae as indicated by carbon isotope ratios. Am. J. Botany **60**, 505–513 (1973)

SMITH, F.A.: Transport of solutes during C_4 photosynthesis: assessment. In: Photosynthesis and Photorespiration (HATCH, M.D., OSMOND, C.B., SLATYER, R.O., eds.), pp. 502–506. New York: Wiley-Interscience 1971

SPANSWICK, R.M.: Symplastic transport in plants. In: Transport at the Cellular Level. (Symposia of the Society of Experiment Botany.) Vol. XXVIII, pp. 127–137. London: Cambridge University Press 1974

SUGIYAMA, T.: Purification, molecular and catalytic properties of pyruvate, phosphate dikinase from the maize leaf. Biochemistry **12**, 2862–2867 (1973)

TING, I.P., OSMOND, C.B.: Photosynthetic phosphoenolpyruvate carboxylases. Characteristics of alloenzymes from C_3 and C_4 plants. Plant Physiol. **51**, 439–447 (1973)

TOLBERT, N.E.: Microbodies—peroxisomes and glyoxysomes Ann. Rev. Plant Physiol. **22**, 45–74 (1971)

TROUGHTON, J.H.: The lack of carbon dioxide evolution in maize leaves in the light. Planta **100**, 87–92 (1971)

TYREE, M.T.: The symplast concept a general theory of symplast transport according to the thermodynamics of irreversible processes. J. Theoret. Biol. **26**, 181–214 (1970)

UMBREIT, W.W., BURRIS, R.H., STAUFFER, J.F.: Manometric Techniques, 4th ed. Minneapolis: Burgess Publ. Co. 1964

VIEWEG, G.H., FEKETE, M.A.R. DE: The regulation of starch metabolism in *Zea mays* leaves. Ber. Deut. Botan. Ges. **86**, 233–239 (1973)

WALKER, D.A., BROWN, J.M.: Physiological studies on acid metabolism. V. Effects of carbon dioxide concentration on phosphoenolpyruvic carboxylase activity. Biochem. J. **67**, 79–83 (1957)

WAYGOOD, E.R., MACHE, R., TAN, C.K.: Carbon dioxide, the substrate for phosphoenopyruvate carboxylase from leaves of maize. Can. J. Botany **47**, 1455–1458 (1969)

WEAST, R.C. (ed.): Handbook of Chemistry and Physics, 49th ed. Cleveland: Chemical Rubber Company 1963

WHELAN, T., SACKETT, W.M., BENEDICT, C.R.: Enzymatic fractionation of carbon isotopes by phosphoenolpyruvate carboxylase from C_4 plants. Plant Physiol. **51**, 1051–1054 (1973)

WISKICH, J.T.: Substrate transport into plant mitochondria: swelling studies. Australian J. Plant Physiol. **1**, 177–181 (1974)

WOO, K.C., ANDERSON, J.M., BOARDMAN, N.K., DOWNTON, W.J.S., OSMOND, C.B., THORNE, S.W.: Deficient photosystem II in agranal bundle sheath chloroplasts of C_4 plants. Proc. Natl. Acad. Sci. US **67**, 18–25 (1970)

ZELITCH, I.: Alternative pathways of glycolate synthesis in tobacco and maize leaves in relation to rates of photorespiration. Plant Physiol. **51**, 299–305 (1973)

5. Interactions among Organelles Involved in Photorespiration

C. SCHNARRENBERGER and H. FOCK

1. Introduction

Illuminated leaves of C_3 plants evolve considerable amounts of CO_2 during carbon dioxide assimilation. This continuous production of CO_2 in the light (photorespiration) shifts the CO_2 compensation concentration, where the rates of CO_2 uptake and CO_2 evolution are equal, towards 30 to 70 vpm CO_2 and is still manifest as an apparent CO_2 outburst at the beginning of a dark period when photosynthesis has already ceased. Photorespiration differs from dark respiration in its response to light, temperature, chemical inhibitors, CO_2 and especially oxygen. The oxygen effect on net photosynthetic CO_2 uptake as well as the inhibition of net photosynthetic O_2 evolution by oxygen can be reduced or eliminated by decreasing the oxygen tension, by reducing the temperature, or by increasing the CO_2 concentration (JACKSON and VOLK, 1970, where most of the gas exchange literature has been extensively reviewed.)

There were three main points which led toward an understanding of the molecular events leading to photorespiration. First, glycolate was recognised being rapidly labeled during photosynthetic CO_2 reduction (BENSON and CALVIN, 1950). Its occurrence appeared to be related to photorespiration, although its origin and its further metabolism were rather obscure at that time. In 1963, TOLBERT proposed the so-called glycolate pathway by which glycolate is metabolized to glyoxylate, glycine, serine, hydroxypyruvate, glycerate, and finally to carbohydrates. In this pathway light-respiratory CO_2 is generated during the conversion of two molecules of glycine to one molecule of serine. Second, in 1968, a new subcellular organelle, the peroxisome, was discovered in green leaves which contained, among others, many enzymes of the glycolate pathway and is involved in photorespiration (TOLBERT, 1971a). Finally, in 1971, a reaction leading to glycolate was discovered (OGREN and BOWES, 1971; ANDREWS et al., 1973; LORIMER et al., 1973). An enzyme displaying ribulose bisphosphate oxygenase activity and associated with the fraction I protein was shown to generate phosphoglycerate and phosphoglycolate from ribulose bisphosphate and oxygen. A phosphoglycolate phosphatase then hydrolyses phosphoglycolate to glycolate and phosphate. By these reactions many regulatory properties of photorespiration could be explained.

Two components of the overall oxygen effect have been recognized, (a) a direct inhibition of photosynthetic CO_2 fixation and (b) an increase in photorespiratory CO_2 evolution by oxygen. The first effect may be explained by a competitive inhibition of the ribulose bisphosphate carboxylase by oxygen, and the second effect by an increase in glycolate formation accompanied by an increased flux of carbon through the glycolate pathway from which photorespiratory CO_2 is derived. This would provide a linkage between photorespiratory gas exchange and substrate metabolism of photorespiration. Substantial evidence in support of this notion has been accumulated and quantitative models have been proposed by OSMOND and BJÖRKMAN (1972) and LAING et al. (1974), even though they do not yet fit all experimental results.

Enzymes involved in glycolate metabolism, and thus in photorespiration, have been studied with respect to their intracellular distribution. It turned out that they were distributed in several cell compartments: the chloroplasts, the peroxisomes, the mitochondria, and the cytosol. This suggests a cooperation of several cell compartments in photorespiration and, in addition, the transport of several metabolites between cell compartments. Direct evidence for such transport is still largely missing. The cooperation argument rests mostly on the compartmentation of the enzymes of glycolate metabolism in different cell organelles, and on the establishment of a route for glycolate metabolism by enzyme inhibitor and radioactive tracer studies.

The purpose of this article is to discuss the metabolism of glycolate in relation to compartmentation and, within the limits of present knowledge, to reconcile biochemical work with gas exchange phenomena. Excessive recapitulations of problems which have already been discussed in recent reviews (ZELITCH, 1964; GOLDSWORTHY, 1970; FOCK, 1970; JACKSON and VOLK, 1970; WOLF, 1970; TOLBERT, 1971; ZELITCH, 1971; TOLBERT, 1973; BLACK, 1973; ZELITCH, 1975) have been avoided.

2. Methodology

2.1 Measurement of Photorespiration

2.1.1 Definitions

Current evidence suggests that dark respiration of C_3-type leaves (see Chap. II, 3) may be inhibited in the light and replaced by photorespiration (FORRESTER et al., 1966a, b; TREGUNNA et al., 1966; HOLMGREN and JARVIS, 1967). In this article we shall use the term "photorespiration" or "CO_2 evolution in the light" for all CO_2 releasing processes in the light regardless of the mechanisms involved and regardless of their localization inside the green cell. Where the processes can be separated, we shall distinguish between "glycolate pathway" or "light respiration" for carbon dioxide produced through the operation of the glycolate pathway, and "mitochondrial" or "dark respiration" for carbon dioxide released during Krebs cycle activity.

Not only mitochondrial respiration but also photorespiration is accompanied by O_2 consumption (HOCH et al., 1963; FOCK et al., 1969; FOCK et al., 1971; MULCHI et al., 1971). Normally photorespiration is characterized through the carbon dioxide evolved in the light. In some cases, however, O_2 consumption, which is more difficult to measure against the high concentration of oxygen in air (21% O_2) than CO_2 evolution in the light, is used to describe photorespiration.

The flux of carbon through the glycolate pathway has been used as another parameter to analyze photorespiration.

2.1.2 CO_2 Gas Exchange

Apparent photosynthesis (APS) is defined as the net rate of CO_2 uptake in the light and is the difference between photosynthetic and photorespiratory processes. In 21% O_2:

$$APS_{21\% O_2} = TPS'_{21\% O_2} - PR'_{21\% O_2}. \quad (1)$$

TPS′ is the rate of CO_2 fixation from the external medium by photosynthesis. PR′ is the amount of photorespiratory CO_2 released into the medium, while the remaining photorespiratory CO_2 is recycled before it can escape into the medium (Fig. 1).

The true rate of CO_2 fixation (TPS) may be higher than TPS′ because of reassimilated carbon (RA) from photorespiratory sources:

$$RA_{21\% O_2} = TPS_{21\% O_2} - TPS'_{21\% O_2}. \quad (2)$$

$$RA_{21\% O_2} = PR_{21\% O_2} - PR'_{21\% O_2}. \quad (3)$$

If one substitutes the $TPS_{21\% O_2}$ and $PR_{21\% O_2}$ from Eqs. (2) and (3) for the terms $TPS'_{21\% O_2}$ and $PR'_{21\% O_2}$ in Eq. (1):

$$APS_{21\% O_2} = TPS_{21\% O_2} - PR_{21\% O_2}. \quad (4)$$

In C_3-type plants the rate of true photosynthesis in normal oxygen ($TPS_{21\% O_2}$) is lower than in a medium with low oxygen partial pressure ($TPS_{1\% O_2}$) because oxygen inhibits photosynthesis ($IPS_{21\% O_2}$):

$$TPS_{1\% O_2} = TPS_{21\% O_2} + IPS_{21\% O_2}. \quad (5)$$

Combining Eqs. (4) and (5) yields Eq. (6):

$$TPS_{1\% O_2} = APS_{21\% O_2} + PR_{21\% O_2} + IPS_{21\% O_2}. \quad (6)$$

Using the measurable parameters $TPS'_{1\% O_2}$ and $PR'_{21\% O_2}$, Eq. (6) becomes:

$$TPS'_{1\% O_2} = APS_{21\% O_2} + PR'_{21\% O_2} + IPS_{21\% O_2}. \quad (6a)$$

APS can easily be determined as apparent or net CO_2 uptake by infrared gas analysis or other means (SESTAK et al., 1971). However, it is an experimental challenge to measure the true rates of CO_2 fixation (TPS) and of CO_2 evolution in the light (PR), as the rate of reassimilation (RA) cannot precisely be determined (Fig. 1). Most methods that have been used to determine photorespiration by gas exchange attempt to restrict TPS or eliminate PR by changing the environment of the photosynthesizing tissue, or by applying chemical inhibitors, so that the CO_2 gas exchange of the remaining single process can be measured and Eqs. (1) and (4) can be solved. The accuracy of measurements of TPS′ and PR′ depends on two conditions which, because of the close relationship between photosynthesis and photorespiration, are not easily met: the inhibition of one process in Eq. (1) must be complete, and the manipulation must not effect the other process in any way (LUDWIG, 1968). Therefore, the current measurements of the rates of photorespiration and true photosynthesis can only approximate PR′ and TPS′ and often underestimate total PR and TPS to a considerable extent (JACKSON and VOLK, 1970; ZELITCH, 1971). Within the limitations outlined, photorespiration (normally only PR′ or a fraction of it) can be estimated by measuring:

(a) CO_2 evolution into CO_2 free air (FOCK and EGLE, 1966; MOSS, 1966; EL-SHARKAWY et al., 1967; HOLMGREN and JARVIS, 1967; HEW and KROTKOV, 1968; JACKSON and VOLK,

1970; LUDLOW and JARVIS, 1971); (b) CO_2 compensation concentration (FORRESTER et al., 1966a; TREGUNNA et al., 1966; HOLMGREN and JARVIS, 1967; HEATH and ORCHARD, 1968; LUDLOW and JARVIS, 1971; FOCK et al., 1971; ZELITCH, 1971); (c) the postillumination CO_2 burst (DECKER, 1955, 1959; TREGUNNA et al., 1961, 1964; SEMENENKO, 1964; MOSS, 1966; EGLE and FOCK, 1967; JACKSON and VOLK, 1970; LUDLOW and JARVIS, 1971; ZELITCH, 1971; LAWLOR, 1975); (d) extrapolation of apparent photosynthesis (APS) to zero CO_2 concentration (HEW et al., 1969a and b; JACKSON and VOLK, 1970; ZELITCH, 1971); (e) use of isotopic CO_2 (KROTKOV et al., 1958; LISTER et al., 1961; OZBUN et al., 1964; GOLDSWORTHY, 1966; LUDWIG, 1968; ZELITCH, 1968; FOCK et al., 1970; JACKSON and VOLK, 1970); (f) use of isotopic O_2 (HOCH et al., 1963; OZBUN et al., 1964; JACKSON and VOLK, 1970; MULCHI et al., 1971); (g) modifying metabolism by applying chemical inhibitors (ZELITCH, 1966, 1971; MOSS, 1968; LUDLOW and JARVIS, 1971).

If only relative rates of photorespiration have to be determined, any one of these methods may be standardized to the desired accuracy. However, all the methods mentioned here are unsatisfactory in the determination of the true rates of photorespiration (PR).

At the present time the best system for the determination of photorespiration (PR′) is the flow-through gas exchange system developed by LUDWIG (1968), which includes an infrared gas analyzer for the measurement of CO_2 and an ionization chamber for the determination of $^{14}CO_2$. This system has been slightly modified by others (ATKINS and CANVIN, 1971; LUDWIG and CANVIN, 1971a; CANVIN and FOCK, 1972; D'AOUST and CANVIN, 1972). The advantages of this system are that it is possible to determine the rate of photorespiration (PR′) and the specific activity of the $^{14}CO_2$ evolved in the light with respect to the $^{14}CO_2$ fixed. Moreover, at different CO_2 and O_2 concentrations and on the same biological material several other parameters may be measured simultaneously (e.g. transpiration rates, boundary layer and stomatal resistances, amount and specific activity of ^{14}C-labeled metabolites).

2.1.3 Estimates of Photorespiration by Leaf Models

Some of the difficulties involved in the estimation of TPS and PR (e.g. determination of reassimilated CO_2) may be partly overcome, if the different gas streams into and out of the leaf are considered as diffusion phenomena driven by concentration gradients and hampered by anatomical and morphological barriers and by the biochemistry of the tissue. These gas streams have often been described with leaf models (MOSS, 1966; LAKE, 1967; BRAVDO, 1968; WAGGONER, 1969; JACKSON and VOLK, 1970).

Here Waggoner's model, slightly modified, has been used for the analysis of photosynthetic and photorespiratory CO_2 fluxes (Fig. 1).

The flux of carbon dioxide moving from the ambient air to the reaction centers in the chloroplasts (c) is controlled by a series of diffusive resistances and by the rate of the opposing stream of photorespiratory CO_2, which joins the main stream within the leaf at "k". The flux of carbon dioxide from the air into the leaf (F_{ai}) is expressed by:

$$F_{ai}=\frac{[CO_2]_a-[CO_2]_i}{r_a+r_s} \tag{7}$$

where the CO_2 gradient (ng CO_2/cm^3) is the difference between the concentration of CO_2 in the atmosphere $[CO_2]_a$ and the intercellular space $[CO_2]_i$, r_a(s/cm) is the resistance of the air over the leaf surface, and r_s(s/cm) the stomatal resistance with respect to $CO_2 \cdot F_{ai}$ is equivalent to APS and 2.78 ng CO_2/cm^2s equals 1 mg CO_2/dm^2h.

From the flux (E) of water vapor from the leaf (JACKSON and VOLK, 1970; LAWLOR and FOCK, 1975) r_a and r_s are calculated:

$$E=\frac{[e]_i-[e]_a}{r_{aw}+r_{sw}} \tag{8}$$

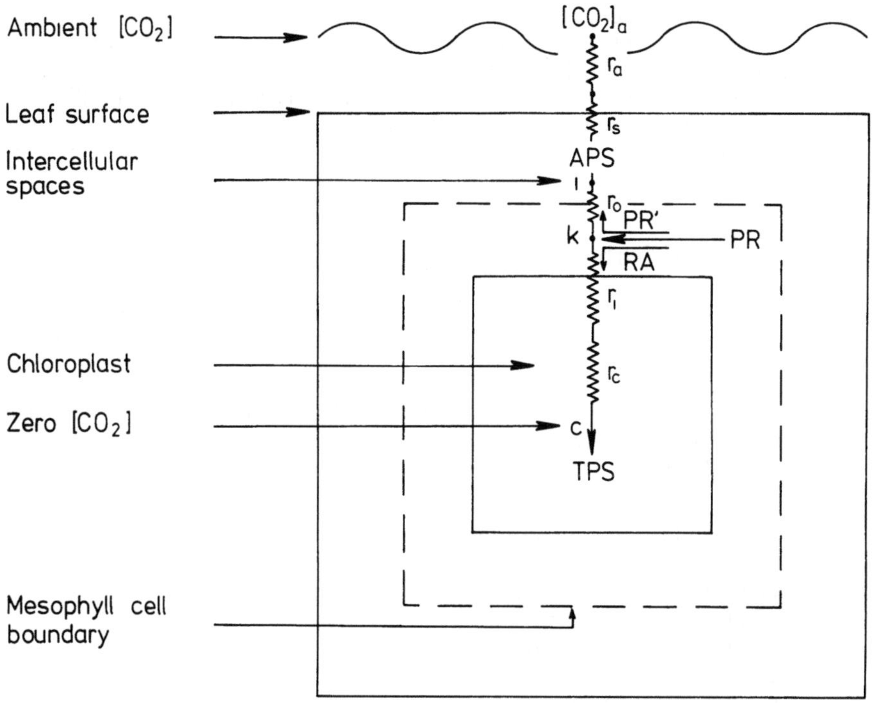

Fig. 1. Leaf model showing the pathway of CO_2 exchange during photosynthesis (adapted from WAGGONER, 1969; ZELITCH, 1971). APS: rate of apparent CO_2 uptake; TPS: rate of true CO_2 uptake at reaction center c. PR: rate of photorespiration joining main gas stream at k inside the mesophyll cell; PR′: rate of CO_2 released into intercellular spaces i; RA: rate of reassimilated CO_2; r_a and r_s are boundary layer and stomatal resistances; the mesophyll resistance r_m is the sum of r_o, r_i, and r_c

where $[e]_i$ is the water vapor content of the air at the evaporating surface assumed to be the saturation vapor content at leaf temperature, and $[e]_a$ is the water vapor content of the air around the leaf. The boundary layer and stomatal resistances to water vapor diffusion are r_{aw} and r_{sw} respectively The corresponding resistances r_a and r_s for CO_2 diffusion are calculated by multiplying r_{aw} and r_{sw} by 1.6, the ratio of the diffusion coefficients between water vapor and CO_2 in air (GAASTRA, 1959; SAMISH and KOLLER, 1968). If r_a plus r_s are known, $[CO_2]_i$ can be calculated from Eq. (7), since F_{ai} is determined through APS by infrared gas analysis.

Under rate limiting CO_2 concentrations and at high light intensity, the CO_2 partial pressure is assumed to be practically zero at the reactions center in the chloroplasts $[CO_2]_c$. Therefore, the total flux of CO_2 from the air into the chloroplasts (F_{ac}), which is equivalent to TPS′, equals:

$$F_{ac}=\frac{[CO_2]_a-[CO_2]_c}{r_a+r_s+r_m}=\frac{[CO_2]_a}{r_a+r_s+r_m} \tag{9}$$

where r_m, the so-called mesophyll resistance, is the sum of r_o (the effective resistance of the mesophyll boundary), r_i (that inside the cell), and r_c (resistance caused by limitations of photochemistry and biochemistry of CO_2 fixation within the chloroplast) (Fig. 1).

The flux of CO_2 per unit time to the carboxylation site in the chloroplast (F_{ic}), which is equivalent to the rate of true photosynthesis (TPS), is given by:

$$F_{ic}=\frac{[CO_2]_i-[CO_2]_c}{r_m}=\frac{[CO_2]_i}{r_m} \tag{10}$$

where $[CO_2]_i$ is the concentration of carbon dioxide inside the cell at "i" (Fig. 1), if the direct CO_2 fluxes from photorespiratory sources into the chloroplast are negligible.

Depending on how precisely such a model describes the actual CO_2 fluxes in a photosynthesizing leaf and on how closely the requirements for the validity of Eqs. (9) and (10) are fulfilled, F_{ac} may be determined through TPS′ and Eq. (9) solved for r_m. Eq. (10) then estimates F_{ic} and the difference between F_{ic} and F_{ac} should give an estimate of the rate of reassimilated carbon [RA in Eqs. (2 and 3)].

2.1.4 Estimates of Carbon Flux through the Glycolate Pathway

The flux of carbon per unit time (turnover rate) through the glycolate pathway (Fig. 6) may be obtained from experiments with isotopic $^{14}CO_2$ (ARONOFF, 1956).

Here only a simple case is described, assuming that the sequence of intermediates of the pathway is real, that the flux is unidirectional without branchpoints, and that each metabolite is only located in a single pool.

We follow REINER'S (1953) mathematical analysis which has been slightly modified:

T = flux rate of carbon (turnover rate)
A = number of carbon atoms in A species
B = number of carbon atoms in B species
A^* = number of isotopic carbon atoms in A species
B^* = number of isotopic carbon atoms in B species
s_A = A^*/A = specific activity of A
s_B = B^*/B = specific activity of B
dt = a small time interval
dn_{AB} = number of carbon atoms transported from A to B during dt
dn_{BX} = number of carbon atoms transported from B to another compound during dt

In the reaction sequence

$$A \rightarrow B \rightarrow X \rightarrow \ldots$$

the rate of carbon flux through compound B is equal to the number of carbon atoms entering B from A during dt or leaving B to X during dt, whichever is smaller. For the changes in B and B* one obtains:

$$dB = dn_{AB} - dn_{BX} \tag{11}$$

$$dB^* = A^*/A \cdot dn_{AB} - B^*/B \cdot dn_{BX} \tag{12}$$

$$dB^* = s_A \cdot dn_{AB} - s_B \cdot dn_{BX}. \tag{12a}$$

Combining Eqs. (11) and (12) and dividing the solutions by dt yields:

$$T_{AB} = \frac{dn_{AB}}{dt} = \frac{(dB^*/dt - s_B \cdot dB/dt)}{s_A - s_B} \tag{13}$$

$$T_{BX} = \frac{dn_{BX}}{dt} = \frac{(dB^*/dt - s_A \cdot dB/dt)}{s_A - s_B}. \tag{14}$$

These two carbon fluxes into and out of the compound B, which are equal under steady rate conditions, should be experimentally determined. For example, if the B species were serine of the glycolate pathway, then one has to measure the changes in the amount of serine (dB/dt) in kinetic $^{14}CO_2$ fixation experiments together with the changes in the radioactivity (dB^*/dt) and the specific activity of serine (s_B) and the precursor glycine (s_A). The measured values are then inserted into Eqs. (13) and (14) for the estimation of the carbon flux through serine.

However, very few standard procedures are available for the determination of specific activities (ZELITCH, 1965; HESS and TOLBERT, 1966; ROBINSON and GIBBS, 1974), which are required for the estimation of carbon flux rates. Only recently methods have been developed to measure the specific activity of ^{14}C-labeled glycolic acid (MAHON et al., 1975) and other intermediates of the glycolate pathway and the photosynthetic reduction cycle in leaves (MAHON et al., 1974a, b, c). Further difficulties in the estimation of turnover rates may arise, if the experiments reveal a more complicated flux scheme rather than the assumed unidirectional transfer. In that case the actual turnover rates have to be determined by a series of flux equations derived from REINER'S (1953) comprehensive analysis. If multiple pools of a substance are located in different cell organelles, then these cell organelles must be separated by nonaqueous isolation procedures (STOCKING, 1971) after being labeled with radioactive carbon, and the specific activity of the compound measured in each of the isolated cell fractions.

2.2 Isolation of Leaf Peroxisomes and Other Cell Organelles

All present methods for the separation of plant peroxisomes from other cell organelles such as chloroplasts and mitochondria involve gentle homogenisation of the leaf tissue, differential centrifugation, and isopycnic centrifugation in a sucrose density gradient. In the last step peroxisomes band at a characteristic density of about 1.25 g/cm^3 and are separated from other cell organelles. During the isolation in hypertonic sucrose solutions, cell organelles retain most of their enzymes. However, some permeability properties of their envelopes may be altered and metabolites may be lost or metabolized.

Details of the isolation procedure for peroxisomes have been published by TOLBERT (1971b). However, optimal conditions for the isolation of peroxisomes may be different for different tissues or species (see PARISH and RICKENBACHER, 1971). Other methods for the isolation of cell organelles such as the aqueous or nonaqueous procedures used for the isolation of chloroplasts (JENSEN and BASSHAM, 1966; STOCKING, 1971), are not yet sufficiently developed to permit a satisfying isolation of peroxisomes.

For the isolation of cell organelles the blending of the leaf tissue has to be done as gently as possible. Grinding for 5–10 s with a Waring blendor is most convenient. Sheering forces and extensive grinding should be reduced to a minimum to avoid organelle damage. The grinding medium should include 0.4 to 1.0 M sucrose and an appropriate buffer like 20 mM glycylglycine pH 7.5. In order to remove unbroken cells or tissue pieces, the cell organelles are either filtered through some layers of cheese cloth or Miracloth, or they are freed from these contaminants by a low speed centrifugation. The resultant suspension is called "crude particulate homogenate". Sine this preparation contains only part of the total enzymes of a leaf, a separate and vigorous homogenization of the same leaf tissue has to be done for a complete extraction of enzymes in order to determine the total activity in that tissue.

The crude particulate homogenate can directly be applied to a sucrose density gradient for the separation of cell organelles by isopycnic centrifugation. Alternatively, a differential centrifugation for 3 min at 3,000 g may be performed at that stage to separate chloroplasts from peroxisomes and mitochondria, the sedimentation velocity of which decreases from whole chloroplasts to broken chloroplasts, peroxisomes, and mitochondria (ROCHA and TING, 1970a). Then, peroxisomes and mitochondria may be pelleted by centrifugation for 20 to 30 min at 6,000 × g.

The final step in isolating peroxisomes is an isopycnic centrifugation in a sucrose density gradient ranging from about 25–60% (w/w) sucrose in 20 mM glycylglycine. The gradients can be prepared as linear or as step gradients. Step gradients require special care of the density of the layers, since the isopycnic density of some cell organelles may vary between different tissues and in tissues at different developmental stages (SCHNARRENBERGER et al.,

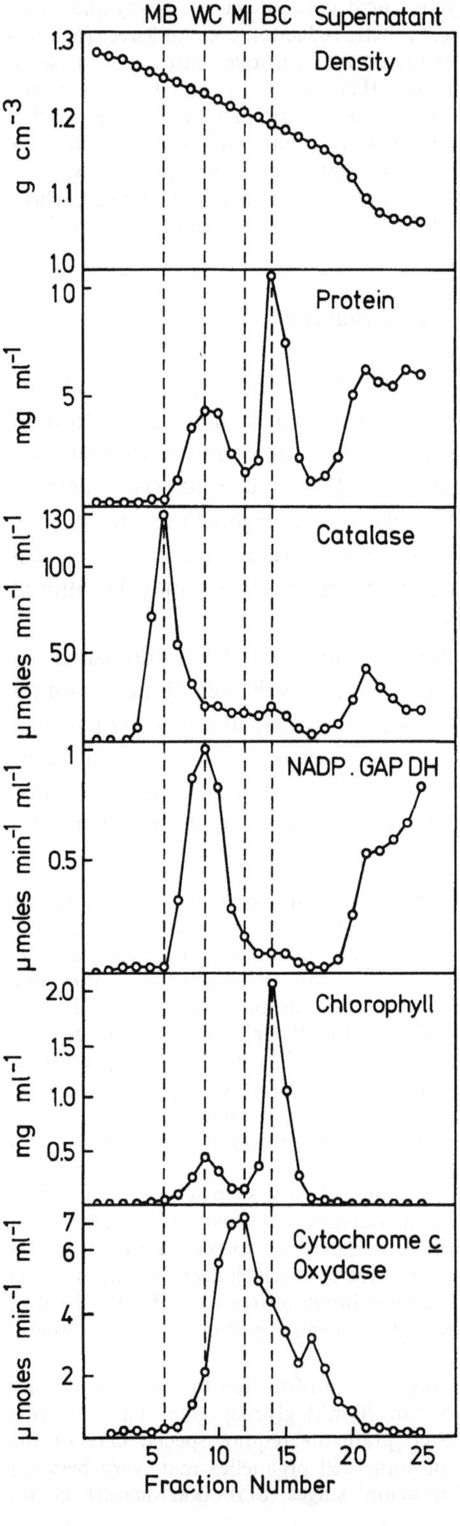

Fig. 2. Distribution of peroxisomes and other cell organelles from spinach leaves in a sucrose gradient after isolation by isopycnic centrifugation. Markers presented are catalase for peroxisomes (MB); NADP, glyceraldehydephosphate dehydrogenase for whole chloroplasts (WC); the main peak of chlorophyll for broken chloroplasts (BC); and cytochrome c ocidase for mitochondria (MI). *Dashed lines*: position of the cell organelles on the gradients (SCHNARRENBERGER, unpublished)

Table 1. Convenient markers and isopycnic densities for cell organelles after isolation by isopycnic centrifugation in sucrose density gradients

Type of cell organelle Marker	Isopycnic density (g/cm³)	References
Peroxisomes	1.25–1.27	
Catalase (EC 1.11.1.6)		TOLBERT (1971)
Glycolate oxidase (EC 1.1.3.1)		TOLBERT (1971)
Hydroxypyruvate dehydrogenase (EC 1.1.1.29)		TOLBERT (1971)
Whole chloroplasts	1.20–1.25	
Chlorophyll peak at high density		ARNON (1949)
NADP: glyceraldehyde-phosphate dehydrogenase (EC. 1.2.1.13)		HEBER et al. (1963)
NADPH: glyoxylate reductase (EC 1.1.1.79)		TOLBERT (1971)
Triosephosphate isomerase (EC 5.3.1.1.)		BEISENHERZ (1955)
Mitochondria	1.18–1.22	
Cytochrome *c* oxidase (EC 1.11.1.5)		TOLBERT et al. (1968)
Broken chloroplasts	1.14–1.18	
Chlorophyll peak at low density		ARNON (1949)
Endoplasmatic reticulum	1.14–1.17	
NADH: cytochrome *c* reductase (EC 1.6.99.3)		DONALDSON et al. (1972)

1971; EYTAN and OHAD, 1972). After centrifugation for 3 h at 50,000 × g, cell organelles have reached their isopycnic density. The gradients are fractionated and the fractions tested for cell organelle markers (see Table 1), sucrose density, and protein. A typical result is given in Figure 2.

If the intracellular location of an enzyme is to be investigated, activity profiles of marker enzymes for as many organelles as possible should be measured on the gradients. For some purposes it is necessary to estimate the absolute purity of cell organelle, for instance of the isolated peroxisomes. Maximal values for the cross contamination with other cell organelles can be obtained by comparing the specific activity (*U*/mg protein) of a marker enzyme from another cell organelle in the peroxisomal band with the specific activity of this marker in its own cell organelle.

3. Glycolate Metabolism and Its Intracellular Compartmentation

3.1 Biosynthesis of Glycolate in Chloroplasts

Glycolate is early labeled during photosynthesis in $^{14}CO_2$ (BENSON and CALVIN, 1950) and often contains a sizeable percentage of the total carbon fixed. The conditions favoring glycolate formation are: low p_{CO_2} (WILSON and CALVIN, 1955; WARBURG and KRIPPAHL, 1960; ZELITCH, 1965), high p_{O_2} (WARBURG and KRIPPAHL, 1960; BASSHAM and KIRK, 1962), high light intensity (ZELITCH, 1958; TOLBERT, 1963; COOMBS and WHITTINGHAM, 1966a, b), and elevated temperatures (ZELITCH, 1966; LAING et al., 1975). ^{14}C-label was first incorporated into Calvin cycle intermediates and then into glycolate (HESS and TOLBERT, 1966; ORTH et al., 1966). Degradation experiments showed that the two carbon atoms of glycolate are already equally labeled after the first few seconds of photosynthetic $^{14}CO_2$ fixation (SCHOU et al., 1950; BASSHAM et al., 1953; ZELITCH, 1965; HESS and TOLBERT, 1966). At the same time P-glycerate is predominantly carboxyl labeled. The C_2 and C_3 carbons of P-glycerate and the C-atoms of other Calvin cycle intermediates which are derived from these two atoms, are equally labeled. The sequence of labeling and the quantitative distribution of the ^{14}C-label in Calvin cycle intermediates and glycolate suggest that glycolate is derived from a Calvin cycle intermediate. On changing from optimal conditions for photosynthesis (high p_{CO_2}, low p_{O_2}) to optimal conditions for photorespiration (low p_{CO_2}, high p_{O_2}) the light dependent carbon metabolism is dramatically altered and the glycolate pathway is emphasized. This shift in metabolism is also shown by feeding experiments with ^{14}C-labeled glucose (WHITTINGHAM et al., 1963; COOMBS and WHITTINGHAM, 1966a, b; WHITTINGHAM and PRITCHARD, 1963; WHITTINGHAM et al., 1967).

Glycolate can also be synthesized by isolated chloroplasts (TOLBERT, 1958; KEARNEY and TOLBERT, 1962; WALKER, 1964; JENSEN and BASSHAM, 1966; BRADBEER and ANDERSON, 1967; ELLYARD and GIBBS, 1969; PLAUT and GIBBS, 1970; SHAIN and GIBBS, 1971). Favorable conditions for glycolate formation in isolated chloroplasts are the same as for glycolate formation in whole leaves, i.e. low p_{CO_2} (ELLYARD and GIBBS, 1969; PLAUT and GIBBS, 1970; ROBINSON and GIBBS, 1974), high p_{O_2} (ELLYARD and GIBBS, 1969), and high light intensity (SHAIN and GIBBS, 1971). From $^{14}CO_2$ glycolate becomes labeled more slowly than the sugar phosphates. It is formed at the expense of sugar phosphates (CHAN and BASSHAM, 1967; PLAUT and GIBBS, 1970; HEBER and KIRK unpubl.). Thus, features of glycolate formation by isolated chloroplasts are virtually the same as in whole leaves.

Glycolate cannot be further metabolized by isolated chloroplasts and represents an end product of photosynthesis in chloroplasts. This is evidenced by the accumulation of glycolate during photosynthesis of isolated chloroplasts (HEBER et al., 1974; HEBER and KIRK, unpubl.), by the rapid excretion of glycolate by isolated chloroplasts (TOLBERT, 1958; KEARNEY and TOLBERT, 1962) and by incubating chloroplasts with glycolate-1-^{14}C or glycolate-2-^{14}C (KEARNEY and TOLBERT, 1962; CHAN and BASSHAM 1967; THOMPSON and WHITTINGHAM, 1968). Traces of glycine, serine, and glyoxylate formed from labeled glycolate, as reported by some authors, may have come from the contamination of the chloroplast preparations with other cell fragments, especially peroxisomes (see below).

For the formation of glycolate several possibilities have been discussed. The reaction of ribulose bisphosphate with oxygen appears to be the predominant source of glycolate. This reaction is catalyzed by the same enzyme which in the presence of an excess of CO_2 carboxylates ribulose bisphosphate (BOWES et al., 1971; ANDREWS et al., 1973) and is the primary enzyme reaction of CO_2 fixation in photosynthetic tissues (MCFADDEN, 1973). The "ribulose bisphosphate carboxylase/oxygenase" (Fig. 3) either uses ribulose bisphosphate and CO_2 as substrate to form 2 molecules of phosphoglycerate, or it uses ribulose bisphosphate and oxygen to form one molecule each of phosphoglycerate and phosphoglycolate. The isolated enzyme incorporates $^{14}CO_2$ into the C-1 position of one of the two phosphoglycerates in the carboxylase reaction. In the oxygenase reaction label from $^{18}O_2$ is only incorporated into the carboxyl group of phosphoglycolate and

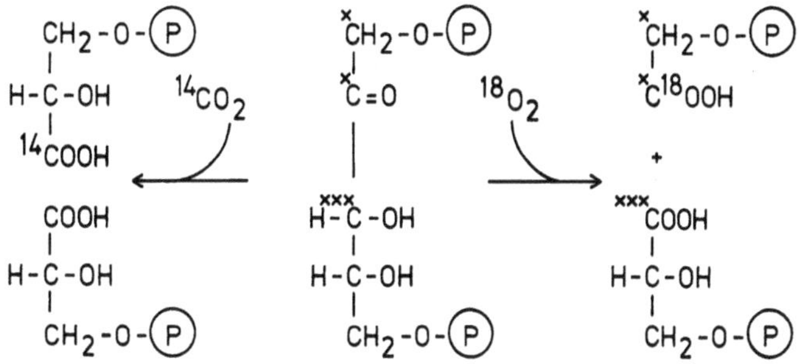

Fig. 3. Labeling pattern as obtained with $^{14}CO_2$ in the ribulose bisphosphate carboxylase reaction (left part of the scheme) and with $^{18}O_2$ in the ribulose bisphosphate oxygenase reaction (right part of the scheme). The asterisks (*) show the distribution of radioactivity in ribulose bisphosphate from $^{14}CO_2$ (after one turn in the Calvin cycle) and in phosphoglycolate and phosphoglycerate after the subsequent ribulose bisphosphate oxygenase reaction

not into that of phosphoglycerate, whereas $H_2{}^{18}O$ is incorporated into both products of this reaction (LORIMER et al., 1973). Phosphoglycerate and phosphoglycolate are the only products of the oxygenase reaction (ANDREWS et al., 1973). The labeling results are consistent with the ^{18}O-labeling pattern of glycine and serine in the subsequent glycolate pathway during in vivo photosynthesis in $^{18}O_2$ (ANDREWS et al., 1971b).

The most important feature in the properties of the ribulose bisphosphate carboxylase/oxygenase is the competition of CO_2 and O_2 for the reaction with ribulose bisphosphate. The reaction with CO_2 is inhibited competitively by oxygen and vice versa, while the affinity for ribulose bisphosphate is the same in the presence of both CO_2 and oxygen (OGREN and BOWES, 1971; BOWES et al., 1971; BOWES and OGREN, 1972; LAING et al., 1974; RYAN and TOLBERT, 1975a, b). In vivo the rate of photosynthetic CO_2 fixation and the extent of glycolate formation, and thus of photorespiration, may be regulated by competition between CO_2 and oxygen. A quantitative model, which has also been tested experimentally, has been proposed by LAING et al. (1974). Further regulatory properties of the ribulose bisphosphate carboxylase/oxygenase have been investigated recently (BAHR and JENSEN, 1974a, b; CHU and BASSHAM, 1974, 1975; RYAN and TOLBERT, 1975a, b; RYAN et al., 1975; LAING et al., 1975).

In a subsequent step, a phosphoglycolate phosphatase splits phosphoglycolate into glycolate and phosphate. This enzyme was first discovered and characterized by RICHARDSON and TOLBERT (1961). It is located in the chloroplasts (YU et al., 1964; THOMPSON and WHITTINGHAM, 1967; RANDALL et al., 1971). Time course studies for the appearance of phosphoglycolate and glycolate in algae have shown that first phosphoglycolate and then glycolate is formed (BASSHAM and KIRK, 1973), even though the rate of phosphoglycolate formation may not account for all the glycolate synthesized.

Another mechanism for glycolate formation involves the oxidation of α, β-dihydroxyethyl-thiamine pyrophosphate (Fig. 4), which is formed in the transketolase reaction (WILSON and CALVIN, 1955; DELAWAN and BENSON, 1958; COOMBS and WHITTINGHAM, 1966a; BRADBEER and ANDERSON, 1967; ELLYARD and GIBBS, 1969; PLAUT and GIBBS, 1970; SHAIN and GIBBS, 1971). The oxidation and the hydrolytic cleavage of the complex would yield glycolate. Ferricyanide and DCPIP

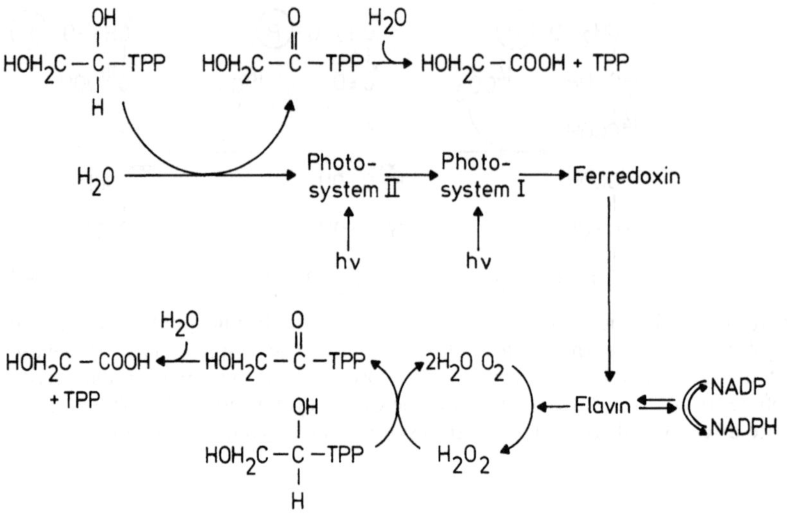

Fig. 4. Scheme of glycolate formation from α, β-dihydroxyethyl thiamine pyrophosphate

can be used as artificial electron acceptors. As natural oxidant H_2O_2 has been proposed (COOMBS and WHITTINGHAM, 1966a), which would be produced during the reduction of oxygen by reduced ferredoxin in a MEHLER (1951) type reaction. Later, PLAUT and GIBBS (1970) and SHAIN and GIBBS (1971) assigned the site of H_2O_2 formation to the ferredoxin, NADP reductase, which is a flavin enzyme. During photosynthesis under low CO_2, NADPH and reduced flavin enzyme would accumulate. Autoxidation of the latter would yield H_2O_2. They also considered a second site of the photosynthetic electron transport chain which is on the donor side of photosystem 2 to act as oxidant of the α, β-dihydroxyethyl-thiamine pyrophosphate in a nitrogen atomsphere. Similarly, EICKENBUSCH and BECK (1973) and EICKENBUSCH et al. (1975) proposed an oxygen dependent and another oxygen independent mechanism for glycolate formation. However, the rates are slow and there is little enzymatic evidence for the mechanisms proposed. TOLBERT and RYAN (1975) argue that during the oxidation of the α, β-dihydroxyethyl-thiamine pyrophosphate no $^{18}O_2$ would be incorporated into the carboxyl group of glycolate. As mentioned above, such incorporation has been observed.

Glycolate formation from acetate has also been studied by MERRETT and GOULDING (1967, 1968). However, a significant metabolic path from acetate to glycolate in green leaves is unlikely and the rates cannot account for photorespiratory glycolate formation. ZELITCH (1965) proposed that glycolate is derived directly from two molecules of CO_2, since in his experiments the specific activity of carbon in glycolate was even higher than in the carboxyl carbon atom of phosphoglycerate during photosynthesis in the presence of $^{14}CO_2$. However, this has never been confirmed. In all other cases the labeling pattern was just the other way around. Finally, glycolate could also be formed in the chloroplasts from glyoxylate by the action of the NADPH-specific glyoxylate reductase (ZELITCH and GOTTO, 1962; THOMPSON and WHITTINGHAM, 1967; TOLBERT et al., 1970). However, the activity of this enzyme is too low to account for the rate of glycolate formation (TOLBERT and RYAN, 1975). In addition, the relationship between glyoxylate and Calvin cycle intermediates would still be an unsolved problem.

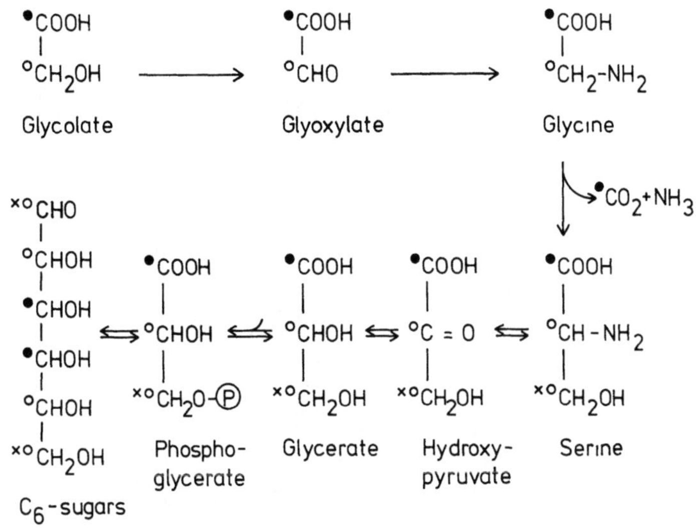

Fig. 5. Labeling pattern in radio tracer experiments showing the path of carbon atoms in the glycolate pathway from glycolate to C_6-sugars

3.2 The Glycolate Pathway and Peroxisomes

The subsequent metabolism of glycolate was first investigated by feeding experiments with glycolate, glyoxylate, glycine, and serine, which were ^{14}C-labeled in the C_1-, C_2-, and C_3-position (TOLBERT and COHAN, 1953; WANG and WAYGOOD, 1962; RABSON et al., 1962; JIMENEZ et al., 1962; TOLBERT, 1963). By these experiments a metabolic route was revealed form glycolate via glyoxylate, glycine, serine, hydroxypyruvate and glycerate to sugars (Fig. 5). This is commonly refered to as the "glycolate pathway". The flow of carbon seemed to be unidirectional from glycolate to glycine, and even the conversion of glycine to serine was much more efficient than the reaction in the reverse direction (WANG and BURRIS, 1963). A useful tool in the study of glycolate metabolism proved to be the use of certain inhibitors, among them α-hydroxysulfonates, inhibitors of glycolate oxidase (ZELITCH, 1958), and isonicotyl hydrazide, an inhibitor of the conversion of glycine to serine (COOMBS and WHITTINGHAM, 1966a, b). If applied to whole leaves or algae, these inhibitors cause a large accumulation of glycolate or glycine.

When intermediates of the glycolate pathway were investigated for their internal distribution of radioactivity during $^{14}CO_2$ incorporation in higher plants, glycine and serine were found uniformly labeled. This labeling pattern is compatible with an origin of the two amino acids from glycolate, since glycolate is equally labeled in both carbon atoms. However, glycerate was predominantly carboxyl labeled. This rather relates the origin of glycerate to phosphoglycerate but not to serine. A similar conclusion was also implied by the observation that $^{18}O_2$ was incorporated rapidly into the carboxyl group of glycine and serine, but not of glycerate (LORIMER et al., 1973).

Most of the enzyme activities involved in the glycolate pathway were found in a newly discovered cell organelle of green leaves, the "peroxisome" (TOLBERT et al., 1968; TOLBERT et al., 1969; TOLBERT and YAMAZAKI, 1969). The function of peroxisomes in photorespiration has been extensively reviewed by TOLBERT

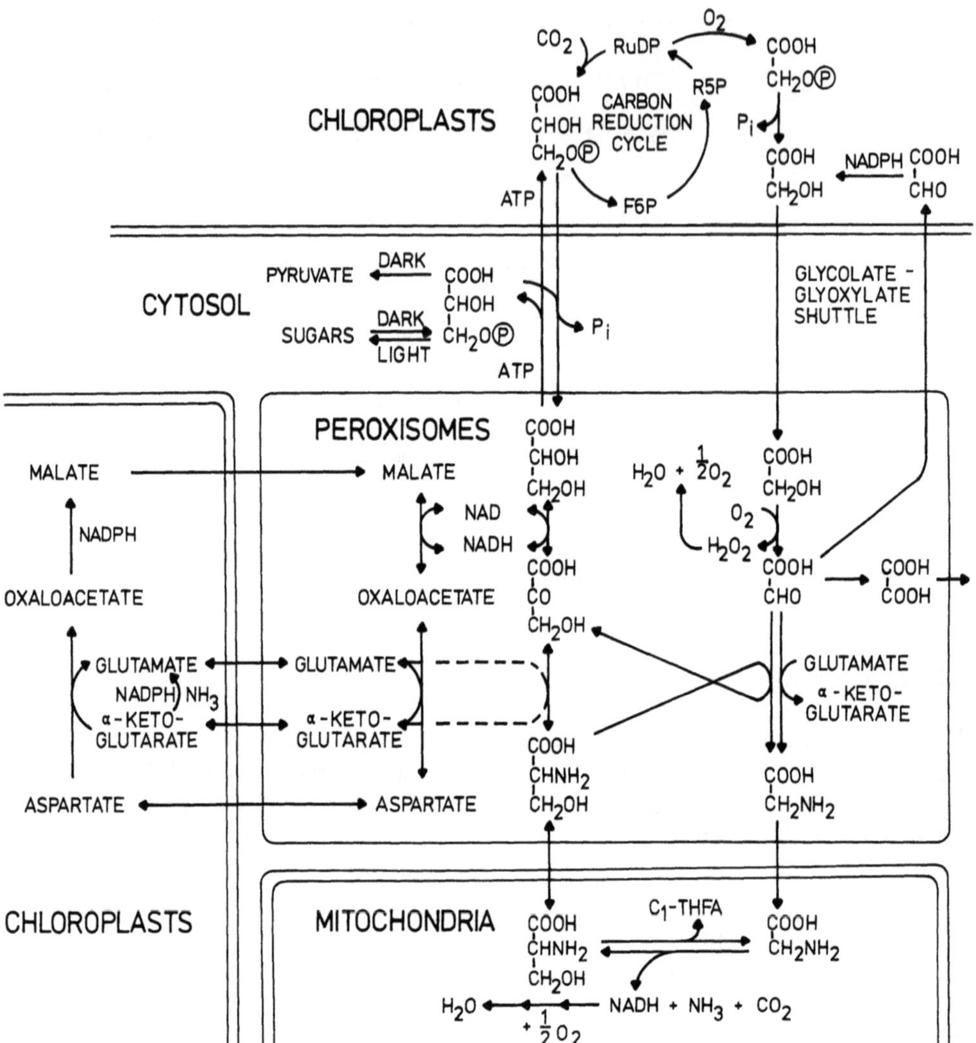

Fig. 6. Scheme of glycolate metabolism and its intracellular compartmentation according to TOLBERT (1973)

(1971 a) and cytological properties have been summarized by VIGIL (1973) and FREDERIK et al. (1975). The enzyme reactions occuring in the peroxisomes or related to the glycolate pathway are presented in Figure 6. They include the conversion of glycolate to glycine and of serine to glycerate in the peroxisomes, and the conversion of glycine to serine in the mitochondria.

Peroxisomes may be grouped in a major class of subcellular organelles, the "microbodies", which are defined in cytologic, biophysic and biochemic terms. In the electron microscope these organelles appear as spheric particles surrounded by a single unit membrane in the cytoplasm. In diameter they vary between 0.2 and 1.5 μm. The inside consists of a coarsely granular matrix, in which cristalline material is sometimes deposited (see HRUBAN and RECHCIGL, 1969; VIGIL, 1973; FREDERIK et al., 1975). Microbodies contain a number of flavoprotein oxidases which generate H_2O_2. This H_2O_2 is decomposed to H_2O and oxygen by the high activity of catalase present in these organelles. The degradation of H_2O_2 generated by flavin oxidases, in turn, may be understood as a general function of this class of cell organelles. Catalase itself is one of the few microbody enzymes which can be stained for electron microscopy. The stain with 3,3-diaminobenzidine indicated that catalase activity occurs exclusively in the peroxisomes of plant leaf cells (FREDERICK and NEWCOMB, 1969b; VIGIL, 1969; HILLIARD et al., 1971; SMAOUI, 1972).

Depending on the type of tissue, microbodies contain specific enzyme complements which are related to their physiological function. So, in green leaf cells, they contain many enzymes of the glycolate pathway and are called peroxisomes (TOLBERT et al., 1968, 1969; YAMAZAKI and TOLBERT, 1970; TOLBERT and YAMAZAKI, 1969; TOLBERT, 1971a). In nongreen plant cells activities of glycolate pathway enzymes in microbodies are low if present at all. Microbodies of germinating, fat-storing seeds, i.e. the glyoxysomes, have been investigated best and contain enzymes for the conversion of fats to sugars (BREIDENBACH and BEEVERS, 1967; BREIDENBACH et al., 1968; COOPER and BEEVERS, 1969).

An important peroxisomal enzyme (TOLBERT et al., 1968, 1969) is the glycolate oxidase which oxidizes glycolate to glyoxylate. The enzyme was first reported by CLAGETT et al. (1949), TOLBERT et al. (1949), and ZELITCH and OCHOA (1953). It is a flavoprotein using FMN as coenzyme. During glycolate oxidation FMN is reduced to $FMNH_2$ which is reoxidized nonenzymatically by molecular oxygen, thereby generating H_2O_2. The glycolate oxidase reaction is irreversible. In a subsequent reaction the H_2O_2 formed is decomposed to water and oxygen by catalase. As glycolate oxidase, catalase was found only in perosxisomes during cell fractionation, but not in chloroplasts where it was previously believed to occur (TOLBERT et al., 1968, 1969).

A number of aminotransferases catalyse some of the following steps in the glycolate pathway. Two of them, glutamate, glyoxylate aminotransferase and serine, glyoxylate aminotransferase, occur only in the peroxisomes (KISAKI and TOLBERT, 1969; YAMAZAKI and TOLBERT, 1970; REHFELD and TOLBERT, 1972). The two reactions catalyzed by these enzymes are irreversible. The enzymes could be separated from each other, but could not yet be freed from an acitivity with alanine as amino donor (REHFELD and TOLBERT, 1972). Both aminotransferases contribute to the conversion of glyoxylate to glycine in the peroxisomes. However, glycine cannot be further metabolized by these organelles and represents a peroxisomal end product.

Two molecules of glycine are then converted to one molecule each of serine and CO_2 in the mitochondria (see below). No information exists on the transfer of glycine from the peroxisomes to the mitochondria. Several biomembrane barriers have to be crossed in this transfer. It may be worthwhile to mention that the chloroplast envelope has only a low permeability for glycine and does not permit passage of more than at most a few micromoles per mg chlorophyll per hour (GIMMLER et al., 1974). Serine can again enter the peroxisomal metabolism. Again its mode of transfer is unknown. In the glycolate pathway serine is converted to hydroxypyruvate by the serine, glyoxylate aminotransferase (YAMAZAKI and TOLBERT, 1970; REHFELD and TOLBERT, 1972). During the conversion of two molecules of glycine to one molecule of serine, one has to postulate—for stoichiometric reasons—, that the reaction of the glutamate, glyoxylate aminotransferase, has to take place twice when the reaction of the serine, glyoxylate aminotransferase takes place once. Enzyme preparations of the serine, glyoxylate aminotransferase also show activity with pyruvate as amino acceptor which might be due to a hydroxypyruvate, alanine aminotransferase, activity reported previously (WILLIS and SALLACH, 1963; CHENG et al., 1968). Such an activity would be needed if the serine were to be formed from glycerate, as is especially the case in algae under some conditions.

Table 2. The intracellular distribution and the occurrence of glycolate metabolizing enzymes

	Intracellular distribution	Catalyzed reactions	Remarks
Glycolate oxidase	Peroxisomes	Glycolate + O_2 → glyoxylate + H_2O_2 L-Lactate + O_2 → pyruvate + H_2O_2	In higher plants and some algae; insensitive to cyanide; strongly inhibited by O_2
Glycolate dehydrogenase	Peroxisomes or mitochondria	Glycolate + X → glyoxylate + XH_2 D-Lactate + X → pyruvate + XH_2	In some green algae; sensitive to cyanide; little or not inhibited by O_2; X is unknown
NADH: Hydroxypyruvate reductase	Peroxisomes	Hydroxypyruvate + NADH ⇌ NAD + glycerate Glyoxylate + NADH ⇌ NAD + glycolate	High substrate affinity for hydroxypyruvate, low substrate affinity for glyoxylate
NADPH: Glyoxylate reductase	Chloroplasts	Glyoxylate + NADPH ⇌ NADP + glycolate	
NAD: D-Lactate dehydrogenase	Cytosol	D-Lactate + NADH ⇌ NAD + pyruvate	In algae having a glycolate dehydrogenase
NAD: L-Lactate dehydrogenase	Cytosol	L-Lactate + NADH ⇌ NAD + pyruvate	In algae having a glycolate oxidase; in lower land plants; in higher plants

The last step in the conversion of serine to glycerate in the peroxisomal part of the glycolate pathway is catalysed by the NADH, hydroxypyruvate reductase (TOLBERT et al., 1970). The enzyme is usually designated as NADH, glyoxylate reductase, and was discovered by ZELITCH (1953, 1955). It can use both glyoxylate and hydroxypyruvate as substrate. However, its affinity for hydroxypyruvate is higher by two orders of magnitude than that for glyoxylate. Its maximal velocity is far higher with hydroxypyruvate than in the reverse direction with glycerate (KOHN and WARREN, 1970; TOLBERT et al., 1970). The enzyme (see Table 2) is different from the NADPH, glyoxylate reductase, in the chloroplasts (ZELITCH and GOTTO, 1962; THOMPSON and WHITTINGHAM, 1967; TOLBERT et al., 1970).

3.3 The Interconversion of Glycine and Serine in the Mitochondria

One of the most important steps of the glycolate pathway is the reaction of two glycines to form one serine and CO_2 which is evolved during light respiration. KISAKI and TOLBERT (1970) showed that ^{14}C-carboxyl-labeled glycolate and glycine are by far the best substrates for this CO_2. Results of $^{14}CO_2$ feeding experiments also imply that the light-respiratory CO_2 is derived directly from glycine, or from a closely related compound (MAHON et al., 1974c).

The conversion of glycine to serine involves two enzyme activities, glycine decarboxylase and hydroxymethyl-transferase activity. Both were detected in the mitochondrial, but not in the peroxisomal cell fraction (KISAKI et al., 1971 a; KISAKI et al., 1971 b; KISAKI et al., 1972; CLANDININ and COSSINS, 1975). The glycine decarboxylase is stimulated by the addition of NAD, pyridoxal phosphate, tetrahydrofolic acid, and oxygen, and is inhibited by NADH, NADPH, and methionine. The activity is highly dependent on the integrity of the mitochondria. A formation of ATP has also been observed during the conversion of glycine to serine in the mitochondria (BIRD et al., 1972).

There is also the possibility that serine is converted back to glycine. This would allow complete oxidation of glycine to CO_2 or glycine formation from serine, when no glycolate is available for glycine formation. However, glycine formation from serine seems to be very small when compared with the conversion of glycine to serine (WANG and BURRIS, 1963). The back reaction was first demonstrated by WILKINSON and DAVIES (1958). CLANDININ and COSSINS (1975) showed that glycine formation from serine also occurred in the mitochondria.

In addition to the mitochondrial enzyme, hydroxymethyltransferase seems also to occur in the chloroplasts (SHAH and COSSINS, 1970; KISAKI et al., 1971 b). KISAKI et al. (1971 b) proposed that the enzyme in the mitochondria catalyses the conversion of glycine to serine, and the enzyme in the chloroplasts takes part in the synthesis of glycine from serine derived from phosphoglycerate. However, there is no evidence that the carbon skeleton of serine can be produced in the chloroplasts.

3.4 Glycerate and Its Relationship to Phosphoglycerate

Feeding studies with ^{14}C-labeled glycolate, glyoxylate, glycine, and serine, clearly demonstrated that carbon moved from these compounds through glycerate and phosphoglycerate into C_6-sugars (RABSON et al., 1962; JIMENEZ et al., 1962; WANG and WAYGOOD, 1962; TOLBERT, 1963). This proved a gluconeogenic function of the glycolate pathway. At the same time the intramolecular distribution of ^{14}C-label in glycerate during the first seconds of photosynthetic $^{14}CO_2$ fixation revealed most of the ^{14}C-label in the carboxyl group of glycerate. This demonstrated that flow of carbon is also possible in the other direction, i.e. from phosphoglycerate to glycerate.

There are two enzyme reactions which catalyse the interconversion of glycerate and phosphoglycerate: glycerate kinase forms phosphoglycerate from glycerate and ATP, while phosphoglycerate phosphatase splits phosphoglycerate into glycerate and phosphate. A glycerate kinase is located in the chloroplasts, at least of corn leaves (HATCH and SLACK, 1969). In the C_3 plants spinach and swiss chard, 50–65% of the glycerate kinase activity is found inside the chloroplasts and the rest outside (HEBER et al., 1974). Glycerate can enter the chloroplasts both as anion and as acid thus mediating an indirect proton flow across the chloroplast envelope, which has a very low permeability for protons. In the chloroplasts it is metabolized to phosphoglycerate and, thereafter, to other Calvin cycle intermediates (HEBER et al., 1974). The total activity of the glycerate kinase is about one fifth of the photosynthetic CO_2 fixation rate in spinach. This rather low activity might constitute a bottleneck in the flow of carbon from glycolate back to the Calvin cycle (HEBER et al., 1974).

Phosphoglycerate phosphatase is an active enzyme in many plants examined (RANDALL and TOLBERT, 1971 a, b; RANDALL et al., 1971). The main activity was found in the soluble cell fraction. About 20% of the activity was associated with starch grains of spinach chloroplasts. The function of the soluble phosphoglycerate phosphatase is certainly related to the formation of glycerate from phosphoglycerate and to serine formation in the peroxisomes. Flow of carbon in this direction must take place whenever the carboxyl groups of glycerate and serine are predominantly labeled during short time $^{14}CO_2$ photosynthesis and labeling thus reflects the labeling pattern of phosphoglycerate.

4. Glycolate Metabolism in C_4-plants, CAM Plants, and Algae

4.1 Glycolate Metabolism in C_4-plants

Higher plants have been divided into C_3 plants, which evolve large amounts of photorespiratory CO_2, and C_4 plants, which do not evolve considerable amounts of photorespiratory CO_2 (HATCH and SLACK, 1970; BLACK, 1973; JACKSON and VOLK, 1970). In C_4 plants photorespiration could only be observed by a direct measurement of light-dependent O_2 uptake (OZBUN et al., 1964). There, the question was as to whether and to what extent glycolate metabolism takes place in C_4 plants despite the lack of apparent CO_2 evolution in the light.

Leaves of C_4 plants have a so-called Kranz anatomy: one layer of green bundle sheath cells surrounds the vascular tissue and is itself surrounded by another layer of chloroplast containing cells, the so-called mesophyll cells. During photosynthesis the cooperation of the two types of cells is required. The mesophyll cells incorporate CO_2 through PEP carboxylase into oxalacetate which is reduced to malate or transaminated to aspartate. These C_4 acids are transferred to the bundle sheath cells. There, the CO_2 is released and refixed by the action of ribulose bisphosphate carboxylase to yield phosphoglycerate, which is reduced and further metabolized along the metabolic route of the Calvin cycle.

CHOLLET and OGREN (1972) showed that the photosynthesis of the bundle sheath cells is inhibited by oxygen and that the inhibition can be reversed by high CO_2 concentrations. Feeding experiments with ^{14}C-labeled sugar phosphates suggest that glycolate is formed in the bundle sheath cells of C_4 plants from Calvin cycle intermediates as in C_3 plants (OSMOND and HARRIS, 1971). Conditions for glycolate formation in the bundle sheath cells are low p_{CO_2} and high p_{O_2} (CHOLLET, 1974). This suggests that glycolate is synthesized in C_4 plants as in C_3 plants by the action of the ribulose bisphosphate carboxylase/oxygenase and phosphoglycolate phosphatase. In a recent study, ZELITCH (1973) compared glycolate formation in the C_3 plant tobacco and in the C_4 plant maize. From his observations he concluded that there exists a rapid pathway for glycolate biosynthesis in tobacco and a different, slow pathway in maize. However, the evidence for another pathway in maize is disputable and the results could possibly also be explained by the difference between the carbon metabolism in C_3 and C_4 plants.

Incubation studies with ^{14}C-labeled glycolate and glycine proved the operation of the glycolate pathway in the bundle sheath cells and revealed a flow of carbon from these compounds to sucrose (OSMOND and HARRIS, 1971). Photorespiratory $^{14}CO_2$ is evolved from glycolate-1-^{14}C and glycine-1-^{14}C by bundle sheath cells.

It is efficiently refixed in the presence of mesophyll cells. Rapid refixation of photorespiratory CO_2 explains the apparent lack of photorespiration in C_4 plants (LIU and BLACK, 1972).

The distribution of glycolate pathway enzymes between the mesophyll cells and the bundle sheath cells has been investigated by several authors (REHFELD et al., 1970; EDWARDS and BLACK, 1971; OSMOND and HARRIS, 1971; HUANG and BEEVERS, 1972; LIU and BLACK, 1972). Using PEP carboxylase as a marker for mesophyll cells, and ribulose bisphosphate carboxylase as a marker for bundle sheath cells, the glycolate pathway enzymes phosphoglycolate phosphatase, glycolate oxidase, glutamate, glyoxylate aminotransferase, serine transmethylase, and NADH, hydroxypyruvate reductase, were found in both cell types, but were two to five times more active in the bundle sheath cells than in the mesophyll cells. Only the serine, pyruvate aminotransferase, in crabgrass (LIU and BLACK, 1972), the glycerate kinase in *Atriplex spongiosa* (OSMOND and HARRIS, 1971), and the phosphoglycerate phosphatase in maize (REHFELD et al., 1970) are more active in the mesophyll cells than in the bundle sheaths cells. In sorghum lower levels of glycolate oxidase and NADH, hydroxypyruvate reductase, but not of catalase were observed in the mesophyll cells, which led HUANG and BEEVERS (1972) to propose a nonspecialized type of microbodies in these cells.

Peroxisomes could be observed in C_4 plants by electron microscopy in both the mesophyll and the bundle sheath cells. The bundle sheath cells contained about 2 to 5 times more peroxisomes than the mesophyll cells depending on the species under investigation. The presence of catalase associated with the peroxisomes could be demonstrated by the stain with 3,3′-diaminobenzidine (DAB) (FREDERICK and NEWCOMB, 1971; BLACK and MOLLENHAUER, 1971; HILLIARD et al., 1971; LIU and BLACK, 1972).

In summary, radio tracer studies and the presence of glycolate pathway enzymes emphasize the existence of both glycolate metabolism and photorespiration in C_4 plants. Both events are mostly restricted to the bundle sheath cells. The lack of light-respiratory CO_2 evolution cannot be attributed to lower activities of glycolate pathway enzymes in C_4 plants if compared with C_3 plants, but is most likely due to a very efficient refixation of photorespired CO_2 by the PEP carboxylase in the mesophyll cells, which surround the bundle sheath cells.

4.2 Glycolate Metabolism in CAM Plants

Very little information is available on glycolate metabolism in CAM (crassulacean acid metabolism) plants. In an electron microscopic study KAPIL et al. (1975) reported numerous microbodies in cells of CAM plants, which can easily be stained for catalase activity with 3,3′-diaminobenzidine. An isolation of peroxisomes from CAM plants has not yet been reported.

4.3 Glycolate Metabolism in Algae

Algae represent another group of photosynthetic organisms in which photorespiration and glycolate metabolism have been studied. Like C_4 plants, algae have a low CO_2 compensation point (EGLE and SCHENK, 1952; SCHAUB and EGLE, 1965; FOCK and EGLE, 1966; TREGUNNA and BROWN, 1967). However, O_2 evolution is inhibited by high p_{O_2} (FOCK et al., 1971; WARBURG, 1920) which indicates photorespiratory processes. Glycolate metabolism and the interaction of subcellular components are generally believed to be similar to those in higher plant cells, i.e. the formation of glycolate takes place in the chloroplasts, the conversion of glycolate to glycine in the peroxisomes, the conversion of glycine to serine in the mitochondria, the interconversion of serine and glycerate again in the peroxisomes, and the further metabolism of glycerate in the cytosol or the chloroplasts (TOLBERT et al., 1971; MERRETT and LORD, 1973; TOLBERT, 1974). However, in algae some significant differences from higher plants have been discovered. They include the excretion of glycolate, a limited conversion of glycolate to glycine and serine, and an increased tendency to form serine from phosphoglycerate.

Glycolate formation by algae became of particular interest when TOLBERT and ZILL (1956) discovered that in *Chlorella* glycolate would be excreted if three conditions were met: presence of low p_{CO_2}, high oxygen content, and high light intensity. The excretion of glycolate, and other organic compounds in some instances represents an interesting aspect of interactions between the cytosol of the algal cell and the surrounding medium. Glycolate excretion occured only if, after growth under high p_{CO_2} (0.2–5% CO_2), algae were transferred to low p_{CO_2}. Thereafter, excretion soon ceased. For this reason glycolate excretion might be understood as an adaptation phenomenon. Although much carbon is lost by glycolate excretion, the excreted glycolate does not seem to be important as a substrate for growth of other algae (DROOP and MCGILL, 1966).

A regulatory mechanism to explain glycolate excretion was deduced from specific changes of some enzyme activities involved in glycolate metabolism. In cells grown at high CO_2 concentrations the level of glycolate dehydrogenase activity is low. After transfer to low p_{CO_2} glycolate is synthesized, accumulates because of a repressed glycolate dehydrogenase activity, and is finally excreted. Adaptation of the cells to low p_{CO_2} causes derepression of the glycolate dehydrogenase. As a consequence glycolate is further metabolized and no longer excreted (NELSON and TOLBERT, 1969; CODD and MERRETT, 1971a). Glycolate excretion can also be induced by the addition of inhibitors of the glycolate pathway such as hydroxypyridine methanesulfonate (LORD and MERRET, 1969; LORD et al., 1970) and isonicotyl hydrazide (PRITCHARD et al., 1962; WHITTINGHAM and PRITCHARD, 1963). Obviously, glycolate must first accumulate before it is excreted. Still it is formed even under conditions where excretion does not take place although this has also been questioned because a difference of glycolate dehydrogenase activity in CO_2 and air grown cells could not be observed (CODD et al., 1969; COLMAN et al., 1974).

The operation of the glycolate pathway has also been demonstrated in algae by ^{14}C-tracer studies (BRUIN et al., 1970; LORD and MERRETT, 1970; MURRAY et al., 1971). In contrast to the situation in higher plants, glycolate was more labeled in the C_2 than in the C_1 position during the first few seconds of photosynthetic $^{14}CO_2$ fixation but rapidly became uniformly labeled thereafter (HESS and TOLBERT, 1967; BRUIN et al., 1970). Under the same conditions phosphoglycerate was also more labeled in the C_3 than in the C_2 position, even though more than 80% of the total label was in the carboxyl group. The labeling of glycolate is consistent with an origin from Calvin cycle intermediates. Glycine was found uniformly labeled, but serine was predominantly carboxyl labeled. This indicates the formation of glycine from glycolate while serine is synthesized from phosphoglycerate (ZAK and NICHIPOROVICH, 1964; HESS and TOLBERT, 1967; BRUIN et al., 1970).

Enzymic evidence for an operative glycolate pathway has been reported for several algae (HESS and TOLBERT, 1967; LORD and MERRETT, 1970; CODD and MERRETT, 1971a; MURRAY et al., 1971). The substrate specificity of glycolate pathway enzymes in algae seems to be similar to that in higher plants with one exception. In higher plants reduced FMN of the glycolate oxidase transfers electrons directly to oxygen. This was originally also reported for algae (LORD and MERRETT, 1968; ZELITCH and DAY, 1968; LORD and MERRETT, 1969). Later, it was recognized that the algal enzyme does not always couple to oxygen but reduces an unknown internal substrate (NELSON and TOLBERT, 1969, 1970; CODD et al., 1969; CODD and MERRETT, 1971a). The enzyme was designated as glycolate dehydrogenase (see Table 2). An evolutionary survey showed that it occurred only in algae, but not in all of them (FREDERIK et al., 1973). The glycolate dehydrogenase differs from the glycolate oxidase by its capability to react with D-lactate and in its inhibition by cyanide and oxygen as summarized in Table 2.

Other than by glycolate dehydrogenase or glycolate oxidase, lactate can also be oxidized to pyruvate by NAD-dependent lactate dehydrogenases. These use either L- or D-lactate as substrate. Interestingly, algae possessing a glycolate dehydrogenase have a D-lactate dehydrogenase, while algae and lower land plants with a glycolate oxidase have a L-lactate dehydrogenase (GRUBER et al., 1974). In higher plants which possess only a glycolate oxidase L-lactate dehydrogenase can be induced by anaerobic growth conditions (DAVIES and DAVIES, 1972). In potato tubers a D,L-lactate dehydrogenase had been reported (ROTHE, 1974).

Using synchronized *Euglena* cell cultures CODD and MERRETT (1971a, b) found that young cells, at the end of the dark phase, accumulate and excrete more glycolate than do mature cells at the end of the light phase. Glycerate was predominantly carboxyl labeled by young cells and uniformly labeled by mature cells. This indicates an origin of serine from phosphoglycerate in young cells and from glycolate in mature cells. Activities of ribulose bisphosphate carboxylase and phosphoglycolate phosphatase did not change greatly over the cell cycle. However, glycolate dehydrogenase activity was low in young cells and high in mature cells. The opposite was observed for phosphoglycerate phosphatase. The enzymic evidence is thus consistent with the labeling data.

Cytologically, all subcellular organelles involved in glycolate metabolism in higher plants also occur in algae. Microbodies were observed electron-microscopically in several species of green algae (GERGIS, 1971; GERHARDT and BERGER, 1971; GIRAND and CZANINSKI, 1971; TOURTE, 1972; SILVERBERG and SAWA, 1974; SILVERBERG, 1975), brown algae (BOUCK, 1965; BISALPUTRA et al., 1971), red algae (OAKLEY and DODGE, 1974), dinoflagellatae (BIBBY and DODGE, 1973) and Euglena (GRAVES et al., 1971; BRODY and WHITE, 1972). In algae the microbodies seemed to be somewhat smaller than in higher plant cells, and in some instances, less numerous. A close association with chloroplasts and mitochondria was also noticed (SILVERBERG and SAWA, 1974; SILVERBERG, 1975). A cytological stain with 3,3′-diaminobenzidine (DAB) for catalase in algal microbodies was positive in some species (STEWARD et al., 1972; TOURTE, 1972; SILVERBERG and SAWA, 1974; SILVERBERG, 1975), but negative in others even though catalase was present. If in the latter case H_2O_2 is not generated by these microbodies because of the presence of a glycolate dehydrogenase instead of a glycolate oxidase, catalase activity might not be required. However, in some instances the staining technique may not have been performed under optimal conditions (Silverberg, 1975).

The isolation of algal microbodies proved to be difficult. A peroxisomal type of microbody could be isolated from some autotrophically grown algae by isopycnic centrifugation (Codd et al., 1972; Stabenau, 1974; Stabenau and Beevers, 1974; Collins and Merrett, 1975). A major problem appeared to be that in many cases peroxisomal enzymes banded partly or completely with the mitochondria. The significance of this is still to be explored. A glyoxysomal type of microbody has also been isolated form algae grown on acetate, which induces glyoxylate cycle enzymes (Graves et al., 1971; Graves et al., 1972).

4.4 Glycolate Metabolism in Blue-Green Algae

Little information exists on the operation of the glycolate pathway in blue-green algae. Glycolate-1-^{14}C and 2-^{14}C can be metabolized, but only little glycine and serine are formed (Codd and Stewart, 1973). Instead, the formation of glycerate from glycolate via glyoxylate and tartronic semialdehyde has been observed. Glycolate assimilation is inhibited by α-hydroxypyridinemethanesulfonate.

Excretion of glycolate is also known from blue-green algae, but glycolate represents only a minor part of the total organic matter excreted (Cheng et al., 1972). Since isonicotyl hydrazide does not much inhibit glycolate excretion, it was also concluded that glycine and serine were not derived from glycolate. A 3-fold increase was produced by α-hydroxypyridinemethanesulfonate in glycolate excretion. Excretion of glycolate from blue-green algae is also dependent on the quality of the light (Döhler and Koch, 1972). In red light at 25° C glycolate is excreted, but in blue or far-red light as well as at a temperature of 35° C it is only excreted in the presence of hydroxypyridinemethanesulfonate. Since blue-green algae do not contain cell organelles glycolate metabolism would have to occur in the cytosol.

5. The Transport of Metabolites between Cell Organelles during Glycolate Metabolism

5.1 Transport of Glycolate Pathway Intermediates between Peroxisomes, Chloroplasts, and Mitochondria

Since glycolate biosynthesis takes place in the chloroplasts, the subsequent glycolate metabolism in the peroxisomes, the interconversion of glycine and serine in the mitochondria, and the interconversion of glycerate and hosphoglycerate in the cytosol or in the chloroplasts, the glycolate pathway involves the cooperation of four compartments and requires a rapid transfer of several metabolites between these cell compartments, i.e. of glycolate from the chloroplasts to the peroxisomes, of glycine from the peroxisomes to the mitochondria, of serine from the mitochondria back to the peroxisomes, and of glycerate to the cytosol or the chloroplasts. This has been pictured in Figure 6.

The cooperation of different cell compartments and the transfer of metabolites between them is deduced from radiotracer and inhibitor studies establishing a metabolic route for glycolate metabolism, and from enzyme compartmentation studies showing that the enzyme reactions for this metabolic route are located at different cellular sites. The magnitude of the transfers involved are estimated from fluxes of photosynthetically fixed CO_2 through certain Calvin cycle and glycolate pathway intermediates. This will be discussed in Section 6.2. Knowledge about the permeability of cell organelle membranes to allow such transfers is scarce. It is known that glycolate, glyoxylate, glycerate, and phosphoglycerate can cross the chloroplast envelope, the former probably after protonation (HEBER et al., 1974). Phosphoglycerate is transported by a carrier mediating specific and obligatory anion exchange in the inner membrane of the chloroplast envelope (see Chaps. II.2 and II.3 of this vol.).

Electron microscopic studies have revealed that chloroplasts, peroxisomes, and mitochondria are often appressed to each other. This did not appear to be due just to lack of space (FREDERICK and NEWCOMB, 1969a). It is tempting to speculate that there is a relationship between this observation and an efficient transport of metabolites between the three cell organelles during glycolate metabolism.

5.2 Sites of CO_2 Evolution and O_2 Uptake during Photorespiration

According to Tolbert's scheme of glycolate metabolism (Fig. 6) lightrespiratory CO_2 is produced during the conversion of glycine to serine in the mitochondria. The stoichiometry requires two glycines for one CO_2 produced. Sites of O_2 uptake are the ribulose bisphosphate oxygenase reaction which consumes one mole of oxygen per one mole of phosphoglycolate produced and the reactions of glycolate oxidase and catalase which together consume half a mole of oxygen per one mole of glycolate oxidized.

The ratio of O_2 uptake to CO_2 evolution within this part of photorespiration is 3:0. Assuming a ratio of CO_2 uptake to O_2 evolution during Calvin cycle photosynthesis of one, photorespiration could, depending on the fate of the serine formed from glycolate, increase the overall CO_2/O_2 ratio during photosynthesis. There are only a few quantitative measurements of both CO_2 uptake and O_2 evolution in the light under nonphotorespiratory and photorespiratory conditions. They indeed indicate a transient increase of the CO_2/O_2 ratio from about 1:0 under low p_{O_2} to almost 2:0 under high p_{O_2} (FOCK et al., 1969; FOCK et al., 1971; FOCK et al., 1972).

In Tolbert's scheme of photorespiration (Fig. 6) serine is metabolized to phosphoglycerate, which again enters the Calvin cycle. Alternatively, TOLBERT and RYAN (1975) proposed that during photosynthesis near CO_2 compensation point the serine formed would be converted back to glycine and the resultant C_1-unit be oxidized to CO_2. Thereby, both carbon atoms of glycine would eventually be released as photorespiratory CO_2. An enzyme system for the oxidation of the C_1-unit has been reported to be present in the mitochondria (CLANDININ and COSSINS, 1975). Under physiologic conditions only little $^{14}CO_2$ was evolved from serine-3-^{14}C or glycine-2-^{14}C as compared with glycine-1-^{14}C, (KISAKI and TOLBERT, 1970). Thus, complete oxidation of glycolate does not normally seem to occur during photorespiration.

Another possibility for photorespiratory CO_2 formation involves the oxidation of glyoxylate to formate and CO_2. This could be accomplished by the H_2O_2 that is formed during the glycolate oxidase reaction (TOLBERT et al., 1949; ZELITCH and OCHOA, 1953; ZELITCH, 1972a). When the exclusive location of glycolate oxidase in the peroxisomes became apparent, an oxidation of glyoxylate to formate and CO_2 by H_2O_2 in the peroxisomes had to be excluded because H_2O_2 would be decomposed much faster to H_2O and $^1/_2$ O_2 by the high peroxisomal catalase activity than it would react nonenzymatically in an oxidation of glyoxylate. Instead, ZELITCH (1972b) reported that glyoxylate is oxidized to formate and CO_2 in the chloroplasts. His assay system included glyoxylate-1-^{14}C, 10 mM $MnCl_2$, envelope-free chloroplasts, and light. However, the "nonenzymatic" control rate without envelope-free chloroplasts increased between pH 7.6 and 8.2 to almost full "enzymatic" activity and the nonenzymatic activity was only shifted by about one pH unit towards acidity in the enzymatic reaction. The significance of the observations is unclear and the system would need some further manifestation of its enzymatic character and also further information about the fate of the formate formed. An oxidation of formate to CO_2 had been observed by the peroxidative action of catalase in the peroxisomes or by a formate dehydrogenase in the mitochondria, but the contribution of these reactions to light-respiratory CO_2 evolution did not seem to be important (LEEK et al., 1972; HALLIWELL, 1974).

Photorespiratory O_2 uptake during glycolate biosynthesis in the chloroplasts by the oxidation of the α, β-dihydroxy-ethyl-thiamine pyrophosphate via H_2O_2 would be as high as in the ribulose bisphosphate oxygenase reaction. Additional O_2 uptake may arise from a glycolate-glyoxylate shuttle between the chloroplasts and the peroxisomes (see Sect. 5.3). This shuttle could theoretically cause an unlimited O_2 uptake. However, under steady state conditions, net O_2 gas exchange by the shuttle is zero because of the O_2 evolution during photosynthetic production of the NADPH needed to reduce glyoxylate. Furthermore, the shuttle might be limited by the low activity of the NADPH-dependent glyoxylate reductase in the chloroplasts (HEBER and KRAUSE, 1971; TOLBERT and RYAN, 1975). Finally, oxygen could be consumed during the reoxidation of NADH formed during the conversion of glycine to serine in the mitochondria. Its reoxidation by the mitochondrial electron transport chain has been deduced from in vitro experiments with isolated mitochondria (BIRD et al., 1972). However, in vivo the ammonia formed during the conversion of glycine to serine has to be reassimilated. This may happen in the mitochondria by the action of glutamate dehydrogenase (RITENOUR et al., 1967). Thus, a reoxidation of NADH by the mitochondrial electron transport chain would be unnecessary.

5.3 Insignificance of the Glycolate-Glyoxylate Shuttle

The glycolate oxidase reaction is an energy wasting process because the reaction is directly coupled to oxygen. This results in an unidirectional conversion of glycolate to glyoxylate, which could not occur if a pyridine nucleotide were electron acceptor. In the latter case the equilibrium of the reaction would be far on the side of glycolate and the oxidized cosubstrate.

With the discovery of a NADH-dependent glyoxylate reductase, ZELITCH (1953) discussed a possible mechanism by which glycolate oxidase and the NADH, glyoxylate reductase, could act together and oxidize NADH by oxygen. When it was recognized that isolated peroxisomes contained glycolate oxidase, catalase, and NADH, glyoxylate reductase, TOLBERT et al. (1968) proposed these three enzymes to function in a terminal oxidizing system for NADH. Later, however, the low affinity for glyoxylate ($Km \approx 2 \cdot 10^{-2}$M) (TOLBERT et al., 1970; KOHN and WARREN, 1970) left doubts about the efficiency of such a peroxisomal system. The affinity of the NADH, glyoxylate reductase, for hydroxypyruvate is higher by two orders of magnitude ($Km \approx 10^{-4}$M) than that for glyoxylate, so that the enzyme would rather act as a hydroxypyruvate reductase in the glycolate pathway than as a glyoxylate reductase in a terminal oxidizing system for NADH.

Another enzyme capable of functioning with glycolate oxidase and catalase as a terminal oxidizing system would be the NADPH-dependent glyoxylate reductase of the chloroplasts (ZELITCH and GOTTO, 1962; THOMPSON and WHITTINGHAM, 1967, 1968; TOLBERT et al., 1970). This enzyme has a high affinity for NADPH and glyoxylate, but not for NADH and hydroxypyruvate. The equilibrium of the reaction catalyzed by the enzyme is far on the side of NADP and glycolate. The oxidation of NADPH is thought to proceed via a shuttle of glycolate and glyoxylate between the chloroplasts and the peroxisomes (see also Fig. 6). The argument in favor of a glycolate glyoxylate shuttle between the chloroplasts and the peroxisomes is that ^{14}C-glyoxylate is taken up and reduced, albeit at a low rate, to glycolate by isolated chloroplasts in the light, whereas ^{14}C-glycolate is not metabolized by isolated chloroplasts (KEARNEY and TOLBERT, 1962; CHAN and BASSHAM, 1967; THOMPSON and WHITTINGHAM, 1968) but oxidized to glyoxylate in the peroxisomes. The glyoxylate can be returned to the chloroplasts. The chloroplast envelope is permeable to both glycolate and glyoxylate and would in principle, allow such a shuttle (TOLBERT, 1958; KEARNEY and TOLBERT, 1962; JENSEN and BASSHAM, 1966; BASSHAM et al., 1968; POINCELOT, 1975). The shuttle transfer of glycolate and glyoxylate would provide a possibility to oxidize some excess NADPH produced by the chloroplasts under photorespiratory conditions where less NADPH for photosynthetic CO_2 reduction would be needed. Taking into account the O_2 evolution during photosynthetic NADP reduction, the net O_2 balance of the shuttle would be zero. Nevertheless, the activity of the NADPH-specific glyoxylate reductase and the intracellular concentration of glyoxylate are too low to allow an extensive glycolate-glyoxylate shuttle (HEBER and KRAUSE, 1971; TOLBERT and RYAN, 1975). Indeed, there is no evidence for the existence of the shuttle under in vivo conditions.

5.4 The Malate-Aspartate Shuttle

When carbon flows through the glycolate pathway the glutamate, glyoxylate aminotransferase, has to be steadily supplied with glutamate and produces α-ketoglutarate. Both compounds have to be balanced with other reactions in a continuous flow system. Also, in the glycerate dehydrogenase reaction, NADH is oxidized in the peroxisomes and has to be recycled. There are two other enzymes of high activity in the peroxisomes, which appear to have a function related to the glycolate

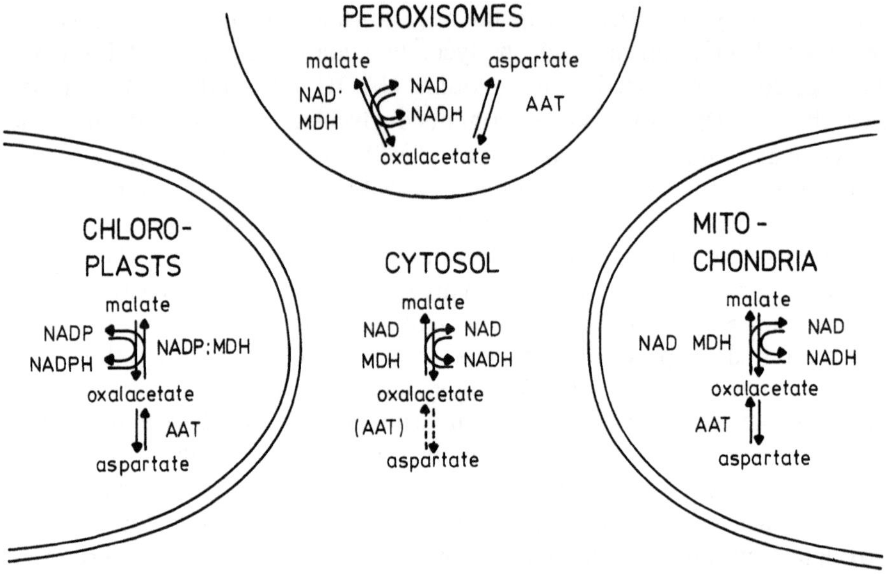

Fig. 7. The intracellular distribution of malate dehydrogenase (MDH) and aspartate aminotransferase (AAT) in C_3-plants. There is no evidence for a cytosolic aspartate aminotransferase in C_3-plants yet, but in some C_4-plants. From this scheme it can be deduced, that the malate-aspartate shuttle may play an important role in the indirect transfer of reducing equivalents between several cell compartments

pathway in the peroxisomes. These are malate dehydrogenase and aspartate aminotransferase. Activities of these enzymes occur also in other cellular compartments. There, they are involved in the shuttle transfer of reducing equivalents across the envelopes of chloroplasts and mitochondria. The envelopes of these organelles are not permeable for NAD(P) and NAD(P)H, (CHAPPEL, 1969; KREBS and VEECH, 1969; VAN DAM and MEYER, 1971; HEBER, 1974). Although nothing is known about the permeability of the peroxisomal membrane for NAD, NADH, malate, aspartate, α-ketoglutarate, and glutamate, it was proposed that the peroxisomal malate dehydrogenase and aspartate aminotransferase are involved in a similar shuttle (YAMAZAKI and TOLBERT, 1970; SCHNARRENBERGER et al., 1971). In Figure 7 enzyme localizations are recorded which form the basis of the proposed shuttle systems.

In leaf cells four isoenzymes of NAD-specific malate dehydrogenase exist, one in the peroxisomes, one in the mitochondria, and one or two in the cytosol (YAMAZAKI and TOLBERT, 1969; ROCHA and TING, 1970a, b, 1971). In the chloroplasts a NADP-specific malate dehydrogenase is present, which is only active in illuminated leaves or can be activated by sulfhydryl reducing agents in vitro, if inactive (HATCH and SLACK, 1969; SLACK et al., 1969; JOHNSON and HATCH, 1970; JOHNSON, 1971; TING and ROCHA, 1971). A chloroplast NAD-specific dehydrogenase has also been reported (HEBER and KRAUSE, 1971). Aspartate aminotransferase activity occurs in three cellular compartments of C_3 plants, the peroxisomes, the mitochondria and the chloroplasts. Different isoenzymes appear to be responsible for this activity (YAMAZAKI and TOLBERT, 1970; REHFELD and TOLBERT, 1972). In C_4 plants isoenzymes of aspartate aminotransferase exist in mesophyll and bundle sheath cells (ANDREWS et al., 1971a; HATCH, 1973; HATCH and MAU, 1973). At least in some C_4 plants, aspartate aminotransferase activity appears to occur also in the cytosol, not only in chloroplasts and mitochondria. In some species of C_4 plants aspartate aminotransferase is involved in the C_4-dicarboxylic carbon metabolism (see Chap. II,5).

Between chloroplasts and the cytosol a shuttle involving transfer of oxalacetate and malate has been experimentally demonstrated (HEBER and KRAUSE, 1971, 1972). Evidence for a malate-aspartate shuttle transfer across the mitochondrial membrane has mainly been obtained for tissues other than plants (see Chap. II,6).

6. The Magnitude of Photorespiration with Respect to the Glycolate Pathway Metabolism and Metabolite Transfer

6.1 Gas Exchange

The substrates and the mechanism of photorespiration by C_3-plants can be clearly distinguished from those of dark respiration (ZELITCH, 1966; ATKINS, 1969; LUDWIG and CANVIN, 1971a, b; MAHON et al., 1974b, c). Most of the information that is based on enzymological experiments already considered in previous sections and on gas exchange, radio tracer, and inhibitor experiments applied to different plant systems, shows that CO_2 evolved in the light is derived from glycolate pathway metabolism (TOLBERT, 1971a; ZELITCH, 1975). Here some effects of external factors on CO_2 evolution in the light, are discussed with the respect to the flux of carbon through the glycolate pathway (see also JACKSON and VOLK, 1970). As has been pointed out, this can be translated into fluxes of metabolites between chloroplasts and peroxisomes and peroxisomes and mitochondria. According to Figure 6 for each CO_2 evolved, two molecules of glycolate should be transferred from the chloroplasts to the peroxisomes and two molecules of glycine from there to the mitochondria. The stoichiometry of CO_2 evolution and serine or glycerate transfers is envisaged to be unity.

6.1.1 Effects of Oxygen and Carbon Dioxide

When the gas exchange of attached C_3-type leaves (e.g. sunflower, tobacco, radish, bean) is studied with the $^{14}CO_2/^{12}CO_2$ isotope technique in an open system (CANVIN and FOCK, 1972), the steady rate of photorespiration (PR′) (300 vpm CO_2, 21% O_2, 25°, moderate light intensity) is at least twice the rate of dark respiration, but more often exceeds it three- fourfold (LUDWIG and CANVIN, 1971b; D'AOUST and CANVIN, 1972). PR′ is usually 4–6 mg CO_2/dm^2 h or 30–45 µmoles CO_2/mg chlorophyll h under normal photosynthetic conditions (Fig. 8). The high rates of PR′ are almost independent of CO_2 concentrations from very low CO_2 partial pressures to at least 300 vpm CO_2 (LUDWIG and CANVIN, 1971b; D'AOUST and CANVIN, 1973; MAHON et al., 1974c), however, PR′ increases between 400 and 900 vpm CO_2 from 6 to 9 mg CO_2/dm^2 h (Fig. 8).

According to these authors, the percentage of refixed photorespiratory CO_2 is constant over a range of low CO_2 concentrations where first order kinetics apply, but decreases when

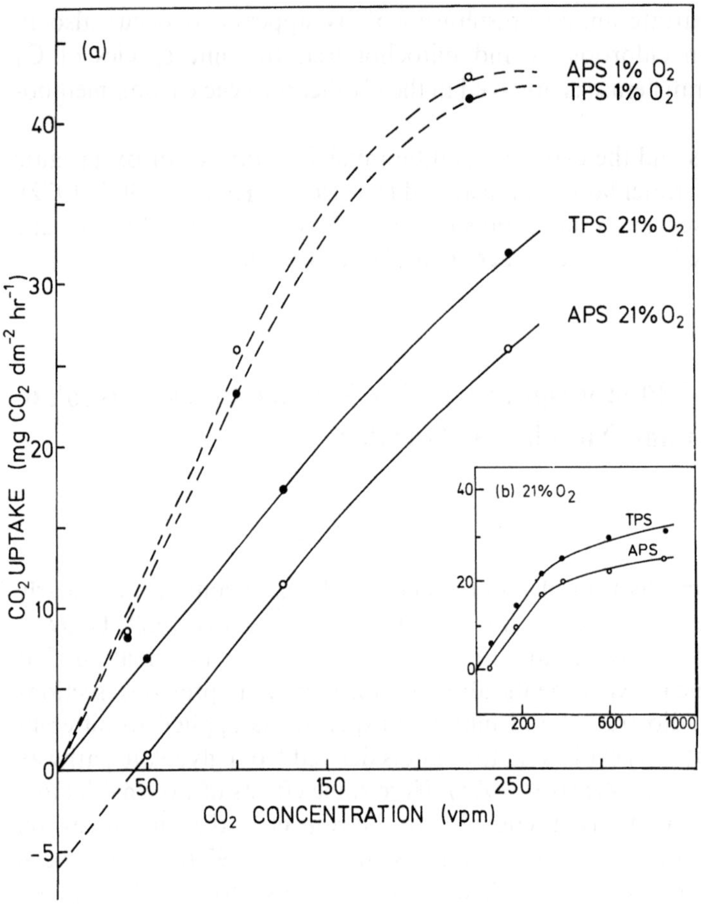

Fig 8a and b. Effects of CO_2 (average in the leaf chamber) and O_2 concentration on the rates of true photosynthesis (TPS′), apparent photosynthesis (APS), and CO_2 evolution in the light (PR′) of photosynthesizing sunflower leaves at 25°. Light intensity=3,500 ft-c. (a) Data from LUDWIG and CANVIN (1971b) measured in an open gas exchange system. (b) Data from BRAVDO and CANVIN, 1974

CO_2 is no longer rate-limiting for photosynthesis. Some 30–35% of the CO_2 evolved in photorespiration is refixed under normal and low CO_2 concentrations in illuminated sunflower leaves. Furthermore, from these results it was concluded that not only the measurable rate (PR′) but also the total process of CO_2 evolution in the light (PR) is independent of low CO_2 concentrations (D'AOUST and CANVIN, 1973).

Others have reported photorespiration to be CO_2 dependent, although the measured fraction of photorespiration (PR′) was not clearly defined. For instance, BIDWELL et al. (1969) found a higher rate of CO_2 evolution in illuminated bean leaves in 30–40 vpm CO_2 than in 300 vpm CO_2. In CO_2 concentrations saturating for photosynthesis, CO_2 evolution in the light was restricted or completely inhibited (BLACK, 1973).

Low photorespiratory rates have also been observed (JACKSON and VOLK, 1970), e.g. at low light intensity (HOCH et al., 1963; HOLMGREN and JARVIS, 1967; HEW et al., 1969a) or in photosynthetically efficient C_3-plant species (ZELITCH, 1975). More often, however, photorespiration may be underestimated if the measuring system determines only a small fraction of the total CO_2 formed and most of the generated CO_2 is immediately refixed (MANGAT et al., 1974). Generally, with CO_2 concentrations limiting the rate of photosynthesis the percentage of refixed photorespiratory CO_2 may vary widely (SAMISH and KOLLER, 1968; HEW et al., 1969a; ZELITCH, 1971; SAMISH et al., 1972). For intact C_3-plants, RAVEN (1972) reported reassimilation of photorespiratory CO_2 between zero and 63%, depending on the resistances to CO_2 diffusion inside the leaf and the characteristics of the measuring system (Sect. 2.1.3, Fig. 1).

Increasing the oxygen concentration in the atmosphere results in a proportional increase in the rate of PR′ and a proportional decrease of APS and TPS′ (LUDWIG and CANVIN, 1971b; D'AOUST and CANVIN, 1973). The inhibition of apparent photosynthesis (APS) by oxygen is due to both an inhibition of true photosynthesis (TPS) and a stimulation of CO_2 evolution in the light (PR) (FORRESTER et al., 1966a; TREGUNNA et al., 1966). These conclusions are in agreement with Eq. (4) to (6) (Sect. 2.1.2). The inhibition of TPS relative to PR is most predominant at low concentrations of carbon dioxide but is small at high CO_2 (LUDWIG and CANVIN, 1971b; D'AOUST and CANVIN, 1973).

Low oxygen concentrations (1–2% O_2) inhibit CO_2 evolution of C_3-plant leaves in the light and allows optimal rates of true photosynthesis (Fig. 8). The terms PR′ and IPS then approach zero and Eq. (6a) is simplified to:

$$TPS'_{1\%O_2} = APS_{1\%O_2}. \tag{15}$$

Therefore apparent photosynthesis (APS) measured by infrared gas analysis and true photosynthesis (TPS′) estimated with the $^{14}CO_2$ isotope technique are virtually identical at low oxygen.

6.1.2 Effects of Light and Temperature

When photorespiration (HOLMGREN and JARVIS, 1967; HEW et al., 1969a) or O_2 uptake (HOCH et al., 1963) of C_3-plants are measured at different light intensity, the amount of CO_2 evolved first decreases at low light intensity as compared with dark CO_2 evolution and then increases with increasing light intensity. At high light intensity the rates are much higher than those of dark respiration. These phenomena have been interpreted to mean that low light is sufficient to restrict dark respiration. Higher light intensity then stimulates a different light dependent respiratory process which, with dark respiration inhibited, is light respiration.

However, dark respiration may also continue in the light under certain experimental conditions (RAVEN, 1972; CHAPMAN and GRAHAM, 1974). For instance, while dark respiration appeared to be inhibited in well-watered illuminated sunflower leaves and replaced by light respiration, sunflower leaves under water stress evolved CO_2 in the light with a substantial dark-respiratory component (LAWLOR and FOCK, 1975).

In C_3-plants the rate of photorespiration progressively increases with increasing temperature. This has been shown by measuring either CO_2 evolution into CO_2-free air (HOLMGREN and JARVIS, 1967; HEW et al., 1969b; HOFSTRA and HESKETH, 1969), CO_2 compensation concentration (ZELITCH, 1966; WHITEMAN and KOLLER, 1967; JOLLIFFE and TREGUNNA, 1968), the postillumination CO_2 burst (DECKER, 1959; HEW et al., 1969b), or by extrapolating apparent photosynthesis to zero CO_2 (HEW et al., 1969b; LAING et al., 1974). In tracer experiments, the release of $^{14}CO_2$ by tobacco leaf discs infiltrated with glycolate-1-^{14}C was greatly enhanced at 35° in comparison to 25° (ZELITCH, 1966). The Q_{10}-value for CO_2 evolution in the light appears to be lower than that in the dark. Above 35° most C_3-plants evolve more carbon dioxide in darkness than in the light, although photorespiration usually exceeds dark respiration below 35° (HEW et al., 1969b; HOFSTRA and HESKETH, 1969).

6.1.3 Effects of External Factors on the Gas Exchange of C_4-plants

Photosynthesizing plants can be separated into those which, when kept in a sealed chamber in 21% O_2 and illuminated with moderate light intensity, decrease the carbon dioxide concentration to less than 10 vpm CO_2 (low compensation concentration plants) and those which decrease the CO_2 concentration to approximately 50 vpm CO_2 (high compensation concentration plants). The phenomenon of a low CO_2 compensation concentration was first noted in the unicellular green alga *Chlorella* (EGLE and SCHENK, 1952) and later in the higher plant species maize (MEIDNER, 1962; MOSS, 1962). Low compensation concentration in higher plants is associated with the C_4-dicarboxylic acid pathway of photosynthesis; high compensation ability is associated with Calvin's photosynthetic reduction cycle or C_3 pathway (BLACK, 1973). The two types of higher plants have therefore been classified as C_4- and C_3-plants. Photorespiratory CO_2 evolution (PR′) by leaves of C_4-plants is very small and less than 10% of dark respiration (JACKSON and VOLK, 1970; VOLK and JACKSON, 1972).

The data available indicate that leaves of the C_4-type are able, upon illumination, to reduce their internal CO_2 concentration to a very low level, because the carboxylation efficiency (FORRESTER et al., 1966a) in this group is high. Therefore the mesophyll resistance (between i and c in Fig. 1) is low compared with the boundary layer (r_a) and stomatal resistances (r_s). The total CO_2 produced in the light (PR) would then immediately be refixed and would be undetected by conventional gas analysis. Alternatively, the rate of photorespiration (PR) is small and the flux of photorespiratory CO_2 into the intercellular spaces would not significantly raise the internal CO_2 concentration. It seems that a high carboxylation efficiency, low mesophyll resistances, and low photorespiratory rates together account for the low CO_2 compensation concentration. Verification of these possibilities depends on the C_4-species tested and on the experimental conditions.

6.1.4 The Specific Activity of the $^{14}CO_2$ Evolved in the Light and in the Dark

Photorespiration of C_3-plants can be characterized by the specific activity of the $^{14}CO_2$ released during or after $^{14}CO_2$ fixation (Fig. 9; GOLDSWORTHY, 1966; LUDWIG, 1968; D'AOUST, 1970). In low O_2 photorespiratory CO_2 is not released into the surroundings (Fig. 8), but under normal conditions (ca. 300 vpm CO_2, 21% O_2, moderate light intensity) $^{14}CO_2$ is evolved from the leaf within 30 s of its supply. The rate of $^{14}CO_2$ evolution increases rapidly with time and becomes constant within 10–15 min; at that time its specific activity reaches an almost constant level of approximately 70–100% relative to the $^{14}CO_2$ offered (D'AOUST and CANVIN, 1972, 1974). The steady state value of this so-called relative specific activity (RSA) of the $^{14}CO_2$ evolved in the light and the rate of labeling are both affected by the CO_2 concentration around the leaves (LUDWIG and CANVIN, 1971b) and by other external parameters. Factors which decrease photosynthesis, e.g. a low concentration of carbon dioxide or water stress (LAWLOR and FOCK, 1975), tend to lower the RSA of the $^{14}CO_2$ evolved in the light and reduce the rate of ^{14}C labeling.

In the few instances where the relative specific activity of photorespiration reaches 100% (LUDWIG and CANVIN, 1971a, b), dark respiration is completely inhibited in the light and the substrate for photorespiration is only derived from the early products of photosynthesis.

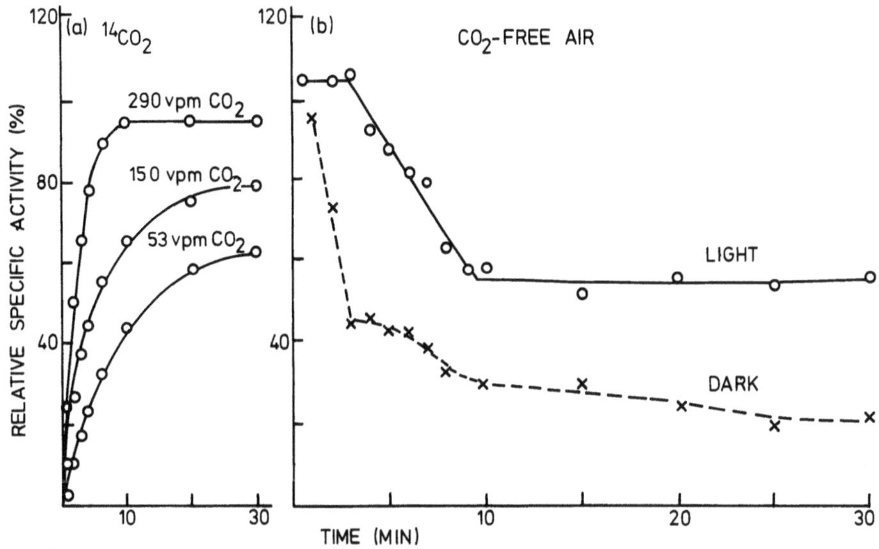

Fig. 9. (a) Effect of CO_2 concentration (average in the leaf chamber) on the relative specific activity (RSA) of the $^{14}CO_2$ evolved from attached sunflower leaves photosynthesizing in $^{14}CO_2$ at 25° C and 3,500 ft-c. RSA is the specific activity of the $^{14}CO_2$ evolved from the leaf as a percentage of the specific activity of the supplied $^{14}CO_2$. Data from LUDWIG and CANVIN (1971b) measured in an open gas exchange system. (b) The relative specific activity (RSA) of the $^{14}CO_2$ evolved from sunflower leaves into CO_2-free air in the light and in the dark after 60 min photosynthesis in $^{14}CO_2$ at 350 vpm CO_2 (inlet) and 25° C. Light intensity=9,000 ft-c (1,700 μE/m²s in the range from 400–700 nm). Data from CANVIN et al. (1975)

When the specific activity (RSA) during steady state photosynthesis is below 100%, photorespiratory CO_2 may also be derived completely from intermediates of the glycolate pathway, if this pathway is supplied both by recent ^{14}C labeled photosynthate and unlabeled storage product(s); then dark respiration would be inhibited. Alternatively, the Krebs cycle may operate in the light, use unlabeled or weakly labeled reserve materials (ATKINS, 1969), and produce carbon dioxide of low specific activity. At the same time, recent assimilate is metabolized through the glycolate pathway and high specific activity $^{14}CO_2$ released. Inside the mesophyll cell (Fig. 1 at k) the two fluxes of carbon dioxide with high and low specific activity mix and yield a total outgoing CO_2 stream (PR) of intermediate specific activity. A substantial fraction of the CO_2 is reassimilated (RA) thus reducing the specific activity of recent photosynthate; the remainder escapes into the intercellular spaces and the rate (PR′) and specific activity is estimated (LUDWIG and CANVIN, 1971a; CANVIN and FOCK, 1972. For information on dark respiration in the light see JACKSON and VOLK, 1970; RAVEN, 1972; CHAPMAN and GRAHAM, 1974; MANGAT et al., 1974; LAWLOR and FOCK, 1975).

After a period of $^{14}CO_2$ fixation, little $^{14}CO_2$ of low specific activity is released by ^{14}C labeled leaves during photosynthesis in 300 vpm $^{12}CO_2$ and 21% O_2, more in 150 vpm $^{12}CO_2$, and even more in 50 vpm $^{12}CO_2$. The specific activity of the $^{14}CO_2$ evolved in darkness by ^{14}C labeled leaves, however, is independent of the CO_2 concentration in the atmosphere (GOLDSWORTHY, 1966; LUDWIG and CANVIN, 1971b). In a CO_2-free atmosphere, the specific activity of the $^{14}CO_2$ evolved during the first minute of illumination and during the post-illumination CO_2 burst on darkening is equivalent to that measured during the preceding period of $^{14}CO_2$ fixation (Fig. 9b; D'AOUST and CANVIN, 1972). Later, the specific activity of the $^{14}CO_2$ released rapidly falls but is significantly higher in the light than in darkness. This indicates that the CO_2-free conditions induce a change in the substrate of photorespiration, recent photosynthate being preferentially oxidized in 300 vpm CO_2, reserve material in CO_2-free air. Moreover, the substrates and mechanisms for CO_2 evolution in the light and in the dark are different, as compounds degraded during illumination contain a different percentage of assimilate from the previous fixation period than those oxidized in the dark (ATKINS, 1969).

6.2 Carbon Flux through the Glycolate Pathway

In the previous sections evidence has been discussed which indicates that most of the CO_2 evolved in the light is formed during glycolate pathway operation. Therefore, intermediates of the glycolate pathway and the $^{14}CO_2$ evolved in the light should have similar specific activities and the immediate substrate of light respiration should be identical in specific activity with that of the $^{14}CO_2$ evolved. This prediction provides an experimental tool for the identification of the light-respiratory substrate and for an estimate of carbon flux through the glycolate pathway: one compares the specific activity of the evolved $^{14}CO_2$ with that of possible precursors and then uses the experimental data to solve Eq. (13) (Sect. 2.1.4), e.g. for an estimate of the flux of carbon from glycine to serine. The limitations of this method have been discussed elsewhere (Sect. 2.1.4; CANVIN et al., 1975).

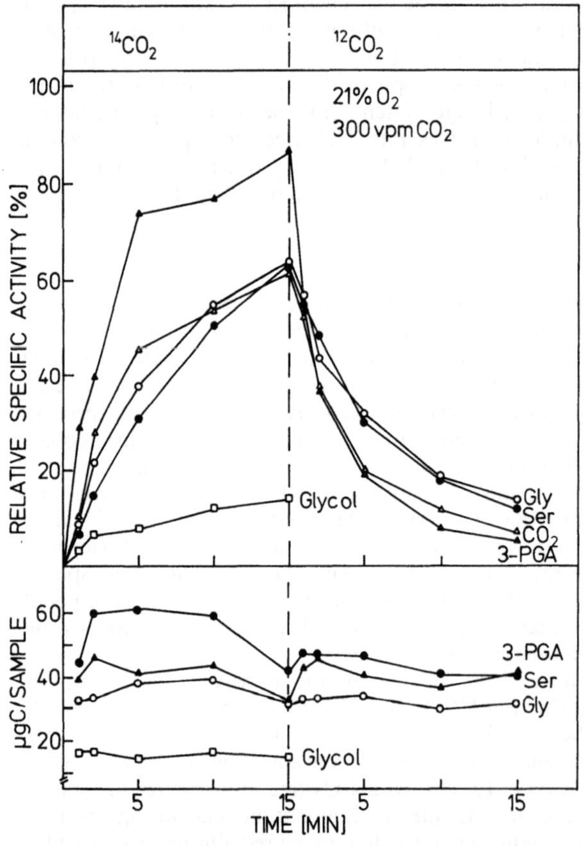

Fig. 10. The relative specific activity (RSA) and the total amounts of metabolites labeled during photosynthesis of sunflower leaf discs in 300 vpm $^{14}CO_2$ and 21% O_2 (average in leaf chamber) and during photosynthesis in 300 vpm CO_2 and 21% O_2 after 15 min exposure to $^{14}CO_2$. Every measured point corresponds to a sample of 9 leaf discs (55.4 cm^2). Light intensity = 109 W/m^2 (ca. 2,800 ft-c), temperature = 25° C. Data from MAHON et al. (1974c). Gly = glycine; Ser = serine; 3-PGA = 3-phosphoglyceric acid; Glycol = glycolate; CO_2 = CO_2 evolved in the light

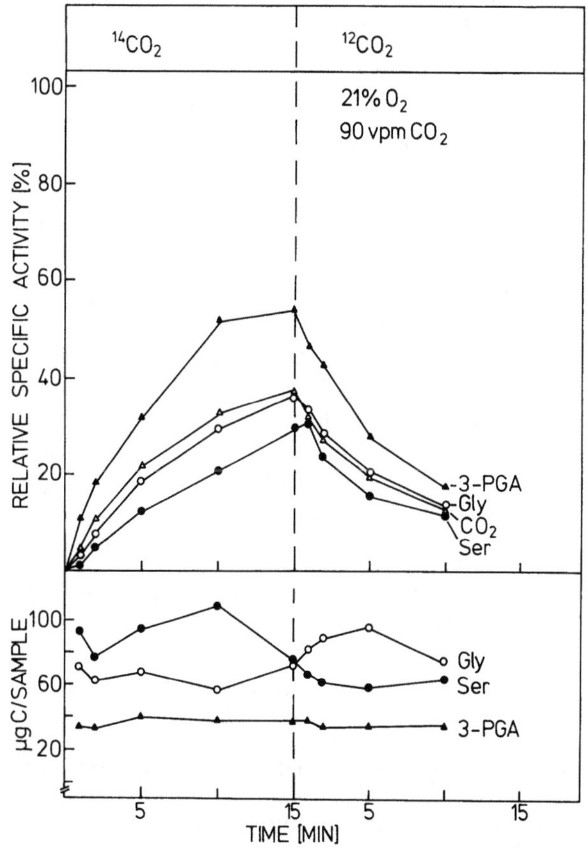

Fig. 11. The relative specific activity (RSA) and the total amounts of metabolites labeled during photosynthesis of sunflower leaf discs in 90 vpm $^{14}CO_2$ and 21% O_2 (average in leaf chamber) and during photosynthesis in 90 vpm CO_2 and 21% O_2 after 15 min exposure to $^{14}CO_2$. Every measured point corresponds to a sample of 9 leaf discs (55.4 cm^2). Light intensity = 109 W/m^2 (ca. 2,800 ft-c); temperature = 25° C. Data from MAHON et al. (1974c). Symbols as in Fig. 10

6.2.1 C_3-plants in Air

The specific activity of glycine, serine, and CO_2 of sunflower leaf discs illuminated in air (ATKINS and CANVIN, 1971) increased rapidly when $^{14}CO_2$ was supplied and declined equally rapid when $^{14}CO_2$ was replaced by CO_2, but did not become the same as that of the $^{14}CO_2$ fed to the leaves even after 15 min of $^{14}CO_2$ incorporation (Fig. 10). The specific activities of the three compounds were always similar despite the fluctuation in the amount of glycine and serine. In air of low CO_2 concentration (90 vpm CO_2, 21% O_2), the specific activities were lower than in air containing 300 vpm CO_2 and the rate of change was slower (Fig. 11). The results suggest that glycine and serine are both in single pools in leaf discs photosynthesizing in air but not necessarily in air containing little CO_2. The specific activities indicate that both glycolate pathway amino acids are closely related to CO_2 evolution in the light (see also Sects. 3.3 and 4.1).

The flux of carbon into serine can be estimated from the amount of radioactivity in the precursor and product and their specific activities according to Eq. (13), assuming this flux is unidirectional and mainly supplied by glycine:

$$T_{AB} = \frac{dn_{AB}}{dt} = \frac{dB^*/dt - s_B \cdot dB/dt}{s_A - s_B} \tag{13}$$

where T_{AB} is the flux rate of carbon from glycine to serine (μgC/dm^2 min); dB^*/dt is the incorporation of ^{14}C (dpm) into serine during dt; dt is the $^{14}CO_2$ feeding time interval between 1 and 2, 2 and 5, 5 and 10, or 10 and 15 min; s_A is the average specific activity of glycine during dt (dpm/μgC); s_B is the average specific activity of serine during dt (dpm/μgC); dB/dt is the change in the amount of serine during dt (dpm/μgC).

The estimates of carbon flux into serine during different periods of $^{14}CO_2$ incorporation (Table 3), which should be constant under the conditions of steady state photosynthesis, decreased continuously during $^{14}CO_2$ uptake in 90 vpm CO_2. Therefore, the estimates in low CO_2, which are subject to considerable uncertainty, do not, if taken at face value fully support the proposed reaction sequence from glycine to serine (Fig. 6). In air (300 vpm CO_2; 21% O_2), however, the estimated rates of carbon flux from glycine to serine were similar between 2 and 15 min of ^{14}C incorporation, averaging 71 μgC/min for 1 dm^2 of leaf discs or 61% of TPS′ (MAHON et al., 1974c).

Table 3. Flux of carbon (μgC/dm^2 min) from glycine to serine estimated according to Eq. (13) from the radioactivity and total amount of compound during $^{14}CO_2$ photosynthesis in 90 and 300 vpm CO_2 (21% O_2). APS= 19 (in 90 vpm CO_2) and 96 μgC/dm^2 min (in 300 vpm CO_2); TPS′=41 (in 90 vpm CO_2) and 116 μgC/dm^2 min (in 300 vpm CO_2); PR′=22 (in 90 vpm CO_2) and 20 μgC/dm^2 min (in 300 vpm CO_2); leaf sample=0.55 dm^2. Data from Figs. 10 and 11

Time of $^{14}CO_2$ fixation (min)	90 vpm CO_2 C-flux (μgC/dm^2 min)	300 vpm CO_2 C-flux (μgC/dm^2 min)
1 to 2	195	153
2 to 5	72	70
5 to 10	27	61
10 to 15	25	83
Mean		71

An independent method of estimating carbon transfer through the glycolate pathway considers the main fluxes entering 3-PGA (Figs. 10, 11). The average specific activity of carbon entering 3-PGA which is assumed to be homogeneous and totally involved in photosynthetic carbon metabolism (GALMICHE, 1973), can be determined from the specific activity of the supplied $^{14}CO_2$, the rate of carbon flux into the leaf (TPS′), and the total flux of carbon through the glycolate pathway including reassimilated carbon dioxide:

$$S_a = \frac{\text{TPS}' S_{CO_2} + F S_{gs}}{\text{TPS}' + F} \tag{16}$$

where S_a is the average specific activity of carbon entering 3-PGA (dpm/μgC); TPS′ is the rate of true CO_2 uptake (μgC/dm^2 min); S_{CO_2} is the specific activity of the $^{14}CO_2$ feeding gas (dpm/μgC); F is the flux of carbon from serine and reassimilated CO_2 into 3-PGA (μgC/dm^2 min); and S_{gs} is the average specific activity of carbon in the glycolate pathway calculated as mean of glycine and serine (dpm/μgC).

Because of the large flux into the relatively small 3-PGA pool, the average specific activity of the carbon entering 3-PGA soon becomes similar to that of 3-PGA itself and the flux from serine and reassimilated CO_2 (F) can be estimated from:

$$F = \frac{\text{TPS}'(S_{CO_2} - S_{PGA})}{S_{PGA} - S_{gs}} \tag{17}$$

where S_{PGA} is the specific activity of 3-PGA and the other symbols are as in Eq. (16). The estimates (Table 4) show that Eq. (17) is not applicable after short periods of $^{14}CO_2$ incorporation because of the initially low specific activity of the Calvin cycle intermediates (FOCK et al., 1974). However, the relatively constant values after 5 min photosynthesis in 300 vpm CO_2 and after 10 min in 90 vpm CO_2 indicate large fluxes of carbon from the glycolate pathway into 3-PGA. Addition of the rate of CO_2 release in light (PR′) gives an estimate of the total fluxes of carbon. They are approximately 106 and 96 μgC/dm^2 min or 91% and 234% of TPS′ in air and in air with low CO_2 respectively, and appear to be CO_2 independent as predicted from gas exchange measurements (Sect. 6.1.1). These fluxes are comparable with those estimated from the glycine to serine conversion (average 71 μgC/dm^2 min, Table 3) by which additional carbon is lost as light-respiratory CO_2. Moreover, the estimated fluxes through the glycolate pathway are sufficient to account for photorespiration (PR) with the stoichiometry of one carbon atom released as CO_2 for every four carbons traversing the pathway.

Table 4. Flux of carbon (μgC/dm^{2}min) into 3-PGA from sources other than supplied CO_2 estimated at 90 and 300 vpm CO_2 (21% O_2) according to Eq. (17). Rates of CO_2 gas exchange as in Table 3. Leaf sample = 0,55 dm^{-2}. Data from Figs. 10 and 11

Time of $^{14}CO_2$ incorporation (min)	90 vpm CO_2		300 vpm CO_2	
	C-flux	% TPS′	C-flux	% TPS′
1	356		404	
2	240		325	
5	157		79	68
10	65	159	116	100
15	83	202	63	54
Mean	74	180	86	74
+PR′	22	54	20	17
Total flux of carbon through glycolate pathway	96	234	106	91

There are no other direct estimates of carbon traffic through the glycolate pathway. However, results obtained by various indirect methods, e.g. ^{14}C-labeling of glycine and serine, accumulation of glycolate by inhibiting glycolate oxidase in leaf discs or from photorespiration, although not directly comparable, also suggest that much of the carbon fixed in photosynthesis passes through the glycolate pathway (BIRD et al., 1974; ZELITCH, 1975).

If the flow of carbon through the glycolate pathway is as great as estimated (Tables 3, 4), the specific activities suggest that glycine and serine could not be formed solely from intermediates of the Calvin cycle. Thus a second source remote from the immediate photosynthetic products is supplying carbon atoms to the pathway in photosynthesizing sunflower leaf discs; one possible intermediate through which reserve material enters is glycolate. The average specific activity of glycolate is relatively low (Fig. 10) suggesting two independent pathways for glycolate formation, one supplied by high specific activity RuBP during $^{14}CO_2$ uptake (Sect. 3.1) and the other by unknown but almost unlabeled storage products. However, some aspects of the labeling data could also be explained by compartmentation of the metabolites with low exchange between labeled and unlabeled pools.

6.2.2 C_3-plants in a Medium with Low Oxygen

In an atmosphere of low oxygen (1–2% O_2) CO_2 is not released from sunflower leaf discs in the light (Sect. 6.1.1) but glycine and serine still incorporate carbon-14 during exposure to $^{14}CO_2$ and the label may be partly flushed out with CO_2 (Fig. 12). The initial rate of glycine labeling, when $^{14}CO_2$ is supplied, and the rate of flushing in CO_2 are much lower than in air (Fig. 10), although there is still appreciable glycolate formation in low oxygen (FOCK and EGLE, 1967). The initially greater specific activity of glycine compared to serine is reversed after longer feeding periods indicating that glycine and serine are not closely related metabolically. Therefore only a limited conversion of glycine to serine occurs. Substantial amounts of both metabolites are formed and further metabolized by independent mechanisms. This suggests a decreased flux of carbon through the glycolate pathway. Eqs. (13) and (14) are invalid under these conditions and carbon flux through intermediates cannot be determined.

6.2.3 C_4-plants in Normal and Low Oxygen

The changes in specific activity of glycine and serine in maize leaf discs during $^{14}CO_2$ uptake in air (300 vpm CO_2, 21% O_2) and in low oxygen (300 vpm CO_2, 1% O_2) are similar to but much slower than in sunflower discs (Figs. 10, 12; MAHON et al., 1974a) suggesting that the glycolate pathway is operating. However, the magnitude of carbon transfer through the pathway cannot be determined, because multiple pathways for the metabolism of glycine and serine may occur.

6.2.4 Carbon Fluxes and Enzyme Activities

The rate of photosynthetic CO_2 fixation can be assumed to be in the order of 600 μmol/dm^2 h at saturating light intensity, 300 vpm CO_2, and 21% O_2. The total rate of light-respiratory CO_2 evolution (PR) is in the order of 200 μmol/dm^2 h

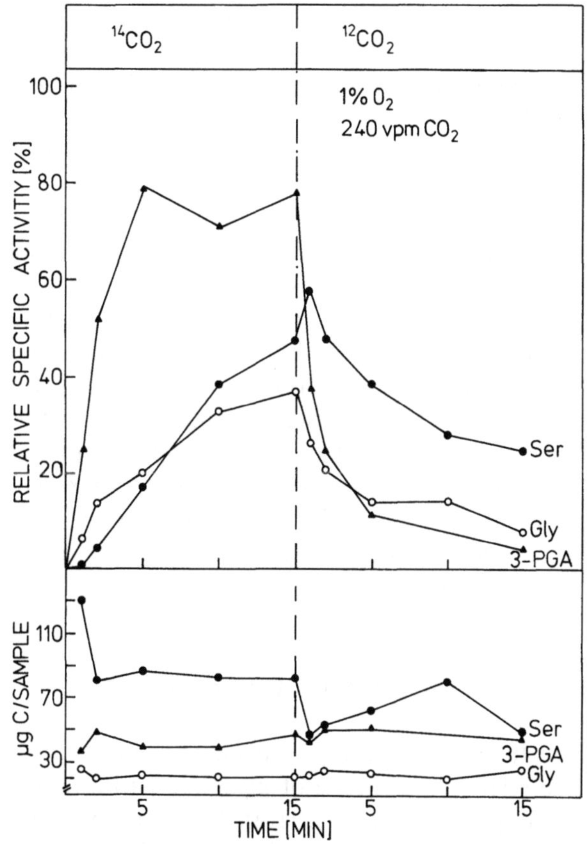

Fig. 12. The relative specific activity (RSA) and the total amounts of metabolites labeled during photosynthesis of sunflower leaf discs in 240 vpm $^{14}CO_2$ and 1% O_2 (average in leaf chamber) and during photosynthesis in 240 vpm CO_2 and 1% O_2 after 15-min exposure to $^{14}CO_2$. Every measured point corresponds to a sample of 9 leaf discs (55.4 cm^2). Light intensity = 109 W/m^2 (ca. 2,800 ft-c), temperature = 25° C. Data from MAHON et al. (1974b). Symbols as in Fig. 10

for intact leaves of C_3 plants or two-thirds of these figures for leaf discs. Carbon fluxes from glycine to serine are approximately 70 µgC/dm^2 min for leaf discs and are thus of the same magnitude as the rate of light-respiratory CO_2 evolution. For reasons of comparison the different units used in the literature can approximately be interconverted according to ZELITCH (1971): 1 dm^2 projected leaf area = 2 mg fresh weight lamina = 3.0 mg chlorophyll = 6.0 mg protein N = 40 mg protein.

Carbon fluxes for light-respiratory CO_2 evolution have to be accomplished by the enzyme activities of the glycolate pathway. According to ANDREWS et al. (1973) glycolate formation can eventually be accounted for by the ribulose bisphosphate carboxylase/oxygenase. LAING et al. (1974) also argue that the oxygenase activity is high enough to account for postulated rates. Phosphoglycolate phosphatase activity seems to be sufficiently available (RANDALL et al., 1971). Glycolate oxidase activity is about half of true photosynthesis (FOCK and KROTKOV, 1969) and seems to be just sufficient for postulated rates (FOCK and KROTKOV, 1969; TOLBERT and YAMAZAKI, 1969). Glutamate, glyoxylate aminotransferase, and serine, glyoxylate aminotransferase, show activities one half and one third of that for glycolate oxidase in the isolated peroxisomes (TOLBERT and YAMAZAKI, 1969) so that these activities are at the lower edge of being sufficient for postulated rates and fluxes.

The rate of CO_2 evolution by isolated mitochondria is estimated to be equivalent to 64 µmol/mg chlorophyll h assuming that 0.4 to 0.6 mg protein are equivalent to 0.2 g of green leaves (KISAKI et al., 1971a). However, the activities of glycine decarboxylase and hydroxymethyl transferase appear to be rather low. Hydroxypyruvate reductase activity is present in leaves in large excess (TOLBERT and YAMAZAKI, 1969) if compared with the activity of glycolate oxidase. Glycerate kinase activity is insufficient to permit high rates of carbon flow through the glycolate pathway back to Calvin cycle intermediates (HEBER et al., 1974). Phosphoglycerate phosphatase activity ranges between 1 and 10 µmol/min mg chlorophyll and the activity is therefore in the range of photorespiration (RANDALL et al., 1971).

In summary, for most enzymes the maximal activities reported appear to be sufficient to account for measured rates of photorespiration and carbon fluxes. Some other reported enzyme activities appear to be too low. Obviously, further work is required.

6.3 The Function of Light Respiration

The function of leaf peroxisomes is in glycolate metabolism (TOLBERT, 1971a). The relative magnitude of glycolate pathway metabolism and of light respiration within plant species (ZELITCH, 1975) and between species (BLACK, 1973) may vary widely and the selective advantage offered by it is not clear. LAING et al. (1974) estimated the approximate stoichiometry of apparent photosynthesis (APS), true photosynthesis (TPS), and photorespiration (PR) of soybean leaves in atmospheres containing 0% and 21% O_2 at 300 vpm CO_2. This model appears to be compatible with CO_2 fluxes during photosynthesis (Sect. 6.1), with the changes in specific activity of metabolites, with the estimated flux of carbon through the glycolate pathway (Sect. 6.2), and with several enzyme activities involved in the pathway (Sects. 3–5). However, the model may have to be modified to include the flux of carbon from storage compounds, e.g. through glycolate or other metabolites, into the stream of recent assimilate.

From an energetic point of view, light respiration seems to be a wasteful process by which glycolate as an unavoidable by-product of photosynthesis (LORIMER and ANDREWS, 1973) is diverted to produce carbohydrates or other cell components. CO_2 release in the light may be essential for most plants as it provides a mechanism to degrade excessive light energy absorbed under conditions of restricted CO_2 supply and thus, to protect the cell against photooxidative destruction of the photosynthetic mechanism (TOLBERT, 1971a; ZELITCH, 1971). Carbon flux models have been proposed by OSMOND and BJÖRKMAN (1972), TOLBERT and RYAN (1975), and LAING et al. (1974). The view of light respiration being a mechanism against photooxidation under conditions of restricted CO_2 supply is supported by Heber's (unpublished) observation that leaves with little CO_2 gas exchange have surprisingly high electron transport activity as judged by fluorescence and light scattering measurements.

7. Concluding Remarks

Photosynthetic CO_2 reduction leads to the synthesis of several compounds. Starch is stored as an insoluble product inside the chloroplasts. Other end products such as triosephosphate and glycolate are soluble and must be exported if photosynthesis is to proceed. They are further metabolized outside the chloroplasts. An ultimate product is sucrose, which may leave the photosynthetic cell. The biochemical path to sucrose is direct and short, if triosephosphate is the substrate. It is involved and complex in the case of glycolate. Several cellular compartments take part in glycolate metabolism. Intermediates on the path from glycolate to sugar must reenter the chloroplasts for further processing. While it is likely that for the transfer of glycolate and glycerate across the envelopes of chloroplasts and peroxisomes unspecific diffusion in the protonated form or unspecific anion exchange are sufficient, it remains unknown whether transport of the amino acids involved in the glycolate pathway is carrier-mediated or proceeds via unspecific diffusion. The significance of the complex cooperation between different cell organelles in the reactions of the glycolate pathway is poorly understood. Biochemical details will be discussed in another volume of the Encyclopedia of Plant Physiology. One purpose of glycolate metabolism may be to supply the cell with the amino acids formed en route to sugar. Such an explanation does not consider the high capacity of the glycolate pathway and its large carbon fluxes. Some of the reactions of the pathway require energy, either in the form of ATP or of reducing equivalents. Although the formation of glycolate in the ribulose bisphosphate oxygenase reaction is exergonic, the complete conversion of sugar phosphate into glycolate and phosphate is also energy-dependent, since energy-requiring reactions of the Calvin cycle have to regenerate sugar phosphate from the phosphoglycerate formed together with phosphoglycolate in the oxygenase reaction. Taking this into account, the sequence of events leading to the complete oxidation of ribulose bisphosphate into glycolate and phosphate may be abbreviated as follows:

$$2\,\mathrm{RuBP}^{4-} + 5\,\mathrm{O}_2 + 8\,\mathrm{OH} \;\; + 8\,\mathrm{ATP}^{4-} + 5\,\mathrm{NADPH} \rightarrow$$

$$5\,\mathrm{glycolate}^{-} + \mathrm{H_2O} + 8\,\mathrm{ADP}^{3-} + 12\,\mathrm{Pi}^{2-} + 5\,\mathrm{NADP}^{+}$$

Considerable consumption of ATP and NADPH is apparent. It appears promising to consider photorespiration and its associated reactions and fluxes in the context of regulated degradation of excess energy. Whether this view covers its main aspects, only future work can show.

References

ANDREWS, T.J., JOHNSON, H.S., SLACK, C.R., HATCH, M.D.: Malic enzyme and aminotransferases in relation to 3-phosphoglycerate formation in plants with the C_4-dicarboxylic acid pathway of photosynthesis. Phytochemistry **10**, 2005–2013 (1971 a)

ANDREWS, T.J., LORIMER, G.H., TOLBERT, N.E.: Incorporation of molecular oxygen into glycine and serine during photorespiration in spinach leaves. Biochemistry **10**, 4777–4782 (1971 b)

ANDREWS, T.J., LORIMER, G.H., TOLBERT, N.E.: Ribulose diphosphate oxygenase. I. Synthesis of phosphoglycolate by fraction-1 protein of leaves. Biochemistry **12**, 11–18 (1973)

ARNON, D.I.: Copper enzymes in isolated chloroplasts. Polyphenoloxidase in Beta vulgaris. Plant Physiol. **24**, 1–15 (1949)

ARONOFF, S.: Techniques of radiobiochemistry. Iowa: The Iowa State College Press 1956

ATKINS, C.A.: Intermediary metabolism of photosynthesis in relation to CO_2 evolution in the light. Ph.D. Thesis, Queen's University, Kingston, Ontario, Canada, 1969

ATKINS, C.A., CANVIN, D.T.: Photosynthesis and CO_2 evolution by leaf discs: gas exchange, extraction, and ion exchange fractionation of ^{14}C-labeled photosynthetic products. Can. J. Botany **49**, 1225–1234 (1971)

BAHR, J.T., JENSEN, R.G.: Ribulose bisphosphate oxygenase activity from freshly ruptured spinach chloroplasts. Arch. Biochem. Biophys. **164**, 408–413 (1974a)

BAHR, J.T., JENSEN, R.G.: Ribulose diphosphate carboxylase from freshly ruptured spinach chloroplasts having an in vivo *Km* [CO_2]. Plant Physiol. **53**, 39–44 (1974b)

BASSHAM, J.A., BENSON, A.A., CALVIN, M.: Isotope studies in photosynthesis. J. Chem. Educ. **30**, 274–283 (1953)

BASSHAM, J.A., KIRK, M.: The effect of oxygen on the reduction of CO_2 to glycolic acid and other products during photosynthesis by Chlorella. Biochem. Biophys. Res. Commun. **9**, 376–380 (1962)

BASSHAM, J.A., KIRK, M.: Sequence of formation of phosphoglycolate and glycolate in photosynthesizing Chlorella pyrenoidosa. Plant Physiol. **52**, 407–411 (1973)

BASSHAM, J.A., KIRK, M., JENSEN, R.G.: Photosynthesis by isolated chloroplasts. I. Diffusion of labeled photosynthetic intermediates between isolated chloroplasts and suspending medium. Biochim. Biophys. Acta **153**, 211–218 (1968)

BEISENHERZ, G.: Triosephosphate isomerase from calf muscle. Methods in Enzymol. **1**, 387–391 (1955)

BENSON, A.A., CALVIN, M.: The path of carbon in photosynthesis. VII. Respiration and photosynthesis. J. Exptl. Botany **1**, 63–68 (1950)

BIBBY, B.T., DODGE, J.D.: The ultrastructure and cytochemistry of microbodies in dinoflagellates. Planta **112**, 7–16 (1973)

BIDWELL, R.G.S., LEVIN, W.B., SHEPHARD, D.C.: Photosynthesis, photorespiration and respiration of chloroplasts from Acetabularia mediterranea. Plant Physiol. **44**, 946–954 (1969)

BIRD, I.F., CORNELIUS, M.J., KEYS, A.J., KUMARASINGHE, S., WHITTINGHAM, C.P.: The rate of metabolism by the glycolate pathway in wheat leaves during photosynthesis. In: Proceedings of the Third International Congress on Photosynthesis. AVRON, M. (ed.), pp 1291–1301. Amsterdam: Elsevier Scientific Publishing Company 1974

BIRD, I.F., CORNELIUS, M.J., KEYS, A.J., WHITTINGHAM, C.P.: Oxidation and phosphorylation associated with the conversion of glycine to serine. Phytochemistry **11**, 1587–1594 (1972)

BISALPUTRA, T., SHIELDS, C.M., MARKHAM, J.W.: In situ observation of the fine structure of Laminaria gametophytes and embryos in culture. I. Methods and the ultrastructure of the zygote. J. Microscopie **10**, 83–98 (1971)

BLACK, C.C.: Photosynthetic carbon fixation in relation to net CO_2 uptake. Ann. Rev. Plant Physiol. **24**, 253–286 (1973)

BLACK, C.C., MOLLENHAUER, H.H.: Structure and distribution of chloroplasts and other organelles in leaves with various rates of photosynthesis. Plant Physiol. **47**, 15–23 (1971)

BOUCK, G.R.: Fine structure and organelle associations in brown algae. J Cell Biol **26**, 523–537 (1965)

BOWES, G., OGREN, W.L.: Oxygen inhibition and other properties of soybean ribulose 1,5-diphosphate carboxylase. J. Biol. Chem. **247**, 2171–2176 (1972)

BOWES, G., OGREN, W.L., HAGEMAN, R.H.: Phosphoglycolate production catalyzed by ribulose diphosphate carboxylase. Biochem. Biophys. Res. Commun. **45**, 716–722 (1971)

BRADBEER, J.W., ANDERSON, C.M.A.: Glycolate formation in chloroplast preparations. In: Biochemistry of Chloroplasts. Vol. II (GOODWIN, T.W., ed.), pp. 175–179. New York: Academic Press 1967

BRAVDO, B.A.: Decrease in net photosynthesis caused by respiration. Plant Physiol. **43**, 479–483 (1968)

BRAVDO, B.A., CANVIN, D.T.: Evaluation of the rate of CO_2 production by photorespiration. Proceedings of the Third International Congress on Photosynthesis (AVRON, M., ed.), pp. 1277–1283. Amsterdam: Elsevier 1974

BREIDENBACH, R.W., BEEVERS, H.: Association of the glyoxylate cycle enzymes in a novel subcellular particle from castor bean endosperm. Biochem. Biophys. Res. Commun. **27**, 462–469 (1967)

BREIDENBACH, R.W., KAHN, A., BEEVERS, H.: Characterization of glyoxysomes from castor bean endosperm. Plant Physiol. **43**, 705–713 (1968)

BRODY, M., WHITE, J.E.: Environmental factors controlling enzymatic activity in microbodies and mitochondria of Euglena gracilis. FEBS Lett. **23**, 149–152 (1972)

BRUIN, W.J., NELSON, E.B., TOLBERT, N.E.: Glycolate pathway in green algae. Plant Physiol. **46**, 386–391 (1970)

CANVIN, D.T., FOCK, H.: Measurement of photorespiration. In: Methods in Enzymology, Vol. XXIV, Part B, pp. 246–260 (SAN PIETRO, A., ed.). New York: Academic Press 1972

CANVIN, D.T., FOCK, H., PRZYBYLLA, K., LLOYD, N.D.H.: Glycine and serine metabolism and photorespiration. Proc. 5th Steenbock Symp. Madison-Wisconsin. In press 1975

CHAN, H.W.-S., BASSHAM, J.A.: Metabolism of ^{14}C-labeled glycolic acid by isolated spinach chloroplasts. Biochim. Biophys. Acta **141**, 426–429 (1967)

CHAPMAN, E.A., GRAHAM, D.: The effect of light on the tricarboxylic acid cycle in green leaves. I. Relative rates of the cycle in the dark and light. Plant Physiol. **53**, 879–885 (1974)

CHAPPELL, J.B.: Transport and exchange of anions in mitochondria. In: Inhibitors: Tools in Cell Research (BÜCHNER, TH., SIES, H., eds.), pp. 335–350. Berlin-Heidelberg-New York: Springer 1969

CHENG, G.P., ROSENBLUM, I.Y., SALLACH, H.J.: Comparative studies of enzymes related to serine metabolism in higher plants. Plant Physiol. **43**, 1813–1820 (1968)

CHENG, K.G., MILLER, A.G., COLMAN, B.: An investigation of glycolate excretion in two species of blue-green algae. Planta **103**, 110–116 (1972)

CHOLLET, R.: $^{14}CO_2$ fixation and glycolate metabolism in the dark in isolated maize (Zea mays L.) bundle sheath strands. Arch. Biochem. Biophys. **163**, 521–529 (1974)

CHOLLET, R., OGREN, W.L.: Oxygen inhibits maize bundle sheath photosynthesis. Biochem. Biophys. Res. Commun. **46**, 2062–2066 (1972)

CHU, D.K., BASSHAM, J.A.: Activation of ribulose 1,5-diphosphate carboxylase by nicotinamide adenine dinucleotide phosphate and other chloroplast metabolites. Plant Physiol. **54**, 556–559 (1974)

CHU, D.K., BASSHAM, J.A.: Regulation of ribulose 1,5-diphosphate carboxylase by substrates and other metabolites. Plant Physiol. **55**, 720–726 (1975)

CLAGETT, C.O., TOLBERT, N.E., BURRIS, R.H.: Oxidation of α-hydroxy acids by enzymes from plants. J. Biol. Chem. **178**, 977–987 (1949)

CLANDININ, M.T., COSSINS, E.A.: Regulation of mitochondrial glycine decarboxylase from pea mitochondria. Phytochemistry **14**, 387–391 (1975)

CODD, G.A., LORD, J.M., MERRETT, M.J.: The glycolate oxidising enzyme of algae. FEBS Lett. **5**, 341–342 (1969)

CODD, G.A., MERRETT, M.J.: Photosynthetic products of division synchronized cultures of Euglena. Plant Physiol. **47**, 635–639 (1971a)

CODD, G.A., MERRETT, M.J.: The regulation of glycolate metabolism in division synchronized cultures of Euglena. Plant Physiol. **47**, 640–643 (1971b)

CODD, G.A., SCHMID, G.H., KOWALLIK, W.: Enzymic evidence for peroxisomes in a mutant of Chlorella vulgaris. Arch. Mikrobiol. **81**, 264–272 (1972)

CODD, G.A., STEWART, W.D.P.: Pathways of glycolate metabolism in the blue-green alga Anabaena cylindrica. Arch. Mikrobiol. **94**, 11–28 (1973)

COLLINS, N., MERRETT, M.J.: The localization of glycolate-pathway enzymes in Euglena. Biochem. J. **148**, 321–328 (1975)

COLMAN, B., MILLER, A.G., GRODZINSKI, B.: A study on the control of glycolate excretion in Chlorella. Plant Physiol. **53**, 395–397 (1974)

COOMBS, J., WHITTINGHAM, C.P.: The mechanism of inhibition of photosynthesis by high partial pressures of oxygen in Chlorella. Proc. Roy. Soc. B. Lond. **164**, 511–520 (1966a)

COOMBS, J., WHITTINGHAM, C.P.: The effect of high partial pressures of oxygen on photosynthesis in Chlorella-I. The effect on end products of photosynthesis. Phytochemistry **5**, 643–651 (1966b)

COOPER, T.G., BEEVERS, H.: Mitochondria and glyoxysomes from castor bean endosperm: Enzyme constitues and catalytic capacity. J. Biol. Chem. **244**, 3507–3513 (1969)

DAM, K. VAN, MEYER, A.J.: Oxidation and energy conservation by mitochondria. Ann. Rev. Biochem. **40**, 115–160 (1971)

D'AOUST, A L.: The effect of oxygen concentration on the carbon dioxide gas-exchange in some higher plants. Ph.D. Thesis, Queen's University, Kingston, Ontario, Canada, 1970

D'AOUST, A.L., CANVIN, D.T.: The specific activity of $^{14}CO_2$ evolved in CO_2-free air in the light and darkness by sunflower leaves following periods of photosynthesis in $^{14}CO_2$. Photosynthetica **6**, 150–157 (1972)

D'AOUST, A.L., CANVIN, D.T.: Effect of oxygen concentration on the rates of photosynthesis and photorespiration of some higher plants. Can. J. Botany **51**, 457–464 (1973)

D'AOUST, A.L., CANVIN, D.T.: Caractéristiques due $^{14}CO_2$ dégagé à la lumière et à l'obscurité par des feuilles de haricot, de radis, de tabac et de tournesol pendant et après une photosynthèse en présence de $^{14}CO_2$. Physiol. Vég. **12**, 545–560 (1974)

DAVIES, D.D., DAVIES, S.: Purification and properties of L(+)-lactate dehydrogenase from potato tubers. Biochem. J. **129**, 831–839 (1972)

DECKER, J.P.: A rapid, postillumination deceleration of respiration in green leaves. Plant Physiol. **30**, 82–84 (1955)

DECKER, J.P.: Comparative responses of carbon dioxide outburst and uptake in tobacco. Plant Physiol. **34**, 100–102 (1959)

DELAWAN, L.A., BENSON, A.A.: Light stimulation of glycolic acid oxidation in chloroplasts. In: The Photochemical Apparatus. Structure and Function. Brookhaven Symp. Biol. **11**, 271–275 (1958)

DÖHLER, G., KOCH, R.: Die Wirkung monochromatischen Lichts auf die extracelluläre Glykolsäure-Ausscheidung und die Lichtatmung bei der Blaualge Anacystis nidulans. Planta **105**, 352–359 (1972)

DONALDSON, R.P., TOLBERT, N.E., SCHNARRENBERGER, C.: A comparison of microbody membranes with microsomes and mitochondria from plant and animal tissue. Arch. Biochem. Biophys. **152**, 199–215 (1972)

DROOP, M.R., MCGILL, S.: The carbon nutrition of some algae: The inability to utilize glycollic acid for growth. J. Mar. Biol. Ass. U.K. **46**, 679–684 (1966)

EDWARDS, G.E., BLACK, C.C.: Photosynthesis in mesophyll cells and bundle sheath cells isolated from Digitaria sanguinalis L. Scap. leaves. In: Photosynthesis and Photorespiration (HATCH, M.D., OSMOND, C.B., SLAYTER, R.O., eds.), pp. 153–168. New York-London-Sydney-Toronto: Wiley-Interscience 1971

EGLE, K., FOCK, H.: Light respiration-correlations between CO_2 fixation, O_2 pressure, and glycolate concentration. In: Biochemistry of Chloroplasts (GOODWIN, T.W., ed.), Vol. II, pp. 79–87. London and New York: Academic Press 1967

EGLE, K., SCHENK, W.: Untersuchungen über die Reassimilation der Atmungskohlensäure bei der Photosynthese der Pflanzen. Beitr. Biol. Pflanzen **29**, 75–105 (1952)

EICKENBUSCH, J.D., BECK, E.: Evidence for involvement of two types of reaction in glycolate formation during photosynthesis in isolated spinach chloroplasts. FEBS Lett. **31**, 225–228 (1973)

EICKENBUSCH, J.D., SCHEIBE, R., BECK, E.: Activated glycol aldehyde and ribulose diphosphate as carbon sources for oxidative glycolate formation in chloroplasts. Z. Pflanzenphysiol. **75**, 375–380 (1975)

ELLYARD, P.W., GIBBS, M.: Inhibition of photosynthesis by oxygen in isolated spinach chloroplasts. Plant Physiol. **44**, 1115–1121 (1969)

EL-SHARKAWY, M.A., LOOMIS, R.S., WILLIAMS, W.A.: Apparent reassimilation of respiratory carbon dioxide by different plant species. Physiol. Plantarum **20**, 171–186 (1967)

EYTAN, G., OHAD, I.: Biogenesis of chloroplast membranes. VII. The observation of membrane homogeneity during development of the photosynthetic lamellar system in an algal mutant (Chlamydomonas reinhardi y-1). J. Biol. Chem. **247**, 112–121 (1972)

FOCK, H : Die Lichtatmung der grünen Pflanzen. Eine kritische Darstellung der bisher erarbeiteten Ergebnisse. Biol. Zbl. **89**, 545–572 (1970)

FOCK, H., BECKER, J.D., EGLE, K.: Use of labeled carbon dioxide for separation of CO_2 evolution from true CO_2 uptake by photosynthesizing Amaranthus and sunflower leaves. Can. J. Botany **48**, 1185–1189 (1970)

FOCK, H., CANVIN, D.T., GRANT, B.R.: Effects of oxygen and carbon dioxide on photosynthetic O_2 evolution and CO_2 uptake in sunflower and Chlorella. Photosynthetica **5**, 389–394 (1971)

FOCK, H., EGLE, K.: Über die „Lichtatmung" bei grünen Pflanzen. I. Die Wirkung von Sauerstoff und Kohlendioxyd auf den CO_2-Gaswechsel während der Licht- und Dunkelphase. Beitr. Biol. Pflanzen **42**, 213–239 (1966)

FOCK, H., EGLE, K.: Über die Beziehungen zwischen dem Glykolsäure-Gehalt und dem Photosynthese-Gaswechsel von Bohnenblättern. Z. Pflanzenphysiol. **57**, 389–397 (1967)

FOCK, H., HILGENBERG, W., EGLE, K.: Kohlendioxid- und Sauerstoff-Gaswechsel belichteter Blätter und die CO_2/O_2-Quotienten bei normalen und niedrigen O_2-Partialdrucken. Planta **106**, 355–361 (1972)

FOCK, H., KROTKOV, G.: Relation between photorespiration and glycolate oxidase activity in sunflower and red kidney bean leaves. Can. J. Botany **47**, 237–240 (1969)

FOCK, H., MAHON, J.D., CANVIN, D.T., GRANT, B.R.: Estimation of carbon fluxes through photosynthetic and photorespiratory pathways. In: Mechanisms of Regulation of Plant Growth (BIELESKI, A.L., FERGUSON, A.R., CRESWELL, M.M., eds.), Bulletin 12, pp. 235–242. Wellington: Roy. Soc. New Zealand 1974

FOCK, H., SCHAUB, H., HILGENBERG, W., EGLE, K.: Über den Einfluß niedriger und hoher O_2-Partialdrucke auf den Sauerstoff- und Kohlendioxidumsatz von Amaranthus und Phaseolus während der Lichtphase. Planta **86**, 77–83 (1969)

FORRESTER, M.L., KROTKOV, G., NELSON, C.D.: Effect of oxygen on photosynthesis, photorespiration and respiration in detached leaves. I. Soybean. Plant Physiol. **41**, 422–427 (1966a)

FORRESTER, M.L., KROTKOV, G., NELSON, C.D.: Effect of oxygen on photosynthesis, photorespiration and respiration in detached leaves. II. Corn and other monocotyledons. Plant Physiol. **41**, 428–431 (1966b)

FREDERICK, S.E., GRUBER, P.J., NEWCOMB, E.H.: Plant microbodies. Protoplasma **84**, 1–29 (1975)

FREDERICK, S.E., GRUBER, P.J., TOLBERT, N.E.: The occurrence of glycolate dehydrogenase and glycolate oxidase in green plants. An evolutionary survey. Plant Physiol. **52**, 318–323 (1973)

FREDERICK, S.E., NEWCOMB, E.H.: Microbody-like organelles in leaf cells. Science **163**, 1353–1355 (1969a)

FREDERICK, S.E., NEWCOMB, E.H.: Cytochemical localization of catalase in leaf microbodies (peroxisomes). J. Cell Biol. **43**, 343–353 (1969b)

FREDERICK, S.E., NEWCOMB, E.H.: Ultrastructure and distribution of microbodies in leaves of grasses with and without CO_2-photorespiration. Planta **96**, 152–174 (1971)

GAASTRA, P.: Photosynthesis of crop plants as influenced by light, carbon dioxide, temperature and stomatal diffusion resistance. Mededel. Landbouwhogeschool Wageningen **59**, 1–68 (1959)

GALMICHE, J.M.: Studies on the mechanism of glycerate 3-phosphate synthesis in tomato and maize leaves. Plant Physiol. **51**, 512–519 (1973)

GERHARDT, B., BERGER, C.: Microbodies und Diaminobenzidin-Reaktion in den Acetat-Flagellaten Polytomella caeca und Chlorogonium elongatum. Planta **100**, 155–166 (1971)

GERGIS, M.S.: The presence of microbodies in three strains of Chlorella. Planta **101**, 180–184 (1971)

GIMMLER, H., SCHÄFER, G., KRAMINER, A., HEBER, U.: Amino acid permeability of the chloroplast envelope as measured by light scattering, volumetry and aminoacid uptake. Planta **120**, 47–61 (1974)

GIRAUD, G., CZANINSKI, Y.: Localisation ultrastructurale d'activités oxydasiques chez le Chlamydomonas reinhardi Compt. Rend. **273**, 2500–2503 (1971)

GOLDSWORTHY, A.: Experiments on the origin of CO_2 released by tobacco leaf segments in the light. Phytochemistry **5**, 1013–1019 (1966)

GOLDSWORTHY, A.: Photorespiration. Botan. Rev. **36**, 321–340 (1970)

GRAVES, L.B., HANZELY, L., TRELEASE, R.N.: The occurrence and fine structural characterisation of microbodies in Euglena gracilis. Protoplasma **72**, 141–152 (1971)

GRAVES, L.B., TRELEASE, R.N., GRILL, A., BECKER, W.M.: Localisation of glyoxylate cycle enzymes in glyoxysomes in Euglena. J. Protozool. **19**, 527–532 (1972)

GRUBER, P.J., FREDERICK, S.E., TOLBERT, N.E.: Enzymes related to lactate metabolism in green algae and lower land plants. Plant Physiol. **53**, 167–170 (1974)

HALLIWELL, B.: Oxidation of formate by peroxisomes and mitochondria from spinach leaves. Biochem. J. **138**, 77–85 (1974)

HATCH, M.D.: Separation and properties of leaf aspartate aminotransferase and alanine aminotransferase isoenzymes operative in the C_4 pathway of photosynthesis. Arch. Biochem. Biophys. **156**, 207–214 (1973)
HATCH, M.D., MAU, S.-L.: Activity location and role of aspartate aminotransferase and alanine aminotransferase isoenzymes in leaves with C_4 pathway of photosynthesis. Arch. Biochem. Biophys. **156**, 195–206 (1973)
HATCH, M.D., SLACK, C.R.: NADP-specific malate dehydrogenase and glycerate kinase in leaves and evidence for their location in chloroplasts. Biochem. Biophys. Res. Commun. **34**, 589–593 (1969)
HATCH, M.D., SLACK, C.R.: Photosynthetic CO_2-fixation pathways. Ann. Rev. Plant Physiol. **21**, 141–162 (1970)
HEATH, O.V.S., ORCHARD, B.: Carbon assimilation at low carbon dioxide levels. II. The processes of apparent assimilation. J. Exptl. Botany **19**, 176–192 (1968)
HEBER, U.: Metabolite exchange between chloroplasts and cytoplasm. Ann. Rev. Plant Physiol. **25**, 393–421 (1974)
HEBER, U., KIRK, M.R., GIMMLER, H., SCHÄFER, G.: Uptake and reduction of glycerate by isolated chloroplasts. Planta **120**, 31–46 (1974)
HEBER, U., KRAUSE, G.H.: Transfer of carbon phosphate energy and reducing equivalents across the chloroplast envelope. In: Photosynthesis and photorespiration (HATCH, M.D., OSMOND, C.B., SLATYER, R.O., eds.), pp. 218–225. New York: Wiley-Interscience 1971
HEBER, U., KRAUSE, G.H.: Hydrogen and proton transfer across the chloroplast envelope. In: Proc. IInd. Intern. Congr. Photosynthesis Res. (FORTI, G., AVRON, M., MELANDRI, A., eds.), Vol. II, pp. 1023–1033 (1972)
HEBER, U., PON, N.G., HEBER, M.: Localization of carboxydismutase and triosephosphate dehydrogenases in chloroplasts. Plant Physiol. **38**, 355–360 (1963)
HESS, J.L., TOLBERT, N.E.: Glycolate, glycine, serine and glycerate formation during photosynthesis by tobacco leaves. J. Biol. Chem. **241**, 5707–5711 (1966)
HESS, J.L., TOLBERT, N.E.: Glycolate pathway in algae. Plant Physiol. **42**, 371–379 (1967)
HEW, C.-S., KROTKOV, G.: Effect of oxygen on the rates of CO_2 evolution in light and in darkness by photosynthesizing and nonphotosynthesizing leaves. Plant Physiol. **43**, 464–466 (1968)
HEW, C.-S., KROTKOV, G., CANVIN, D.T.: Determination of the rate of CO_2 evolution by green leaves in light. Plant Physiol. **44**, 662–670 (1969a)
HEW, C.-S., KROTKOV, G., CANVIN, D.T.: Effects of temperature on photosynthesis and CO_2 evolution in light and darkness by green leaves. Plant Physiol. **44**, 671–677 (1969b)
HILLIARD, J.H., GRACEN, V.E., WEST, S.H.: Leaf microbodies (peroxisomes) and catalase localization in plants differing in their photosynthetic carbon pathways. Planta **97**, 93–105 (1971)
HOCH, G., OWENS, O.H., KOK, B.: Photosynthesis and respiration. Arch. Biochem. Biophys. **101**, 171–180 (1963)
HOFSTRA, G., HESKETH, J.D.: Effects of temperature on the gas exchange of leaves in the light and dark. Planta **85**, 228–237 (1969)
HOLMGREN, P., JARVIS, P.: Carbon dioxide efflux from leaves in light and darkness. Physiol. Plantarum **20**, 1045–1051 (1967)
HRUBAN, Z., RECHCIGL, M.: Microbodies and related particles. Morphology, Biochemistry and Physiology. In: International Rev. Cyt. Suppl. 1. New York: Academic Press 1969
HUANG, A.H., BEEVERS, H.: Microbody enzymes and carboxylases in sequential extracts from C_4 and C_3 leaves. Plant Physiol. **50**, 242–248 (1972)
JACKSON, W.A., VOLK, R.J.: Photorespiration. Ann. Rev. Plant Physiol. **21**, 385–432 (1970)
JENSEN, R.G., BASSHAM, J.A.: Photosynthesis by isolated chloroplasts. Proc. Natl. Acad. Sci. US **56**, 1095–1101 (1966)
JIMENEZ, E., BALDWIN, R.L., TOLBERT, N.E., WOOD, W.A.: Distribution of C^{14} in sucrose from glycolate-C^{14} and serine-3-C^{14} metabolism. Arch. Biochem. Biophys. **98**, 172–175 (1962)
JOHNSON, H.S.: NADP-malate dehydrogenase: Photoactivation in leaves of plants with Calvin cycle photosynthesis. Biochem. Biophys. Res. Commun. **43**, 703–709 (1971)
JOHNSON, H.S., HATCH, M.D.: Properties and regulation of leaf nicotinamide adenine dinucleotide phosphate malate dehydrogenase and malic enzyme in plants with the C_4-dicarboxylic acid pathway of photosynthesis. Biochem. J. **119**, 273–280 (1970)

JOLLIFFE, P.A., TREGUNNA, E.B.: Effect of temperature, CO_2 concentration, and light intensity on oxygen inhibition of photosynthesis in wheat leaves. Plant Physiol. **43**, 902–906 (1968)

KAPIL, R.N., PUGH, T.D., NEWCOMB, E.H.: Microbodies and an anomalous "microcylinder" in the ultrastructure of plants with crassulacean acid metabolism. Planta **124**, 231–244 (1975)

KEARNEY, P.C., TOLBERT, N.E.: Appearance of glycolate and related products of photosynthesis outside of chloroplasts. Arch. Biochem. Biophys. **98**, 164–171 (1962)

KISAKI, T., IMAI, A., TOLBERT, N.E.: Intracellular localisation of enzymes related to photorespiration in green leaves. Plant Cell Physiol. **12**, 267–273 (1971a)

KISAKI, T., TOLBERT, N.E.: Glycolate and glyoxylate metabolism by isolated peroxisomes or chloroplasts. Plant Physiol. **44**, 242–250 (1969)

KISAKI, T., TOLBERT, N.E.: Glycine as a substrate for photorespiration. Plant Cell Physiol. **11**, 247–258 (1970)

KISAKI, T., YANO, N., HIRABAYASHI, S.: Photorespiration, stimulation of glycine decarboxylation by oxygen in tobacco leaf segments. Plant Cell Physiol. **13**, 581–584 (1972)

KISAKI, T., YOSHIDA, N., IMAI, A.: Glycine decarboxylase and serine formation in spinach leaf mitochondrial preparation with reference to photorespiration. Plant Cell Physiol. **12**, 275–288 (1971b)

KOHN, L.D., WARREN, W.A.: The kinetic properties of spinach leaf glyoxylic acid reductase. J. Biol. Chem. **245**, 3831–3839 (1970)

KREBS, H.A., VEECH, R.L.: Interrelations between diphospho- and triphosphopyridine nucleotides. FEBS Symposium, Vol. 17, pp. 101–109 (1969)

KROTKOV, G., RENUCKLES, V.C., THIMANN, K.V.: Effect of light on the CO_2 absorption and evolution by Kalanchoe, wheat and pea leaves. Plant Physiol. **33**, 289–292 (1958)

LAING, W.A., OGREN, W.L., HAGEMAN, R.H.: Regulation of soybean net photosynthetic CO_2 fixation by the interaction of CO_2, O_2, and ribulose 1,5-diphosphate carboxylase. Plant Physiol. **54**, 678–685 (1974)

LAING, W.A., OGREN, W.L., HAGEMAN, R.H.: Bicarbonate stabilization of ribulose 1,5-diphosphate carboxylase. Biochemistry **14**, 2269–2275 (1975)

LAKE, J.V.: Respiration of leaves during photosynthesis. I. Estimates from an electrical analogue. Australian J. Biol. Sci. **20**, 487–493 (1967)

LAWLOR, D.W.: Water stress induced changes in photosynthesis, in photorespiration and CO_2 compensation concentration of wheat. Photosynthetica, in press (1975)

LAWLOR, D.W., FOCK, H.: Photosynthesis and photorespiratory CO_2 evolution of water-stressed sunflower leaves. Planta **126**, 247–258 (1975)

LEEK, A.E., HALLIWELL, B., BUTT, V.S.: Oxidation of formate and oxalate in peroxisomal preparations from leaves of spinach beet (Beta vulgaris L.). Biochim. Biophys. Acta **286**, 299–311 (1972)

LISTER, G.R., KROTKOV, G., NELSON, C.D.: A closed-circuit apparatus with an infrared CO_2 analyzer and a Geiger tube for continuous measurement of CO_2 exchange in photosynthesis and respiration. Can. J. Botany **39**, 581–591 (1961)

LIU, A.Y., BLACK, C.C.: Glycolate metabolism in mesophyll cells and bundle sheath cells isolated from crabgrass, Digitaria sanguinalis (L.) Scop., leaves. Arch. Biochem. Biophys. **149**, 269–280 (1972)

LORD, J.M., CODD, G.A., MERRETT, M.J.: The effect of light quality on glycolate formation and excretion in algae. Plant Physiol. **46**, 855–856 (1970)

LORD, J.M., MERRETT, M.J.: Glycolate oxidase in *Chlorella pyrenoidosa.* Biochim. Biophys. Acta **159**, 543–544 (1968)

LORD, J.M., MERRETT, M.J.: The effect of hydroxymethanesulphonate on photosynthesis in Chlorella pyrenoidosa. J. Exptl. Botany **20**, 743–750 (1969)

LORD, J.M., MERRETT, M.J.: The pathway of glycolate utilization in Chlorella pyrenoidosa. Biochem. J. **117**, 929–937 (1970)

LORIMER, G.H., ANDREWS, T.J.: Plant photorespiration. An inevitable consequence of the existence of atmospheric oxygen. Nature **243**, 359–360 (1973)

LORIMER, G.H., ANDREWS, T.J., TOLBERT, N.E.: Ribulose diphosphate oxygenase. II. Further proof of reaction products and mechanism of action. Biochemistry **12**, 18–23 (1973)

LUDLOW, M.M., JARVIS, P.G.: Methods of measuring photorespiration in leaves. In: Plant Photosynthetic Production (SESTAK, CATSKY, JARVIS, eds.), pp. 294–315. The Hague: Dr. W. Junk, N.V. 1971

LUDWIG, L.J.: The relationship between photosynthesis and respiration in sunflower leaves. Ph.D. Thesis, Queen's University, Kingston, Ontario, Canada, 1968
LUDWIG, L.J., CANVIN, D.T.: An open gas-exchange system for the simultaneous measurement of the CO_2 and $^{14}CO_2$ fluxes from leaves. Can. J. Botany **49**, 1299–1313 (1971a)
LUDWIG, L.J., CANVIN, D.T.: The rate of photorespiration during photosynthesis and the relationship of the substrate of light respiration to the products of photosynthesis in sunflower leaves. Plant Physiol. **48**, 712–719 (1971b)
MAHON, J.D., EGLE, K., FOCK, H.: A radio-gas chromatographic method for determining the specific radioactivity of glycolic acid in ^{14}C-labeled leaf tissue. Can. J. Biochem. **53**, 609–614 (1975)
MAHON, J.D., FOCK, H., CANVIN, D.T.: Changes in specific radioactivities of sunflower leaf metabolites during photosynthesis in $^{14}CO_2$ and $^{12}CO_2$ at normal and low oxygen. Planta **120**, 125–134 (1974b)
MAHON, J.D., FOCK, H., CANVIN, D.T.: Changes in specific radioactivity of sunflower leaf metabolites during photosynthesis in $^{14}CO_2$ and $^{12}CO_2$ at three concentrations of CO_2. Planta **120**, 245–254 (1974c)
MAHON, J.D., FOCK, H., HÖHLER, T., CANVIN, D.T.: Changes in specific radioactivities of corn-leaf metabolites during photosynthesis in $^{14}CO_2$ and $^{12}CO_2$ at normal and low oxygen. Planta **120**, 113–123 (1974a)
MANGAT, B.S., LEVIN, W.B., BIDWELL, R.G.S.: The extent of dark respiration in illuminated leaves and its control by ATP levels. Can. J. Botany **52**, 673–681 (1974)
MCFADDEN, B.A.: Autotrophic CO_2 assimilation and the evolution of ribulose diphosphate carboxylase. Bacteriol. Rev. **37**, 289–319 (1973)
MEHLER, A.H.: Studies on reactions of illuminated chloroplasts I. Mechanism of the reduction of oxygen and other Hill reagents. Arch. Biochem. Biophys. **33**, 65–77 (1951)
MEIDNER, H.: The minimum intercellular space CO_2 concentration (T) of maize leaves and its influence on stomatal movements. J. Exptl. Botany **13**, 284–293 (1962)
MERRETT, M.J., GOULDING, K.H.: Glycolate formation during the photoassimilation of acetate by Chlorella. Planta **75**, 275–278 (1967)
MERRETT, M.J., GOULDING, K.H.: The glycolate pathway during the photoassimilation of acetate by Chlorella. Planta **80**, 321–327 (1968)
MERRETT, M.J., LORD, J.M.: Glycolate formation and metabolism by algae. New Phytologist **72**, 751–767 (1973)
MOSS, D.N.: The limiting carbon dioxide concentration for photosynthesis. Nature **193**, 587 (1962)
MOSS, D.N.: Respiration of leaves in light and darkness. Crop Sci. **6**, 351–354 (1966)
MOSS, D.N.: Photorespiration and glycolate metabolism in tobacco leaves. Crop Sci. **8**, 71–76 (1968)
MULCHI, C.L., VOLK, R.J., JACKSON, W.A.: Oxygen exchange of illuminated leaves at carbon dioxide compensation. In: Photosynthesis and Photorespiration (HATCH, M.D., OSMOND, C.B., SLATYER, R.O., eds.), pp. 35–50. New York: Wiley-Interscience 1971
MURRAY, D.R., GIOVANELLI, J., SMILLIE, R.M.: Photometabolism of glycollate by Euglena gracilis. Australian J. Biol. Sci. **24**, 23–33 (1971)
NELSON, E.B., TOLBERT, N.E.: The regulation of glycolate metabolism in Chlamydomonas rheinhardtii Biochim. Biophys. Acta **184**, 263–270 (1969)
NELSON, E.B., TOLBERT, N.E.: Glycolate dehydrogenase in green algae. Arch. Biochem. Biophys. **141**, 102–110 (1970)
OAKLEY, B.R., DODGE, J.D.: The ultrastructure and cytochemistry of microbodies in Porphyridium. Protoplasma **80**, 233–244 (1974)
OGREN, W.L., BOWES, G.: Ribulose diphosphate carboxylase regulates soybean photorespiration. Nature New Biol. **230**, 159–160 (1971)
ORTH, G.M., TOLBERT, N.E., JIMENEZ, E.: Rate of glycolate formation during photosynthesis at high pH. Plant Physiol. **41**, 143–147 (1966)
OSMOND, C.B., BJÖRKMAN, O.: Simultaneous measurements of oxygen effects on net photosynthesis and glycolate metabolism in C_3 and C_4 species. Carnegie Inst. Year Book **71**, 141–148 (1972)
OSMOND, C.B., HARRIS, B.: Photorespiration during C_4 photosynthesis. Biochim. Biophys. Acta **234**, 270–282 (1971)

OZBUN, J.L., VOLK, R.J., JACKSON, W.A.: Effects of light and darkness on gaseous exchange of bean leaves. Plant Physiol. **39**, 523–527 (1964)

PARISH, R.W., RICKENBACHER, R.: The isolation of peroxisomes, mitochondria and chloroplasts from leaves of spinach beet (Beta vulgaris L. ssp. vulgaris). Eur. J. Biochem. **22**, 423–429 (1971)

PLAUT, Z., GIBBS, M.: Glycolate formation in intact spinach chloroplasts. Plant Physiol. **45**, 470–474 (1970)

POINCELOT, R.: Transport of metabolites across isolated envelope membranes of spinach chloroplasts. Plant Physiol. **55**, 847–852 (1975)

PRITCHARD, G.G., GRIFFIN, W.J., WHITTINGHAM, C.P.: The effect of carbon dioxide concentration, light intensity and isonicotyl hydrazide on the photosynthetic production of glycollic acid by Chlorella. J. Exptl. Botany **13**, 176–184 (1962)

RABSON, R., TOLBERT, N.E., KEARNEY, P.C.: Formation of serine and glyceric acid by the glycolate pathway. Arch. Biochem. Biophys. **98**, 154–163 (1962)

RANDALL, D.D., TOLBERT, N.E.: 3-phosphoglycerate phosphatase in plants I. Isolation and characterization from sugarcane leaves. J. Biol. Chem. **246**, 5510–5517 (1971a)

RANDALL, D.D., TOLBERT, N.E.: 3-phosphoglycerate phosphatase in plants. III. Activity associated with starch particles. Plant Physiol. **48**, 488–492 (1971b)

RANDALL, D.D., TOLBERT, N.E., GREMEL, D.: 3-phosphoglycerate phosphatase in plants. II. Distribution, physiological considerations, and comparison with P-glycolate phosphatase. Plant Physiol. **48**, 480–487 (1971)

RAVEN, I.A.: Endogenous inorganic carbon sources in plant photosynthesis. II. Comparison of total CO_2 production in the light with measured CO_2 evolution in the light. New Phytologist **71**, 995–1014 (1972)

REHFELD, D.W., RANDALL, D.D., TOLBERT, N.E.: Enzymes of the glycolate pathway in plants without CO_2-photorespiration. Can. J. Botany **48**, 1219–1226 (1970)

REHFELD, D.W., TOLBERT, N.E.: Aminotransferases in peroxisomes from spinach leaves. J. Biol. Chem. **247**, 4803–4811 (1972)

REINER, J.M.: The study of metabolic turnover rates by means of isotopic tracers. Arch. Biochem. Biophys. **46**, 53–79 (1953)

RICHARDSON, K.E., TOLBERT, N.E.: Phosphoglycolic acid phosphatase. J. Biol. Chem. **236**, 1285–1290 (1961)

RITENOUR, G.L., JOY, K.W., BUNNING, G., HAGEMAN, R.H.: Intracellular location of nitrate reductase, nitrite reductase and glutamic acid dehydrogenase in green leaf tissue. Plant Physiol. **42**, 233–237 (1967)

ROBINSON, J.M., GIBBS, M.: Photosynthetic intermediates, the Warburg effect and glycolate synthesis in isolated spinach chloroplasts. Plant Physiol. **53**, 790–797 (1974)

ROCHA, V., TING, I.P.: Preparation of cellular plant organelles from spinach leaves. Arch. Biochem. Biophys. **140**, 398–407 (1970a)

ROCHA, V., TING, I.P.: Tissue distribution of microbody, mitochondrial, and soluble malate dehydrogenase isoenzymes. Plant Physiol. **46**, 754–756 (1970b)

ROCHA, V., TING, I.P.: Malate dehydrogenases of leaf tissue from Spinacia oleracea: properties of three isoenzymes. Arch. Biochem. Biophys. **147**, 114–122 (1971)

ROTHE, G.M.: Catalytic properties of three lactate dehydrogenases from potato tubers (Solanum tuberosum). Arch. Biochem. Biophys. **162**, 17–21 (1974)

RYAN, F.J., BARKER, R., TOLBERT, N.E.: Inhibition of ribulose diphosphate carboxylase oxygenase by xylitol 1,5-diphosphate. Biochem. Biophys. Res. Commun. **65**, 39–46 (1975)

RYAN, F.J., TOLBERT, N.E.: Ribulose diphosphate carboxylase/oxygenase III. Isolation and properties. J. Biol. Chem. **250**, 4229–4233 (1975a)

RYAN, F.J., TOLBERT, N.E.: Ribulose diphosphate carboxylase/oxygenase. IV. Regulation by phosphate esters. J. Biol. Chem. **250**, 4234–4238 (1975b)

SAMISH, Y., KOLLER, D.: Photorespiration in green plants during photosynthesis estimated by use of isotopic CO_2. Plant Physiol. **43**, 1129–1132 (1968)

SAMISH, Y.B., PALLAS, J.E., DORNHOFF, G.M., SHIBLES, R.M.: A re-evaluation of soybean leaf photorespiration. Plant Physiol. **50**, 28–30 (1972)

SCHAUB, H., EGLE, K.: Über die CO_2-Kompensationslage bei der Photosynthese von Algensedimenten. Beitr. Biol. Pflanzen **41**, 5–10 (1965)

SCHNARRENBERGER, C., OESER, A., TOLBERT, N.E.: Development of microbodies in sunflower cotyledons and castor bean endosperm during germination. Plant Physiol. **48**, 466–474 (1971)

SCHNARRENBERGER, C., OESER, A., TOLBERT, N.E.: Isolation of plastids from sunflower cotyledons during germination. Plant Physiol. **50**, 55–59 (1972)

SCHOU, L., BENSON, A.A., BASSHAM, J.A., CALVIN, M.: The path of carbon photosynthesis, XI. The role of glycolic acid. Physiol. Plantarum **3**, 487–495 (1950)

SEMENENKO, V.E.: Characteristics of carbon dioxide gas exchange in the transition states of photosynthesis upon changing from light to darkness. Light induced evolution of CO_2. Fiziol. Rast. **11**, 375–484 (1964)

SESTAK, Z., CATSKY, J., JARVIS, P.G.: Plant photosynthetic production; manual of methods. The Hague: Dr. W. Junk N.V. 1971

SHAH, S P.J., COSSINS, E.A.: The biosynthesis of glycine and serine by isolated chloroplasts. Phytochemistry **9**, 1545–1551 (1970)

SHAIN, Y., GIBBS, M.: Formation of glycolate by a reconstituted spinach chloroplasts preparation. Plant Physiol. **48**, 325–330 (1971)

SILVERBERG, B.A.: Ultrastructural and cytochemical characterisation of microbodies in the green algae. Protoplasma **83**, 269–295 (1975)

SILVERBERG, B.A., SAWA, T.: Cytochemical localisation of oxydase activities with diaminobenizidine in the green alga Chlamydomonas dysosmos. Protoplasma **81**, 177–188 (1974)

SLACK, C.R., HATCH, M.D., GOODCHILD, D.J.: Distribution of enzymes in mesophyll and parenchyma-sheath chloroplasts of maize leaves in relation to the C_4-dicarboxylic acid pathway of photosynthesis. Biochem. J. **114**, 489–498 (1969)

SMAOUI, A.: Distribution des peroxysomes dans les feuilles de deux Chenopodiacies differant par leur structure et leur metabolisme photosynthetique: Atriplex halimus L. et Chenopodium album L. Compt. Rend. **275**, 1031–1034 (1972)

STABENAU, H.: Verteilung von Microbody-Enzymen aus Chlamydomonas in Dichtegradienten. Planta **118**, 35–42 (1974)

STABENAU, H., BEEVERS, H.: Isolation and characterization of microbodies from the alga Chlorogonium elongatum. Plant Physiol. **53**, 866–869 (1974)

STEWART, K.D., FLOYD, G.L., MATTOX, K.R., DAVIS, M.E.: Cytochemical demonstration of a single peroxisome in a filamentous green alga. J. Cell Biol. **54**, 431–434 (1972)

STOCKING, C.R.: Chloroplasts: Nonaqueous. Methods in Enzymol. **23**, 221–228 (1971)

THOMPSON, C.M., WHITTINGHAM, C.P.: Intracellular localization of phosphoglycolate phosphatase and glycolate reductase Biochim. Biophys. Acta **143**, 642–644 (1967)

THOMPSON, C.M., WHITTINGHAM, C.P.: Glycolate metabolism in photosynthesising tissue. Biochim. Biophys. Acta **153**, 260–269 (1968)

TING, I.P., ROCHA, V.: NADP-specific malate dehydrogenase of green spinach leaf tissue. Arch. Biochem. Biophys. **147**, 156–164 (1971)

TOLBERT, N.E.: Secretion of glycolic acid by chloroplasts. In: The Photochemical Apparatus–Its Structure and Function. Brookhaven Symposia in Biology, No. 11, pp. 271–275 (1958)

TOLBERT, N.E.: Glycolate pathway. In: Photosynthetic Mechanisms in Green Plants. Publication 1145. Natl. Acad. Sci. USA, Natl. Res. Council, pp. 648–662 (1963)

TOLBERT, N.E.: Microbodies- peroxisomes and glyoxysomes. Ann. Rev. Plant Physiol. **22**, 45–74 (1971a)

TOLBERT, N.E.: Isolation of leaf peroxisomes. Methods in Enzymol. **23**, 665–682 (1971b)

TOLBERT, N.E.: Glycolate biosynthesis. In: Current Topics in Cellular Regulation (HORECKER, B.L., STADTMAN, E.R., eds.), Vol. 7, pp. 21–50. New York: Academic Press 1973

TOLBERT, N.E.: Photorespiration. In: Algal Physiology and Biochemistry (STEWARD, W.D.P., ed.), pp. 474–504. Oxford: Blackwell Scientific Public 1974

TOLBERT, N.E., CLAGETT, C.O., BURRIS, R.H.: Products of the oxidation of glycolic acid and l-lactic acid by enzymes from tobacco leaves. J. Biol. Chem. **181**, 905–914 (1949)

TOLBERT, N.E., COHAN, M.S.: Products formed from glycolic acid in plants. J. Biol. Chem. **204**, 649–654 (1953)

TOLBERT, N.E., NELSON, E.B., BRUIN, W.J.: Glycolate pathway in algae. In: Photosynthesis and Photorespiration (HATCH, M.D., OSMOND, C.B., SLATYER, R.O., eds.), pp. 506–513. New York: Wiley Interscience 1971

TOLBERT, N.E., OESER, A., KISAKI, T., HAGEMAN, R.H., YAMAZAKI, R.K.: Peroxisomes from spinach leaves containing enzymes related to glycolate metabolism. J. Biol. Chem. **243**, 5179–5184 (1968)

TOLBERT, N.E., OESER, A., YAMAZAKI, R.K., HAGEMAN, R.H., KISAKI, T.: A survey of plants for leaf peroxisomes. Plant Physiol. **44**, 135–147 (1969)

TOLBERT, N.E., RYAN, F.J.: Glycolate biosynthesis by ribulose diphosphate carboxylase/oxygenase. Proc. 3rd. Intern. Congr. Photosynth. Res. Vol. II, pp. 1303–1319 (1975)

TOLBERT, N.E., YAMAZAKI, R.K.: Leaf peroxisomes and their relation to photorespiration and photosynthesis. In: Nature and Function of Peroxisomes. Ann. N.Y. Acad. Sci. **168**, 325–341 (1969)

TOLBERT, N.E., YAMAZAKI, R.K., OESER, A.: Localization and properties of hydroxypyruvate and glyoxylate reductases in spinach leaf particles. J. Biol. Chem. **245**, 5129–5136 (1970)

TOLBERT, N.E., ZILL, L.P.: Excretion of glycolic acid by algae during photosynthesis. J. Biol. Chem. **222**, 895–906 (1956)

TOURTE, M.: Mise en evidence d'une activité catalasique dans les peroxisomes de Micrasterias fimbriata (Ralfs). Planta **105**, 50–59 (1972)

TREGUNNA, E.B., BROWN, D.L.: Inhibition of respiration during photosynthesis by some algae. Can. J. Botany **45**, 1135–1143 (1967)

TREGUNNA, E.B., KROTKOV, G., NELSON, C.D.: Evolution of carbon dioxide by tobacco leaves during the dark period following illumination with light of different intensities. Can. J. Botany **39**, 1045–1056 (1961)

TREGUNNA, E.B., KROTKOV, G., NELSON, C.D.: Further evidence on the effects of light on respiration during photosynthesis. Can. J. Botany **42**, 989–997 (1964)

TREGUNNA, E.B., KROTKOV, G., NELSON, C.D.: Effect of oxygen on the rate of photorespiration in detached tobacco leaves. Physiol. Plantarum **19**, 723–733 (1966)

VIGIL, E.L.: Intracellular localization of catalase (peroxidase) activity in plant microbodies. J. Histochem. Cytochem. **17**, 425–428 (1969)

VIGIL, E.L.: Structure and function of plant microbodies. Subcell. Biochem. **2**, 237–285 (1973)

VOLK, R.J., JACKSON, W.A.: Photorespiratory phenomena in maize. Oxygen uptake, isotope discrimination, and carbon dioxide efflux. Plant Physiol. **49**, 218–223 (1972)

WAGGONER, P.E.: Predicting the effect upon net photosynthesis of changes in leaf metabolism and physics. Crop Sci. **9**, 315–321 (1969)

WALKER, D.A.: Improved rates of carbon dioxide fixation by illuminated chloroplasts. Biochem. J. **92**, 22c–23c (1964)

WALKER, D.A.: Photosynthetic induction phenomena and the light activation of ribulose diphosphate carboxylase. New Phytologist **72**, 209–235 (1973)

WANG, D., BURRIS, R.H.: Carbon metabolism of C^{14}-labeled amino acids in wheat leaves II. Serine and its role in glycine metabolism. Plant Physiol. **38**, 430–439 (1963)

WANG, D., WAYGOOD, E.R.: Carbon metabolism of C^{14}-labelled amino acids in wheat leaves I. A pathway of glyoxylate-serine metabolism. Plant Physiol. **37**, 826–832 (1962)

WARBURG, O.: Über die Geschwindigkeit der photochemischen Kohlensäurezersetzung in lebenden Zellen II. Biochem. Z. **103**, 188–217 (1920)

WARBURG, O., KRIPPAHL, G.: Glykolsäurebildung in Chlorella. Z. Naturforsch. **15b**, 197–199 (1960)

WHITEMAN, P.C., KOLLER, D.: Interactions of carbon dioxide concentration, light intensity and temperature on plant resistances to water vapour and carbon dioxide diffusion. New Phytologist **66**, 463–473 (1967)

WHITTINGHAM, C.P., BIRMINGHAM, M., HILLER, R.G.: The photometabolism of glucose in Chlorella. Z. Naturforsch. **18b**, 701–706 (1963)

WHITTINGHAM, C.P., COOMBS, J., MARKER, A.F.H.: The role of glycolate in photosynthetic carbon fixation. In: Biochemistry of Chloroplasts, Vol. II, pp. 155–173 (GOODWIN, T.W., ed.), London and New York: Academic Press 1967

WHITTINGHAMM, C.P., PRITCHARD, G.G.: The production of glycolate during photosynthesis in Chlorella. Proc. Roy. Soc. B **157**, 366–380 (1963)

WILKINSON, A.P., DAVIS, D.D.: Serine-glycine interconversion in plant tissues. Nature **181**, 1070–1071 (1958)

WILLIS, J.E., SALLACH, H.J.: Serine biosynthesis from hydroxypyruvate in plants. Phytochemistry **2**, 23–28 (1963)

WILSON, A.T., CALVIN, M.: The photosynthetic cycle. CO_2 dependent transients. J. Am. Chem. Soc. **77**, 5948–5957 (1955)

WOLF, F.T.: Photorespiration, the C_4 pathway of photosynthesis and related phenomena. Advanc. Frontiers Plant Sci. **26**, 161–231 (1970)

YAMAZAKI, R.K., TOLBERT, N.E.: Malate dehydrogenase in leaf peroxisomes. Biochim. Biophys. Acta **178**, 11–20 (1969)

YAMAZAKI, R.K., TOLBERT, N.E.: Enzymic characterization of leaf peroxisomes. J. Biol. Chem. **245**, 5137–5144 (1970)

YU, Y.L., TOLBERT, N.E., ORTH, G.M.: Isolation and distribution of phosphoglycolate phosphatase. Plant Physiol. **39**, 643–647 (1964)

ZAK, E.G., NICHIPOROVICH, A.A.: The pathway of acid synthesis during photosynthesis. Fiziol. Rast. **11**, 945–950 (1964)

ZELITCH, I.: Oxidation and reduction of glycolic and glyoxylic acids in plants II. Glyoxylic acid reductase. J. Biol. Chem. **201**, 719–726 (1953)

ZELITCH, I.: The isolation and action of crystalline glyoxylic acid reductase from tobacco leaves. J. Biol. Chem. **216**, 553–575 (1955)

ZELITCH, I.: The role of glycolic acid oxidase in the respiration of leaves. J. Biol. Chem. **233**, 1299–1303 (1958)

ZELITCH, I.: Organic acids and respiration in photosynthetic tissues. Ann. Rev. Plant. Physiol. **15**, 121–142 (1964)

ZELITCH, I.: The relation of glycolic acid synthesis to the primary photosynthetic carboxylation reaction in leaves. J. Biol. Chem. **240**, 1869–1876 (1965)

ZELITCH, I.: Increased rate of net photosynthetic carbon dioxide uptake caused by the inhibition of glycolate oxidase. Plant Physiol. **41**, 1623–1631 (1966)

ZELITCH, I.: Water and CO_2 transport in the photosynthetic process. In: Harvesting the Sun (SAN PIETRO, A., GREER, F.A., ARMY, T.J., eds.), pp. 231–248. New York: Academic Press 1967

ZELITCH, I.: Investigations on photorespiration with a sensitive ^{14}C-assay. Plant Physiol. **43**, 1829–1837 (1968)

ZELITCH, I.: Photosynthesis, photorespiration and plant productivity. New York: Academic Press 1971

ZELITCH, I.: Comparison of the effectiveness of glycolic acid and glycine as substrates for photorespiration. Plant Physiol. **50**, 109–113 (1972a)

ZELITCH, I.: The photooxidation of glyoxylate by envelope free spinach chloroplasts and its relation to photorespiration. Arch. Biochem. Biophys. **150**, 698–707 (1972b)

ZELITCH, I.: Alternate pathways of glycolate synthesis in tobacco and maize leaves in relation to rates of photorespiration. Plant Physiol. **51**, 299–305 (1973)

ZELITCH, I.: Pathway of carbon fixation in green plants. Ann. Rev. Biochem. **44**, 123–145 (1975)

ZELITCH, I., DAY, P.R.: Glycolate oxidase activity in algae. Plant Physiol. **43**, 289–291 (1968)

ZELITCH, I., GOTTO, A.M.: Properties of a new glyoxylate reductase from leaves. Biochem. J. **84**, 541–546 (1962)

ZELITCH, I., OCHOA, S.: Oxidation and reduction of glycolic and glyoxylic acids in plants. I. Glycolic acid oxydase. J. Biol. Chem. **201**, 707–718 (1953)

6. Transport of Metabolites between Cytoplasm and the Mitochondrial Matrix

H.W. HELDT

1. Introduction

In all aerobic eucaryotic cells the mitochondria are the site of cellular respiration. In the mitochondria, substrates are split into CO_2 and hydrogen, and the latter is oxidized to water. The energy gained from this oxidation is utilized to synthesize ATP, which drives the cells' many energy consuming processes (e.g. mechanical, chemical, or osmotic work). Mitochondria are found in leaf cells, but they are specially abundant in nonphotosynthetic plant tissue, e.g. in the root and in the shoot.

Mitochondria of plant cells and of animal cells are essentially alike. Nearly all systematic studies on mitochondrial structure and function have been performed with mammalian mitochondria, and in particular with mitochondria from rat liver tissue. For reviews on this subject see KLINGENBERG (1970a, b, 1976), HELDT (1972), MUNN (1974), MEIJER and VAN DAM (1974), and WILLIAMSON (1976). Ion transport in plant mitochondria has been reviewed by PACKER et al. (1970). Most of the data given here will be based on mammalian mitochondria, and special reference will be made to relevant work on plant mitochondria.

2. Structure of Mitochondria

Mitochondria contain two types of membranes, (see Chap. I) which differ in their fine structure, lipid composition, and protein content (Fig. 1). The *outer membrane*, surrounding the mitochondrion, has a low protein content and contains,

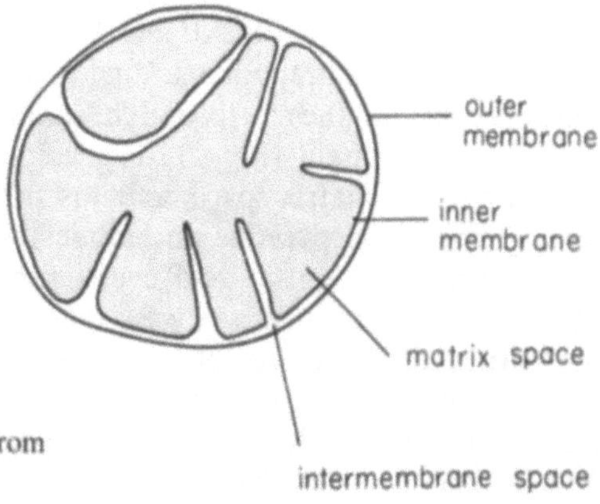

Fig. 1. Schematic diagram of a mitochondrion from corn. (From MALONE et al., 1974)

beside phospholipids, some cholesterol. The *inner membrane* contains much more protein than lipids and it also contains cardiolipin, which appears to be lacking in the outer membrane (PARSON and YANO, 1967; PARSON et al., 1967; STOFFEL and SCHIEFER, 1968). The inner membrane extends, densely packed in endlessly folded or tubular arrangements, into the interior of the mitochondria, thus forming a very large surface. The two membranes divide the mitochondria into two spaces. The *matrix space* is the space surrounded by the inner membrane, and the *intermembrane space* is the space between the outer and the inner membrane.

Investigations with isolated mitochondria showed that the intermembrane space is freely accessible to substances of low molecular weight, such as inorganic ions, nucleotides, amino acids, and sucrose, but it is not accessible to large molecules like polymeric sugars or proteins (PFAFF et al., 1968). This led to the conclusion that the outer membrane is freely permeable to all low molecular weight substances. Thus, from a functional standpoint, the intermembrane space may be regarded as part of the extramitochondrial compartment. The partition between the mitochondrial and extramitochondrial compartment is the inner membrane. This view is supported by the observation that mitochondria can be deprived of their outer membrane without losing essential metabolic functions (SCHNAITMAN and GREENAWALT, 1968).

3. Functional Organization of Mitochondrial Metabolism

The mitochondria represent a distinct metabolic compartment within the cell. The inner membrane, surrounding this compartment, is permeable to monocarboxylates passing, after protonation, through the membrane as undissociated acids, e.g. acetate (WILSON et al., 1969; KLINGENBERG, 1970a). It is in principle impermeable to nucleotides, inorganic phosphate, and intermediates of the metabolic cycles in the mitochondria. This impermeability of the inner membrane is bypassed by specific translocators, enabling transport of ADP, ATP, inorganic phosphate, pyruvate, dicarboxylates, and tricarboxylates, which will be considered later. The organization of mitochondrial metabolism is revealed from the enzyme distribution. Table 1 shows the localization of a number of mitochondrial enzymes as found in mammalian mitochondria. In those cases tested in plant mitochondria, the distribution of enzymes between the inner membrane and the matrix was found to be essentially the same (MOREAU and LANCE, 1972).

The inner membrane has a dual function: it is the site of specific metabolite transport between the matrix and the extramitochondrial space; and it also contains the respiratory chain (Fig. 2). The associated ATP synthase is directed toward the matrix. The matrix space contains the mitochondrial nucleotides (mainly adenine, guanine, and pyridine nucleotides), and the soluble enzymes of the substrate breakdown cycles (citric acid cycle, β oxydation of fatty acids). The reduction equivalents formed in these cycles (NADH, succinate, and acylCoA) diffuse from the matrix to the inside of the inner membrane and become oxidized by the dehydrogenases of the respiratory chain. ADP and inorganic phosphate also reach the site of ATP synthesis from the matrix, and the ATP thus formed is released into the matrix again. Therefore the phosphorylation of extramitochondrial ADP

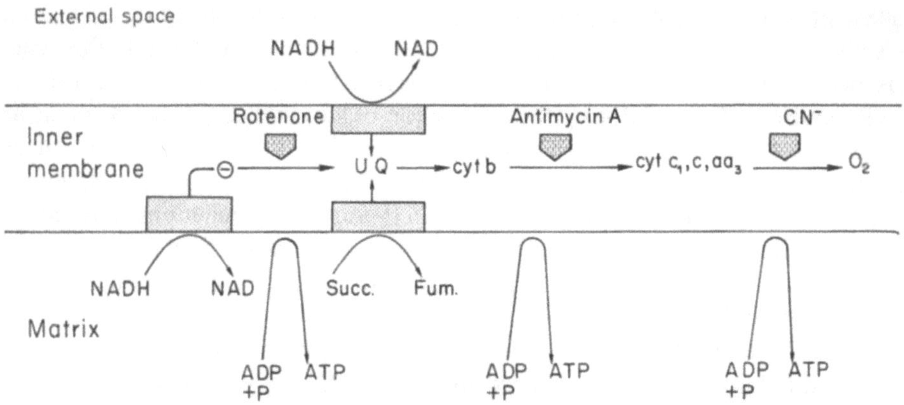

Fig. 2. Schematic diagram of the reduction of mitochondrial and extramitochondrial NADH by the respiratory chain. *UQ* ubiquinone, *cyt* cytochrome, *succ* succinate, *Fum* fumarate

Table 1. Localization of mitochondrial enzymes

Site	Enzyme
Inner membrane	Respiratory chain NADH dehydrogenase Succinate dehydrogenase Fatty acylCoA dehydrogenase ATP synthetase Metabolite carriers
Matrix	Dehydrogenases of citric acid cycle β-Hydroxyacyl CoA dehydrogenase Succinate thiokinase Transaminases Pyruvate carboxylase Activation of short chain fatty acids GTP-AMP-P transferase
Intermembrane space	Adenylate Kinase Nucleoside diphosphate kinase
Outer membrane	Activation of long chain fatty acids Phospholipase Enzymes of lecithin synthesis

The localization of these enzymes has been established with mammalian mitochondria. For references see HELDT (1972) and MUNN (1974, pp. 142–147).

involves a transport of ADP and inorganic phosphate across the mitochondrial membranes into the matrix, and a transport of ATP in the other direction. This will be studied in detail below.

AMP, pyridine, and guanine nucleotides are unable to pass the inner membrane. There is no direct communication between the pools of these nucleotides inside and outside the inner membrane. In rat liver mitochondria, AMP is formed from ATP during the activation of short chain fatty acids. Since there is no adenylate kinase present in the matrix, this AMP cannot be phosphorylated by ATP. Instead, it is phosphorylated by GTP. The GTP-AMP-P-transferase catalyzing this reaction, has been found exclusively in the matrix (HELDT and SCHWALBACH, 1967). The GTP required is formed during the conversion of succinyl coenzyme A to succinate. Besides its function in phosphorylating AMP, the GTP can be also utilized in the matrix for the phosphorylation of ADP or the formation of phosphoenolpyruvate from oxaloacetate. However, these GTP-consuming reactions may not occur in plant mitochondria, since the succinic thiokinase catalyzing the conversion of succinyl coenzyme A was found in plant tissue to be specific for adenine nucleotides (KAUFMANN and ALIVISATOS, 1955).

Since the inner membrane is impermeable to pyridine nucleotides, extramitochondrial NADH cannot be oxidized by the NADH dehydrogenase of the respiratory chain which is directed toward the matrix. Mammalian mitochondria are unable to oxidize extramitochondrial NADH directly. This is different in mitochondria from fungi and plants, where a second dehydrogenase of the respiratory chain has been found reacting with extramitochondrial NADH (BONNER and VOSS, 1961; CUNNINGHAM 1964; ONISHI et al., 1966; v. JAGOW and KLINGENBERG, 1970; COLEMAN and PALMER, 1972; KOEPPE and MILLER, 1972; DOUCE et al., 1973; DAY and WISKICH, 1974). However, there is a marked difference between these two dehydrogenases. Whereas oxidation by the respiratory chain of 1 mol of mitochondrial NADH yields 3 mol of ATP, only 2 ATP are formed from the oxidation of extramitochondrial NADH (Fig. 2). Rotenone, which inhibits the oxidation of mitochondrial NADH does not inhibit the oxidation of extramitochondrial NADH, but in each case respiration is blocked by the addition of Antimycin A, which is known to be an inhibitor of electron transport between ubiquinone and cytochrom c. Apparently, reducing equivalents from extramitochondrial NADH are fed into the respiratory chain in the region of ubiquinone. Similarly, in mammalian mitochondria extramitochondrial α-glycerophosphate was found to be oxidized and yielded only 2 ATP per mol of reducing equivalent (KLINGENBERG and BUCHHOLZ, 1970).

The dehydrogenase for external NADH may not occur in all plant mitochondria. It was observed by WILLENBRINK (personal communication) that in intact potato mitochondria there was almost no oxidation of external NADH, unless the mitochondria were damaged by osmotic shock. On the other hand, it has been found with mitochondria from hypocotyls from mung bean (*Phaseolus aureus*) that external NADH is not only oxidized by the external NADH dehydrogenase of the inner membrane, but also by an NADH dehydrogenase located in the outer mitochondrial membrane. Oxidation of NADH by the latter system is not linked to oxidative phosphorylation and is insensitive to antimycin A (DOUCE et al., 1973). The physiological function of this NADH oxidation remains to be discovered.

4. Transport of Adenine Nucleotides

Isolated mitochondria contain a certain number of adenine nucleotides (HELDT and KLINGENBERG, 1965; ONISHI et al., 1967). These are not released by normal careful washing procedures, but they are exchanged with external adenine nucleotides (KLINGENBERG and PFAFF, 1966). The sum of the mitochondrial adenine nucleotides (AMP+ADP+ATP) remains constant during this exchange. Thus, for each nucleotide molecule transported into the mitochondria, another one is transported out. The transport process causing this 1:1 exchange has been named *adenine nucleotide translocation* (HELDT et al., 1965). It is limited to ADP and ATP only; AMP, guanine-, cytidine-, uracil- and pyridine nucleotides are not transported (Table 2). There is evidence that the adenine nucleotide transport is facilitated by a carrier (KLINGENBERG et al., 1973, 1974a). The properties of the carrier are summarized in Table 3.

Table 2. Transport of nucleotides (200 μM) into rat liver mitochondria. Temp. 18° C, uncoupled state. (KLINGENBERG, 1976)

Nucleotide	Uptake activity (%)
ADP	100
ATP	70
AMP	2
GDP, CDP, UDP, IDP	0
3,5-ADP	0

Table 3. Properties of adenine nucleotide translocation in rat liver mitochondria. (From KLINGENBERG, 1976)

Km for ADP (18°)	1–4 μM
Maximal translocation activity with ADP (18°)	200 μmol/g protein/min
Number of specific binding sites for ADP	~1 mol/mole cyt aa_3
Turnover number per binding site (18°)	500/min
Activation energy (18–30°)	11 Kcal/mol
Dissociation constant of carboxyatractyloside (K_D)	0.02 μM
Carrier protein	dimer, composed of 2 units of 29,000 MW

Several potent inhibitors of adenine nucleotide transport are known, and all these compounds are of biological origin. The inhibitors atractyloside and carboxyatractyloside (HELDT, 1969; VIGNAIS et al., 1973; KLINGENBERG, 1974) are glycosides contained in the rhizomes of *Atractylis gummifera*, a thistle growing in the southern Mediterranean area. The highly toxic material from this plant has been known since ancient times. Pedanios Dioscurides reported it as "Chamaileon Leukos" in "De Materia Medica" in about 80 A.D. Of the two, carboxyatractyloside is the stronger inhibitor. Decarboxylation of carboxyatractyloside yields atractyloside. Another inhibitor of adenine nucleotide transport is bongkrekic acid, a very toxic unsaturated long chain fatty acid with 3 carboxylic groups from *Pseudomonas cocovenenans* (HENDERSON and LARDY, 1970; KLINGENBERG et al., 1970; KEMP et al., 1970; KLINGENBERG and BUCHHOLZ, 1973).

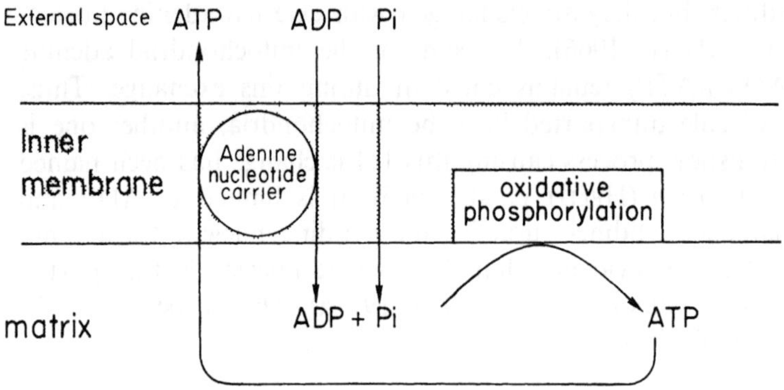

Fig. 3. Schematic diagram of the role of the adenine nucleotide carrier in the overall reaction of oxidative phosphorylation

It appears that these inhibitors react at different sides of the inner membrane. Whereas atractyloside and carboxyatractyloside seem to be bound to the carrier from the external side, bongkrekic acid appears to be bound from the matrix side (KLINGENBERG and BUCHHOLZ, 1973; SCHERER and KLINGENBERG, 1974). Besides these substances, long chain acylCoA esters were also shown to be inhibitors of adenine nucleotide translocation (PANDE and BLANCHAER, 1971; LERNER et al., 1972; KLINGENBERG et al., 1973; VAARTJES et al., 1972; MOREL et al., 1974).

The adenine nucleotide translocation enables the export of the ATP, generated by oxidative phosphorylation at the matrix side of the inner membrane, to the cytoplasm (Fig. 3). Therefore the transport can be regarded as a partial step of the overall reaction of oxidative phosphorylation. This explains why inhibitors of this transport are so highly toxic.

The specificity of the adenine nucleotide carrier is different for the two directions of transport. For transport into the mitochondria, ADP is favored more than ATP (PFAFF et al., 1969), whereas for the outward transport ATP is preferred (KLINGENBERG, 1976). However, this is true only of intact mitochondria, in which there is coupling between the electron transport of the respiratory chain and ATP synthesis. If this "coupling" is eliminated by "uncouplers", e.g. carbonylcyanide phenylhydrazone, or if the mitochondria become anaerobic, these differences in specificity disappear, in which case both adenine nucleotides are about equally well transported in either direction. Apparently the consumption of energy, as supplied from electron transport of the respiratory chain, is required to maintain the differences in specificity. These differences in specificity are reflected by the *Km* for ADP and ATP. In the uncoupled state the *Km* for ATP is very similar to the *Km* of ADP. Energization of the mitochondrial membrane by electron transport has no major effect on the *Km* of ADP, but increases the *Km* of ATP by about one order of magnitude (KLINGENBERG, 1975, 1976). This increase of the *Km* for ATP seems to be caused by a membrane potential, which is built up by electron transport (KLINGENBERG, 1970a). Since ATP carries one negative charge more than ADP, a membrane potential could allow a discrimination against the transport of ATP in one direction and of ADP in the other.

The asymmetry in the specificity of transport of adenine nucleotides leads to an asymmetric distribution of ADP and ATP on the two sides of the membrane.

Table 4. Phosphorylation potential $\Delta G'$ of ATP in rat liver mitochondria and in the medium in the state of respiratory control (20° C).
Data from HELDT et al. (1972)
$\Delta G' = \Delta G'_0 - RT \ln (ATP)/(ADP) \times (Pi)$
$\Delta G'_0 = -8.8$ Kcal/mol (ALBERTI, 1968)

	ATP/ADP	Pi (mM)	G' (Kcal/mol)
Mitochondrial	3.9	3.5	−12.9
Extramitochondrial	29	0.48	−15.2

When rat liver mitochondria are carrying out oxidative phosphorylation, the ATP/ADP ratio in the matrix is found to be lower than in the extramitochondrial space (Table 4). Very similar results have been obtained in whole tissue from fractionation of freeze stop liver material, using nonaqueous solvents (ELBERS et al., 1974).

With the aid of the measured phosphate concentrations it is possible to calculate the free energy of hydrolysis of ATP, known as the phosphorylation potential, for both spaces. Taking the experimental values from Table 4, the phosphorylation potential of the ATP in the external space is calculated to be about 2Kcal more negative than in the matrix space. Thus, the energy which is required for the release of ATP from the mitochondria may be considerable. Since this energy is derived from the same intermediate energy pool which supplies the ATP synthesis, the ratio of ATP formed to oxygen consumed (P/O ratio) is decreased below the theoretical value (KLINGENBERG, 1970a).

The function of the energy control of the adenine nucleotide translocation appears to ensure an early response of ATP synthesis in the mitochondria to ATP consumption in the cytoplasm. A relatively small increase in the level of ADP in the cytoplasm leads to a much higher increase of the mitochondrial ADP concentration, thus stimulating a relatively high rate of ATP synthesis. The energy consumption required for the adenine nucleotide transport could therefore be regarded as a contribution to the maintenance of a constant cellular phosphorylation potential of the ATP.

The data on adenine nucleotide transport discussed so far have been obtained mainly with rat liver mitochondria. Specific transport of ADP and ATP, having very similar properties to the adenine nucleotide carrier in mammalian mitochondria, has been demonstrated in mitochondria from *Saccharomyces cerevisiae* (ONISHI et al., 1967) and *Neurospora crassa* (v. JAGOW, 1974). Atractyloside was also inhibitory. Furthermore, there is indirect evidence that atractyloside also inhibits adenine nucleotide transport in mitochondria from maize (BERTAGNOLLI and HANSON, 1973) and from potato (WILLENBRINK, personal communication). On the other hand, it has been reported that in mitochondria from Jerusalem artichoke (*Helianthus tuberosus*), which are insensitive to atractyloside, this carrier may be absent (PASSAM et al., 1973). Further investigations will be necessary in order to exclude the possibility that the latter observation does not reflect damage to the mitochondria investigated.

5. Transport of Inorganic Phosphate

Inorganic phosphate is transported across the inner membrane either in exchange with hydroxyl ions or with dicarboxylates (FONYO, 1968; TYLER, 1968; CHAPPEL, 1969; PAPA et al., 1969). These two types of phosphate transport are facilitated by two different carriers. The first carrier (phosphate carrier), facilitating a counterexchange with hydroxyl ions (or a cotransport with protons, which cannot be distinguished experimentally) enables the uptake of phosphate, as required for oxidative phosphorylation (Fig. 3). At 0° the *Km* for phosphate was found to be 1.6 mM, and the V_{max} 205 µmol/min (COTY and PEDERSEN, 1974). Thus, phosphate transport by the phosphate carrier is over ten times more active than phosphate transport by the so called dicarboxylate carrier (cf. Table 5, note temperature differences of measurements).

Phosphate transport *via* counterexchange with hydroxyl ions would lead to an acidification of the matrix space. In the course of oxidative phosphorylation this generation of protons in the matrix is compensated by the consumption of protons due to the formation of ATP from ADP and phosphate, and by the extrusion of protons from the matrix due to the ATP/ADP counterexchange. In contrast to this, active proton transport is required for an accumulation of phosphate in the matrix. In this way the respiration-driven proton pump acts as the driving force for the active uptake of phosphate by mitochondria. Consequently, the accumulation of phosphate is abolished when uncoupler is added to respiring mitochondria.

Very low concentrations of SH-reagents, e.g. N-ethyl-maleinimide or mersalyl, selectively inhibit phosphate transport (MEIJER et al., 1970; KLINGENBERG et al., 1974b). This indicates that the carrier involved contains free SH groups.

6. Transport of the Intermediates of the Tricarboxylic Acid Cycle

Three different carriers have been identified (CHAPPEL, 1969; PAPA et al., 1971a; PALMIERI et al., 1971, 1972a, b), which catalyze a counterexchange of anions (Fig. 4):

1. *The dicarboxylate carrier* specifically transports either phosphate or dicarboxylates, e.g. malate and succinate. The carrier has two separate binding sites (PALMIERI et al., 1971), one specific for phosphate and one for dicarboxylates. It has its highest affinity for malate and is less active with succinate (Table 5). The maximal velocities are similar with all three compounds, and all these dicarboxylates were shown to compete with each other for transport. This clearly indicates that the dicarboxylates mentioned are all transported by the same carrier. Furthermore, evidence has been presented that oxaloacetate is also transported by this carrier (GIMPEL et al., 1973). Glutamate, aspartate, and monocarboxylates do not react with the carrier. The fivefold higher affinity of the carrier for malate, as compared with succinate, indicates that the main function of the carrier is the transfer of reducing equivalents between the matrix and the extramitochondrial space by a malate-linked shuttle.

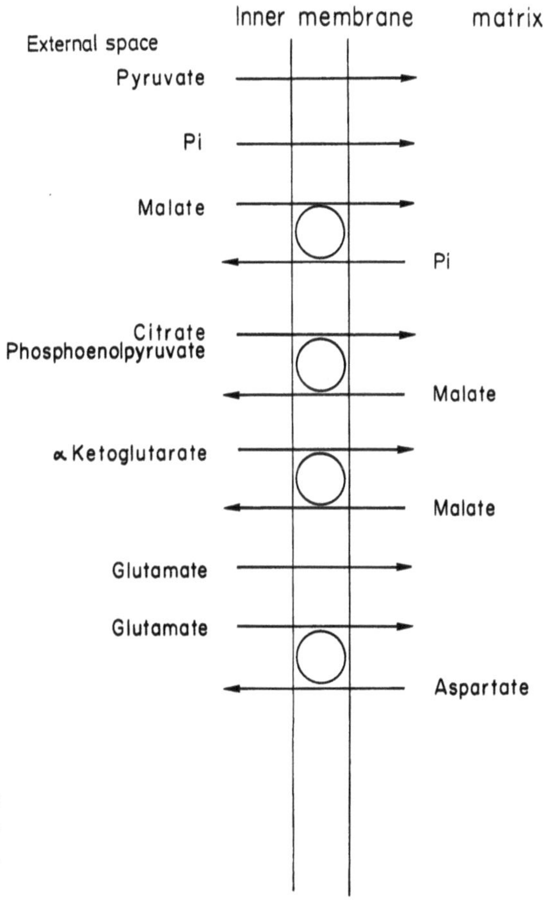

Fig. 4. Schematic diagram of specific transport of phosphate and of carboxylates across the inner mitochondrial membrane

2. *The tricarboxylate carrier* transports citrate, isocitrate, cis aconitate, phosphoenolpyruvate, and also the dicarboxylates malate and succinate. The carrier has only a single binding site for all substrates mentioned (PALMIERI et al., 1972a). Its affinity for tricarboxylates is much higher than for dicarboxylates (Table 6).

3. *The α-ketoglutarate carrier* has a single binding site for α ketoglutarate and other dicarboxylates, e.g. malate (DE HAAN and TAGER, 1968; PALMIERI et al., 1972b; SLUSE and LIEBECQ, 1973). This carrier also reacts with aspartate, glutamate, and oxaloacetate, but the affinity of the carrier is rather low (Table 7).

All three carriers are inhibited by SH reagents, but higher concentrations are required than for inhibition of the phosphate carrier (MEIJER et al., 1970; QUAGLIARIELLO and PALMIERI, 1972; PALMIERI et al., 1972a). Thus, by careful addition of SH reagents, the phosphate carrier can be selectively inhibited without major inhibition of the dicarboxylate, tricarboxylate- and ketoglutarate carriers. These three carriers, but not the phosphate carrier, are competitively inhibited to various extents by butylmalonate, phenylsuccinate, and pentylmalonate (QUAGLIARIELLO and PALMIERI, 1972). A more specific inhibitor for the tricarboxylate carrier is 1.2.3.-benzene tricarboxylate (ROBINSON et al., 1972).

Furthermore, all three carriers, but again not the phosphate carrier, are inhibited by the metal complexing reagent bathophenanthrolin (QUAGLIARIELLO and PAL-

Table 5. Specificity of the dicarboxylate carrier in rat liver mitochondria. Temp. 9° C. (From PALMIERI et al., 1971)

	Km (mM)	V_{max} (μmol/min/g protein)
Malate	0.23	69
Malonate	0.37	77
Succinate	1.17	64
Phosphate	1.50	similar to malate

Activation energy (succinate, 0–20° C): 22 Kcal (QUAGLIARIELLO et al., 1969)

MIERI, 1972). From this it is concluded that a metal ion is located at the substrate binding site of all three carriers.

A comparison of the transport of inorganic phosphate, malate, α-ketoglutarate, and citrate reveals a sequential linkage of these processes in a cascade-like manner (Fig. 4) (MCGIVAN and KLINGENBERG, 1971). Inorganic phosphate is the key substrate, since it can be taken up into the matrix by the phosphate carrier without exchange with other substrates. An accumulation of dicarboxylates requires such an uptake of phosphate first, and then an exchange of this phosphate in the matrix with dicarboxylates in the extramitochondrial space, as facilitated by the dicarboxylate carrier. Likewise, the accumulation of citrate involves an uptake of phosphate, followed by exchange of mitochondrial phosphate with dicarboxylates, and finally the exchange of mitochondrial dicarboxylates with extramitochondrial tricarboxylates. The accumulation of α-ketoglutarate proceeds accordingly. This explains why the accumulation of citrate in the matrix requires that dicarboxylates and phosphate are present. Therefore, inhibition of the phosphate carrier may also result in an inhibition of the uptake of di- and tricarboxylates. In this cascade, the proton pump of the respiratory chain is also the driving force for the active uptake of di- and tricarboxylates.

Table 6. Properties of the tricarboxylate carrier in rat liver mitochondria. Temp. 9° C. (From PALMIERI et al., 1972a)

Citrate:		
Km	0.12 mM	
V_{max}	22.5 μmol/min/g protein	
Activation energy (0–14° C): 20.1 Kcal		
Competitive inhibition of citrate transport by:		
phosphoenolpyruvate	Ki	0.11 mM
malate	Ki	0.7 mM
succinate	Ki	>3 mM

Table 7. Properties of the α-ketoglutarate carrier in rat liver mitochondria. Temp. 9° C. (From PALMIERI et al., 1972b)

α-Ketoglutarate:		
Km	0.046 mM	
V_{max}	43 μmol/min/g protein	
Activation energy (0–12° C): 20.5 Kcal		
Competitive inhibition of α-ketoglutarate transport by:		
malate	Ki	0.12 mM
succinate	Ki	1.6 mM
oxaloacetate	Ki	1.1 mM
aspartate	Ki	2.7 mM
glutamate	Ki	2.5 mM

Table 8. Maximal activities of metabolite carriers at 25°. Values extrapolated from the activation energies and the maximal activities in Tables 3, 5–7

Carrier for	μmol/min/g protein (25° C)
ADP	310
Phosphate	very high
Dicarboxylates	570
α-Ketoglutarate	320
Citrate	156

For comparison: maximal rate of respiration approx. 150 μ atoms 0/min/g protein (25° C).

In order to compare the activities of these carriers, the maximal activities have been extrapolated to 25° (Table 8). At this temperature the maximal activities of all carriers mentioned are higher than the maximal rate of respiration. Because the activation energies may change with increasing temperature, as has been shown for the adenine nucleotide carrier (KLINGENBERG, 1976), an extrapolation to the temperature of 37° would be rather speculative.

7. Transport of Glutamate and Aspartate

Like phosphate, glutamate can be transported into the mitochondria without exchanging with other metabolites (AZZI et al., 1967; McGIVAN and CHAPPEL, 1969; MEIJER et al., 1972; MEIJER and VIGNAIS, 1973). This transport is inhibited by lipid-soluble SH reagents and also with the antibiotic Avenaciolide. Recently, a proteolipid has been isolated by GAUTHERON et al. (1974) from pig heart mitochondria, showing specific binding of glutamate and stimulating the transfer of glutamate across artificial lipid membranes in an inhibitor-sensitive manner. The findings indicate the existence of a specific glutamate carrier in the mitochondria.

Glutamate is also transported in exchange with aspartate (KING and DIWAN, 1972; LA NOUE et al., 1973, 1974). This transport system shows an asymmetric specificity. When the mitochondria are in the energized state, the efflux of aspartate coupled to influx of glutamate is much faster than the influx of aspartate. When the mitochondria are uncoupled, the efflux of aspartate is prevented and the influx promoted. This asymmetric transport could be explained by assuming that glutamate is transported as the undissociated acid, whereas aspartate is transported as anion. A membrane potential, generated by electron transport of the respiratory chain, would enable an unidirectional efflux of aspartate from the mitochondria. In this way, aspartate may be transported against a concentration gradient with energy provided by the respiratory chain.

8. Transport of Pyruvate

Pyruvate may be regarded as one of the most important substrates of mitochondria. Because of its very low pK (2.5), it is unlikely to pass the inner mitochondrial membrane by unfacilitated diffusion of the undissociated acid, as seems to be the case with acetate. Recent investigations indicate the existence of a specific translocator for pyruvate (PAPA et al., 1971 b; BROUWER et al., 1973). This transport may be a cotransport of pyruvate with a proton; there appears to be no exchange with other metabolites involved. One indication for the existence of this transport system is the observation that uptake of pyruvate into the mitochondria is specifically inhibited by cyano-4-hydroxycinnamate (HALESTRAP and DENTON, 1974).

9. Transport of Phosphate and of Carboxylates in Plant Mitochondria

The few data available on transport in plant mitochondria indicate that the transport reactions are very similar to those known in mammalian mitochondria. It was concluded from swelling experiments that mitochondria from potato (*Solanum tuberosum* L.) possess a phosphate-hydroxyl carrier, a dicarboxylate carrier catalyzing an exchange of dicarboxylates with inorganic phosphate, and a tricarboxylate carrier exchanging tricarboxylates with dicarboxylates (PHILLIPS and WILLIAMS, 1973). The transport of phosphate was inhibited by the SH-reagent N-ethylmaleinimide, and of dicarboxylates with pentylmalonate. Furthermore, oxidation of malate, occuring in the matrix, was shown to be inhibited by butylmalonate in mitochondria from Jerusalem artichoke (*Helianthus tuberosus*) (COLEMAN and PALMER, 1972) and from cauliflower (*Brassica oleracea* L.) (DAY and WISKICH, 1974).

10. Uptake of Fatty Acids

In rat liver mitochondria short chain fatty acids are activated in the matrix space (AAS and BREMER, 1968). The uptake of these fatty acids may proceed by diffusion of the undissociated fatty acids across the inner membrane. In contrast to this, long chain fatty acids are shown to be activated on the outer mitochondrial membrane or the membrane of the endoplasmic reticulum (BREMER et al., 1967; AAS, 1971). Since the inner membrane is impermeable to coenzyme A, and their acylderivatives, the mitochondria are thus faced with the problem of transferring the long chain acylcoenzyme A esters from the external space into the matrix (Fig. 5).

With isolated mitochondria carnitine is required for the oxidation of palmitate (FRITZ, 1959), indicating that carnitine is involved in the transfer of the activated palmitate across the inner membrane. Furthermore, two different acyl coenzyme A carnitine transferases have been discovered (WEST et al., 1971), one of which is bound to the inner membrane. From these findings, RAMSEY and TUBBS (1975)

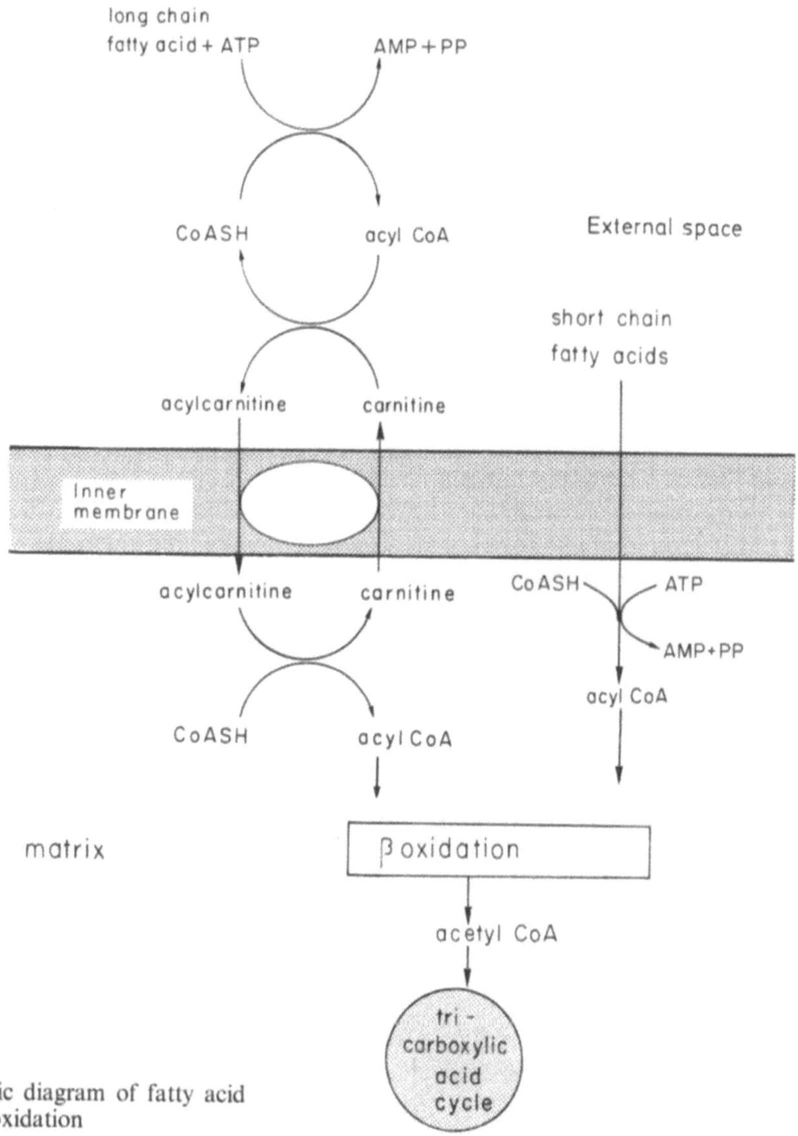

Fig. 5. Schematic diagram of fatty acid activation and oxidation

proposed a pathway for the uptake of the activated fatty acids, as shown in Figure 5. The acylcoenzyme A outside the inner membrane is converted into acyl carnitine, which is transported by a specific carrier across the inner membrane in exchange for carnitine. It then reacts with the coenzyme A in the matrix, as catalyzed by the acyl coenzyme A carnitine transferase located in the inner membrane.

Since carnitine has been found in several plant tissues, especially in seedlings, (FRAENKEL and FRIEDMAN, 1957; PANTER and MUDD, 1969) it is likely that in plant mitochondria also an acyl carnitine transferase may be involved in the uptake of activated fatty acids.

11. Oxidation of Extramitochondrial NADH by Mitochondria

Since the inner membrane is impermeable to pyridine nucleotides, the extramitochondrial NADH arising from glycolysis cannot directly be oxidized by the respiratory chain of mammalian mitochondria. Oxidation of extra mitochondrial NADH requires an indirect transfer of reducing equivalents by metabolite shuttles.

In the glycerophosphate shuttle, as proposed by Bücher (ZEBE et al., 1959), the extramitochondrial NADH is utilized for the formation of α-glycerophosphate from dihydroxyacetone phosphate, as catalyzed by the soluble α-glycerophosphate dehydrogenase of the extramitochondrial compartment. The α-glycerophosphate is oxidized again by the membrane-bound α-glycerophosphate dehydrogenase of the respiratory chain, which is accessible from the outside of the inner membrane. Since this is a flavin enzyme, in this way 2 mol of ATP are obtained from the oxidation of 1 mol extramitochondrial NADH.

In the malate-aspartate shuttle, the extramitochondrial NADH is first utilized to reduce extramitochondrial oxaloacetate to malate, and the malate is transported into the mitochondria (Fig. 6). There it is reoxidized, yielding NADH and oxaloacetate. The NADH is oxidized by the respiratory chain, and the oxaloacetate is transaminated by glutamate, forming aspartate and α-ketoglutarate. The latter two compounds are transported across the inner membrane in exchange with glutamate and malate. By transamination in the extramitochondrial compartment, glutamate and oxaloacetate are regenerated there.

It may be noted that the redoxpotential of the mitochondrial NADH is much more negative than of the cytoplasmic NADH (BÜCHER and KLINGENBERG, 1958; WILLIAMSON et al., 1967). Therefore, transfer of reducing equivalents from the cytoplasm to the mitochondrial NADH pool is an uphill transport, requiring energy. In the case of the α-glycerophosphate shuttle, this is avoided, and the reducing equivalents are fed into the respiratory chain, yielding only 2 mol of ATP per 2 H oxidized. With the malate-aspartate shuttle, 3 mol of ATP are obtained from oxidation of mitochondrial NADH, but energy has to be spent for the uphill transport of reducing requivalents. It is assumed that this energy is provided from the respiratory chain for energy requiring aspartate efflux from the mitochondria (see above). Stimulation of aspartate efflux accompanied by an inhibition of aspartate influx would make the malate-aspartate shuttle an unidirectional process, thus enabling uphill transport of reducing equivalents.

In contrast to mammalian mitochondria, plant mitochondria are able to oxidize extramitochondrial NADH directly. Similarly to the oxidation of α-glycerophosphate in mammalian mitochondria, there are only 2 mol of ATP obtained per 2 H oxidized (see Sect. 3). The question arises whether the malate/aspartate shuttle may also contribute in plant tissue to the oxidation of cytoplasmic NADH. The oxidation of extramitochondrial NADH could be especially important for germinating seedlings containing fat as storage. COOPER and BEEVERS (1969) reported that in the endosperm of germinating castor bean the enzymes for β-oxidation of fatty acids are localized primarily in the glyoxysomes, and not in the mitochondria, but that the NADH produced by such oxidation is not utilized in the glyoxysomes. Here the external NADH oxidase of the mitochondria seems to be responsible for the oxidation of the NADH generated in the glyoxysomes.

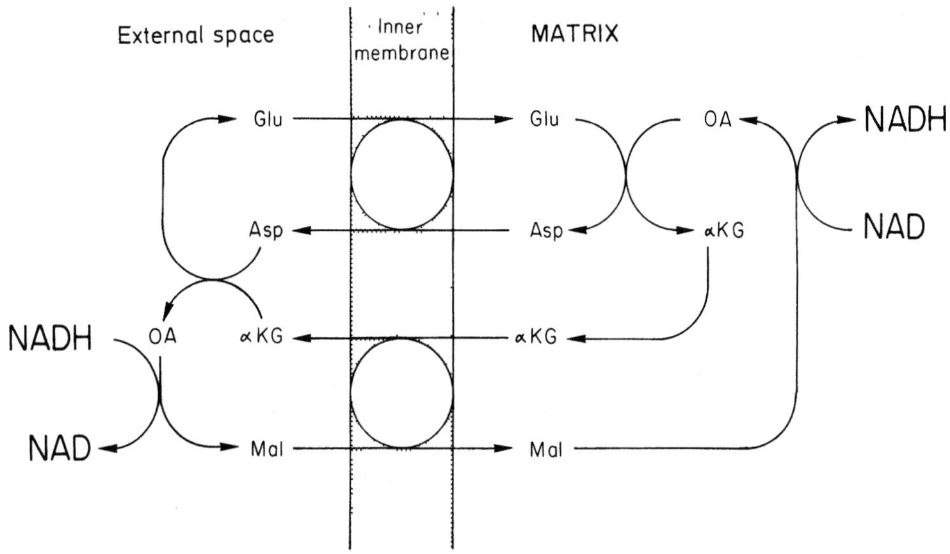

Fig. 6. Schematic diagram of the malate-aspartate cycle

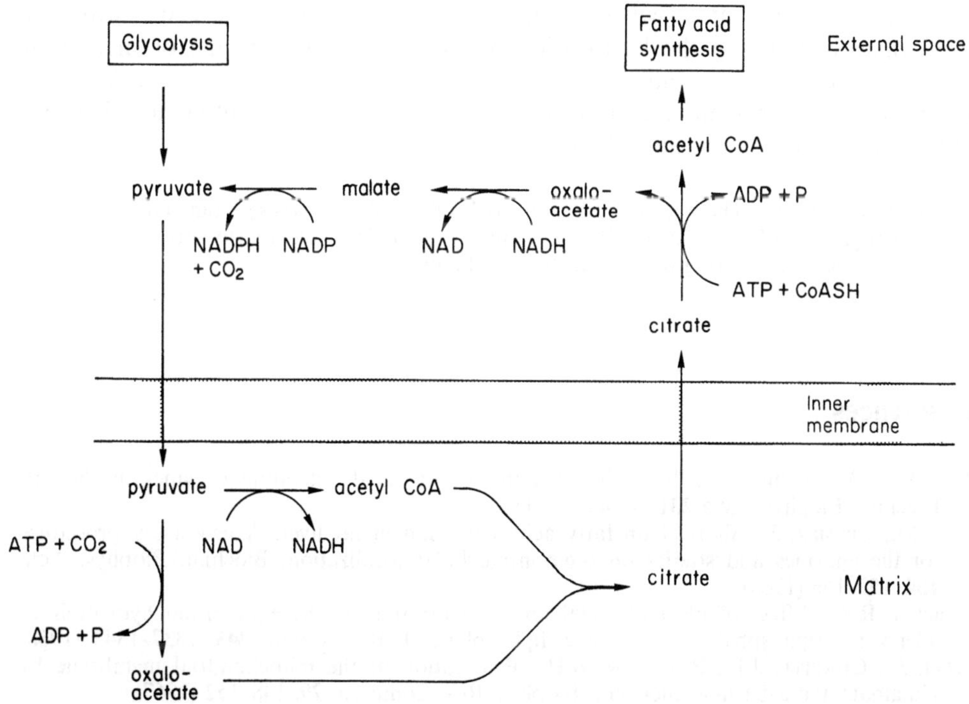

Fig. 7. Schematic diagram of lipogenesis

12. On a Possible Role of Mitochondrial Transport in Lipogenesis

In mammalian lipogenesis, the synthesis of fatty acids takes place in the extramitochondrial compartment, whereas the precursor acetyl coenzyme A is formed in the mitochondrial matrix. Thus, acetyl coenzyme A has to be brought from the matrix to the extramitochondrial space. In mammalian mitochondria there is no direct transfer of acetyl coenzyme A across the inner mitochondrial membrane, as the transfer proceeds in an indirect way (Fig. 7). The acetyl coenzyme A condenses in the matrix with oxaloacetate, forming citrate, and the latter is transported by the tricarboxylate carrier to the extramitochondrial compartment (LOWENSTEIN, 1968). There it is split again to acetyl coenzyme A and oxaloacetate at the expense of coenzyme A and ATP. It is feasible that the tricarboxylate carrier in plant mitochondria may also play a role in fatty acid synthesis.

13. Conclusion

This paper has made it very obvious that our knowledge of plant mitochondria is fragmentary. This may in part be due to the fact that intact plant mitochondria are more difficult to prepare than mammalian mitochondria. In order to understand the metabolic role of the mitochondria in the plant cell, we are still dependent on extrapolation of our knowledge of mammalian mitochondria to plant mitochondria. Investigations on the localization of enzymes, measurements of metabolite transport etc. will be required in order to gain more insight into the functions of mitochondria in plant metabolism, and to reveal possible differences between mitochondria of animals and plants.

Acknowledgments. The author is grateful to Prof. J. WILLENBRINK, Köln, Dr. R. LILLEY, Wollongong, and to Dr. G. KELLY, Weihenstephan, for their advice and to Mrs. D. MARONDE for her technical help during the preparation of the manuscript.

References

AAS, M.: Organ and subcellular distribution of fatty acid activating enzymes in the rat. Biochim. Biophys. Acta **231**, 32–47 (1971)

AAS, M., BREMER, J.: Short chain fatty acid activation in rat liver. A new assay procedure for the enzymes and studies on their intracellular localization. Biochim. Biophys. Acta **164**, 157–166 (1968)

ALBERTI, R.A.: Effect of pH and metal ion concentration on the equilibrium hydrolysis of adenosine triphosphate to adenosine diphosphate. J. Biol. Chem. **243**, 1337–1343 (1968)

AZZI, A., CHAPPEL, J.B., ROBINSON, B.H.: Penetration of the mitochondrial membrane by glutamate and aspartate. Biochem. Biophys. Res. Commun. **29**, 148–152 (1967)

BERTAGNOLLI, B.L., HANSON, J.B.: Functioning of the adenine nucleotide transporter in the arsenate uncoupling of corn mitochondria. Plant Physiol. **52**, 431–435 (1973)

BONNER, W.D., VOSS, D.O.: Some characteristics of mitochondria extracted from higher plants. Nature **191**, 682–684 (1961)

BREMER, J., NORUM, K.R., FARSTAD, M.: Intracellular distribution of some enzymes involved in the metabolism of fatty acids. In: Mitochondrial Structure and Compartmentation (QUAGLIARIELLO, E., PAPA, S., SLATER, E.C., TAGER, J.M., eds.), pp. 380–384. Bari: Adriatica Editrice 1967

BROUWER, A., SMITS, G.G., TAS, J., MEIJER, A.J., TAGER, J.M.: Substrate anion transport in mitochondria. Biochimie **55**, 717–725 (1973)

BÜCHER, TH., KLINGENBERG, M.: Wege des Wasserstoffs in der lebendigen Organisation. Angew. Chem. **70**, 552–570 (1958)

CHAPPEL, J.B.: Transport and exchange of anions in mitochondria. In: Inhibitors, Tools of Cell Research (BÜCHER, TH., SIES, H., eds.), pp. 335–350. Berlin-Heidelberg-New York: Springer 1969

COLEMAN, J.O.D., PALMER, J.M.: The oxidation of malate by isolated plant mitochondria. Eur. J. Biochem. **26**, 499–509 (1972)

COOPER, T.G., BEEVERS, H.: β-Oxidation in glyoxysomes from castor bean endosperm. J. Biol. Chem. **244**, 3514–3520 (1969)

COTY, W.A., PEDERSEN, P.L.: Phosphate transport in rat liver mitochondria. J. Biol. Chem. **249**, 2593–2598 (1974)

CUNNINGHAM, W.B.: The oxidation of externally added NADH by corn root mitochondria. Plant Physiol. **39**, 699–703 (1964)

DAY, D.A., WISKICH, I.T.: The oxidation of malate and exogenous NADH by isolated plant mitochondria. Plant Physiol. **53**, 104–109 (1974)

DOUCE, R., MANNELLA, C.A., BONNER, W.D.: The external NADH dehydrogenases of intact plant mitochondria. Biochim. Biophys. Acta **292**, 105–116 (1973)

ELBERS, R., HELDT, H.W., SCHMUCKER, P., SOBOLL, S., WIESE, H.: Measurement of the ATP/ADP ratio in mitochondria and in the extramitochondrial compartment by fractionation of freeze stopped liver tissue in nonaqueous media. Z. Physiol. Chem. **355**, 378–393 (1974)

FONYO, A.: Phosphate carrier of rat liver mitochondria: Its role in phosphate outflow. Biochem. Biophys. Res. Commun. **32**, 624–628 (1968)

FRAENKEL, G., FRIEDMAN, S.: Carnitine. In: Vitamins and Hormones (HARRIES, R.S., MARRIAN, G.F., THIMAN, K.V., eds.), Vol. 15, pp. 73–118. New York: Academic Press 1957

FRITZ, I.B.: Action of carnitine on long chain fatty acid oxidation by liver. Am. J. Physiol. **197**, 297–304 (1959)

GAUTHERON, D.C., JULLIARD, J.H., GODINOT, C.: Protein-lipid associations and glutamate transport across mitochondrial membranes. In: Membrane Proteins in Transport and Phosphorylation (AZZONE, G.F., KLINGENBERG, M.E., QUAGLIARIELLO, E., SILIPRANDI, N., eds.), pp. 91–96. Amsterdam-London: North-Holland Publ. Co. 1974

GIMPEL, J.A., DE HAAN, E.J., TAGER, J.M.: Permeability of isolated mitochondria to oxaloacetate. Biochim. Biophys. Acta **292**, 582–591 (1973)

HAAN, E.J. DE, TAGER, J.M.: Evidence for a permeability barrier for α-oxoglutarate in rat liver mitochondria. Biochim. Biophys. Acta **153**, 98–112 (1968)

HALESTRAP, A.P., DENTON, R.M.: Specific inhibition of pyruvate transport in rat liver mitochondria and in human erythrocytes by α-cyano-4-hydroxycinnamate. Biochem. J. **138**, 313–316 (1974)

HELDT, H.W.: The inhibition of adenine nucleotide translocation by atractyloside. In: Inhibitors, Tools in Cells Research (BÜCHER, TH., SIES, H., eds.), pp. 301–317. Berlin-Heidelberg-New York: Springer 1969

HELDT, H.W.: Energiestoffwechsel in Mitochondrien. Angew. Chem. **84**, 792–798 (1972)

HELDT, H.W., JACOBS, H., KLINGENBERG, M.: Endogenous ADP of mitochondria, an early phosphate acceptor of oxidative phosphorylation as disclosed by kinetic studies with C^{14} labelled ADP and ATP and with atractyloside. Biochem. Biophys. Res. Commun. **18**, 174–179 (1965)

HELDT, H.W., KLINGENBERG, M.: Endogenous nucleotides of mitochondria participating in phosphate transfer reactions as studied with ^{32}P labelled orthophosphate and ultramicroscale ion exchange chromatography. Biochem Z. **343**, 433–451 (1965)

HELDT, H.W., KLINGENBERG, M., MILOVANCEV, M.: Differences between the ATP/ADP ratios in the mitochondrial matrix and in the extramitochondrial space. Eur. J. Biochem. **30**, 434–440 (1972)

HELDT, H.W., SCHWALBACH, K.: The participation of GTP-AMP-P transferase in substrate level phosphate transfer of rat liver mitochondria. Eur. J. Biochem. **1**, 199–206 (1967)
HENDERSON, P.J.F., LARDY, H.A.: Bongkrekic acid, an inhibitor of the adenine nucleotide translocase of mitochondria. J. Biol. Chem. **245**, 1319–1326 (1970)
JAGOW, G. V.: Die Atmungskette der niederen Eukarionten Saccaromyces carlsbergensis und Neurospora crassa. Habilitationsschrift Medizinische Fakultät der Universität (1974)
JAGOW, G. V., KLINGENBERG, M.: Pathways of hydrogen in mitochondria of Saccharomyces carlsbergensis. Eur. J. Biochem. **12**, 583–592 (1970)
KAUFMAN, S., ALIVISATOS, S.G.A.: Purification and properties of the phosphorylating enzyme from spinach. J. Biol. Chem. **216**, 141–152 (1955)
KEMP, J.A., OUT, T.A., GUIOT, H.F.L., SOUVERIJN, J.H.M.: The effect of adenine nucleotides and pH on the inhibition of oxidative phosphorylation by bongkrekic acid. Biochim. Biophys. Acta **223**, 460–462 (1970)
KING, M.J., DIWAN, J.J.: Transport of glutamate and aspartate across the membranes of rat liver mitochondria. Arch. Biochem. Biophys. **152**, 670–676 (1972)
KLINGENBERG, M.: Metabolite transport in mitochondria: an example for intracellular membrane function. Essays of Biochemistry **6**, 119–159 (1970a)
KLINGENBERG, M.: Mitochondria Metabolite Transport. FEBS Lett. **6**, 145–154 (1970b)
KLINGENBERG, M.: The mechanism of the mitochondrial ADP, ATP carrier as studied by kinetics of ligand binding. In: Dynamics of Energy Transducing Membranes (ERNSTER, L., ESTABROOK, R., SLATER, E.C., eds.), pp. 511–527. Amsterdam: Elsevier Publ. Co. 1974
KLINGENBERG, M.: Energetic aspect of transport of ADP and ATP through the mitochondrial membrane. Ciba Found. Symp. Energy Transformation Biological Systems. London in print (1975)
KLINGENBERG, M.: The ADP, ATP carrier in mitochondrial membranes. In: The Enzymes of Biological Membranes, Vol. 3 (MARTONOSI, A., ed.), pp. 383–438. New York: Plenum Press 1976
KLINGENBERG, M., BUCHHOLZ, M.: Localization of the glycerol phosphate dehydrogenase in the outer phase of the mitochondrial inner membrane. Eur. J. Biochem. **13**, 247–252 (1970)
KLINGENBERG, M., BUCHHOLZ, M.: On the mechanism of bongkrekate effect on the mitochondrial adenine nucleotide carrier as studied through the binding of ADP. Eur. J. Biochem. **38**, 346–358 (1973)
KLINGENBERG, M., DURAND, R., GUERIN, B.: Analysis of the reactivity of SH-Reagents with the mitochondrial phosphate carrier. Eur. J. Biochem. **42**, 135–150 (1974b)
KLINGENBERG, M., GREBE, K., HELDT, H.W.: On the inhibition of the adenine nucleotide translocation by bongkrekic acid. Biochem. Biophys. Res. Commun. **39**, 344–351 (1970)
KLINGENBERG, M., PFAFF, E.: Structural and functional compartmentation in mitochondria. In: Regulation of Metabolic Processes in Mitochondria (TAGER, J.M., PAPA, S., QUAGLIARIELLO, E., SLATER, E.C., eds.), pp. 180–201. Amsterdam: Elsevier Publ. Co. 1966
KLINGENBERG, M., RICCIO, P., AQUILA, H., SCHMIEDT, B., GREBE, K., TOPITSCH, P.: Characterisation of the ATP/ADP carrier in mitochondria. In: Membrane Proteins in Transport and Phosphorylation (AZZONE, G.F., KLINGENBERG, M.E., QUAGLIARIELLO, E., SILIPRANDI, N., eds.), pp. 229–243. Amsterdam: North Holland Publ. Co. 1974a
KLINGENBERG, M., SCHERER, B., STENGEL-RUTKOWSKI, L., BUCHHOLZ, M., GREBE, K.: Experimental demonstration of the reorienting (mobile) carrier mechanism exemplified by the mitochondrial adenine nucleotide translocator. In: Mechanisms in Bioenergetic. (AZZONE, G.F., ERNSTER, L., PAPA, S., QUAGLIARIELLO, E., SILIPRANDI, N., eds.), pp. 257–284. New-York-London: Academic Press 1973
KOEPPE, D.E., MILLER, R.J.: Oxidation of reduced nicotinamide adenine dinucleotide phosphate by isolated corn mitochondria. Plant Physiol. **49**, 353–357 (1972)
LA NOUE, K.F., MEIJER, A.J., BROUWER, A.: Evidence for electrogenic aspartate transport in rat liver mitochondria. Arch. Biochem. Biophys. **161**, 544–550 (1974)
LA NOUE, K.F., WALAJTYS, E.I., WILLIAMSON, J.R.: Regulation of glutamate metabolism and interactions with the citric acid cycle in rat heart mitochondria. J. Biol. Chem. **248**, 7171–7183 (1973)
LERNER, E., SHUG, A.L., ELSON, CH., SHRAGO, E.: Reversible inhibition of adenine nucleotide translocation by long chain fatty acyl coenzyme A esters in liver mitochondria of diabetic and hibernating animals. J. Biol. Chem. **247**, 1513–1519 (1972)

LOWENSTEIN, J.M.: Citrate and the conversion of carbohydrate into fat. In: The Metabolic Role of Citrate. (GOODWIN, T.W., ed.), pp. 61–86. London-New York: Academic Press 1968

MALONE, C., KOEPPE, D.E., MILLER, R.I.: Corn mitochondrial swelling and contraction. Plant Physiol. **53**, 918–927 (1974)

MCGIVAN, J.D., CHAPPEL, J.B.: Avenaciolide: a specific inhibitor of glutamate transport in rat liver mitochondria. Biochem. J. **116**, 37P (1969)

MCGIVAN, J.D., KLINGENBERG, M.: Correlation between H^+ and anion movement in mitochondria and the key role of the phosphate carrier. Eur. J. Biochem. **20**, 392–399 (1971)

MEIJER, A.J., BROUWER, A., REIJNGOUD, D.J., HOEK, J.B., TAGER, J.M.: Transport of glutamate in rat liver mitochondria. Biochim. Biophys. Acta **283**, 421–429 (1972)

MEIJER, A.J., DAM, K. VAN: The metabolic significance of anion transport in mitochondria. Biochim. Biophys. Acta **346**, 213–244 (1974)

MEIJER, A.J., GROOT, G.S.P., TAGER, J.M.: Effect of sulfhydrylblocking reagents on mitochondrial anion-exchange reactions involving phosphate. FEBS Lett. **8**, 41–44 (1970)

MEIJER, A.J., VIGNAIS, P.M.: Kinetic study of glutamate transport in rat liver mitochondria. Biochim. Biophys. Acta **325**, 375–384 (1973)

MOREAU, F., LANCE, C.: Isolement et propriétés des membranes externes et internes de mitochondries végétales. Biochimie **54**, 1335–1348 (1972)

MOREL, F., LAUQUIN, G., LUNARDI, J., DUSZYNSKI, J., VIGNAIS, P.V.: An appraisal of the functional significance of the inhibitory effect of long chain acyl CoAs on mitochondrial transports. FEBS Lett. **39**, 133–138 (1974)

MUNN, E.A.: The Structure of Mitochondria. London-New York: Academic Press 1974

ONISHI, T., KAWAGUCHI, K., HAGIHARA, B.: Preparation and some properties of yeast mitochondria. J. Biol. Chem. **241**, 1797–1806 (1966)

ONISHI, T., KRÖGER, A., HELDT, H.W., PFAFF, E., KLINGENBERG, M.: The response of the respiratory chain and adenine nucleotide system to oxidative phosphorylation in yeast mitochondria. Eur. J. Biochem. **1**, 301–311 (1967)

PACKER, L., MURAKAMI, S., MEHARD, C.W.: Ion transport in chloroplasts and in plant mitochondria. Ann. Rev. Plant Physiol. **21**, 271–304 (1970)

PALMIERI, F., PREZIOSO, G., QUAGLIARIELLO, E., KLINGENBERG, M.: Kinetic study of the dicarboxylate carrier in rat liver mitochondria. Eur. J. Biochem. **22**, 66–74 (1971)

PALMIERI, F., QUAGLIARIELLO, E., KLINGENBERG, M.: Kinetics and specificity of the oxoglutarate carrier in rat liver mitochondria. Eur. J. Biochem. **29**, 408–416 (1972b)

PALMIERI, F., STIPANI, I., QUAGLIARIELLO, E., KLINGENBERG, M.: Kinetic study of the tricarboxylate carrier in rat liver mitochondria. Eur. J. Biochem. **26**, 587–594 (1972a)

PANDE, S.V., BLANCHAER, M.C.: Reversible inhibition of mitochondrial adenosine diphosphate phosphorylation by long chain acyl coenzyme A esters. J. Biol. Chem. **246**, 402–411 (1971)

PANTER, R.A., MUDD, J.B.: Carnitine levels in some higher plants. FEBS Lett. **5**, 169–170 (1969)

PAPA, S., FRANCAVILLA, A., PARADIES, G., MEDURI, B.: The transport of pyruvate in rat liver mitochondria. FEBS Lett. **12**, 285–288 (1971b)

PAPA, S., LOFRUMENTO, N.E., KANDUC, D., PARADIES, G., QUAGLIARIELLO, E.: The transport of citric acid cycle intermediates in rat liver mitochondria. Eur. J. Biochem. **22**, 134–143 (1971a)

PAPA, S., LOFRUMENTO, N.E., LOGLISCI, M., QUAGLIARIELLO, E.: On the transport of inorganic phosphate and malate in rat liver mitochondria. Biochim. Biophys. Acta **189**, 311–314 (1969)

PARSON, D.F., WILLIAMS, G.R., THOMPSON, D., WILSON, D., CHANCE, B.: Improvement in the procedure for purification of mitochondrial outer and inner membrane. Comparison of the outer membrane with smooth endoplasmic reticulum. In: Mitochondrial Structure and Compartmentation (QUAGLIARIELLO, E., PAPA, S., SLATER, E.C., TAGER, J.M., eds.), pp. 29–68. Bari: Adriatica Editrice 1967

PARSON, D.F., YANO, Y.: The cholesterol content of the outer and the inner membranes of guinea pig liver mitochondria. Biochim. Biophys. Acta **135**, 362–364 (1967)

PASSAM, H.C., SOUVERIJN, J.H.M., KEMP, A., JR.: Adenine nucleotide translocation in Jerusalem artichoke mitochondria. Biochim. Biophys. Acta **305**, 88–94 (1973)

PFAFF, E., HELDT, H.W., KLINGENBERG, M.: Kinetics of the adenine nucleotide exchange. Eur. J. Biochem. **10**, 484–493 (1969)

PFAFF, E., KLINGENBERG, M., RITT, E., VOGELL, W.: Korrelation des unspezifisch permeablen mitochondrialen Raumes mit dem Intermembran-Raum. Eur. J. Biochem. **5**, 222–232 (1968)

PHILLIPS, M.L., WILLIAMS, G.R.: Anion transporters in plant mitochondria. Plant Physiol. **51**, 667–670 (1973)

QUAGLIARIELLO, E., PALMIERI, F.: Kinetics of substrate uptake by mitochondria. Identification of carrier sites for substrates and inhibitors. In: Biochemistry and Biophysics of Mitochondrial Membranes (AZZONE, G.F., CARAFOLI, E., LEHNINGER, A.L., QUAGLIARIELLO, E., SILIPRANDI, N., eds.), pp. 659–680. New York and London: Academic Press 1972

QUAGLIARIELLO, E., PALMIERI, F., PREZIOSO, G., KLINGENBERG, M.: Kinetics of succinate uptake by rat liver mitochondria. FEBS Lett. **4**, 251–254 (1969)

RAMSEY, R.R., TUBBS, P.K.: The mechanism of fatty acid uptake by heart mitochondria: An acylcarnitine—carnitine exchange. FEBS Lett. **54**, 21–25 (1975)

ROBINSON, B.H., WILLIAMS, G.R., HALPERIN, M.L., LEZNOFF, C.C.: Inhibitors of the dicarboxylate and tricarboxylate transporting systems of rat liver mitochondria. J. Membrane Biol. **7**, 391–401 (1972)

SCHERER, B., KLINGENBERG, M.: Demonstration of the relationship between the adenine nucleotide carrier and the structural changes of mitochondria as induced by ADP. Biochemistry **13**, 161–170 (1974)

SCHNAITMAN, C.A., GREENAWALT, J.W.: Enzymatic properties of the inner and outer membranes of rat liver mitochondria. J. Cell Biol. **38**, 158–175 (1968)

SLUSE, F.E., LIEBECQ, C.: Kinetics and mechanism of the exchange reactions catalyzed by the oxoglutarate translocator of rat heart mitochondria. Biochimie **55**, 747–754 (1973)

STOFFEL, W., SCHIEFER, H.G.: Biosynthesis and composition of phosphatides in outer and inner mitochondrial membranes. Z. Physiol. Chem. **349**, 1017–1026 (1968)

TYLER, D.D.: The inhibition of phosphate entry into rat liver mitochondria by organic mercurials and by formaldehyde. Biochem. J. **107**, 121–123 (1968)

VAARTJES, W.J., KEMP, A., JR., SOUVERIJN, J.H.M., VAN DEN BERGH, S.G.: Inhibition by fatty acyl esters of adenine nucleotide translocation in rat liver mitochondria. FEBS Lett. **23**, 303–308 (1972)

VIGNAIS, P.V., VIGNAIS, P.M., LAUQUIN, G., MOREL, F.: Binding of adenosine diphosphate and of antagonist ligands to the mitochondrial ADP carrier. Biochimie **55**, 763–778 (1973)

WEST, D.W., CHASE, J.F.A., TUBBS, P.K.: The separation and properties of two form of carnitine palmitoyl transferase from ox liver mitochondria. Biochem. Biophys. Res. Commun. **42**, 912–918 (1971)

WILLIAMSON, D.H., LUND, P., KREBS, H.A.: The redox state of free nicotinamide-adenine dinucleotide in the cytoplasm and mitochondria of rat liver. Biochem. J. **103**, 514–527 (1967)

WILLIAMSON, J.R.: Mitochondrial anion translocation and its role in the regulation of hepatic metabolism. In: Gluconeogenesis: Its regulation in mammalian species (HANSON, R.W., MEHLMAN, M.A., eds.), pp. 165–220. John Wiley and Sons: New York 1976

WILSON, R.H., HANSON, J.B., MOLLENHAUER, H.H.: Active swelling and acetate uptake in corn mitochondria. Biochemistry **8**, 1203–1213 (1969)

ZEBE, E., DELBRÜCK, A., BÜCHER, TH.: Über den Glycerin 1P Cyclus im Flugmuskel von Locusta migratoria. Biochem. Z. **331**, 254–272 (1959)

7. Interactions between Cytoplasm and Vacuole

PH. MATILE and A. WIEMKEN

1. Introduction

The emergence of ultrastructural plant cytology in the past twenty-five years entailed much uncertainy in the use of the term vacuole. It is therefore necessary to introduce the subject of this Chapter with some terminological considerations.

Originally the term vacuole designated structureless spaces within the plant protoplasm. In the case of parenchyma cells it unequivocally referred to the large central space containing an aequeous fluid, the *cell sap*. It was established that a membrane boundary exists between the vacuolar fluid and the cytoplasm. Since osmotic properties of plant cells were attributed to solutes accumulated in the cell sap and to the semipermeability of the vacuolar membrane, the latter was termed *tonoplast* (GUILLIERMOND et al., 1933; ZIRKLE, 1937; KÜSTER, 1951).

However, in the era of electron microscopy numerous additional cell spaces separated from the cytoplasm by a single membrane were discovered. The cytoplasm appeared to be interspersed with membrane-confined multiform enclaves, and the vacuoles turned out to be but comparatively voluminous components of an elaborate system of endomembranes. In Figure 1 this system is diagrammatically represented and the various developmental relationships between its morphologically discernible parts are shown. Evidently the meaning of the term vacuole has become vague and sometimes even controverisal. It is therefore necessary to adopt an adequate definition for both vacuole-like cell spaces and cytoplasm.

Such a definition can proceed from the fact that the protoplast of an eucaryotic cell contains in principle two aequeous phases which are separated by a single membrane. The *matrix* phase comprises nucleoplasm, cytoplasm, and the inner compartments of mitochondria and plastids; important properties of these matrices include the presence of protein synthesizing systems (ribosomes) and a normally granular and electron-dense appearance on electron micrographs. The other phase, the *vacuome,* contains the enchylema (the sap within the ER-cisternae) and the cell sap. It is composed of the spaces encircled by the membranes of the nuclear envelope, rough and smooth endoplasmic reticulum (rER, sER), Golgi membranes, microbody membrane, and the membranes of large central vacuoles and their small precursors or successors. In addition, the vacuome includes the spaces enclosed by the inner and outer membranes of mitochondria and plastids and the thylakoid compartment of chloroplasts. The vacuome can communicate with the space outside the protoplast, i.e. outside the plasmalemma. It is characterized by the absence of ribosomes and by its usually electron-transparent appearance. Spherosomes and oil bodies are outstanding in that they originate by inflation of the inner hydrophobic phase of ER-derived membranes.

Membranes surrounding the various components of the vacuome are ontogenetically related. Differentiations of this membrane system appear to follow several

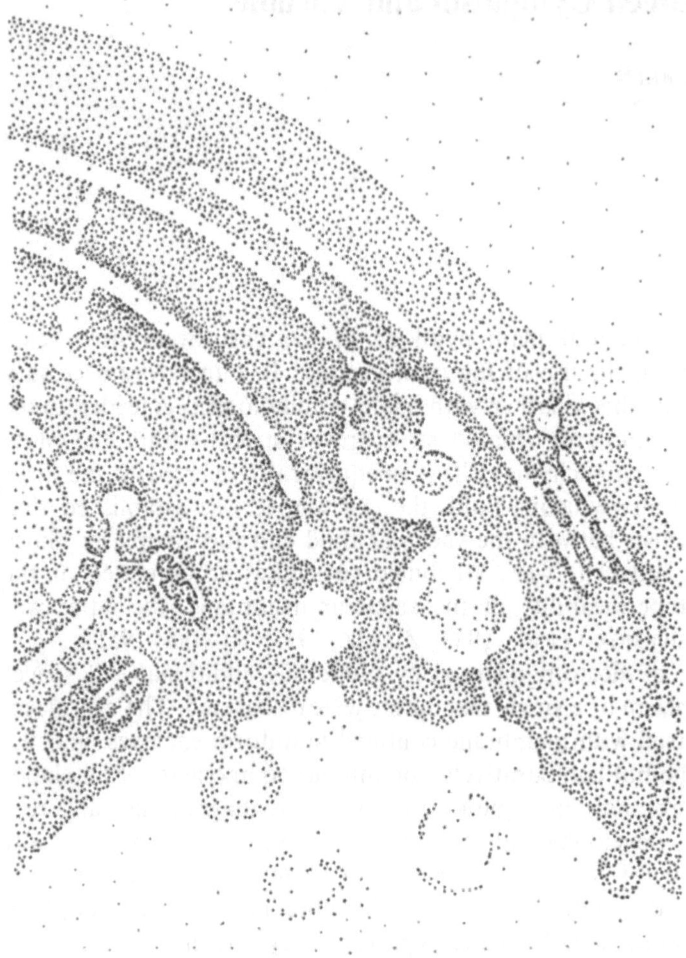

Fig. 1. Diagrammatic representation of the vacuome. It shows the developmental and dynamic features of the endomembrane system which forms the boundaries between the matrices and the vacuolar spaces. Examples of various interconnections within the vacuome are given. (See SCHNEPF, 1966; DE DUVE, 1969; BUVAT, 1971; BRACKER and GROVE, 1971)

distinct pathways, leading e.g. from the ER to Golgi membranes and plasmalemma, to microbody membranes and to tonoplasts. As a result of membrane differentiation the Golgi apparatus, lysosomes, microbodies (see Chap. II, 8) and various structurally less characteristic vacuoles assume distinct functional properties. It should be emphasized that the term vacuole as it is subsequently used is only operational and descriptive. It always implies the relevant interassociations and conversions within the system of endomembranes.

As comparatively little is known about the biochemical properties of tonoplasts, the functional features of the central vacuole and ontogenetically related components of the vacuolar system must be outlined on the basis of the chemical composition of cell saps. On the one hand, the cell sap is characterized by the presence of a variety of substances that are temporarily or definitively removed from the cytoplasm. On the other hand, the presence of a number of hydrolytic enzymes points to a function of vacuoles in intracellular digestive processes. Aleurone (protein) bodies which belong to the vacuolar system demonstrate the dual function of vacuoles quite strikingly: they represent the site of accumulation of storage proteins which, upon seed germination, are digested within the same compartment. Both functions, accumulation and digestion, suggest a passive nature of the vacuome. The active principles which control the function of vacuoles in the frame of cell metabolism are undoubtedly located in the cytoplasm. It will be seen that although a number of biochemical properties of vacuoles have been investigated, the mechanisms of interaction between cytoplasm and vacuole remain largely to be elucidated.

2. Vacuoles as Repositories of Substances

2.1 The Internal Environment of the Cytoplasm

The cytoplasm must be regarded as the truly living entity of the cell. Its structure and activity depend on a finely balanced interplay of macromolecules, micromolecules, and inorganic ions. Moreover, it is an open system in that it depends on a continuous exchange of substances and energy with the environment. Sudden changes of the conditions of the external environment threaten the subtle balances within the cytoplasm. Therefore, the vacuome constituting an internal environment under the control of the cell is most important for the maintenance of cytoplasmic activities under ever-changing external conditions.

The function of the vacuome in protecting the cytoplasm from possible stresses is evident when the special position held by vacuolar substances is considered. On the one hand they are separated by the vacuolar membranes from the main sites of synthesis and metabolic control and cannot therefore interfere directly with cytoplasmic processes. On the other hand the vacuolar substances are under the control of the cytoplasm. Independent of the external environment they can be stored or consumed according to the requirements of the cytoplasmic activities. Toxic secondary products of plants such as alkaloids or polyphenols can be immobilized within the vacuoles. The vacuome has, therefore, an important function for detoxifying the cytoplasm. The ability to store toxic products of metabolism and coloured compounds such as anthocyanins may also have important ecological functions. Last but not least, the accumulation of nutritive substances within the vacuome plays an important role in the physiology of germination of seeds.

2.2 Methods Used for Evaluating Vacuolar Compartmentation of Substances

The information available to date on the chemical composition of cell saps is largely based on a variety of attempts to evade the difficulties of isolating vacuoles. Direct analysis of cell saps is restricted either to coenocytic vesicles (e.g. *Chara, Nitella, Valonia*) the cell sap of which is comparatively easily sucked out by means of inserted microsyringes, or to vacuoles which can be isolated (see Vol. 2, Part A: Chap. 6.4.1.2).

2.2.1 Isolation of Vacuoles

Vacuoles that are liberated from osmotically lysed yeast spheroplasts are suitable for investigating the vacuolar enzymes (see Sect. 3.1.1); the stretching of the tonoplast which is associated with the osmotic lysis results, however, in a release of small molecules. Therefore, the rupture of yeast spheroplasts must be performed under isotonic conditions in order to preserve the vacuolar substances. Corresponding techniques based on metabolic lysis (INDGE, 1968a, b) or on mechanical distintegration (WIEMKEN and DÜRR, 1974) of spheroplasts have been developed. The isolation and purification of the vacuoles liberated from spheroplasts is achieved either by flotation or by sedimentation in media of appropriate densities (WIEMKEN and DÜRR, 1974).

The isolation of vacuoles from certain higher plant tissues will be discussed later (see Sect. 3.1.1). As far as compartmental analysis of micromolecules and inorganic ions is concerned, the latex from laticifers of certain species which are characterized by the presence of numerous, easily isolated vacuoles represents a convenient material (RIBAILLIER et al., 1971; MATILE et al., 1970). It has the disadvantage of originating from an extremely specialized tissue.

2.2.2 Stepwise Extraction of Cells

The plasmalemma of yeast cells (SCHLENK et al., 1970) as well as of higher plant cells (SIEGEL and DALY, 1966; RUESINK, 1971) is disrupted in the presence of macromolecular polycations. Under isotonic conditions the vacuolar membrane is unaffected. This difference in the susceptibilities of plasmalemma and vacuolar membrane can be used for extracting cells sequentially (SCHLENK et al., 1970; WIEMKEN and NURSE, 1973). In a first step of extraction yeast cells are treated with a macromolecular polycation, such as cytochrome c, in the presence of an osmotic stabilizer and at low ionic strength. This treatment causes the release of cytoplasmic pools of soluble substances. Upon subsequent hypotonic treatment (distilled water) the vacuoles are disrupted and the vacuolar pools are released.

2.2.3 Indirect Evidence from Isotope Kinetics

Overwhelming evidence for an uneven distribution of metabolites and inorganic ions within plant cells has been obtained from the analysis of kinetics of isotope labeling. These studies led to the view of cell compartmentation. Correlations between kinetic compartments and morphological cell spaces have been widely postulated (reviews by STEWARD and BIDWELL, 1966; BEEVERS et al., 1966; Oaks and BIDWELL, 1970; DAVIS, 1972; see also Vol. 2, Part A: Chaps. 5, 13).

The basic observation is that labeled metabolites are not in equilibrium with the total cellular pools. On the basis of isotope kinetics the existence of essentially two pools of metabolites can generally be distinguished. One of them is characterized by its small size and rapid turnover. In contrast to this *active metabolic pool,* a large pool appears to be remote from the sites of metabolic activities. This *inactive storage pool* was repeatedly thought to be localized in vacuoles; in the case of amino acids this could be verified directly with isolated vacuoles (see Sect. 2.3.1). Since the analysis of isotope kinetics or regulatory mechanisms has yielded similar results with regard to organic acids and sugars, as well as to inorganic ions such as nitrate (FERRARI et al., 1973; MARTIN, 1973) and phosphate (ULLRICH et al., 1965; see also BIELESKI, 1973), the storage pools of these substances are likely to be localized in vacuoles as are the storage pools of amino acids.

Compartmentation has also been inferred from the kinetics of transport of inorganic ions (review by LÜTTGE, 1973). Phenomena such as the biphasic Michaelis-Menten kinetics of ion uptake were interpreted to indicate the existence of two transporting systems differing in affinity. The high-affinity system is thought by some workers to be localized in the plasmalemma, the low-affinity system in the tonoplast. However, there is no general agreement about this interpretation (NISSEN, 1974). Concentrations of inorganic ions present in compartments which exhibit different rates of isotope exchange were calculated on the basis of efflux kinetics. The slowly exchanging pool is thought to be localized in the vacuoles (see LÜTTGE, 1973; and Vol. 2, Part A: Chaps. 5, 13; Part B: Chap. 3.2).

2.2.4 Electron Microscopy, Autoradiography, Electron Microprobe

Vacuolar depositions of various substances have been demonstrated using specific staining procedures for electron microscopy. This is exemplified by the following selected studies in which heavy metals were employed for obtaining the required contrast. Vacuolar Cl^- was detected in the form of precipitated AgCl (STEVENINCK and CHENOWETH, 1972), phosphate as lead phosphate (POUX, 1965), alkaloids as complexed with hexachloroplatinic acid (NEUMANN and MÜLLER, 1967) and tannins as precipitates with ferrous ion (DIERS et al., 1973; CHAFE and DURZAN, 1973). An outstanding inorganic vacuolar deposit is represented by barium sulfate in the statoliths of *Chara* rhizoids; its chemical identification was performed using electron diffraction (SCHRÖTER et al., 1975). Electron microscope work has also covered such crystalline vacuolar inclusions as Ca-oxalate (WATTENDORFER, 1969; SCHÖTZ et al., 1970).

An example of autoradiographical demonstration of vacuolar localization of alkaloids is provided by the work of NEUMANN and MÜLLER (1972). Methods used for the autoradiographical localization of water soluble substances have been extensively discussed by LÜTTGE (1972).

Promising possibilities fõr investigating the subcellular compartmentation of inorganic ions are provided by the technique of electron probe analysis (see LÄUCHLI, 1972). It has been used by PALLAGHY (1973) for assessing vacuolar and cytoplasmic concentrations of K^+ and Cl^- in freeze-substituted leaf sections of corn. This work shows that the scanning technique may be suited for the study of interactions between cytoplasm and vacuole.

2.2.5 Direct Microscopic Observations

It should be mentioned that the vacuolar location of a number of naturally coloured compounds can be directly observed in the light microscope (see KÜSTER, 1951). In addition, UV-absorbing compounds such as purine derivatives can be detected in yeast vacuoles using UV microscopy (SVIHLA et al., 1963). The use of vital stains has played an important role in investigation of the vacuome (see GUILLIERMOND, 1941). Cationic (basic) stains like neutral red accumulate in vacuoles in a purely physicochemical fashion and provide information about corresponding binding sites and the pH in cell saps (STADELMANN and KINZEL, 1972).

2.3 Substances Present in Vacuoles

2.3.1 Metabolic Intermediates

Intermediary products of metabolism such as organic acids, amino acids and sugars are frequently accumulated in large quantities. The acidity of saps squeezed from plant tissues is due to the presence of TCA-cycle intermediates and other acids. Coexistence of these acids and acid-sensitive cytoplasmic enzymes in the same cell space appears to be impossible. It is, therefore, generally assumed that acids formed in excess, as well as inhibitors such as malonic acid encountered in many species, are stored in a cell space remote from the sites of metabolic activity. In fact, discrete pools of intermediates which are not in free equilibrium with metabolic pools have repeatedly been described on the basis of isotope-labeling experiments (see BEEVERS et al., 1966; OAKS and BIDWELL, 1970). These pools are conceived as vacuolar storage pools. In the case of organic acids the acidity of cell saps, as demonstrated by natural indicators (anthocyanins) or vital indicator stains, strongly supports this view. Moreover, in rootlets the relative sizes of storage pools of organic acids are larger in the highly vacuolated proximal tissues than in the less vaculated tips (MACLENNAN et al., 1963). Kinetic studies on the release of ^{14}C-labeled malate from leaf slices of *Bryophyllum* provided indirect evidence for the vacuolar storage of the accumulated acid (KLUGE and HEININGER, 1973). The accumulation of malate in the dark, characteristic of the specific organization of photosynthesis in CAM plants, is a conspicuous feature of *Bryophyllum*. Problems of compartmentation bearing on the rhythmic accumulation and consumption of acids in leaf cells of CAM plants were discussed recently by QUEIROZ (1974). A direct demonstration of citric acid accumulation in the vacuolar compartment of laticifers of *Hevea* has been presented by RIBAILLIER et al. (1971); its concentration in isolated vacuoles is 24 times higher than in the cytoplasm, whereas about equal concentrations of malate were found in the two compartments.

Storage pools of amino acids undoubtedly exist in higher plant cells; their location in vacuoles has so far not been demonstrated directly (see e.g. AACH and HEBER, 1967). In the *Hevea* latex the bulk of free amino acids is present in the cytoplasmic fraction, however, basic amino acids are enriched in the vacuoles (BRZOZOWSKA et al., 1974). Abundant data is available with regard to vacuolar amino acid pools in fungal cells. INDGE (1968c) showed that crude preparations of vacuoles obtained from yeast spheroplasts contain a considerable proportion of the total amino acids. Using the technique of stepwise extraction, WIEMKEN and NURSE (1973) recovered large vacuolar storage pools. Moreover, the composition (accumulated exogenous and nitrogen-rich species of amino acids!) and isotope kinetics revealed the function as a storage pool. Storage of amino acids was also shown after isolation of the vacuoles from yeast spheroplasts (WIEMKEN and DÜRR, 1974). A similar storage pool is present in hyphae of *Neurospora,* from which vesicles containing more than 90% of the total ornithine and arginine were isolated by WEISS (1973).

Purines represent another group of metabolic intermediates which may be accumulated in yeast cells under certain nutritional conditions. If *Saccharomyces cerevisiae* is grown in the presence of methionine, S-adenosyl-methionine is synthesized; its vacuolar deposition can be observed in the UV microscope (SVIHLA et al., 1963) or assessed by sequential extraction of cells (SCHLENK et al., 1970). The low water solubility of compounds such as uric acid (Fig. 2C) and isoguanine explains the fact that their accumulation in *Candida utilis* results in the formation of large crystals (ROUSH, 1961).

On the problem of compartmentation of sugars in plant cells little information is available. In *Hevea* latex sugars were found to be localized mainly in the cytoplasm (RIBAILLIER et al., 1971). On the other hand there is considerable indirect evidence for the vacuolar deposition of sugars, particularly in tissues specialized in sugar accumulation (HEBER, 1957; HAWKER and HATCH, 1965; HUMPHREYS, 1973; see also OAKS and BIDWELL, 1970; Vol. 2, Part B: Chap. 5.3.1.1).

2.3.2 Reserve Substances

The most conspicuous among vacuolar reserve substances are the seed proteins which are stored in aleurone bodies. A fascinating further example of accumulation in vacuoles of a specific protein, chymotrypsin inhibitor protein I, has been detected in leaves of *Solanaceae* (tomato, potato). This protein is synthesized in large amounts under certain environmental conditions, and this is correlated with the appearance of electron-dense bodies in the vacuoles (RYAN and SHUMWAY, 1971; Fig. 2B). As exemplified by EDELMAN and JEFFORD (1968) fructosans (inulin) represent reserve substances which are deposited in vacuoles. Finally, lipids were detected in fungal vacuoles (e.g. PFEIFFER, 1963; SMITH and MARCHANT, 1968; BAUER and SIGARLAKIE, 1973) as well as in vacuoles of certain higher plant cells; at least some of these observations concern spherosomes that were taken up into vacuoles (see MATILE, 1975).

2.3.3 Secondary Products of Metabolism

The obvious vacuolar accumulation of coloured phenolic glycosides has already been mentioned. As demonstrated cytochemically at the level of electron microscopy, colourless phenolic compounds such as tannins are also perfectly compartmented in the vacuome (CHAFE and DURZAN, 1973; DIERS et al., 1973; BAUR and WALKINSHAW, 1974; see Fig. 2A). A special type of tannin vacuole is present in the contractile motor cells of the pulvini of *Mimosa pudica* (TORIYAMA and SATÔ, 1971; TORIYAMA and JAFFE, 1972).

Vacuolar localizations have also been reported for alkaloids. The observations concern specifically cells or tissues that are specialized in the accumulation of these compounds. In *Macleaya cordata* tissue cultures the yellow colour of idioblasts is caused by the presence of sanguinarine which, according to NEUMANN and MÜLLER (1967), is localized in the vacuoles of these specialized cells. Small vacuoles (alkaloidal vesicles) isolated from the capsule latex of *Papaver somniferum* contain almost all the morphine (FAIRBAIRN et al., 1974). It should, however, be pointed

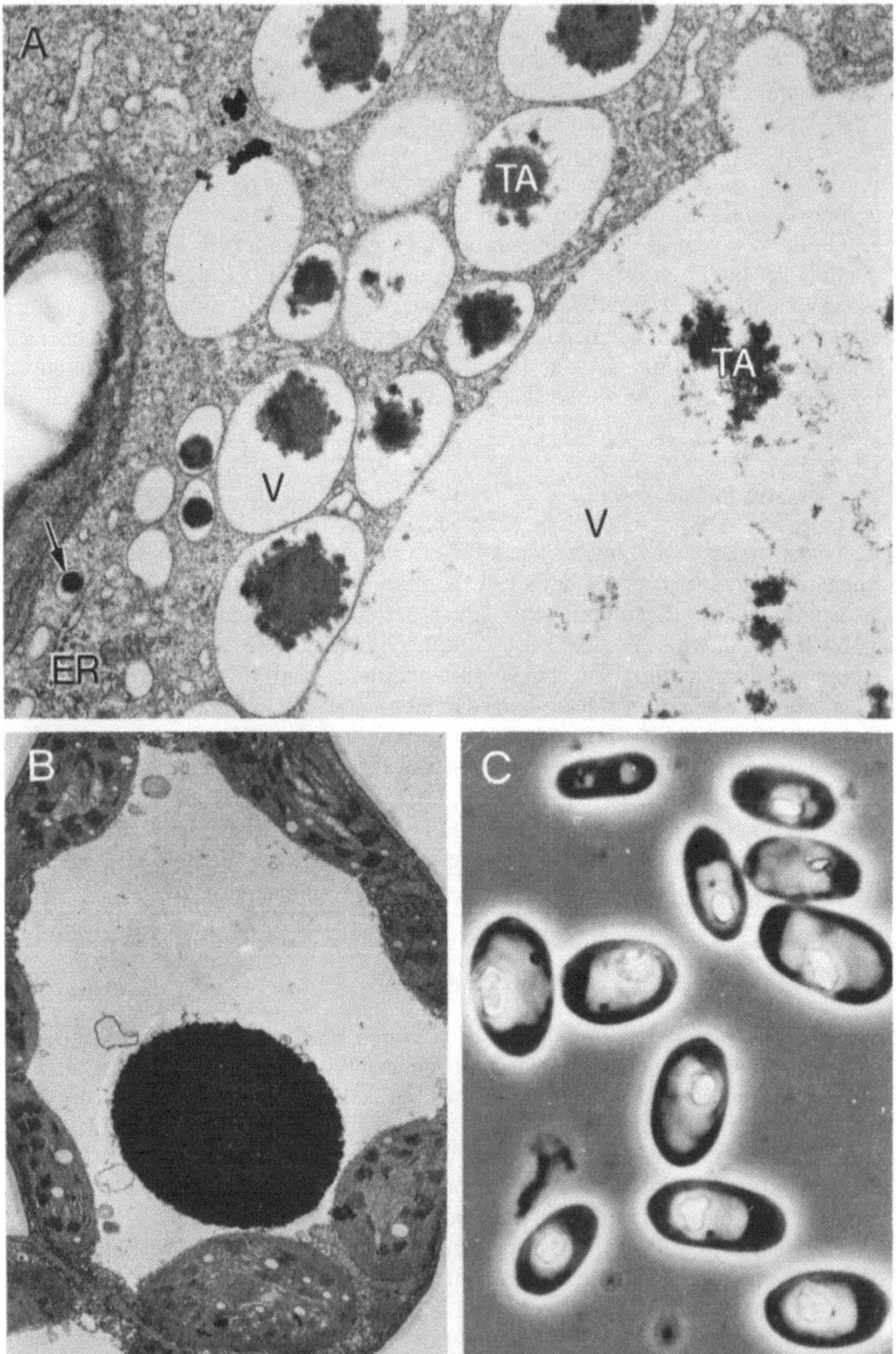

Fig. 2A–C

out that alkaloids may not always be localized exclusively in vacuoles; cell walls and plastids may represent additional sites of deposition (MÜLLER et al., 1971; NEUMANN and MÜLLER, 1972, 1974). Vacuoles containing isochinoline alkaloids can easily be isolated from latex of *Chelidonium majus* laticifers (MATILE et al., 1970). Of the four major alkaloids, sanguinarine and chelerythrine appear to be particularly enriched in these vacuoles (JANS, 1973). The corresponding data presented in Table 1 demonstrate that the accumulation of alkaloids in latex vacuoles of *Chelidonium* shows some specificity.

2.3.4 Inorganic Substances

The coenocytic vesicles of giant algal cells like *Valonia* and the internodes of *Nitella* lend themselves to direct analysis of mineral salts present in cell saps (review by STEWARD and SUTCLIFFE, 1959). The composition of these saps reveals a marked specificity of accumulation if vacuolar concentrations of K^+, Na^+, Mg^{2+}, Ca^{2+} and Cl^- are compared with the media. Similar specificities of vacuolar accumulation of K^+, Na^+ and Cl^- have been calculated on the basis of isotope kinetics in a variety of tissues (reviews by LÜTTGE, 1973 and in Vol. 2, Parts A and B).

RIBAILLIER et al. (1971) have determined quantitatively the distribution of some inorganic ions between the cytoplasm (serum) and vacuoles (lutoids) of *Hevea* latex. The ratios of apparent concentrations, based on the volumes of the two subcellular fractions concerned, exhibit considerable differences from one species of ion to another. In contrast to a 90-fold accumulation of Mg^{2+} in vacuoles, Ca^{2+} and Cu^{2+} are much less enriched; about equal apparent concentrations of K^+ were found in cytoplasm and vacuoles. Since *Hevea* vacuoles remain in the serum during cell fractionation, i.e. in their natural environment, the above data most likely correspond with the in vivo situation. Other inorganic constituents of *Hevea* vacuoles showing a conspicuous accumulation are acid soluble inorganic phosphates (RIBAILLIER et al., 1971). In yeast about 40% of this phosphorous fraction is present in a crude fraction of vacuoles prepared from spheroplasts (INDGE, 1968c). This vacuolar phosphorous is largely present in the form of polyphosphates (volutin) and is probably responsible for the binding of neutral red by yeast vacuoles (see GUILLIERMOND, 1941). The appearance of polyphosphate granules in vacuoles of *Scenedesmus* is correlated with phosphate uptake during the dark period of the growth cycle (SUNDBERG and NILSHAMMAR-HOLMVALL, 1975). The location of large storage pools of phosphate in vacuoles has also been inferred from indirect compartmental analysis (ULLRICH et al., 1965; BIELESKI, 1973). Although it is a self-evident fact it should be mentioned that the most abundant inorganic constituent of cell saps is water, the solvent of vacuolar solutes.

Fig. 2A–C. Vacuolar deposits. (A) Portion of a cultured white spruce cell showing electron dense tannin materials *TA* in vacuoles *V*. Arrow marks tiny tannin containing vacuole which may have arisen from the endoplasmic reticulum *ER*. ×26,500. (Courtesy of D.J. DURZAN.) (B) Vacuolar protein body in a tomato leaf cell. The appearance of vacuolar electron dense inclusion bodies is associated with the accumulation of chymotrypsin inhibitor protein *I* in the leaves. ×5,500. (Courtesy of L.K. SHUMWAY and C.A. RYAN.) (C) Crystals of uric acid in vacuoles of *Candida utilis* cells grown in the presence of uric acid. ×5,000

2.4 Mechanisms of Transport in Vacuoles

2.4.1 Transport of Ions

A vast and explosively expanding literature is concerned with the uptake mechanisms of inorganic ions in plant tissues. This topic is considered in detail in Vol. 2, Parts A and B. As the cytoplasm of a differentiated plant cell forms only a thin layer around the large vacuolar compartment, plant physiologists were always aware of the necessity to consider the influence of fluxes across the tonoplast upon the overall fluxes into and out of the cells. Therefore, the complex kinetic data obtained in experiments on ion uptake or isotope exchange were analyzed and interpreted with regard to plant cells representing a system of essentially two membrane-bound compartments. Consequently, two transport mechanisms were assumed to be located in the plasmalemma and tonoplast. It is surprising that the possible influence of the vacuoles on transport in primitive plants such as fungi is generally not considered (but see Vol. 2, Part A: Chap. 7).

As several excellent reviews appeared recently on this subject (Steward and Mott, 1970; MacRobbie, 1971; Epstein, 1973; Lüttge, 1973; Nissen, 1974; Vol. 2, Part A: Chap. 5; Part B: Chap. 3.2) only the most important facts and viewpoints of these indirect kinetic assessments of possible transport mechanisms across the tonoplast will be summarized here. If Michaelis-Menten formalism is applied to the process of ion absorption or ion exchange in plant tissues, it appears that at least two systems for ion transport are present. One system seems to involve transport mechanisms (carriers) with a high affinity for the ions ($Km < 1$ mM) and a low maximal velocity, whereas the mechanisms of the other system are characterized by a comparatively low affinity ($Km > 1$ mM) and a high maximal velocity. The two systems also differ in their specificity for the ions. The ideas about the location of these carriers are controversial. One group of researchers favors the idea that the two systems operate in series; the high-affinity system is thought to be located in the plasmalemma and the low-affinity system in the tonoplast (see Laties, 1969; Lüttge, 1973). This hypothesis is based on the assumption that ions can passively diffuse into the cytoplasm, bypassing the high-affinity system (plasmalemma), if the external concentration is high. Consequently, the low-affinity system residing in the tonoplast can be recognized at high external concentrations because, under these conditions, it is rate-limiting for the measured overall uptake by the cells. A comparison of the characteristics of ion uptake by highly vacuolated and less vacuolated root cells as well as a number of other investigations and considerations appear to support this hypothesis (see Lüttge, 1973).

Another group of investigators advocates the idea of a parallel operation of both transport systems across the plasmalemma (see Epstein, 1973). Evidence is presented that the plasmalemma is not permeable to inorganic ions to any marked degree even at high concentrations, and that ions are taken up from the medium only via transport mechanisms. Hence vacuolar transport mechanisms are never rate-limiting for the entry of ions into the cells and cannot therefore be explored by analyzing the kinetics of ion uptake.

Careful analysis of a large amount of old and new data has recently provided evidence for the existence of a single but multiphasic transport mechanism at the plasmalemma and probably a similar mechanism at the tonoplast (Nissen, 1974). This short outline of the debate about different models of uptake mechanisms for inorganic ions serves merely to show the difficulties encountered in the indirect exploration of vacuolar transport by the analysis of uptake kinetics assessed in whole cells or tissues.

In the giant coenocytic algae the ion concentrations of the vacuolar sap as well as the fluxes and electropotentials across the tonoplast can be directly measured (GUTKNECHT and DAINTY, 1968; MACROBBIE, 1971). On the basis of these data it is possible to distinguish between passive flux of ions across the membrane and active flux which is directed against an electrochemical gradient (potential). It turned out that several ions can be transported actively into vacuoles. Thus the tonoplast must be equipped with some sort of pumps which are driven by metabolic energy. Several principles of transformation of metabolic energy into the pumping of solutes against gradients across biological membranes are currently being discussed: electrogenic pumps (HIGINBOTHAM and ANDERSON, 1974), proton transport (RAVEN and SMITH, 1974), and the involvement of pinocytic (vesicular) transport (BAKER and HALL, 1973; see also reviews in Vol. 2, Parts A and B).

It is doubtful whether the overall concentrations of ions measured in the cytoplasmic compartment are meaningful with regard to concentration gradients, as a large proportion of the water present in the cytoplasmic matrix is probably structurized and therefore not available as a solvent. Current studies show that the cytoplasm must be conceived as an integrated structure of both macromolecules and micromolecules including inorganic ions and water (LING, 1969; WIGGINS, 1971; DAMADIAN, 1974). The high degree of organization within the cytoplasm is also indicated by the phenomenon of metabolic channeling which must be caused by an uneven distribution of metabolites (see DAVIES, 1972). The presence of cytoplasm therefore introduces enormous difficulties with regard to studies on transport across the tonoplast. Hence isolated vacuoles, that is the replacement of the cytoplasmic "black box" by a simple solution, may provide interesting experimental possibilities.

Vacuoles with intact, tight tonoplasts can be isolated from yeast spheroplasts (see Sect. 2.2.1). They provide direct access to the tonoplast from the cytoplasmic side and were recently used for the study of arginine transport (BOLLER et al., 1975). Arginine represents one of those amino acids which in yeast cells can be highly accumulated within vacuoles (WIEMKEN and DÜRR, 1974). These large storage pools of arginine are retained in the isolated vacuoles. Upon the addition of arginine to the medium, a specific exchange reaction between vacuolar and external arginine takes place. This reaction exhibits saturation kinetics with an apparent *Km* of 30 μM. It is inhibited competitively by D-arginine, L-histidine and L-canavanine but not by L-lysine, L-ornithine and other amino acids. In contrast, the arginine uptake system of the spheroplasts (presumably located in the plasmalemma) is characterized by a lower *Km* (1 μM) and by its inhibition in the presence of L-ornithine and L-lysine. Whilst the tonoplast system in vitro mediates an exchange of arginine without the need for energy, the plasmalemma system catalyzes an energy-dependent vectorial transport across the membrane. From the pattern of inhibition caused by a variety of arginine analogs it was concluded that the vacuolar transport system recognizes more specifically the guanidino group, whereas the plasmalemma system is more specific for the L-α-amino group (BOLLER et al., 1975). These results demonstrate that specific systems (carriers) mediate the transport across the tonoplasts. However, they do not explain the accumulation of compounds such as arginine within the yeast vacuole.

Ions can be accumulated to a certain extent without being pumped actively into vacuoles. As vacuoles often contain high molecular weight polyanionic substances such as polyphosphates or tannins (pp. 261–263), cations can be accumulated by ionic binding to these cation-exchangers (which cannot permeate the tonoplast). Furthermore, if pH differences exist between the cell sap and the cytoplasm, the vacuoles can act as ion traps. The permeability of biological membranes is generally higher for the undissociated form of weak electrolytes than for the ionized form. As the vacuoles are normally more acid than the cytoplasm they can act as traps for weak bases. This mechanism appears to be responsible for vital staining of vacuoles with the basic stain neutral red. This compound permeates the tonoplast at neutral pH as an uncharged hydrophobic molecule. It is ionized in the more acid cell sap and trapped in this form. Consequently, accumulation of neutral red proceeds until the pH on both sides of the tonoplast is equalized.

A key problem to be solved in this context concerns the maintenance of pH gradients between cytoplasm and cell sap and the mechanisms responsible for the vacuolar deposition of high molecular weight ion exchanger molecules. In the case of ion exchangers a synthesis in the tonoplast and vectorial release into the cell sap could be involved. A similar process has also been proposed for the vacuolar accumulation of uncharged macromolecules. For example, the enzyme which catalyzes the transfer of fructose from a trisaccharide to a growing chain of inulin was said to be located in the tonoplast (EDELMAN and JEFFORD, 1968). As to the maintenance of pH gradients, it is likely that active pumping mechanisms, e.g. proton-translocating ATPases, are involved (see RAVEN and SMITH, 1974; Vol. 2, Part A: Chap. 12).

2.4.2 Accumulation of Products of Secondary Metabolism

The mechanisms involved in the vacuolar deposition of secondary products of metabolism appear to depend on the class of excretory product concerned. To begin with the isoprenoids, it is interesting to note that rubber globules present in mature laticifers may be localized exclusively in the cytoplasm (*Papaver, Taraxacum, Hevea*: DICKENSON, 1964; SCHULZE et al., 1967) or e.g. in the genus *Euphorbia*, both in vacuoles and cytoplasm (SCHULZE et al., 1967; MARTY, 1968; SCHNEPF, 1964). It is obvious that there is no need for a specific compartmentation of the metabolically inert polyterpene. The transfer of rubber particles, which originate in the cytoplasm, to the vacuolar cell space must therefore be interpreted as a rather incidental event. In fact, penetrating studies on the differentiation of laticifers of *Euphorbia characias* undertaken by MARTY (1971a) revealed that the sequestration of rubber globules in the large vacuole is the result of an extensive autophagic activity associated with vacuolation.

As for the mechanism involved in the vacuolar accumulation of alkaloids, it is noteworthy that isolated vacuoles of *Chelidonium* latex absorb sanguinarine which is added to the suspending medium (Fig. 3; MATILE et al., 1970). This absorption is associated with the simultaneous release of dihydrocoptisine (JANS, 1973). As neither a requirement for metabolic energy nor a temperature dependency which would typify a catalyzed transport of sanguinarine across the tonoplast could be detected, it appears that a purely physical exchange mechanism is responsible for the sanguinarine accumulation observed. Such an ion exchange mechanism

Table 1. Localization of some alkaloids in vacuoles of *Chelidonium majus* latex. (Data from JANS, 1973)

Alkaloid	μg/ml latex			Ratio A/B
	Complete latex	Vacuoles (A)	Latex minus vacuoles (B)	
Sanguinarine	332	365	64	5.71
Chelerythrine	174	139	21	6.62
Berberine	218	128	77	1.66
Dihydrocoptisine	1,100	704	611	1.15

would require the existence of vacuolar anions which form non-diffusible salts with alkaloid cations. According to JANS (1973) the isolated *Chelidonium* vacuoles are, in fact, rich in chelidonic acid and phenolic compounds; in vitro these substances form precipitates with isochinoline alkaloids. If a precipitate of berberine and tannin is wrapped in a dialysis bag and exposed to a solution of sanguinarine, the latter alkaloid is readily accumulated in the model vacuole and this is accompanied by the release of berberine (MATILE, unpublished results). Hence sanguinarine is able to displace berberine. Different affinities of alkaloids for the vacuolar anions appear to be responsible for the different degrees of vacuolar accumulation of alkaloids as observed in the latex of *Chelidonium* (Table 1). It is interesting to note that the highly accumulated alkaloids, sanguinarine and chelerythrine, are poisonous in the sense that tonoplasts are destroyed in the presence of free alkaloid. As shown in Figure 3, isolated vacuoles are lysed as soon as their capacity to absorb the alkaloid is exceeded. In contrast, vacuoles are perfectly stable in the presence of berberine and dihydrocoptisine, which are characterized by a comparatively low degree of vacuolar accumulation (JANS, 1973).

The movement of exogeneously supplied alkaloids to the sites of accumulation is largely independent of metabolic energy, though the cellular concentration eventually exceeds the external concentration (NEUMANN and MÜLLER, 1972). In conclusion, detoxification in the case of alkaloids appears to be due to the physicochemical properties of alkaloids and their vacuolar counterions, rather than to specific transporting functions associated with vacuolar membranes.

The problem to be considered in the context of an ion exchange mechanism for alkaloid accumulation in vacuoles obviously concerns the mechanism responsible for the proper compartmentation of alkaloid-binding vacuolar compounds.

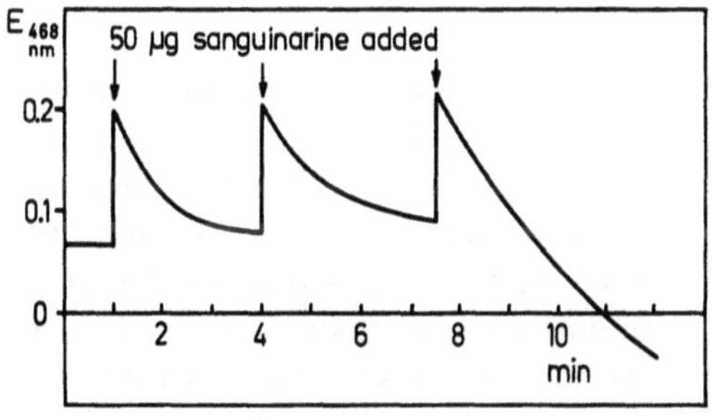

Fig. 3. Absorption of sanguinarine by isolated latex vacuoles of *Chelidonium majus*. Cuvettes of double beam spectrophotometer contained 3.0 ml of a suspension of vacuoles and 2.9 ml with same amount of vacuoles. 50 μg of sanguinarine dissolved in 0.1 ml sorbitol medium were added as indicated. After third addition of alkaloid, vacuoles begin to lyse and decreasing turbidity in sample cuvette causes rapid drop in optical density. (From MATILE et al., 1970)

Cytochemical investigations performed with cultured cells of conifers suggest that the synthesis of tannins is localized in the ER and in small vacuoles derived therefrom, the products being released directly into the vacuome (Fig. 2A; CHAFE and DURZAN, 1973; BAUR and WALKINSHAW, 1974). Biochemical data on the association of enzymes involved in the synthesis of phenolic compounds with microsomal membranes (e.g. AMRHEIN and ZENK, 1971; ALIBERT et al., 1972) support these morphological observations. Hence the accumulation of phenolic compounds in the vacuoles may be a consequence of the specific localization of their synthesis in the endomembrane system of plant cells.

3. Vacuoles as Digestive Compartments

The detection of latent, membrane-bound hydrolases in homogenates of rat liver has led to the concept of *lysosomes* (see DE DUVE, 1969). This concept states that unspecific hydrolases such as proteolytic enzymes, nucleases, and phosphatases are localized in a distinct cell compartment, intracellular digestive processes thus being spatially separated from the cytoplasm. In animal cells lysosomes represent a polymorphous class of small vacuole-like organelles with diameters in general not exceeding 1 μm. Attempts to investigate the existence of analogous organelles in plant cells have yielded considerable evidence favoring the lysosomal nature of vacuoles. (Reviews on the lysosomal system of plant cells by GAHAN, 1973; PITT, 1975; MATILE, 1969, 1975.)

3.1 Hydrolases Localized in Vacuoles

3.1.1 Remarks on Methods Employed

At first sight, cytochemical techniques appear to be ideally suited for investigating the presence or absence of enzymes in cell compartments or membranes. Unfortunately, safe cytochemical methods, particularly at the level of electron microscopy, are available for only a few hydrolases.

Using the Gomori technique for detecting phosphatases, cytochemists have studied mainly the cellular localizations of these hydrolases. At the level of light microscopy a granular, occasionally reticular distribution of acid-unspecific phosphatase as well as of certain other hydrolases was observed in various higher plant and fungal cells (see reviews by GAHAN, 1967; REISS, 1973; PITT, 1975; MATILE, 1975).

The identity of such hydrolase-positive granules with vacuoles is evident from various cytochemical investigations at the ultrastructural level (e.g. POUX, 1963a, 1970; CATESSON and CZANINSKI, 1968; BERJAK, 1968; HALPERIN, 1969; VINTÉJOUX, 1970; COULOMB, 1971a; HALL and DAVIE, 1971; SEXTON et al., 1971; FIGIER, 1972; CRONSHAW and CHARVAT, 1973).

Immunocytochemistry may provide a promising extension of cytochemical hydrolase localizations to enzymes whose activity is difficult to transmute specifically into electron dense precipitates. A corresponding attempt to localize RNase (an acid endonuclease) in a highly vacuolated tissue has recently yielded evidence for the presence of this enzyme in the cell sap (BAUMGARTNER, 1975). Immunocytochemical methods have the considerable disadvantage of requiring the laborious purification of the enzyme proteins in question. In any case, a more detailed analysis can be carried out using preparations of isolated and purified vacuoles.

The size and fragility of large vacuoles poses considerable technical problems with regard to their liberation from cells. It is therefore not surprising that successful isolations of higher plant vacuoles are restricted either to small vacuoles of meristematic cells (MATILE, 1966, 1968) or to laticifers of certain species which contain large numbers of small vacuoles instead of a single large one (PUJARNISCLE, 1968). Differential and isopycnic gradient centrifugation in suitable media were employed for separating vacuoles from the other protoplasmic constituents. Isolations of vacuoles from parenchyma cells are possible, on principle, as shown by COCKING (1960); large-scale isolations with the aim of investigating their lysosomal nature have not been carried out so far.

The circumstance which renders the liberation of large vacuoles more difficult is the rigidity of plant cell walls. The drastic mechanical procedures required to disintegrate cells inevitably result in the disrupture of central vacuoles. Protoplasts (spheroplasts) obtained through the action of cell wall lytic enzymes appear to represent, therefore, a convenient material for liberating vacuoles by the use of more gentle procedures. Indeed, osmotically stabilized yeast spheroplasts can be lysed, e.g. by hypotonic treatments, so that intact vacuoles are liberated with a high yield (MATILE and WIEMKEN, 1967, 1974; WIEMKEN and NURSE, 1973). Alternatively, metabolic lysis of yeast spheroplasts has been employed (INDGE, 1968a; MATERN et al., 1974; HASILIK et al., 1974). For separating vacuoles from the other cell constitutents centrifugations in appropriate media (sorbitol, sucrose, Ficoll) are generally used. For example, yeast vacuoles obtained from osmotically lysed spheroplasts are isolated by flotation in Ficoll media.

Yet another technique for investigating hydrolase localizations in vacuoles is stratification of organelles, which is achieved by centrifuging whole cells or tissues. In subcells prepared from stalks of *Acetabularia* a separation of vacuole and cytoplasm can be observed after centrifugation at low speed; the subcellular distribution of hydrolases can subsequently be determined by analyzing the vacuolar and cytoplasmic moieties separated from each other (LÜSCHER and MATILE, 1974).

3.1.2 Vacuoles as Lysosomes

Hydrolases that were discovered in preparations of vacuoles isolated from various tissues and cells are compiled in Table 2. One of the most thoroughly investigated plant lysosomes is the yeast cell vacuole (Table 3). It appears that all classes of hydrolases, peptidases, esterases and glycosidases, are present in isolated plant vacuoles. Hence vacuoles are equipped with the enzymic tools for the potential self-digestion of cells.

The postulation of the lysosomal nature of vacuoles requires the demonstration of *latent* hydrolase activities, i.e. the substrates should not be attacked by the hydrolases unless the membranes of isolated vacuoles are disrupted, e.g. in the presence of detergent or after ultrasonication. If the isolated structures are stable

Table 2. Lysosomal nature of isolated vacuoles

Object	Hydrolases localized	References
Zea mays, root meristem	Proteinase, exopeptidases, RNase, DNase, phosphatase, phosphodiesterase, acetylesterase, β-amylase, α- and β-glucosidase, β-galactosidase	MATILE (1965, 1968), unpublished results
Pisum sativum, rootlets	Phosphatase, RNase, α-amylase	NAKANO and ASAKI (1972) HIRAI and ASAKI (1973)
Cucurbita pepo, root meristem	Phosphatase	COULOMB (1968)
Asplenium fontanum, meristematic fronds	Proteinase, RNase, DNase, phosphatase, phosphodiesterase, β-galactosidase	COULOMB (1971 b)
Solanum tuberosum, young dark grown shoots	RNase, phosphatase, phosphodiesterase, acetylesterase	PITT and GALPIN (1973)
Lycopersicon esculentum, fruit	Proteinase, RNase, phosphatase	HEFTMANN (1971)
Hevea brasiliensis, latex	Proteinase, RNase, DNase, phosphatase, phosphodiesterase, β-glucosidase, β-galactosidase, β-N-acetyl-glucosaminidase	PUJARNISCLE (1968)
Chelidonium majus, latex	Proteinase, RNase, phosphatase	MATILE et al. (1970)
Polysphondylium pallidum	Phosphatase, RNase, DNase, proteinase, amylase, β-N-acetyl-glucosaminidase	O'DAY (1973)
Dictyostelium discoideum, starved myxamebae	Phosphatase, β-N-acetyl-glucosaminidase, α-mannosidase	WIENER and ASHWORTH (1970)
Acetabularia mediterranea, subcells, prepared from stalks	RNase, phosphatase (Stratification)	LÜSCHER and MATILE (1974)
Nitella axilliformis, *Chara braunii*, internodes	Phosphatase, carboxypeptidase (Separation of cell sap)	DOI et al. (1975)
Coprinus lagopus, vegetative hyphae, fruiting bodies	Proteinases, RNase, phospatase, β-glucosidase, chitinase	ITEN and MATILE (1970)
Saccharomyces cerevisiae, spheroplasts	see Table 3	
Neurospora crassa, conidia	Proteinases, RNase, phosphatase, β-glucosidase, chitinase phosphatases, invertase	MATILE (1971)

under the conditions of the enzyme assay, the latency of vacuolar hydrolases can, in fact, be demonstrated (e.g. MATILE, 1971; PITT and GALPIN, 1973).

It should be pointed out that technical difficulties do not allow a quantitative analysis of hydrolase distributions in plant cells. The presence of these enzymes in the soluble fraction of tissue homogenates or of lysates of yeast spheroplasts

Table 3. Activities of enzymes (per unit protein) present in isolated yeast vacuoles compared with activities in total lysate of spheroplasts. Activities in lysate taken to be 1. (Compiled from MATILE and WIEMKEN, 1967; WIEMKEN, 1969; VAN DER WILDEN et al., 1973; MEYER and MATILE, 1975; LENNEY et al., 1974)

Protease A		20.7
Protease B		40.2
Carboxypeptidase		24.0
Aminopeptidase		7.7
Ribonuclease		19.2
Invertase		20.1
α-Mannosidase		20.0
Exo-β-Glucanase		27.8
Alk. Phosphatase (Mg^{2+})	(derepressed cells)	>40.0
Acid Phosphatase (Co^{2+})	(derepressed cells)	15.0
ATPase (Mg^{2+})	(membrane associated enzyme)	9.0
$NADH_2$-dichlorophenol-indophenol-oxidoreductase	(mitochondrial and microsomal enzyme)	0.638
$NADH_2$-cytochrome c-oxidoreductase	(mitochondrial and microsomal enzyme)	0.016
Cytochrome-c-oxidase	(mitochondrial enzyme)	< 0.01
Succinate-dehydrogenase	(mitochondrial enzyme)	< 0.01
Glucose-6-phosphate-dehydrogenase	(soluble enzyme)	< 0.01
Ethanol-dehydrogenase	(soluble enzyme)	< 0.01
α-Glucosidase	(soluble enzyme, derepressed cells)	< 0.01

may suggest their occurrence in the cytoplasmic matrix. However, the liberation of vacuoles is always associated with the rupture of a fraction of the population of these fragile organelles, causing the release of enzymes which actually were dissolved in the cell sap. In addition, a number of hydrolases, especially glycosidases and acid phosphatase, have a dual localization in vacuoles and in the space outside the plasmalemma; upon homogenization of tissues these secreted enzymes may be released from the cell walls and appear in the soluble fraction. The dual location of hydrolases such as acid phosphatase in vacuoles and in cell walls is documented by a large number of cytochemical observations (see MATILE, 1975). On the other hand, the lysosomal concept is convincingly confirmed by the fact that reaction products of acid phosphatase were never observed in the cytoplasmic matrix. However, intracellular hydrolases were localized cytochemically in vacuoles which belong to various different parts of the endomembrane system encircling the vacuome. It is, therefore, not surprising that these enzymes are also present in microsomal fractions of tissue homogenates which contain various small components and fragments of this membrane system (e.g. MATILE, 1968; COULOMB, 1971b; NAKANO and ASAKI, 1972; PITT and GALPIN, 1973). Hence the view that vacuoles represent the principal intracellular location of hydrolases appears to be justified.

Hydrolases may be located differentially within vacuoles. The reaction product of acid phosphatase appears to be deposited frequently toward the tonoplast. Indeed, a considerable fraction of this enzyme is associated with membranes obtained from preparations of root cell vacuoles (PARISH, 1975). Structural latency of vacuolar acid phosphatase, as demonstrated by PITT and GALPIN (1973), strengthens this point. RNase is almost completely solubilized upon the disrupture of isolated yeast vacuoles, whereas α-mannosidase seems to be a constituent of the tonoplast (VAN DER WILDEN et al., 1973).

Another interesting aspect of differential localization concerns the specific inhibitor proteins of proteases present in yeast and other fungal cells. In baker's yeast these inhibitors are absent from purified vacuoles (LENNEY et al., 1974; MATERN et al., 1974; HASILIK et al., 1974); they are probably located in the cytoplasmic matrix. It appears then that proteases, and probably the other vacuolar hydrolases as well, are present in plant cells in active form.

3.2 Autophagy

3.2.1 Turnover

Since the hydrolases of the vacuome are separated from their potential substrates by membranes, it might be envisaged that these enzymes are destined eventually to break down the protoplasm after the decay of tonoplasts upon cell death. *Autolysis* of cells is, however, only one aspect of the lytic compartment. An important reference to the action of vacuolar hydrolases in the living cells can be deduced from the well-known metabolic lability of cellular proteins and cytoplasmic nucleic acids. Turnover of these biopolymers must be regarded as the result of simultaneous synthesis and degradation within living cells. The demonstration of protein catabolism implies the existence of processes that result in the exposure of cellular protein to the vacuolar peptidases. *Autophagy* is the adequate term for this phenomenon.

Assessments of protein turnover were carried out employing various isotope-labeling techniques (review by HUFFAKER and PETERSON, 1974). Attempts to measure rates of protein synthesis and degradation individually are complicated by unknown interactions between distinct pools of amino acids, e.g. shifting of amino acids from protein catabolism to the protein precursor pool. A most sophisticated technique for estimating net protein synthesis was based by TREWAVAS (1972a) on the assessment of radio-activities of an exogeneously supplied amino acid in the immediate protein precursor pool, amino-acyl-tRNA. This strategy allowed the calculation of fluxes of methionine in and out of protein. In *Lemna* apparent half-lifes of total protein of about seven days were found when the plants were growing fast; turnover rates were increased when the plants were kept on a suboptimal medium, and rates of protein degradation were particularly high when growth was stopped by placing *Lemna* on distilled water (TREWAVAS, 1972b).

Similar correlations between protein turnover rates and developmental conditions (e.g. STEWARD and BIDWELL, 1966; HALVORSON, 1960) suggest that protein degradation is chiefly associated with cell differentiation. This aspect is particularly obvious in senescent plant organs. In the fading corolla of the morning glory *Ipomoea* the synthesis of certain enzymes occurs during a period of rapid decline of the total protein content (MATILE and WINKENBACH, 1971; WIEMKEN and WIEMKEN, 1975; BAUMGARTNER et al., 1975). Senescence appears, therefore, as a cell differentiation process in which protein degradation exceeds protein synthesis.

Penetrating studies on the turnover of individual species of biopolymers are rare. They reveal, however, that the rates of turnover may be different for individual proteins. For example, in cultured tobacco cells two metabolically closely related enzymes, nitrate and nitrite reductase, are characterized by remarkably different half-lifes: nitrate reductase appears to be a short-lived enzyme with a half-life of only 6 h, whereas nitrite reductase has a half-life as long as 124 h (ZIELKE and FILNER, 1971; KELKER and FILNER, 1971). These studies also reveal the significance of protein turnover in the regulation of metabolic activity according to the nutritional conditions.

It is obvious that the digestive processes which are touched on in studies on turnover cannot be understood completely in terms of substrate (digested proteins) and enzymes (peptidases) involved. The decisive element responsible for the organized breakdown within the living cell is undoubtedly the membrane separating cytoplasmic matrix and vacuome. The importance of membranes is convincingly shown in a study by PAYNE and BOULTER (1974) in which degradation of cytoplasmic ribosomal RNA in senescent broad-bean cotyledons is compared with degradation in the homogenized tissue. The abolition of compartmentation associated with tissue grinding, i.e. the liberation of RNase from lysosomal structures, results in a remarkably accelerated and unorganized digestion of ribosomal RNA as compared with the autophagic breakdown.

3.2.2 Mechanisms of Autophagy

Electron micrographs of plant cells demonstrate as a rule that the vacuoles are not empty in the sense that they are devoid of structured material. A number of authors have explicitly pointed out the fact that membranes and even recognizable organelles such as mitochondria appear to be suspended in the cell sap (e.g. SIEVERS, 1966; THORNTON, 1968; WARDROP, 1968; HALPERIN, 1969; ZANDONELLA, 1970; MATILE and WINKENBACH, 1971; GEZELIUS, 1972; CRONSHAW and CHARVAT, 1973). This phenomenon was first interpreted in terms of degradative processes taking place in vacuoles by POUX (1963b). There is little doubt that the vacuolar inclusions represent the morphological counterpart of what can be biochemically described as intracellular digestion. This view is strongly supported by the fact that isolated vacuoles contain the hydrolases required for the degradation of cytoplasmic material which eventually is sequestered in vacuoles.

Obviously the tonoplast must play a prominent role in the autophagic process. The corresponding intracellular transport can be compared with phagocytosis, that is, with the uptake of extracellular particles in animal cells through the invagination and vesiculation of the plasmalemma. Electron micrographs of plant cells, particularly freeze-etchings, suggest that phagocytosis-like invaginations of the tonoplast result in the formation of intravacuolar vesicles which contain a portion of cytoplasm (Fig. 4A, B). Indeed, FINERAN (1970) was able to demonstrate that in freeze-etched specimens of root tip cells the membrane surrounding vacuolar inclusions has the typical structure of the tonoplast.

It is unlikely that the tonoplast invaginations represent artifacts of chemical fixation. They were observed in freeze-etched yeast and other fungal cells without chemical prefixation (WIEMKEN, 1969; ITEN and MATILE, 1970; GRIFFITHS, 1971). In chemically fixed cells, tonoplast invaginations grown into myelin-like whorls of membranes were repeatedly observed (THOMAS and ISAAC, 1967; COULOMB and BUVAT, 1968; BOWES, 1969; MESQUITA, 1972; REID and MEIER, 1972; COULOMB, 1973a). Although chemical fixation does not necessarily provide adequate pictures

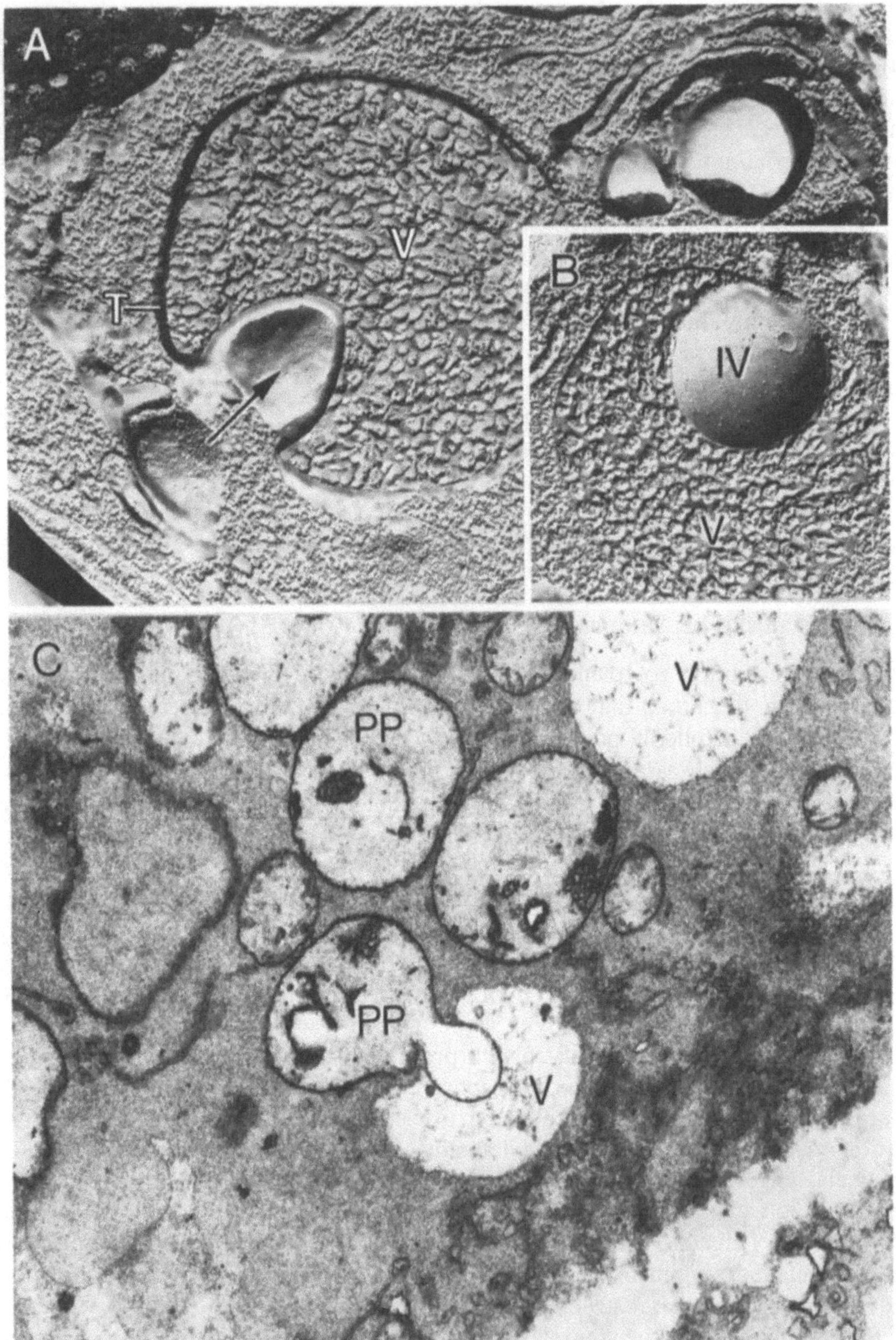

Fig. 4A–C. Autophagic activity of vacuoles. (A) Invagination (*arrow*) of tonoplast *T* in root meristem cell of *Zea mays*. ×30,000. (Courtesy of H. MOOR.) (B) Intravacuolar vesicle *IV* in vacuole *V* of root meristem cell of *Zea mays*. ×27,600. (From MATILE and MOOR, 1968.) (C) Autophagic activity of vacuoles in embryo of *Fraxinus excelsior* seed recovering from pathological changes accumulated over long period of enforced dormancy. Specific engulfment of proplastid *PP* by invaginating tonoplast. ×15,800. (Courtesy of T.A. VILLIERS)

of tonoplast invaginations, it nevertheless suggests that membrane growth may occur at distinct points in this membrane. Invaginating regions of tonoplasts exhibit a fine structure which is remarkably different from that of noninvaginating regions; the convex fracture face of the invaginated membranes contains a much higher population of globular particles, suggesting an active involvement of the membrane in the engulfment of cytoplasm to be transported into vacuoles (FINERAN, 1970).

The autophagic activity of vacuoles is far from being understood. In contrast to phagocytosis in animal cells, in which cytoplasmic microfilaments appear to be involved in the formation of plasmalemma invaginations (ALLISON et al., 1971), corresponding observations are lacking completely in the case of tonoplast invaginations. In any case, the invaginations imply the alteration of the thermodynamically most probable and stable spherical shape of vacuoles and, therefore, must represent an active, energy-dependent process.

It can be deduced from electron micrographs that intravacuolar vesicles that have been formed through tonoplast invagination are unstable. Cytoplasmic material can be seen in vacuoles with and without a wrapping membrane. Upon the rupture of the membrane envelope the cytoplasmic content of the vesicles is exposed to attack by the vacuolar digestive enzymes. It may be speculated that the stability of the tonoplast is provided by the cytoplasmic matrix; the contact with the cytoplasm is interrupted after the fission of invaginated portions of the tonoplast, and this may be responsible for the decay of intravacuolar vesicles.

A number of morphological observations suggest that specific cytoplasmic constituents may be engulfed by tonoplasts. FINERAN (1971) presented micrographs showing invaginating tonoplasts of root tip cells which appear to lead to the sequestration of ER, nuclear envelope, ribosomes, Golgi vesicles, extruded portions of mitochondria and to plastids. According to GIFFORD and STEWART (1968) large inclusions are formed within proplastids of apical shoot cells of *Bryophyllum* and *Kalanchöe*; these are eventually extruded and transferred to vacuoles by tonoplast invaginations. In imbibed *Fraxinus* seeds that were not allowed to germinate for prolonged periods of time, the specific engulfment of proplastids and mitochondria was observed after the onset of germination (Fig. 4C). VILLIERS (1971) interprets this phenomenon in terms of autophagic elimination of organelles that have been damaged by the artificial ageing of embryonic cells. MARTY (1971b) reported the sequestration of leucoplasts in the central vacuole of *Euphorbia characias* laticifers. Hence the morphological descriptions of autophagy reveal a certain specificity with regard to the cellular material grasped by the invaginating tonoplast. However, the biochemical counterpart which would be necessary for a complete description of the above examples of autophagic activity is lacking.

The autophagic processes described so far take place in existing vacuoles. A different mechanism, which is characterized by the coincidence of vacuolation and sequestration of cytoplasmic material, represents an alternative to tonoplast invagination. VILLIERS (1967) reported the occurrence of so-called "cytolysomes", i.e. portions of cytoplasm wrapped up by a double membrane envelope. This envelope develops from endomembranes (VILLIERS, 1972) and similar structures were observed in a variety of objects (e.g. BUVAT, 1968; MARTY, 1970, 1973; COULOMB and COULOMB, 1972; CRESTI et al., 1972; MESQUITA, 1972, Fig. 5A). The membranes may also form several envelopes in such a fashion that a series of concentric shells of cytoplasm is sequestered (COULOMB and COULOMB, 1972;

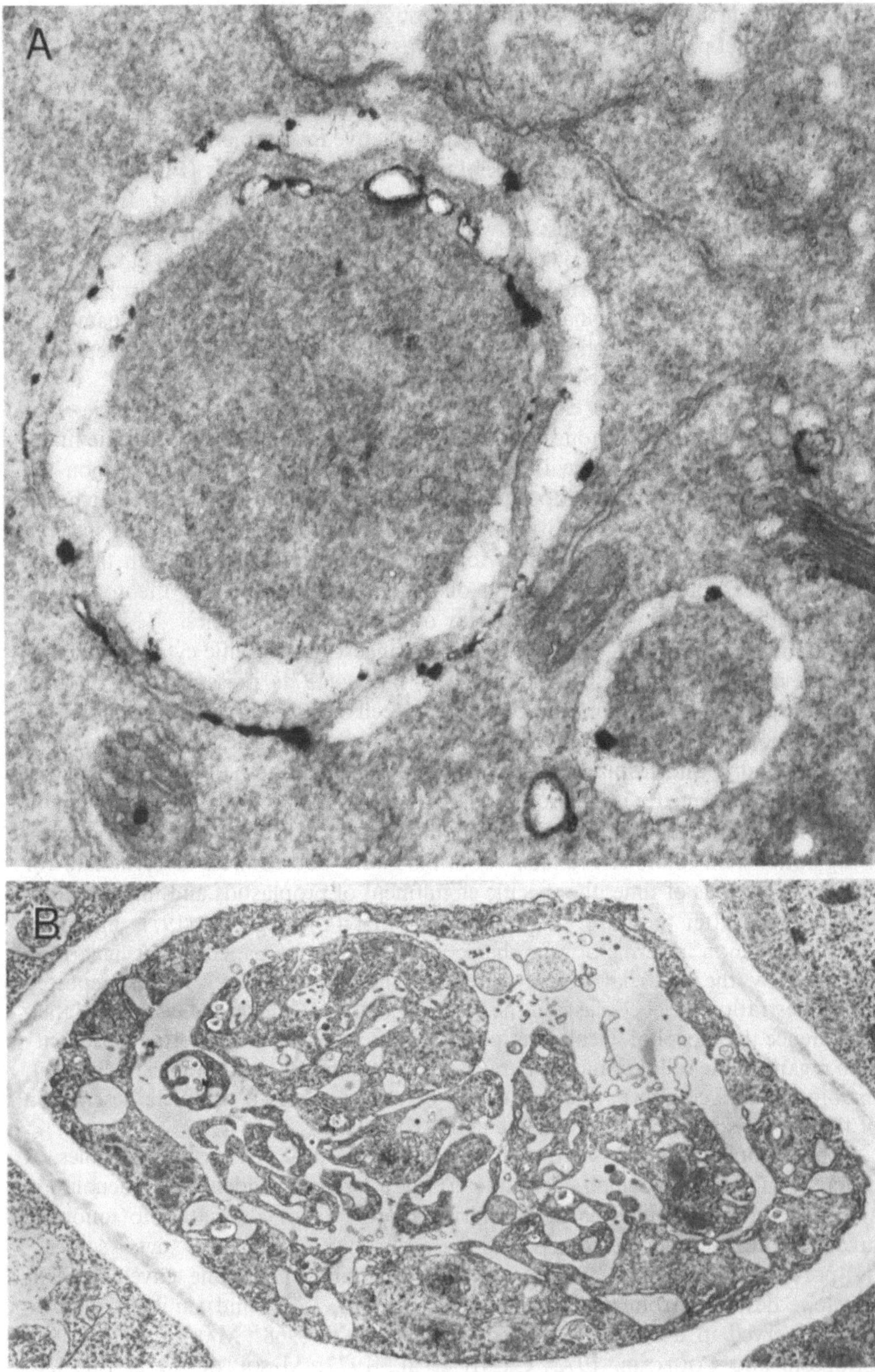

Fig. 5A and B

CRESTI et al., 1972; MESQUITA, 1972). BUVAT (1968) described the lining up of numerous vesicles around a cytoplasmic portion; the subsequent fusion of these vesicles results again in the formation of a cytolysome. In the course of all of these processes the endomembrane cisterna begins to inflate and, after completion of the envelopes, the cytolysomes closely resemble vacuoles which contain large intravacuolar vesicles.

The difference between the two mechanisms of autophagy described obviously concerns only the sequence of events. In the case of sequestration of cytoplasm by endomembranes, autophagy and vacuolation coincide, whereas autophagy associated with tonoplast invagination occurs after vacuoles have developed. It is, therefore, not surprising that CARROLL and CARROLL (1973) observed a mechanism of autophagy which appears to represent an intermediate stage between tonoplast invagination and sequestration by endomembranes.

Autophagic processes are summarized in the diagram of Figure 6. From the viewpoint of compartmentation of digestive enzymes it is important to recognize that the mechanisms of autophagy guarantee the continuity of the membrane system which separates the cytoplasmic matrix from the vacuome. This continuity of membrane barriers appears to represent a prerequisite for the organized (controlled) catabolism of cellular constituents.

3.2.3 Heterophagy

The narrow, submicroscopic pores of plant cell walls are likely to prevent the phagocytic uptake of large exogeneous particles which is a common phenomenon in protozoa, amebae and other animal cells. Exceptions such as the uptake of *Rhizobium* bacteria in cortical root cells of legumes involve a corresponding modification of the cell walls. They demonstrate, however, that phagocytosis is in principle possible in plant cells. In contrast to heterophagy in animal cells the phagocytic uptake of the bacteria is not immediately followed by intracellular digestion. The plasmalemma-homologous vacuoles in which the bacteroids stay are conversed into heterophagic (digestive) vacuoles only when the system has gone through a symbiotic stage of development. Necrosis and digestion of the bacteroids take place in the senescent host cells and are followed by their autolysis (TRUCHET and COULOMB, 1973).

Endocytic processes resulting in the formation of multivesicular bodies that are eventually transferred to vacuoles were observed in meristematic root cells (COULOMB, 1973b) and in cultured plant cells (MAHLBERG et al., 1974). The biochemical counterpart and functional significance of these cases of endocytosis is unknown.

Fig. 5A and B. Autophagy in differentiating cells of *Euphorbia characias*. (A) Development of autophagic vacuoles in root meristem cell. Formation of membrane envelopes wrapping up portions of cytoplasm. Note reaction product of acid phosphatase in cisternae of membrane envelopes. ×45,000. (Courtesy of F. MARTY.) (B) Cross section of laticifer, with central cytoplasm fragmented. Autophagy associated with formation of large central vacuole. ×7,350. (Courtesy of F. MARTY)

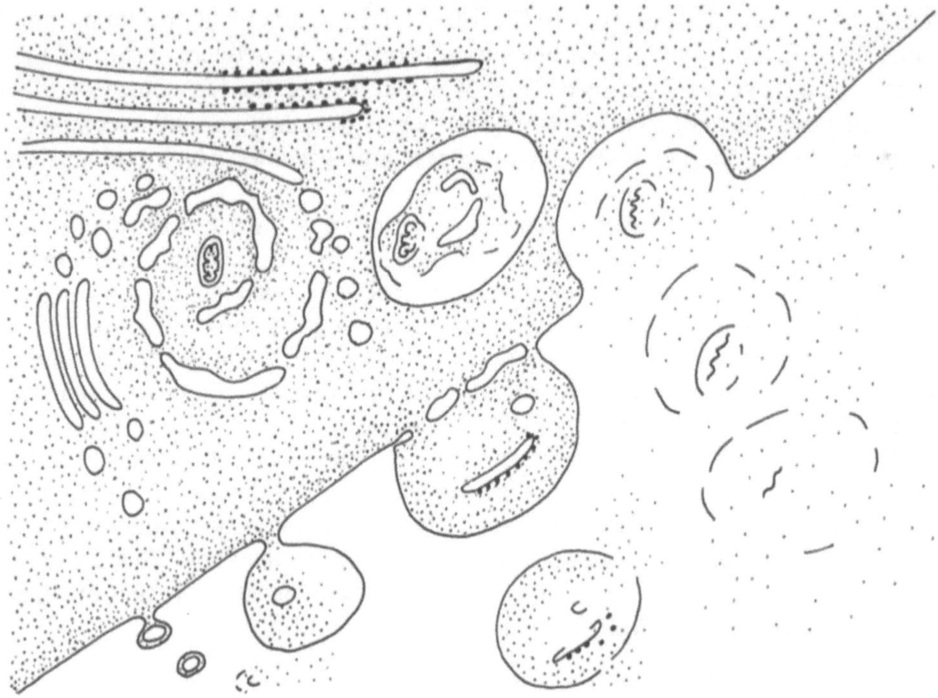

Fig. 6. Diagrammatic representation of mechanisms of autophagy observed in plant cells

3.2.4 Autolysis

Whilst autophagy concerns digestive activity within living cells, the irreversible chaotic breakdown which occurs upon cell death is termed autolysis. It may be conceived as the consequence of annihilation of the continuity of the endomembrane system which separates cytoplasmic matrix and vacuome and is responsible for the organized interactions between these two cell spaces. This view is supported by the observation of a diffused distribution of acid phosphatase in senescent cells which have reached the final stage of development (BERJAK and VILLIERS, 1970, 1972). Morphological observations in mesophyll cells of petals of the ephemeral flowers of *Ipomoea* suggest that severe changes of tonoplast permeability precede the final phase of cell ageing, which appears to be induced by the rupturing of the tonoplast (MATILE and WINKENBACH, 1971, Fig. 7). The breakdown of tonoplasts was, in fact, repeatedly observed in senescent cells (see MATILE, 1975). It results in the free mixing of vacuolar hydrolases and cytoplasmic constituents. In *Ipomoea* petals it is only after abolition of hydrolase compartmentation that the progressive degradation of nuclei can be observed. This illustrates the irreversibility of autolysis.

In flax cotyledons various fine structural changes are induced by the herbicide Paraquat; if the treated organs are illuminated, tonoplast rupture followed by disorganization in the protoplasm is evident after a few hours (HARRIS and DODGE, 1972). This observation raises the question of how autolysis is triggered under normal conditions. A likely cause of tonoplast rupture is the cessation of metabolic activity at the end of senescence, which comprises cessation of the supply of membrane building material or energy necessary for the maintenance of membrane

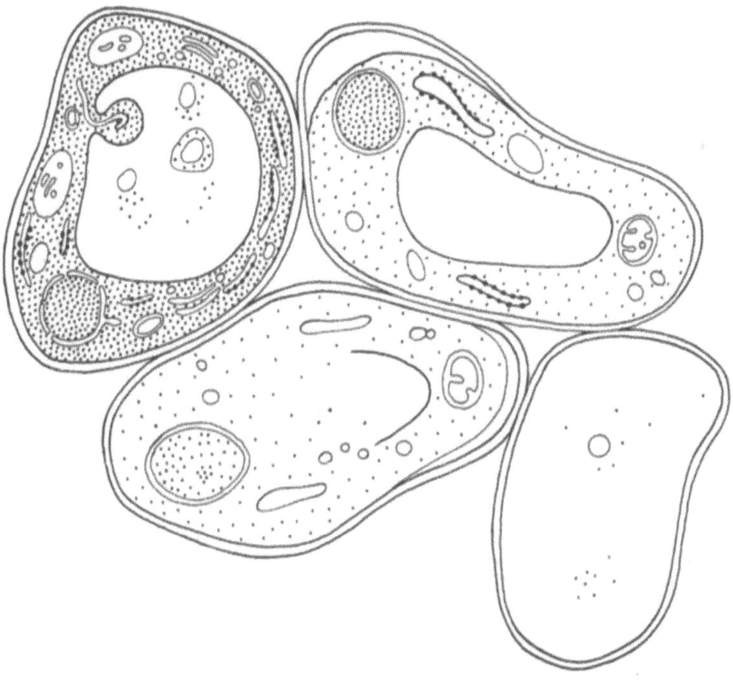

Fig. 7. Ageing and autolysis in mesophyll cells of *Ipomoea* corolla. Adapted from MATILE and WINKENBACH (1971). Reproduced by permission of McGraw-Hill Book Company (UK)

integrity. In *Coprinus* the well-known phenomenon of autolysis of gills coincides indeed with the sharp decline of the activities of respiratory enzymes (ITEN and MATILE, 1970).

3.2.5 Digestive Processes in Plant Development

Interactions between the cytoplasm and the digestive cell compartment have important functions in the differentiation of cells and development of plant organs. A few selected examples may illustrate this point.

Morphological correlates of autophagy, i.e. vacuoles containing cytoplasmic material, are particularly conspicuous in differentiating cells. Extensive autophagic activity was observed for instance in differentiating laticifers (MARTY, 1970, Fig. 5B), in developing sporangiophores (THORNTON, 1968), and in sporogenous cells (CARROLL and CARROLL, 1973) of fungi. Upon the differentiation of meristem cells into tracheids the cells are cleared of their cytoplasm by autophagy (WODZICKI and BROWN, 1973).

The early senescence and eventual autolysis in developing tracheids exemplifies the nonsynchronous differentiation of cells within plant tissues and organs. It is, therefore, impossible to correlate the overall biochemical changes associated with the development of organs with the cytological phenomena that can be observed in individual cells.

In the fading corolla of *Ipomoea tricolor* the rapid decline of protein and other biopolymers is associated with autophagy and autolysis which can be observed in the mesophyll cells and, at later stages of senescence, also in the epidermal cells (MATILE and WINKENBACH, 1971). The reuse of breakdown products in the developing organs of the plant is suggested by the disappearance of amino acids and other micromolecules from the ageing corolla (WINKENBACH, 1970). Hence the products of protein and nucleic acid breakdown formed presumably in the vacuoles appear to be subsequently metabolized in the cytoplasm and eventually exported via the sieve tubes (WIEMKEN, 1975).

A convincing example of autophagic activity which is correlated with cell differentiation is provided by *Euglena*. If this photoautotrophic organism is starved in the dark on an inorganic medium, it adapts to autophagic metabolism (BLUM and BUETOW, 1963). Under these conditions the flagellate digests 45% of its cellular protein, 55% of the RNA, and 30% of the DNA after 13 days without losing its viability. Starvation is accompanied by a marked increase in protease activity (BERTINI et al., 1965), and intracellular digestion of cytoplasmic constituents such as mitochondria could be observed in the electron microscope (MALKOFF and BUETOW, 1964; BRANDES and BERTINI, 1964). This example is outstanding because the investigation covered the major aspects, biochemical and morphological, of autophagy. A somewhat less complete study concerns the correlation between biochemical differentiation and autophagic activity in baker's yeast. This facultative anaerobic organism exhibits diauxic growth if cultured on glucose media. An extensive biochemical differentiation takes place when the cells switch from fermentative to respirative metabolism or vice versa. During the respective transition phases the pattern of enzyme activities changes dramatically (BECK and v. MEYENBURG, 1968). On the one hand these changes are evoked by regulatory mechanisms of protein synthesis such as glucose repression of enzymes involved in the oxidative metabolism. On the other hand, a prominent role of protein catabolism is suggested by the increased hydrolase activities which are present in yeast when biochemical differentiations take place (WIEMKEN, 1969; LENNEY et al., 1974).

4. Concluding Remarks

Research on the interactions between cytoplasm and vacuoles reveals quite strikingly the fundamental difficulties encountered in cell physiology. As long as intact cells are observed, the mechanisms involved in the operation of the vacuome can only be postulated on the basis of indirect conclusions. In turn, attempts to directly investigate these mechanisms using isolated vacuoles can never be fully successful as the cytoplasm, the truly living entity is eliminated. In particular, phenomena such as autophagy which involve the action of both membranes and cytoplasm cannot be studied with isolated vacuoles. It is, therefore, necessary to combine data obtained from methodologically quite divergent investigations in order eventually to obtain a concept of the interactions between cytoplasm and vacuole.

In general terms it can be recognized that the vacuome has an important function in what is conceived as *homeostasis*. A homeostatic external environment of cells as it is provided by the body fluids of animals is absent in plants and unicellular organisms. It is obvious that the vacuome of plant cells represents an internal environment which guarantees the maintenance of homeostasis. This subcellular analog of body fluid of animals (WIEMKEN and NURSE, 1973) serves as a device for coping with the environmental changes to which plant cells are more or less directly exposed. With regard to the digestive functions a striking similarity between the lysosomal systems of plant and animal cells is obvious. In this context homeostatic functions may be described in terms of turnover reactions which contribute to the maintenance of metabolic activities according to the external and developmental conditions.

References

AACH, H.G., HEBER, U.: Kompartimentierung von Aminosäuren in der Blattzelle. Z. Pflanzenphysiol. **57**, 317–328 (1967)

ALIBERT, G., RANJEVA, R., BOUDET, A.: Recherches sur les enzymes catalysant la formation des acides phénoliques chez *Quercus pedunculata*. II. Localisation intracellulaire de la phenylalanine ammoniaquelyase, de la cinnamate 4-hydroxylase, et de la «benzoate synthase». Biochim. Biophys. Acta **279**, 282–289 (1972)

ALLISON, A.C., DAVIES, P., DE PETRIS, S.: Role of contractile microfilaments in macrophage movement and endocytosis. Nature New Biol. **232**, 153–155 (1971)

AMRHEIN, N., ZENK, M.H.: Untersuchungen zur Rolle der Phenylalanin-ammonium-lyase (PAL) bei der Regulation der Flavonoidsynthese im Buchweizen (Fagopyrum esculentum, Moench). Z. Pflanzenphysiol. **64**, 145–168 (1971)

BAKER, D.A., HALL, J.L.: Pinocytosis, ATP-ase and ion uptake by plant cells. New Phytologist **72**, 1281–1291 (1973)

BAUER, H., SIGARLAKIE, E.: Cytochemistry on ultrathin frozen sections of yeast cells. J. Microscopie **99**, 205–218 (1973)

BAUMGARTNER, B.: Alterungsprozesse in der Korolle von *Ipomoea tricolor*. Thesis No. 5432, ETH Zürich, 1975

BAUMGARTNER, B., KENDE, H., MATILE, P.: RNase in ageing Japanese morning glory. Plant Physiol. **55**, 734–737 (1975)

BAUR, P.S., WALKINSHAW, C.H.: Fine structure of tannin accumulations in callus cultures of *Pinus elliotti*. Can. J. Botany **52**, 615–620 (1974)

BECK, C., MEYENBURG, H.K. VON: Enzyme pattern and aerobic growth of *Saccharomyces cerevisiae* and various degrees of glucoselimitation. J. Bacteriol. **96**, 479–486 (1968)

BEEVERS, H., STILLER, M.L., BUTT, V.S.: Metabolism of the organic acids. In: Plant Physiology. STEWARD, F.C. (ed.), Vol. IVB, pp. 119–242. New York-London: Academic Press 1966

BERJAK, P.: A lysosome-like organelle in the root cap of *Zea mays*. J. Ultrastruct. Res. **23**, 233–242 (1968)

BERJAK, P., VILLIERS, T.A.: Ageing in plant embryos. I. The establishment of the sequence of development and senescence in the root cap during germination. New Phytologist **69**, 929–938 (1970)

BERJAK, P., VILLIERS, T.A.: Ageing in plant embryos. V. Lysis of the cytoplasm in non-viable embryos. New Phytologist **71**, 1075–1079 (1972)

BERTINI, F., BRANDES, D., BUETOW, D.E.: Increased acid hydrolase activity during carbon starvation in *Euglena gracilis*. Biochim. Biophys. Acta **107**, 171–173 (1965)

BIELESKI, R.L.: Phosphate pools, phosphate transport and phosphate availability. Ann. Rev. Plant Physiol. **24**, 225–252 (1973)

BLUM, J.J., BUETOW, D.E.: Biochemical changes during acetate deprivation and repletion in *Euglena*. Exptl. Cell Res. **29**, 407–421 (1963)

BOLLER, T., DÜRR, M., WIEMKEN, A.: Characterization of a specific transport system for arginine in isolated yeast vacuoles. Eur. J. Biochem. **54**, 81–91 (1975)

BOWES, B.G.: Electron microscopic observations on Myelinlike bodies and related membranous elements in *Glechoma hederacea L.* Z. Pflanzenphysiol. **60**, 414–417 (1969)
BRACKER, C.E., GROVE, S.N.: Continuity between cytoplasmic endomembranes and outer mitochondrial membranes in fungi. Protoplasma **73**, 15–34 (1971)
BRANDES, D., BERTINI, F.: Role of Golgi apparatus in the formation of cytolysomes. Exptl. Cell Res. **35**, 194–217 (1964)
BRZOZOWSKA, J., HANOWER, P., CHEZEAU, R.: Free amino acids of *Hevea brasiliensis* latex. Experientia **30**, 894–895 (1974)
BUVAT, R.: Diversité des vacuoles dans les cellules de la racine d'orge (*Hordeum sativum*). Compt. Rend. **267**, 296–298 (1968)
BUVAT, R.: Origin and continuity of cell vacuoles. In: Results and Problems in Cell Differentiation. REINERT, J., URSPRUNG, H. (eds.), Vol. II. Berlin-Heidelberg-New York: Springer 1971
CARROLL, F.E., CARROLL, G.C.: Senescence and death of the conidiogenous cell in *Stemphylium botryosum* Wallroth. Arch. Mikrobiol. **94**, 109–124 (1973)
CATESSON, A.-M., CZANINSKI, Y.: Localisation ultrastructurale de la phosphatase acide et cycle saisonnier dans les tissus conducteurs de quelques arbres. Bull. Soc. Franç. Physiol. Végétale **14**, 165–173 (1968)
CHAFE, S.C., DURZAN, D.J.: Tannin inclusions in cell suspension cultures of white spruce. Planta **113**, 251–262 (1973)
COCKING, E.C.: A method for the isolation of plant protoplasts and vacuoles. Nature **187**, 962–963 (1960)
COULOMB, C.: Phénomènes d'autophagie liés à la différentiation cellulaire dans les jeunes racines de scorsonère (*Scorzonera hispanica*). Compt. Rend. **276**, 1161–1164 (1973a)
COULOMB, C.: Diversité des corps multivésiculaires et notion d'hétérophagie dans le méristème radiculaire de scorsonère (*Scorzonera hispanica*). J. Microscopie **16**, 345–360 (1973b)
COULOMB, C., BUVAT, R.: Processus de dégénérence cytoplasmique partielle dans les cellules de jeunes racines de *Cucurbita pepo*. Compt. Rend. **267**, 843–844 (1968)
COULOMB, P.: Etude préliminaire sur l'activité phosphatasique acide des particules analogues aux lysosomes des cellules radiculaires jeunes de la Courge (*Cucurbita Pepo L.* Cucurbitacée). Compt. Rend. **267**, 2133–2136 (1968)
COULOMB, P.: Phytolysosomes dans le meristème radiculaire de la Courge (*Cucurbita pepo L.* Cucurbitacée). Activité phosphatasique acide et activité peroxydasique. Compt. Rend. **272**, 48–51 (1971a)
COULOMB, P.: Phytolysosomes dans les frondes d'*Asplenium fontanum* (Filicinées, Polypodiacées). Isolement sur gradient dosages de quelques hydrolases et contrôle des culots obtenus par la microscopie électronique. J. Microscopie **11**, 299–318 (1971b)
COULOMB, P., COULOMB, C.: Processus d'autophagie cellulaire dans les cellules de méristèmes radiculaires en état d'anoxie. Compt. Rend. **274**, 214–217 (1972)
CRESTI, M., PACINI, E., SARFATTI, G.: Ultrastructural studies on the autophagic vacuoles in *Eranthis hiemalis endosperm*. J. Submicrosc. Cytol. **4**, 33–44 (1972)
CRONSHAW, J., CHARVAT, I.: Localization of β-glycerophosphatase activity in the myxomycete *Perichaena vermicularis*. Can. J. Botany **51**, 97–101 (1973)
DAMADIAN, R.: Biological ion exchanger resins. Ann. N.Y. Acad. Sci. **204**, 211–248 (1974)
DAVIS, R.H.: Metabolite distribution in cells. Science **178**, 835–840 (1972)
DICKENSON, P.B.: The ultrastructure of the latex vessels of *Hevea brasiliensis*. Proc. Nat. Rubber Prod. Res. Assoc., Jubilee Conf., Cambridge, MULLINS, L. (ed.), pp. 52–66. London: Maclaren and Sons Ltd. 1964
DIERS, L., SCHÖTZ, F., MEYER, B.: Über die Ausbildung von Gerbstoffvakuolen bei Oenotheren. Cytobiology **7**, 10–19 (1973)
DOI, E., OHTSURU, C., MATOBA, T.: Lysosomal nature of plant vacuoles II. Acid hydrolases in the central vacuole of internodal cells of Charophyta. Plant Cell Physiol. (Tokyo) **16**, 581–588 (1975)
DUVE, C. DE: The lysosome in retrospect. In: Lysosomes in Biology and Pathology. DINGLE, J.T., FELL, H.B. (eds.), Vol. 1, pp. 3–40. Amsterdam-London: North Holland Publ. Co. 1969
EDELMAN, J., JEFFORD, T.G.: The mechanism of fructosan metabolism in higher plants as exemplified in *Helianthus tuberosus*. New Phytologist **67**, 517–531 (1968)

EPSTEIN, E.: Mechanisms of ion uptake through plant cell membranes. Intern. Rev. Cytol. **34**, 123–168 (1973)
FAIRBAIRN, J.W., HAKIM, F., EL KHEIR, Y.: Alkaloidal storage, metabolism and translocation in the vesicles of *Papaver somniferum* latex. Phytochemistry **13**, 1133–1139 (1974)
FERRARI, T.E., YODER, O.C., FILNER, P.: Anerobic nitrite production by plant cells and tissues: evidence for two nitrate pools. Plant Physiol. **51**, 423–431 (1973)
FIGIER, J.: Localisation intrastructurale de la phosphatase acide dans les glandes pétiolaires d'*Impatiens holstii*. Planta **108**, 215–226 (1972)
FINERAN, B.A.: Organization of the tonoplast in frozen-etched root tips. J. Ultrastruct. Res. **35**, 574–586 (1970)
FINERAN, B.A.: Ultrastructure of vacuolar inclusions in root tips. Protoplasma **72**, 1–18 (1971)
GAHAN, P.B.: Histochemistry of lysosomes. Intern. Rev. Cytol. **21**, 1–63 (1967)
GAHAN, P.B.: Plant lysosomes. In: Lysosomes in Biology and Pathology. DINGLE, J.T. (ed.), Vol. III, pp. 69–85. Amsterdam-London: North-Holland Publ. Co. 1973
GEZELIUS, K.: Acid phosphatase localization during differentiation in the cellular slime mold *Dictyostelium discoideum*. Arch. Microbiol. **85**, 51 (1972)
GIFFORD, E.M., STEWART, K.D.: Inclusions of the proplastids and vacuoles in the shoot apices of *Bryophyllum* and *Kalanchöe*. Am.. J. Botany **55**, 269–279 (1968)
GRIFFITHS, D.A.: Hyphal structure in *Fusarium oxysporum* (Schlecht) revelaed by freeze-etching. Arch. Microbiol. **79**, 93–101 (1971)
GUILLIERMOND, A.: The Cytoplasm of the Plant Cell. Waltham, Mass. USA: Chronica Botanica Company, 1941
GUILLIERMOND, A., MANGENOT, G., PLANTEFOL, L.: Traité de cytologie végétale. Paris: Le François 1933
GUTKNECHT, J., DAINTY, J.: Ionic relations in marine algae. Oceanogr. Mar. Biol. Ann. Rev. **6**, 163–200 (1968)
HALL, J.L., DAVIE, C.A.M.: Localization of acid hydrolase activity in *Zea mays* L. root tips. Ann. Botany (London) **35**, 849–855 (1971)
HALPERIN, W.: Ultrastructural localization of acid phosphatase cultured cells of *Daucus carota*. Planta **88**, 91–102 (1969)
HALVORSON, H.O.: The induced synthesis of proteins. Advan. Enzymol. **22**, 99–156 (1960)
HARRIS, N., DODGE, A.D.: The effect of paraquat on flax cotyledon leaves: Changes in fine structure. Planta **104**, 201–209 (1972)
HASILIK, A., MÜLLER, H., HOLZER, H.: Compartmentation of the Tryptophone-Synthetase-Proteolysing system in *Saccharomyces cerevisiae*. Eur. J. Biochem. **48**, 111–117 (1974)
HAWKER, J.S., HATCH, M.D.: Mechanism of sugar storage by mature stem tissue of sugar cane. Physiol. Plantarum **18**, 444–453 (1965)
HEBER, U.: Zur Frage der Lokalisation von löslichen Zuckern in der Pflanzenzelle. Ber. Deut. Botan. Ges. **70**, 371–382 (1957)
HEFTMANN, E.: Lysosomes in tomatoes. Cytobios **3**, 129–136 (1971)
HIGINBOTHAM, N., ANDERSON, W.P.: Electrogenic pumps in higher plant cells. Can. J. Botany **52**, 1011–1021 (1974)
HIRAI, M., ASAKI, T.: Membranes carrying acid hydrolases in pea seedling roots. Plant Cell Physiol. **14**, 1019–1030 (1973)
HUFFAKER, R.C., PETERSON, L.W.: Protein turnover in plants and possible means of its regulation. Ann. Rev. Plant Physiol. **25**, 363–392 (1974)
HUMPHREYS, T.E.: Sucrose transport at the tonoplast. Phytochemistry **12**, 1211–1219 (1973)
INDGE, K.J.: Metabolic lysis of yeast protoplasts. J. Gen. Microbiol. **51**, 433–440 (1968a)
INDGE, K.J.: The isolation and properties of the yeast cell vacuole. J. Gen. Microbiol. **51**, 441–446 (1968b)
INDGE, K.J.: Polyphosphates of the yeast cell vacuole. J. Gen. Microbiol. **51**, 447–455 (1968c)
ITEN, W., MATILE, P.: Role of chitinase and other lysosomal enzymes of *Coprinus lagopus* in the autolysis of fruiting bodies. J. Gen. Microbiol. **61**, 301–309 (1970)
JANS, B.: Untersuchungen am Milchsaft des Schöllkrautes (*Chelidonium majus* L.). Ber. Schweiz. Botan. Ges. **83**, 306–344 (1973)
KELKER, H.C., FILNER, P.: Regulation of nitrite reductase and its relationship to the regulation of nitrate reductase in cultured tobacco cells. Biochim. Biophys. Acta **252**, 69–82 (1971)

KLUGE, M., HEININGER, B.: Untersuchungen über den Efflux von Malat aus den Vacuolen der assimilierenden Zellen von *Bryophyllum* und mögliche Einflüsse dieses Vorganges auf den CAM. Planta **113**, 333–344 (1973)
KÜSTER, E.: Die Pflanzenzelle, 2nd ed. Jena: Fischer, 1951
LATIES, G.G.: Dual mechanisms of salt uptake in relation to compartmentation and long-distance transport. Ann. Rev. Plant Physiol. **20**, 89–116 (1969)
LÄUCHLI, A.: Electron probe analysis. In: Microautoradiography and Electron Probe Analysis. Their Application to Plant Physiology. LÜTTGE, U. (ed.), pp. 191–236. Berlin-Heidelberg-New York: Springer 1972
LENNEY, J.F., MATILE, P., WIEMKEN, A., SCHELLENBERG, M., MEYER, J.: Activities and cellular localization of yeast proteases and their inhibitors. Biochem. Biophys. Res. Commun. **60**, 1378–1383 (1974)
LING, G.N.: A new model for the living cell: a summary of the theory and recent experimental evidence in its support. Intern. Rev. Cytol. **26**, 1–62 (1969)
LÜSCHER, A., MATILE, P.: Distribution of acid RNase and other enzymes in stratified *Acetabularia*. Planta **118**, 323–332 (1974)
LÜTTGE, U.: Microautoradiography of water-soluble inorganic ions. In: Microautoradiography and Electron Probe Analysis. Their Application to Plant Physiology. LÜTTGE, U. (ed.), pp. 61–98. Berlin-Heidelberg-New York: Springer 1972
LÜTTGE, U.: Stofftransport der Pflanzen. Berlin-Heidelberg-New York: Springer 1973
MACLENNAN, D.H., BEEVERS, H., HARLEY, J.L.: Compartmentation of acids in plant tissues. Biochem. J. **89**, 316–327 (1963)
MACROBBIE, E.A.C.: Fluxes and compartmentation in plant cells. Ann. Rev. Plant Physiol. **22**, 75–96 (1971)
MAHLBERG, P.G., TURNER, F.R., WALKINSHAW, C., VENKETESWARAN, S.: Ultrastructural studies in plasma membrane related secondary vacuoles in cultured cells. Am. J. Botany **61**, 730–738 (1974)
MALKOFF, D.B., BUETOW, D.E.: Ultrastructural changes during carbon starvation in *Euglena gracilis*. Exptl. Cell Res. **35**, 58–68 (1964)
MARTIN, P.: Nitratstickstoff in Buschbohnenblättern unter dem Gesichtspunkt der Kompartimentierung der Zellen. Z. Pflanzenphysiol. **70**, 158–165 (1973)
MARTY, F.: Infrastructure des lacticifères différenciés d'*Euphorbia characias*. Compt. Rend. **267**, 299–302 (1968)
MARTY, F.: Rôle du système membranaire vacuolaire dans la différenciation des lacticifères d'*Euphorbia characias* L. Compt. Rend. **271**, 2301–2304 (1970)
MARTY, F.: Vésicules autophagiques des laticifères différenciés d'*Euphorbia characias* L. Compt. Rend. **272**, 399–402 (1971a)
MARTY, F.: Différenciation des plastes dans les laticifères d'*Euphorbia characias* L. Compt. Rend. **272**, 223–226 (1971b)
MARTY, F.: Mise en évidence d'un appareil provacuolaire et de son rôle dans l'autophagie cellulaire et l'origine des vacuoles. Compt. Rend. **276**, 1549–1552 (1973)
MATERN, H., BETZ, H., HOLZER, H.: Compartmentation of inhibitors of proteinases A and B and carboxypeptidase in yeast. Biochem. Biophys. Res. Commun. **60**, 1051–1097 (1974)
MATILE, P.: Enzyme der Vakuolen aus Wurzelzellen von Maiskeimlingen. Ein Beitrag zur funktionellen Bedeutung der Vakuole bei der intrazellulären Verdauung. Z. Naturforsch. **21** b, 871–878 (1966)
MATILE, P.: Lysosomes of root tip cells in corn seedlings. Planta **79**, 181–196 (1968)
MATILE, P.: Plant lysosomes. In: Lysosomes in Biology and Pathology. DINGLE, J.T., FELL, H.B. (eds.), pp. 406–430. Amsterdam-London: North Holland Publ. Co. 1969
MATILE, P.: Vacuoles, lysosomes of *Neurospora*. Cytobiologie **3**, 324–330 (1971)
MATILE, P.: The lytic compartment of plant cells. Cell Biology Monographs. Vol. I. Wien-New York: Springer 1975
MATILE, P., JANS, B., RICKENBACHER, R.: Vacuoles of *Chelidonium* latex: Lysosomal property and accumulation of alkaloids. Biochem. Physiol. Pflanzen **161**, 447–458 (1970)
MATILE, P., MOOR, H.: Vacuolation: origin and development of the lysosomal apparatus in root tip cells. Planta **80**, 159–175 (1968)
MATILE, P., WIEMKEN, A.: The vacuole as the lysosome of the yeast cell. Arch. Mikrobiol **56**, 148–155 (1967)

MATILE, P., WIEMKEN, A.: Vacuoles and spherosomes. Methods Enzymol. **31**, 572–578 (1974)
MATILE, P., WINKENBACH, F.: Function of lysosomes and lysosomal enzymes in the senescing corolla of the morning glory (*Ipomoea purpurea*). J. Exptl. Botany **22**, 759–771 (1971)
MESQUITA, J.F.: Ultrastructure de formations comparables aux vacuoles autophagiques dans les cellules des racines de l'*Allium cepa* L. et du *Lupinus albus* L. Cytologia **37**, 95–110 (1972)
MEYER, J., MATILE, P.: Subcellular distribution of yeast invertase isoenzymes. Arch. Microbiol. **103**, 51–55 (1975)
MÜLLER, E., NELLES, A., NEUMANN, D.: Beiträge zur Physiologie der Alkaloide. I. Aufnahme von Nikotin in Gewebe von *Nicotianes rustica* L. Biochem. Physiol. Pflanzen **162**, 272 (1971)
NAKANO, M., ASAKI, T.: Subcellular distribution of hydrolase in germinating cotyledons. Plant Cell Physiol. **13**, 101–110 (1972)
NEUMANN, D., MÜLLER, E.: Intrazellulärer Nachweis von Alkaloiden in Pflanzenzellen im licht- und elektronenmikroskopischen Maßstab. Flora, A **158**, 479–491 (1967)
NEUMANN, D., MÜLLER, E.: Beiträge zur Physiologie der Alkaloide. III. *Chelidonium majus* L. und *Sanguinaria canadensis* L.: Ultrastruktur der Alkaloidbehälter, Alkaloidaufnahme und -verteilung. Biochem. Physiol. Pflanzen **163**, 375–391 (1972)
NEUMANN, D., MÜLLER, E.: Beiträge zur Physiologie der Alkaloide. IV. Alkaloidbildung in Kalluskulturen von *Macleaya*. Biochem. Physiol. Pflanzen **165**, 271–282 (1974)
NISSEN, P.: Uptake mechanisms: inorganic and organic. Ann. Rev. Plant Physiol. **25**, 53–79 (1974)
OAKS, A., BIDWELL, R.G.S.: Compartmentation of intermediary metabolites. Ann. Rev. Plant Physiol. **21**, 43–66 (1970)
O'DAY, D.H.: Intracellular localization and extracellular release of certain lysosomal enzyme activities from amoebae of the cellular slime mold *Polysphondylium pallidum*. Cytobios **7**, 223–232 (1973)
PALLAGHY, C.K.: Electron probe microanalysis of K^+ and Cl^- in freeze-substituted leaf sections of *Zea mays*. Australian J. Biol. Sci. **26**, 1015–1034 (1973)
PARISH, R.W.: The lysosome-concept in plants II. Location of acid hydrolases in maize root tips. Planta **123**, 15–31 (1975)
PAYNE, P.I., BOULTER, D.: Catabolism of plant cytoplasmic ribosomes: RNA breakdown in senescent cotyledons of germinating broad-bean seedlings. Planta **117**, 251–258 (1974)
PFEIFFER, H.H.: Cytotopochemische Beiträge zum Lipidvorkommen in Vakuolen. Protoplasma **57**, 636–642 (1963)
PITT, D.: Lysosomes and Cell Function. London-New York: Longman 1975
PITT, D., GALPIN, M.: Isolation and properties of lysosomes from dark-grown potato shoots. Planta **109**, 233–258 (1973)
POUX, N.: Localisation de la phosphatase acide dans les cellules meristématiques de blé (*Triticum vulgare* Vill.). J. Microscopie **2**, 485–489 (1963a)
POUX, N.: Sur la présence d'enclaves cytoplasmiques en voie de dégénerescence dans les vacuoles des cellules végétales. Compt. Rend. **257**, 736–738 (1963b)
POUX, N.: Mise en évidence des phosphates en microscopie électronique dans les cellules embryonnaires de quelques dicotyledones. Compt. Rend. **261**, 1064–1066 (1965)
POUX, N.: Localisation d'activités enzymatiques dans le méristème radiculaire de *Cucumis sativus* L. III. Activité phosphatasique acide. J. Microscopie **9**, 407–434 (1970)
PUJARNISCLE, S.: Caractère lysosomal des lutoïdes du latex d'*Hevea brasiliensis*. Physiol. Vég. **6**, 27–46 (1968)
QUEIROZ, O.: Circadian rhythms and metabolic patterns. Ann. Rev. Plant Physiol. **25**, 115–134 (1974)
RAVEN, J.A., SMITH, F.A.: Significance of hydrogen ion transport in plant cells. Canad. J. Botany **52**, 1035–1047 (1974)
REID, J.S.G., MEIER, H.: The function of the aleurone layer during galactomannan mobilization in germinating seeds of fenugreek (*Trigonella foenum-graecum* L.) crimson clover (*Trifolium incarnatum* L.) and lucerne (*Nedicago sativa* L.): a correlative biochemical and ultrastructural study. Planta **106**, 44–60 (1972)
REISS, J.: Enzyme cytochemistry of fungi. Progr. Histochem. Cytochem. Vol. V, No. 4 (1973)
RIBAILLIER, D., JACOB, J.L., D'AUZAC, J.: Sur certains caractères vacuolaires des lutoïdes du latex d'*Hevea brasiliensis*. Physiol. Vég. **9**, 423–437 (1971)

ROUSH, A.H.: Crystallization of purines in the vacuole of *Candida utilis*. Nature **190**, 449 (1961)

RUESINK, A.W.: The plasma membrane of Avena coleoptile protoplasts. Plant Physiol. **47**, 192–195 (1971)

RYAN, C.A., SHUMWAY, L.K.: Studies on the structure and function of chymotrypsin inhibitor I in the *Solanaceae* family. Proc. Inter. Res. Conf. Proteinase Inhibitors. FRITZ, H., TSCHESCHE, H. (eds.), pp. 175–188. Berlin-New York: Walter de Gruyter 1971

SCHLENK, F., DAINKO, J.L., SVIHLA, G.: The accumulation and intracellular distribution of biological Sulfonium compounds in yeast. Arch. Biochem. Biophys. **140**, 228–236 (1970)

SCHNEPF, E.: Zur Cytologie und Physiologie pflanzlicher Drüsen. 5. Teil: Elektronenmikroskopische Untersuchungen an Cyathialnektarien von *Euphorbia pulcherima* in verschiedenen Funktionszuständen. Protoplasma **58**, 193–219 (1964)

SCHNEPF, E.: Organellen-Reduplikation und Zell-Kompartimentierung. In: Probleme der Biologischen Reduplikation. SITTE, P. (ed.), pp. 372–390. Berlin-Heidelberg-New York: Springer 1966

SCHÖTZ, F., DIERS, L., BATHELT, H.: Zur Feinstruktur der Raphidenzellen. I. Die Entwicklung der Vakuolen und der Raphiden. Z. Pflanzenphysiol. **63**, 91–113 (1970)

SCHRÖTER, K., LÄUCHLI, A., SIEVERS, A.: Mikroanalytische Identifikation von Bariumsulfat-Kristallen in den Statolithen der Rhizoide von *Chara fragilis*, Desv. Planta **122**, 213–225 (1975)

SCHULZE, C., SCHNEPF, E., MOTHES, K.: Über die Lokalisation der Kautschukpartikel in verschiedenen Typen von Milchröhren. Flora, A **158**, 458–460 (1967)

SEXTON, R., CRONSHAW, J., HALL, J.L.: A study of the biochemistry and cytochemical localization of β-glycerophosphatase activity in root tips of maize and pea. Protoplasma **73**, 417–441 (1971)

SIEGEL, S.M., DALY, O.: Regulation of betacyanin efflux from beet root by poly L-Lysine, Ca-ion and other substances. Plant Physiol. **41**, 1429–1434 (1966)

SIEVERS, A.: Lysosomen-ähnliche Kompartimente in Pflanzenzellen. Naturwissenschaften **53**, 334–335 (1966)

SMITH, D.G., MARCHANT, R.: Lipid inclusions in the vacuoles of *Saccharomyces cerevisiae*. Arch. Mikrobiol. **60**, 340–347 (1968)

STADELMANN, E.J., KINZEL, H.: Vital staining of plant cells. In: Methods in Cell Physiology, PRECOTT, D.M. (ed.), Vol. V, pp. 325–372. New York–London: Academic Press 1972

STEVENINCK, R.F.M., CHENOWETH, A.R.F.: Ultrastructural localization of ions. I. Effect of high external sodium chloride concentrations on the apparent distribution of chloride in leaf *parenchyma* cells of barley seedlings. Australian J. Biol. Sci. **25**, 499–516 (1972)

STEWARD, F.C., BIDWELL, R.G.S.: Storage pools and turnover systems in growing and non-growing cells: experiments with C^{14}-sucrose, ^{14}C-glutamine and ^{14}C-asparagine. J. Exptl. Botany **17**, 726–774 (1966)

STEWARD, F.C., MOTT, R.L.: Cells, solutes and growth: salt accumulation in plants reexamined. Intern. Rev. Cytol. **28**, 275–370 (1970)

STEWARD, F.C., SUTCLIFFE, J.F.: Plants in relation to inorganic salts. In: Plant Physiology. STEWARD, C.F. (ed.), Vol. II, pp. 253–478. New York-London: Academic Press 1959

SUNDBERG, I., NILSHAMMAR-HOLMVALL, M.: The diurnal variation of phosphate uptake and ATP level in relation to deposition of starch, lipid, and polyphosphate in synchronized cells of *Scenedesmus*. Z. Pflanzenphysiol. **76**, 270–279 (1975)

SVIHLA, G., DAINKO, J.L., SCHLENK, F.: Ultraviolet microscopy of purine compounds in the yeast vacuole. J. Bacteriol. **85**, 399–409 (1963)

THOMAS, P.L., ISAAC, P.K.: An electron microscope study of intravacuolar bodies in the uredia of wheat stem rust and in hyphae of other fungi. Can. J. Botany **45**, 1473–1478 (1967)

THORNTON, R.M.: The fine structure of Phycomyces. I. Autophagic vesicles. J. Ultrastruct. Res. **21**, 269–280 (1968)

TORIYAMA, H., JAFFE, M.J.: Migration of calcium and its role in the regulation of seismonasty in the motor cell of *Mimosa pudica* L. Plant Physiol. **49**, 72–81 (1972)

TORIYAMA, H., SATÔ, S.: On the contents of the central vacuole in the *Mimosa* motor cell. Cytologia (Tokyo) **36**, 359–375 (1971)

TREWAVAS, A.: Determination of the rates of protein synthesis and degradation in *Lemna minor*. Plant Physiol. **49**, 40–46 (1972a)

TREWAVAS, A.: Control of the protein turnover rates in *Lemna minor*. Plant Physiol. **49**, 47–51 (1972b)

TRUCHET, G., COULOMB, P.: Mise en évidence et évolution du système phytolysosomal dans les cellules des différentes zones de nodules radiculaires de pois (*Pisum sativum* L.). Notion d'hétérophagie. J. Ultrastruct. Res. **43**, 36–57 (1973)

ULLRICH, W., URBACH, W., SANTARIUS, K.A., HEBER, U.: Die Verteilung des Orthophosphates auf Plastiden, Cytoplasma und Vakuole in der Blattzelle und ihre Veränderung im Licht-Dunkel-Wechsel. Z. Naturforsch. **20**b, 905–910 (1965)

VILLIERS, T.A.: Cytolysomes in long-dormant plant embryo cells. Nature **214**, 1356–1357 (1967)

VILLIERS, T.A.: Lysosomal activities of the vacuole in damaged and recovering plant cells. Nature New Biol. **233**, 57–58 (1971)

VILLIERS, T.A.: Cytological studies in dormancy. II. Pathological ageing changes during prolonged dormancy and recovery upon dormancy release. New Phytologist **71**, 145–152 (1972)

VINTÉJOUX, C.: Localisation d'une activité phosphatasique acide dans des inclusions cytoplasmiques des feuilles d'hibernacles chez l'*Utricularia neglecta* L. (Lentibulariacées). Compt. Rend. **270**, 3213–3216 (1970)

WARDROP, A.B.: Occurrence of structures with lysosome-like function in plant cells. Nature **218**, 978–980 (1968)

WATTENDORFER, J.: Feinbau und Entwicklung der verkorkten Calcium-Oxalat-Kristallzellen in der Rinde von *Larix decidua* Mill. Z. Pflanzenphysiol. **60**, 307–347 (1969)

WEISS, R.L.: Intracellular localization of ornithine and arginine pools in *Neurospora*. J. Biol. Chem. **248**, 5409–5413 (1973)

WIEMKEN, A.: Eigenschaften der Hefevacuole. Thesis 4340, ETH Zürich, 1969

WIEMKEN, A., DÜRR, M.: Characterization of amino acid pools in the vacuolar compartment of *Saccharomyces visiae*. Arch. Microbiol. **101**, 45–57 (1974)

WIEMKEN, A., NURSE, P.: The vacuole as a compartment of amino acid pools in yeast. Proc. Third Intern. Symp. Yeasts, Otaniemi, Helsinki, 1973. Part II, pp. 331–347

WIEMKEN, V.: Der Abbau von Zellwänden bei den Blüten von *Ipomoea tricolor* und beim Samen von *Phoenix dactylifera*. Thesis 5501 ETH Zürich, 1975

WIEMKEN, V., WIEMKEN, A.: Dichtemarkierung von β-Glycosidasen in welkenden Blüten von *Ipomoea tricolor* (Cav. Z. Pflanzenphysiologie **75**, 186–190 (1975)

WIENER, E., ASHWORTH, J.M.: The isolation and characterization of lysosomal particles from myxamoebae of the cellular slime mold *Dictyostelium discoideum*. Biochem. J. **118**, 505–512 (1970)

WIGGINS, P.M.: Water structure as a determinant of ion distribution in living tissue. J. Theoret. Biol. **32**, 131–146 (1971)

WILDEN, W. VAN DER, MATILE, P., SCHELLENBERG, M., MEYER, J., WIEMKEN, A.: Vacuolar membranes: isolation from yeast cells. Z. Naturforsch. **28**c, 416–421 (1973)

WINKENBACH, F.: Zum Stoffwechsel der aufblühenden und welkenden Korolle der Prunkwinde *Ipomoea purpurea*. I. Beziehungen zwischen Gestaltwandel, Stofftransport, Atmung und Invertaseaktivität. Ber. Schweiz. Botan. Ges. **80**, 374–390 (1970)

WODZICKI, T.J., BROWN, C.L.: Organization and breakdown of the protoplast during maturation of pine tracheids. Am. J. Botany **60**, 631–640 (1973)

ZANDONELLA, P.: Infrastructure du tissu nectarigène floral de *Beta vulgaris* L.: le vacuome et la dégradation du cytoplasme dans les vacuoles. Compt. Rend. Sci. Nat. **271**, 70–73 (1970)

ZIELKE, H.R., FILNER, P.: Synthesis and turnover of nitrate reductase induced by nitrate in cultured tobacco cells. J. Biol. Chem. **246**, 1772–1779 (1971)

ZIRKLE, C.: The plant vacuole. Botan. Rev. **3**, 1–30 (1937)

8. Interactions among Cytoplasm, Endomembranes, and the Cell Surface

D.J. MORRÉ and H.H. MOLLENHAUER

1. Introduction

Protoplasts of mature cells of higher plants are delimited by a cellulosic wall which overlies the plasma membrane (cell membrane or plasmalemma) at the external or environmental surface. Internally, these cells contain a watery central vacuole surrounded by the tonoplast (vacuole membrane). These external and internal extraprotoplasmic compartments, and the membranes which surround them, arise through biosynthetic activities of the cytoplasm. Coordinated increases in their surface area accompany increases in cell volume and account for the elongation phase of plant growth. An understanding of the mechanisms and control of formation of these cellular components constitutes an important study area of plant physiology.

In this chapter, information on the formation of cell surfaces in plants is reviewed. Emphasis is on correlative aspects, especially those which pertain to interactions among constituents of the soluble cytoplasm (cytosol) and internal membranes that lead to formation and control of biosynthesis of cell walls and other components of cell surfaces.

2. Cell Surfaces and Expansion Growth

Growth of plant cells occurs by division followed by expansion or elongation. Both phases involve net increases in surfaces. At cytokinesis a new cell wall, the cell plate, is formed (WHALEY and MOLLENHAUER, 1963; FREY-WYSSLING et al., 1964; WHALEY et al., 1966). Cell expansion is, by definition, a period of increase in cell surface (BONNER, 1961).

The basic concepts of how plant cells expand have remained unchanged for many years (HEYN, 1940). An irreversible increase in the surface is accompanied by an uptake of water. Implicit in the former is an increase in surface of both wall and plasma membrane. Implicit in the latter is an increase in the surface of the tonoplast.

Surface growth may be restricted to the cell apex, as in tip growth (ROELOFSEN, 1959; ROSEN, 1968; GROVE et al., 1970; BARTNICKI-GARCIA, 1973), or distributed more or less uniformly over the entire area of the cell (ROELOFSEN, 1959). In general, both patterns of surface growth share common structural and biosynthetic features, the major differences being that the sites of tip growth are determined ultimately by systems of "guide elements" (FRANKE et al., 1972a) which direct the flow of precursors and biosynthetic activities to sites where growth occurs.

3. Role of the Soluble Cytoplasm (Cytosol)

The role of the cytosol in the biosynthesis of surface components is generally considered to be primarily as a source of enzymes, precursors, and activated intermediates for synthesis of macromolecules and macromolecular complexes. Other roles may involve the system of guide elements assumed to function in delivery

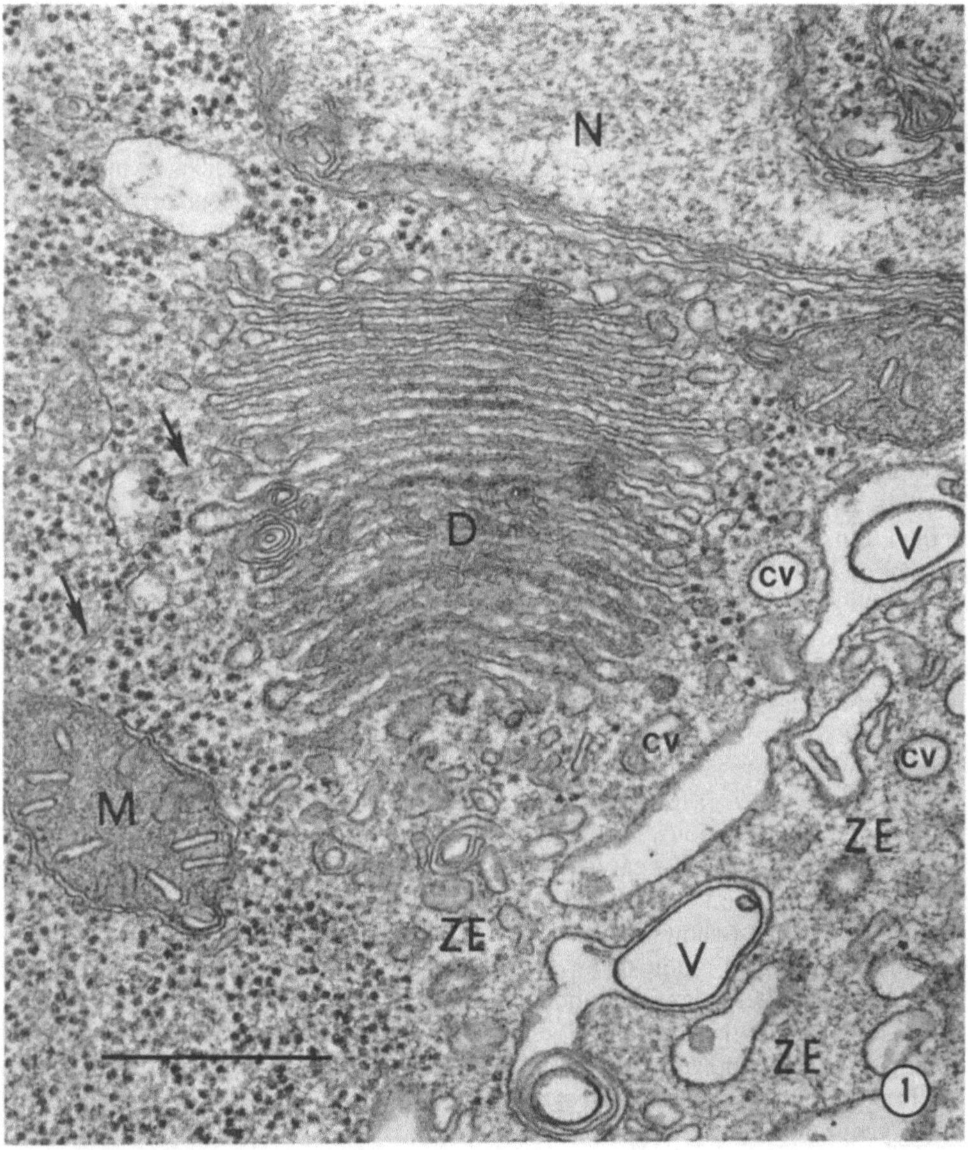

Fig. 1. Thin section of *E. gracilis* showing a dictyosome *D* within a zone of exclusion *ZE* adjacent to and continuous with the zone of exclusion surrounding coated vesicles *cv* and the thick membranes of the water expulsion vacuole *V*. Several microtubules (*arrows*) extend from the dictyosome to the peripheral cytoplasm. *N* nucleus. *M* mitochondrion. Glutaraldehyde-paraformaldehyde-osmium tetroxide fixation (MOLLENHAUER, 1974a). Scale line = 0.5 μ

of secretory products to specified regions of the surface (FRANKE et al., 1972a), or as a suitable milieu for the transformations required for membrane assembly (MORRÉ, 1975) and/or membrane differentiation (GROVE et al., 1968; MORRÉ and MOLLENHAUER, 1974; Fig. 1).

3.1 Biosynthetic Role of the Cytosol

Most biosynthetic pathways include "soluble" enzymes, especially those enzymes involving molecular rearrangements of small molecules. Familiar examples include enzymes of the cytosol per se, e.g. enzymes of glycolysis, of the pentose phosphate pathway, and of amino acid metabolism. "Soluble" enzymes are present also in the cytosol of organelles, e.g. carbon fixing reactions of photosynthesis of the chloroplast stroma and enzymes of the Krebs cycle of the mitochondrial matrix (OAKS and BIDWELL, 1970). In contrast, assembly of macromolecules or macromolecular complexes most frequently involves enzymes and/or precursors structured as part of membranes, or with identifiable regions of the cytoplasm such as ribosomes or portions of chromosomes.

Interactions among the cytosol and the morphologically distinguishable cell components are expressed or implied in all phases of cellular activity. A few examples are given to underscore the ubiquity and complexity of these interactions.

In lecithin biosynthesis, the terminal enzyme, the CDP-choline-1,2-diglyceride transferase, is found exclusively within endomembranes (MORRÉ et al., 1970), chiefly endoplasmic reticulum (LORD et al., 1973; MOORE et al., 1973) and Golgi apparatus (MORRÉ et al., in press). The enzymes of CDP-choline formation, the P-choline cytidyl transferases, are distributed, at least in *Allium cepa,* almost equally between cytosol and endomembranes (MORRÉ et al., 1970). The choline kinase is chiefly in the cytosol (MORRÉ et al., 1970; JOHNSON and KENDE, 1971).

Enzymes of polysaccharide formation, e.g. cellulose synthetase, and presumably those for the formation of lipid intermediates as well (VILLEMEZ and CLARK, 1969), are membrane associated in eukaryotes (HASSID, 1969; LELOIR, 1971; NORTHCOTE, 1969b; see COLVIN and BEER, 1960, for bacterial cellulose) but may be solubilized from the membrane in active form (HASSID, 1969; TSAI and HASSID, 1971; LARSEN and BRUMMOND, 1974). Enzymes of the sugar nucleotide pathway, and other enzymes for the generation of soluble activated intermediates, are largely assumed to reside in the cytosol, but rat liver microsomal membranes contain a biosynthetic pathway for formation of UDP-glucose (BERTHILLIER and GOT, 1974; FROT-COUTAZ et al., 1975). Table 1 shows the distribution of membrane-bound enzymes of β-1,4-glucan formation for onion stem (*Allium cepa*).

A similar pattern of distribution of enzymes, i.e. those early in the pathway being soluble and those late in the pathway being structure-associated, are encountered in protein synthesis (HENDLER, 1974), steroidogenesis (DENNICK, 1972; SCALLEN et al., 1974), nucleic acid synthesis (KIT, 1970), and synthesis of glycolipids and glycoproteins (ROSEMAN, 1970).

The regulated transfer in situ of low molecular weight precursors from cytoplasm to endomembrane compartments has been little studied. This is due in large measure to a lack of applicable experimental techniques. Of greatest importance in such studies is the need to determine local concentrations of substrates, cofactors, ions, and enzymes in the vicinity of various membranes and organelles, rather than

Table 1. Distribution of polysaccharide-synthesizing enzymes among cell-free fractions of onion (*Al. cepa*) stem (C.A. LEMBI and L. ORDIN, unpublished)

Fraction[a]	PGM[b] µmol/h/ mg protein	UDPG-PPase[c] µmol/h/ mg protein	Glucan synthetase[d]		
			CHCl₃-MeOH Sol (mµmol/h/ mg protein)	NaOH Sol (mµmol/h/ mg protein)	NaOH Insol (mµmol/h/ mg protein)
Smooth membranes[e]	0.62	1.57	15.8	2.40	0.40
Dictyosomes	1.36	4.11	102.4	14.64	5.52
Plasma membranes	1.00	0.57	105.2	10.40	3.44

[a] Fractions were obtained from stem axes of onion according to LEMBI et al. (1971).
[b] PGM = phosphoglucomutase.
[c] UDPG-PPase = Uridine diphosphoglucose pyrophosphorylase.
[d] Procedure of ORDIN and HALL (1968) and PINSKY and ORDIN (1969). The UDP-glucose concentration was 6.5 µmolar.
[e] Fraction consisting of membranes not stained after PTA-chromic acid (PACP procedure) for electron microscopic identification of plasma membrane vesicles (ROLAND et al., 1972). For details of fraction compositions, see VAN DER WOUDE et al. (1974).

the total or average cytosol distribution. In practice, this has been difficult to achieve. Use of autoradiography is limited because of extraction and displacement of low molecular weight substances during specimen preparation. With existing cell fractionation procedures, problems of mixing are intensified and, at best, an average cytoplasmic distribution is measured. Usually, interactions are inferred from findings with in vitro incubations or from experiments in which radioactive precursors, provided externally to living cells, are incorporated via metabolic pathways known to be localized within internal compartments.

3.2 Guide Elements of the Cytosol

A function of microtubules and microfilaments as "guide elements" for vectorial movements and transport of cytoplasmic components has been indicated for mitotic chromosomes (PICKETT-HEAPS, 1967b), as well as for translocations of endocytotic and exocytotic vesicles (NEWCOMB, 1969; KAMIYA, 1971; MOLLENHAUER, 1974a; MOLLENHAUER and MORRÉ, 1976). Microtubules are implicated in the translocations of pigment granules in melanophores (BIKLE et al., 1966) and vesicle migrations in neurons (SCHMITT, 1968; DAHLSTRÖM, 1969; KREUTZBERG, 1969; SMITH, 1970; SMITH et al., 1970), during cnidoblast formation in *Hydra* (LENTZ, 1965), rabdite formation in *Planaria* (LENTZ, 1967), lipoprotein and protein secretion in liver (ORCI et al., 1973; STEIN et al., 1974; MORRÉ et al., in press), insulin secretion in pancreatic beta cells (MALAISSE-LAGAE et al., 1971), cellulase secretion in *Achlya* (THOMAS et al., 1974), cell plate formation in cultured cells of endosperm of *Haemanthus katerinae* (HEPLER and JACKSON, 1968), directed migration of cytoplasmic components and wall formation in the coenocytic marine alga *Caulerpa prolifera* (SABNIS and JACOBS, 1967), and in vesicle movements during mitosis (MCINTOSH et al., 1969). In higher plants, microtubules have been discussed in regard to localization of wall deposition during secondary thickenings of xylem elements

(HEPLER and NEWCOMB, 1964; WOODING and NORTHCOTE, 1964; CRONSHAW and BOUCK, 1965; PICKETT-HEAPS, 1966; PICKETT-HEAPS and NORTHCOTE, 1966; ROBARDS, 1968; NORTHCOTE, 1969a, 1971; HEPLER and FOSKETT, 1971; MAITRA and DE, 1971; see O'BRIEN, 1972, for a critical review).

Microfilaments (PARTHASARATHY and MÜHLETHALER, 1972) appear also to be involved with the flow of vesicles in maize root cap and epidermis (MOLLENHAUER and MORRÉ, 1976), neurons (SCHMITT, 1968), and pollen tubes (FRANKE et al., 1972a), and with protoplasmic streaming in slime molds (KOMNICK et al., 1970; WOOLLEY, 1970), *Amoeba* (POLLARD and KORN, 1971), *Nitella* (NAGAI and REBHUN, 1966), *Chara* (PICKETT-HEAPS, 1967c), *Avena* coleoptiles (O'BRIEN and THIMANN, 1966), and cultured nerve cells (YAMADA et al., 1971, and references cited).

Movements thought to be mediated by microtubules are usually characterized by saltatory motion (FREED and LEBOWITZ, 1970; TILNEY, 1971). Cytoplasmic streaming, on the other hand, appears to involve microfilaments (KAMIYA, 1971; WESSELS et al., 1971; see, however, SABNIS and JACOBS, 1967). For example, in pollen tubes of *Lilium* and *Cliva,* concentrations of colchicine that caused the complete disappearance of microtubules did not inhibit either tip growth or cytoplasmic streaming. However, both these processes are sensitive to cytochalasin B and vinblastine (FRANKE et al., 1972a). Similar findings are obtained with migration of secretory vesicles derived from Golgi apparatus of outer cap and epidermal cells of the maize root (MOLLENHAUER and MORRÉ, 1976). Here vesicle migration is unaffected by colchicine, but completely inhibited by cytochalasin B. Treatment with cytochalasin B caused vesicles to accumulate in large numbers at their sites of origin near the Golgi apparatus (Fig. 2). These findings, plus those of the effect of cytochalasin B on other tip growing systems such as root hairs of *Brassica* and rhizoids of the coenocytic green alga *C. prolifera* (HERTH et al., 1972a), are consistent with a role of microfilamentous structures in these processes. However, possible effects of cytochalasin B on lateral cross-link elements (ESTENSEN and PLAGEMANN, 1972; FRANKE et al., 1972a) and direct effects on membranes (FRANKE et al., 1972a; KLETZIEN and PERDUE, 1973; MIRANDA et al., 1974b; POLLARD and WEIHING, 1974) or regions of membrane-microfilament association (STAEHELIN et al., 1972), emphasize the need for caution in specifically associating all processes inhibited by cytochalasin B only with microfilaments.

3.3 Milieu for Biosynthesis and Transformation (Zones of Exclusion)

Often cell-free systems, especially of protein and cellulose synthesis as examples, are only a few percent as efficient in vitro as they are in situ. Some of the loss in efficiency may be attributed to damage incurred during isolation. However, it is also possible that critical cytoplasmic factors have been lost or that some special relationship of cell component to cytosol is required for optimum activity. Examples include stimulating factors or noncatalytic "carrier" proteins necessary for steroid biogenesis (DENNICK, 1972; SCALLEN et al., 1974), biosynthesis of triglycerides (MANLEY and and SKRDLANT, 1974; RONCARI, 1974), and the metabolism of kaurine (MOORE et al., 1972) by endoplasmic reticulum membranes.

An example of an organized region of the cytosol which may facilitate either membrane biogenesis or membrane transformations (including flow and transloca-

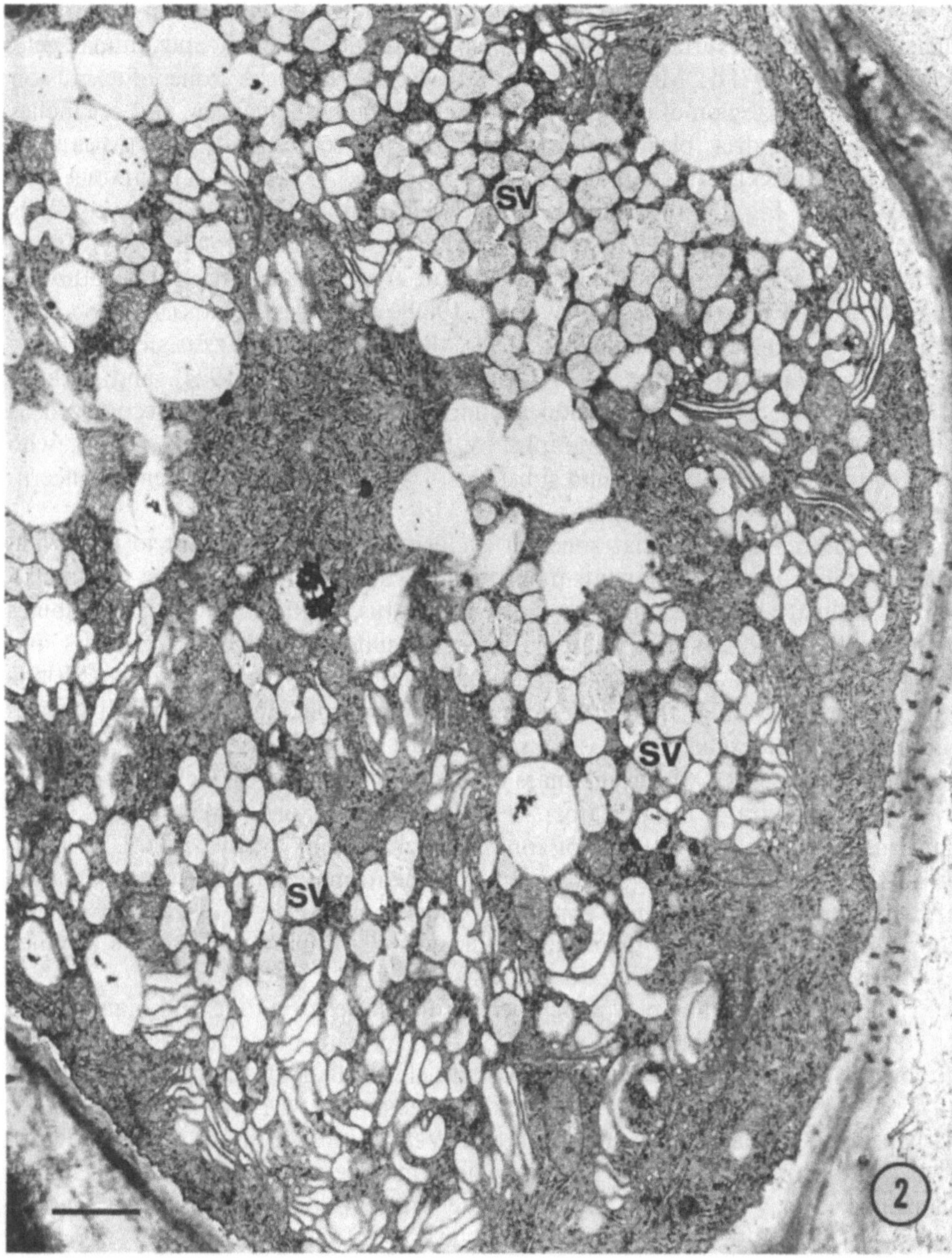

Fig. 2. A portion of an outer cap cell of the root tip from a maize (*Z. mays*) seedling treated with 100 μg/ml cytochalasin B in 1% dimethylsulfoxide for 2 h. Note the heavy concentration of secretory vesicles *sv* in the inner portions of the cytoplasm and the lack of secretory vesicles near the cell surface. This distribution confirms an origin of the secretory vesicles in the slime-producing cells from the Golgi apparatus and demonstrates that cytochalasin B prevents the vectorial migration of mature secretory vesicles to the cell surface (MOLLENHAUER and MORRÉ, 1976). Glutaraldehyde-paraformaldehyde-picric acid-osmium tetroxide fixation. Scale line = 1 μ

tion), or both, are the zones of exclusion which surround the Golgi apparatus, nuclei, and cortical surface of the plasma membrane of plant and animal cells (MORRÉ et al., 1971d; MOLLENHAUER and MORRÉ, 1972). A zone of exclusion is a differentiated region of cytoplasm in which ribosomes, glycogen, and organelles such as mitochondria, plastids, or microbodies are scarce or absent (MORRÉ et al., 1971d; FRANKE et al., 1972b; MIRANDA et al., 1974a; MOLLENHAUER and MORRÉ, 1975; Fig. 1). In addition to the cell components listed above, zones of exclusion surround Golgi apparatus equivalents (BRACKER, 1967, 1968; MORRÉ et al., 1971d), water expulsion vacuoles (MORRÉ et al., 1971d; Fig. 1), microtubules (LEDBETTER and PORTER, 1963; NEWCOMB, 1969), centrioles (SOROKIN, 1968; FULTON, 1971; STRAPRANS and DIRKSEN, 1974), the cell surface (WESSELS et al., 1971; MORRÉ and VANDERWOUDE, 1974; POLLARD and WEIHING, 1974), and perhaps other organelles. Zones of exclusion around Golgi apparatus were recognized as early as 1954 in electron micrographs by SJÖSTRAND and HANZON (1954), who referred to them as "Golgi ground substance." Yet there is little evidence concerning their function or composition.

It has been postulated that zones of exclusion may be important to the origin and continuity of cell components by providing a suitable milieu for the multiplication of cellular structures such as Golgi apparatus, centrioles, and microtubules (MORRÉ et al., 1971d). For example, in early stages of oögenesis (ADAMS and HERTIG, 1964; YAMAMOTO, 1964; WARD and WARD, 1968), and in seeds (MORRÉ et al., 1971d), the Golgi apparatus seems to develop from clusters of small vesicles within a zone of exclusion. WARD and WARD (1968) favored the view that the vesicles arose de novo within the mass of fibers in the zone of exclusion, possibly by the combination of the fine fibers with lipid. KARTENBECK and FRANKE (1971) have suggested that constituents of zones of exclusion may provide pools of intracellular lipoprotein aggregates to be utilized in the formation and/or transformation of dictyosome cisternae. Thus, zones of exclusion might serve as repositories for constituents removed from membranes during transformation or provide a milieu in which membrane transformations and assembly take place.

Fibrous elements may play a significant role in the formation and maintenance of zones of exclusion (FRANKE et al., 1972b). As with other zones of exclusion, the one at the cell periphery often appears finely fibrillar or amorphous (WESSELS, 1971; WESSELS et al., 1971; AXLINE and REAVEN, 1974; MIRANDA et al., 1974a; MORRÉ and BRACKER, 1974a; POLLARD and WEIHING, 1974; SINGH, 1974; MOLLENHAUER and MORRÉ, 1972). The fibrous regions may be organized as distinct microfilaments or may appear as dense felt-like masses along the inner cell surface (WESSELS, 1971; WESSELS et al., 1971; LE BEAUX, 1972; SCHROEDER, 1973; SPOONER, 1973; AXLINE and REAVEN, 1974; MIRANDA et al., 1974a; SINGH, 1974). The felt-like masses may represent depolymerized microfilaments (WESSELS, 1971; AXLINE and REAVEN, 1974; MIRANDA et al., 1974a). In some cells at least, both the microfilaments and the felt-like masses are contractile (SCHROEDER, 1973) and are capable of binding heavy meromyosin (MIRANDA et al., 1974a). Therefore, these components of the zone of exclusion may contain actin or actin-like proteins (ISHIKAWA et al., 1969).

Fuzzy, alveolate, or coated membranes and vesicles are a consistent feature of all zones of exclusion (FRANKE et al., 1971a; MORRÉ et al., 1971d; MOLLENHAUER and MORRÉ, 1972; MOLLENHAUER et al., in press; Fig. 1). The fuzzy or coated appearance of the membranes is due to the presence of a nap-like, electron-dense coating. The coat appears to be derived from irregular hexagonal subunits which often give coated membranes and vesicles an alveolate (honeycomb) appearance (BONNETT and NEWCOMB, 1966; NEWCOMB, 1967; SLAUTTERBACK, 1967). Contiguous parallel sides of the hexagons appear in sections as bristles 150 to 200 Å in length. The coated vesicles of the Golgi apparatus, nuclear envelope, and endoplasmic reticulum are distinct from the coated vesicles of pinocytosis, but both types of vesicles have been suggested to function in intracellular transport (FRIEND and FARQUHAR, 1967).

Coated membranes and vesicles seem to be a nearly-exclusive feature of zones of exclusion (MORRÉ et al., 1971d). The specialized coatings may reflect some property of membrane transfer processes (vesicle formation, vesicle migration, tubule extension), or the differentiation of membranes or both. As such, coated membranes are characteristic of cellular regions which function in membrane flow and differentiation. Zones of exclusion may provide the milieu favoring membrane differentiation and expansion, and the formation of alveolate coats may facilitate either one or both processes.

The extent to which the rest of the cytoplasm is structured or differentiated is unknown. The colloidal nature of the fluid cytoplasm is generally accepted from cytological and physiological studies, as is the existence of microfilaments, multienzyme complexes and other organized aggregates in this region. Similarly, membrane lumina (the inner spaces of Golgi apparatus and ER cisternae), and the plastid matrix surrounding the thylakoids (FRANKE et al., 1971b), or the mitochondrial matrix, may be structured. Bacterial cells with a minimum of internal membrane organization might be expected to rely heavily on a structured cytosol to achieve compartmentation. Microorganization of the cytoplasm may be of a degree of importance equal to, or greater than, the more familiar organization afforded by cellular membranes (MOLLENHAUER and MORRÉ, 1972).

4. Role of Internal Membranes

Three aspects of the role of internal membranes in the intracellular transport and exchange of constituents and components of the protoplast surface will be discussed, first as separate topics, and then integrated in the final sections of the chapter.

4.1 Role in the Biosynthesis, Assembly, Transformation, and Transport of Membranes

Potential origins of surface membranes (plasma membranes and tonoplast from endoplasmic reticulum *via* well-established secretory routes are extensively documented (FRANKE et al., 1971a; MORRÉ et al., 1971d; MORRÉ and MOLLENHAUER, 1974; NORTHCOTE, 1974). The participating system of membranes, the endomembrane system, consists of generating elements, transition elements, and endproducts (MORRÉ and MOLLENHAUER, 1974). Each form a part of a functionally interconnected and interrelated system of internal membrane compartments (Fig. 3) to facilitate and insure the ordered transport of materials from sites of synthesis to sites of utilization. Processes of membrane flow and differentiation are combined to account for the derivation of diverse membrane types from common progenitor membranes, endoplasmic reticulum, and nuclear envelope (MORRÉ and MOLLENHAUER, 1974).

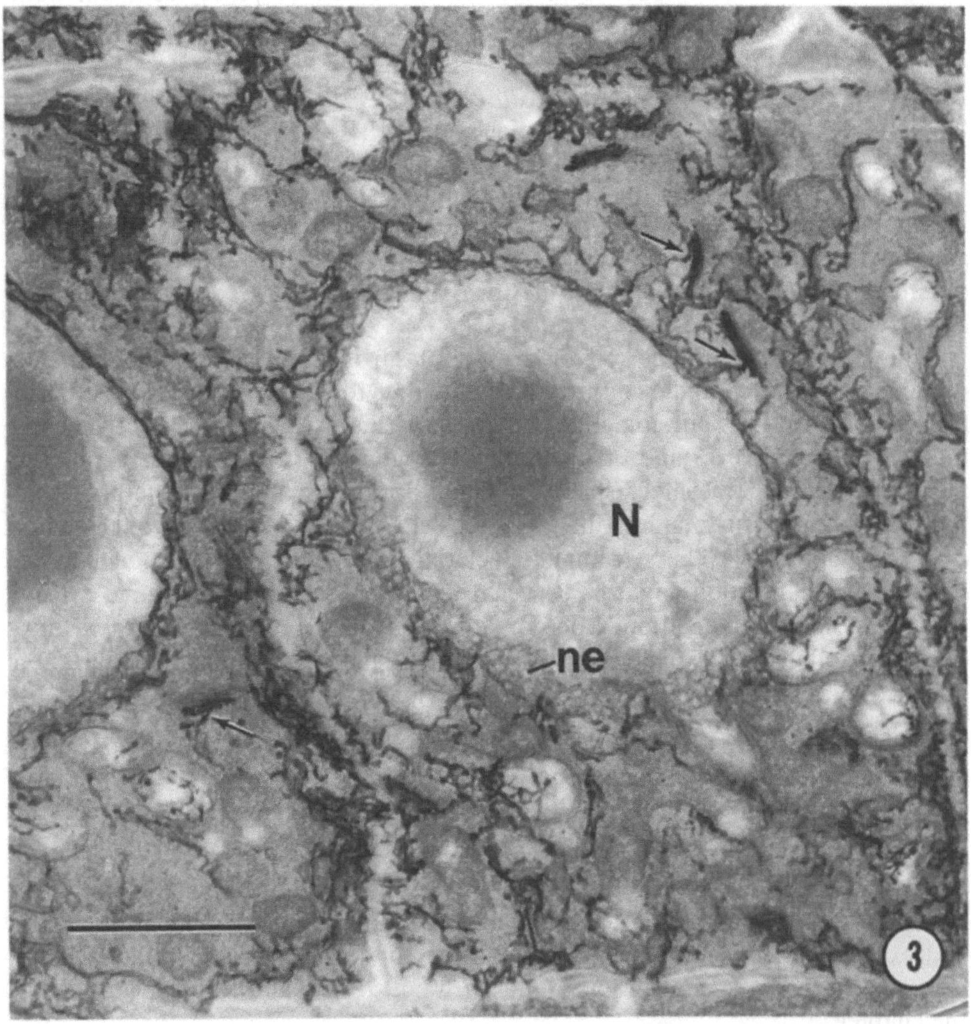

Fig. 3. The endomembrane system of a plant cell as visualized using thick sections and high voltage electron microscopy. The section, 0.5 μ thick, is of a cortical cell of the root of *Ricinus* (castor bean) impregnated with osmium tetroxide (POUX, 1973). Deposits of reduced osmium fill the lumens of the nuclear envelope, the endoplasmic reticulum, and the dictyosome cisternae nearest the forming face. The nuclear envelope and its pores *ne* can be seen through the thickness of the section as well as numerous points of continuity between the nuclear envelope and the endoplasmic reticulum. Apparent connections between endoplasmic reticulum and dictyosomes are also numerous (*arrows*). Information from thick sections reinforces the concept of an endomembrane system, as opposed to the view that each cell component is independent of the others, as is sometimes gained from conventional thin sections. *N* nucleus. Electron micrograph courtesy of N. POUX, Ecole Normale Supérieure, Paris, and N. CARASSO and P. FAVARD, Centre de Recherches d'Ivry, C.N.R.S., Ivry, France. Scale line = 2 μ

4.1.1 The Endomembrane System

The concept of an endomembrane system (MORRÉ et al., 1971d; MORRÉ and MOLLENHAUER, 1974) was proposed to explain the functional continuum within which processes of membrane differentiation and physical transfer of membrane from one compartment to another (membrane flow) could account for the biogenesis of the cell's internal and external membranes (Fig. 4). The endomembrane system is a term for the structural and developmental continuum of internal membranes which characterize the cytoplasm of eukaryotic cells. Included within the endomembrane system are nuclear envelope, rough and smooth endoplasmic reticulum, Golgi apparatus, and various cytoplasmic vesicles. Plasma membrane, vacuole membranes, and lysosomes, are regarded as endproducts of the system. Endomembrane components are distinguished biochemically from the semiautonomous organelles (with definite inner and outer membrane systems), such as chloroplasts and mitochondria, by (1) the absence of DNA and cytochrome oxidase; (2) the inability to generate ATP through respiratory or photosynthetic chain-linked phosphorylation; and (3) a high degree of functional interdependence.

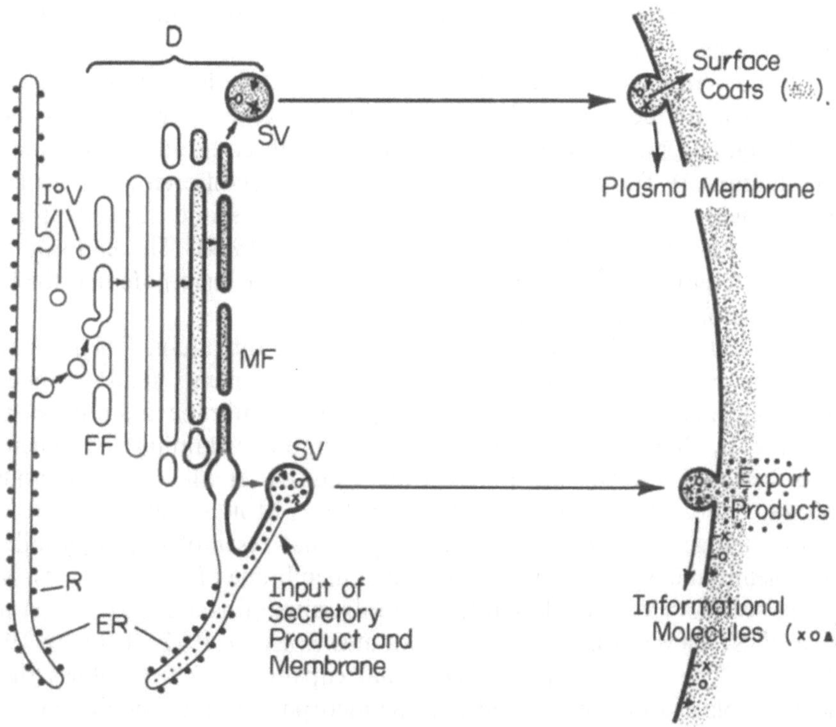

Fig. 4. Diagrammatic representation of endomembrane functioning in membrane flow and differentiation

Various types of evidence for the operation of an endomembrane system within eukaryotic cells have been summarized (FRANKE et al., 1971a, c; MORRÉ et al., 1971d; MORRÉ and MOLLENHAUER, 1974; MORRÉ et al., 1974a; NORTHCOTE, 1974; MORRÉ, 1975) and only major aspects and new information will be emphasized here. The evidence is as follows:

(a) Morphological: Membrane connections and associations. Endomembrane components are organized in the cell as a morphological continuum: nuclear envelope --→ rough endoplasmic reticulum --→ smooth endoplasmic reticulum --→ Golgi apparatus and other transition elements --→ secretory vesicles --→ endproducts.

Direct membrane continuities and potential ontogenetic relationships between each of the separate components is well documented (BRACKER et al., 1971; FRANKE et al., 1971a; CARASSO et al., 1971, 1974; MORRÉ et al., 1971d; CAROTHERS, 1972; CHARDARD, 1973; FAVARD and CARASSO, 1973; FINERAN, 1973b; MARTY, 1973b, c; PEEL et al., 1973; POUX, 1973; MORRÉ and MOLLENHAUER, 1974). Additionally, continuities occur between endomembrane components and the outer envelope systems of mitochondria (BRACKER and GROVE, 1971; FRANKE and KARTENBECK, 1971; MORRÉ et al., 1971c) and plastids (CRAN and DYER, 1973; CROTTY and LEDBETTER, 1973). The potential for interconnection is more readily visualized from thick electron microscope sections of plant tissues, where lumina of internal membranes are impregnated with heavy metals (CARASSO et al., 1974; Fig. 3). Here, multiple direct nuclear envelope-endoplasmic reticulum-Golgi apparatus connections are visible.

(b) Patterns of membrane differentiation parallel those of connections and associations.

Endomembrane characteristics display a morphological, structural, chemical, and enzymatic continuum beginning with the generating elements (endoplasmic reticulum and nuclear envelope) and ending with endproducts (plasma membrane and tonoplast). The generating elements are the least differentiated of the endomembranes with the greatest potential for biosynthesis and capacity for membrane transformation (MORRÉ et al., 1971d; FRANKE, 1974). The endproducts are the most differentiated of the endomembranes with the least potential for biosynthesis and transformations (MORRÉ, 1975).

The transition elements (smooth endoplasmic reticulum, Golgi apparatus, vesicles) are intermediate between the generating elements (endoplasmic reticulum and nuclear envelope) and endproducts (plasma membrane and tonoplast). Electron microscopy shows progressive changes in membrane staining and dimensions from one face of a dictyosome to the other (GROVE et al., 1968; SCHNEPF, 1969b; MORRÉ et al., 1971d; VIAN and ROLAND, 1972; MORRÉ and MOLLENHAUER, 1974). These changes are from a membrane type that resembles endoplasmic reticulum to a membrane type that resembles plasma membrane (Table 2; Fig. 5). Membranes of the secretory vesicles elaborated by Golgi apparatus are also plasma membrane-like (VAN DER WOUDE et al., 1971; VIAN and ROLAND, 1972; ROLAND, 1973; Fig. 5; Table 2). This imparts special functional capabilities which allows them to fuse with the plasma membrane and become incorporated into the existing cell surface.

This feature of Golgi apparatus functioning, the differentiation of membranes, emerges as a major function of Golgi apparatus. It is found in Golgi apparatus regardless of what type of contents characterizes the interiors of the vesicles. It is found in Golgi apparatus where no conspicuously large secretory vesicles are produced. It occurs in smooth membranes, other than Golgi apparatus, which appear to function as transition elements in the transfer and transformation of cellular membranes (FRANKE et al., 1971a; FRANKE and ECKERT, 1971; MORRÉ et al., 1971d). In *Cyanidium,* a unicellular alga, changes in structure of the plasma membrane were demonstrated by freeze-fracture-etching (STAEHELIN, 1968). In old cells, the plasma membrane is characterized by arrays of hexagonally-packed particles or by long striated folds, but, just prior to cell division, all folds and differentiations of the membrane disappear so that the membrane is undifferentiated and "embryonic" in appearance again. Differentiation is an aspect of membranes that may be universal, especially among Golgi apparatus and other transition elements. Thus Golgi apparatus or equivalent systems of transition elements (BRACKER, 1967, 1968; GIRBARDT, 1969; FRANKE et al., 1971a; FRANKE and ECKERT, 1971; MORRÉ et al., 1971d; MORRÉ and MOLLENHAUER, 1974) are a potential major supplier of certain types of specialized membranes to the cell, especially in terms of contributions to the cell surface.

The biochemical basis for membrane differentiation has been sought through studies that compare endoplasmic reticulum, Golgi apparatus, and plasma membrane fractions isolated from rat liver (MORRÉ et al., 1971b and ref. cit.; MORRÉ et al., 1971a; MORRÉ et al., 1974a; MORRÉ and MOLLENHAUER, 1974; MORRÉ et al., in press). If Golgi apparatus function in the conversion of endoplasmic reticulum membranes to plasma membranes, the composition of Golgi apparatus membranes should reflect this transformation (KEENAN and MORRÉ, 1970). Similarly, a comparison of endoplasmic reticulum membranes and plasma membranes will indicate the biochemical changes required to effect the transformation.

Table 2. Dimensions of membranes from plant stems. Tissue was fixed with glutaraldehyde-osmium tetroxide and sections were post-stained either with alkaline lead citrate or by the PACP procedure (ROLAND et al., 1972) (D.J. MORRÉ and F.M. TWOHIG, unpublished)

Membrane type	Membrane thickness (Å)	
	Soybean (*Glycine max*) hypocotyl Lead-stained	Onion (*Al. cepa*) stem PACP-stained
Nuclear envelope		
Outer leaflet	56	–
Inner leaflet	56	–
Endoplasmic reticulum	56	64
Golgi apparatus		
Forming face cisterna 1	56	64
Intercalary cisterna 2	60	73
Intercalary cisterna 3	64	78
Maturing face cisternae 4 (5)	69	91
Secretory vesicle	78	106
Plasma membrane	88	112
Tonoplast	80	106
Microbody	56	69
Mitochondria		
Outer membrane	56	62
Inner membranes	56	56
Etioplast		
Outer membrane	56	–
Inner membranes	56	–

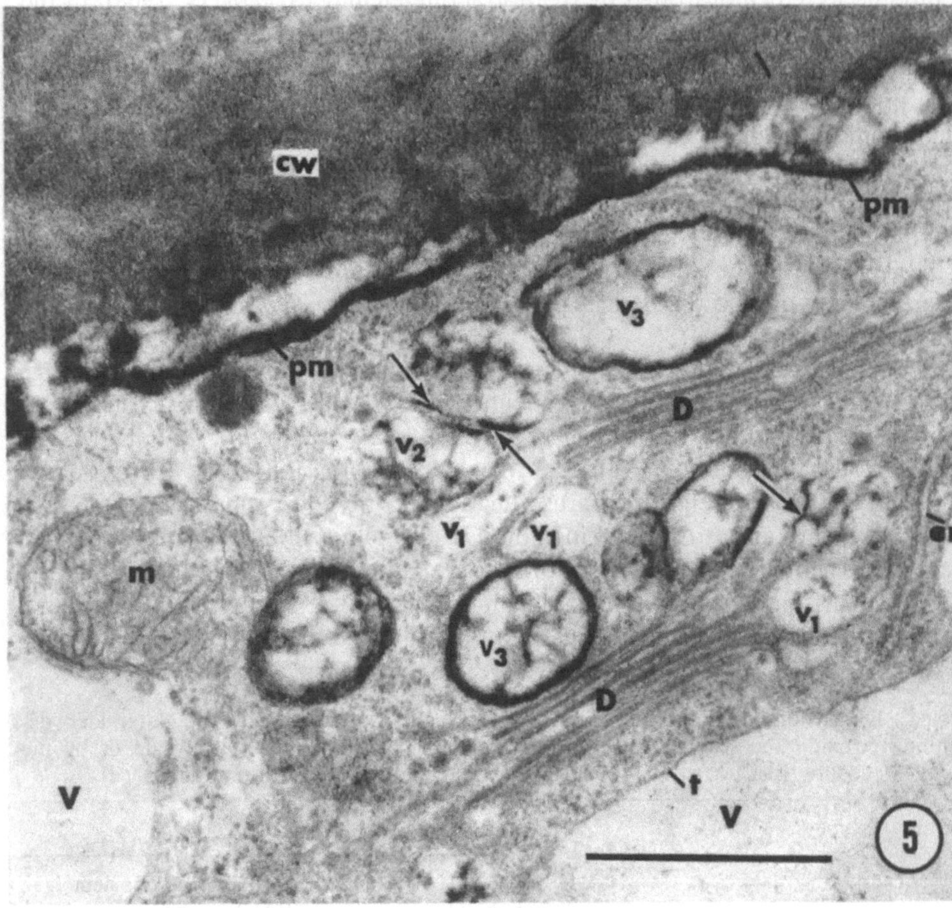

Fig. 5. Differentiation of membranes of pea (*Pisum sativum*) root cap as demonstrated using the PACP procedure for plasma membranes of plants (ROLAND, 1969; ROLAND et al., 1972). Membranes of mature secretory vesicles v_3 of dictyosomes D stain intensely in the same manner as plasma membranes *pm*, while membranes of immature secretory vesicles v_1, endoplasmic reticulum *er*, and tonoplast *t* are unstained. Staining of vesicles of intercalary cisternae appears in patches (*arrows*). *m* mitochondrion. *V* vacuole. *cw* cell wall. (From VIAN and ROLAND, 1972.) Scale line=0.5 μ

The transitional nature of Golgi apparatus membranes, revealed first from morphological studies (GROVE et al., 1968), is reflected in the lipid and protein composition of the membranes for liver (KEENAN and MORRÉ, 1970; YUNGHANS et al., 1970; MORRÉ et al., 1974a) and in the organization of lipids and proteins within the membrane for liver (MORRÉ et al., 1974a), and other tissues (STAEHELIN and KIERMAYER, 1970; VIAN, 1972, 1974; FINERAN, 1973b). Phospholipid and fatty acids of the major lipid classes found in Golgi apparatus are intermediate between those of the endoplasmic reticulum (or nuclear envelope) and plasma membrane (KEENAN and MORRÉ, 1970; FRANKE, 1974). At present levels of resolution, all endomembrane fractions (rough endoplasmic reticulum, smooth endoplasmic reticulum, Golgi apparatus, and plasma membrane) have major protein bands in common, based on analyses by polyacrylamide disc gel electrophoresis comparing apparent molecular weights (MORRÉ et al., 1974a; ELDER and MORRÉ, 1976). Enzymatic activities characteristic of plasma membranes, i.e. plasma membrane marker enzymes, appear to be acquired at the Golgi apparatus, whereas enzyme activities characteristic of endoplasmic reticulum membranes appear to be lost. Biochemical findings with pancreas (MELDOLESI et al., 1971a, b, c) and bovine and rat mammary gland (KEENAN et al., 1974b) are consistent with a transitional nature for Golgi apparatus in these tissues as well. For a discussion of the opposite view, involving limited intermixing of membranes, see FLEISCHER and FLEISCHER (1971), MELDOLESI and COVA (1972), BERGERON et al. (1973a), and DAUWALDER and WHALEY (1974b).

Both the plasma membrane and the membranes of Golgi apparatus are asymmetrically substituted with specific carbohydrate groups (ROLAND, 1969, 1973; PINTO DA SILVA et al., 1971; VAN DER WOUDE et al., 1971; ROUGIER et al., 1973). These carbohydrate-containing constituents of membranes are only beginning to be studied in plants (ROBERTS et al., 1971; ROBERTS and POLLARD, 1975; CLARK et al., 1975; glycolipids of plastids being an exception, MAZLIAK, 1973), but in animal cells, these molecules form a significant pool of informational molecules potentially important to immune responses and to the control of many of the diverse activities of the cell (COOK and STODDART, 1973; KEENAN and MORRÉ, 1973). Their precise locations are visualized with the electron microscope using carbohydrate complexing reagents combined with heavy metals (THIÉRY, 1969; VAN DER WOUDE et al., 1971; OVTRACHT and THIÉRY, 1972; ROLAND, 1973; ROUGIER et al., 1973), lectins coupled with ferritin or other markers (HIRANO et al., 1972), or other stains (ROLAND and VIAN, 1971; ROLAND, 1973; ROUGIER et al., 1973). Because of the manner in which the vesicles fuse with the plasma membrane, the inner surfaces of the cisternae or vesicles of the Golgi apparatus are equivalent to the external surfaces of the plasma membranes. The biosynthetic enzymes (ROSEMAN, 1970) which attach the carbohydrate residues to glycoproteins (MORRÉ et al., 1969; SCHACHTER et al., 1970; SCHACHTER, 1974) and glycolipids (KEENAN et al., 1974b) are localized primarily or even exclusively in the Golgi apparatus. Thus, Golgi apparatus appears to be the origin of specific membrane components, some of which may be of considerable importance in determining surface characteristics of cells (see Sect. 6).

(c) Biosynthetic machinery for major protein and lipid constituents of membranes is localized exclusively in the generating elements (rough endoplasmic reticulum and nuclear envelope).

In rat liver, at least, the majority of the hydrophobic membrane proteins of endoplasmic reticulum appear to be synthesized on polyribosomes associated with rough endoplasmic reticulum (and nuclear envelope?) (ELDER and MORRÉ, 1976). Isolated plasma membranes are unable to synthesize membrane proteins. Polysomes of the Golgi apparatus zone (FRANKE et al., 1972b; MOLLENHAUER and MORRÉ, 1974; MORRÉ et al., in press), as well as free polysomes of the cytosol, may participate also in the synthesis of membrane proteins (ELDER and MORRÉ, 1976). Finally, the terminal enzymes of glycerolipid biosynthesis are localized in endoplasmic reticulum of both plant and animal cells (VAN GOLDE et al., 1971, 1974; LORD et al., 1973; MOORE et al., 1973; MORRÉ et al., in press). Although contributions to phospholipid biosynthesis by Golgi apparatus have been clearly shown (MORRÉ, 1970; MORRÉ et al., 1971b; MORRÉ et al., in press), plasma membranes do not synthesize glycerolipids (VAN GOLDE et al., 1974; MORRÉ et al., in press; JELSEMA and MORRÉ, in prep.).

Initial steps in glycosylations of lipids and proteins to form glycolipids and glycoproteins also may occur in endoplasmic reticulum. Terminal reactions are more clearly localized in the Golgi apparatus. The potential contribution of plasma membrane to protein and lipid glycosylation is less certain in animal cells (KEENAN and MORRÉ, 1975) and in plants (VAN DER WOUDE et al., 1974).

(d) Kinetics of in vivo labeling of different endomembrane fractions from membrane precursors supplied exogenously suggest the following sequence: (1.) initial incorporation into membranes of rough endoplasmic reticulum, (2.) subsequent transfer to transition elements and endproducts, (3.) accumulation of the synthesized products in endproducts of the endomembrane system, and (4.) gradual turnover and replacement of endproducts with a $t_{\frac{1}{2}}$ of degradation of about 2 days (FRANKE et al., 1971c; MORRÉ, 1975).

In vivo experiments with rats show that isotopically labeled amino acids are rapidly incorporated into proteins of the rough endoplasmic reticulum and nuclear envelope without discernable lag (less than 1 min). After labeling of rough endoplasmic reticulum and nuclear envelope, radioactive proteins subsequently appear in Golgi apparatus and plasma membranes (MORRÉ, 1975). The lag between labeling of rough endoplasmic reticulum and other cell components varies from a few minutes for smooth endoplasmic reticulum to more than 20 min for plasma membrane.

In studies with plants (Fig. 6), incorporation of ^{14}C-leucine into microsomes (endoplasmic reticulum) and nuclear fractions was linear, following a lag in uptake of the radioactive amino acids. Incorporation into dictyosomes and a mixed plasma membrane-tonoplast fraction lagged 30–60 min behind incorporation into microsomes.

Less is known about the kinetics of incorporation of lipid constituents into different endomembrane components. Kinetics of incorporation of ^{14}C-choline show that labeled lecithin appears first in a rough microsome (=endoplasmic reticulum) fraction (KAGAWA et al., 1973) and later appears in mitochondria and glyoxysomes. The relative specific activity of lecithin is microsomes > dictyosomes > plasma membrane + tonoplast (MORRÉ, 1970).

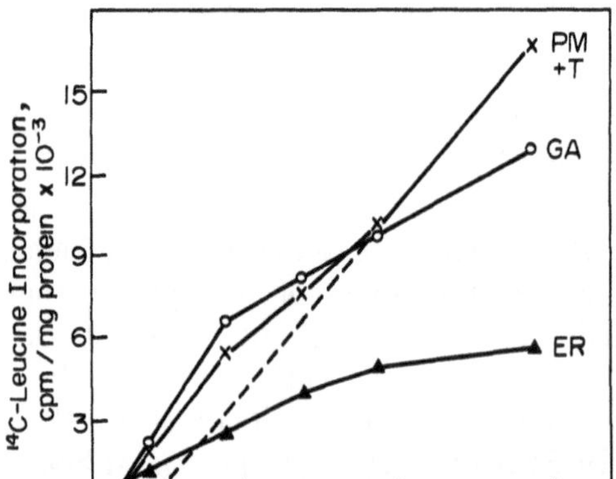

Fig. 6. Kinetics of incorporation of radioactivity from U-^{14}C-leucine into membrane proteins of onion stem (MORRÉ, 1970). Data are corrected for a lag in uptake of the amino acid and expressed on a protein basis. $PM+T$ = plasma membrane + tonoplast. GA = Golgi apparatus. ER = endoplasmic reticulum. (Adapted from MORRÉ and VAN DER WOUDE, 1974)

In the biosynthesis of glycolipids and glycoproteins of membranes, the attachment of the sugars is probably a late event in both plants (CHRISPEELS, 1970) and animals (MORRÉ, 1975). Thus a "multistep" mechanism is indicated for membrane biogenesis within the endomembrane system. Primary synthesis of a "basic" membrane of lipid and protein is accomplished by endoplasmic reticulum and nuclear envelope, to which additional components, e.g. enzymes and sugar moieties, are added sequentially by Golgi apparatus and other transition elements (MORRÉ, 1975 and ref. cit., see also RICHARD and LOUISOT, 1972).

4.1.2 Membrane Differentiation

Membrane differentiation is defined as the progressive change in the appearance, composition, organization, and functional specialization observed among endomembranes. These changes are reflected in measurements of membrane thickness (Table 2). As emphasized in section 4.1.1, differences shown from membrane measurements are reflected in biochemical and other parameters. Differentiation follows established export pathways and occurs from endoplasmic reticulum-like to plasma membrane-like or from endoplasmic reticulum-like to tonoplast-like depending on the pathway (MORRÉ and MOLLENHAUER, 1974). To establish the concept of membrane differentiation in plants, a special staining procedure (RAMBOURG, 1969; ROLAND, 1969; ROLAND and VIAN, 1971) specific for the plasma membrane of plants has been used. The procedure involves destaining of thin sections of glutaraldehyde-osmium tetroxide-fixed tissues with periodic acid, followed by staining with 1% phosphotungstic acid in 10% chromic acid (PACP). Only plasma membranes and membranes resembling plasma membranes are stained. The staining is sufficiently specific to identify vesicles of plasma membranes in cell fractions (ROLAND et al., 1972).

ROLAND and colleagues (ROLAND, 1969, 1973; ROLAND and VIAN, 1971; ROLAND et al. 1972; VIAN and ROLAND, 1972; FRANTZ et al., 1973) have used PACP staining to establish that secretory vesicles formed by Golgi apparatus acquire plasma membrane characteristics in advance of their fusion with the plasma membrane (Fig. 5). The transformation begins in patches and the appearance of PACP staining coincides with the accumulation of fibrillar materials within the vesicle interiors (Fig. 5).

This one aspect of membrane differentiation, the acquisition by Golgi apparatus of PACP staining reactivity characteristic of plasma membrane, has been demonstrated in vitro. FRANTZ et al. (1973) showed that dictyosomes of soybean hypocotyls did not stain (except for vesicles) with the PACP procedure, either in situ or when freshly isolated. However, after incubation of pellets of isolated dictyosomes for either 2 h at 25° or 8 h at 0°, the membranes of dictyosomes acquired PACP reactivity. Acquisition of staining reactivity began at the mature pole or face of the dictyosome and progressed toward the forming pole or face of the dictyosome.

Differentiation of membranes is probably a general phenomenon of transition elements and endproducts and may be a major feature of the mechanism whereby surface membranes acquire specific characteristics (FALK, 1969; FRANKE and ECKERT, 1971; FRANKE et al., 1971a; MORRÉ et al., 1971d; DOBBERSTEIN and KIER-

MAYER, 1972; WHALEY et al., 1972; KIERMAYER and DOBBERSTEIN, 1973; MORRÉ and VAN DER WOUDE, 1974).

4.1.3 Membrane Flow

Membrane flow, the physical transfer of membrane from one compartment to another within the cell (FRANKE et al., 1971c), is implicit in most discussions of secretion and surface growth by vesicular additions. Secretory vesicles which originate at the Golgi apparatus migrate to and fuse with the plasma membrane. They provide the clearest example of a flow mechanism. An origin of secretory vesicles at Golgi apparatus has been visualized recently for plant roots in experiments where cytochalasin B was used to block vesicle migration (MOLLENHAUER and MORRÉ, 1976) so that vesicles accumulate in large numbers at the mature face around each dictyosome (Fig. 2).

Membrane flow from nuclear envelope to endoplasmic reticulum and from endoplasmic reticulum to Golgi apparatus has been more difficult to establish. Progressive appearance of membrane proteins labeled with amino acids (RAY et al., 1968; FRANKE et al., 1971c), labeled membrane lipids (MORRÉ et al., 1974a), and drug-induced NADPH-mixed function oxidases (MORRÉ et al., 1974a) in nuclear envelope, rough endoplasmic reticulum, smooth endoplasmic reticulum, and Golgi apparatus are consistent with a flow mechanism. Yet, the argument always remains that the actual movement is through the cytosol from one membrane piece to the next (MORRÉ et al., in press). A visual marker attached to the membrane is needed to resolve this question. The marker should be of a type that can be recognized both when attached to the membrane and when free in the cytoplasm. Such a marker is provided by the mastigonemes or flimmer of flagella. These extracellular adornments are firmly based in the plasma membrane surrounding the flagella of a variety of motile cells (BOUCK, 1969). The mastigonemes are assembled initially within internal endomembranes and are then transported to the ultimate site of flagellar attachment. The precise function of the mastigonemes in unknown, but in some organisms the presence of mastigonemes correlates with an ability to locomote in the same direction as the wave propagation along the flagellum (BOUCK, 1969, 1971).

In the green alga *Ochromonas* (BOUCK, 1969, 1971; HILL, 1973; HILL and OUTKA, 1974), mastigonemes are arranged on the flagella, either as singlets or in tufts, attached in two unbalanced files (one of singles, one of tufts) on nearly opposite sides of the flagellum. In this organism, flagella released mechanically are replaced; the reappearance of mastigonemes provides a definitive marker to monitor the secretory route of the plasma membrane of the flagellum.

Flagella released mechanically are replaced over the course of several hours (Fig. 7). During flagellar replacement, tubular mastigonemes appear early within cytoplasmic compartments. In the work of HILL (1973; HILL and OUTKA, 1974), mastigonemes were most concentrated in the perinuclear space between the inner and outer membranes of the nuclear envelope, attached to the membrane, in the first 10 min following deflagellation. Mastigonemes appeared also in endoplasmic reticulum cisternae adjacent to the nuclear envelope and in endoplasmic reticulum cisternae adjacent to dictyosomes.

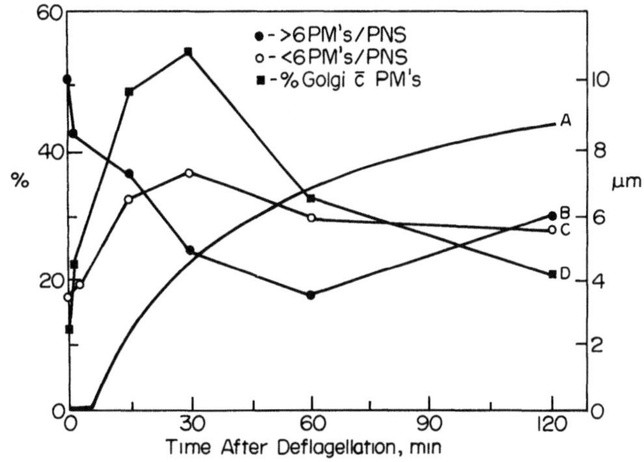

Fig. 7. Intracytoplasmic distribution of perinuclear mastigonemes *PMs* from thin sections of *Ochromonas minute* fixed at intervals after deflagellation. Curve A: Reflagellation reference curve determined from light microscopy of living cells. Curve B: % mid-nuclear sections with greater than 6 mastigonemes/perinuclear space. Curve C: % mid-nuclear sections with less than 6 mastigonemes/perinuclear space. Curve D: % Golgi apparatus with mastigonemes. (From HILL, 1973 and HILL and OUTKA, 1974 through the courtesy of D.E. OUTKA, Iowa State University, Ames, Iowa)

Before the mastigonemes reached the Golgi apparatus, they were intermixed as tufts and single mastigonemes. However, between 15 and 30 min after deflagellation, mastigonemes appeared in secretory vesicles of the Golgi apparatus, which occupied a position near the new flagellar base, and the singles and tufts were segregated (HILL, 1973; HILL and OUTKA, 1974). The single mastigonemes were concentrated in vesicles on one aspect of the dictyosome, while tufts of mastigonemes were concentrated in vesicles of the opposite aspect. Apparently, upon fusion of the vesicles with the plasma membrane, the new flagellar membranes were delivered with the mastigonemes in the correct final orientation to yield the two unbalanced files. Thus, the Golgi apparatus seems to serve as a "sorting center" to yield the final asymmetric arrangement of mastigonemes on the flagellum (HILL, 1973; HILL and OUTKA, 1974).

Although mastigonemes represent only one component of the membrane surface in these organisms, an origin at or near the nucleus is clearly shown with subsequent attachment to the nuclear envelope and migration via a directed flow mechanism to a specific region of the cell surface. The time course of appearance of mastigonemes in various cell components is summarized in Figure 7. At no time were mastigonemes observed as being free in the cytosol and a mechanism involving migration of mastigonemes through the cytosol seems unlikely (BOUCK, 1969, 1971; HILL, 1973).

BOUCK (1971) calculated the number of mastigonemes required from the Golgi apparatus per unit time to account for flagellar regeneration. During the most active phase of regeneration, the flagellum elongated approximately 1 μ in 20 min, and each micron of flagellum averaged about 17 attached mastigonemes or about 1 mastigoneme per min.

Mastigonemes are common within Golgi apparatus vesicles only during periods of active flagellar regeneration. Few remain after the flagellum is fully formed.

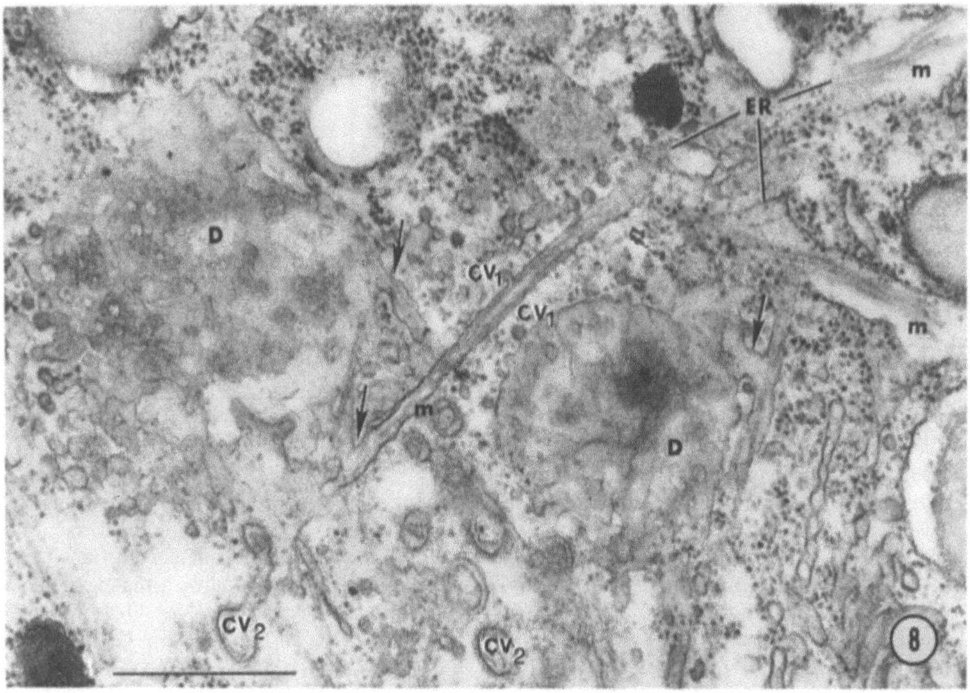

Fig. 8. Portion of a developing sporangia of *Pythium middletonii*. Intercisternal mastigonemes *m* occur in rough endoplasmic reticulum *ER* which is continuous with smooth tubules at the peripheries of two dictyosomes *D* seen in face view. *Arrows*: sites of continuity between the mastigoneme-containing endoplasmic reticulum and the dictyosome cisternae. Also illustrated are two classes of coated vesicles *cv* associated with the dictyosomes. One has a nap-like coat cv_1. The other has a spiny coat cv_2. Electron micrograph courtesy of C.E. HEINTZ and C.E BRACKER, in prep. See also BRACKER et al. (1970). Scale line = 0.5 μ

Thus, a type of feedback mechanism was suggested to operate between endomembrane activity, mastigoneme production, and flagellar growth (BOUCK, 1971). A similar type of control is observed in the production of secretory vesicles during tip growth of pollen tubes and fungal hyphae (MORRÉ and VANDERWOUDE, 1974; C.E. BRACKER, personal communication).

An endomembrane pathway of mastigoneme migration is indicated for other algae (MANTON et al., 1965; BOUCK, 1969; LEADBEATER, 1969, 1971; SCHNEPF and DEICHGRÄBER, 1969, 1972a, b; HEATH et al., 1970; LEEDALE et al., 1970; HEYWOOD, 1972) and fungi (BRACKER et al., 1970). In the latter, transfer between endoplasmic reticulum and Golgi apparatus was shown to involve direct membrane connections (Fig. 8).

4.2 Role in the Biosynthesis, Assembly, Transformation, and Transport of Products for Secretion to the Cell's Exterior

While mammalian cells are known to produce and export a wide variety of macromolecular products (e.g. digestive enzymes, hormones, connective elements, bone, egg shells, surface coats, circulating lipoproteins, blood clotting factors, etc.)

through activities of the endomembrane system, much less is known about the diversity of macromolecular export products of plant cells.

A variety of secretory products are recognized for plants in addition to preformed structural elements such as mastigonemes. These include cell wall components, slimes and mucilages, digestive enzymes, oils, and the dilute secretions such as nectars, guttation fluids, and stigmatic exudates.

4.2.1 Cell Walls

The cell surface of plants in characterized by the biphasic cell wall (ROELOFSEN, 1959; WILSON, 1964; MÜHLETHALER, 1967; PRESTON, 1974). The microfibrillar or discontinuous phase is assembled from predominantly one type of polysaccharide (β-1,4-glucan, β-1,3-glucan, β-1,3-xylan, β-1,4-mannan, or chitin) via a process involving plasma membrane-bound enzymes and surface-based assembly and orientation mechanisms (see Sect. 5, 3). The matrix or continuous phase is a mixture of pectins and hemicelluloses derived from predominantly mixed polymers of uronic acids, pentoses, and hexoses (SHAFIZADEH and MCGINNIS, 1971) and must be distinguished from extraneous wall components of similar composition such as slimes and mucilages (MOLLENHAUER, 1974b) which pass through the wall. Matrix materials are initially secreted, perhaps via endomembrane vesicles derived either from Golgi apparatus, endoplasmic reticulum, or both, but there is no proof of this except for dividing and tip-growing cells. Continued synthesis of matrix polysaccharides by vesicle membranes after incorporation into plasma membrane at the cell surface remains a possibility (VANDERWOUDE et al., 1969, 1971; ROLAND, 1973; ROUGIER et al., 1973; MORRÉ and VANDERWOUDE, 1974).

Some algal walls are rich in protein (ROBERTS et al., 1972; HILLS, 1973; HILLS et al., 1973; up to 40%, HERTH, personal communication); much of it distinct hydroxyproline-rich glycoproteins. All cell walls contain variable amounts of polypeptides rich in hydroxyproline and covalently linked to polysaccharide (LAMPORT, 1965, 1970). Functions of the hydroxyproline-rich wall component in both cell extension (LAMPORT, 1965, 1970) and the termination of cell extension (CLELAND and KARLSNES, 1967; RIDGE and OSBORNE, 1970; WINTER et al., 1971; SADAVA and CHRISPEELS, 1973; SADAVA et al., 1973) have been discussed.

ALBERSHEIM and coworkers (see KEEGSTRA et al., 1973) present a model of molecular arrangements and cross-linkages in primary cell wall of growing sycamore cells. Based on methylation analyses, they show the primary wall to consist of 10% araban, 2% 3,6-linked arabinogalactan, 23% cellulose, 4% oligo-arabinosides (bound to hydroxyproline), 8% 4-linked galactan, 10% protein rich in hydroxyproline, 16% rhamnogalactan and 21% xyloglucan. Covalent bonds link the xyloglucan and pectic components, and wall protein is bound to oligoarabinosides via hydroxyproline. In the model, only the cellulosic fibrils are not covalently bound, but are attached to the xyloglucan by hydrogen bonds. The authors propose that lowering of the pH facilitates sliding of the xyloglucan past the cellulose as a non-enzymatic mechanism for control of extension growth, and that auxin acts in this manner via an effect on the excretion of hydrogen ions (CLELAND, 1971, 1973; reviewed by ROLAND and PILET, 1974). More recently, LABAVITCH and RAY (1974) report liberation of a xyloglucan polymer from cell walls of elongating pea (*Pisum sativum*) stem segments following auxin treatment.

1. Direct export of cell wall materials via secretory vesicles derived from Golgi apparatus or other endomembrane components is evident in the following situations:

Cell Plate Formation. Here vesicles of endomembrane origin fuse to form the partition or cell wall which separates the two daughter cells (cell plate). Vesicle membranes yield new plasma membrane while vesicle contents contribute to the new cell wall or middle lamella (WHALEY and MOLLENHAUER, 1963; FREY-WYSSLING et al., 1964; WHALEY et al., 1966; HEPLER and JACKSON, 1968; PICKETT-HEAPS, 1967d, 1968a; DAUWALDER and WHALEY, 1974a; see O'BRIEN, 1972, for a critical review). An involvement of coated vesicles in this process has been shown (FRANKE and HERTH, 1974).

Tip Growth. Walls of cells that elongate by tip growth (ROELOFSEN, 1959) are derived in large measure from contents of secretory vesicles. The vesicle membranes form the new plasma membrane. Examples are pollen tubes (ROSEN et al., 1964; SASSEN, 1964; ROSEN, 1968; VANDERWOUDE et al., 1971; ENGELS, 1974), rhizoids (SIEVERS, 1965, 1967; RAWLENCE and TAYLOR, 1972; CHEN, 1973), fungal hyphae (GIRBARDT, 1969; GROVE and BRACKER, 1970; GROVE et al., 1970; BARTNICKI-GARCIA, 1973), neurons (YAMADA et al., 1971), and certain plant hairs (SIEVERS, 1963; BONNETT and NEWCOMB, 1966; SCHRÖTER and SIEVERS, 1971; WATSON and BERLIN, 1973; CHAMBERLIN, 1974) of which there is great variety (UPHOF, 1962). Fiber growth has also been indicated to involve vesicles prior to deposition of secondary layers (BERLIN and RAMSEY, 1970; WATSON and BERLIN, 1973).

Cell Wall Protein. The mechanism of synthesis and secretion of cell wall proteins is unknown, although participation of endomembranes has been implicated (RAY et al., 1969; DASHEK, 1970). Studies of GARDINER and CHRISPEELS (1975) show Golgi apparatus to be a major potential site of transport and glycosylation of hydroxyproline-rich cell wall glycoprotein in phloem parenchyma of carrot root.

Algal Walls. Many alagal walls are fabricated, at least in part, within cisternae and vesicles derived from the Golgi apparatus or other endomembranes (DODGE, 1973). More specialized cell surfaces which deviate in detail from the usual pattern include the coccoliths of the Chrysophycean algae, the silicaceous walls of the diatoms, and the complex pellicles of the Euglenoid flagellates.

(a) Scale formation and secretion in chrysophycean algae. A stepwise assembly of a complex cell wall subunit in Golgi apparatus cisternae has been shown for the surface scales of Chrysophytes, first by MANTON and colleagues (see MANTON and LEEDALE, 1961; MANTON, 1966, 1967; DODGE, 1973 for additional references) and detailed for *Pleurochrysis scherfelii* by BROWN and coworkers (BROWN et al., 1969, 1970, 1973; most recently through use of the silver-methenamine-periodic acid staining procedure, BROWN and ROMANOVICS, unpubl.). *P. scherfelii* (and other Chrysophytes) grow in two major, morphologically distinct phases. Different types of scales are produced in each phase. In the Pleurochrysis phase of the growth cycle of the alga, the scales consist of: (1) a radial system of non-cellulosic microfibrils; (2) a cellulosic (alkali-insoluble β-1,4-glucan) microfibrillar system spirally arranged and covalently linked to protein (HERTH et al., 1972b); and (3) an amorphous matrix deposited upon and within the fibrillar network. In the Cricosphaera phase, two different scale types are produced. One has a peripheral network of calcium carbonate crystals deposited on the rim. These calcified scales are known as coccoliths. In all three types of scales, the radial microfibrils are assembled on the rim. The radial microfibrils are assembled first, unfold, and are followed by deposition of the spiral bands of cellulosic microfibrils. Finally, the covering of amorphous material is added, followed by the calcification of the rim if the scale is one of those destined to become coccoliths. The entire process takes place within the confines of a single cisterna of the Golgi apparatus (see MANTON and LEEDALE, 1961; MANTON, 1966, 1967; OUTKA and WILLIAMS, 1971; DODGE, 1973; MOESTRUP and THOMSEN, 1974 for additional examples). Work of BROWN (1969) indicates that the complex surface scales are synthesized and discharged along with Golgi apparatus cisternae at a rate of one in less than every 2 min. In this group of organisms, single intact cisternae are discharged from the dictyosome and function as secretory vesicles and fuse with the plasma membrane to discharge the scales at the cell surface (Fig. 9). The kinetics of the appearance of scales at the cell surface has been monitored (Fig. 10). Additionally, the scale-containing vesicles are sufficiently large so that their formation within and discharge from the cytoplasm has been observed and recorded by cinematography (BROWN, 1969). Since both the origin of the vesicle from the Golgi apparatus and its final fusion with the plasma membrane has been viewed in living cells, the total contribution of the Golgi apparatus to cell wall formation and the direction and rapidity of membrane flow in this system can hardly be disputed.

(b) Wall formation in diatoms and desmids. The formation of the silica wall of the diatom frustule occurs in a manner whereby each daughter cell takes one

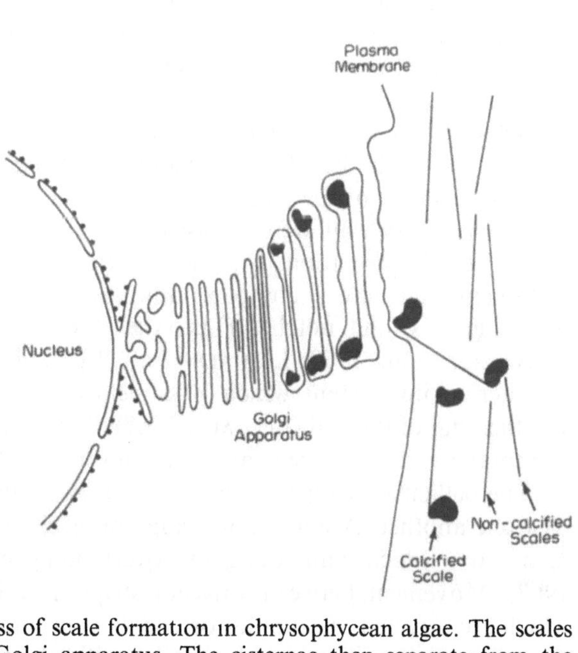

Fig. 9. Diagram illustrating the process of scale formation in chrysophycean algae. The scales are formed within cisternae of the Golgi apparatus. The cisternae then separate from the stack and the cisternal membranes fuse with the plasma membrane during discharge of individual scales. Calcification occurs at the scale margin (*solid black projections*) prior to discharge of the scale to the cell surface (OUTKA and WILLIAMS, 1971). Adapted from BROWN et al. (1970) and ELDER et al. (1972)

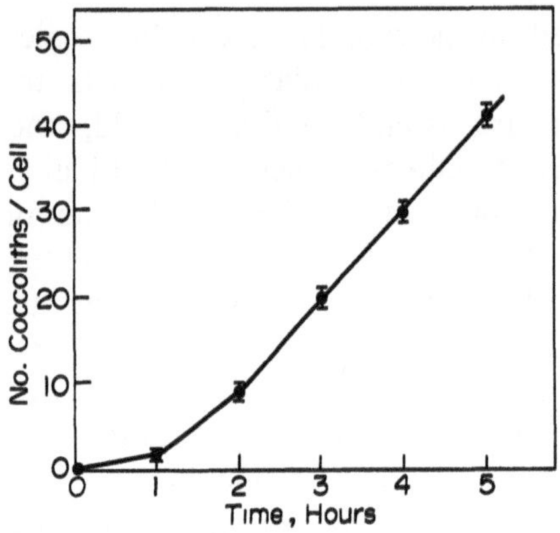

Fig. 10. Kinetics of secretion of calcified scales (coccoliths) in *Hymenomonas carterae*. Calcium-depleted cells were transferred to a medium favoring coccolith formation. After an initial lag of one to two h, coccoliths were produced at an average rate of 10–11/cell/h. From WILLIAMS (1974)

of the original overlapping valves of the frustule and each has to make a new valve (DODGE, 1973) i.e. semiconservative replication of cell walls. REIMANN (1964), DRUM and PANKRATZ (1964), STOERMER et al. (1965), LAURITIS et al. (1968) have shown that silicone deposition vesicles first collect beneath the plasma membrane at the center of the site of a new valve. As the vesicles fuse, a continuous sac, the silicalemma, is formed which eventually surrounds the area where the new valve will form. STOERMER et al. (1965) suggest that the silica precursors are formed near the Golgi apparatus but definitive evidence is lacking (DODGE, 1973). Later, the portion of the plasma membrane outside the new valve is abandoned and a new one is formed beneath the completed frustule (DODGE, 1973).

During secondary wall formation in the desmid *Micrasterias denticulatus*, an unusual disc-like vesicle is observed, apparently produced by dictyosomes (DOBBERSTEIN and KIERMAYER, 1972; KIERMAYER and DOBBERSTEIN, 1973). These vesicles are characterized by unusually thick membranes (160–200 Å) and carry globular particles of about 200 Å diameter on the inner membrane surface. As the unusual vesicles fuse with the plasma membrane, the globular particles are deposited at the outer surface of the plasma membrane at sites where the cellulose microfibrils of the secondary wall are synthesized.

(c) Pellicle formation in euglenoid flagellates. The cell surface of euglenoids is differentiated into a complex, multilayered pellicle with a repeating system of ridges, groves, and notches (LEEDALE, 1964, 1967; ARNOTT and WALNE, 1967; SCHWELITZ et al., 1970; LEEDALE et al., 1970). The resulting pellicular strips form an interlocking system which pass helically along the cell and eventually fuse at either end of the cell (LEEDALE, 1967). The entire surface is covered by a plasma membrane and a surface coat of mucin (SCHWELITZ et al., 1970).

The pellicular strips stretch and relax with euglenoid movements and can slide past one another. A continuous ridge on each strip faces outward and articulates in a grove which runs along the overlapping edge of the next strip (LEEDALE, 1967). Movement between adjacent strips may be aided by the mucin deposits, which are secreted via tubules or canals continuous with a system of endoplasmic reticulum cisternae oriented parallel with the cell surface (MOLLENHAUER and EVANS, 1970). Four microtubles form part of each pellicular ridge within a zone of exclusion also occupied by a second fibrous component, as yet incompletely characterized (MOLLENHAUER and EVANS, 1970).

In *Euglena gracilis,* microtubules have been reported within the endoplasmic reticulum and secretory vesicles of the Golgi apparatus (MOLLENHAUER, 1974a) but a role for the endomembrane system in formation of the complex pellicle remains to be elucidated.

2. Direct export of cell wall materials via secretory vesicles derived from Golgi apparatus or other endomembrane components is less clear for the following:

Primary Walls of Elongating Cells of Higher Plants. Unequivocal evidence is lacking for participation of contributions from Golgi apparatus or other endomembrane vesicles to the formation of primary walls during the grand phase of cell expansion in higher plants (O'BRIEN, 1972; Fig. 11) or in cell wall regeneration in protoplasts (BURGESS and FLEMING, 1974; FOWKE et al., 1974; see, however, PRAT and ROLAND, 1971; ROLAND and PRAT, 1972). The vesicles reported by HEYN (1971) in the cell wall region may represent lomasomes or paramural bodies (MARCHANT and ROBARDS, 1968; MARCHANT and MOORE, 1973; see ROLAND,

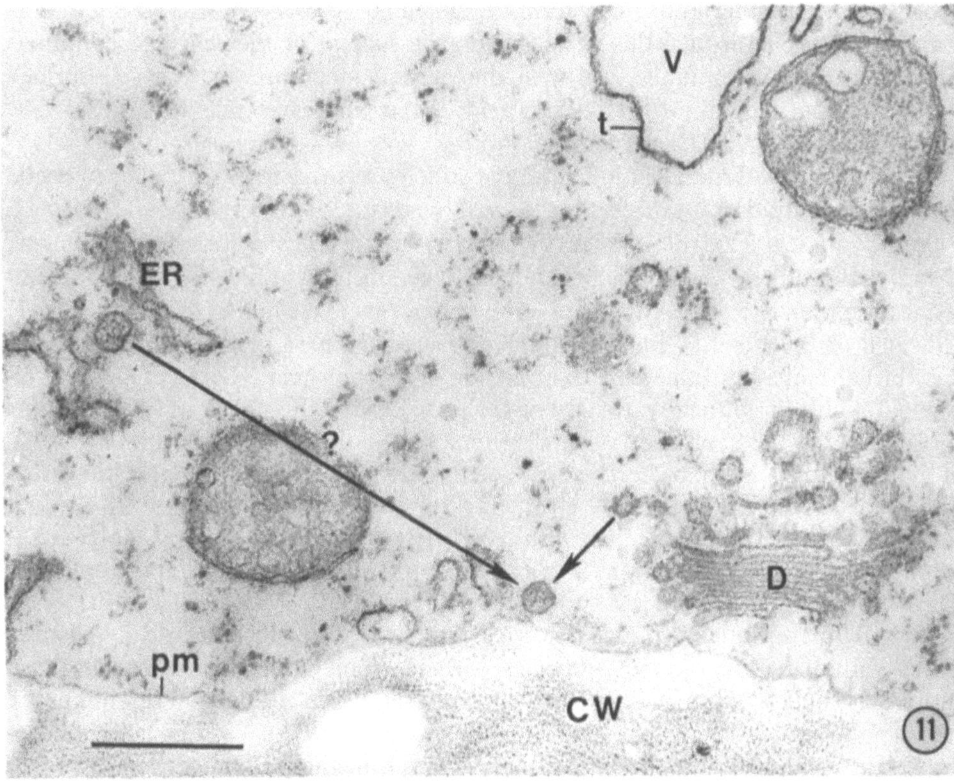

Fig. 11. Portion of the cytoplasm of a cell of tumor callus of *Vinca rosea,* Armin Braun strain, obtained through the courtesy of ANNE MATTHYSEE, Indiana University, Indianapolis. Vesicles with thick membranes of the plasma membrane type and containing electron-dense fibrillar contents are found associated with both Golgi apparatus *D* dictyosome and with endoplasmic reticulum *ER*. However, in the absence of more specific markers and an appropriate time dimension (i.e. as with mastigonemes, Fig. 7; coccoliths, Fig. 10), potential involvement of such vesicles in wall formation or surface growth remains unproven. Glutaraldehyde-osmium tetroxide fixation. *pm* plasma membrane. *CW* cell wall. *t* tonoplast. *V* vacuole. From a study with KRISTINE HESS. Scale line = 0.5 μ

1973, for literature) derived from the plasma membrane rather than vesicles derived from Golgi apparatus.

In those examples where evidence favors endomembrane participation in the formation of primary cell walls, the mechanism may be viewed as follows:

(a) Secretory vesicles as identified from electron dense cores for root cortex (RISUENO et al., 1968) or by carbohydrate staining reactions for stem collenchyma and parenchyma (ROLAND and SANDOZ, 1969; LANDRE, 1970; see however, PICKETT-HEAPS, 1968a) are formed by Golgi apparatus or other transition elements. Some, but not all, of the vesicles appear to accumulate radioactive sugars (PICKETT-HEAPS, 1967d, for sieve tube cells of wheat; COULOMB and COULON, 1971, for meristematic cells of *Cucurbita* root, and FOWKE and PICKETT-HEAPS, 1972, for fiber cells of *Marchantia* thalli; see, however, O'BRIEN, 1972).

(b) Based on studies with pollen tubes, MORRÉ and VAN DER WOUDE (1974) suggested that the glycosyl transferases of polysaccharide formation were activated or added to the membrane as the membrane of the secretory vesicle differentiates from endoplasmic reticulum-like to plasma membrane-like. In their scheme, formation of cell wall material begins within the forming secretory vesicle during migration of the vesicle through the cytoplasm to the surface of the cell, and continues after the vesicle membranes fuse with the plasma membrane at the cell surface (VAN DER WOUDE et al., 1969, 1971, 1974; VIAN and ROLAND, 1972; ROLAND, 1973; MORRÉ and VAN DER WOUDE, 1974).

(c) Synthesis within the endomembrane vesicles seems largely restricted to pectic and hemicellulosic substances of the wall matrix (MOLLENHAUER and MORRÉ, 1966a; MÜHLETHALER, 1967; NORTHCOTE, 1968). Only in the scale-forming algae, where wall synthesis is essentially completed within the confines of an endomembrane cisterna, is a role for the Golgi apparatus in cellulose synthesis clearly indicated (BROWN et al., 1969, 1970). However, even here, recent evidence shows the fibrillar material of the scale to be cellulosic glycoprotein (HERTH et al., 1972b). The β-1,4-glucan synthetase activity of Golgi apparatus (RAY et al., 1969; VAN DER WOUDE et al., 1974) or endoplasmic reticulum (SHORE and MACLACHLAN, 1974) of higher plants is perhaps best interpreted as either a non-cellulosic glucan synthetase or as cellulose synthetase en route to its site of primary action at the cell surface (NORTHCOTE, 1970). Even with secretory vesicles purified from Golgi apparatus of rat liver, the fractions are enriched in the glycosyltransferase activities characteristic of this tissue (MERRITT and MORRÉ, 1973; MORRÉ et al., in press).

A polymerization or condensation of materials into morphologically distinct fibrils within mature secretory vesicles derived from Golgi apparatus has been shown for pollen tubes of lily (VAN DER WOUDE et al., 1971) and in immature carposporangia of the red alga, *Polysiphona* (HAWKINS, 1974). Various high molecular weight polysaccharides, including polygalacturonic acid (LEPPARD and COLVIN, 1972), may appear fibrillar in the electron microscope. Apparently linear polysaccharides may form paracrystalline fibrils and it is not justified morphologically to distinguish a fibrillar component as cellulosic unless confirmed by chemical and X-ray analyses (FRANKE et al., 1974).

Secondary Walls. There are no unequivocal examples of involvement of endomembranes or endomembrane vesicles in secondary wall formation in higher plants. The possibility of secondary thickenings of the xylem being derived from vesicles

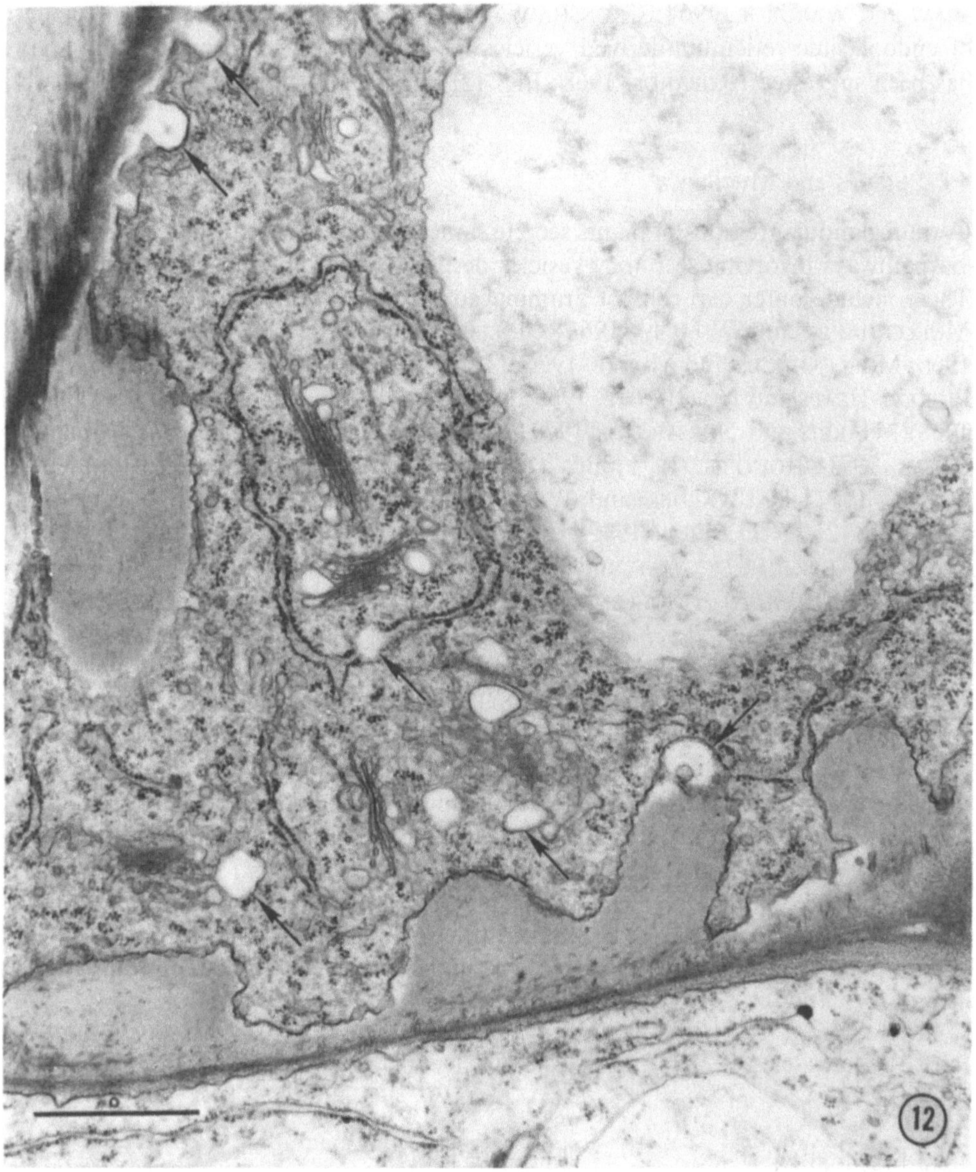

Fig. 12. Portion of developing xylem of okra (*Hibiscus esculentus*) illustrating the appearance of the cytoplasm during secondary wall formation. Dilations resembling secretory vesicles are associated with Golgi apparatus cisternae, endoplasmic reticulum, and the cell periphery (*arrows*). Yet the nature of these dilations or their potential contributions to the cell surface, if any, are unknown. Glutaraldehyde-osmium tetroxide fixation. Block-stained with uranyl acetate. Scale line = 1 μ

of the Golgi apparatus has been noted (WOODING and NORTHCOTE, 1964; MOLLENHAUER, 1967b; Fig. 12). WOODING (1968) provides autoradiographic evidence for polysaccharide formation in Golgi apparatus of developing sieve tubes. Other examples are less clear (O'BRIEN, 1972) or evidence is lacking entirely (e.g. CRON-

SHAW and WARDROP, 1964; CRONSHAW and BOUCK, 1965). Similarly, involvement of endoplasmic reticulum-derived vesicles in secondary wall formation in xylem has been suggested (ROBARDS, 1968; Fig. 12) but remains unproven.

4.2.2 Slimes and Mucilages

Certain glandular regions of plants secrete slimes and mucilages to the cell's exterior via pathways involving secretory vesicles derived primarily from Golgi apparatus. These include outer cap cells of graminaceous roots (MOLLENHAUER et al., 1961; MOLLENHAUER and WHALEY, 1963; MOLLENHAUER, 1965; JUNIPER and ROBERTS, 1966; MOLLENHAUER and MORRÉ, 1966a, b; NORTHCOTE and PICKETT-HEAPS, 1966; PICKETT-HEAPS and NORTHCOTE, 1966; MORRÉ et al., 1967; PICKETT-HEAPS, 1967a, 1968a; HARRIS and NORTHCOTE, 1970, 1971; KIRBY and ROBERTS, 1971; MOLLENHAUER, 1971; ROUGIER, 1971; BOWLES and NORTHCOTE, 1972, 1974; JONES and MORRÉ, 1973; DAUWALDER and WHALEY, 1974a, b; WRIGHT and NORTHCOTE, 1974; Fig. 2; see O'BRIEN, 1972, for critical review) and roots of other species (MOLLENHAUER, 1967a; VIAN and ROLAND, 1972), mucilage secreting cells of okra (*Hibiscus esculentus*) (MOLLENHAUER, 1967b), and *Plantago* (HYDE, 1970), secretory trichomes of *Psychotria bacteriophila* (HORNER and LERNSTEN, 1968) and *Pharbitus* (UNZELMAN and HEALY, 1974), in the oil ducts of *Heracleum* which secrete mucilagenous components (SCHNEPF, 1969d), trapping slimes of insectivorous plants (SCHNEPF, 1961, 1968b, 1969d), gland hairs of ochrea of *Rumex* and *Rheum* (SCHNEPF, 1968a), and in the secretion of the slime which constitutes the outer layers of *Spirogyra* (JORDAN, 1970) and other algae (DODGE, 1973). [See, however, FAHN and EVERT (1974) for an alternative pathway for mucilage secretion in secretory ducts of the secondary phloem of *Rhus glabra*.] Even though the compositions of slimes and cell walls may be similar, slimes do not normally appear to make permanent contributions to cell walls. Their elaboration by Golgi apparatus does not constitute proof of a similar involvement of Golgi apparatus in wall formation.

The topic of secretion of slimes and mucilages is considered in Volume 2, Part B: Chaps. 5.1, p. 222; and 5.3, p. 244. Details are provided there.

4.2.3 Essential Oils

Secretory cells of the oil ducts of many species (see SCHNEPF, 1969a, c, d, e, 1972; HEINRICH, 1970; WOLLENWEBER and SCHNEPF, 1970; SCHNEPF and KALSOVA, 1972) characteristically contain prominent arrays of smooth, tubular endoplasmic reticulum. This type of endoplasmic reticulum is also prevalent in some mammalian cells involved in steroid biogenesis (PORTER, 1961). The morphological pattern may be indicative of a general involvement of smooth endoplasmic reticulum in isoprenoid metabolism such as synthesis of steroid hormones in animals and essential oils in plants (see also Vol. 2, Part B: Chap. 5.3.2.2, p. 267).

In glandular hairs of *Mentha piperita*, oils accumulate in vacuoles (AMELUNXEN et al., 1969). In other secretory cells of the holocrine type, plastids, or even the ground cytoplasm, may be the sites of synthesis and/or accumulation of oils (HEINRICH, 1966, 1970; AMELUNXEN and ARBEITER, 1967).

4.2.4 Encrusting Materials

Little is known concerning the secretion of lignin precursors and other secondary products into cell walls or vacuoles. Available information is summarized by PICKETT-HEAPS (1968b), NORTHCOTE (1969a, b) and SCHNEPF (1973).

4.2.5 Dilute Secretions Containing Lipids, Sugars, Mucilages, and Salts

Included in this category are nectar secretions of nectaries (SCHNEPF, 1969a, 1973), various stigmatic secretions (MARTIN and TELEK, 1971), exudates of hydathodes (SCHNEPF, 1969a, 1973), and salts secreted by special salt glands (CAMPBELL et al., 1974). Additionally, plant cells, especially roots, may secrete a variety of low molecular weight substances (see Vol. 2, Part B: Chaps. 5.2 and 5.3).

According to KROH (1967), the production of the lipid-containing stigmatic secretion of *Petunia* is related to endoplasmic reticulum. RACHMILEVITZ and FAHN (1973) noted an increase in endoplasmic reticulum in developing nectaries of *Vinca* and *Citrus* and suggested that sugar was transported via vesicles derived from endoplasmic reticulum. However, both nectaries (SCHNEPF, 1969a, 1973; RACHMILEVITZ and FAHN, 1973) and the canal cells of *Lilium* pistils (ROSEN and THOMAS, 1970), as well as hydathodes (SCHNEPF, 1969a, 1973) and salt glands (THOMPSON and LIU, 1967; THOMSON et al., 1969; CAMPBELL et al., 1974) appear to fit well ultrastructurally into the category which GUNNING and PATE (1969) have termed transfer cells. These are thought to participate in an eccrine (active transport across the plasma membrane) type of secretion (SCHNEPF, 1973) adapted to intensive, short-distance transport (absorption and secretion by vascular parenchyma, tapetum of anthers, pericycle of root nodules, haustorial connections, embryo sacs and embryos).

Transfer cells are characterized by labyrinthine (with many ingrowths) cell walls, a correspondingly increased surface area of plasma membrane, and, frequently, a well-developed endoplasmic reticulum. This type of cell exemplifies developmental specialization favoring interactions among cytoplasm, endomembranes, and the cell surface (see Vol. 1, 17BII and 19; Vol. 2, Part B: Chaps. 1.4.2.2, 2.4, 5.3.1.1.1.1, 5.3.1.2.2.4, 6.2.2, 6.5.2, 10.2).

4.2.6 Digestive Enzymes

Plants secrete extracellular digestive enzymes including amylase (JONES, 1969, 1971; HESLOP-HARRISON and KNOX, 1971, also for references; VARNER and MENSE, 1972), pectinases (MORRÉ, 1968), cellulases (FAN and MACLACHLAN, 1967; THOMAS et al., 1974) and other carbohydrases (JONES, 1971; CORTAT et al., 1972; REID and MEIER, 1973b; STIEGLITZ and STERN, 1973), nucleases (JONES and PRICE, 1970; HESLOP-HARRISON and KNOX, 1971), and proteases (SCHWAB et al., 1969; HESLOP-HARRISON and KNOX, 1971) as examples. Secretion via vesicles derived from endoplasmic reticulum has been suggested for α-amylase (VIGIL and RUDDAT, 1973), glucanase (CORTAT et al., 1972), wall-degrading enzymes (BAL and PAYNE, 1972), and other extracellular materials (see FRANKE et al., 1972a). Secretion via vesicles derived from Golgi apparatus has been suggested for ribonuclease (JONES and PRICE, 1970), proteases (SCHWAB et al., 1969), and certain phosphatases (POUX, 1970; PALISANO and WALNE, 1972; see DAUWALDER et al., 1969, 1972, for reviews).

However, such secretory pathways for extracellular digestive enzymes in plants have not been corroborated by biochemical or cytochemical studies as with animal cells (e.g. Jamieson and Palade, 1967a, b, 1971) so their operation is largely speculative (Heslop-Harrison and Knox, 1971).

Treatment of barley aleurone layers with gibberellic acid to stimulate synthesis and secretion of α-amylase does appear to increase membrane synthesis and levels of certain phospholipid biosynthetic enzymes (Jones, 1969; Evins and Varner, 1971; Johnson and Kende, 1971; Koehler and Varner, 1973; Vigil and Ruddat, 1973; Ben-Tal and Varner, 1974), but probably not through a direct effect of the hormone on membrane synthesis (Firn and Kende, 1974; Jelsema and Ruddat, personal communication).

4.2.7 Water Expulsion

In *Vacuolaria virescens* and *Glaucocystis* studied by Schnepf and Koch (1966a, b), water is extruded via vesicles derived from the Golgi apparatus. Within 10 min, an amount of Golgi apparatus membranes equivalent to the total surface of the cell is estimated to become incorporated into the cell membrane by exocytosis. According to their estimates, the plasma membrane must turn over approximately 6 times each hour during active water secretion.

4.3 Role in the Biosynthesis, Assembly, Transformation, and Transport of Products among Internal Compartments

The cytoplasmic vesicle or primary vesicle provides one form of functional continuity between systems of transition elements and provides a structural link between smooth-surfaced portions of the endoplasmic reticulum and the Golgi apparatus. Bounding membranes of these vesicles are sometimes coated with an electron-dense material having the alveolate or honey-comb pattern in the electron microscope which is characteristic of the so-called coated vesicles (Newcomb, 1967).

A derivation of spherosomes or lipid bodies from endoplasmic reticulum (Frey-Wyssling et al., 1963) and/or vice versa (Jelsema et al., 1975a) has been suggested. Membranes of protein bodies of seeds (e.g. aleurone grains) appear to originate from tonoplast during seed development and then give rise to tonoplast during germination (Öpik, 1968; Jones and Price, 1970; Mollenhauer, unpublished observations).

The remainder of the discussion of internal surfaces and secretions will be limited to the formation of the vacuole membrane and contents. By analogy with lysosomes of mammalian cells, the internal compartments of the vacuolar apparatus of plant cells conform to an external milieu inside the cell. Other forms of internal secretions might be identified, such as the transfer of sugars from the chloroplast to the cytosol (see Chap. II,2). This type of process is probably best regarded as transport or secretion of the eccrine type. Special membrane-bounded transport vesicles or compartments do not appear to be involved.

4.3.1 Vacuoles

One of the most characteristic features of the mature plant cell is the large central vacuole, which occupies 70–90% of the cell volume and restricts the cytoplasm to a narrow band against the cell wall. The membrane surrounding the vacuole is the tonoplast. A special mechanical function is attributed to vacuoles for maintaining the turgidity of plant cells, which, in turn, is important to cell expansion mechanisms. Additionally, the vacuole constitutes an intracytoplasmic compartment for waste products (endproducts of metabolism, products of detoxification) and also for storage (protein, polyphosphates, inulin), and as a major repository for inorganic ions and organic acids (see Chap. II,7). Plant vacuoles occasionally react positively to detection criteria for hydrolases at acid pH and may identify with lysosomes as an important part of the plant's digestive apparatus.

In addition to water and dissolved mineral salts, vacuoles are thought to contain proteins, organic acids, amino acids, peptides, sugars and other carbohydrates, nucleic acids in various stages of decomposition, alkaloids (chiefly in dicotyledonous plants), tannins, phenols, flavinols, anthocyanins, mucilages, and glycosides. Some are species specific like the alkaloids and glycosides of various pigments but others, such as organic acids, are more general. The pH of the vacuole varies from 5.0 to 6.5 but may be as low as 1.0 in *Begonia* (2.0 in *Oxalis* and 2.0 to 2.5 in *Citrus*) and subject to considerable diurnal variation. In land plants, cations are generally balanced with organic acids, whereas chlorides are said to predominate in algae. Details of the composition of vacuolar contents are given by VOELLER et al. (1964).

4.3.2 Vacuole Membranes

An origin of plant vacuoles has been suggested from endoplasmic reticulum cisternae (POUX, 1962; BOWES, 1965; MATILE, 1968; MATILE and MOOR, 1968; MESQUITA, 1969; ROBINSON et al., 1969; BERJAK and VILLIERS, 1970; RISUENO et al., 1970; BUVAT, 1971, also for references; BERJAK, 1972; COULOMB et al., 1972; FIGIER, 1973; BRACKER, 1974; MATILE, 1974) or vesicles or cisternae derived from Golgi apparatus (MARINOS, 1963; UEDA, 1966; MATILE and MOOR, 1968; COULOMB et al., 1972; MCBRIDE and COLE, 1972), and plasma membrane (MAHLBERG, 1972; MAHLBERG et al., 1974). All are perhaps correct in certain developmental stages. In general, the bounding membranes of vacuoles seem to derive from parts of pre-existing membrane systems of the cytoplasm. All of these pre-existing membranes may ultimately trace their origins to endoplasmic reticulum. Direct continuities of plant vacuoles with rough endoplasmic reticulum have been reported (MESQUITA, 1969; BERJAK, 1972; MORRÉ and MOLLENHAUER, 1974; Fig. 13). In some fungi, membranes of immature vacuoles are frequently continuous with rough endoplasmic reticulum (BRACKER, 1974). In spite of these many suggestions and observations, a definitive account of vacuole formation is lacking and their mode of origin is largely still a matter of speculation.

If vacuole membranes do trace their origins to endoplasmic reticulum, the continuity between endoplasmic reticulum and tonoplast is likely bridged by a type of vacuole-forming transition element or provacuolar structure (MARTY, 1973a; MORRÉ and MOLLENHAUER, 1974). These provacuoles or provacuolar bodies

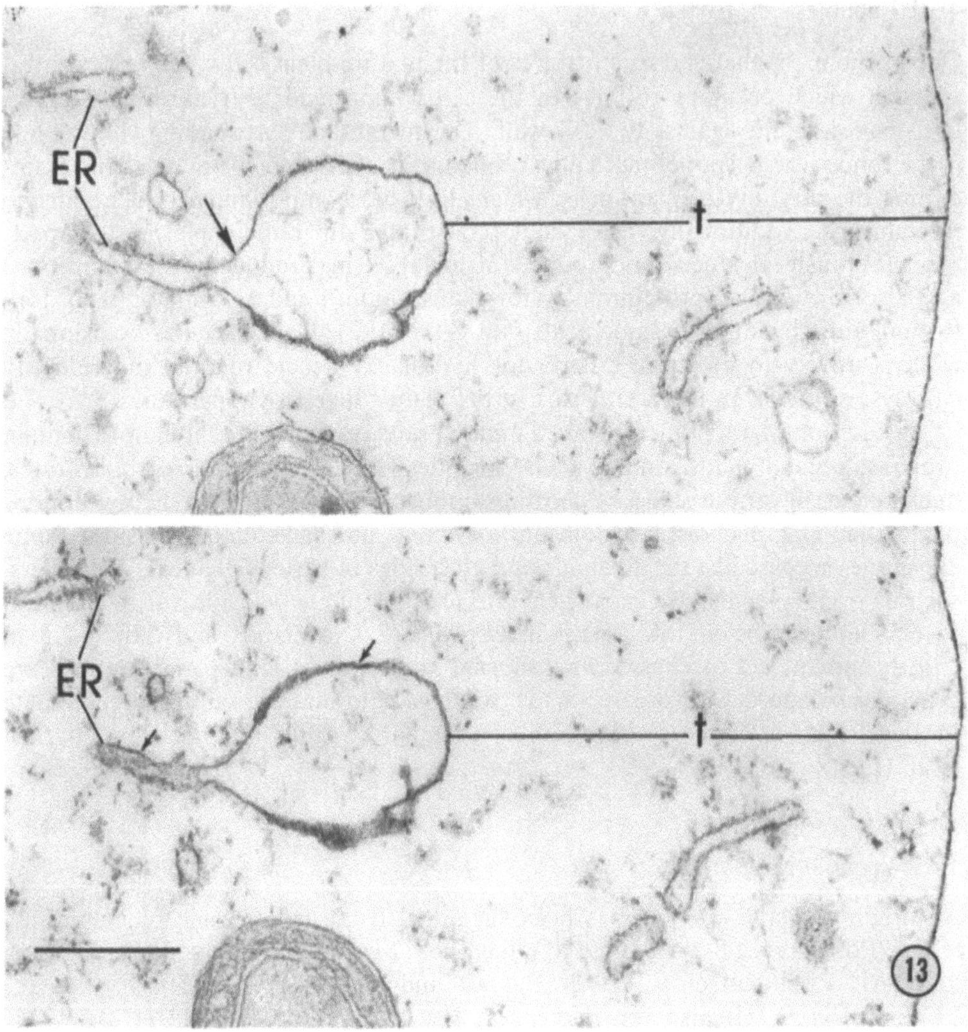

Fig. 13. Tumor callus of *V. rosea* as in Fig. 11 but showing skipped serial sections through a connection between endoplasmic reticulum and a tonoplast-like vesicle. Membrane types were differentiated on the basis of staining characteristics with alkaline lead citrate (MORRÉ and MOLLENHAUER, 1974). Tonoplast *t* membranes were thick and darkly stained while endoplasmic reticulum *ER* was thin and lightly stained (*small arrows*, see also Fig. 11). In addition to clearline membrane and luminal continuity, these electron micrographs show membrane differentiation from endoplasmic reticulum-like to tonoplast-like within the smooth (lacking ribosomes) transition elements (*large arrow*) joining the two structures. From a study with KRISTINE HESS. Scale line = 0.5 μ

are formed directly from other endomembranes, such as endoplasmic reticulum, and later coalesce to form the large central vacuole of the mature cell (MARTY, 1973a). Possible candidates include the aleurone grains and protein bodies of seeds (MOLLENHAUER, unpublished), the provacuolar bodies of roots (WHALEY et al., 1964), and the provacuolar apparatus of meristematic cells (MARTY, 1973a). In elongating plant cells of leaves, roots, and stems, formation of a central vacuole is preceeded by formation of many small vacuoles which have been suggested to function as a provacuolar apparatus (MARTY, 1973a). A similar pattern of vacuole formation has been deduced from studies of thick sections using high voltage electron microscopy (CARASSO et al., 1974; NICOLE POUX, personal communication).

4.3.3 Vacuole Contents

Movement of water into vacuoles by diffusion as well as active transport of salts and other solutes into and out of vacuoles (MACROBBIE, 1970) is frequently assumed and supported by a variety of indirect evidence (EPSTEIN, 1960; MACLENNAN et al., 1963; DAINTY, 1968; HOOYMANS, 1971; POOLE, 1971; HANSON et al., 1973; HUMPHREYS, 1973). These movements are essential to the maintenance of turgor, for vacuole formation, and in the space filling function of vacuoles during growth (HEYN, 1940; LOESCHER and NEVINS, 1973; HETTIARATCHI and O'CALLAGHAN, 1974; Vol. 2, Parts A and B, especially Part A: Chap. 11).

Only in a few instances have vesicular contributions to vacuole contents been reported. For example, from freeze-fracture-etch studies of pea roots, MATILE and MOOR (1968) report the appearance in vacuoles of vesicles apparently derived from Golgi apparatus. The membranes of the vesicles do not seem to fuse with the tonoplast in the same manner as for external secretions via the plasma membrane. Rather, the vesicles seem to be incorporated into the vacuole via a process analogous to phagocytosis.

5. Role of the Cell Surface

Both cell walls and vacuolar contents may be regarded as secretory products when secretion is defined as the transport of products of synthesis from sites of formation in internal compartments to sites of utilization or release at the cell surface (MOLLENHAUER and MORRÉ, 1966a). Because of their location exterior to the protoplast, cell walls are secreted in the general sense of the term including those components synthesized at or near the plasma membrane.

5.1 Role of the Plasma Membrane in Cell Wall Formation

In vivo and in vitro studies of cell wall biogenesis indicate that cellulose synthesis occurs at the cell surface, possibly at the surface of the plasma membrane (MÜHLETHALER, 1967; RAY, 1967; VILLEMEZ et al., 1968; WOODING, 1968; NORTHCOTE, 1969a, b; VILLEMEZ, 1970; VAN DER WOUDE et al., 1971; WILLISON and COCKING, 1972; ROBINSON and PRESTON, 1972b; ROBINSON and RAY, 1973; SMITH and STONE, 1973; MORRÉ and VAN DER WOUDE, 1974), while pectic and hemicellulosic substances are synthesized by Golgi apparatus (MOLLENHAUER and MORRÉ, 1966a; HARRIS and NORTHCOTE, 1971; VAN DER WOUDE et al., 1971; BOWLES and NORTHCOTE, 1972; EISINGER and RAY, 1972; ROBINSON and RAY, 1973; MORRÉ and

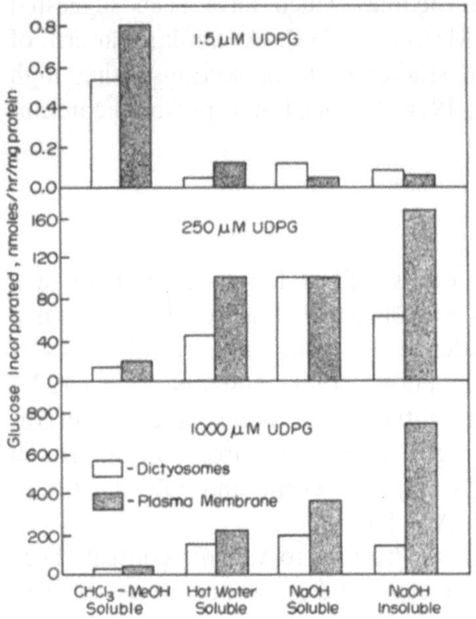

Fig. 14. Incorporation of ^{14}C-glucose from UDP-^{14}C-glucose into materials of various solubilities at three concentrations of UDP-glucose for dictyosome and plasma membrane fractions from onion stem (VAN DER WOUDE et al., 1974). The results show graphically the shift in a preponderance of synthesis of NaOH soluble and insoluble glucans from dictyosomes to the plasma membrane with increasing concentrations of UDP-glucose. *Open bars*: dictyosomes. *Shaded bars*: plasma membrane. Unpublished data of W.J. VAN DER WOUDE

VAN DER WOUDE, 1974) or, perhaps more correctly, by secretory vesicles derived from Golgi apparatus (VAN DER WOUDE et al., 1969, 1971; MORRÉ and VAN DER WOUDE, 1974) or endoplasmic reticulum (Fig. 11). Results of VAN DER WOUDE et al. (1972, 1974) demonstrate synthesis of β-1,3-glucans and β-1,4-glucans by plasma membranes from onion stem (Fig. 14), soybean hypocotyl, and oat root. As cautioned by ROBINSON and PRESTON (1972a), however, it is not possible to equate the in vitro synthesis of β-1,4-glucans with the formation of the same type of polymers that give rise to paracrystalline microfibrils of cellulose I.

β-1,3-glucans are normal wall constituents, although their amounts may increase as a result of injury or disturbance to the cell (CURRIER, 1957). In this regard, isolated plasma membranes may be regarded as parts of injured cells so that an increased propensity toward β-1,3-glucan formation by cell fractions should not be unexpected.

The proportion of β-1,3- and β-1,4-glucans synthesized by isolated membranes depends on the concentration of UDP-glucose. At low concentrations of UDP-glucose (1.5 µM) the number of β-1,3-linked residues increases disproportionately to the number of β-1,4-linked residues (ORDIN and HALL, 1968; PÉAUD-LENÖEL and AXELOS, 1970; TSAI and HASSID, 1971, 1973; VAN DER WOUDE et al., 1974). The activities may result from differential responses of two different enzymes (TSAI and HASSID, 1971). At 6.5 µM UDP-glucose, LEMBI and ORDIN observed that dictyosome and plasma membrane fractions from onion stem synthesized only β-1,4-glucans (Table 3).

In a fungus with chitinous cell walls, *Mucor rouxii*, the chitin synthetase particle has been solubilized by gentle procedures either from the cell wall, the plasma membrane, or the cell wall-plasma membrane interface (RUIZ-HERRERA and BARTNICKI-GARCIA, 1974). Thus, a site of synthesis at or near the cell surface is indicated for chitin microfibrils. Other types of polysaccharides involved in elaboration of

Table 3. Relative radioactivity in products of cellulase digestion of alkali-insoluble (NaOH Insol) polysaccharides. Data are from experiments described in Table 1 by the method of ORDIN and HALL (1968) (C.A. LEMBI and L. ORDIN, unpublished)

Fraction	Origin	Cello-dextrin	Laminari-biosyl glucose	Cello-biose	Glucose
Smooth membranes	tr	tr	0	2.5	0
Dictyosomes	2.0	4.4	0	5.1	0
Plasma membranes	tr	tr	0	3.0	0

wall microfibrils include xylans, mannans, and glucomannans. With glycomannans, at least, participation of Golgi apparatus-derived secretory vesicles has been indicated (REID and MEIER, 1973a).

Accumulating evidence suggests that glycolipids function as cofactors or intermediates in the synthesis of cellulose and other cell wall polysaccharides in plants (COLVIN, 1961; KAUSS, 1969; PINSKY and ORDIN, 1969; VILLEMEZ and CLARK, 1969; VILLEMEZ, 1970; FORSEE and ELBEIN, 1973; BEN-ARIE et al., 1973; ELBEIN and FORSEE, 1973; JUNG and TANNER, 1973). Based on studies of wall formation in bacteria (GARCIA et al., 1974) and of glycoprotein formation in mammals (LELOIR, 1971; LENNARZ and SCHER, 1972; BEHRENS et al., 1973; WAECHTER et al., 1973; MARTIN and THORNE, 1974), lipid intermediates emerge as potentially common features of surface coat formation in a variety of cell types. Glycolipid synthesis is associated with plasma membrane fractions in plants (VAN DER WOUDE et al., 1974) as well as membranes of dictyosomes. In mammalian cells, synthesis of glycosphingolipids resides principally in Golgi apparatus (KEENAN et al., 1974a; KEENAN and MORRÉ, 1975), but the site of formation of the glycolipid intermediates of coat biosynthesis is unknown.

Particles on plasma membranes visible in freeze-fracture-etch preparations are suggested to be multienzyeme complexes of cellulose microfibril synthesis (MOOR and MÜHLETHALER, 1963; BARNETT and PRESTON, 1970; ROBINSON and PRESTON, 1972b; see MÜHLETHALER, 1967 and SHAFIZADEH and MCGINNIS, 1971, for additional references) and provide one basis for the intriguing template transfer model of microfibril orientation postulated by KIERMAYER and DOBBERSTEIN (1973) for *Micrasterias* (Sect. 4.2.1). Yet, freeze-fracture particles occur in a variety of membrane types not involved in cellulose biogenesis (PINTO DA SILVA et al., 1971; TILLACK et al., 1972). Moreover, particles may be absent from some plasma membranes during the phase of greatest cellulose synthesis (WILLISON and COCKING, 1972). Statistical studies of the surfaces of different species and different tissues within a species show no correlation between particle distribution and microfibril orientation (CHAFE and WARDROP, 1970).

Attachment of fibrillar material to the plasma membrane has been observed (PRAT and ROLAND, 1971; PRAT, 1973; see, however, WILLISON and COCKING, 1972), but a relationship between fibrils and freeze-fracture particles in support of the enzyme complex theory of cellulose biosynthesis (PRESTON, 1964; PRESTON and GOODMAN, 1968) remains to be established.

5.2 Plasmodesmata

An aspect of the cell surface unique to plants is the presence of plasmodesmata. Plasmodesmata provide channels for symplastic transport of substances as large as virus particles, as well as the transmission of stimuli from cell to cell (ROBARDS, 1975). It has long been suggested that plasmodesmata are formed during cytokinesis. Trapping of endoplasmic reticulum strands within forming cell plates is confirmed by electron microscopy, so that the desmotubules of the plasmodesmata may develop from endoplasmic reticulum strands that traverse the cell plate following division. Additionally, plasmodesmata also must develop secondarily by somehow "penetrating" an existing wall (ROBARDS, 1975). Even small meristematic plant cells are estimated to contain 10^3 to 10^5 connections with their neighbors (CLOWES and JUNIPER, 1968); frequencies commonly exceed 10^6 per mm^2 (ROBARDS, 1975).

Although plasma membranes of adjacent cells are contiguous at plasmodesmata (BURGESS, 1971; VIAN and ROUGIER, 1974; ROBARDS, 1975, also for references), plasmodesmata do not constitute open cytoplasmic channels. Each is occupied almost entirely by a desmotubule which is either closed at both ends or continuous with endoplasmic reticulum. Thus, plasmodesmata provide, at best, continuity of surface membranes, of endomembranes, and of the cavities of the endoplasmic reticulum (endoplasmic space) among cells, but do not provide for direct exchange of soluble cytoplasm (see also Vol. 2, Part B: Chaps. 2 and 3.4).

The potential for flow of membrane, or membrane associated stimuli, as well as translocation of solutes or particles from cell to cell via plasmodesmata is great. Yet, except for transfer of viruses within the endoplasmic space (ESAU et al., 1967; see ROBARDS, 1975 for additional references), technical problems have thus far precluded any direct demonstration of such a translocatory function.

5.3 Role of Cell Walls in Cell Wall Formation

Suggestions of wall bound or soluble enzymes (e.g. COLVIN and BEER, 1960; COLVIN, 1961) which function in wall formation have failed to receive confirmation from analysis of isolated wall fractions, except possibly for chitin synthetase (RUIZ-HERRERA and BARTNICKI-GARCIA, 1974). Yet, many indications, especially the formation of corner thickenings in walls of collenchyma (WILSON, 1964), show that wall synthesis or alterations in cell wall architecture occur at points distant from the plasma membrane. Additionally, even if glucan units of cellulose chains are polymerized by enzymes bound to the plasma membrane, some mechanism must exist for their assembly into microfibrils and for the subsequent ordered orientation of microfibrils into defined layers (PRESTON, 1974). The problem of microfibril orientation is complicated by uncertainties as to whether the orientation of cellulose chains is parallel or antiparallel (PRESTON, 1974).

Recent findings of ROLAND and collaborators (in preparation) are contrary to the concept of multinet growth and suggest that orientation of microfibrils is a directed process rather than a mechanical orientation during cell expansion. In spite of much speculation by various authors (see PRESTON, 1974), the mechanisms whereby microfibril assembly and orientation are achieved are unknown. One possibility, that of self-assembly through crystallization of preformed polysaccharide chains, remains attractive in view of the lack of evidence for enzymes of cell wall synthesis within the cell wall proper. Still, a directing force must be supplied whether it be microtubules, flow of particles within the plasma membrane, internal stress vectors within the wall, or some other determinant of patternization. Even the polymerization process has characteristics of a structure-controlled (template mechanism) process (MARX-FIGINI, 1969). Clearly, the extent to which the wall is responsible for its own formation remains as a most challenging subject area of cytoplasma-membrane-surface coat interactions.

6. Specializations in the Formation of Plant Cell Surfaces

In addition to the carbohydrates of the conventional cellulosic cell wall, other components of the plant surface must also be synthesized and secreted. These include various types of receptor molecules and environmental receptors of the cell surface.

Plant cells respond to a variety of hormonal and environmental stimuli, at least some of which may be associated with the cell surface or expressed as altered surface (especially cell wall) properties (WILSON, 1964; OLSON et al., 1965; CLELAND, 1971; ALBERSHEIM and ANDERSON-PROUTY, 1975). With animal and bacterial cells, a number of such receptors have been localized on the plasma membrane or associated surface coats (CUATRECASAS, 1974). Included for mammalian cells are antigens, hormone receptors, toxin-binding proteins, aggregation factors, and receptors to initiate endocytosis, and, for bacteria, antigens and toxin receptors (ROSEMAN, 1971; PINTO DA SILVA et al., 1971; KEENAN et al., 1972a, b; CUATRECASAS, 1973, 1974; VAN HEYNINGEN, 1974; ALBERSHEIM and ANDERSON-PROUTY, 1975). Presumably, such receptors originate via a biosynthetic route similar to that for other surface constituents (ROSEMAN, 1970; LEBLOND and BENNETT, 1974; MORRÉ, 1975).

A secretory route within the endomembrane system is indicated for the glucagon-stimulated adenylate cyclase of rat liver (MORRÉ et al., 1974b; YUNGHANS and MORRÉ, in prep.). Insulin binding activity has been reported for the plasma membrane and Golgi apparatus of rat liver (BERGERON et al., 1973b). Additionally, the Golgi apparatus of rat liver is the site of completion of a variety of glycolipids (KEENAN et al., 1974a) and glycoproteins (SCHACHTER, 1974, also for references) which may function as surface antigens.

In plants, little is known concerning the distribution and/or origin of surface molecules which function as environmental receptors. Dictyosomes are differentially distributed upon geostimulation (SHEN-MILLER and MILLER, 1972). Plasma membranes of sugarcane cells bind the host-specific toxin, helminthosporoside (STROBEL and HESS, 1974). The inducer of sexuality in *Volvox* appears to be a glycoprotein (PALL, 1974). Auxin binding activity is membrane-associated (HERTEL et al., 1972) and concentrated in isolated fractions of plant plasma membranes (Fig. 15). To what extent this binding reflects a transport form of the hormone is unknown. Plasma membranes from plants specifically bind the inhibitor of auxin transport N-1-naphthylphthalamic acid (NPA) (LEMBI et al., 1971); this binding to membranes is competitively inhibited by morphactins (THOMSON and LEOPOLD, 1974).

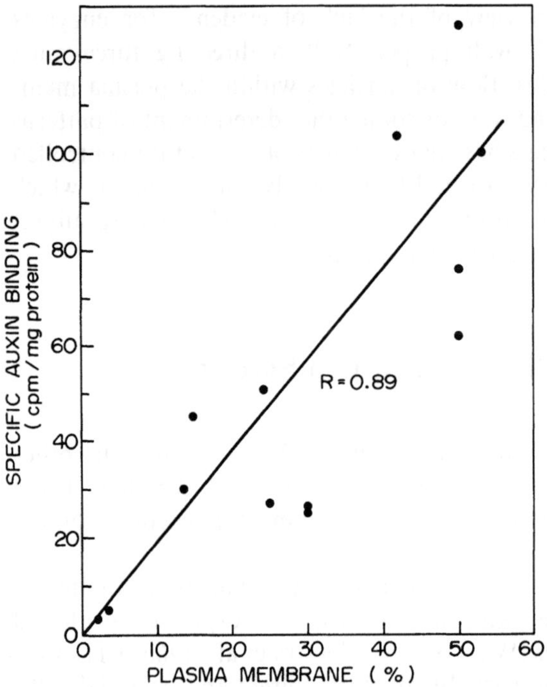

Fig. 15. Auxin binding by the method of HERTEL et al. (1972) as a function of plasma membrane content of cell fractions from soybean (*Glycine max*) hypocotyls by the method of HARDIN et al. (1972). Plasma membrane content was estimated according to ROLAND et al. (1972). From a study with F.A. WILLIAMSON and KRISTINE HESS

Auxin binding, however, may not be restricted to plasma membranes; fractions enriched in Golgi apparatus and/or endoplasmic reticulum also bind auxin, although to a lesser extent than do plasma membranes. Mitochondria do not bind auxin (WILLIAMSON et al., in prep.).

Auxin stimulations of β-glucan synthetase activities have been reported both in vivo (SPENCER et al., 1971; RAY, 1973a, b; SHORE and MACLACHLAN, 1973) and with plasma membrane fractions (VAN DER WOUDE et al., 1972). Other lines of investigation show more clearly a direct interaction between auxin and the plasma membrane (CLARK et al., 1965; HARDIN et al., 1972; MORRÉ and BRACKER, 1974b; MORRÉ et al., in prep.).

Attempts to localize binding sites for gibberellic acid on isolated plasma membranes have not been successful (JELSEMA et al., 1975b). Cytokinins seem to bind to ribosomes (BERRIDGE et al., 1970).

Attention has focused recently on the role of plant lectins (SHARON and LIS, 1972) in the regulation of plant growth and development. Fractions with lectin activity have been extracted from both cell walls and, to a lesser extent, from cytoplasmic compartments (KAUSS and GLASER, 1974). However, a clear regulatory role for plant lectins in plants remains to be established.

Another type of regulatory agent for plant growth and development which seems to act at the cell surface is calcium ions (MORRÉ and BRACKER, 1974a). Calcium interacts with either the plasma membrane or a fibrillar layer of the cortical cytoplasm just underlying the cytoplasm to cause a thickening of the plasma membrane (MORRÉ and BRACKER, 1974a), a contraction of the surface of the protoplast (ROLAND and MORRÉ, unpubl.), and the cessation of cell expansion (Table 4). Isolated plasma membranes are enriched in calcium ions relative to other membrane fractions (LAMANT and ROLAND, 1972). Interestingly, the wall

Table 4. Effect of $CaCl_2$ and indole-3-acetic acid (IAA) on cell expansion and wall extensibility of living and non-living pea internode sections. Tissue was incubated for 12 h at 25°. Non-living tissue was prepared by freezing and thawing or by boiling the tissue in methanol. These methods gave identical results. Extensibility was determined using an Instron stress-stain analyzer (OLSON et al., 1965). Values are averages from 15 determinations. Solutions were unbuffered (COARTNEY, 1967)

Treatment	Increase in section length (mm)		Relative wall extensibility	
	Living	Non-living	Living	Non-living
Water	0.9	0	5.7	8.3
$CaCl_2$ (10^{-2} M)	0.2	0	4.9	8.2
IAA (10^{-4} M)	4.7	0	10.0	8.3
IAA + $CaCl_2$	0.6	0	5.4	8.4

loosening associated with auxin-induced growth is also inhibited by calcium (COARTNEY, 1967; Table 4). This latter response of wall stiffening is not exhibited by walls of cells killed by freezing and thawing (Table 4) and may be related to the membrane response. Clearly, the mechanism whereby plant cells respond to a changing environment affords one of the most challenging opportunities for the study of cytoplasm-endomembrane-cell surface interactions.

7. Summary

Interactions among cytoplasm (cytosol, guide elements) and endomembranes (nuclear envelope, endoplasmic reticulum and Golgi apparatus and other transition elements) and the protoplasmic surface (plasma membrane, tonoplast, cell wall, and vacuole contents) are among the more complex of the intracellular transport and exchange processes of plants. Aspects include formation and transport of membranes, biosynthesis and secretion of low molecular weight compounds, and the assembly, transport, and secretion of macromolecules and macromolecular complexes. Each is facilitated by a system of internal membrane compartments, the endomembrane system. The result is growth along with manifestations of differentiation that involve replacement or modification of existing cell surfaces, plus the continual replacement of surfaces during homeostasis to balance degradation during turnover.

Work supported in part by grants from the NIH CA 13145 and HD 06624 and from the NSF BMS75-15527.

References

ADAMS, E.C., HERTIG, A.T.: Studies on guinea pig oocytes. I. Electron microscopic observations on the development of cytoplasmic organelles in oocytes of primordial and primary follicles. J. Cell Biol. **21**, 397–427 (1964)

ALBERSHEIM, P., ANDERSON-PROUTY, A.J.: Carbohydrates, proteins, cell surfaces, and the biochemistry of pathogenesis. Ann. Rev. Plant Physiol. **26**, 31–52 (1975)

AMELUNXEN, F., ARBEITER, H.: Untersuchungen an den Spritzdrüsen von *Dictamnus albus* L. Z. Pflanzenphysiol. **58**, 49–69 (1967)

AMELUNXEN, F., WAHLIG, T., ARBEITER, H.: Über den Nachweis des ätherischen Öls in isolierten Drüsenhaaren und Drüsenschuppen von *Mentha piperita* L. Z. Pflanzenphysiol. **61**, 68–72 (1969)

ARNOTT, H.J., WALNE, P.L.: Observations on the fine structure of the pellicle pores of *Euglena granulata*. Protoplasma **64**, 330–344 (1967)

AXLINE, S.G., REAVEN, E.P.: Inhibition of phagocytosis and plasma membrane mobility of the cultivated macrophage by cytochalasin B. Role of subplasmalemmal microfilaments. J. Cell Biol. **62**, 647–659 (1974)

BAL, A.K., PAYNE, J.F.: Endoplasmic reticulum activity and cell wall breakdown in quiescent root meristems of *Allium cepa* L. Z. Pflanzenphysiol. **66**, 265–272 (1972)

BARNETT, J.R., PRESTON, R.D.: Arrays of granules associated with the plasmalemma in swarmers of *Cladophora*. Ann. Botany **34**, 1011–1017 (1970)

BARTNICKI-GARCIA, S.: Fundamental aspects of hyphal morphogenesis. In: Microbial Differentiation. ASHWORTH, J.M., SMITH, J.E. (eds.), pp. 245–267. London: Cambridge University Press 1973

BEAUX, Y.J. LE: An ultrastructural study of a cytoplasmic filamentous body, termed nematosome, in the neurons of the rat and cat substantia nigra. Z. Zellforsch. **133**, 289–325 (1972)

BEHRENS, N.H., CARMINATTI, H., STANELONI, R.J., LELOIR, L.F., CANTARELLA, A.I.: Formation of lipid-bound oligosaccharides containing mannose. Their role in glycoprotein synthesis. Proc. Natl. Acad. Sci. US **70**, 3390–3394 (1973)

BEN-ARIE, R., ORDIN, L., KINDINGER, J.I.: A cell-free xylan synthesizing enzyme system from *Avena sativa*. Plant Cell Physiol. **14**, 427–434 (1973)

BEN-TAL, Y., VARNER, J.E.: Gibberellic acid dependent activation of phosphorylcholine glyceride transferase in barley aleurone layers. Plant Physiol. **53**, 54 (Abstr.) (1974)

BERGERON, J.J.M., EHRENREICH, J.H., SIEKEVITZ, P., PALADE, G.E.: Golgi fractions prepared from rat liver homogenates. II. Biochemical characterization. J. Cell Biol. **59**, 73–88 (1973a)

BERGERON, J.J.M., EVANS, W.H., GESCHWIND, I.I.: Insulin binding to rat liver Golgi fractions. J. Cell Biol. **59**, 771–776 (1973b)

BERJAK, P.: Lysosomal compartmentation: Ultrastructural aspects of the origin, development, and function of vacuoles in root cells of *Lepidium sativum*. Ann. Botany **36**, 73–81 (1972)

BERJAK, P., VILLIERS, T.A.: Aging in plant embryos. I. The establishment of the sequence of development and senescence in the root cap during germination. New Phytologist **69**, 929–938 (1970)

BERLIN, J.D., RAMSEY, J.C.: Electron microscopy of the developing cotton fiber. In: 28th Ann. Proc. Electron microscop. Soc. Am. ARCENEAUX, C.J. (ed.), pp. 128–129, 1970

BERRIDGE, M.V., RALPH, R.K., LETHAM, D.S.: The binding of kinetin to plant ribosomes. Biochem. J. **119**, 75–84 (1970)

BERTHILLIER, G., GOT, R.: Biosynthèse de l'UDPglucose par les microsomes des hépatocytes de rat. Biochim. Biophys. Acta **362**, 390–402 (1974)

BIKLE, D., TILNEY, L.G., PORTER, K.R.: Microtubules and pigment migration in the melanophores of *Fundulus heteroclitus* L. Protoplasma **61**, 322–345 (1966)

BONNER, J.: On the mechanics of auxin-induced growth. In: Plant Growth Regulation. KLEIN, R.M. (ed.), pp. 307–328. Ames: Iowa State University Press 1961

BONNETT, H.T., NEWCOMB, E.H.: Coated vesicles and other cytoplasmic components of growing root hairs of radish. Protoplasma **62**, 59–75 (1966)

BOUCK, G.B.: Extracellular microtubules. The origin, structure, and attachment of flagellar hairs in *Fucus* and *Ascophyllum* antherozoids. J. Cell Biol. **40**, 446–460 (1969)

BOUCK, G.B.: The structure, origin, isolation, and composition of the tubular mastigonemes of the *Ochromonas* flagellum. J. Cell Biol. **50**, 362–384 (1971)

BOWES, B.G.: The origin and development of vacuoles in *Glechoma hederacea* L. Cellule **65**, 359–364 (1965)

BOWLES, D.J., NORTHCOTE, D.H.: The sites of synthesis and transport of extracellular polysaccharides in the root tissues of maize. Biochem. J. **130**, 1133–1145 (1972)

BOWLES, D.J., NORTHCOTE, D.H.: The amounts and rates of export of polysaccharides found within the membrane system of maize root cells. Biochem. J. **142**, 139–144 (1974)

BRACKER, C.E.: Ultrastructure of fungi. Ann. Rev. Phytopathol. **5**, 343–374 (1967)

BRACKER, C.E.: The ultrastructure and development of sporangia in *Gilbertella persecaria*. Mycologia **60**, 1016–1067 (1968)

BRACKER, C.E.: The endomembrane system of fungi. In: Proceedings VIII Intern. Cong. Electron Microscopy, Canberra. SANDERS, J.V. and GOODCHILD, D.J. (eds.), Vol. 2, pp. 558–559. Canberra: Aust. Acad. Sci. 1974

BRACKER, C.E., GROVE, S.N.: Continuity between cytoplasmic endomembranes and outer mitochondrial membranes in fungi. Protoplasma **73**, 15–34 (1971)

BRACKER, C.E., GROVE, S.N., HEINTZ, C.E., MORRÉ, D.J.: Continuity between endomembrane components in hyphae of *Pythium spp*. Cytobiologie **4**, 1–8 (1971)

BRACKER, C.E., HEINTZ, C.E., GROVE, S.N.: Structural and functional continuity among endomembrane organelles in fungi. In: Microscopie Électronique. 7th Congr. Intern. Microscop. Electron., Grenoble. Soc. Franç. Microscop. Électron., Paris, 1970, pp. 103–104

BROWN, R.M., JR.: Observations on the relationship of the Golgi apparatus to wall formation in the marine chrysophycean alga, *Pleurochrysis scherffelii* Pringsheim. J. Cell Biol. **41**, 109–123 (1969)

BROWN, R.M., JR., FRANKE, W.W., KLEINIG, H., FALK, H., SITTE, P.: Cellulosic wall component produced by the Golgi apparatus of *Pleurochrysis scherffelii*. Science **166**, 894–896 (1969)

BROWN, R.M., JR., FRANKE, W.W., KLEINIG, H., FALK, H., SITTE, P.: Scale formation in chrysophycean algae. I. Cellulosic and noncellulosic wall components made by the Golgi apparatus. J. Cell Biol. **45**, 246–271 (1970)

BROWN, R.M., JR., HERTH, W.W., FRANKE, W., ROMANOVICZ, D.: The role of the Golgi apparatus in the biogenesis and secretion of a cellulosic glycoprotein in *Pleurochrysis:* A model system for the synthesis of structural polysaccharides. In: Biogenesis of Plant Cell Wall Polysaccharides. LOEWUS, F. (ed.), pp. 207–257. New York: Academic Press 1973

BURGESS, J.: Observations on structure and differentiation in plasmodesmata. Protoplasma **73**, 83–95 (1971)

BURGESS, J., FLEMING, E.N.: Ultrastructural observations of cell wall regeneration around isolated tobacco protoplasts. J. Cell Sci. **14**, 439–449 (1974)

BUVAT, R.: Origin and continuity of cell vacuoles. In: Results and Problems in Cell Differentiation. II. Origin and Continuity of Cell Organelles. REINERT, J., URSPRUNG, H. (eds.), pp. 127–157. Berlin-Heidelberg-New York: Springer 1971

CAMPBELL, N., THOMSON, W.W., PLATT, K.: The apoplastic pathway of transport to salt glands. J. Exptl. Botany **25**, 61–69 (1974)

CARASSO, N., FAVARD, P., MENTRÉ, P., POUX, N.: High voltage transmission electron microscopy of cell compartments in araldite thick sections. In: High Voltage Electron Microscopy. SWANN, P.R., HUMPHRIES, C.J., GORINE, M.J. (eds.), pp. 414–418. New York: Academic Press 1974

CARASSO, N., OVTRACHT, L., FAVARD, P.: Observation, en microscopie électronique haute tension, de l'appareil de Golgi sur coupes de 0.5 A 5 μ d'épaisseur. Compt. Rend. **273**, 876–879 (1971)

CAROTHERS, Z.B.: Studies of spermatogenesis in the hepaticae. III. Continuity between plasma membrane and nuclear envelope in androgonial cells of *Blasia*. J. Cell Biol. **52**, 273–282 (1972)

CHAFE, S.C., WARDROP, A.B.: Microfibril orientation in plant cell walls. Planta **92**, 13–24 (1970)

CHAMBERLIN, A.H.L.: Preliminary observations on the unicellular hairs of *Ceramium rubrum*. Brit. Phycol. J. **9**, 216 (Abstr.) (1974)

CHARDARD, R.: Observations aux microscopes électroniques à bas et haut voltages, de coupes fines et épaisses de cellules méristématiques radiculaires de *Zea mays* L. imprégnées par des sels métalliques. Compt. Rend. **276**, 2155–2158 (1973)

CHEN, J.C.W.: The kinetics of tip growth in the *Nitella* rhizoid. Plant Cell Physiol. **14**, 631–640 (1973)

CHRISPEELS, M.J.: Biosynthesis of cell wall protein: Sequential hydroxylation of proline, glycosylation of hydroxyproline and secretion of the glycoprotein. Biochem. Biophys. Res. Commun. **39**, 732–737 (1970)

CLARK, J.E., MORRÉ, D.J., CHERRY, J.H.: RNA polymerase stimulated by non-protein constituents released from plasma membranes by auxins. Plant Physiol. **56** Suppl., Abstr. 2 (1975)

CLELAND, R.: Cell wall extension. Ann. Rev. Plant Physiol. **22**, 197–222 (1971)

CLELAND, R.: Auxin-induced hydrogen ion excretion from *Avena* coleoptiles. Proc. Natl. Acad. Sci. US **70**, 3092–3093 (1973)

CLELAND, R., KARLSNES, A.M.: A possible role of hydroxyproline-containing proteins in the cessation of cell elongation. Plant Physiol. **42**, 669–671 (1967)

CLOWES, F.A.L., JUNIPER, B.E.: Plant Cells. Oxford: Blackwell, 1968

COARTNEY, J.S.: Physical and chemical analyses of auxin-induced cell wall loosening. PhD. Thesis, Purdue University 1967

COLVIN, J.R.: Synthesis of cellulose from ethanol-soluble precursors in green plants. Can. J. Biochem. Physiol. **39**, 1921–1926 (1961)

COLVIN, J.R., BEER, M.: The formation of cellulose microfibrils in suspensions of *Acetobacter xylinum*. Can. J. Microbiol. **6**, 631–637 (1960)

COOK, G.M.W., STODDART, R.W.: Surface Carbohydrates of the Eukaryotic Cell. New York and London: Academic Press, 1973

CORTAT, M., MATILE, P., WIEMKEN, A.: Isolation of glucanase-containing vesicles from budding yeast. Arch. Mikrobiol. **82**, 189–205 (1972)

COULOMB, P., COULOMB, C., COULON, J.: Origine et fonctions des phytolysosomes dans le méristème radiculaire de la courge (*Cucurbita pepo* L., Cucurbitaceaée). I. Origine des phytolysosomes. Relations reticulum endoplasmique-dictyosomes-phytolysosomes. J. Microscop. **13**, 263–280 (1972)

COULOMB, P., COULON, J.: Fonctions de l'appareil de Golgi dans les méristèmes radiculaires de la courge (*Cucurbita pepo* L., Cucurbitaceaée). J. Microscopie **10**, 203–214 (1971)

CRAN, D.G., DYER, A.F.: Membrane continuity and associations in the fern *Dryopteris borreri*. Protoplasma **76**, 103–108 (1973)

CRONSHAW, J., BOUCK, G.B.: The fine structure of differentiating xylem elements. J. Cell Biol. **24**, 415–431 (1965)

CRONSHAW, J., WARDROP, A.B.: The organization of cytoplasm in differentiating xylem. Aust. J. Botany **12**, 15–23 (1964)

CROTTY, W.J., LEDBETTER, M.C.: Membrane continuities involving chloroplasts and other organelles in plant cells. Science **182**, 839–841 (1973)

CUATRECASAS, P.: Gangliosides and membrane receptors for cholera toxin. Biochemistry **12**, 3558–3566 (1973)

CUATRECASAS, P.: Membrane receptors. Ann. Rev. Biochem. **43**, 169–214 (1974)

CURRIER, H.B.: Callose substances in plant cells. Am. J. Botany **44**, 478–488 (1957)

DAHLSTRÖM, A.: Synthesis, transport, and life-span of amine storage granules in sympathetic adrenergic neurons. In: Cellular Dynamics of the Neuron. BARONDES, S.H. (ed.), Vol. VIII, pp. 153–174. New York and London: Academic Press 1969

DAINTY, J.: The structure and possible function of the vacuole. In: Plant Cell Organelles. PRIDHAM, J.B. (ed.), pp. 40–46. New York and London: Academic Press 1968

DASHEK, W.V.: Synthesis and transport of hydroxyproline-rich components in suspension cultures of sycamore-maple cells. Plant Physiol. **46**, 831–838 (1970)

DAUWALDER, M., WHALEY, W.G.: Patterns of incorporation of [^{3}H] galactose by cells of *Zea mays* root tips. J. Cell Sci. **14**, 11–27 (1974a)

DAUWALDER, M., WHALEY, W.G.: Staining of cells of *Zea mays* root apices with the osmium-zinc iodide and osmium impregnation techniques. J. Ultrastruct. Res. **45**, 279–296 (1974b)

DAUWALDER, M., WHALEY, W.G., KEPHART, J.E.: Phosphatases and differentiation of the Golgi apparatus. J. Cell Sci. **4**, 455–497 (1969)

DAUWALDER, M., WHALEY, W.G., KEPHART, J.E.: Functional aspects of the Golgi apparatus. Sub-Cell. Biochem. **1**, 225–276 (1972)
DENNICK, R.G.: The intracellular organization of cholesterol biosynthesis. Steroids Lipids Res. **3**, 236–256 (1972)
DOBBERSTEIN, B., KIERMAYER, O.: Das Auftreten eines besonderen Typs von Golgivesikeln während der Sekundärwandbildung von *Micrasterias denticulata* Bréb. Protoplasma **75**, 185–194 (1972)
DODGE, J.D.: The Fine Structure of Algal Cells. New York: Academic Press 1973
DRUM, R.W., PANKRATZ, H.S.: Pyrenoids, raphes, and other fine structure in diatoms. Am. J. Botany **51**, 405–418 (1964)
EISINGER, W., RAY, P.M.: Function of Golgi apparatus in synthesis and transport of cell wall polysaccharides. Plant Physiol. **49**, S-2 (Abstr.) (1972)
ELBEIN, A.D., FORSEE, W.T.: Studies on the biosynthesis of cellulose. In: Biogenesis of Plant Cell Wall Polysaccharides. LOEWUS, F. (ed.), pp. 259–295. New York: Academic Press 1973
ELDER, J.H., LEMBI, C.A., ANDERSON, L., MORRÉ, D.J.: Scale calcification in a chrysophycean alga: A test system for the effects of DDT on biological calcification. Proc. Indiana Acad. Sci. **81**, 106–113 (1972)
ELDER, J.H., MORRÉ, D.J.: Synthesis *in vitro* of intrinsic membrane proteins by free, membrane-bound and Golgi apparatus-associated polyribosomes from rat liver. J. Biol. Chem. **251**, 5054–5068 (1976)
ENGELS, F.M.: Function of Golgi vesicles in relation to cell wall synthesis in germinating petunia pollen tube wall. II. Chemical composition of Golgi vesicles and pollen tube wall. Acta Botan. Neerl. **23**, 81–89 (1974)
EPSTEIN, E.: Spaces, barriers, and ion carriers: Ion absorption by plants. Am. J. Botany **47**, 393–399 (1960)
ESAU, K., CRONSHAW, J., HOEFERT, L.L.: Relation of beet yellows virus to the phloem and to movement in the sieve tube. J. Cell Biol. **32**, 71–87 (1967)
ESTENSEN, B.D., PLAGEMANN, P.G.W.: Cytochalasin B: Inhibition of glucose and glucosamine transport. Proc. Natl. Acad. Sci. US **69**, 1430–1434 (1972)
EVINS, W.H., VARNER, J.E.: Hormone-controlled synthesis of endoplasmic reticulum in barley aleurone cells. Proc. Natl. Acad. Sci. US **68**, 1631–1633 (1971)
FAHN, A., EVERT, R.F.: Ultrastructure of the secretory ducts of *Rhus glabra*. Am. J. Botany **61**, 1–14 (1974)
FALK, H.: Fusiform vesicles in plant cells. J. Cell Biol. **43**, 167–174 (1969)
FAN, D.F., MACLACHLAN, G.A.: Massive synthesis of ribonucleic acid and cellulase in the pea epicotyl in response to indoleacetic acid, with and without concurrent cell division. Plant Physiol. **42**, 1114–1122 (1967)
FAVARD, P., CARASSO, N.: The preparation and observation of thick biological sections in the high voltage electron microscope. J. Microscop. **97**, 59–81 (1973)
FIGIER, J.: Etude infrastructurale des glandes pétiolaire d'*Impatiens holstii*. Le Botaniste **56**, 311–338 (1973)
FINERAN, B.A.: Association between endoplasmic reticulum and vacuoles in frozen-etched root tips. J. Ultrastruct. Res. **43**, 75–87 (1973a)
FINERAN, B.A.: Organization of the Golgi apparatus in frozen-etched root tips. Cytobiologie **8**, 175–193 (1973b)
FIRN, R.D., KENDE, H.: Some effects of applied gibberellic acid on the synthesis and degradation of lipids in isolated barley aleurone layers. Plant Physiol. **54**, 911–915 (1974)
FLEISCHER, B., FLEISCHER, S.: Comparison of cellular membranes of liver with emphasis on the Golgi complex as a discrete organelle. Biomembranes **2**, 75–94 (1971)
FORSEE, W.T., ELBEIN, A.D.: Biosynthesis of mannosyl- and glucosyl-phosphoryl-polyprenols in cotton fibers. J. Biol. Chem. **248**, 2858–2867 (1973)
FOWKE, L.C., BECH-HANSEN, C.W., GAMBOURG, O.L.: Electron microscopic observations of cell regeneration from cultured protoplasts of *Ammi visnaga*. Protoplasma **79**, 235–248 (1974)
FOWKE, L.C., PICKETT-HEAPS, J.D.: A cytochemical and autoradiographic investigation of cell wall deposition in fiber cells of *Marchantia beteroana*. Protoplasma **74**, 19–32 (1972)
FRANKE, W.W.: Nuclear envelopes. Structure and biochemistry of the nuclear envelope. Phil. Trans. Roy. Soc. London Ser. B **268**, 67–93 (1974)

FRANKE, W.W., ECKERT, W.A.: Cytomembrane differentiation in a ciliate, *Tetrahymena pyriformis*. II. Bifacial cisternae and tubular formations. Z. Zellforsch. **122**, 244–253 (1971)
FRANKE, W.W., ECKERT, W.A., KRIEN, S.: Cytomembrane differentiation in a ciliate, *Tetrahymena pyriformis*. I. Endoplasmic reticulum and dictyosomal equivalents. Z. Zellforsch. **119**, 577–604 (1971a)
FRANKE, W.W., HERTH, W.: Morphological evidence for de novo formation of plasma membrane from coated vesicles in exponentially growing cultured plant cells. Exptl. Cell Res. **89**, 447–451 (1974)
FRANKE, W.W., HERTH, W., VANDERWOUDE, W.J., MORRÉ, D.J.: Tubular and filamentous structures in pollen tubes: Possible involvement as guide elements in protoplasmic streaming and vectorial migration of secretory vesicles. Planta **105**, 317–341 (1972a)
FRANKE, W.W., KARTENBECK, J.: Outer mitochondrial membrane continuous with endoplasmic reticulum. Protoplasma **73**, 35–41 (1971)
FRANKE, W.W., KARTENBECK, J., KRIEN, S., VANDERWOUDE, W.J., SCHEER, U., MORRÉ, D.J.: Inter- and intracisternal elements of the Golgi apparatus: A system of membrane-to-membrane cross-links. Z. Zellforsch. **132**, 365–380 (1972b)
FRANKE, W.W., KARTENBECK, J., ZENTGRAF, H., SCHEER, U., FALK, H.: Membrane to membrane cross-bridges. A means to orientation and interaction of membrane faces. J. Cell Biol. **51**, 881–888 (1971b)
FRANKE, W.W., MORRÉ, D.J., DEUMLING, B., CHEETHAM, R.D., KARTENBECK, J., JARASCH, E.-D., ZENTGRAF, H.W.: Synthesis and turnover of membrane proteins in rat liver: An examination of the membrane flow hypothesis. Z. Naturforsch. **26b**, 1031–1039 (1971c)
FRANKE, W.W., SCHEER, U., HERTH, W.: General and molecular cytology. In: Progress in Botany. ELLENBERG, H., ESSER, K., MERXMÜLLER, H., SCHNEPF, E., ZEIGLER, H. (eds.), Vol. XXXVI, pp. 1–20, 1974
FRANTZ, C., ROLAND, J.-C., WILLIAMSON, F.A., MORRÉ, D.J.: Différenciation *in vitro* des membranes des dictyosomes. Compt. Rend. **277**, 1471–1474 (1973)
FREED, J.J., LEBOWITZ, M.M.: The association of a class of saltatory movements with microtubules in cultured cells. J. Cell Biol. **45**, 334–354 (1970)
FREY-WYSSLING, A., GRIESHABER, E., MÜHLETHALER, K.: Origin of spherosomes in plant cells. J. Ultrastruct. Res. **8**, 506–516 (1963)
FREY-WYSSLING, A., LÓPEZ-SÁEZ, J.F., MÜHLETHALER, K.: Formation and development of the cell plate. J. Ultrastruct. Res. **10**, 422–432 (1964)
FRIEND, D.S., FARQUHAR, M.G.: Functions of coated vesicles during protein absorption in the rat vas deferens. J. Cell Biol. **35**, 357–376 (1967)
FROT-COUTAZ, J., BERTHILLIER, G., GOT, R.: Mise en evidence d'un recepteur d'UDP-glucose dans les membranes microsomiques des hepatocytes de rat. FEBS Lett. **52**, 81–84 (1975)
FULTON, C.: Centrioles. In: Results and Problems in Cell Differentiation, II. Origin and Continuity of Cell Organelles. REINERT, J., URSPRUNG, H. (eds.), pp. 170–221. Berlin-Heidelberg-New York: Springer 1971
GARCIA, R.C., RECONDO, E., DANKERT, M.: Polysaccharide biosynthesis in *Acetobacter xylinum*. Enzymatic synthesis of lipid diphosphate and monophosphate sugars. Eur. J. Biochem. **43**, 93–105 (1974)
GARDINER, M., CHRISPEELS, M.J.: Involvement of the Golgi apparatus in the synthesis and secretion of hydroxyproline-rich cell wall glycoproteins. Plant Physiol. **55**, 536–541 (1975)
GIRBARDT, M.: Die Ultrastruktur der Apikalregion von Pilzhyphen. Protoplasma **67**, 413–441 (1969)
GROVE, S.N., BRACKER, C.E.: Protoplasmic organization of hyphal tips among fungi: Vesicles and Spitzenkörper. J. Bacteriol. **104**, 989–1009 (1970)
GROVE, S.N., BRACKER, C.E., MORRÉ, D.J.: Cytomembrane differentiation in the endoplasmic reticulum-Golgi apparatus-vesicle complex. Science **161**, 171–173 (1968)
GROVE, S.N., BRACKER, C.E., MORRÉ, D.J.: An ultrastructural basis for hyphal tip growth in *Pythium ultimum*. Am. J. Botany **57**, 245–266 (1970)
GUNNING, B.E.S., PATE, J.S.: "Transfer cells": Plant cells with wall ingrowths, specialized in relation to short distance transport of solutes—their occurrence, structure, and development. Protoplasma **68**, 107–133 (1969)
HANSON, J.B., LEONARD, R.T., MOLLENHAUER, H.H.: Increased electron density of tonoplast membranes in washed corn root tissue. Plant Physiol. **52**, 298–300 (1973)

HARDIN, J.W., CHERRY, J.H., MORRÉ, D.J., LEMBI, C.A.: Enhancement of RNA polymerase activity by a factor released by auxin from plasma membrane. Proc. Natl. Acad. Sci. US **69**, 3146–3150 (1972)

HARRIS, P.J., NORTHCOTE, D.H.: Patterns of polysaccharide biosynthesis in differentiating cells of maize root-tips. Biochem. J. **120**, 479–491 (1970)

HARRIS, P.J., NORTHCOTE, D.H.: Polysaccharide formation in plant Golgi bodies. Biochim. Biophys. Acta **237**, 56–64 (1971)

HASSID, W.Z.: Biosynthesis of oligosaccharides and polysaccharides in plants. Science **165**, 137–144 (1969)

HAWKINS, E.K.: Golgi vesicles of uncommon morphology and wall formation in the red alga, *Polysiphonia*. Protoplasma **80**, 1–14 (1974)

HEATH, I.B., GREENWOOD, A.D., GRIFFITHS, H.B.: The origin of flimmer in *Saprolegnia, Dictyuchus, Synura* and *Cryptomonas*. J. Cell Sci. **7**, 445–461 (1970)

HEINRICH, G.: Licht- und elektronenmikroskopische Untersuchungen zur Genese der Exkrete in den lysigenen Exkreträumen von *Citrus medica*. Flora **156A**, 451–456 (1966)

HEINRICH, G.: Elektronmikroskopische Beobachtungen an den Drüsenzellen von *Poncirus trifoliata*; zugleich ein Beitrag zur Wirkung ätherischer Öle auf Pflanzenzellen und eine Methode zur Unterscheidung flüchtiger von nichtflüchtigen lipophilen Komponenten. Protoplasma **69**, 15–36 (1970)

HENDLER, R.W.: Protein synthesis by membrane-bound polysomes. Biomembranes **5**, 147–211 (1974)

HEPLER, P.K., FOSKETT, D.E.: The role of microtubules in vessel member differentiation in *Coleus*. Protoplasma **72**, 213–236 (1971)

HEPLER, P.K., JACKSON, W.T.: Microtubules and early stages of cell-plate formation in the endosperm of *Haemanthus katherinae* Baker. J. Cell Biol. **38**, 437–446 (1968)

HEPLER, P.K., NEWCOMB, E.H.: Microtubules and fibrils in the cytoplasm of *Coleus* cells undergoing secondary wall deposition. J. Cell Biol. **20**, 529–532 (1964)

HERTEL, R., THOMSON, K.-ST., RUSSO, V.E.A.: *In vitro* auxin binding to particulate cell fractions from corn coleoptiles. Planta **107**, 325–340 (1972)

HERTH, W., FRANKE, W.W., STADLER, J., BITTIGER, H., KEILICH, G., BROWN, R.M., JR.: Further characterization of the alkali-stable material from the scales of *Pleurochrysis scherffelii*: A cellulosic glycoprotein. Planta **105**, 79–92 (1972b)

HERTH, W., FRANKE, W.W., VAN DER WOUDE, W.J.: Cytochalasin stops tip growth in plants. Naturwissenschaften **59**, 38–39 (1972a)

HESLOP-HARRISON, Y., KNOX, R.B.: A cytochemical study of the leafgland enzymes of insectivorous plants of the genus *Pinguicula*. Planta **96**, 183–211 (1971)

HETTIARATCHI, D.R.P., O'CALLAGHAN, J.R.: A membrane model of plant cell expansion. J. Theoret. Biol. **45**, 459–465 (1974)

HEYN, A.N.J.: The physiology of cell elongation. Bot. Rev. **6**, 515–574 (1940)

HEYN, A.N.J.: Observations on the exocytosis of secretory vesicles and their products in coleoptiles of *Avena*. J. Ultrastruct. Res. **37**, 69–81 (1971)

HEYNINGEN, W.E. VAN: Gangliosides as membrane receptors for tetanus toxin, cholera toxin, and serotonin. Nature **249**, 415–417 (1974)

HEYWOOD, P.: Structure and origin of flagellar hairs in *Vacuolaria virescens*. J. Ultrastruct. Res. **39**, 608–623 (1972)

HILL, F.G.: The structure, arrangement, and intracellular origin of tubular mastigonemes in *Ochromonas minute* and *Monas* sp. (Chrysophyceae). Ph.D. Thesis, Iowa State University 1973

HILL, F.G., OUTKA, D.E.: The structure and origin of mastigonemes in *Ochromonas minute* and *Monas* sp. J. Protozool. **21**, 299–312 (1974)

HILLS, G.J.: Cell wall assembly in vitro from *Chamydomonus reinhardi*. Planta **115**, 7–23 (1973)

HILLS, G.J., GURNEY-SMITH, M., ROBERTS, K.: Structure, composition, and morphogenesis of the cell wall of *Chlamydomonas reinhardi*. II. Electron microscopy and optical diffraction analysis. J. Ultrastruct. Res. **43**, 179–192 (1973)

HIRANO, H., PARKHOUSE, B., NICOLSON, G.L., LENNOX, E.S., SINGER, S.J.: Distribution of saccharide residues on membrane fragments from a myeloma-cell homogenate. Its implications for membrane biogenesis. Proc. Natl. Acad. Sci. US **69**, 2945–2949 (1972)

HOOYMANS, J.J.M.: Effect of pretreatment on the location of the limiting step in vacuolar chloride accumulation by barley roots. Z. Pflanzenphysiol. **65**, 309–314 (1971)

HORNER, H.T., LERSTEN, N.R.: Development, structure, and function of secretory trichomes in *Psychotria bacterophila* (Rubiaceae). Am. J. Botany **55**, 1089–1099 (1968)

HUMPHREYS, T.E.: Sucrose transport at the tonoplast. Phytochemistry **12**, 1211–1219 (1973)

HYDE, B.B.: Mucilage-producing cells in the seed coat of *Plantago ovata*: Developmental fine structure. Am. J. Botany **57**, 1197–1206 (1970)

ISHIKAWA, H., BISCHOFF, R., HOLTZER, H.: Formation of arrowhead complexes with heavy meromyosin in a variety of cell types. J. Cell Biol. **43**, 312–328 (1969)

JAMIESON, J.D., PALADE, G.E.: Intracellular transport of secretory proteins in the pancreatic exocrine cell. I. Role of the peripheral elements of the Golgi complex. J. Cell Biol. **34**, 577–596 (1967a)

JAMIESON, J.D., PALADE, G.E.: Intracellular transport of secretory proteins in the pancreatic exocrine cell. II. Transport to condensing vacuoles and zymogen granules. J. Cell Biol. **34**, 597–615 (1967b)

JAMIESON, J.D., PALADE, G.E.: Synthesis, intracellular transport, and discharge of secretory proteins in stimulated pancreatic exocrine cells. J. Cell Biol. **50**, 135–158 (1971)

JELSEMA, C.L., MORRÉ, D.J.: The distribution of phospholipid biosynthetic enzymes among cell components of rat liver. (In Preparation)

JELSEMA, C.L., MORRÉ, D.J., RUDDAT, M.: Isolation and characterization of spherosomes from aleurone layers of wheat. Proc. Indiana Acad. Sci. **84**, 166–178 (1975a)

JELSEMA, C.L., RUDDAT, M., MORRÉ, D.J., WILLIAMSON, F.A.: Specific binding of gibberellic acid (GA_1) to defined fractions from aleurone layers of wheat. Plant Physiol. **56** (Suppl.), 45 (Abstr.) (1975b)

JOHNSON, K.D., KENDE, H.: Hormonal control of lecithin synthesis in barley aleurone cells: Regulation of the CDP-choline pathway by gibberellin. Proc. Natl. Acad. Sci. US **68**, 2674–2677 (1971)

JONES, D.D., MORRÉ, D.J.: Golgi apparatus mediated polysaccharide secretion by outer root cap cells of *Zea mays*. III. Control by exogenous sugars. Physiol. Plantarum **29**, 68–75 (1973)

JONES, R.L.: Gibberellic acid and fine structure of barley aleurone cells. II. Changes during the synthesis and secretion of α-amylase. Planta **88**, 73–86 (1969)

JONES, R.L.: Gibberellic acid-enhanced release of β-1,3-glucanase from barley aleurone cells. Plant Physiol **47**, 412–416 (1971)

JONES, R.L., PRICE, J.M.: Gibberellic acid and the fine structure of barley aleurone cells. III. Vacuolation of the aleurone cell during the phase of ribonuclease release. Planta **94**, 191–202 (1970)

JORDAN, E.G.: Ultrastructural aspects of cell wall synthesis in *Spirogyra*. Protoplasma **69**, 405–416 (1970)

JUNG, P., TANNER, W.: Identification of the lipid intermediate in yeast mannan biosynthesis. Eur. J. Biochem. **37**, 1–6 (1973)

JUNIPER, B.E., ROBERTS, R.M.: Polysaccharide synthesis and the fine structure of root cells. J. Roy. Microscop. Soc. **85**, 63–72 (1966)

KAGAWA, T., LORD, J.M., BEEVERS, H.: The origin and turnover of organelle membranes in castor bean endosperm. Plant Physiol. **51**, 61–65 (1973)

KAMIYA, N.: Motilität des Plasmas. In: Die Zelle, Struktur und Funktion. METZNER, H. (ed.), pp. 373–404. Stuttgart: Wissenschaftliche Verlagsgesellschaft 1971

KARTENBECK, J., FRANKE, W.W.: Dense cytoplasmic aggregates associated with Golgi apparatus cisternae of rat hepatocytes. Protoplasma **72**, 49–53 (1971)

KAUSS, H.: A plant mannosyl-lipid acting in reversible transfer of mannose. FEBS Lett. **5**, 81–84 (1969)

KAUSS, H., GLASER, C : Carbohydrate-binding proteins from plant cell walls and their possible involvement in extension growth. FEBS Lett. **45**, 304–307 (1974)

KEEGSTRA, K., TALMADGE, K.W., BAUER, W.D., ALBERSHEIM, P.: The structure of plant cell walls. III. A model of the walls of suspension-cultured sycamore cells based on the interconnections of the macromolecular components. Plant Physiol. **51**, 188–197 (1973)

KEENAN, T.W., HUANG, C.M., MORRÉ, D.J.: Gangliosides: Nonspecific localization in the surface membranes of bovine mammary gland and rat liver. Biochem. Biophys. Res. Commun. **47**, 1277–1283 (1972a)

KEENAN, T.W., MORRÉ, D.J.: Phospholipid class and fatty acid composition of Golgi apparatus isolated from rat liver and comparison with other cell fractions. Biochemistry **9**, 19–25 (1970)

KEENAN, T.W., MORRÉ, D.J.: Mammary carcinoma: Enzymatic block in disialoganglioside biosynthesis. Science **182**, 935–937 (1973)

KEENAN, T.W., MORRÉ, D.J.: Glycosyltransferases: Do they exist on the surface of mammalian cells? FEBS Lett. **55**, 8–13 (1975)

KEENAN, T.W., MORRÉ, D.J., BASU, S.: Ganglioside biosynthesis: Concentration of glycosphingolipid glycosyltransferases in Golgi apparatus from rat liver. J. Biol. Chem. **29**, 310–315 (1974a)

KEENAN, T.W., MORRÉ, D.J., HUANG, C.M.: Distribution of gangliosides among subcellular fractions from rat liver and bovine mammary gland. FEBS Lett. **24**, 204–208 (1972b)

KEENAN, T.W., MORRÉ, D.J., HUANG, C.M.: Membranes of the mammary gland. In: Lactation: A Comprehensive Treatise. LARSON, B.L., SMITH, V.R. (eds.), Vol. II, pp. 191–233. New York: Academic Press 1974b

KIERMAYER, O., DOBBERSTEIN, B.: Membrankomplexe dictyosomaler Herkunft als „Matrizen" für die extraplasmatische Synthese und Orientierung von Mikrofibrillen. Protoplasma **77**, 437–451 (1973)

KIRBY, E.G., ROBERTS, R.M.: The localized incorporation of ^{3}H-L-fucose into cell-wall polysaccharides of the cap and epidermis of corn roots. Autoradiographic and biosynthetic studies. Planta **99**, 211–221 (1971)

KIT, S.: Nucleotides and nucleic acids. In: Metabolic Pathways. GREENBERG, D.M. (ed.), Vol. IV, pp. 69–275. New York: Academic Press 1970

KLETZIEN, R.F., PERDUE, J.F.: The inhibition of sugar transport in chick embryo fibroblasts by cytochalasin B. J. Biol. Chem. **248**, 711–719 (1973)

KOEHLER, D.E., VARNER, J.E.: Hormonal control of orthophosphate incorporation into phospholipids of barley aleurone layers. Plant Physiol. **52**, 208–214 (1973)

KOMNICK, H., STOCKEM, W., WOHLFARTH-BOTTERMANN, K.E.: Weitreichende fibrilläre Protoplasmadifferenzierungen und ihre Bedeutung für die Protoplasmaströmung. VII. Experimentelle Induktion, Kontraktion und Extraktion der Plasmafibrillen von *Physarum polycephalum*. Z. Zellforsch. **109**, 420–430 (1970)

KREUTZBERG, G.W.: Neuronal dynamics and axonal flow. IV. Blockage of "intraaxonal" enzyme transport by colchicine. Proc. Natl. Acad. Sci. US **62**, 722–728 (1969)

KROH, M.: Bildung und Transport des Narbensekrets von *Petunia hybrida*. Planta **77**, 250–260 (1967)

LABAVITCH, J.M., RAY, P.M.: Relationship between promotion of xyloglucan metabolism and induction of elongation by indoleacetic acid. Plant Physiol. **54**, 499–502 (1974)

LAMANT, A., ROLAND, J.C.: Isolement et analyse cationique d'une fraction membranaire enriche en plasmalemme, à partir de racines de Lupin (*L. luteus* L.). Compt. Rend. **274**, 3391–3394 (1972)

LAMPORT, D.T.A.: The protein components of primary cell walls. Advan. Bot. Res. **2**, 151–218 (1965)

LAMPORT, D.T.A.: Cell wall metabolism. Ann. Rev. Plant Physiol. **21**, 235–270 (1970)

LANDRE, P.: Activité golgienne en liaison avec celle du plasmalemme dans les cellules stomatiques de la moutarde (*Sinapis alba* L.) lors de la formation de l'ostiole. Compt. Rend. **271**, 904–907 (1970)

LARSEN, G.L., BRUMMOND, D.O.: β-(1→4)-D-glucan synthesis from UDP-[^{14}C]-D-glucose by a solubilized enzyme from *Lupinus albus*. Phytochem. **13**, 361–365 (1974)

LAURITIS, J.A., COOMBS, J., VOLCANI, B.E.: Studies on the biochemistry and fine structure of silica shell formation in diatoms. IV. Fine structure of the apochlorotic diatom *Nitzschia alba* Lewin and Lewin Arch. Mikrobiol. **62**, 1–16 (1968)

LEADBEATER, B.S.C.: A fine structural study of *Olisthodiscus luteus* Carter. Br. Phycol. J. **4**, 3–17 (1969)

LEADBEATER, B.S.C.: The intracellular origin of flagellar hairs in the dinoflagellate *Woloszynskia micra* Leadbeater and Dodge. J. Cell Sci. **9**, 443–451 (1971)

LEBLOND, C.P., BENNETT, G.: Elaboration and turnover of cell coat glycoproteins. In: The Cell Surface in Development. MOSCONA, A.A. (ed.), pp. 29–49. New York: John Wiley and Sons 1974

LEDBETTER, M.C., PORTER, K.R.: A "microtubule" in plant cell fine structure. J. Cell Biol. **19**, 239–250 (1963)
LEEDALE, G.F.: Pellicle structure in *Euglena*. Br. Phycol. Bull. **2**, 291–306 (1964)
LEEDALE, G.F.: Euglenoid Flagellates. Inglewood Cliffs/N.J.: Prentice-Hall 1967.
LEEDALE, G.F., LEADBEATER, B.S.C., MASSALSKI, A.: The intracellular origin of flagellar hairs in the Chrysophyceae and Xanthophyceae. J. Cell Sci. **6**, 701–719 (1970)
LELOIR, L.F.: Two decades of research on the biosynthesis of oligosaccharides. Science **172**, 1299–1303 (1971)
LEMBI, C.A., MORRÉ, D.J., THOMSON, K.S., HERTEL, R.: N-1-napthylphthalamic-acid-binding-activity of a plasma membrane-rich fraction from maize coleoptiles. Planta **99**, 37–45 (1971)
LENNARZ, W.J., SCHER, M.G.: Metabolism and function of polyisoprenol sugar intermediates in membrane-associated reactions. Biochim. Biophys. Acta **265**, 417–441 (1972)
LENTZ, T.L.: The fine structure of differentiating interstitial cells in *Hydra*. Z. Zellforsch. **67**, 547–560 (1965)
LENTZ, T.L.: Rhabdite formation in Planaria: The role of microtubules. J. Ultrastruct. Res. **17**, 114–126 (1967)
LEPPARD, G.G., COLVIN, J.R.: Electron-opaque fibrils and granules in and between the cell walls of higher plants. J. Cell Biol. **53**, 695–703 (1972)
LOESCHER, W.H., NEVINS, D.J.: Turgor-dependent changes in *Avena* coleoptile cell wall composition. Plant Physiol. **52**, 248–251 (1973)
LORD, J.M., KAGAWA, T., MOORE, T.S., BEEVERS, H.: Endoplasmic reticulum as the site of lecithin formation in castor bean endosperm. J. Cell Biol. **57**, 659–667 (1973)
MACLENNAN, D.H., BEEVERS, H., HARLEY, J.L.: Compartmentation of acids in plant tissues. Biochem. J. **89**, 316–327 (1963)
MACROBBIE, E.A.C.: The active transport of ions in plant cells. Quart. Rev. Biophys. **3**, 251–294 (1970)
MAHLBERG, P.: Further observations on the phenomenon of secondary vacuolation in living cells. Am. J. Botany **59**, 172–179 (1972)
MAHLBERG, P.G., TURNER, F.R., WALKINSHAW, C., VENKETESWARAN, S.: Ultrastructural studies plasma membrane related secondary vacuoles in cultured cells. Am. J. Botany **61**, 730–738 (1974)
MAITRA, S.C., DE, D.N.: Role of microtubules in secondary thickening of differentiating xylem element. J. Ultrastruct. Res. **34**, 15–22 (1971)
MALAISSE-LAGAE, F., GREIDER, M.H., MALAISSE, W.J., LACY, P.E.: The stimulus-secretion coupling of glucose-induced insulin release. IV. The effect of vincristine and deuterium oxide on the microtubular system of the pancreatic beta cell. J. Cell Biol. **49**, 530–535 (1971)
MANLEY, E.R., SKRDLANT, H.B.: Conversion of diglyceride to triglyceride by rat liver microsomes: A requirement for the 105,000 × g supernatant. Federation Proc. **33**, 1525 (Abstr.) (1974)
MANTON, I.: Observations on scale production in *Pyramimonas amylifera* Conrad. J. Cell Sci. **1**, 429–438 (1966)
MANTON, I.: Further observations on the fine structure of *Chrysochromulina chiton* with special reference to the haptonema, "peculiar" Golgi structure and scale production. J. Cell Sci. **2**, 265–272 (1967)
MANTON, I., LEEDALE, G.F.: Observations on the fine structure of *Paraphysomonas vestita*, with special reference to the Golgi apparatus and the origin of scales. Phycologia **1**, 37–57 (1961)
MANTON, I., RAYNS, D.G., ETTL, H., PARKE, M.: Further observations on green flagellates with scaly flagella: the genus *Heteromastix* Korshikov. J. Marine Biol. Assoc. U.K **45**, 241–255 (1965)
MARCHANT, R., MOORE, R.T.: Lomasomes and plasmalemmasomes in fungi. Protoplasma **76**, 235–247 (1973)
MARCHANT, R., ROBARDS, A.W.: Membrane systems associated with the plasmalemma of plant cells. Ann. Botany (London) **32**, 457–471 (1968)
MARINOS, N.G.: Vacuolation in plant cells. J. Ultrastruct. Res. **9**, 177–185 (1963)
MARTIN, F.W., TELEK, L.: The stigmatic secretion of the sweetpotato. Am. J. Botany **58**, 317–322 (1971)

MARTIN, H.G., THORNE, K.J.I.: The involvement of endogenous dolichol in the formation of lipid-linked precursors of glycoprotein in rat liver. Biochem. J. **138**, 281–289 (1974)
MARTY, F.: Mise en évidence d'un appareil provacuolaire et de son rôle dans l'autophagie cellulaire et l'origine des vacuoles. Compt. Rend. **276**, 1549–1552 (1973a)
MARTY, F.: Sites réactifs à l'iodure de zinc-tétroxyde d'osmium dans les cellules de la racine d'*Euphorbia characias*. Compt. Rend. **277**, 1317–1320 (1973b)
MARTY, F.: Observation au microscope électronique à haute tension (3 MeV) de cellules végétales en coupes épaisses de 1 à 5 μ. Compt. Rend. **277**, 2681–2684 (1973c)
MARX-FIGINI, M.: On the biosynthesis of cellulose in higher and lower plants. J. Polymer Sci. C-**28**, 57–67 (1969)
MATILE, P.: Lysosomes of root tip cells in corn seedlings. Planta **79**, 181–196 (1968)
MATILE, P.: Lysosomes. In: Dynamic Aspects of Plant Ultrastructure. ROBARDS, A.W. (ed.), pp. 178–218. London: McGraw-Hill 1974
MATILE, P., MOOR, H.: Vacuolation: Origin and development of the lysosomal apparatus in root-tip cells. Planta **80**, 159–175 (1968)
MAZLIAK, P.: Lipid metabolism in plants. Ann. Rev. Plant Physiol. **24**, 287–310 (1973)
MCBRIDE, D.L., COLE, K.: Ultrastructural observations on germinating monospores in *Smithora naiadum* (Rhodphyceae, Bangiophycidae). Phycologia **11**, 181–191 (1972)
MCINTOSH, J.R., HEPLER, P.K., VAN WIE, D.G.: Model for mitosis. Nature **224**, 659–663 (1969)
MELDOLESI, J., COVA, D.: Composition of cellular membranes in the pancreas of the guinea pig. IV. Polyacrylamide gel electrophoresis and amino acid composition of membrane proteins. J. Cell Biol. **55**, 1–18 (1972)
MELDOLESI, J., JAMIESON, J.D., PALADE, G.E.: Composition of cellular membranes in the pancreas of the guinea pig. I. Isolation of membrane fractions. J. Cell Biol. **49**, 109–129 (1971a)
MELDOLESI, J., JAMIESON, J.D., PALADE, G.E.: Composition of cellular membranes in the pancreas of the guinea pig. II. Lipids. J. Cell Biol. **49**, 130–149 (1971b)
MELDOLESI, J., JAMIESON, J.D., PALADE, G.E.: Composition of cellular membranes in the pancreas of the guinea pig. III. Enzymatic activities. J. Cell Biol. **49**, 150–158 (1971c)
MERRITT, W.D., MORRÉ, D.J.: A glycosyl transferase of high specific activity in secretory vesicles derived from isolated Golgi apparatus of rat liver. Biochim. Biophys. Acta **304**, 397–407 (1973)
MESQUITA, J.F.: Electron microscope study of the origin and development of the vacuoles in root-tip cells of *Lupinus albus* L. J. Ultrastruct. Res. **26**, 242–250 (1969)
MIRANDA, A.F., GODMAN, G.C., DEITSCH, A.D., TANENBAUM, S.W.: Action of cytochalasin D on cells of established lines. I. Early events. J. Cell Biol. **61**, 481–500 (1974b)
MIRANDA, A.F., GODMAN, G.C., TANENBAUM, S.W.: Action of cytochalasin D on cells of established lines. II. Cortex and microfilaments. J. Cell Biol. **62**, 406–423 (1974a)
MOESTRUP, O., THOMSEN, H.A.: An ultrastructural study of the flagellate *Pyramimonas orientalis* with particular emphasis on Golgi apparatus activity and the flagellar apparatus. Protoplasma **81**, 247–269 (1974)
MOLLENHAUER, H.H.: Transition forms of Golgi apparatus secretion vesicles. J. Ultrastruct. Res. **12**, 439–446 (1965)
MOLLENHAUER, H.H.: A comparison of root cap cells of epiphytic, terrestrial, and aquatic plants. Am J. Botany **54**, 1249–1259 (1967a)
MOLLENHAUER, H.H.: The fine structure of mucilage secreting cells of *Hibiscus esculentus* pods. Protoplasma **63**, 353–362 (1967b)
MOLLENHAUER, H.H.: Fragmentation of mature dictyosome cisternae. J. Cell Biol. **49**, 212–214 (1971)
MOLLENHAUER, H.H.: Distribution of microtubules in the Golgi apparatus of *Euglena gracilis*. J. Cell Sci. **15**, 89–97 (1974a)
MOLLENHAUER, H.H.: Role of Golgi apparatus in the formation of plant slimes, cuticles, and extraneous wall materials. Proc. Electron Microscop. Soc. Amer. **32**, 86–87 (1974b)
MOLLENHAUER, H.H., EVANS, W.: Organization of the pellicle and the muciferous glands in *Euglena gracilis*. 28th Ann. Proc. Electron Microscop. Soc. Am. 1970
MOLLENHAUER, H.H., HASS, B.S., MORRÉ, D.J.: Evidence for membrane transformations in Golgi apparatus of rat spermatids: A role for thick cisternae. J. Microscop. (In press)

MOLLENHAUER, H.H., MORRÉ, D.J.: Golgi apparatus and plant secretion. Ann. Rev. Plant Physiol. **17**, 27–46 (1966a)
MOLLENHAUER, H.H., MORRÉ, D.J.: Tubular connections between dictyosomes and forming secretory vesicles in plant Golgi apparatus. J. Cell Biol. **29**, 373–376 (1966b)
MOLLENHAUER, H.H., MORRÉ, D.J.: Intercisternal substances of the Golgi apparatus, zones of exclusion and other 'invisible' structures that contribute to subcellular compartmentalization. What's New in Plant Physiol. **4** (No. 5) 1–4 (1972)
MOLLENHAUER, H.H., MORRÉ, D.J.: Polyribosomes associated with the Golgi apparatus. Protoplasma **79**, 333–336 (1974)
MOLLENHAUER, H.H., MORRÉ, D.J.: A possible role for intercisternal elements in the formation of secretory vesicles in plant Golgi apparatus. J. Cell Sci. **19**, 231–237 (1975)
MOLLENHAUER, H.H., MORRÉ, D.J.: Cytochalasin B, but not colchicine, inhibits migration of secretory vesicles in root tips of maize. Protoplasma **87**, 39–48 (1976)
MOLLENHAUER, H.H., WHALEY, W.G.: An observation on the functioning of the Golgi apparatus. J. Cell Biol. **17**, 222–225 (1963)
MOLLENHAUER, H.H., WHALEY, W.G., LEECH, J.H.: A function of the Golgi apparatus in outer rootcap cells. J. Ultrastruct. Res. **5**, 193–200 (1961)
MOOR, H., MÜHLETHALER, K.: Fine structure in frozen-etched yeast cells. J. Cell Biol. **17**, 609–628 (1963)
MOORE, T.C., BARLOW, S.A., COOLBAUGH, R.C.: Participation of noncatalytic 'carrier' protein in the metabolism of kaurene in cell-free extracts of pea seeds. Phytochemistry **11**, 3225–3233 (1972)
MOORE, T.S., LORD, J.M., KAGAWA, T., BEEVERS, H.: Enzymes of phospholipid metabolism in the endoplasmic reticulum of castor bean endosperm. Plant Physiol. **52**, 50–53 (1973)
MORRÉ, D.J.: Cell wall dissolution and enzyme secretion during leaf abscission. Plant Physiol. **43**, 1545–1559 (1968)
MORRÉ, D.J.: *In vivo* incorporation of radioactive metabolites by Golgi apparatus and other cell fractions of onion stem. Plant Physiol. **45**, 791–799 (1970)
MORRÉ, D.J.: Membrane biogenesis. Ann. Rev. Plant Physiol. **26**, 441–481 (1975)
MORRÉ, D.J., BRACKER, C.E.: Influence of calcium ions on the plant cell surface: Membrane fusions and conformational changes. Proc. Electron Microscop. Soc. Am. **32**, 154–155 (1974a)
MORRÉ, D.J., BRACKER, C.E.: Conformational alteration of soybean plasma membranes induced by auxin and calcium ions. Plant Physiol. **53**, 42 (Abstr.) (1974b)
MORRÉ, D.J., ELDER, J.H., JELSEMA, C.L., KEENAN, T.W., MERRITT, W.D.: Sites of membrane biogenesis within the endomembrane system of rat liver. In: Membrane Composition and Biogenesis. BARRNETT, R.J. (ed.). New York: Wiley Interscience (In press)
MORRÉ, D.J., FRANKE, W.W., DEUMLING, B., NYQUIST, S.E., OVTRACHT, L.: Golgi apparatus function in membrane flow and differentiation: origin of plasma membranes from endoplasmic reticulum. Biomembranes **2**, 95–104 (1971a)
MORRÉ, D.J., JONES, D.D., MOLLENHAUER, H.H.: Golgi apparatus mediated polysaccharide secretion by outer root cap cells of *Zea mays*. I. Kinetics and secretory pathway. Planta **74**, 286–301 (1967)
MORRÉ, D.J., KEENAN, T.W., HUANG, C.M.: Membrane flow and differentiation: Origin of Golgi apparatus membranes from endoplasmic reticulum. In: Advances in Cytopharmacology. CECCARELLI, B., CLEMENTI, F., MELDOLESI, J. (eds.), Vol. II, pp. 107–125. New York: Raven Press 1974a
MORRÉ, D.J., KEENAN, T.W., MOLLENHAUER, H.H.: Golgi apparatus function in membrane transformations and product compartmentalization: Studies with cell fractions isolated from rat liver. In: Advances in Cytopharmacology. CLEMENTI, F., CECCARELLI, B. (eds.), Vol. I, pp. 159–182. New York: Raven Press 1971b
MORRÉ, D.J., MERLIN, L.M., KEENAN, T.W.: Localization of glycosyl transferase activities in a Golgi apparatus-rich fraction isolated from rat liver. Biochem. Biophys. Res. Commun. **37**, 813–819 (1969)
MORRÉ, D.J., MERRITT, W.D., LEMBI, C.A.: Connections between mitochondria and endoplasmic reticulum in rat liver and onion stem. Protoplasma **73**, 43–49 (1971c)
MORRÉ, D.J., MOLLENHAUER, H.H.: The endomembrane concept: a functional integration of endoplasmic reticulum and Golgi apparatus. In: Dynamic Aspects of Plant Ultrastructure. ROBARDS, A.W. (ed.), pp. 84–137. New York and London: McGraw-Hill 1974

MORRÉ, D.J., MOLLENHAUER, H.H., BRACKER, C.E.: The origin and continuity of Golgi apparatus. In: Results and Problems in Cell Differentiation. II. Origin and Continuity of Cell Organelles. REINERT, J., URSPRUNG, H. (eds.), pp. 82–126. Berlin-Heidelberg-New York: Springer 1971d

MORRÉ, D.J., NYQUIST, S., RIVERA, E.: Lecithin biosynthetic enzymes of onion stem and the distribution of phosphorylcholine-cytidyl transferase among cell fractions. Plant Physiol. **45**, 800–804 (1970)

MORRÉ, D.J., VAN DER WOUDE, W.J.: Origin and growth of cell surface components. In: Macromolecules Regulating Growth and Development. HAY, E.D., KING, T.J., PAPACONSTANTINOU, J. (eds.), pp. 81–111. New York: Academic Press 1974

MORRÉ, D.J., YUNGHANS, W.N., VIGIL, E.L., KEENAN, T.W.: Isolation of organelles and endomembrane components from rat liver; Biochemical markers and quantitative morphometry. In: Methodological Developments in Biochemistry. REID, E. (ed.), Vol. IV, pp. 195–236. London: Longmans 1974b

MÜHLETHALER, K.: Ultrastructure and formation of plant cell walls. Ann. Rev. Plant Physiol. **18**, 1–24 (1967)

NAGAI, R., REBHUN, L.I.: Cytoplasmic microfilaments in streaming *Nitella* cells. J. Ultrastruct. Res. **14**, 571–589 (1966)

NEWCOMB, E.H.: A spiny vesicle in slime-producing cells of the bean root. J. Cell Biol. **35**, C17–C22 (1967)

NEWCOMB, E.H.: Plant microtubules. Ann. Rev. Plant Physiol. **20**, 253–288 (1969)

NORTHCOTE, D.H.: Structure and function of plant-cell membranes. Brit. Med. Bull. **24**, 107–112 (1968)

NORTHCOTE, D.H.: Fine structure of cytoplasm in relation to synthesis and secretion in plant cells. Proc. Roy. Soc. (London), Ser. B **173**, 21–30 (1969a)

NORTHCOTE, D.H.: The synthesis and metabolic control of polysaccharides and lignin during differentiation of plant cells. Essays in Biochem. **5**, 89–137 (1969b)

NORTHCOTE, D.H.: The Golgi apparatus. Endeavour **30**, 26–33 (1970)

NORTHCOTE, D.H.: Organization of structure, synthesis and transport within the plant during cell division and growth. Symp. Soc. Exptl. Biol. **25**, 51–69 (1971)

NORTHCOTE, D.H.: Complex envelope system. Phil. Trans. Roy. Soc. London Ser. B **268**, 119–128 (1974)

NORTHCOTE, D.H., PICKETT-HEAPS, J.D.: A function of the Golgi apparatus in polysaccharide synthesis and transport in the root-cap cells of wheat. Biochem. J. **98**, 159–167 (1966)

OAKS, A., BIDWELL, R.G.S.: Compartmentation of intermediary metabolites. Ann. Rev. Plant Physiol. **21**, 43–66 (1970)

O'BRIEN, T.P.: The cytology of cell-wall formation in some eukaryotic cells. Botan. Rev. **38**, 87–118 (1972)

O'BRIEN, T.P., THIMANN, K.V.: Intracellular fibers in oat coleoptile cells and their possible significance in cytoplasmic streaming. Proc. Natl. Acad. Sci. US **56**, 888–894 (1966)

OLSON, A.C., BONNER, J., MORRÉ, D.J.: Force extension analysis of *Avena* coleoptile cell walls. Planta **66**, 126–134 (1965)

ÖPIK, H.: Development of cotyledon cell structure in ripening *Phaseolus vulgaris* seeds. J. Exptl. Botany **19**, 64–76 (1968)

ORCI, L., LEMARCHAND, Y., SINGH, A., ASSIMACOPOULOS-JEANNET, F., ROUILLER, C., JEANRENAUD, B.: Role of microtubules in lipoprotein secretion by the liver. Nature **244**, 30–32 (1973)

ORDIN, L., HALL, M.A.: Cellulose synthesis in higher plants from UDP glucose. Plant Physiol. **43**, 473–476 (1968)

OUTKA, D.E., WILLIAMS, D.C.: Sequential coccolith morphogenesis in *Hymenomonas carterae*. J. Protozool **18**, 285–297 (1971)

OVTRACHT, L., THIÉRY, J.P.: Mise en évidence par cytochimie ultrastructurale de compartiments physiologiquement différents dans un même saccule golgien. J. Microscop. **15**, 135–170 (1972)

PALISANO, J.R., WALNE, P.L.: Acid phosphatase activity and ultrastructure of aged cells of *Euglena granulata*. J. Phycol. **8**, 81–88 (1972)

PALL, M.L.: Evidence for the glycoprotein nature of the inducer of sexuality in *Volvox*. Biochem. Biophys. Res. Commun. **57**, 683–688 (1974)

Parthasarathy, M.V., Mühlethaler, K.: Cytoplasmic microfilaments in plant cells. J. Ultrastruct. Res. **38**, 46–62 (1972)

Péaud-Lenöel, C., Axelos, M.: Structural features of the β-glucans enzymatically synthesized from uridine diphosphate glucose by wheat seedlings. FEBS Lett. **8**, 224–228 (1970)

Peel, M.C., Lucas, I.A.N., Duckett, J.G., Greenwood, A.D.: Studies of sporogenesis in the Rhodophyta. I. An association of the nuclei with endoplasmic reticulum in post-meiotic tetraspore mother cells of *Corallina officinalis* L. Z. Zellforsch. **147**, 59–74 (1973)

Pickett-Heaps, J.D.: Incorporation of radioactivity into wheat xylem walls. Planta **71**, 1–14 (1966)

Pickett-Heaps, J.D.: Preliminary attempts at ultrastructural polysaccharide localization in root tip cells. J. Histochem. Cytochem. **15**, 442–455 (1967a)

Pickett-Heaps, J.D.: The effects of colchicine on the ultrastructure of dividing plant cells, xylem wall differentiation and distribution of cytoplasmic microtubules. Develop. Biol. **15**, 206–236 (1967b)

Pickett-Heaps, J.D.: Ultrastructure and differentiation in *Chara* sp. I. Vegetative cells. Aust. J. Biol. Sci. **20**, 539–551 (1967c)

Pickett-Heaps, J.D.: Further observations on the Golgi apparatus and its functions in cells of the wheat seedling. J. Ultrastruct. Res. **18**, 287–303 (1967d)

Pickett-Heaps, J.D.: Further ultrastructural observations on polysaccharide localization in plant cells. J. Cell Sci. **3**, 55–64 (1968a)

Pickett-Heaps, J.D.: Xylem wall deposition. Radioautographic investigations using lignin precursors. Protoplasma **65**, 181–205 (1968b)

Pickett-Heaps, J.D., Northcote, D.H.: Relationship of cellular organelles to formation and development of the plant cell wall. J. Exptl. Botany **17**, 20–26 (1966)

Pinsky, A., Ordin, L.: Role of lipid in the cellulose synthetase enzyme system from oat seedlings. Plant Cell Physiol. **10**, 771–785 (1969)

Pinto Da Silva, P., Douglas, S.D., Branton, D.: Localization of A antigen sites on human erythrocyte ghosts. Nature **232**, 194–196 (1971)

Pollard, T.D., Korn, E.D.: Filaments of *Amoeba proteus*. II. Binding of heavy meromyosin by thin filaments in motile cytoplasmic extracts. J. Cell Biol. **48**, 216–219 (1971)

Pollard, T.D., Weihing, R.R.: Actin and myosin and cell movement. CRC Crit. Rev. in Biochem. **2**, 1–65 (1974)

Poole, R.J.: Effect of sodium on potassium fluxes at the cell membrane and vacuole membrane of red beet. Plant Physiol. **47**, 731–734 (1971)

Porter, K.R.: The ground substances: Observations from electron microscopy. In: The Cell. Brachet, J., Mirsky, A.E. (eds.), Vol. II, pp. 621–675. New York: Academic Press 1961

Poux, N.: Nouvelles observations sur la nature et l'origine de la membrane vacuolaire des cellules végétales. J. Microscop. **1**, 55–66 (1962)

Poux, N.: Localization d'activités enzymatiques dans le méristème radiculaire de *Cucumis sativus* L. III Activité phosphatasique acide. J. Microscop. **9**, 407–434 (1970)

Poux, N.: Observation en microscopie électronique de cellules végétales imprégnées par l'osmium. Compt. Rend. **276**, 2163–2166 (1973)

Prat, R.: Contribution à l'étude des protoplastes végétaux. II. Ultrastructure du protoplaste isolé et régénération de sa paroi. J. Microscop. **18**, 65–86 (1973)

Prat, R., Roland, J.C.: Étude ultrastructurale des premier stades de néoformation d'une enveloppe par les protoplasts végétaux séparés méchaniquement de leur paroi. Compt. Rend. **273**, 165–168 (1971)

Preston, R.D.: Structural and mechanical aspects of plant cell walls with particular reference to synthesis and growth. In· Formation of Wood in Forest Trees Zimmerman, M.H. (ed.), pp. 169–188. New York: Academic Press 1964

Preston, R.D.: The Physical Biology of Plant Cell Walls. London: Chapman and Hall 1974

Preston, R.D., Goodman, R.N.: Structural aspects of cellulose microfibril biosynthesis. J. Roy. Microscop. Soc. **88**, 513–527 (1968)

Rachmilevitz, T., Fahn, A.: Ultrastructure of nectaries of *Vinca rosea* L., *Vinca major* L., and *Citrus sinensis* Osbeck cv. Valencia and its relation to the mechanism of nectar secretion. Ann. Botany **37**, 1–9 (1973)

Rambourg, A.: Localisation ultrastructurale et nature du matériel colore au niveau de la surface cellulaire par le mélange chromique-phosphotungstique. J. Microscop. **8**, 325–342 (1969)

RAWLENCE, D.J., TAYLOR, A.R.A.: A light and electron microscopic study of rhizoid development in *Polysiphonia lanosa* L. Tandy. J. Phycol. **8**, 15–24 (1972)
RAY, P.M.: Radioautographic study of cell wall deposition in growing plant cells. J. Cell Biol. **35**, 659–674 (1967)
RAY, P.M.: Regulation of β-glucan synthetase activity by auxin in pea stem tissue. I. Kinetic aspects. Plant Physiol. **51**, 601–608 (1973a)
RAY, P.M.: Regulation of β-glucan synthetase activity by auxin in pea stem tissue. II. Metabolic requirements. Plant Physiol. **51**, 609–614 (1973b)
RAY, P.M., SHININGER, T.L., RAY, M.M.: Isolation of β-glucan synthetase particles from plant cells and identification with Golgi membranes. Proc. Natl. Acad. Sci. US **64**, 605–612 (1969)
RAY, T.K., LIEBERMAN, I., LANSING, A.I.: Synthesis of the plasma membrane of the liver cell. Biochem. Biophys. Res. Commun. **31**, 54–58 (1968)
REID, J.S.G., MEIER, H.: Formation of the endosperm galactomannan in leguminous seeds: Preliminary communications. Caryologia **25**, 219–222 (1973a)
REID, J.S.G., MEIER, H.: Enzymatic activities and galactomannan mobilisation in germinating seeds of Fenugreek (*Trigonella foenum-graecum* L. Leguminosae). Planta **112**, 301–308 (1973b)
REIMANN, B.E.F.: Desposition of silica inside a diatom cell. Exptl. Cell Res. **34**, 605–608 (1964)
RICHARD, M., LOUISOT, P.: Localization des glycosyl-transférases dans les membranes endoplasmiques des splénocytes de rat. Experientia **28**, 516–517 (1972)
RIDGE, I., OSBORNE, D.J.: Hydroxyproline and peroxidases in cell walls of *Pisum sativum*: Regulation by ethylene. J. Exptl. Botany **21**, 843–856 (1970)
RISUENO, M.C., GIMENÉZ-MARTÍN, G., LÓPEZ-SÁEZ, J.F.: Role of Golgi vesicles in plant cell elongation. Experientia **24**, 926 (1968)
RISUENO, M.C., SOGO, J.M., GIMENÉZ-MARTÍN, G., R-GARCIÁ, M.I.: Vacuolation in the cytoplasm of plant cell. Cytologia **35**, 609–621 (1970)
ROBARDS, A.W.: On the ultrastructure of differentiating secondary xylem in willow. Protoplasma **65**, 449–464 (1968)
ROBARDS, A.W.: Plasmodesmata. Ann. Rev. Plant Physiol. **26**, 13–29 (1975)
ROBERTS, K., GURNEY-SMITH, M., HILLS, G.J.: Structure, composition and morphogenesis of the cell wall of *Clamydomonas Reinhardi*. I. Ultrastructure and preliminary chemical analysis. J. Ultrastruct. Res. **40**, 599–613 (1972)
ROBERTS, R.M., CONNOR, A.B., CETORELLI, J.J.: The formation of glycoproteins in tissues of higher plants. Specific labelling with D-[1-^{14}C] glucosamine. Biochem. J. **125**, 999–1008 (1971)
ROBERTS, R.M., POLLARD, W E.: The incorporation of D-glucosamine into glycolipids and glycoproteins of membrane preparations from *Phaseolus aureus* hypocotyls. Plant Physiol. **55**, 431–436 (1975)
ROBINSON, P.M., PARK, D., MCCLURE, W.K.: Observations on induced vacuoles in fungi. Trans. Brit. Mycol. Soc. **52**, 447–450 (1969)
ROBINSON, D.G., PRESTON, R.D.: Polysaccharide synthesis in mung bean roots: An X-ray investigation. Biochim. Biophys. Acta **273**, 336–345 (1972a)
ROBINSON, D.G., PRESTON, R.D.: Plasmalemma structure in relation to microfibril biosynthesis in *Oocystis*. Planta **104**, 234–246 (1972b)
ROBINSON, D.G., RAY, P.M.: Separation of the synthesis of cellulose and non-cellulosic wall polysaccharides. Plant Physiol. **51**, 59 (Abstr.) (1973)
ROELOFSEN, P.A.: The plant cell-wall. Hand. Pflanzenanatomie **2**, 4 (1959)
ROLAND, J.C.: Mise en évidence sur coupes ultrafines de formations polysaccharidiques directement associées au plasmalemme. Compt. Rend. **269**, 939–942 (1969)
ROLAND, J.-C.: The relationship between the plasmalemma and plant cell wall. Intern. Rev. Cytol. **36**, 45–92 (1973)
ROLAND, J.-C., LEMBI, C.A., MORRÉ, D.J.: Phosphotungstic acid-chromic acid as a selective electron-dense stain for plasma membranes of plant cells. Stain Technol. **47**, 195–200 (1972)
ROLAND, J.-C., PILET, P.-E.: Implications du plasmalemme et de la paroi dans la croissance des cellules végétales. Experientia **30**, 441–451 (1974)
ROLAND, J.-C., PRAT, R.: Les protoplastes et quelques problémes concernant le rôle et l'elaboration des parois. Colloq. Intern. Centre Natl. Rech. Sci. (Paris) **212**, 243–271 (1972)

ROLAND, J.-C., SANDOZ, D.: Détection cytochimique des sites de formation des polysaccharides pré-membranaires dans les cellules végétales. J. Microscop. **8**, 263–268 (1969)
ROLAND, J.-C., VIAN, B.: Réactivité du plasmalemme végétal. Etude cytochimique. Protoplasma **73**, 121–137 (1971)
RONCARI, D.A.K.: Enhancement of triglyceride synthesis in mammalian adipose tissue by soluble factors. Federation Proc. **33**, 1525 (Abstr.) (1974)
ROSEMAN, S.: The synthesis of complex carbohydrates by multiglycosyltransferase systems and their potential function in intracellular adhesion. Chem. Phys. Lipids **5**, 270–297 (1970)
ROSEN, W.G.: Ultrastructure and physiology of pollen. Ann. Rev. Plant Physiol. **19**, 435–462 (1968)
ROSEN, W.G., THOMAS, H.R.: Secretory cells of lily pistils. I. Fine structure and function. Amer. J. Bot. **57**, 1108–1114 (1970)
ROSEN, W.G., GAWLIK, S.R., DASHEK, W.V., SIEGESMUND, K.A.: Fine structure and cytochemistry of Lilium pollen tubes. Am. J. Botany **51**, 61–71 (1964)
ROUGIER, M.: Étude cytochimique de la sécrétion des polysaccharides végétaux à l'aide d'un matériel de choix: les cellules de la coiffe de *Zea mays*. J. Microscop. **10**, 67–82 (1971)
ROUGIER, M., VIAN, B., GALLANT, D., ROLAND, J.C.: Aspects cytochimiques de l'étude ultrastructurale des polysaccharides végétaux. Ann. Biol. **12**, 44–75 (1973)
RUIZ-HERRERA, J., BARTNICKI-GARCIA, S.: Synthesis of cell wall microfibrils in vitro by a "soluble" chitin synthetase from *Mucor rouxii*. Science **186**, 357–359 (1974)
SABNIS, D.D., JACOBS, W.P.: Cytoplasmic streaming and microtubules in the coenocytic marine alga *Caulerpa prolifera*. J. Cell Sci. **2**, 465–472 (1967)
SADAVA, D., CHRISPEELS, M.J.: Hydroxyproline-rich cell wall protein (Extensin): Role in the cessation of elongation in excised pea epicotyls. Develop. Biol. **30**, 49–55 (1973)
SADAVA, D., WALKER, F., CHRISPEELS, M.J.: Hydroxyproline-rich cell wall protein (Extensin): Biosynthesis and accumulation in growing pea epicotyls. Develop. Biol. **30**, 42–48 (1973)
SASSEN, M.M.A.: Fine structure of *Petunia* pollen grain and pollen tube. Acta Botan. Neerl. **13**, 175–181 (1964)
SCALLEN, T.J., SRIKANTAIAH, M.V., SEETHARAM, B., HANSBURY, E., GAVEY, K.L.: Sterol carrier protein hypothesis. Federation Proc. **33**, 1733–1746 (1974)
SCHACHTER, H.: Glycosylation of glycoproteins during intracellular transport of secretory products. In: Advances in Cytopharmacology. CECCARELLI, B., CLEMENTI, F., MELDOLESI, J. (eds.), Vol. II, pp. 207–218. New York: Raven Press 1974
SCHACHTER, H., JABBAL, I., HUDGIN, R.L., PINTERIC, L., MCGUIRE, E.J., ROSEMAN, S.: Intracellular localization of liver sugar nucleotide glycoprotein glycosyltransferases in a Golgi-rich fraction. J. Biol. Chem. **245**, 1090–1100 (1970)
SCHMITT, F.O.: Fibrous proteins-neuronal organelles. Proc. Natl. Acad. Sci. US **60**, 1092–1101 (1968)
SCHNEPF, E.: Quantitative Zusammenhänge zwischen der Sekretion des Fangschleimes und den Golgi-Strukturen bei *Drosophyllum lustianicum*. Z. Naturforsch. **16b**, 605–610 (1961)
SCHNEPF, E.: Zur Feinstruktur der schleimsezernierenden Drüsenhaare auf der Ochrea von *Rumex* and *Rheum*. Planta **79**, 22–34 (1968a)
SCHNEPF, E.: Transport by compartments. In: Transport and distribution of matter in cells of higher plants. MOTHES, K., MULLER, E., NELLES, A., NEUMANN, D. (eds.), pp. 39–49. Berlin: Akademie-Verlag 1968b
SCHNEPF, E.: Sekretion und Exkretion bei Pflanzen. Protoplasmatologia, Handbuch der Protoplasmaforschung **8**, 1–181 (1969a)
SCHNEPF, E.: Membranfluß und Membrantransformation. Ber. Deut. Botan. Ges. **82**, 407–413 (1969b)
SCHNEPF, E.: Über den Feinbau von Öldrüsen. I. Die Drüsenhaare von *Arctium lappa*. Protoplasma **67**, 185–194 (1969c)
SCHNEPF, E.: Über den Feinbau von Öldrüsen. II. Die Drüsenhaare in *Calceolaria*-Blüten. Protoplasma **67**, 195–203 (1969d)
SCHNEPF, E.: Über den Feinbau von Öldrüsen. IV. Die Ölgänge von Umbelliferen: *Heracleum spondylinum* und *Dorema ammoniacum*. Protoplasma **67**, 375–390 (1969e)
SCHNEPF, E.: Tubuläres endoplasmatisches Reticulum in Drüsen mit lipophilen Ausscheidungen von *Ficus, Ledum* und *Salvia*. Biochem. Physiol. Pflanzen **163**, 113–125 (1972)
SCHNEPF, E.: Sezernierende und exzernierende Zellen bei Pflanzen. In: Grundlagen der Cytologie. HIRSCH, G.C. (ed.), pp. 461–477. Jena: Gustav Fischer 1973

SCHNEPF, E., DEICHGRÄBER, G.: Über die Feinstruktur von *Synura petersenii* unter besonderer Berücksichtigung der Morphogenese ihrer Kieselschuppen. Protoplasma **68**, 85–106 (1969)

SCHNEPF, E., DEICHGRÄBER, G.: Tubular inclusions in the endoplasmic reticulum of the gland hairs of *Ononis repens* L. (Fabaceae). J. Microscop. **14**, 361–364 (1972a)

SCHNEPF, E., DEICHGRÄBER, G.: Über den Feinbau von Theka, Pusule und Golgi-Apparat bei dem Dinoflagellaten *Gymnodium* spec. Protoplasma **74**, 411–425 (1972b)

SCHNEPF, E., KALSOVA, A.: Zur Feinstruktur von Öl- und Flavon-Drüsen. Ber. Deut. Botan. Ges. **85**, 249–258 (1972)

SCHNEPF, E., KOCH, W.: Golgi-Apparat und Wasserausscheidung bei *Glaucocystis*. Z. Pflanzenphysiol. **55**, 97–109 (1966a)

SCHNEPF, E., KOCH, W.: Über die Entstehung der pulsierenden Vacuolen von *Vacuolaria virescens* (Chloromonadophyceae) aus dem Golgi-Apparat. Arch. Mikrobiol. **54**, 229–236 (1966b)

SCHROEDER, T.E.: Cell constriction: Contractile role of microfilaments in division and development. Am. Zool. **13**, 949–960 (1973)

SCHRÖTER, K., SIEVERS, A.: Wirkung der Turgorreduktion auf den Golgi-Apparat und die Bildung der Zellwand bei Wurzelhaaren. Protoplasma **72**, 203–211 (1971)

SCHWAB, D.W., SIMMONS, E., SCALA, J.: Fine structure changes during function of the digestive gland of Venus's-flytrap. Am. J. Botany **56**, 88–100 (1969)

SCHWELITZ, F.D., EVANS, W.R., MOLLENHAUER, H.H., DILLEY, R.A.: The fine structure of the pellicle of *Euglena gracilis* as revealed by freeze-etching. Protoplasma **69**, 341–349 (1970).

SHAFIZADEH, F., MCGINNIS, G.D.: Morphology and biogenesis of cellulose and plant cell-walls. Advan. Carbohydrate Chem. **26**, 297–349 (1971).

SHARON, N., LIS, H.: Lectins: Cell-agglutinating and sugar-specific proteins. Science **177**, 949–959 (1972)

SHEN-MILLER, J., MILLER, C.: Distribution and activation of the Golgi apparatus in geotropism. Plant Physiol. **49**, 634–639 (1972)

SHORE, G., MACLACHLAN, G.A.: Indoleacetic acid stimulates cellulose deposition and selectively enhances certain β-glucan synthetase activities. Biochim. Biophys. Acta **329**, 271–282 (1973)

SHORE, G., MACLACHLAN, G.: Changes in subcellular location of cellulose synthetases during growth. Plant Physiol. **53**, 16 (Abstr.) (1974)

SIEVERS, A.: Beteiligung des Golgi-Apparates bei der Bildung der Zellwand von Wurzelhaaren. Protoplasma **56**, 188–192 (1963)

SIEVERS, A.: Elektronenmikroskopische Untersuchungen zur geotropischen Reaktion. I. Über Besonderheiten im Feinbau der Rhizoide von *Chara foetida*. Z. Pflanzenphysiol. **53**, 193–213 (1965)

SIEVERS, A.: Elektronenmikroskopische Untersuchungen zur geotropischen Reaktion. II Die polare Organization des normal wachsenden Rhizoids von *Chara foetida*. Protoplasma **64**, 225–253 (1967)

SINGH, A.: The subplasmalemmal microfilaments in Kupffer cells. J. Ultrastruct. Res. **48**, 67–68 (1974).

SJÖSTRAND, F.S., HANZON, V.: Ultrastructure of Golgi apparatus of exocrine cells of mouse pancreas. Exptl. Cell Res. **7**, 415–429 (1954)

SLAUTTERBACK, D.B.: Coated vesicles in absorptive cells of *Hydra*. J. Cell Sci. **2**, 563–572 (1967)

SMITH, D.S.: Bridges between vesicles and axoplasmic mictrotubules. J. Cell Biol. **47**, 195a (Abstr.) (1970)

SMITH, D.S., JÄRLFORS, U., BERÁNEK, R.: The organization of synaptic axoplasm in the lamprey (*Petromyzon marinus*) central nervous system. J. Cell Biol. **46**, 199–219 (1970)

SMITH, M.M., STONE, B.A.: β-Glucan synthesis by cell-free extracts from *Lolium multiflorum* endosperm. Biochim. Biophys. Acta **313**, 72–94 (1973)

SOROKIN, S.P.: Reconstructions of centriole formation and ciliogenesis in mammalian lungs. J. Cell Sci. **3**, 207–230 (1968)

SPENCER, F.S., ZIOLA, B., MACLACHLAN, G.A.: Particulate glucan synthetase activity: dependence on acceptor, activator, and plant growth hormone. Can. J. Biochem. **49**, 1326–1332 (1971)

SPOONER, B.S.: Microfilaments, cell shape changes, and morphogenesis of salivary epithelium. Am. Zool. **13**, 1007–1022 (1973)

STAEHELIN, L.A.: Ultrastructural changes of the plasmalemma and the cell wall during the life cycle of *Cyanidium caldarium*. Proc. Roy. Soc. B **171**, 249–259 (1968)
STAEHELIN, L.A., CHLAPOWSKI, F.J., BONNEVILLE, M.A.: Lumenal plasma membrane of the urinary bladder. I. Three-dimensional reconstruction from freeze-etch images. J. Cell Biol. **53**, 73–91 (1972)
STAEHELIN, L.A., KIERMAYER, O.: Membrane differentiation in the Golgi complex of *Micrasterias denticulata* Bréb. visualized by freeze-etching. J. Cell Sci. **7**, 787–792 (1970)
STEIN, O., SANGER, L., STEIN, Y.: Colchicine-induced inhibition of lipoprotein and protein secretion into the serum and lack of interference with secretion of biliary phospholipids and cholesterol by rat liver *in vivo*. J. Cell Biol. **62**, 90–103 (1974)
STIEGLITZ, H., STERN, H.: Regulation of β-1,3-glucanase activity in developing anthers of *Lilium*. Develop. Biol. **34**, 169–173 (1973)
STOERMER, E.F., PANKRATZ, H.S., BOWEN, C.C.: Fine structure of the diatom *Amphipleura pellucida*. II. Cytoplasmic fine structure and frustule formation. Am. J. Botany **52**, 1067–1078 (1965)
STRAPRANS, I., DIRKSEN, E.R.: Microtubule protein during ciliogenesis in the mouse oviduct. J. Cell Biol. **62**, 164–174 (1974)
STROBEL, G.A., HESS, W.M.: Evidence for the presence of the toxin-binding protein on the plasma membrane of sugarcane cells. Proc. Natl. Acad. Sci. US **71**, 1413–1417 (1974)
THIÉRY, J.-P.: Role de l'appareil de Golgi dans la synthèse des mucopolysaccharides étude cytochimique. I. Mise en évidence de mucopolysaccharides dans les vésicules de transition entre l'ergastoplasme et l'appareil de Golgi. J. Microscop. **8**, 689–708 (1969)
THOMAS, D. DES S., LUTZAC, M., MANAVATHU, E.: Cytochalasin selectively inhibits synthesis of a secretory protein, cellulase, in *Achlya*. Nature **249**, 140–142 (1974)
THOMSON, K.-S., LEOPOLD, A.C.: *In vitro* binding of morphactins and 1-N-naphthylphthalamic acid in corn coleoptiles and their effects on auxin transport. Planta **115**, 259–270 (1974)
THOMSON, W.W., BERRY, W.L., LIU, L.L.: Localization and secretion of salt by the salt glands of *Tamarix aphylla*. Proc. Natl. Acad. Sci. US **63**, 310–317 (1969)
THOMSON, W.W., LIU, L.L.: Ultrastructural features of the salt gland of *Tamarix aphylla* L. Planta **73**, 201–220 (1967)
TILLACK, T.W., SCOTT, R.E., MARCHESI, V.T.: The structure of erythrocyte membranes studied by freeze-etching. II. Localization of receptors for phytohemagglutinin and influenza virus to the intramembranous particles. J. Exptl. Med. **135**, 1209–1227 (1972)
TILNEY, L.G.: Origin and continuity of microtubules. In: Origin and Continuity of Cell Organelles. REINERT, J., URSPRUNG, H. (eds.), pp. 222–260. Berlin-Heidelberg-New York: Springer 1971
TSAI, C.M., HASSID, W.Z.: Solubilization and separation of uridine diphospho-D-glucose: β-(1→4)-glucan and uridine diphospho-D-glucose: β-(1→3)-glucan glucosyltransferases from coleoptiles of *Avena sativa*. Plant Physiol. **47**, 740–744 (1971)
TSAI, C.M., HASSID, W.Z.: Substrate activation of β-(1→3)-glucan synthetase and its effect on the structure of β-glucan obtained from UDP-D-glucose and particulate enzyme of oat coleoptiles. Plant Physiol. **51**, 998–1001 (1973)
UEDA, K.: Fine structure of *Chlorogonium elongatum* with special reference to vacuole development. Cytologia **31**, 461–472 (1966)
UNZELMAN, J.M., HEALY, P.L.: Development, structure, and occurrence of secretory trichomes of *Pharbitis*. Protoplasma **80**, 285–303 (1974)
UPHOF, J.C.TH.: Plant hairs. Hand. der Pflanzenanatomie **4**, 5 (1962)
VANDERWOUDE, W.J., LEMBI, C.A., MORRÉ, D.J.: Auxin (2,4-D) stimulation (*in vivo* and *in vitro*) of polysaccharide synthesis in plasma membrane fragments isolated from onion stems. Biochem. Biophys. Res. Commun. **46**, 245–253 (1972)
VANDERWOUDE, W.J., LEMBI, C.A., MORRÉ, D.J., KINDINGER, J.I., ORDIN, L.: β-glucan synthetases of plasma membrane and Golgi apparatus from onion stem. Plant Physiol. **54**, 333–340 (1974)
VANDERWOUDE, W.J., MORRÉ, D.J., BRACKER, C.E.: A role for secretory vesicles in polysaccharide biosynthesis. Proc. 11th Intern. Botan. Congr. (Seattle/Wash.), p. 226 (Abstr.) (1969)
VANDERWOUDE, W.J., MORRÉ, D.J., BRACKER, C.E.: Isolation and characterization of secretory vesicles in germinated pollen of *Lilium longiflorum*. J. Cell Sci. **8**, 331–351 (1971)

VAN GOLDE, L.M.G., FLEISCHER, B., FLEISCHER, S.: Some studies on the metabolism of phospholipids in Golgi complex from bovine and rat liver in comparison to other subcellular fractions. Biochim. Biophys. Acta **249**, 318–330 (1971)

VAN GOLDE, L.M.G., RABEN, J., BATENBURG, J.J., FLEISCHER, B., ZAMBRANO, F., FLEISCHER, S.: Biosynthesis of lipids in Golgi complex and other subcellular fractions from rat liver. Biochim. Biophys. Acta **360**, 179–192 (1974)

VARNER, J.E., MENSE, R.M.: Characteristics of the process of enzyme release from secretory plant cells. Plant Physiol. **49**, 187–189 (1972)

VIAN, B.: Aspects, en cryodécapage, de la fusion des membranes des vésicules cytoplasmiques et du plasmalemme lors des phénomènes de sécrétion végétale. Compt. Rend. **275**, 2471–2474 (1972)

VIAN, B.: Précisions fournies par le cryodécapage sur la restructuration et l'assimilation au plasmalemme des membranes des dérivés golgiens. Compt. Rend. **278**, 1483–1486 (1974)

VIAN, B., ROLAND, J.-C.: Différenciation des cytomembranes et renouvellement du plasmalemme dans les phénomènes de sécrétions végétales. J. Microscop. **13**, 119–136 (1972)

VIAN, B., ROUGIER, M.: Ultrastructure des plasmodesmes après cryo-ultramicrotomie. J. Microscop. **20**, 307–312 (1974)

VIGIL, E.L., RUDDAT, M.: Effect of gibberellic acid and actinomycin D on the formation and distribution of rough endoplasmic reticulum in barley aleurone cells. Plant Physiol. **51**, 549–558 (1973)

VILLEMEZ, C.L.: Characterization of intermediates in plant cell wall biosynthesis. Biochem. Biophys. Res. Commun. **40**, 636–641 (1970)

VILLEMEZ, C.L., CLARK, A.F.: A particle bound intermediate in the biosynthesis of plant cell wall polysaccharides. Biochem. Biophys. Res. Commun. **36**, 57–63 (1969)

VILLEMEZ, C.L., MCNAB, J.M., ALBERSHEIM, P.: Formation of plant cell wall polysaccharides. Nature **218**, 878–880 (1968)

VOELLER, B.R., LEDBETTER, M.C., PORTER, K.R.: The plant cell: Aspects of its form and function. In: The Cell. ALLFREY, V., MIRSKY, A.E. (eds.), Vol. 6, pp. 245–312. New York: Academic Press 1964

WAECHTER, C.J., LUCAS, J.J., LENNARZ, W.J.: Membrane glycoproteins. I. Enzymatic synthesis of mannosyl phosphoryl-polyisoprenol and its role as a mannosyl donor in glycoprotein synthesis. J. Biol. Chem. **248**, 7570–7579 (1973)

WARD, R.T., WARD, E.: The multiplication of Golgi bodies during maturation in the oocytes of *Rana pipiens*. J. Microscop. **7**, 1007–1020 (1968)

WATSON, M.W., BERLIN, J.D.: Differentiation of lint and fuzz fibers on the cotton ovule. J. Cell Biol. **59**, 360a (Abstr.) (1973)

WESSELS, N.K.: How living cells change shape. Sci. Am. **225**, 77–82 (1971)

WESSELS, N.K., SPOONER, B.S., ASH, J.F., BRADLEY, M.O., LUDUENA, M.A., TAYLOR, E.L., WRENN, J.T., YAMADA, K.M.: Microfilaments in cellular and developmental processes. Science **171**, 135–143 (1971)

WHALEY, W.G., DAUWALDER, M., KEPHART, J.E.: The Golgi apparatus and an early stage in cell plate formation. J. Ultrastruct. Res. **15**, 169–180 (1966)

WHALEY, W.G., DAUWALDER, M., KEPHART, J.E.: Golgi apparatus: influence on cell surfaces. Science **175**, 596–599 (1972)

WHALEY, W.G., KEPHART, J.E., MOLLENHAUER, H.H.: The dynamics of cytoplasmic membranes during development. In: Cellular Membranes in Development. LOCKE, M. (ed.), pp. 135–173. New York: Academic Press 1964

WHALEY, W.G., MOLLENHAUER, H.H.: The Golgi apparatus and cell plate formation—a postulate. J. Cell Biol. **17**, 216–221 (1963)

WILLIAMS, D.C.: Studies of protistan mineralization. I. Kinetics of coccolith secretion in *Hymenomonas carterae*. Calcif. Tissue Res. **16**, 227–237 (1974)

WILLISON, J.H.M., COCKING, E.C.: The production of microfibrils at the surface of isolated tomato-fruit protoplasts. Protoplasma **75**, 397–403 (1972)

WILSON, K.: The growth of plant cell walls. Intern. Rev. Cytol. **17**, 1–49 (1964)

WINTER, H., MEYER, L., HENGEVELD, E., WIERSEMA, P.K.: The role of wall-bound hydroxyproline-rich protein in cell extension. Acta Botan. Neerl. **20**, 489–497 (1971)

WOLLENWEBER, E., SCHNEPF, E.: Vergleichende Untersuchungen über die flavonoiden Exkrete von „Mehl-" und „Öl-" Drüsen bei Primeln und die Feinstruktur der Drüsenzellen. Z. Pflanzenphysiol. **62**, 216–227 (1970)

WOODING, F.B.P.: Radioautographic and chemical studies of incorporation into sycamore vascular tissue walls. J. Cell Sci. **3**, 71–80 (1968)

WOODING, F.B.P., NORTHCOTE, D.H.: The development of the secondary wall of the xylem in *Acer pseudoplatanus*. J. Cell Biol. **23**, 327–337 (1964)

WOOLLEY, D.E.: An actin-like protein from amoebae of *Dictyostelium discoideum*. Federation Proc. **29**, 667 (Abstr.) (1970)

WRIGHT, K., NORTHCOTE, D.H.: The relationship of root-cap slimes to pectins. Biochem. J. **139**, 525–534 (1974)

YAMADA, K.M., SPOONER, B.S., WESSELS, N.K.: Ultrastructure and function of growth cones and axons of cultured nerve cells. J. Cell Biol. **49**, 614–635 (1971)

YAMAMOTO, M.: Electron microscopy of fish development. III. Changes in the ultrastructure of the nucleus and cytoplasm of the oocyte during its development in *Oryzias latipes*. J. Fac. Sci. Univ. Tokyo **10**, 335–346 (1964)

YUNGHANS, W.N., KEENAN, T.W., MORRÉ, D.J.: Isolation of Golgi apparatus from rat liver. III. Lipid and protein composition. Exptl. Mol. Path. **12**, 36–45 (1970)

III. Intracellular Transport in Relation to Energy Conservation

1. Ion Transport and Energy Conservation in Chloroplasts

R.E. McCarty

1. Introduction

Photosynthetic phosphorylation, the light-driven synthesis of ATP from ADP and Pi, in chloroplasts was discovered over twenty years ago by Arnon et al. (1954). Within a few years, it was firmly established that photosynthetic phosphorylation (photophosphorylation in short) is coupled to light-dependent electron flow through an electron transport chain rather similar to that in mitochondria (Jagendorf, 1958; Arnon, 1958). However, the realization that ion fluxes might have something to do with photophosphorylation is only a decade old.

In an attempt to uncover high-energy intermediates Hind and Jagendorf (1963) found that chloroplasts illuminated in the presence of a mediator of electron flow were able to synthesize substantial amounts of ATP when they were injected into a solution containing ADP and Pi in the dark. Similar observations were made by Shen and Shen (1962). Since the yield of ATP was more than one hundred times greater than the content of electron transport carriers such as cytochrome f, plastocyanin or P_{700}, it was clear that the intermediate state responsible for ATP synthesis in this two-stage system could not have been an electron transport component in some high energy form. The impetus for the discovery of light-dependent, reversible H^+ uptake in chloroplasts by Jagendorf and Hind (1963) came from Mitchell's (1961) first proposal that a transmembrane H^+ gradient might provide the driving force for ATP synthesis in mitochondria and chloroplasts. Subsequently the extent of H^+ uptake was correlated to the extent of postillumination ATP formation (Jagendorf and Uribe, 1966a), providing the first indication that a gradient in H^+ may be involved in phosphorylation.

Dramatic proof that H^+ gradients can serve as the driving force for ATP synthesis in chloroplasts was then published by Jagendorf and Uribe (1966b). Chloroplasts were found to synthesize ATP in the dark as a result of a rapid shift in the pH of the suspending medium from around 4 to 8. Large amounts of ATP were formed especially when the capacity of the chloroplasts to store H^+ was increased by the addition of a buffer with the appropriate characteristics (Uribe and Jagendorf, 1967a, 1967b).

Thus, the broad outline of a plausible pathway of energy conservation in chloroplasts was completed by the mid 1960s.

$$\text{light} \rightarrow \text{electron flow} \rightarrow H^+ \text{ uptake} \rightarrow \text{ATP synthesis.}$$

This pathway is consistent with the basic tenets of Mitchell's chemiosmotic hypothesis (1961, 1966). Not all investigators were ready to accept the possibility that an electrochemical gradient in H^+ was the high energy state. An alternate pathway for energy conservation was proposed:

$$\begin{array}{c} H^+ \text{ uptake} \\ \downarrow\uparrow \\ \text{light} \rightarrow \text{electron flow} \rightarrow \sim \rightarrow \text{ATP} \end{array}$$

where ~ is "squiggle" and represents some kind of high energy chemical bond or state. Thus, H^+ uptake and the gradient formed were considered to be on a side path, not on the main route toward ATP formation.

Since 1966, the evidence accumulated seems to weigh heavily in favor of the chemiosmotic interpretation of energy conservation. First, no evidence for a high-energy intermediate other than the electrochemical gradient in H^+ has been presented. Second, the results of more detailed analyses of H^+ uptake, the magnitude of the electrochemical gradient in H^+ and of the ATPase system of chloroplasts performed in several laboratories provide, in general, more support for the involvement of the H^+ gradient in photophosphorylation. In the absence of evidence to the contrary, it seems logical to conclude that electron transport and phosphorylation are coupled to each other through an electrochemical gradient in H^+.

This article is not intended to be a comprehensive review on the relation of ion fluxes to energy conservation in chloroplasts. Instead some of the more recent aspects of this problem will be discussed in some detail. To fill in the gaps, the reader is directed to recent review articles on this subject (Walker and Crofts, 1970; Packer et al., 1970; Schwartz, 1971; Dilley, 1971; Greville, 1969; Witt, 1971; Hind and McCarty, 1973; Trebst, 1974) as well as to a book (Govindjee, 1975) and to various symposia volumes (Forti et al., eds., 1971; Avron, ed., 1974).

2. Mechanisms of H^+ Uptake in Chloroplasts

It is important to clarify at the onset of this article the nature of the chloroplast preparations used in most of the investigations reported. Although chloroplasts in situ have an outer double membrane envelope surrounding an internal membrane system, the outer membranes of chloroplasts are readily lost during isolation of chloroplasts unless special precautions are taken. The inner membranes contain most, if not all, of the chlorophyll and catalyze electron transport, H^+ uptake and phosphorylation. The outer membrane system is devoid of these activities but contains specific transport systems for a number of metabolites. The transport of ions referred to in this article occurs across the inner or thylakoid membranes of the chloroplasts. Most investigators study these processes in chloroplast preparations which are essentially naked inner membranes. These preparations have been given various names including thylakoid membranes, class II chloroplasts, broken chloroplasts, chloroplast membranes, and lamellar membranes.

2.1 H^+ Uptake in the Absence of Permeating Electron Donors or Acceptors

Two mechanisms to explain light-dependent H^+ uptake in chloroplasts have been proposed. MITCHELL (1966) suggested that the translocation of H^+ in chloroplasts was directly coupled to electron flow and proposed a tentative scheme for H^+ uptake coupled to noncyclic electron flow which is shown in Figure 1. The appearance of H^+ inside the thylakoids was postulated to occur by two mechanisms. First, if the oxidation of water were to be catalyzed by an enzyme system bound to the inner surface of the thylakoid membranes, H^+ could be liberated inside. Second, he proposed that a redox loop, similar to those postulated for mitochondrial electron flow, existed between the two photosystems. In this scheme, a mobile hydrogen carrier (for example, plastoquinones) and electron carriers (for example, cytochromes) are required. When plastoquinone is reduced it picks up H^+ from the medium. If oxidized plastoquinone were in contact with the external medium, it would accept H^+ from the outside when it is reduced. It could then diffuse across the membrane where it is oxidized by an electron acceptor. The H^+ formed as a result of the oxidation would be liberated inside the thylakoids.

The other mechanism, proposed by DILLEY and VERNON (1967), involved the operation of a H^+-cation exchange carrier. They proposed that such a carrier could undergo a reversible change in conformation which was linked to changes in the redox state of an electron carrier.

Although there is no compelling reason to reject either hypothesis, the properties of H^+ uptake in chloroplasts seem to be consistent with the simpler (and, therefore, more acceptable) model of MITCHELL. MITCHELL'S suggested mechanism of H^+ uptake requires that the thylakoid membrane be anisotropic. This has been shown to be the case. Coupling factor 1 (MCCARTY and RACKER, 1966), ferredoxin-$NADP^+$ reductase (BERZBORN, 1968), and ferredoxin (HIEDEMAN-VAN WYK and KANNANGARA, 1971), are accessible to antibodies and are therefore localized on the outer surface of thylakoids. In contrast, plastocyanin (HAUSKA et al., 1971) and cytochrome f (RACKER et al., 1971) do not react with antibodies unless the chloroplasts were disrupted in the presence of the antibody. Furthermore, added

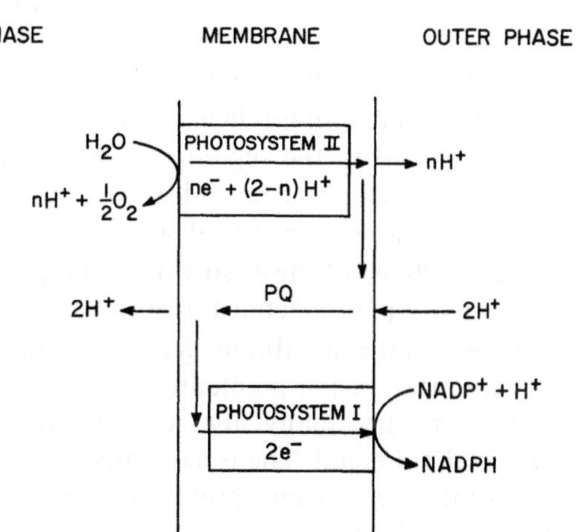

Fig. 1. Possible H^+ translocating oxido-reduction system for noncyclic phosphorylation in chloroplasts suggested by MITCHELL (1966). For an n of 2, 1 or 0 an $H^+/2e^-$ ratio of 4, 3 or 2, respectively, was predicted. It now appears that the $H^+/2e^-$ is 4 and that the H^+ formed upon the oxidation of water are released inside. *PQ* plastoquinone

plastocyanin enhanced phosphorylation in plastocyanin-deficient chloroplast preparations only when the chloroplasts were exposed to sonic oscillation in the presence of plastocyanin which allows the plastocyanin to come into contact with the inside of the thylakoid vesicles (Hauska et al., 1971).

Work with artificial electron donors and acceptors, which is summarized in Section 2,2, supports the notion that the electron donation to the photosystems occurs toward the inner surface of the membrane whereas electron acceptance from the photosystems occurs more toward the outside. Moreover, reactions catalyzed by photosystem II are sensitive to the treatment of chloroplasts with either trypsin (Selman and Bannister, 1971; Selman et al., 1973), diazonium benzene sulfonate (Giaquinta et al., 1973) and lactoperoxidase-catalyzed iodination (Arntzen et al., 1974). These reagents are not likely to cross the membrane and therefore exert their effects on components which are accessible from the outside. However, oxygen evolution probably occurs inside the thylakoids (Fowler and Kok, 1974; Junge and Ausländer, 1974) and, hence at least some components of photosytem II are also likely to be localized more toward the inside. The conclusion that the water splitting reaction is located toward the inside of the thylakoids is also supported by the finding (Harth et al., 1974b; Cohn et al., 1975) that the inactivation of oxygen evolution at alkaline pH values is markedly enhanced by uncouplers. By decreasing the pH differential across thylakoid membranes, uncouplers would allow the internal pH in illuminated chloroplasts to approach that of the external pH. Thus, inactivation of oxygen evolution occurs only when the internal pH is high, indicating that a component(s) of the water splitting system is localized toward the inside of the membrane.

It should be pointed out that the treatment of chloroplasts with either trypsin (Selman and Bannister, 1971; Selman et al., 1973) or diazonium benzene sulfonate (Giaquinta et al., 1973) inhibits oxygen evolution, but not the photosystem II-dependent oxidation of diphenylcarbazide. These findings are in apparent conflict with the notion that the water splitting system is located toward the inside. However, as discussed by Trebst (1974), these treatments could result in disorganization of the membrane and an exposure of the water splitting system to reagents which do not cross the membrane.

In view of the strong possibility that electron donation to the photosystems occurs toward the inside whereas electron acceptance occurs more toward the outside, it would be expected that the primary photochemical reactions take place across the membrane. The extensive studies of Witt and his collaborators (see Witt, 1971 for a comprehensive review) on the light-induced increase of absorption of chloroplasts in the region of 518 nm have given support to this concept. This absorption change is probably a shift in the spectrum of chloroplast pigments caused by the generation of an electric field. In view of the very fast rise time (less than 20 ns) of the absorption change, it can be considered to be a consequence of a primary photochemical reaction. Thus, photochemical charge separation could occur across the membrane generating an electric field which influences the absorption spectrum of the pigments.

In summary, many lines of evidence give strong support to the notion that the thylakoid membrane is asymmetric and that electron transport can occur across the membrane. An interpretation of the localization of components in thylakoids is given in Figure 2.

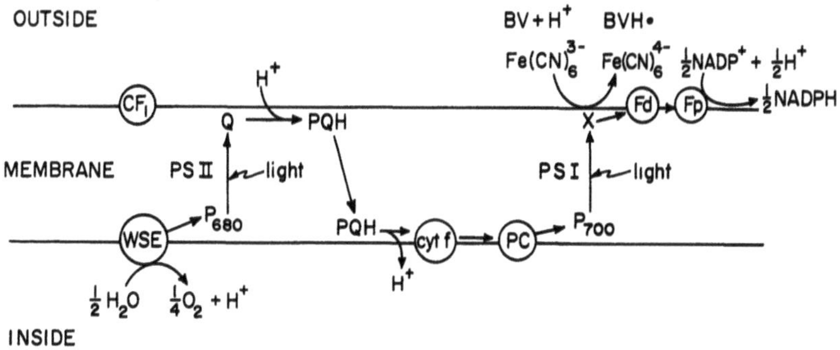

Fig. 2. A schematic representation of the distribution of some chloroplast components within the thylakoid membrane. Key: CF_1 coupling factor 1; *WSE* water splitting enzyme; P_{680} reaction center chlorophyll of photosystem II (*PS II*); *Q* the primary electron acceptor of *PS II*; *PQH* half-reduced plastoquinone; *cyt f* cytochrome f; *PC* plastocyanin; P_{700} the reaction center chlorophyll of photosystem I (*PS I*); *X* the primary electron acceptor of *PS I*; *Fd* ferredoxin; *Fp* ferredoxin-NADP$^+$ reductase, a flavoprotein; *BV* benzylviologen

MITCHELL's speculations on the nature of H^+ uptake in chloroplasts also require a precise stoichiometry between the number of H^+ translocated and electrons transported down the chain (H^+/e^- ratio). H^+/e^- ratios ranging from over 5 to 1 have been reported. The higher values were obtained by DILLEY (1969) and DILLEY and VERNON (1967), and by KARLISH and AVRON (1968), and led DILLEY to conclude that the H^+ carrier hypothesis was more tenable than MITCHELL's hypothesis. However, both DILLEY and KARLISH and AVRON estimated H^+ uptake using conventional pH meters. The slow response of pH meters and glass electrodes complicates the assessment of the H^+/e^- ratio. Furthermore, DILLEY (1969) and DILLEY and VERNON (1967) calculated H^+/e^- ratios from the steady state rate of electron flow and the *initial* rate of H^+ uptake. It is probable that the true initial rate of electron flow is considerably higher than the steady state rate. In addition, the rate and extent of H^+ uptake can be increased dramatically by reagents such as pyridine and aniline which increase the internal buffering capacity of the chloroplasts without affecting the rate of electron flow. Variations in the internal buffering capacity of chloroplasts with pH could, therefore, also complicate the measurement of H^+/e^- ratios.

In every case where the slow response of the glass electrode has either been corrected for or obviated, a low and constant H^+/e^- ratio has been observed. IZAWA and HIND (1967), who used a flow system to overcome the slow response of the glass electrode, calculated an H^+/e^- ratio of 2 at pH 6.2 using dichlorophenolindophenol as the electron acceptor and 1, using ferricyanide as the acceptor. Initial rates of electron transport and H^+ uptake were used in the estimation of the H^+/e^- ratio. TELFER and EVANS (1972) and RUMBERG et al. (1969) corrected for the response time of the glass electrode and found values of the H^+/e^- ratio near 1 when the steady state rate of electron flow was compared to the initial rate of H^+ efflux. RUMBERG et al. also calculated H^+/e^- ratio from the initial rates of H^+ influx and electron flow and obtained values close to 2. It is possible that the initial rates of H^+ efflux are underestimated due to the formation of a membrane potential positive outside during H^+ efflux. In agreement with this notion, several investigators (cf. KARLISH and AVRON,

1971) have shown that valinomycin plus K^+ which should help to keep an H^+ diffusion potential to a minimum, increases H^+ efflux rates in chloroplasts. It should be pointed out, however, that valinomycin can have secondary effects on phosphorylation and electron flow (McCarty, 1969; Telfer and Barber, 1974).

H^+ uptake in chloroplasts illuminated with repetitive, single turnover flashes has also been estimated by monitoring the absorption or fluorescence of pH indicators. Schliephake et al. (1968) used bromthymol blue but this indicator does not behave as an ideal indicator of the external pH (Cost and Frenkel, 1967). More recently, cresol red has been used by Junge and Ausländer (1974). They report that the absorption changes in cresol red (30 μM) induced by flash illumination were sensitive to the buffering capacity of the medium and that the light-induced shift in the absorption spectrum of the indicator was identical to that observed upon alkalinization.

With benzylviologen, which accepts electrons from photosystem I, $2H^+$ disappeared from the medium as a result of a single turnover flash (Junge and Ausländer, 1974). With 0.3 mM ferricyanide, which also probably accepts electrons from photosystem I, only one H^+ was consumed per electron. In the presence of 10 mM ferricyanide, no change in pH of the medium was detected. If, however, 1 μM carbonylcyanide p-trifluoromethoxyphenylhydrazone (FCCP) was added, no change in the absorption of cresol red was detected with benzylviologen whereas approximately 1 H^+/e^- *appeared* in the medium when either the high or low concentration of ferricyanide was used. These results were interpreted as evidence that H^+ is released inside the thylakoids. In the presence of FCCP, the H^+ permeability of the membranes is enhanced. In the absence of FCCP the half time for H^+ efflux is probably on the order of several seconds. This is too slow to be detected by the method used. Fowler and Kok (1974) who used a sensitive, rapidly responding glass electrode to measure pH charges, found that the flash yield of H^+ release oscillated with the same periodicity as that observed for O_2 evolution. These oscillations could be observed only in the presence of an uncoupler such as gramicidin. Thus, in agreement with the conclusions of Junge and Ausländer (1974), Fowler and Kok suggested that water oxidation and its concomitant release of H^+ occurs inside the thylakoids. Therefore, 1 H^+/e^- would be liberated inside the thylakoids as a direct consequence of the splitting of water.

In the presence of benzylviologen, 2 H^+ disappeared from the medium per electron transferred. One of these H^+ is probably required in the benzylviologen-mediated reduction of oxygen:

$$\mathrm{BV} + e^- + H^+ \rightleftharpoons \mathrm{BVH}\cdot$$

$$\mathrm{BVH}\cdot + {}^1/_2\, O_2 \rightleftharpoons \mathrm{BV} + {}^1/_2\, H_2O_2$$

where BV stands for benzylviologen. Only 1 H^+/e^- was taken up from the medium with flash excitation when a low concentration of ferricyanide was used as electron acceptor. Ferricyanide is a pure electron acceptor at pH 8. The plastoquinone antagonist, dibromomethylisopropyl-p-benzoquinone (DBMIB) (Trebst et al., 1970), blocks electron flow between the two photosystems probably at plastoquinone [see Trebst (1974) for a review]. At high concentrations (>1 μM) DBMIB itself can support photosystem II-dependent electron flow and phosphorylation (Gould and Izawa 1973a) since reduced DBMIB is reoxidized by molecular

Table 1. Summary of Junge's experiments on H^+/e^- (JUNGE and AUSLÄNDER, 1974; AUSLÄNDER et al., 1974)

Electron acceptor	H^+/e^- (observed)[a]		H^+/e^- (calculated)
	−FCCP	+FCCP	
Benzlviologen	−2	0	2
Ferricyanide	−1	+1	2
Benzlviologen+DBMIB	−2	−1	1
Ferricyanide+DBMIB	−1	0	1

[a] In the absence of FCCP the cresol red method probably detects changes in H^+ concentration only in the suspending medium. In the presence of FCCP, which markedly enhances the permeability of thylakoids to H^+, H^+ released inside the thylakoids are also detected. Therefore, the number of H^+ appearing inside per electron transferred may be calculated

oxygen. Ferricyanide may also be used to reoxidize reduced DBMIB. AUSLÄNDER et al. (1974) found that DBMIB inhibits the flash-induced acidification supported by ferricyanide in the presence of 1 μM FCCP. They calculate an H^+/e^- ratio of approximately 1 under these conditions, in contrast to the value of 2 in the absence of DBMIB. The results of JUNGE's work are summarized in Table 1. GOULD and IZAWA (1974) observed an H^+/e^- ratio in the presence of substrate amounts of DBMIB and absence of ferricyanide of about 0.5. Their determinations were based in measurements of H^+ uptake with the glass electrode and oxygen uptake was used to assay electron flow. In view of the sensitivity the H^+/e^- ratio to DBMIB and to high concentrations of ferricyanide which can oxidize plastoquinone directly, it is probable that a "site" for H^+ uptake exists in the plastoquinone region of the chain. This conclusion is in agreement with the observations of AVRON and CHANCE (1966) that a "site" of energy conservation also exists in this region of the chain. Thus, the scheme proposed by MITCHELL for H^+ uptake in chloroplasts is supported. Although the evidence is consistent with plastoquinone acting as a mobile H^+ carrier, other mechanisms have not been ruled out. It should be emphasized, however, that there is good evidence (HINKLE, 1973) that benzoquinone can act as an electron and hydrogen carrier across model membranes.

If H^+ uptake is directly coupled to plastoquinone reduction, the rate of H^+ uptake should be equivalent to that of plastoquinone reduction. However, AUSLÄNDER and JUNGE (1974), FOWLER and KOK (1974), and GRÜNHAGEN and WITT (1970) noted that the half time for H^+ uptake associated with electron flow between the photosystems in chloroplasts was approximately 20–60 ms which is considerably slower than the rate of plastoquinone reduction [0.6 ms, according to STIEHL and WITT (1969)]. This discrepancy could be caused by a diffusion barrier to H^+ in the membrane. For example, plastoquinone could be reduced at a site toward the outer surface of the thylakoids which is shielded by lipids from the external aqueous environment. Plastoquinone reduction within this space could occur quite rapidly resulting in a depletion of H^+ within the space. The reequilibriation of H^+ into the space with H^+ from the outside could occur slowly due to slow permeation of H^+. The structure of chloroplasts could also explain the delay in the disappearance of H^+ from the medium. Plastoquinone could be located within the grana regions of the chloroplast membranes. Diffusion of H^+ between these stacked membrane regions could be slow. AUSLÄNDER and JUNGE (1974) showed that disruption of chloroplast structure either by prolonged grinding of chloroplasts with sand or by treatment with digitonin increased the rate of H^+ uptake from the medium. Uncoupling agents also enhanced the rate of H^+ uptake. However, the fastest rates of H^+ uptake observed were still only one-half those of plastoquinone reduction. It seems reasonable that a barrier to the diffusion of H^+ could still exist even in the presence of uncouplers. It would be of interest to examine H^+ uptake by this method in subchloroplast particles treated with a galactolipid lipase (ANDERSON et al., 1974). Although over 70% of the galactolipid in these particles can be removed by the lipase, presumably from the surface, much of the phosphorylation and electron flow catalyzed by these particles can be retained.

2.2 H^+ Uptake in the Presence of Permeating Electron Donors or Acceptors

Phosphorlylation coupled to cyclic electron flow supported by either pyocyanine or N-methyl phenazonium methosulfate (PMS) is insensitive to DBMIB (BÖHME et al., 1971). Cyclic phosphorylation is a reaction catalyzed by photosystem I. Dichlorophenyl-1,1-dimethylurea (DCMU), which blocks electron flow on the photosystem II side of plastoquinone, also has no effect on PMS-dependent phosphorylation (JAGENDORF and MARGULIES, 1960). Thus, it appears unlikely that the sites of H^+ release inside the thylakoids, determined by the Berlin group, would operate under conditions of cyclic electron flow. Yet, PMS-dependent cyclic electron flow supports rapid and extensive H^+ uptake. These observations led to the proposal that mediators of cyclic electron flow act as artificial H^+ carriers across the membranes.

On the basis of experiments with hydrophilic and hydrophobic electron donors and acceptors (SAHA et al., 1971) and on work with antibodies to components of the electron transport chain, it can be concluded that whereas electron donation to photosystem I occurs within or on the inside of thylakoid membranes, electron acceptance is located at or near the outer surface. The evidence for this conclusion is outlined in detail by TREBST (1974) and will be only briefly considered here.

Charged oxidants such as ferricyanide, viologens, ferredoxin and sulfonated dichlorophenol indophenol are good electron acceptors from photosystem I and are unlikely to readily cross chloroplast membranes. In contrast, HAUSKA et al. (1973) have shown that whereas reduced 2,6-dichlorophenol indophenol (DCPIP) is a good electron donor to photosystem I, reduced DCPIP-sulfonate is a very poor donor. In addition, HAUSKA (1972), demonstrated that PMS-sulfonate and pyocyanine-sulfonate did not catalyze cyclic phosphorylation. NIR and PEASE (1973) used diaminobenzidine to donate electrons to photosystem I. When diaminobenzidine is oxidized it forms an insoluble precipitate. A precipitate was observed by electron microscopy to be inside of thylakoids illuminated in the presence of diaminobenzidine. In contrast, ferricyanide reduction appears to take place on the outer surface of the thylakoids (HALL et al., 1971). Furthermore, the accessibility of ferredoxin-$NADP^+$ reductase (BERZBORN, 1968), and the inaccessibility of plastocyanin (HAUSKA et al., 1971) to antibodies support the concept that the donor site of photosystem I is inside of thylakoids whereas the acceptor site is oriented toward the outside.

HAUSKA et al. (1974) proposed that H^+ uptake in chloroplasts mediated by phenazines such as PMS or pyocyanine occurs as follows: PMS is reduced at the outer surface of the thylakoids and picks up one or more H^+ from the medium. The reduced PMS diffuses inside the chloroplasts where it is oxidized by photosystem I liberating one or more protons (depending on whether PMS cycles through the fully oxidized to fully reduced forms). In view of the insensitivity of cyclic phosphorylation in the presence of high concentrations of PMS to the treatment of chloroplasts with CN^-, which inactivates plastocyanin (IZAWA et al., 1973), it would appear that reduced PMS can be oxidized directly by P_{700}. The idea that PMS acts as a H^+ carrier across thylakoids is supported by the observation (cf. HAUSKA et al., 1973) that the oxidation of reduced N,N,N',N' tetramethyl-phenylene diamine (TMPD) by non-cyclic electron flow through photosystem I is not coupled to phosphorylation. TMPD does not accept H^+ on reduction. A scheme for PMS-supported H^+ uptake in chloroplasts, after HAUSKA et al. (1974), is shown in Figure 3A.

It should be pointed out that mediators need not act as H^+ carriers to support cyclic phosphorylation. For example, it has been known for many years that ferredoxin, especially at high concentrations, can catalyze cyclic phosphorylation (TAGAWA et al., 1963) even though it almost certainly does not cross the membrane. In view of the sensitivity of ferredoxin-supported cyclic phosphorylation to DBMIB (HAUSKA et al., 1974), an involvement of plastoquinone in this kind of phosphorylation seems assured. HAUSKA et al. (1974) postulated that reduced ferredoxin donates electrons to plastoquinone toward the outside of the thylakoids and that plastoquinone acts as the H^+ carrier as it does for non-cyclic electron flow. Other redox

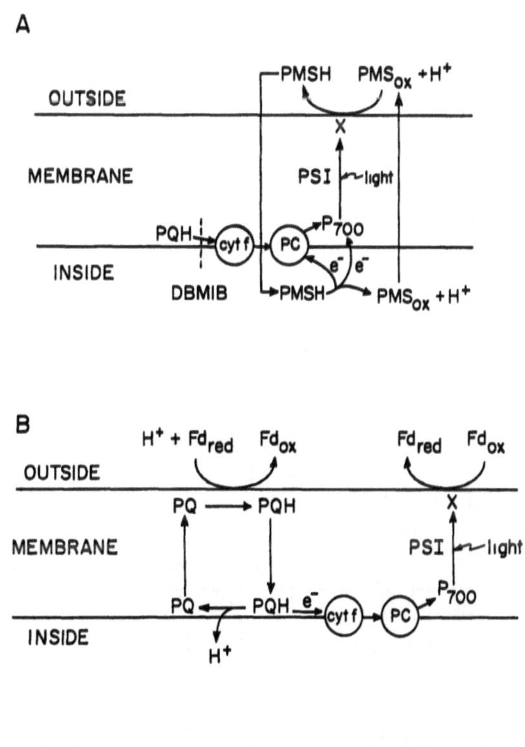

Fig. 3A and B. Possible pathways of cyclic electron flow. (A) in the presence of a permeating electron (and hydrogen) acceptor (*PMS*). (B) with a non-permeating electron acceptor (ferredoxin *Fd*). After HAUSKA et al. (1974)

compounds, especially quinones of E'_0 values at pH 7 of 60 mV or less, supported a DBMIB-sensitive cyclic phosphorylation. A possible mechanism for H^+ uptake supported by electron flow in the presence of ferredoxin or these quinones is shown in Figure 3B.

When water oxidation is blocked by a number of treatments including the incubation of chloroplasts with either high concentrations of Tris buffer (YAMASHITA and BUTLER, 1968) or hydroxylamine (ORT and IZAWA, 1973; CHENIAE and MARTIN, 1971) a number of reductants are oxidized by photosystem II in a DCMU-sensitive manner. In hydroxylamine-treated chloroplasts, the oxidation of benzidine coupled to the reduction of either $NADP^+$ (HARTH et al., 1974a) or methyl viologen (ORT and IZAWA, 1973) is coupled to ATP synthesis with a P/2e ratio near 1. If, however, N,N,N',N'-tretramethyl benzidine is used as the electron donor, the P/2e ratio was close to 0.5. Similar results were obtained using ferrocyanide (at high concentrations) as the electron donor (IZAWA and ORT, 1974). Thus, benzidine, like water, may be oxidized inside the thylakoids resulting in the liberation of H^+ inside. N,N,N',N'-tetramethylbenzidine or ferrocyanide can be an efficient electron donor to photosystem II, but does not liberate H^+ inside. However the H^+ uptake associated with the plastoquinone region of the chain can still operate in the presence of N,N,N',N'-tetramethylbenzidine so some phosphorylation can still occur.

When electron flow is blocked by DBMIB or KCN, various hydrophobic electron acceptors (including DBMIB itself at higher concentrations) can still be reduced, using either water or various artificial electron donors to photosystem II (cf. review by TREBST, 1974). H^+ uptake was found to be associated with electron flow from water to DBMIB (GOULD and IZAWA, 1974). Furthermore, phosphorylation with a P/2e ratio of from 0.3 to 0.7 has been shown to be coupled to electron flow through this abbreviated portion of the electron transport chain (GOULD and IZAWA, 1973b; TREBST and REIMER, 1973a). When oxidized phenylene diamines were used as electron acceptors, uncoupling agents actually *inhibited* electron flow (TREBST and REIMER, 1973b; COHEN et al., 1975). A direct action of the uncoupling agents on electron flow was ruled out by COHEN et al. and, therefore, a role for internal H^+ for phenylene diamine-supported electron flow was proposed. Phenylene diamines may be reduced toward the inside of the thylakoids resulting in the uptake of two H^+ from the inside (see Fig. 4). The reduced phenylene diamine would then diffuse across the membrane where it is oxidized

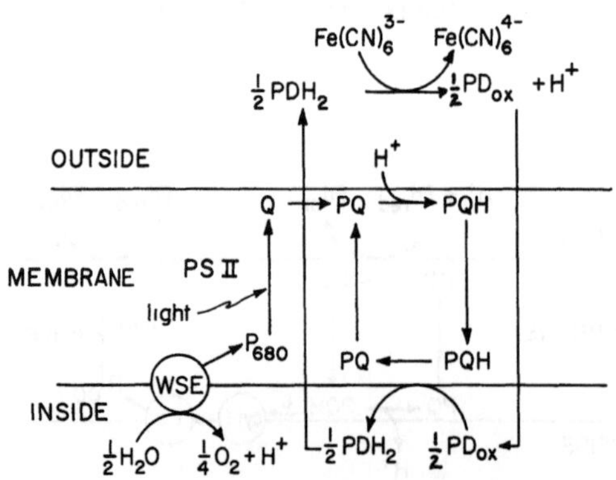

Fig. 4. Trebst's suggestion (TREBST, 1974) for the pathway of electron flow from water to ferricyanide, mediated by phenylene diamine (*PD*). *PD*$_{ox}$ stands for oxidized phenylene diamine

by ferricyanide in the medium resulting in the liberation of H^+ into the medium. Thus, phenylene diamine is actually envisioned as carrying H^+ out of the chloroplasts and the reduction of oxidized phenylene diamines would be dependent on the pH gradient. This could explain why the P/2e ratios with phenylene diamines are low; the phenylene diamines are in essence uncoupling agents!

In summary, the view that H^+ translocation in chloroplasts is a direct consequence of vectoral electron flow has been receiving strong experimental support over the past few years. Although the possibility of an H^+ pump connected to some high energy intermediate cannot be entirely ruled out, there seems to be no need to postulate its existence.

3. Counter Ion Fluxes

It is clear that H^+ uptake in chloroplasts must be associated with the translocation of another ion or ions. DILLEY and VERNON (1965) reported that Mg^{2+} and K^+ efflux appeared to balance H^+ influx in a medium containing Tris-acetate, but no added Mg^{2+} or K^+. DEAMER and PACKER (1969) first showed that some Cl^- uptake could be catalyzed by illuminated chloroplasts but they were unable to quantitate the extent of uptake in relation to H^+ uptake. Amine cations are accumulated by chloroplasts to massive extents (CROFTS, 1968; GAENSSLEN and MCCARTY, 1971). This process provided a method for increasing light-dependent Cl^- uptake. GAENSSLEN and MCCARTY (1971), who used a microcentrifugation method to estimate uptakes, found that approximately 1 Cl^- was accumulated per ethylamine cation taken up when the Cl^- concentration was greater than 10 mM. Similar observations were made by ROTTENBERG et al. (1972) who measured Cl^- uptake by a different microcentrifugation method in chloroplasts illuminated in the absence of amines.

HIND et al. (1974) measured H^+, Cl^-, Na^+, K^+ and Mg^{2+} fluxes in chloroplasts simultaneously through the use of ion-specific electrodes. This is a remarkable feat in view of the fact that even measurements with one electrode at high sensitivities can be difficult. Because of the logarithmic response of the ion-specific electrodes to the concentration (activity) of the ions to which they are sensitive, the uptake assays were run in the presence of low concentrations of the suspected H^+ counter ions. An external pH of 6.6 was used since H^+ uptake is close to maximal at this pH. The extent of Cl^- influx and of Mg^{2+} efflux (or release) in illuminated chloroplasts was nearly equivalent to that of H^+ influx. All ion movements reversed in the dark. Only a very small amount of K^+ was released in the light and there was no change in the concentration of Na^+ in the medium. At Mg^{2+} concentrations greater than 1 mM, K^+ fluxes were abolished. Thus, under these conditions, H^+ uptake appears to be nearly totally balanced by Cl^- influx and Mg^{2+} efflux. HIND et al. (1974) proposed that the Mg^{2+}, rather than crossing thylakoid membranes, may exchange with H^+ on the surface of the membranes in an energy-dependent manner. Alternately, Mg^{2+} could cross the membrane as a counter ion for H^+. A role for Mg^{2+} fluxes in H^+ uptake in intact chloroplasts was also recently suggested by BARBER et al. (1974). An obligate role for Mg^{2+} in H^+ uptake appears to be unlikely. Chloroplasts, repeatedly illuminated at high Cl^- concentrations (50 mM) in the presence of 1 mM EDTA, retain the capacity to take up H^+ (McCarty, unpublished observations). It will be of interest to examine Mg^{2+} fluxes in the presence of high concentrations of Cl^- and at pH values close to those where phosphorylation can take place rapidly.

4. The Magnitude of the Electrochemical Gradient

Although the amount of H^+ accumulated by illuminated chloroplasts may be readily estimated through the use of pH indicating dyes or of pH electrodes, it has been more difficult to develop reliable techniques to determine the pH inside the thylakoids. Furthermore, it has been even more difficult to estimate the magnitude of the membrane potential in illuminated chloroplasts. Unfortunately, the direct approach, that of impaling chloroplasts with microelectrodes, does not appear to be too feasible in view of the size of the electrodes relative to the intrathylakoid space. Therefore, more indirect methods to measure light-induced differentials in pH and changes in membrane potential must be and have been used (cf. also Vol. 2, Part A: Chaps. 4.2.8.1 and 12).

4.1. The H^+ Concentration Gradient

Rumberg and Siggel (1969) were the first to attempt to measure the light-induced H^+ concentration gradient in chloroplasts. This gradient is denoted ΔpH for pH_{out}–pH_{in}. It is important to keep in mind that several investigators use the notation ΔpH for the light-induced rise in the pH of the chloroplast suspending medium measured by the glass electrode. Rumberg and Siggel assumed that internal H^+ concentration was the main factor controlling the rate of electron flow between the photosystems. Avron and Chance (1966) had previously shown that uncoupling agents and phosphorylation, which should (and do) decrease ΔpH, enhance the rate of reduction of cytochrome f, but have little effect on the rate of its oxidation. Therefore, a rate limiting step for electron flow probably exists between plastoquinone and P_{700}. In view of the possibility that plastoquinone releases H^+ inside the thylakoids upon oxidation, it seems reasonable to propose that high concentrations of H^+ inside could exert a backpressure on this oxidation. Rumberg and Siggel (1969) showed that the rate of P^+_{700} reduction in the dark following illumination was sensitive to the pH of the medium. High concentrations of gramicidin were used and it was assumed that the internal pH was the same as the external pH under these conditions. The rate of P^+_{700} reduction in coupled chloroplasts at an external pH of 8.0 was the same as that at pH 5.1 in gramicidin uncoupled chloroplasts, indicating a pH differential of 2.9 units. However, more recent evidence (Bamberger et al., 1973; Portis et al., 1975) suggests that the rate of electron flow in chloroplasts is dependent not only on the internal H^+ concentration but also on ΔpH. These experiments will be discussed in Section 7. Thus, although the approach used by Rumberg and Siggel to estimate ΔpH is certainly ingenious, it is probably not valid under all conditions.

The extensive studies of Crofts during the middle 1960s lead to the conclusion that the light-dependent uptake of NH_4^+ and organic amines was the result of

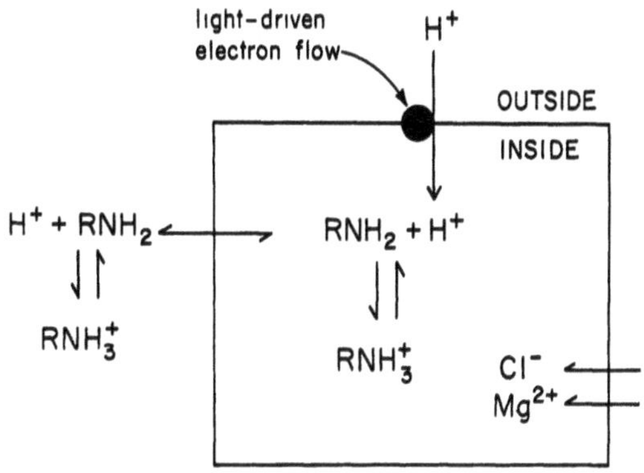

Fig. 5. Mechanism of amine uptake in illuminated chloroplasts (CROFTS, 1968)

acidification of the internal space of chloroplasts (see CROFTS, 1968 for a review of this work). The mechanism of amine uptake proposed by Crofts is shown in Figure 5. The uncharged form of amines was thought to permeate thylakoid membranes quite readily whereas amine cation was thought to be poorly permeating. These assumptions are consistent with the properties of amine uncoupling of phosphorylation in chloroplasts (HIND and WHITTINGHAM, 1963). Upon illumination, H^+ is accumulated inside the thylakoids and the amine base in that space is rapidly protonated resulting in a decreased concentration of the uncharged amine inside. Since the uncharged amine permeates readily, this would result in the movement of more of this species into the thylakoids to maintain concentration equilibrium. The uptake of amine in this manner would continue until the steady state is reached. With amines of high pKa (9 or greater) such as NH_4^+ or alkylamines, uptake of amine by this mechanism results in an *apparent* inhibition of H^+ uptake, measured by the glass electrode, since at pH values of 8 or below, 1 H^+ would be released into the medium for each amine base accumulated. The sensitivity of light-dependent H^+ uptake to amines of high pKa is amply documented. In the presence of amines of low pKa such as pyridine (pKa 5.3) or aniline (pKa 4.6) H^+ uptake is increased dramatically (NELSON et al., 1971; PORTIS and MCCARTY, 1973). The uptake of aniline or pyridine at pH values greater than 6.0 does not result in the liberation of H^+ into the medium since the uncharged form predominates. Although H^+ uptake in the presence of amines of low pKa gives a different result from that in the presence of amines of high pKa with respect to what happens in the external phase, in both cases massive amounts of amine cation can be accumulated inside when high amine concentrations are used. This massive uptake results in marked swelling of the chloroplasts (IZAWA, 1965) and the swelling with associated leakiness to ions may explain the uncoupling of phosphorylation in chloroplasts by amines (GAENSSLEN and MCCARTY, 1971; WALKER, 1975).

The relationship between H^+ uptake and amine cation uptake has been exploited as an approach to the estimation of ΔpH. It may be readily shown (see, for example, HIND and MCCARTY, 1973) that at the steady state in illuminated chloroplasts the amine cation concentration gradient across thylakoid membranes should be equivalent to that in H^+ provided the following conditions hold true: first, low amine concentrations must be used to obivate massive swelling and uncoupling by the amine. Second, the rate of penetration of the amine base must be much faster than that of amine cation. Third, there should be little or no binding of the amine cation.

To estimate ΔpH in chloroplasts from the amine cation concentration gradient, a reliable method for the determination of the extent of amine uptake in the light as well as of the internal osmotic volume of chloroplasts are required. Three different methods have been used to estimate the extent of amine uptake at the steady-state microcentrifugation, electrode measurements and determination of the energy-dependent quenching of fluorescent amines. Since the magnitude of ΔpH is of extreme importance in the energetics of phosphorylation and since the validity of some of the methods used to estimate ΔpH has been questioned, these methods will be described briefly.

Two quite different microcentrifugation techniques have been employed to estimate amine uptake and internal volumes. ROTTENBERG et al. (1971, 1972) suspended chloroplasts (0.4 mg chlorophyll/ml) in a medium which contained ^3H_2O and ^{14}C-methylamine at a nonuncoupling concentration. Samples of these mixtures were illuminated for 1–2 min at a light intensity of 65,000 lux within the microcentrifuge. The samples were then centrifuged with the light still on for 3 min. A high chlorophyll concentration had to be used to allow tight packing of the pellets during centrifugation. Total water in the pellets was estimated by determining the amount of ^3H_2O trapped and water external to the thylakoid inner space was determined through the use of ^{14}C-sorbitol. The ΔpH values in the presence of pyocyanine at pH 8.0 ranged from about 2.1 to 2.5. These values seem to be on the low side compared to those obtained with other methods. It is likely that light intensity was a limiting factor since rather dense chloroplast suspensions had to be used. Furthermore, concentration of the chloroplasts during centrifugation would tend to make light saturation even more difficult. In agreement with this notion, ΔpH, assayed at high pH with pyocyanine, was not saturated at the highest light intensity used. Furthermore, it would be expected that ΔpH formed as a result of pyocyanine-dependent cyclic electron flow would be greater than that formed as a result of non-cyclic electron flow to ferricyanide or other photosystem I electron acceptors since the rate of cyclic phosphorylation is higher than that of noncyclic phosphorylation at high light intensities. This was not observed. However, PORTIS and McCARTY (1974) showed that ΔpH in the presence of pyocyanine was greater than that in the presence of methylviologen at high light intensities but that at lower light intensities, ΔpH values were the same.

The other microcentrifugation technique was developed originally for mitochondria by WERKHEISER and BARTLEY (1957) and was used successfully for mitochondria by PFAFF et al. (1968). This method was modified for the study of light and ATP-dependent amine uptake by Gaensslen in my laboratory (GAENSSLEN, 1971; GAENSSLEN and McCARTY, 1971, 1972). Many refinements to this method have since been made by PORTIS (PORTIS and McCARTY, 1973, 1974). To separate the chloroplasts rapidly from their suspending medium, chloroplasts are centrifuged through a silicone fluid layer. The polyethylene centrifuge tubes used contain three layers of 0.1 ml. The bottom layer consists of 8% glycerol in 2% trichloroacetic acid. The middle layer is a silicone fluid mixture of a density selected so that the chloroplasts can pass through it, but which is light enough to float on the glycerol layer. Chloroplasts (0.01 mg of chlorophyll) suspended in 0.1 ml of a medium containing a ^{14}C-amine (usually hexylamine at 25 μM concentration) and ^{3}H-sorbitol as well as buffer, salts and mediators of electron flow are placed in the tubes with an air space between this layer and the top of the silicone fluid. The tubes are illuminated within the microcentrifuge for 1 to 3 min and then centrifuged for 15 s at an average centrifugal force of 10,000 $\times g$. From high speed photography it is apparent that most of the chloroplasts pass through the silicone fluid within 3 to 5 s after the centrifuge acceleration starts. Radioactivity in aliquots of the glycerol layer is then determined by liquid scintillation counting. As estimated from the amount of ^{3}H-sorbitol which appears in the lower layer, only 0.5 to 0.8% of the suspending medium is carried through the silicone fluid with the chloroplasts. Only a very small correction for amine external to the chloroplasts (less than 1%) must be applied in the calculation of hexylamine uptake. Internal volumes are estimated by comparing the uptake of ^3H_2O to ^{14}C-sorbitol. Since the internal volumes of chloroplasts are on the order of only 10–20 µl/mg chlorophyll under our conditions, the correction for trapped suspending medium is much more critical. Although the error limits of hexylamine uptake are $\pm$1–2%, those for internal volumes are $\pm$10–20%.

Several approaches have been used to test the validity of this method. The extents of uptake of NH_4^+ (GAENSSLEN and MCCARTY, 1972) and of aniline (PORTIS and MCCARTY, 1973) determined by the microcentrifugation method compared favorably with those determined by electrode techniques. Moreover, ΔpH was found to be independent of the ^{14}C-hexylamine concentration over the range of 10 to 100 µM. More recently it was shown (ALEGRE and PORTIS, unpublished observations, 1974) that binding of amine cation probably makes a negligible contribution to the light-dependent uptake of amines. The uptake of ^{14}C-triethylmethyl ammonium ions (10 to 100 µM) to chloroplasts was exceedingly small (less than 0.2% even at 10 µM). Illumination of the chloroplasts or prolonged incubations (up to 30 min) in the dark did not increase the uptake. Finally, the uptake of hexylamine was found to decrease with increasing osmolarity of the suspending medium (PORTIS, unpublished observations, 1974). Since the internal volume also decreased as the osmolarity increased, osmolarity had little effect on the magnitude of ΔpH.

ROTTENBERG and GRUNWALD (1972) used cation-sensitive electrodes to estimate the extent of light-dependent NH_4^+ in chloroplasts. From these measurements made with low NH_4^+ concentrations to avoid uncoupling together with independent measurements of the internal volume of chloroplasts, ΔpH could be calculated. This method suffers from some limitations. The internal volumes cannot be determined under the conditions used for estimating NH_4^+ uptake. Moreover, the ionic composition of the suspending medium is limited due to the sensitivity of most cation-sensitive electrodes to monovalent ions other than NH_4^+. Finally, the sensitivity of the method is such that low values of ΔpH cannot be reliably determined.

An approach to the estimation of ΔpH using a pH-sensitive glass electrode was developed by PORTIS and MCCARTY (1973). The extent of aniline cation uptake in illuminated chloroplasts was determined by measuring the enhancement of H^+ uptake by mM concentrations of aniline. The extents of aniline uptake estimated by this procedure were very similar to those determined by the silicone fluid microcentrifugation technique. Internal volumes had to be determined independently. Although this method works well in the pH range of 6.5–7, the extent of stimulation at pH 8 is rather low because the concentration of aniline cation decreases with increasing pH and less aniline needs to be accumulated to reach the same concentration gradient in aniline cation.

Light-dependent and uncoupler-sensitive quenching of the fluorescence of heterocyclic amines (Kraayenhof, 1970), especially 9-aminoacridine, has been used to estimate ΔpH in chloroplasts. Schuldiner et al. (1972) made the assumption that the quenching of 9-aminoacridine was a result of its uptake into the internal osmotic compartment of the chloroplasts and that the fluorescence of 9-aminoacridine in this space was completely quenched. These assumptions in conjunction with the assumptions that only 9-aminoacridine free base readily permeates and that there is little binding of the acridine allowed the derivation of the following relationship between ΔpH and the fraction of the fluorescence quenched (Q):

$$\Delta \mathrm{pH} = \log\left(\frac{Q}{1-Q}\right) + \log V$$

where V is the ratio between the internal osmotic volume of chloroplasts and the total volume of the assay mixture. Internal volumes had to be estimated by the microcentrifugation method under what were apparently quite different conditions of osmolarity. An internal volume of 50 µl/mg chlorophyll was assumed, but the actual internal volumes could have been considerably lower. Very low quenchings can result in rather high ΔpH values. For example, for only 5% quenching at an assumed internal volume of 50 µl per mg chlorophyll and with 50 µg chlorophyll/3 ml (V=1,200), a ΔpH of 1.80 can be calculated. If however, the internal volume is 12.5 µl (V=4,800) 5% quenching would correspond to a ΔpH of 2.40!

The studies of Deamer et al. (1972) on the quenching of 9-aminoacridine in liposomes gave support to the notion that the fluorescence quenching could be directly related to the magnitude of the pH differential across the liposomal membranes. More recent work (Fiolet et al., 1974), however, casts considerable doubt on the interpretation of Schuldiner et al. (1972) that the fluorescence quenching of 9-aminoacridine is an adequate measure of ΔpH in chloroplasts. If the fluorescence quenching were due to an uptake of 9-aminoacridine, the quenching should be independent of the 9-aminoacridine concentration when non-uncoupling concentrations are used (less than 20 µM; Schuldiner et al., 1972). However, Fiolet et al. (1974) found that the extent of quenching was dependent on the concentration of 9-aminoacridine. Moreover, the dependence of quenching on the concentration of 9-aminoacridine was markedly influenced by the composition of the suspending medium. It was also expected that the quenching would increase in a linear manner with respect to the amount of chloroplasts added, but this was not observed. Furthermore, the quenching was totally unaffected by changing the osmolarity of the suspending medium by addition of sorbitol, 50–500 mM. Rottenberg et al. (1972) previously showed that the internal volume of chloroplasts responds to changes in the sorbitol concentration. According to their data, the internal volume of chloroplasts suspended in 500 mM sorbitol would have been about one-tenth that of chloroplasts in 50 mM sorbitol. In the presence of 50 mM sorbitol, the quenching was about 70%. A ten-fold decrease in the internal volume should, if the model of Schuldiner et al. (1972) were correct, have decreased the quenching to 34%.

Furthermore, the quenching of the fluorescence of 9-amino-6-chloro-2-methoxyacridine by chloroplasts and liposomes was greater than that of 9-aminoacridine fluorescence. This should also not be the case according to the model of SCHULDINER et al. (1972). On the basis of these observations, FIOLET et al. (1974) concluded that the quenching of 9-aminoacridine may be the result of an interaction of the amine with chloroplast or liposome membranes rather than its accumulation in the internal osmotic volume of the chloroplasts. Thus, although the fluorescence quenching method for estimating ΔpH is certainly convenient, it is probably not entirely valid.

It appears that, for the present time at least, the silicone fluid technique for determining ΔpH is the method of choice. Although it is somewhat more laborious than electrode measurements, it is reasonably precise (± 0.05 pH) and can be used to estimate ΔpH over a very wide range. Moreover, internal volumes can be determined under the same conditions as ΔpH. It should also be pointed out that a similar technique was developed independently in Heldt's laboratory (HELDT and RAPLEY, 1968, 1970) for the study of the transport of metabolites across the outer membranes of intact chloroplasts.

The magnitude of ΔpH at the steady state has been found to vary considerably depending on the conditions under which the assays are run and on the quality of the chloroplast preparations used. In view of the uncertainties about the validity of the 9-aminoacridine fluorescence quenching technique and of the probability that the microcentrifugation technique used by ROTTENBERG et al. (1972) underestimates ΔpH, this discussion of the magnitude of ΔpH will be limited to results obtained by the electrode and silicone fluid methods. Under optimal conditions, it would appear that ΔpH can be as high as 3.5 units. ROTTENBERG and GRUNWALD (1972) calculated ΔpH values of about 3.2–3.5 from the uptake of NH_4^+ at an external pH of 8–8.5. High light intensities were used and pyocyanine was used as the mediator of cyclic electron flow. Under similar conditions, the silicone fluid method yielded estimates of ΔpH from the uptake of aniline of 3.1 (PORTIS and MCCARTY, 1973) and 3.0–3.2 from the uptake of hexylamine (PORTIS and MCCARTY, 1974). It should be pointed out that some of the chloroplasts remain in the suspending medium after centrifugation. Approximately 10–15% of the chloroplasts do not appear to penetrate the silicone fluid. We generally do not correct the observed ΔpH or internal volumes for the amount of chloroplasts left in the suspending medium because the amount is not routinely determined in each experiment. Furthermore, the correction is relatively large only at high ΔpH values (2.9 or greater). Assuming that only 85% of the chloroplasts penetrate the silicone fluid layer, the correction for an observed ΔpH of 3.2 would be 0.15 unit, making the corrected ΔpH 3.35.

Low concentrations of either ATP or ADP inhibit electron flow (AVRON et al., 1958) and enhance the extent of H^+ uptake in illuminated chloroplasts (MCCARTY et al., 1971; TELFER and EVANS, 1972) at external pH values of 7.5–8.5. ATP or ADP also increases the magnitude of ΔpH (PORTIS and MCCARTY, 1974). In the presence of ADP, ΔpH values as high as 3.3 were observed. After correction for the chloroplasts which did not enter the silicone fluid, ΔpH approaches 3.5. If Pi (or arsenate) was present in addition to ADP to allow phosphorylation, ΔpH collapsed to about 2.8–2.9, clearly showing that phosphorylation and ΔpH are at the very least competing for a common energy source. This effect of ADP and Pi, but not that of ADP by itself, was reversed by phlorizin or 4′-deoxyphlorizin, which are energy transfer inhibitors (WINGET et al., 1969).

4.2. The Membrane Potential

Estimates of the membrane potential at the steady state in illuminated chloroplasts must be made by even more indirect means than those of ΔpH. These estimates vary over a rather wide range (10 to 100 mV, inside positive). Witt and his collaborators have studied the relationship between the light-induced change in absorbance of chloroplasts in the region of 515 nm and ion transport and phosphorylation. These studies have been reviewed in depth by Witt (1971) and, more superficially, by Hind and McCarty (1973). Therefore, the conclusions of the Berlin group will only be briefly mentioned here. The light-induced change in absorption at 518 nm probably represents a shift in the spectrum of chloroplast pigments (carotenoids and chlorophylls) under the influence of an electric field. Witt refers to this shift as the "field indicating absorption change". That the 518 nm shift is related to ion fluxes is confirmed by the observation that uncouplers, and, especially, ionophorous antibiotics such as valinomycin (in the presence of K^+) remarkably accelerated the dark decay of the absorption change induced by repetitive flash illumination. Based on a number of assumptions, the capacitance of chloroplast membranes was calculated by Schliephake et al. (1968). A membrane potential of about 50 mV was calculated for a single turnover flash and 100 mV at the steady state. The validity of these calculations depends not only on the assumptions used to estimate the capacitance of chloroplast membranes, but also on the assumption that the entire 518 nm shift may be attributed to a membrane potential in H^+.

Strichartz and Chance (1972) attempted to calibrate the 518 nm shift in chloroplasts in the dark by establishing K^+ diffusion potentials across the membranes. Jackson and Crofts (1969) had previously calibrated the carotenoid shift in chromatophores of *Rhodopseudomonas spheroides* in a similar manner. In the presence of valinomycin to enhance K^+ permeation the change in absorption at 518 nm in chloroplasts kept in the dark was linearily related to the log of K^+ concentration, in accordance with the Nernst relationship. Strichartz and Chance (1972) assumed an internal K^+ concentration of 2 mM and calculated the diffusion potentials. The magnitude of the light-induced 515 nm change in chloroplasts corresponded to a K^+ diffusion potential of about 30 mV.

The response of H^+ uptake and ΔpH in chloroplasts in ionophorous antibiotics suggests that the membrane potential component of the H^+ gradient in the steady state is not large. For example, although valinomycin plus K^+ markedly enhance the extent of H^+ uptake in chromatophores (Jackson et al., 1968) and subchloroplast particles (McCarty, 1970), they have little effect on the extent of H^+ uptake in chloroplasts, although the rate of H^+ movement is enhanced (Karlish and Avron, 1971). Moreover, ΔpH in chloroplasts is insensitive to valinomycin and K^+ (Portis, unpublished experiments). By allowing rapid permeation of K^+, valinomycin should dissipate a membrane potential. If a potential of 60 mV existed in the steady state in illuminated chloroplasts, its total collapse should result in an increase of ΔpH of 1 unit. Although the fact that valinomycin has secondary effects on the phosphorylation device (McCarty, 1969; Karlish and Avron, 1971; Telfer and Barber, 1974) may complicate the interpretation of these results, the lack of its effect on ΔpH would seem to indicate that the potential in chloroplasts illuminated to the steady state and suspended in a medium containing high Cl^-

concentrations is rather small. However, the actual magnitude of the potential is still an open question and assumes some importance in the calculation of the total energy stored in the H^+ gradient in chloroplasts.

In summary, the maximum value of the electrochemical gradient in H^+ across illuminated chloroplast membranes at the steady state is probably in the neighborhood of 210 mV. This value is equivalent to an H^+ concentration gradient of over 3,000-fold.

5. Relation of H^+ Movements to Phosphorylation

JAGENDORF and URIBE (1966b) showed that artificially generated pH differentials across chloroplast membranes can serve as the driving force for ATP synthesis. Although it seems reasonable to conclude that the pH differentials formed as a consequence of light-driven electron flow also drive ATP synthesis, it is difficult to prove. However, the evidence supports, in general, the concept that the pH gradient is utilized for ATP formation in a process mediated by an ATPase complex.

Phosphorylation decreases the extent of H^+ uptake in illuminated chloroplasts (SCHWARTZ, 1968; DILLEY and SHAVIT, 1968; SCHRÖDER et al., 1971; GOULD and IZAWA, 1974). PORTIS and MCCARTY (1974) showed that the failure of some investigators to observe this effect was probably due to the fact that they used concentrations of the phosphorylation reagents which were considerably below saturation. The magnitude of ΔpH also is decreased by phosphorylation by 0.4 to 0.5 unit (PORTIS and MCCARTY, 1974; PICK et al., 1973 observed a decrease in 9-aminoacridine fluorescence quenching by phosphorylation). At first glance, the diminution in ΔpH by phosphorylation appears rather small. However, the rate of phosphorylation is apparently quite sharply dependent on ΔpH (PORTIS and MCCARTY, 1974). Phosphorylation rates and ΔpH were altered by variation of the light intensity and the addition of either DCMU or an uncoupler (carbonylcyanide m-chlorophenylhydrazone). Phosphorylation was much more sensitive than ΔpH to either light intensity or the inhibitors. Plots of log of the phosphorylation rate *versus* ΔpH were linear over the entire range (phosphorylation rates of 1 to nearly 1,000 μmol ATP formed per hour/mg chlorophyll). A ten-fold decrease in the phosphorylation rate was accompanied by a decrease in ΔpH of only 0.3 unit. Thus, a decrease in ΔpH of only 0.4 to 0.5 unit by phosphorylation may be explained. Furthermore, the existence of an exponential relationship between the rate of phosphorylation and ΔpH suggests that a "threshold" value (PICK et al., 1974) of ΔpH for phosphorylation has little meaning.

The energy stored in the H^+ gradient must be sufficient to drive ATP formation if a chemiosmotic mechanism occurs. The magnitude of the potential ($\Delta G'$) against which chloroplasts can synthesize ATP can be calculated from the equation:

$$\Delta G' = \Delta G'_0 + 2.303\,RT \log \frac{[ATP]}{[ADP][P_i]}$$

where $\Delta G'_0$ is the standard free energy change for ATP hydrolysis under the conditions used. KRAAYENHOF (1969) found that phosphorylation seemed to approach equilibrium at a [ATP]/[ADP] [Pi] of over 10^4. Using a revised value of $\Delta G_0'$ under these conditions (ROSING and SLATER, 1972), the calculated $\Delta G'$ for ATP synthesis would be approximately 13.5 kcal/mol. These estimations were made with intact (Class I) chloroplasts assayed under hypotonic conditions. To my knowledge, estimates of $\Delta G'$ in washed thylakoid preparations have not been made. Even though there is some uncertainty that the phosphorylation system was in an equilibrium state in Kraayenhof's experiments and intact chloroplasts were used, a $\Delta G'$ for ATP synthesis of about 14 kcal/mol is close to the values obtained for mitochondria or chromatophores under somewhat different conditions. KRAAYENHOF (1969) used $NADP^+$ as the electron acceptor and with mediators of cyclic electron flow at saturating concentrations and at high light intensities, it is possible that $\Delta G'$ could be somewhat higher. Although I will use a value of 14 kcal/mol for the $\Delta G'$ of ATP synthesis, it is clear that $\Delta G'$ must be redetermined in thylakoid preparations under conditions where ΔpH is determined.

In view if the uncertainty of the magnitude of the membrane potential in chloroplasts illuminated to the steady state, calculated values of the energy stored in the H^+ gradient are also rather uncertain. The free energy stored in the gradient (ΔG_{H^+}) can be calculated in calories from the expression

$$\Delta G_{H^+} = 2.303\ RT\Delta pH + F\Delta\Psi$$

where F is the Faraday and $\Delta\psi$ the membrane potential in volts. MITCHELL (1966) expressed this equation in terms of electrical units

$$\frac{\Delta G_{H^+}}{F} = \frac{2.303\,RT}{F}\Delta pH + \Delta\Psi.$$

The quantity $\Delta G_{H^+}/F$ is termed the "proton motive force".

In this discussion, I will assume that $\Delta\Psi$ is small and, therefore, $\Delta G_{H+} \approx 1.36\,\Delta pH$. It should be remembered, however, that estimates of ΔG_{H+} based on ΔpH alone may be too low.

In the comparison of ΔG_{H^+} to $\Delta G'$, maximal values of both quantities should be used. Ideally, ΔpH and $\Delta G'$ should be measured in the same experiment at equilibrium. Measurements of this kind with chloroplasts have not been reported, although preliminary experiments have been carried out in this laboratory (PORTIS and McCARTY, unpublished experiments). However, ΔpH values determined in the presence of ATP and absence of added ADP and Pi should be similar to those under conditions where phosphorylation comes into equilibrium. The ΔpH values obtained in the presence of ATP were around 3.5 (see Sect. 4). The ΔG_{H^+} would be 1.36×3.5 or 4.8 kcal/mol.

Whether a ΔG_{H^+} of 4.8 kcal/mol is sufficient to drive ATP synthesis against a potential of 14 kcal depends, of course, on the stoichiometry of the ATP synthesis reaction with respect to H^+, that is, the number of H^+ which falls down a potential difference of 4.8 kcal per ATP synthesized (the H^+/ATP ratio). Estimates of the H^+/ATP ratio vary from 2 to 4. From comparison of the steady state rates of phosphorylation and the initial rates of H^+ efflux upon darkening of chloroplasts, SCHWARTZ (1968) calculated an H^+/ATP ratio of 2. However, SCHRÖDER et al. (1971), using a similar method, calculated an H^+/ATP ratio of 3 provided valinomycin and K^+ were present in the assay medium. At alkaline pH values valinomycin plus K^+ enhances the rate of H^+ efflux about two-fold, possibly because it helps to prevent the formation of a reverse membrane potential during H^+ efflux. Since SCHWARTZ did not use valinomycin, it is quite possible that his estimates of H^+ efflux were low. CARMELI (1970) calculated that $2H^+$ were translocated per ATP split in chloroplasts carrying out active ATP hydrolysis in the dark. This could also be an underestimation since the initial rate of H^+ influx was determined with the slowly responding glass electrode.

An indication that the H^+/ATP ratio is 3 was obtained from a quite different experimental approach (PORTIS and MCCARTY, 1974). A linear correlation between phosphorylation rates and ΔpH was found. If it is assumed that the rate of phosphorylation at a given, constant external pH is limited by the internal H^+ concentration ($[H^+]_{in}$) the rate of phosphorylation would be proportional to an overall catalytic rate constant k times $[H^+]_{in}^n$ where n is the H^+/ATP ratio. Therefore, the -log phosphorylation rate equals -log $k + n(\mathrm{pH}_{in})$. Since the external pH was constant, plots of log phosphorylation rate versus ΔpH are the same as those of -log phosphorylation rate versus pH_{in}. The slopes of these plots were always close to 3 (3.2 ± 0.2) indicating that if the assumptions are correct, H^+/ATP is 3.

Although an H^+/ATP ratio of 2 and a ΔG_{H^+} of 4.8 kcal falls short of a $\Delta G'_{ATP}$ of 14 kcal, an H^+/ATP ratio of 3 would yield a total of about 14 kcal. Therefore, assuming that the H^+/ATP ratio is 3, a ΔpH of 3.5 units is probably sufficient to drive ATP formation even at high phosphate potentials.

Phosphorylation and H^+ efflux may be represented by

$$n\mathrm{H}_{in}^+ + \mathrm{ADP} + \mathrm{Pi} \rightleftharpoons \mathrm{ATP} + n\mathrm{H}_{out}^+.$$

At equilibrium, it may be shown that n 1.36 $\Delta\mathrm{pH} = \Delta G'_{ATP}$. It should be possible to estimate n from this relationship. For example, ΔpH and $\Delta G'$ could be determined at varying light intensities. From the slopes of plots of ΔpH against $\Delta G'$, n could be found. Preliminary experiments indicate that n must be greater than 2.

Accepting an H^+/e^- ratio of 2 and an H^+/ATP ratio of 3, the maximum ATP/$2e^-$ ratios would be 1.33. In washed thylakoid preparations ATP/$2e^-$ ratios on the order 1.1 to 1.2 can be routinely observed (IZAWA and GOOD, 1968), without "correction" of the electron transport rates by subtraction of the electron transport observed under non-phosphorylating conditions. Unless cyclic phosphorylation contributes, an ATP/$2e^-$ of 1.2 is inconsistent with an H^+/ATP ratio of 4, postulated by RUMBERG and SCHRÖDER (1972).

In intact chloroplasts, assayed under hypotonic conditions, ATP/$2e^-$ ratios as high as 1.7 have been observed without subtraction on non-phosphorylating electron flow rates (REEVES and HALL, 1973). It is possible that the H^+/ATP ratio in these chloroplasts is 2 or that the H^+/e^- ratio differs from that in the thylakoid preparations. Participation of cyclic phosphorylation in these preparations cannot be entirely ruled out. Clearly, estimates of H^+/e^- and H^+/ATP are needed in intact chloroplasts.

In summary, although more work must be done before firm conclusions may be reached, the evidence supports the notion that ΔpH can be the driving force for ATP synthesis in illuminated chloroplasts.

6. Relation of the Rate of Electron Flow to H^+ Movements

Light-dependent electron flow is coupled to the inward translocation of H^+, mediated, quite possibly, by plastoquinone (see Sect. 2). Moreover, the rate limiting step for electron flow between the two photosystems is on the photosystem II side of cytochrome f, or close to the plastoquinone region of the chain. Assuming that the oxidation of plastoquinone is obligatorily linked to the release of H^+ inside the thylakoids, high internal H^+ concentrations should decrease the rate of plastoquinone oxidation (and, therefore, overall electron flow) by exerting a backpressure. Rumberg and Siggel (1969) and Rumberg et al. (1969) presented evidence in support of this hypothesis. The rate of H^+ uptake corresponds to that of electron flow and, at the steady state, the rate of H^+ efflux must equal the influx rate. Therefore, under steady-state conditions, H^+ efflux rates must also correspond to the rate of electron flow. Accordingly, the rate of electron flow should rise in a predictable manner when the H^+ permeability of the thylakoids is increased by uncoupling agents. A good correlation between the rate of electron flow (ferricyanide reduction) and the rate of H^+ efflux, assayed at pH 6 and varied through the use of the uncoupler desaspidin was obtained. Furthermore, the rate of electron flow increased with external pH over the range of 6 to 8 in a manner very similar to the rate of H^+ efflux. These kinds of results lead to the conclusion that internal pH controls the rate of electron flow.

However, more recent work suggests that ΔpH as well as internal pH and external pH are important factors in the control of the rate of electron transport (Bamberger et al., 1973). How ΔpH could be involved was not explained and would not be expected unless ΔpH in some way controls the rate of H^+ efflux. We postulate (Portis et al., 1975) that the magnitude of ΔpH influences the rate at which H^+ leaks through coupling factor 1 and that the dependence of electron flow and of phosphorylation on external pH may be explained on the basis of the phenomenon.

The magnitude of ΔpH determined from the uptake of aniline or hexylamine over the range of pH 7.0 to pH 8.5 was similar (Portis and McCarty, 1973 and 1974). Although ΔpH has been reported to be considerably higher at more alkaline pH values (see for example, Rottenberg and Grunwald, 1972) it is probable that the extent of amine uptake in these experiments was limited by the low concentration of amine base at pH values below 8 (Portis and McCarty, 1973). The fact that ΔpH is the same at pH 7 as it is at pH 8.5 is rather surprising in view of the observations by many investigators (see Good et al., 1966, for a review) that the rate of electron flow at pH 7 is only about one-fourth that at pH 8. It is clear, therefore, that the rate of H^+ efflux must be much greater at pH 8 than at pH 7. Some increase in the rate of H^+ efflux with increasing pH has been detected (Rumberg et al., 1969).

At a given external pH, the rate of electron flow should increase with increasing light intensity in the same manner as H^+ influx. As long as the permeability of thylakoids is independent of the magnitude of ΔpH, plots of the rate of electron flow versus $[H^+]_{in}$ (or log of the rate of electron flow vs ΔpH) obtained by varying the light intensity should be linear. Although this is true at low light intensities, at higher light intensities which generate ΔpH values in the range of 2.7 to 2.8, this relationship breaks down at an external pH of 8.0 and electron flow increases much more rapidly with light intensity than $[H^+]_{in}$. At an external pH of 7.0, however, the breakdown of the relationship between electron flow is much less pronounced. Moreover, ATP nearly totally prevents the breakdown of the relationship between electron flow and $[H^+]_{in}$ at pH 8.0 (PORTIS et al., 1975), indicating that coupling factor 1 (CF_1) may be involved in the efflux of H^+ from chloroplasts.

CF_1, which catalyzes the terminal steps of photophosphorylation, appears to have no *direct* function in H^+ uptake in chloroplasts (MCCARTY and RACKER, 1966). The finding that ATP enhances the extent of H^+ uptake (MCCARTY et al., 1971; TELFER and EVANS, 1972) and that this effect was sensitive to an antiserum to CF_1 (MCCARTY et al., 1971) suggested that the state of CF_1 could regulate the permeability of chloroplasts to H^+. Membrane-bound CF_1 changes its conformation upon illumination (RYRIE and JAGENDORF, 1971; MCCARTY et al., 1972). These changes, detected either by a light-dependent tritiation of CF_1 with ^3H_2O (RYRIE and JAGENDORF, 1971, 1972) or by a light-dependent reaction of CF_1 with N-ethylmaleimide (MCCARTY and FAGAN, 1973), are sensitive to uncoupling agents. Furthermore, the reaction of NEM with CF_1 is also sensitive to ATP (MAGNUSSON and MCCARTY, 1975), indicating that ATP may alter the conformation of CF_1. An irreversible inhibition of phosphorylation by N-ethylmaleimide is caused by the reaction of CF_1 with the inhibitor (MCCARTY and FAGAN, 1973). The light intensity dependence of the development of the inhibition is similar to that of phosphorylation (PORTIS et al., 1975). These results indicate that the change in CF_1 which allows its reaction with N-ethylmaleimide shows a similar sharp dependence on ΔpH as does phosphorylation. Little inhibition occurs at ΔpH values below 2.8. Since this change appears to be minimal at low light intensities and at pH 7, it may be that CF_1 does not undergo dramatic changes in its conformation under these conditions.

If it is postulated that the altered conformation of CF_1 allows the efflux of H^+ from inside the chloroplasts, the dependence of electron flow on ΔpH as well as external pH may be explained. At pH 7, even though ΔpH is high CF_1 cannot assume its altered conformation and the rate of H^+ leakage through CF_1 is low. In agreement with this notion, ATP has little effect on either ΔpH or electron flow at pH 7. In contrast, at pH 8 to 8.5, CF_1 readily assumes a new conformation which opens up a new channel for H^+ efflux only when ΔpH is high. This channel could be through CF_1 itself. By preventing this change, ATP would inhibit H^+ efflux thereby enhancing ΔpH and inhibiting electron flow. In essence, CF_1 may be considered to be a gated translocator for H^+.

Since ΔpH is the same at pH 7 as it is at pH 8 it is difficult to explain why the rate of phosphorylation is lower at pH 7 than at pH 8 unless a pH-dependent conformational change is required or some part of the phosphorylation reaction itself is pH-dependent. The former possibility seems to be supported by the pH-dependence of the reaction of CF_1 with N-ethylmaleimide. Thus even though at pH 7 the internal H^+ concentration in illuminated chloroplasts is ten times that at pH 8, CF_1 could be unable to ultilize these protons to drive phosphorylation because it cannot assume the proper conformation.

7. Conclusion

In my opinion, the evidence presented here gives strong support to the concept that electron flow in chloroplasts is coupled to ATP synthesis through an electrochemical gradient in H^+. Although not reviewed here, work on the reconstitution of oxidative phosphorylation and H^+ translocation in mitochondria (RACKER, 1974) also points to the central role of ion gradients in this process. Because it involves light-driven H^+ uptake, one reconstituted system deserves mention here. Halophilic bacteria such as *Halobactorium halobium*, contain as part of their outer membrane a protein known as bacteriorhodopsin. Illumination of bacteria results in an H^+ efflux and an increase in ATP inside the cells (OESTERHELT and STOECKENIUS, 1973). Further, bacteriorhodopsin has been purified and has been shown to change its H^+ binding capacity upon illumination. The incorporation of bacteriorhodopsin into phospholipid vesicles resulted in vesicles capable of carrying out light-dependent H^+ uptake which was sensitive to uncouplers. Vesicles formed in the presence of bacteriorhodopsin and the oligomycin-sensitive ATPase of beef heart mitochondria catalyzed photophosphorylation (RACKER and STOECKENIUS, 1974).

Now that the electrochemical gradient in H^+ has been identified as a most likely candidate for the position of "high energy intermediate", more attention will undoubtedly be paid to the terminal steps of phosphorylation in chloroplasts, and how H^+ gradients may be coupled to the synthesis of ATP. MITCHELL (1974) proposed a molecular mechanism for H^+-translocating ATPases. Detailed studies of the coupling factor-ATPase system of chloroplasts will hopefully provide some answers to this most important question.

BOYER et al. (1973) and CROSS and BOYER (1975) proposed that conformational changes of the ATPase could provide the driving force for ATP synthesis by providing energy for the dissociation of firmly bound ATP. Although CF_1 does change its conformation in an energy-dependent manner, the relationships between these changes and phosphorylation are far from clear. However, the chemiosmotic hypothesis and conformational hypothesis are not necessarily mutually exclusive. The energy required for conformational changes in the ATPase system could be derived from the electrochemical gradient in H^+ and these changes could then allow the synthesis of ATP.

References

ANDERSON, M.M., McCARTY, R.E., ZIMMER, E.A.: The role of galactolipids in spinach chloroplast lamellar membranes I. Partial purification of a bean leaf galactolipid lipase and its action on subchloroplast particles. Plant Physiol. **53**, 699–704 (1974)

ARNON, D.I.: Chloroplasts and photosynthesis: Brookhaven Symp. Biol. **11**, 181–235 (1958)

ARNON, D.I., ALLEN, M.B., WHATLEY, F.R.: Photosynthesis by isolated chloroplasts. Nature **174**, 394–396 (1954)

ARNTZEN, C.J., VERNOTTE, C., BRIANTIS, J.M., ARMOND, P.: Lactoperoxidase-catalyzed iodination of chloroplast membranes II. Evidence for surface localization of photosystem II reaction centers. Biochem. Biophys. Acta **368**, 39–53 (1974)

AUSLÄNDER, W., HEATHCOTE, P., JUNGE, W.: On the reduction of chlorophyll-A_1 in the presence of the plastoquinone antagonist, dibromothymoquine. FEBS Letters **47**, 229–235 (1974)
AUSLÄNDER, W., JUNGE, W.: The electric generator in the photosynthesis of green plants II. Kinetic correlation between protolytic reactions and redox reactions. Biochim. Biophys. Acta **357**, 285–298 (1974)
AVRON, M. (ed.): Proc. 3rd Intern. Congr. Photosyn. Res., Rehovot, Amsterdam. Elsevier. Vol. 1–3 (1974)
AVRON, M., CHANCE, B.: Relation of phosphorylation to electron transport in isolated chloroplasts. Brookhaven Symp. Biol. **19**, 149–160 (1966)
AVRON, M., KROGMANN, D.W., JAGENDORF, A.T.: The relation of photosynthetic phosphorylation to the Hill reaction. Biochim. Biophys. Acta **30**, 144–153 (1958)
BAMBERGER, E.S., ROTTENBERG, H., AVRON, M.: Internal pH, ΔpH, and the kinetics of electron transport in chloroplasts. Europ. J. Biochem. **34**, 557–563 (1973)
BARBER, J., MILLS, J., NICOLSON, J.: Studies with cation specific ionophores show that within the intact chloroplast Mg^{++} acts as the main exchange cation for H^+ pumping. FEBS Letters **49**, 106–110 (1974)
BERZBORN, R.J.: Nachweis der Ferredoxin-NADP-Reductase in der Oberfläche des chloroplasten-lamellar Systems mit Hilfe spezifischer Antikörper. Z. Naturforsch. **23b**, 1096–1104 (1968)
BÖHME, H., REIMER, S., TREBST, A.: The effect of dibromothymoquinone, an antagonist of plastoquinone, on non cyclic and cyclic electron flow systems in isolated chloroplasts. Z. Naturforsch. **26b**, 341–352 (1971)
BOYER, P.D., CROSS, R.L., MOMSEN, W.: A new concept for energy coupling in oxidative phosphorylation based on a molecular explanation of the oxygen exchange reactions. Proc. Natl. Acad. Sci. US **70**, 2837–2839 (1973)
CARMELI, C.: Proton translocation induced by ATPase activity in chloroplasts. FEBS Letters **7**, 297–300 (1970)
CHENIAE, G.M., MARTIN, I.F.: Effects of hydroxylamine on photosystem II. I. Factors affecting the decay of O_2 evolution. Plant Physiol. **47**, 568–575 (1971)
COHEN, W.S., COHN, D.E., BERTSCH, W.: Acceptor specific inhibition of photosystem II electron transport by uncoupling agents. FEBS Letters **49**, 350–355 (1975)
COHN, D.E., COHEN, W.S., BERTSCH, W.: Inhibition of photosystem II by uncouplers at alkaline pH and its reversal by artificial electron donors. Biochim. Biophys. Acta **376**, 97–105 (1975)
COST, K., FRENKEL, A.W.: Light-induced interactions of *Rhodospirillum rubrum* chromatophores with bromthymol blue. Biochemistry **6**, 663–667 (1967)
CROFTS, A.R.: Ammonium ion uptake by chloroplasts, and the high-energy state. In: (J. JÄRNFELT, ed.) Regulatory Functions of Biological Membranes, pp. 247–263. Amsterdam: Elsevier Publ. Co. 1968
CROSS, R.L., BOYER, P.D.: The rapid labeling of adenosine triphosphate by ^{32}P-labeled inorganic phosphate and the exchange of phosphate oxygens as related to conformational coupling in oxidative phosphorylation. Biochemistry **14**, 392–398 (1975)
DEAMER, D.W., PACKER, L.: Light-dependent anion transport in isolated spinach chloroplasts. Biochim. Biophys. Acta **172**, 539–545 (1969)
DEAMER, D.W., PRINCE, R.C., CROFTS, A.R.: The response of fluorescent amines to pH gradients across liposome membranes. Biochim. Biophys. Acta **247**, 323–335 (1972)
DILLEY, R.A.: Evidence for the requirement of H^+ ion transport in photophosphorylation in spinach chloroplasts. In: (H. METZNER, ed.) Progress in Photosynthesis Research, Vol. III, pp. 1354–1360. Tübingen: Intern. Union of Biol. Sci. 1969
DILLEY, R.A.: Coupling of ion and electron transport in chloroplasts. In: Current Topics in Bioenergetics (D.R. SANADI, ed.), Vol. 4, pp. 237–271. New York-London: Academic Press 1971
DILLEY, R.A., SHAVIT, N.: On the relationship of H^+ transport to photophosphorylation in spinach chloroplasts. Biochim. Biophys. Acta **162**, 86–96 (1968)
DILLEY, R.A., VERNON, L.P.: Ion and water transport processes related to the light-dependent shrinkage of spinach chloroplasts. Arch. Biochem. Biophys. **14**, 365–376 (1965)
DILLEY, R.A., VERNON, L.P.: Quantum requirement of the light-induced proton uptake by spinach chloroplasts. Proc. Natl. Acad. Sci. US **57**, 395–400 (1967)

FIOLET, J.T.W., BAKKER, E.P., VAN DAM, K.: The fluorescent properties of acridines in the presence of chloroplasts or liposomes. On the quantitative relationship between the fluorescence quenching and the transmembrane proton gradient. Biochim. Biophys. Acta **368**, 432–445 (1974)
FORTI, G., AVRON, M., MELANDRI, A. (eds.): Proc. 2nd Intern. Congr. Photosyn. Res., Stresa (Vol. I–III). The Hague: Dr. W. Junk, N.V. 1971
FOWLER, C.F., KOK, B.: Proton evolution associated with the photooxidation of water in photosynthesis. Biochim. Biophys. Acta **357**, 299–307 (1974)
GAENSSLEN, R.E.: Amine uptake in chloroplasts. Ph.D. Thesis, Cornell University. 145 pp. (1971)
GAENSSLEN, R.E., MCCARTY, R.E.: Amine uptake in chloroplasts. Arch. Biochem. Biophys. **147**, 55–65 (1971)
GAENSSLEN, R.E., MCCARTY, R.E.: Determination of solute accumulation in chloroplasts by rapid centrifugal transfer through silicone fluid layers. Anal. Biochem. **48**, 504–514 (1972)
GIAQUINTA, R.T., DILLEY, R.A., ANDERSON, B.J.: Light potentiation of photosynthetic oxygen evolution inhibition by water soluble chemical modifiers. Biochem. Biophys. Res. Comm. **52**, 1410–1417 (1973)
GOOD, N.E., IZAWA, S., HIND, G.: Uncoupling and energy transfer inhibition in photosynthesis. In: (D.R. SANADI, ed.) Current Topics in Bioenergetics, Vol. 1, pp. 75–112. New York-London: Academic Press 1966
GOULD, J.M., IZAWA, S.: Photosystem-II electron transport and phosphorylation with dibromothymoquinone as the electron acceptor. Europ. J. Biochem. **37**, 185–192 (1973a)
GOULD, J.M., IZAWA, S.: Studies on the energy coupling sites of photophosphorylation I. Separation of site 1 and site II by partial reactions of the chloroplast electron transport chain. Biochim. Biophys. Acta **314**, 211–223 (1973b)
GOULD, J.M., IZAWA, S.: Studies on the energy coupling sites of photophosphorylation. IV. The relation of proton fluxes to the electron transport and ATP formation associated with photosystem II. Biochim. Biophys. Acta **333**, 509–524 (1974)
GOVINDJEE (ed.): Bioenergetics of Photosynthesis. New York-London: Academic Press 1975
GREVILLE, G.D.: A scrutiny of Mitchell's chemiosmotic hypothesis of respiratory chain and photosynthetic phosphorylation. In: Current Topics in Bioenergetics (D.R. SANADI, ed.), Vol. 3, pp. 1–78. New York-London: Academic Press. 1969
GRÜNHAGEN, H.H., WITT, H.T.: Primary ionic events in the functional membrane of photosynthesis. Z. Naturforsch. **25b**, 373–386 (1970)
HALL, D.O., EDGE, H., KALINA, M.: The site of ferricyanide photoreduction in the lamellae of isolated spinach chloroplasts: A cytochemical study. J. Cell Sci. **9**, 289–303 (1971)
HARTH, E., OETTMEIER, W., TREBST, A.: Native and artificial energy conserving sites operating in coupled electron donor systems for photosystem II. FEBS Letters **43**, 231–234 (1974a)
HARTH, E., REIMER, S., TREBST, A.: Control of photosynthetic oxygen evolution by the internal pH of the chloroplast thylakoid. FEBS Letters **42**, 165–168 (1974b)
HAUSKA, G.A.: Lipophilicity and catalysis of photophosphorylation I. Sulfonated phenazonium compounds are ineffective in mediating cyclic photophosphorylation in photosystem I subchloroplast vesicles. FEBS Letters **28**, 217–220 (1972)
HAUSKA, G.A., MCCARTY, R.E., BERZBORN, R.J., RACKER, E.: Partial resolution of the enzymes catalyzing photophosphorylation. VII. The function of plastocyanin and its interaction with a specific antibody. J. Biol. Chem. **246**, 3524–3531 (1971)
HAUSKA, G.A., REIMER, S., TREBST, A.: Native and artificial energy-conserving sites in cyclic photophosphorylation systems. Biochim. Biophys. Acta **357**, 1–13 (1974)
HAUSKA, G.A., TREBST, A., DRABER, W.: Lipophilicity and catalysis of photophosphorylation II. Quinoid compounds as artificial carriers in cyclic phosphorylation and photoreductions by photosystem I. Biochim. Biophys. Acta **305**, 632–641 (1973)
HELDT, H.W., RAPLEY, L.: Unspecific permeation and specific uptake of substrances in spinach chloroplasts. FEBS Letters **7**, 139–142 (1968)
HELDT, H.W., RAPLEY, L.: Specific transport of inorganic phosphate and dihydroxyacetone phosphate, and of dicarboxylates across the inner membrane of spinach chloroplasts. FEBS Letters **10**, 143–148 (1970)
HIEDEMAN-VAN WYK, D., KANNANGARA, C.G.: Localization of ferredoxin in the thylakoid membrane with immunological methods. Z. Naturforsch. **26b**, 46–50 (1971)

HIND, G., JAGENDORF, A.T.: Separation of light and dark stages in photophosphorylation. Proc. Natl. Acad. Sci. US **49**, 715–722 (1963)
HIND, G., MCCARTY, R.E.: The role of cation fluxes in chloroplast activity. In: Photophysiology (A.C. GIESE, ed.), Vol. VIII, pp. 113–156. New York-London: Academic Press 1973
HIND, G., NAKATANI, H.Y., IZAWA, S.: Light-dependent redistribution of ions in suspensions of chloroplast thylakoid membranes. Proc. Natl. Acad. Sci. US **71**, 1484–1488 (1974)
HIND, G., WHITTINGHAM, C.P.: Reduction of ferricyanide by chloroplasts in the presence of nitrogenous bases. Biochim. Biophys. Acta **75**, 194–202 (1963)
HINKLE, P.C.: Electron transfer across membranes and energy coupling. Fed. Proc. **32**, 1988–1992 (1973)
IZAWA, S.: The swelling and shrinking of chloroplasts during electron transport in the presence of phosphorylation uncouplers. Biochim. Biophys. Acta **102**, 373–378 (1965)
IZAWA, S., GOOD, N.E.: The stoichiometric relation of phosphorylation to electron flow in chloroplasts. Biochim. Biophys. Acta **162**, 380–391 (1968)
IZAWA, S., HIND, G.: The kinetics of the pH rise in illuminated chloroplast suspensions. Biochim. Biophys. Acta **143**, 377–390 (1967)
IZAWA, S., KRAAYENHOF, R., RUUGE, E.H., DEVAULT, D.: The site of KCN inhibition in the photosynthetic electron transport pathway. Biochim. Biophys. Acta **314**, 328–339 (1973)
IZAWA, S., ORT, D.R.: Photooxidation of ferrocyanide and iodide ions and associated phosphorylation in NH_2OH-treated chloroplasts. Biochim. Biophys. Acta **357**, 127–143 (1974)
JACKSON, J.B., CROFTS, A.R.: The high energy state in chromatophores from *Rhodopseudomonas spheroides*. FEBS Letters **4**, 185–192 (1969)
JACKSON, J.B., CROFTS, A.R., VON STEDINGK, L.-V.: Ion transport induced by light and antibiotics in chromatophores from *Rhodospirillum rubrum*. Europ. J. Biochem. **6**, 41–54 (1968)
JAGENDORF, A.T.: The relationship between electron transport and phosphorylation in spinach chloroplasts. Brookhaven Symp. Biol. **11**, 236–258 (1958)
JAGENDORF, A.T., HIND, G.: Studies on the mechanism of photophosphorylation. In: Photosynthetic Mechanisms of Green Plants, pp. 599–610. National Academy of Sciences, National Research Council Publications 1145 (1963)
JAGENDORF, A.T., MARGULIES, M.: Inhibition of spinach chloroplast reactions by p-chlorophehyl-1,1-dimethylurea. Arch. Biochem. Biophys. **90**, 184–195 (1960)
JAGENDORF, A.T., URIBE, E.G.: Photophosphorylation and the chemisomotic hypothesis. Brookhaven Symp. Biol. **19**, 215–246 (1966a)
JAGENDORF, A.T., URIBE, E.G.: ATP formation caused by acid-base transition of spinach chloroplasts. Proc. Natl. Acad. Sci. US **55**, 170–177 (1966b)
JUNGE, W., AUSLÄNDER, W.: The electric generator in photosynthesis in green plants I. Vectorial and protolytic properties of the electron transport chain. Biochim. Biophys. Acta **333**, 59–70 (1974)
KARLISH, S.J.D., AVRON, M.: The relevance of light-induced proton uptake to the mechanism of energy coupling in photophosphorylation. In: (K. SHIBATA, A. TAKAMYA, A.T. JAGENDORF, R.C. FULLER, eds.) Comparative Biochemistry and Biophysics of Photosynthesis, pp. 214–221. Tokyo-State College. Univ. Tokyo Press/Univ. Park Press 1968
KARLISH, S.J.D., AVRON, M.: Energy transfer inhibition and ion movement in isolated chloroplasts. Europ. J. Biochem. **20**, 51–57 (1971)
KRAAYENHOF, R.: "State 3- State 4 transitions" and phosphate potential in "class I" spinach chloroplasts. Biochim. Biophys. Acta **180**, 213–215 (1969)
KRAAYENHOF, R.: Quenching of uncoupler fluorescence in relation to the "energized state" in chloroplasts. FEBS Letters **6**, 161–165 (1970)
MAGNUSSON, R.P., MCCARTY, R.E.: Influence of adenine nucleotides on the inhibition of photophosphorylation in spinach chloroplasts by N-ethylmaleimide. J. Biol. Chem. **250**, 2593–2598 (1975)
MCCARTY, R.E.: The uncoupling of phosphorylation by valinomycin and ammonium chloride. J. Biol. Chem. **244**, 4292–4298 (1969)
MCCARTY, R.E.: The stimulation of post-illumination ATP synthesis by valinomycin. FEBS Letters **9**, 313–316 (1970)
MCCARTY, R.E., FAGAN, J.: Light-stimulated incorporation of N-ethylmaleimide into coupling factor 1 in spinach chloroplasts. Biochemistry **12**, 1503–1507 (1973)
MCCARTY, R.E., FUHRMAN, J.S., TSUCHIYA, Y.: Effects of adenine nucleotides on hydrogen-ion transport in chloroplasts. Proc. Natl. Acad. Sci. US **68**, 2522–2526 (1971)

McCarty, R.E., Pittman, P.R., Tsuchiya, Y.: Light-dependent inhibition of photophosphorylation by N-ethylmaleimide. J. Biol. Chem. **247**, 3048–3051 (1972)
McCarty, R.E., Racker, E.: Effect of a coupling factor and its antiserum on photophosphorylation and hydrogen ion transport. Brookhaven Symp. Biol. **19**, 202–214 (1966)
Mitchell, P.: Coupling of phosphorylation to electron and hydrogen transfer by a chemiosmotic type of mechanism. Nature **191**, 144–148 (1961)
Mitchell, P.: Chemiosmotic coupling in oxidative and photosynthetic phosphorylation. Biol. Rev. Cambridge **41**, 445–502 (1966)
Mitchell, P.: A chemiosmotic mechanism for proton-translocating adenosine triphosphatases. FEBS Letters **43**, 189–194 (1974)
Nelson, N., Nelson, H., Naim, V., Neumann, J.: Effect of pyridine on the light-induced pH rise and postillumination ATP synthesis in chloroplasts. Arch. Biochem. Biophys. **145**, 263–267 (1971)
Nir, I., Pease, D.C.: Chloroplast organization and the ultrastructural localization of photosystems I and II. J. Ultrastruct. Res. **42**, 534–550 (1973)
Oesterhelt, D., Stoeckenius, W.: Functions of a new photoreceptor membrane. Proc. Natl. Acad. Sci. US **70**, 2853–2857 (1973)
Ort, D.R., Izawa, S.: Studies on the energy-coupling sites of photophosphorylation II. Treatment of chloroplasts with NH_2OH plus ethylenediamine tetraacetate to inhibit water oxidation while maintaining energy-coupling efficiencies. Plant Physiol. **52**, 595–600 (1973)
Packer, L., Murakami, S., Mehard, C.W.: Ion transport in chloroplasts and plant mitochondria. Ann. Rev. Plant Physiol. **21**, 271–304 (1970)
Pfaff, E., Klingenberg, M., Ritt, E., Vogell, W.: Korrelation des unspezifisch permeablen mitochondrialen Raumes mit dem „Intermembran-Raum". Europ. J. Biochem. **5**, 222–232 (1968)
Pick, U., Rottenberg, H., Avron, M.: Effect of phosphorylation on the size of the proton gradient across chloroplast membranes. FEBS Letters **32**, 91–94 (1973)
Pick, U., Rottenberg, H., Avron, M.: The dependence of photophosphorylation in chloroplasts on ΔpH and external pH. FEBS Letters **48**, 32–36 (1974)
Portis, A.R., Jr., Magnusson, R.P., McCarty, R.E.: Conformational changes in coupling factor 1 may control the rate of electron flow in chloroplasts. Biochem. Biophys. Res. Commun. **64**, 877–884 (1975)
Portis, A.R., Jr., McCarty, R.E.: On the pH dependence of the light-induced hydrogen ion gradient in spinach chloroplasts. Arch. Biochem. Biophys. **156**, 621–625 (1973)
Portis, A.R., Jr., McCarty, R.E.: Effects of adenine nucleotides and of photophosphorylation on H^+ uptake and the magnitude of the H^+ gradient in illuminated chloroplasts. J. Biol. Chem. **249**, 6250–6254 (1974)
Racker, E.: Oxidative phosphorylation. In: (O. Hayaski, ed.) Molecular Oxygen in Biology: Topics in Molecular Oxygen Research, pp. 339–361. Amsterdam: North Holland 1974
Racker, E., Hauska, G.A., Lein, S., Berzborn, R.J., Nelson, N.: Resolution and reconstitution of the system of photophosphorylation. Proc. 2nd Intern. Congr. Photosyn. Res., Stresa, pp. 1097–1113, 1971
Racker, E., Stoeckenius, W.: Reconstitution of purple membrane vesicles catalyzing light-driven proton uptake and adenosine triphosphate formation. J. Biol. Chem. **249**, 662–664 (1974)
Reeves, S.G., Hall, D.O.: The stoichiometry (ATP/$2e^-$ ratio) of non-cyclic photophosphorylation in isolated spinach chloroplasts. Biochim. Biophys. Acta **314**, 68–78 (1973)
Rosing, J., Slater, E.C.: The value of $\Delta G°$ for the hydrolysis of ATP. Biochim. Biophys. Acta **267**, 275–290 (1972)
Rottenberg, H., Grunwald, T.: Determination of ΔpH in chloroplasts 3. Ammonium uptake as a measure of ΔpH in chloroplasts and subchloroplast particles. Europ. J. Biochem. **25**, 71–74 (1972)
Rottenberg, H., Grunwald, T., Avron, M.: Direct determination of ΔpH in chloroplasts, and its relation to the mechanisms of photoinduced reactions. FEBS Letters **13**, 41–44 (1971)
Rottenberg, H., Grunwald, T., Avron, M.: Determination of ΔpH in chloroplasts 1. Distribution of [^{14}C] Methylamine. Europ. J. Biochem. **25**, 54–63 (1972)
Rumberg, B., Reinwald, E., Schröder, H., Siggel, U.: Correlations between electron transfer,

proton translocation and phosphorylation in chloroplasts. In: (H. METZNER, ed.) Progress in Photosynthesis Research, Vol. III, pp. 1374–1382. Tübingen: Intern. Union Biol. Sci. 1969

RUMBERG, B., SCHRÖDER, H.: Ion transfer and phosphorylation. Abstracts Inter. Cong. Photobiology. Bochum, Abstract No. 036. 1972

RUMBERG, B., SIGGEL, V.: pH changes in the inner phase of the thylakoids during photosynthesis. Naturwissenschaften **56**, 130–138 (1969)

RYRIE, I.J., JAGENDORF, A.T.: An energy-linked conformational change in the coupling factor protein in chloroplasts. Studies with hydrogen exchange. J. Biol. Chem. **246**, 3771–3774 (1971)

RYRIE, I.J., JAGENDORF, A.T.: Correlation between a conformational change in the coupling factor protein and the high energy state in chloroplasts. J. Biol. Chem. **247**, 4453–4459 (1972)

SAHA, S., OUITRAKUL, R., IZAWA, S., GOOD, N.E.: Electron transport and photophosphorylation in chloroplasts as a function of the electron acceptor. J. Biol. Chem. **246**, 3204–3209 (1971)

SCHLIEPHAKE, W., JUNGE, W., WITT, H.T.: Correlation between field formation, proton translocation and the light reactions of photosynthesis. Z. Naturforsch. **23b**, 1571–1578 (1968)

SCHRÖDER, H., MUHLE, H., RUMBERG, B.: Relationship between ion transport phenomena and phosphorylation in chloroplasts. In: (G. FORTI, M. AVRON, A. MELANDRI, eds.) Proc. 2nd Intern. Congr. Photosyn. Res. Stresa, Vol. III, pp. 919–930. The Hague: Dr. W. Junk N.V. 1971

SCHULDINER, S., ROTTENBERG, H., AVRON, M.: Determination of ΔpH in chloroplasts. 2. Fluorescent amines as a probe for the determination of ΔpH in chloroplasts. Europ. J. Biochem. **25**, 64–70 (1972)

SCHWARTZ, M.: Light induced proton gradient links electron transport and phosphorylation. Nature **219**, 915–919 (1968)

SCHWARTZ, M.: The relation of ion transport to phosphorylation. Ann. Rev. Plant. Physiol. **22**, 469–484 (1971)

SELMAN, B.R., BANNISTER, T.T.: Trypsin inhibition of photosystem II. Biochim. Biophys. Acta **253**, 428–436 (1971)

SELMAN, B.R., BANNISTER, T.T., DILLEY, R.A.: Trypsin inhibition of electron transport. Biochim. Biophys. Acta **292**, 566–581 (1973)

SHEN, Y.K., SHEN, G.M.: Studies on photophosphorylation. II. The light intensity effect and intermediate steps of photophosphorylation. Scientia Sinica **9**, 1097–1106 (1962)

STIEHL, H.H., WITT, M.T.: Quantitative treatment of the function of plastoquinone in photosynthesis. Z. Naturforsch. **24b**, 1588–1599 (1969)

STRICHARTZ, G.R., CHANCE, B.: Absorbance changes at 520 nm caused by salt addition to chloroplast suspensions in the dark. Biochim. Biophys. Acta **256**, 71–84 (1972)

TAGAWA, H., TSUJIMOTO, H.Y., ARNON, D.I.: Analysis of photosynthetic reactions by the use of monochromatic light. Nature **199**, 1247–1252 (1963)

TELFER, A., BARBER, J.: Twofold effect of valinomycin on isolated spinach chloroplasts: uncoupling and inhibition of electron transport. Biochim. Biophys. Acta **333**, 343–352 (1974)

TELFER, A., EVANS, M.C.W.: Evidence for chemi-osmotic coupling of electron transport to ATP synthesis in spinach chloroplasts. Biochim. Biophys. Acta **256**, 625–637 (1972)

TREBST, A.: Energy conservation in photosynthetic electron transport of chloroplasts. Ann. Rev. Plant Physiol. **25**, 423–458 (1974)

TREBST, A., HARTH, E., DRABER, W.: On a new inhibitor of photosynthetic electron transport in isolated chloroplasts. Z. Naturforsch. **25b**, 1157–1159 (1970)

TREBST, A., REIMER, S.: Properties of photoreductions by photosystem II in isolated chloroplasts. An energy-conserving step in the photoreduction of benzoquinones by photosystem II in the presence of dibromothymoquinone. Biochim. Biophys. Acta **305**, 129–139 (1973a)

TREBST, A., REIMER, S.: Properties of photoreductions by photosystem II in isolated chloroplasts III. The effect of uncouplers on phenylenediamine shuttles across the membrane in the presence of dibromothymoquinone. Biochim. Biophys. Acta **325**, 546–557 (1973b)

URIBE, E.G., JAGENDORF, A.T.: Organic Acid specificity for acid-induced ATP synthesis by isolated chloroplasts. Plant Physiol. **42**, 706–711 (1967a)

URIBE, E.G., JAGENDORF, A.T.: On the localization of organic acids in acid-induced ATP synthesis. Plant Physiol. **24**, 697–705 (1967b)

Walker, D.A., Crofts, A.R.: Photosynthesis. Ann. Rev. Biochem. **39**, 389–428 (1970)
Walker, N.A.: Uncoupling in particles and intact chloroplasts by amines and nigericin—a discussion of the role of swelling. FEBS Letters **50**, 98–101 (1975)
Werkheiser, W.C., Bartley, W.: The study of steady-state concentration of internal solutes of mitochondria by rapid centrifugal transfer through a fixation medium. Biochem. J. **66**, 79–87 (1957)
Winget, G.D., Izawa, S., Good, N.E.: The inhibition of photophosphorylation by phlorizin and closely related compounds. Biochem. **8**, 2067–2074 (1969)
Witt, H.T.: Coupling of quanta, electrons, fields, ions and phosphorylation in the functional membrane of photosynthesis. Results by pulse spectroscopic methods. Quart. Rev. Biophys. **4**, 365–477 (1971)
Yamashita, T., Butler, W.L.: Photoreduction and photophosphorylation with tris-washed chloroplasts. Plant Physiol. **43**, 1978–1986 (1968)

2. Ion Transport in Plant Mitochondria

R.H. WILSON and R.J. GRAESSER

1. Introduction

The essentiality of most if not all plant mineral elements was established during the first half of the twentieth century. Research efforts in mineral nutrition, in recent years, have largely been devoted to studies concerning the mechanism of assimilation of mineral elements into plants as well as the subsequent fate of mineral ions once within the plant. At the cellular level, interest has centered on the mechanism of ion movement within different cellular compartments and the role of mineral ions once within each respective compartment.

In this regard, mitochondria have proven to be excellent research tools. They can be rapidly isolated from other cellular materials in relatively pure form and with their energy-generating and ion-transport systems still intact. Mitochondrial ion transport includes some of the most rapid transport systems known in plants. For example, isolated maize (*Zea mays* L.) mitochondria can accumulate over a μmole of Ca^{2+}/mg protein in a 10-min period (HODGES and HANSON, 1965).

Much of the knowledge and considerable theory concerning the mechanism of ion transport has come from studies with mitochondria isolated from animal sources, particularly rat liver. While the results reported and conclusions drawn from studies with plant mitochondria are in general agreement with those found with animal mitochondria, they are not always identical. Moreover, in some instances the physiological role of the transport system in plants and animals may be quite different. Thus the movement of Ca^{2+} across animal mitochondrial membranes has been implicated in the processes of bone deposition (LEHNINGER, 1970) as well as muscle contraction in different animal tissues. Clearly an equivalent role for Ca^{2+} transport in plant mitochondria does not exist and yet a well-defined Ca^{2+} plus Pi transport system is present in plants. Its role remains unclear.

In this chapter the salient features of ion transport in plant mitochondria will be discussed and compared with their counterpart in animal tissues. In addition to this article a number of review articles concerned with ion transport into mitochondria, both in plants and animals, have been written (HANSON and HODGES, 1967; LEHNINGER et al., 1967; PULLMAN and SCHATZ, 1967; LARDY and FERGUSON, 1969; HENDERSON, 1971; SLATER, 1971; VAN DAM and MEYER, 1971; AZZONE et al., 1972).

2. Monovalent Salt Transport

2.1 Monovalent Cation and Anion Transport in Plant Mitochondria

One fundamental difference between the membranes of isolated plant and animal mitochondria is the greater permeability of plant mitochondria to monovalent cations, notable K^+. Maize mitochondria incubated in a 0.1 M KCl medium passively accumulate between 150 to 200 nmoles K^+ per mg protein in a little over 10 min (KIRK and HANSON, 1973). The influx results in osmotic swelling. Similar responses have been observed in isolated bean (*Phaseolus vulgaris* L.) mitochondria (WILSON, 1975). Other monovalent cations which passively penetrate include Na^+, Li^+ and organic cations such as $Tris^+$ [Tris (hydroxymethyl) aminomethane] and tetramethylammonium$^+$ (WILSON et al., 1969). Ammonia accompanied by acetate penetrates the most rapidly of all salts examined, with passive swelling completed in less than a minute. In this latter respect plant mitochondria resemble rat liver mitochondria (CHAPPELL and CROFTS, 1966). While the permeability of isolated plant mitochondria to monovalent cation salts may have some as yet unrealized physiological significance in the living cell, the possibility that it might result from membrane damage incurred during the harsh grinding procedure required to break open plant cells cannot be overlooked. Recently DOUCE et al. (1972) have developed an isolation procedure for plant mitochondria which appears to reduce, if not eliminate, much of the membrane damage which normally occurs during the isolation process. While the procedure is laborious and time-consuming, it preserves mitochondria integrity with unimpaired membranes. Interestingly, these mitochondria do not swell passively in monovalent salt solutions, resembling in that respect rat liver mitochondria (BONNER, 1972).

Passive swelling observed in mitochondria suspended in monovalent cation salt solutions is dramatically altered when an energy source is added. In a medium with KCl as the osmoticant, the addition of an oxidizable substrate or ATP results in mitochondrial contraction (Fig. 1). The extent of the contraction is related to the tightness of coupling with greater contraction occurring in mitochondria with higher P/O ratios and respiratory control values (STONER and HANSON, 1966; EARNSHAW and TRUELOVE, 1968; WILSON et al., 1969).

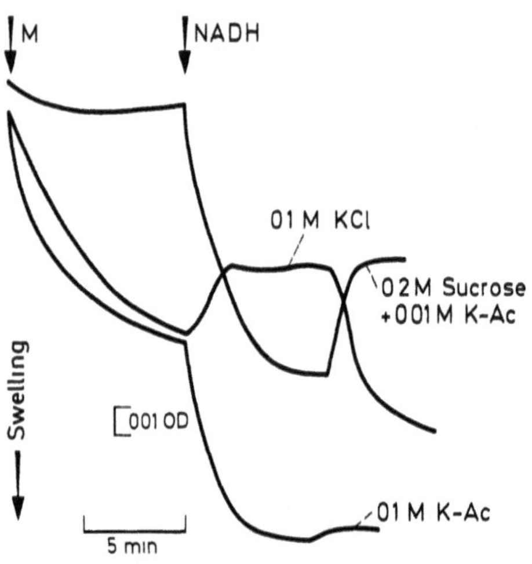

Fig. 1. Volume changes in maize mitochondria in different osmoticants. The complete reaction medium contained: 1 mg/ml bovine serum albumin, 0.025 M Tris-Tricine pH 7.5 and osmoticant as shown. Protein was 0.76 mg per cuvette. Procedures were as outlined in methods of WILSON et al. (1969). Authors' unpublished data

In contrast to the substrate-supported contraction observed with plant mitochondria, a number of external salt environments produce additional swelling when an energy source is added, including Pi, arsenate and several organic acids (WILSON et al., 1969; LEE and WILSON, 1972). In a solution of 0.1 M K-acetate as osmoticant, respiration supports active swelling and the influx of over 150 nmoles K and acetate/mg protein. Electron micrographs show the diameter of acetate-swollen maize mitochondria to be between 0.9 to 1.4 μ which is approximately 40% greater than the diameter of the contracted mitochondria. The swollen mitochondria are characterized by an attenuated matrix with inner and outer membranes pressed together. In many mitochondria the swelling has damaged or broken the outer membrane, and in some cases the outer membrane is no longer present. Active swelling can be inhibited by initially including in the medium an ATP generating system (ADP, Pi and Mg^{2+}) or a Ca^{2+} plus Pi transport system (WILSON et al., 1969; LEE and WILSON, 1972). However, when no competing systems for energy utilization are available, swelling in a K-acetate medium greatly retards or completely inhibits subsequent ATP synthesis or Ca^{2+} plus Pi transport. Moreover, when NADH is depleted and thus respiration terminated in swollen mitochondria, they do not subsequently contract. In contrast, when mitochondria are swollen in a osmoticant composed of a mixture of K-acetate and sucrose, the termination of respiration results in the efflux of K^+ and acetate and the partial restoration of the contracted state (Fig. 1).

In another study with maize mitochondria, active swelling was induced by potassium phosphate with a corresponding stimulation of respiration. Mercurial mersalyl inhibits the swelling but not the respiratory increase (HANSON et al., 1972). These results have been interpreted to indicate that the respiratory stimulation associated with the Pi-induced swelling is due to an internal Pi-dependent acceleration of the turnover of the high energy intermediate state, a so-called "acceptorless respiration", rather than directly to anion transport stimulation of respiration.

While both plant and animal mitochondria appear to behave as simple osmometers (LORIMER and MILLER, 1969; BENTZEL and SOLOMON, 1967), volume changes in vivo similar to those observed with isolated mitochondria have not yet been reported. With chloroplasts, MURAKAMI and PACKER (1970) have shown osmotic swelling in vivo when algal cells are incubated in sodium acetate. From electron micrographs of whole plant cells a wide range of morphological forms of mitochondria are observed. Considering the range of osmotic adjustments to which plant tissues are subjected, it seems probable that changes in osmotic condition in the cell produce corresponding alteration in the volume of the mitochondria and these changes account for some, if not most, observed variations in morphological forms seen in mitochondria in vivo. Irrespective of the physiological significance of the osmotic responses observed, they have provided an important means of studying ion transport and energy conservation processes.

2.2 Chemiosmotic Theory of Ion Transport

Despite the few studies concerned with the chemiosmotic theory of energy coupling in plant mitochondria, its importance to the understanding of ion transport requires a familiarity with its basic tenets. The chemiosmotic theory postulates that the principle function of oxidative phosphorylation is the electrogenic translocation of protons across the mitochondrial membrane (MITCHELL, 1968). The theory further states that protons are prevented from re-equilibrating across the membrane except by stoichiometric exchange with cations or in accompaniment with stoichiometric amounts of anions. The former transport system is called an antiport and the latter is referred to as a symport. If the rapidity of transport approaches that of the respiration-driven proton extrusion, the chemical potential of protons will return to its initial value but the membrane potential retains a maximal value. The flow of charge species down an electrical gradient results in the discharge of the membrane potential and relieves the pressure on the proton gradient so that H^+ translocation will increase. The sum of the electrical and chemical potentials across the membrane is collectively referred to as the proton motive force (PMF) and is equivalent to the high-energy intermediate state in traditional coupling terminology. Thus, reduction of the PMF by ATP generation, uncouplers or ion transport releases the back pressure on the PMF so that respiration is stimulated (HENDERSON, 1971).

Measurements of proton movements associated with energy generation and utilization in animal mitochondria are relatively easy and common whereas electrical potential measurements are not. The minute size of mitochondria prohibits direct quantitative evaluation of electrical gradients with currently available technology. In plant mitochondria, few measurements of H^+ fluxes have been made and these values are relatively small compared with those obtained with animal mitochondria under equivalent conditions (JOHNSON and WILSON, 1972; KIRK and HANSON, 1973; CHEN and LEHNINGER, 1973). For example, when valinomycin is added to bean mitochondria incubated in KCl, a rapid ejection of approximately 80 nmoles H^+/mg protein occurs into the external medium but is followed by an immediate spontaneous decline in the H^+ level (WILSON, 1975). This decline may be due to a leaky inner membrane which prohibits the build up of a larger proton gradient. As previously suggested, the harsh grinding required to disrupt plant tissue during isolation may damage mitochondrial membranes during the isolation and thus limit the build up of the proton gradient. However, chloroplasts which require similar grinding procedures during isolation, exhibit large proton fluxes. Chloroplast experiments, in fact, have contributed to the popularity of the chemiosmotic theory (PACKER et al., 1970; HENDERSON, 1971). Alternatively, the absence of large proton fluxes in plant mitochondria may reflect a greater role of the electrical potential as opposed to the chemical potential in the energy conserving strategy of plant mitochondria.

The lack of means to monitor and quantify the electrical potential in mitochondria, coupled with the small and inconsistent proton flux measurements reported in plant mitochondria, have greatly hindered our understanding of the energetics of ion transport. However, future research with plant mitochondria will require an evaluation of the role of proton fluxes in these organelles.

2.3 Mechanism of Salt Flux

The previously mentioned passive swelling of plant mitochondria suspended in a salt solution is a consequence of the inward flux of salt down a concentration gradient followed by osmotic equivalent amounts of water (Fig. 1). In accordance with the chemiosmotic theory of energy coupling, when an oxidizable substrate or ATP is added to swollen mitochondria an electrochemical gradient builds up across the inner membrane. The electrical component of this gradient supports the efflux of Cl^- against its own concentration gradient. A concomitant efflux of K^+ may occur by exchange diffusion with external H^+ generated by respiration. This diminishes the total proton motive force and thus should produce a stimulation of respiration to restore the loss in PMF. However, no such respiratory stimulation has been reported. An alternative explanation may be that K^+ diffuses outward along with Cl^- to maintain electroneutrality. The efflux of KCl increases the internal water potential and an osmotic equivalent amount of water fluxes outward resulting in mitochondria contraction. When respiration terminates, the electrochemical potential across the mitochondrial membrane dissipates and KCl moves inward back down its own concentration gradient established during respiration. The concomitant decrease in water potential internally results in the inward movement of osmotic amounts of water, producing re-swelling.

When the osmoticant is K-acetate, the observed active swelling can be explained by a similar hypothesis to that applied to animal mitochondria, bearing in mind the greater permeability of isolated plant mitochondrial membranes to K^+. Respiration triggers the external release of H^+ and acetate anions in the medium equilibrate with the external protons to form neutral molecules which influx through the membrane down the concentration gradient. After dissociating internally, acetate anions pull in K^+ to maintain electroneutrality and osmotic water follows, inducing swelling.

If active swelling is extensive, the mitochondrial membranes apparently are damaged, prohibiting them from supporting further coupling responses including salt transport, ATP synthesis or contraction (WILSON et al., 1969; LEE and WILSON, 1972). When the acetate-induced swelling is limited by partially replacing external K-acetate with sucrose (Fig. 1), the termination of respiration permits the efflux of K-acetate from the higher internal salt concentration outward down a chemical gradient. The corresponding loss of water causes the mitochondria to contract. These mitochondrial membranes are undamaged and their coupling responses are unaltered.

2.4 Monovalent Cation Transport with Ionophorous Antibiotics

The movement of monovalent cations across biological membranes or even artificial hydrophobic barriers is greatly enhanced by the presence of ionophores which are antibiotics that facilitate the passive permeability of specific cations through membrane barriers (PRESSMAN et al., 1967). With mitochondria, such ionophores affect energy metabolism as well as membrane permeability. Thus, when valinomycin is added to rat liver mitochondria incubated in a KCl solution, K^+ moves inward, H^+ ions are ejected, respiration is stimulated, and mitochondria swell (PRESSMAN, 1965). These results are interpreted within the frame-work of the chemiosmotic theory to indicate the discharge of an electrical potential as K^+ moves inward down its electrical gradient. An osmotic influx of H_2O results in swelling. The dissipation of the electrical gradient is compensated by the buildup of a proton gradient which requires energy and thus induces a stimulation of respiration. The proton gradient maintains the level of the PMF across the mitochondrial membrane (HENDERSON, 1971).

While relatively few studies are available regarding the response of plant mitochondria to ionophores, those reports where ionophores have been examined support a basically similar pattern of response to that seen with ionophores used with other membrane systems. Since plant mitochondria show some permeability to univalent cations in the absence of ionophores, their presence tends to enhance and accelerate the response of mitochondrial membranes rather than qualitatively alter it. Thus, when valinomycin is added to bean mitochondria in a medium containing K^+ and Pi, the mitochondria swell more rapidly and extensively than when valinomycin is absent (WILSON et al., 1972). In addition, respiration is stimulated. When energy is exhausted, contraction ensues, partially reversing the swelling. These results have been interpreted to indicate an active carrier-mediated transport of Pi inward during respiration and subsequent efflux of Pi down an established salt concentration gradient when respiration terminates. However, the results also fit the chemiosmotic theory which states that an active valinomycin-induced influx of K^+ occurs down an electrical gradient with a concomitant influx of Pi anions either accompanied by protons (symport) or in exchange with an OH^- ions (antiport) to maintain electroneutrality. Respiration is stimulated to maintain the PMF. The termination of respiration dissipates the proton motive force resulting in the efflux of salt down the established concentration gradient and osmotically induced contraction follows. If high concentrations of K-PO_4 are added to the external medium to eliminate the salt concentration gradient, contraction is inhibited. A similar response and explanation applies to volume changes observed with mitochondria in a medium containing K-acetate, except that acetate binds a proton and moves across the membrane as a neutral uncharged molecule (WILSON et al., 1972).

The valinomycin-induced stimulation of respiration is explained in the chemiosmotic theory as a release of back pressure on the proton motive force due to the energy demands of ion transport. However, HANSON et al. (1972) have pointed out that there is no consistent correlation between salt transport and valinomycin-induced respiration. Alternatively, the enhanced respiration may be due to the previously mentioned leakiness of plant mitochondrial membranes to protons and thus to a continuous dissipation of the proton gradient generated in response to the valinomycin induced K^+ influx. The net result is a release of the electrochemical potential gradient and, as a consequence, a stimulation of respiration. The response appears similar to the gramicidin-induced release of respiration in rat liver mitochondria triggered by the capacity of gramicidin to transport both K^+ and protons with equal facility (PRESSMAN, 1965). Such a situation is equivalent to an uncoupled state and is consistent with the suggestion that valinomycin causes uncoupling in maize mitochondria (HENSLEY, 1975b). For valinomycin to cause uncoupling in rat liver mitochondria however, a proton conductor such as nigericin must be present (HENDERSON, 1971).

When KCl replaces potassium phosphate in a medium with valinomycin, similar responses are observed but with one important difference. The penetration of Cl^- inward is limited by the absence of a symport which can transport an accompanying proton inward or an antiport which will support an exchange of

Cl^- with OH^- ions. The result is a Donnan equilibrium in which a chemical and charge potential regulate the rate and extent of the valinomycin supported K^+ and Cl^- penetration and, therefore, active swelling. The consequence is that the rate of salt influx is slow and less extensive than in K-PO_4 or K^+-acetate media. When respiration terminates, no measurable contraction is observed in KCl due to the absence of a buildup of a concentration gradient of salt internally.

In a study with the ionophorous antibiotic, gramicidin, maize mitochondria undergo an accelerated rate as well as an extent of passive swelling when added to a KCl medium containing gramicidin. This result indicates that K^+ penetration normally limits the rate of passive swelling and gramidicin enhances that rate. Metabolic inhibitors do not alter this passive swelling response consistent with the lack of involvement of endogenous respiration. Gramicidin appears to facilitate the passive flux of K^+ down a concentration gradient, and movement of a small amount of K^+ produces an opposing electrical charge or diffusion potential sufficient to pull Cl^- inward to maintain electroneutrality (COCKRELL et al., 1967). The concomitant increase in the internal water potential induces swelling.

In another response of maize mitochondria to gramicidin, energy-dependent contraction in a KCl medium is reversed when an ionophore is added (MILLER et al., 1970). Concomitant with the active swelling, mitochondria show an increase in respiration, and a reduction in both ADP/O ratios and respiratory control. These results indicate that gramicidin produces an uncoupled condition in corn mitochondria and in that respect resemble the situation in rat liver mitochondria (HARRIS et al., 1967).

2.5 Substrate Transport into Mitochondria

Studies with animal mitochondria have shown them to possess a series of exchange diffusion systems which catalyze the uptake of substrate anions in exchange for internal anions. The energy for these exchange reactions comes from electrical or chemical gradients initially established by respiration or other carrier systems. The exchange is followed by an osmotic adjustment in the mitochondrial matrix space. Numerous carrier systems have been reported for animal mitochondria including dicarboxylic, tricarboxylic, amino acid and fatty acid carriers (CHAPPELL and HAARHOFF, 1967; CHAPPELL, 1968; KLINGENBERG, 1970). A discussion of these anion transport systems in animal mitochondria appears elsewhere in this volume (HELDT, in press).

Bean mitochondria show a propensity to swell in a wide range of K^+-anion salts including several which are oxidized by mitochondria (LEE and WILSON, 1972).The swelling can occur both in an energy independent (passive) and energy dependent (active) manner. Monocarboxylic acids including acetate, propionate and β-hydroxybutyrate exhibit both a higher rate and magnitude of active swelling than the dicarboxylic acids. The tricarboxylic acid, citrate, is among the least effective. With the above-mentioned monovalent anions the addition of an ATP generating system or 2,4-DNP partially inhibits the active swelling but oligomycin restores it. These results indicate that entry occurs by a process similar to that previously described for acetate, which involves salt flow down an electrochemical gradient.

Dicarboxylic acid carriers for succinate and malate in animal mitochondria have been shown to be inhibited by a series of analogues of malonic acid (ROBINSON and WILLIAMS, 1969). Succinate respiration is inhibited by these same compounds in intact, but not broken, mung bean (*Phaseolus aureus* Roxb.) mitochondria permitting PHILLIPS and WILLIAMS (1973) to conclude that the entry of succinate and possibly malate is inhibited by these compounds in a manner similar to that reported in rat liver mitochondria (CHAPPELL, 1968; ROBINSON and WILLIAMS, 1969). The results are consistent with a carrier system for these substrates, but certainly do not prove it.

Regarding uptake of other citric acid cycle compounds, oxaloacetic acid (OAA), which serves as a branch point in the pathway of carboxylate, amino acid, fatty acid and oxidative metabolism, readily penetrates mung bean mitochondria. DOUCE and BONNER (1972) demonstrated that OAA rapidly penetrates tightly coupled intact mung bean and potato (*Solanum tuberosum* L.) mitochondria and leads to an inhibition of all citric acid cycle oxidation. The inhibition is thought to occur as a result of the conversion of OAA to malate by malate dehydrogenase which results in the consumption of the common pool of NADH in the mitochondria. The rapid but passive penetration of OAA into these plant mitochondria is in direct contrast to rat liver mitochondria where OAA penetration is severely restricted.

Finally, it has recently been reported that cauliflower (*Brassica aleracea* L.) mitochondria contains an adenine nucleotide translocator similar to that described in animal mitochondria (JUNG and HANSON, 1973). Much more is known concerning this interesting transport system in animal mitochondria and is discussed in this volume (see Chap. II, 6, p. 239).

3. Divalent Cation Transport

3.1 Divalent Cation Transport in Plant Mitochondria

Many observations concerning Ca^{2+} transport in plant mitochondria have resulted from studies on isolated maize mitochondria (HANSON and HODGES, 1967). Maize mitochondria accumulate Ca^{2+} by a process which has an obligatory requirement for Pi and competes for energy used for ATP formation (HANSON et al., 1965). Permeant anions such as acetate substitute very poorly for Pi during Ca^{2+} accumulation (EARNSHAW et al., 1973). It has recently been shown (EARNSHAW and HANSON, 1973) that in the absence of exogenous Pi, a small amount of Ca^{2+} is taken up during respiration and is rapidly released after respiration ceases. If Pi is then added in the substrate depleted, postenergized state, the rate of Ca^{2+} release is greatly reduced, indicating the passive diffusion of Pi into the mitochondrial matrix. Energy for Ca^{2+} plus Pi uptake may be provided by oxidizable substrates or by ATP. An enhanced Ca^{2+} plus Pi uptake results when both ATP *and* substrate are provided (HODGES and HANSON, 1965). Ca^{2+} plus Pi uptake in plant mitochondria produces no burst of respiration. It is inhibited by the presence of exogenous ADP (HODGES and HANSON, 1965; JOHNSON and WILSON, 1973) and by the uncoupler 2,4-DNP, but not by oligomycin (TRUELOVE and HANSON, 1966). From previous data, it had been hypothesized (KENEFICK and HANSON, 1966) that a conserved energy potential for Ca^{2+} plus Pi transports exists in the contracted mitochondrial state. But recent studies have demonstrated that not only is the respiration-linked contraction of corn mitochondria in a KCl medium due to an efflux pumping of the KCl that enters during passive swelling (KIRK and HANSON, 1973), but that the uptake of Pi and retention of Ca^{2+} in the postenergized state also occurs in a nonpenetrating sucrose medium (EARNSHAW and HANSON, 1973), indicating that contraction and reswelling per se are not obligatory for Pi uptake and Ca^{2+} retention.

In another transport study of divalent cations by plant mitochondria, red beet (*Beta vulgaris* L.) root mitochondria were found to actively accumulate Mg^{2+} plus Pi with a Mg^{2+}:Pi stoichiometry of about 1.5:1 (MILLARD et al., 1964, 1965). Like Ca^{2+} transport, substrate-driven Mg^{2+} transport is inhibited by addition of ADP and monovalent cations, while oligomycin stimulates Mg^{2+} accumulation approximately 50%. In contrast, Mg^{2+} is not accumulated in "Kentucky wonder" bean mitochondria (JOHNSON and WILSON, 1973; WILSON and MINTON, 1974), mung bean mitochondria and maize shoot mitochondria.

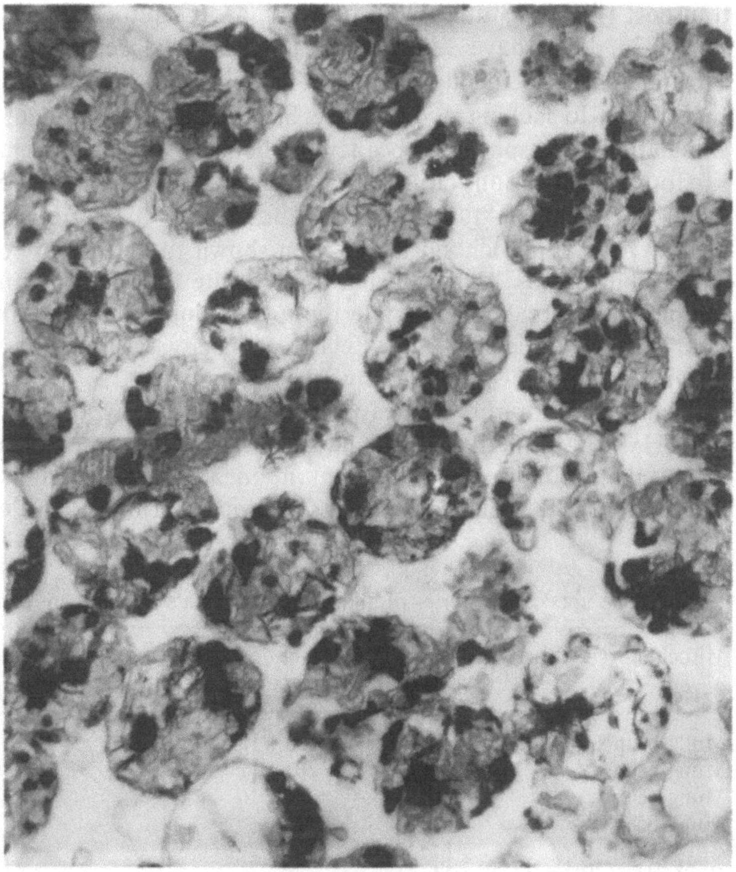

Fig. 2. Sr^{2+} and Pi deposits in bean mitochondria. Mitochondria were prepared for electron microscopy by the procedure in RAMIREZ-MITCHELL et al. (1973). Mag. ×32,640

Table 1. Characteristics of Sr^{2+} uptake into bean mitochondria[a]

Medium[b]	Sr^{2+} uptake (μmol/mg protein)	% of maximum
Complete	0.267	100
Complete + 3 mM ATP	0.259	97
-8 mM succinate	—	0
-8 mM succinate, -Pi	0.000	0
-Pi	0.015	6
+100 μM 2,4-dinitrophenol	0.035	13
+2 μg/ml oligomycin	0.255	95
+196 μM ADP plus hexokinase plus 25 mM glucose	0.057	22

[a] From JOHNSON and WILSON (1973).
[b] The complete reaction mixture of 6.4 ml contained 0.4 M mannitol, 50 mM Tris-Tricine (pH 7.5), 1 mg/ml albumin, 5 μM rotenone, 2.0 mM $SrCl_2$, 2.0 mM Tris phosphate and 8 mM Tris succinate. Protein was 0.80 mg.

Mitochondrial transport of Sr^{2+} and Ba^{2+} is qualitatively similar to Ca^{2+} transport (Fig. 2). Studies with bean hypocotyl mitochondria (JOHNSON and WILSON, 1972, 1973) have shown Sr^{2+} uptake to be dependent on the presence of substrate and Pi. Uncouplers and ATP formation inhibit Sr^{2+} uptake (Table 1). In addition to lacking a detectable Sr^{2+}-induced ejection of H^+ or a transient stimulation of respiration, bean mitochondria also lack a high affinity binding site for divalent cations (JOHNSON and WILSON, 1973; REYNAFARJE and LEHNINGER, 1969). In a similar study (WILSON and MINTON, 1974) Ba^{2+} uptake by mung bean hypocotyl mitochondria has been shown to be supported by oxidizable substrates, but not by ATP, and has an absolute requirement for Pi. ATP synthesis inhibits Ba^{2+} uptake 50% while the uncoupling agent, 2,4-dinitrophenol, inhibits Ba^{2+} uptake over 80%. The relative capacity of mung bean mitochondria to accumulate divalent cations follows the order $Sr^{2+} > Ca^{2+} > Ba^{2+} \gg Mg^{2+}$ (Table 2).

Hanson's laboratory evokes a modified chemical hypothesis for Ca^{2+} accumulation (HANSON et al., 1972), which adheres to the concept of a $X \sim P$ high energy intermediate but is also dependent upon a PMF, i.e., a proton gradient. The dehydration act, which forms the $X \sim I$ with a primary covalent bond from IOH

Table 2. The uptake of a series of alkaline earth metals by plant mitochondria[a]

Mitochondria source	Ion	Net divalent cation uptake (nmol/mg protein)	% of Sr^{2+} uptake
Mung bean	Sr^{2+}	308 ± 18	100
Mung bean	Ca^{2+}	264 ± 11	86
Mung bean	Ba^{2+}	234 ± 16	76
Mung bean	Mg^{2+}	13 ± 5	4
Corn shoot	Mg^{2+}	21	7
Kentucky pole bean	Mg^{2+}	8	3

[a] Data from WILSON and MINTON (1974).

Table 3. Comparison of the Ca^{2+} uptake properties of plant and animal mitochondria

Physiological Observation	Plant		Animal (rat liver)	
	Occurrence	Reference[c]	Occurrence	Reference[c]
Ca^{2+} stimulates state IV respiration	No	4, 7	Yes	2, 3, 11, 14
Ca^{2+} accumulation has an absolute requirement for Pi	Yes[a]	4, 8, 10	No	2, 3, 14
Inner mitochondrial membrane has high affinity Ca^{2+} binding sites	No[b]	4, 10	Yes	2, 11
Accumulation of Ca^{++} results in H^+ ion ejection	Yes	4, 16	Yes	11, 12
Ca^{2+} plus Pi are accumulated in preference to phosphorylation of ADP	No	10	Yes	2, 11
Exogenous NADH can serve as oxidizable substrate to drive Ca^{2+} uptake	Yes	6, 9, 15	No	1, 5, 13

[a] Exception: Arsenate ion is a weak substitute for Pi in the presence of oligomycin (KENEFICK and HANSON, 1966) and acetate substitutes in corn (EARNSHAW et al., 1973) but not bean mitochondria (JOHNSON and WILSON, 1973).
[b] Except for the sweet potato (CHEN and LEHNINGER, 1973).
[c] References
1 BERT and BARTLEY (1960)
2 CARAFOLI and LEHNINGER (1971)
3 CHANCE (1965)
4 CHEN and LEHNINGER (1973)
5 DESHPANDE et al. (1961)
6 DOUCE et al. (1973)
7 HANSON and HODGES (1967)
8 HODGES and HANSON (1965)
9 IKUMA and BONNER (1967)
10 JOHNSON and WILSON (1973)
11 LEHNINGER (1970)
12 LEHNINGER et al. (1967)
13 PAPA et al. (1967)
14 ROSSI and LEHNINGER (1964)
15 WILSON and HANSON (1969)
16 EARNSHAW et al. (1973)

and XH, results in the polar ejection of H^+ and OH^- ions. $I \sim X$ is phosphorylated to $X \sim P$, regenerating IOH. The hydrolysis of $X \sim P$ phosphorylates ADP to ATP, regenerating XH. Simultaneously, the polar ejection of H^+ and OH^- ions creates a proton gradient which drives Pi uptake through a Pi^-OH^- antiporter (CHAPPELL and CROFTS, 1966; CHAPPELL and HAARHOFF, 1967). Then Ca^{2+} accompanies Pi down the electrochemical gradient. Hanson's modified chemical hypothesis requires that respiration generates both a PMF for transport and $I \sim X$ for phosphorylation. This model differs from the ion transport models currently supported in the animal literature, although most ion transport hypotheses involve a respiration-linked PMF. It is possible that the high energy ($I \sim X$) stated in the model of Hanson's may reflect a chemical potential component of the total PMF (HENSLEY, 1975a). In summary, Table 3 outlines the major differences reported between the transport of divalent cations in plants and animal (rat liver) mitochondria.

3.2 Theories of Ca^{2+} Transport in Animal Mitochondria

At least three major theories concerning Ca^{2+} transport in animal mitochondria have been proposed. The first of these involves an energy carrier $X \sim I$ which combines obligatorily and stoichiometrically with Ca^{2+} at the inner mitochondrial membrane and causes an unidirectional transport of Ca^{2+} across the membrane into the mitochondrial matrix where the high energy intermediate becomes discharged (CHANCE, 1965). To account for the experimentally observed simultaneous ejection of protons into the external medium as Ca^{2+} is transported inward, $X \sim I$ is considered to be protonated. This model provides for respiration-dependent transport of Ca^{2+} against a gradient by direct chemical coupling.

In an alternative theory the Ca^{2+} transport system is seen as being composed of a Ca^{2+}-specific electrogenic uniporter which either responds passively to a Ca^{2+} concentration gradient or transmembrane electrical potential (LEHNINGER, 1972), or responds actively to an electron transport (ATPase) generated proton gradient or transmembrane potential (CHEN and LEHNINGER, 1973). Several lines of evidence support this model. When no permeant anions are present, the $Ca^{2+} : \sim$ accumulation ratio may vary greatly, depending on pH conditions or KCl concentrations (LEHNINGER, 1970). This is at odds with the chemical-coupling theory, which demands a stoichiometric chemical interaction of Ca^{2+} and $X \sim I$. However, this evidence does support an independent carrier model assuming that the variations in the $Ca^{2+} : \sim$ accumulation ratio, which are induced by changes in pH or KCl concentration, are caused by fluctuations in the electrochemical gradient across the mitochondrial membrane (LEHNINGER, 1970).

With the discovery of high affinity binding sites (REYNAFARJE and LEHNINGER, 1969) there now appear to be at least two sets of respiration-independent Ca^{2+}-binding sites, an abundance of low affinity sites and much less numerous high affinity sites. The high affinity sites have an affinity for Ca^{2+} comparable to that observed with respiring mitochondria. They bind Ca^{2+}, Sr^{2+} or Mn^{2+}, but not Mg^{2+}, K^{+} or Na^{+} in the absence of electron transport or ATP hydrolysis. The high affinity Ca^{2+}-binding sites are not found in yeast, blowfly flight muscle or plant mitochondria, none of which accumulate Ca^{2+} in the absence of Pi (CARAFOLI and LEHNINGER, 1971).

Studies using ruthenium red, which reacts specifically with mucopolysaccharides, have shown it to specifically inhibit the ability of rat liver mitochondria to transport Ca^{2+} (MOORE, 1971, VASINGTON et al., 1972). Energy conservation, however, is not inhibited. The conclusion drawn is that mucopolysaccharides (in the form of mucoproteins or muco- or glycolipids) are at the active center of the sites mediating mitochondrial Ca^{2+} transport and could be identical with the high affinity binding sites reported by LEHNINGER (1970). It is interesting that PRESTIPINO et al. (1974) recently found that the Ca^{2+}-binding glycoprotein from rat liver mitochondria, in the presence of Ca^{2+}, increases the electrical conductance of lecithin bilayers in a reaction which is sensitive to ruthenium red. Lanthanides have also been found to specifically inhibit the reactions of Ca^{2+} with the mitochondrial membrane and the accumulation of Ca^{2+} into the mitochondria (MELA, 1969). Since no other mitochondrial functions are inhibited, lanthanides appear to block a specific, divalent cation carrier. In another study LEHNINGER and CARAFOLI (1971) show that La^{3+} is rapidly bound to rat liver mitochondria in a respiration-independent process accompanied by loss of H^{+} to the medium. Apparently this lanthanide exceeds Ca^{2+} for binding sites but is not transported by the Ca^{2+} carrier. In plants, both La^{3+} and ruthenium red have been shown to be inhibitory in sweet potato (*Ipomoea batatas* L.) mitochondria (CHEN and

LEHNINGER, 1973), but ruthenium red is without effect in inhibiting Ca^{2+} plus Pi transport in bean mitochondria (JOHNSON and WILSON, 1973).

When Pi, acetate or bicarbonate accompanies Ca^{2+} uptake in rat liver mitochondria, the carrier model postulates that the anion permeates as a protonated species with the H^+ ions generated by electron transport. Once inside, the H^+ is lost to the excess OH^- produced in the matrix by electron transport. This result is a net influx of anions which is postulated to generate a negative gradient internally and pull positively charged Ca^{2+} into the matrix.

A third theory of Ca^{2+} transport in animal mitochondria involves the chemiosmotic hypothesis (MITCHELL, 1968; MITCHELL and MOYLE, 1969). The transport of Ca^{2+} in the presence of an oxidizable substrate but absence of permeant anion, includes a temporary burst of respiration and acidification of the medium, followed by a cessation of respiration (CHANCE, 1965). The hypothesis is that H^+ ions are ejected from the mitochondrial matrix at each of the three phosphorylation sites during respiration, producing an electrical gradient and creating a membrane potential, with the matrix being negatively charged in relation to the exterior. Ca^{2+} crosses the inner membrane and enters the matrix by moving passively down the electrical gradient. The entrance of Ca^{2+} causes a collapse of the membrane potential and uptake ceases. In the presence of an oxidizable substrate and permeant anion (e.g., Pi), respiration remains stimulated and is coupled with the accumulation in the matrix of massive quantities of Ca^{2+} and Pi in the form of hydroxyapatite precipitate (LEHNINGER et al., 1967). In this case, the mechanism is hypothesized as being the same with the addition of an antiport which transports an OH^- outward simultaneously with a Pi anion influx into the matrix.

3.3 Regulation of Ca^{2+} Flux

A number of Ca^{2+} transport regulators, as well as regulatory conditions, have been reported in the animal literature, mainly from studies on isolated rat liver mitochondria in the presence of respiratory substrate. The concentrations of Ca^{2+} and Pi in the medium determine the rate of Ca^{2+} uptake (BRIERLEY et al., 1964; ROSSI and LEHNINGER, 1964). When Pi is present in the medium (in the absence of ATP or oligomycin), it either prevents Ca^{2+}-binding or causes Ca^{2+} efflux. Oligomycin or ATP prevents the deleterious effect of Pi so that massive amounts of Ca^{2+}, stoichiometric with electron transport, can be accumulated in the presence of substrate when Pi and ATP or oligomycin are also present in the medium (ROSSI and LEHNINGER, 1964). However, when accumulated Ca^{2+} is being actively maintained by exogenous ATP in the absence of substrate, oligomycin blocks the maintenance of Ca^{2+} (DRAHOTA et al., 1965). Uncoupling agents such as 2,4-dinitrophenol have been shown to prevent substrate-driven Ca^{2+} accumulation (LEHNINGER et al., 1967) and to discharge half or more of the endogenous Ca^{2+} very rapidly (CARAFOLI and LEHNINGER, 1971). Atractylate also blocks ATP-linked maintenance of accumulated Ca^{2+} by preventing the entrance of ATP into the intact mitochondria (DRAHOTA et al., 1965). Estrogens and other hormones selectively inhibit human myometrial mitochondrial Ca^{2+} uptake (BATRA, 1973a, b) while parathyroid hormone produces a change in osteoblast (bone cell) mitochondrial granules (CAMERON et al., 1967). Finally, a cellular intermediary metabolite, phosphoenolpyruvate (PEP), increases the rate of Ca^{2+} efflux in rat liver and heart mitochondria (CHUDAPONGSE and HAUGAARD, 1973). In rat heart mitochondria, the effect occurs only when the mitochondria are respiring in the presence of NAD-linked substrates. In rat liver mitochondria, however, the effect is also seen when succinate serves as substrate. The PEP inhibition is reversed by ATP and by atractylate. ITP or IMP will not substitute for ATP. The effect is specific to PEP in that no other glycolytic intermediate tested affects Ca^{2+} uptake. Neither state 3 and state 4 respiration nor Pi transport are affected. In another study (PENG et al., 1974), PEP has been shown to induce Ca^{2+} efflux in rat liver mitochondria pre-loaded with Ca^{2+} but only when Pi has also been accumulated during the Ca^{2+}-loading. Mersalyl (which blocks the respiration-linked Pi/OH^- exchange carrier of the inner mitochondrial membrane) and bongkrekic acid (which blocks the adenine nucleotide translocase of the inner mitochondrial membrane) are potent inhibitors of the PEP-induced Ca^{2+} efflux, suggesting that the intramitochondrial ATP/Pi ratio influences the stability of the accumulated Ca^{2+}.

Numerous divalent cation transport regulators and regulatory conditions also appear in the plant literature. ATP and ADP have opposing effects on plant mitochondria. ATP supports Ca^{2+} uptake in maize mitochondria while ADP inhibits Ca^{2+} uptake (HODGES and HANSON, 1965). Accumulated Sr^{2+} in bean mitochondria is released by the initiation of ATP synthesis, i.e., by addition of exogenous ADP (JOHNSON and WILSON, 1973). In both maize and mung bean mitochondria, oligomycin reverses the ADP inhibition of substrate-driven Ca^{2+} transport and also blocks ATP-driven Ca^{2+} uptake but does not inhibit substrate-driven Ca^{2+} uptake (HODGES and HANSON, 1965; JOHNSON and WILSON, 1973). The uncoupler,

Table 4. The effect of phopho-enol-pyruvate and oxalacetate on Ca^{2+} uptake[a]

Treatment	% Inhibition of Ca^{2+} uptake
8.0 mM Succinate	
Complete medium alone (control)	(450)[b]
+1,000 μM PEP	57
+100 μM OAA	98
0.2 mM NADH	
Complete medium alone (control)	(575)
+1,000 μM PEP	30
+100 μM OAA	17
+1,000 μM OAA	46

[a] The reaction mixture contained: 0.3 M mannital, 50 mM tricine buffer pH 7.4, 2 mM Tris-phosphate, 2 mM $CaCl_2$, 1 mg bovine serum albumin in 6.4 ml. Total protein was 0.6 mg. Details of the methods have been given (GRAESSER, 1975).

[b] Parenthesis is the net Ca^{2+} uptake in nmol/mg protein.

2,4-dinitrophenol, also inhibits divalent cation uptake (HODGES and HANSON, 1965; JOHNSON and WILSON, 1973). The initial concentrations of Ca^{2+} and Pi have been shown to affect the rate of Ca^{2+} uptake by maize mitochondria (HODGES and HANSON, 1965). As in animal mitochondria, PEP has been shown to inhibit accumulation of Ca^{2+} by isolated mung bean mitochondria in the presence of oxidizable substrate and Pi (GRAESSER, 1975). PEP is twice as inhibitory to succinate-driven Ca^{2+} uptake as to NADH- or malate/pyruvate-driven Ca^{2+} uptake. Atractylate and ATP, but not ITP, reverse the PEP effect. Pyruvate *plus* cofactors of the pyruvate dehydrogenase reaction also reverse the PEP inhibition of Ca^{2+} uptake. PEP reduces succinate-driven state 3 and state 4 respiratory rates but ADP/O ratios are not affected. Another intermediary metabolite, oxaloacetic acid (OAA), also inhibits Ca^{2+} uptake by mung bean mitochondria (GRAESSER, 1975). Succinate-driven Ca^{2+} uptake is inhibited completely by OAA while NADH-driven Ca^{2+} uptake is only partially inhibited at equivalent concentrations (Table 4). In inhibiting succinate- and NADH-driven Ca^{2+} uptake OAA is more potent than PEP. These results may be due to the general inhibitory effect of OAA on citric acid cycle metabolism as shown by DOUCE and BONNER (1972).

3.4 Role for Mitochondria in the Transport and Regulation of Ca^{2+}

In the animal literature there are several proposed roles for mitochondria in the transport and regulation of Ca^{2+}, often associated with a function specifically linked to the given animal tissue. An example is calcification in the avian shell gland (SCHRAER et al., 1973), which functions to secrete and deposit large quantities of Ca^{2+} at specific time intervals associated with shell formation. Based on previous observations, HOHMAN and SCHRAER (1966) suggest that during shell formation, Ca^{2+} is moved from the mitochondria through the endoplasmic reticulum (ER) to the cell exterior for deposition as calcium carbonate. In the absence of shell formation, Ca^{2+} is moved into the mitochondria from the cytoplasm while the Ca^{2+} level of ER decreases.

Another calcification system is that of bone and cartilage. A gradient of mitochondrial granules in intact chondrocyte cells of rat epiphyseal tissue has been reported (MARTIN and MATTHEWS, 1969), with those mitochondria in the zone of hypertrophy containing granules more dense and numerous than those mitochondria in the provisional calcification zone. The demonstration by micro-incineraton that these granules are mineral in nature (MARTIN and MATTHEWS, 1970) not only supports the theory that these granules might be related to the onset of calcification but also suggests that osteoblast and osteocyte mitochondrial granules (GONZALES and KARMOVSKY, 1961) may serve a similar function for bone. In another study ARSENIS (1972), demonstrated the existence of a gradient of mitochondrial and other enzymatic activities thought to be involved in calcification, in the various zones of calf scapula epiphyseal cartilage, increasing from the resting zone towards the columnar and hypertrophic zones. A hypothesis for Ca^{2+} deposition in bone (LEHNINGER, 1970) suggests that "micropackets" of amorphous calcium phosphate are moved through the mitochondrial membrane either directly or by reverse phagocytosis. After entering the cytosol, the "micropackets" are moved to deposition sites along the extracellular bone matrix by a similar mechanism. Experimental evidence (GONZALES and KARMOVSKY, 1961; MARTIN and MATTHEWS, 1969; LEHNINGER, 1970) supports this view. But, because the evidence is based primarily upon the frequent occurrence of mitochondrial granules at calcification sites, and since these granules also appear in liver where calcification is *not* occurring (REYNOLDS, 1965), or appear only infrequently in the avian shell gland (SCHRAER et al., 1973) where calcification takes place at a high rate, a mitochondrial role for calcification should still be viewed as speculative.

A calcification process has been shown in slime mold mitochondria from the Myxomycete *Didymium squamulosum.* These mitochondria accumulate granules of calcium phosphate just prior to deposition of calcium carbonate ("lime") on the surfaces of the sporangial peridium, stalk, columella and/or capillitium during the process of sporulation (GUSTAFSON and THURSTON, 1974). The granules of calcium phosphate disappear from the mitochondria after "lime" deposition, suggesting that the mitochondria actively function in Ca^{2+} concentration, storage, and deposition. No ultrastructural evidence has been found to suggest that intact granules are transported from the mitochondria through the plasmalemma of the sporangium or incipient spores to the external sporangial surface, leaving open the question as to whether the transport is similar to that of Lehninger's "micropacket" hypothesis. While other calcification processes are known to occur in plant tissues, as yet, no studies have attempted to correlate these deposits with the accumulation of salt in mitochondria.

Several non-calcification systems exist in which a regulatory role for Ca^{2+} by animal mitochondria have been suggested. Ca^{2+} has been shown to be a major factor in excitation-contraction coupling and in relaxation of cardiac, skeletal, and smooth muscle. In cardiac and red skeletal muscle, which contain relatively little sarcoplasmic reticulum but an abundance of mitochondria, it has been suggested that mitochondrial Ca^{2+} uptake serves as a mechanism in controlling cytoplasmic Ca^{2+} in these tissues (CHANCE, 1965; CARAFOLI et al., 1969; LEHNINGER, 1970). It has been suggested that mitochondria also play a significant role in controlling the intracellular Ca^{2+} concentration, and thereby contraction and relaxation, of human myometrium (uterine smooth muscle) (BATRA, 1973a).

Many regulatory enzymes, as studied in vitro, are strongly activated by Ca^{2+} (ATKINSON, 1965). The feasibility of such a role for a metal ion in metabolic control has been discussed (EVANS and SORGER, 1966; BYGRAVE, 1967). Among the factors controlling the concentration of ions in the soluble cytoplasm are the presence of agents capable of chelating ions, the movement of ions between the cell and the extracellular medium, and the uptake or release of the ions by cell organelles (BYGRAVE, 1967; THORNE and BYGRAVE, 1974). The view that it is the mitochondrial accumulation and release of Ca^{2+} that may serve to coordinate metabolic events in the cytoplasm and mitochondrial matrix space has been hypothesized by RASMUSSEN (1966, 1970) and THORNE and BYGRAVE (1974). Cations (particularly Ca^{2+} and Mn^{2+}) translocated through the mitochondrial membrane can serve to directly regulate an NAD-specific glutamic dehydrogenase from *Blastocladiella emersonii,* an aquatic mold, by relieving an inhibition of the enzyme caused by citrate and some members of the citric acid cycle (LE'JOHN, 1968). Ca^{2+} has been found to inhibit glycolysis in extracts from Ehrlich ascites tumor cells (BYGRAVE, 1967). The inhibition mechanism is due to competition between

Ca^{2+} ions, on one hand, and Mg^{2+} and K^+ ions on the other, the latter being obligatory activators of pyruvate kinase. Inhibition of pyruvate kinase by Ca^{2+} leads to a buildup of phosphoenolpyruvate which, itself, has been shown to prevent the accumulation of Ca^{2+} in mitochondria or to trigger its release from inside the mitochondria (CHUDAPONGSE and HAUGAARD, 1973; PENG et al., 1974; GRAESSER, 1975). It has been suggested that the relative rates of pyruvate kinase and pyruvate carboxylase (a mitochondrial enzyme also inhibited by Ca^{2+}) could be partially controlled by the active transport of Ca^{2+} between the mitochondria and the cytoplasm (GEVERS and KREBS, 1966). Thus, if the concentration of Ca^{2+} is high in the cytoplasm, pyruvate kinase activity should be low and pyruvate carboxylase activity high. The reverse situation would exist when Ca^{2+} is transported out of the cytoplasm and into the mitochondria. Recently it has been shown that by increasing the concentration of Ca^{2+} in the mitochondria, the activity of pyruvate dehydrogenase complex is enhanced in beef heart mitochondria (SCHUSTER and OLSON, 1974). In plants, the number of Ca^{2+}-activated enzymes so far reported is limited and these do not appear to be specifically associated with mitochondria (WYN-JONES and LUNT, 1967).

In animal tissues increasing evidence indicates that Ca^{2+} and cyclic-AMP serve as "secondary messengers" functioning by a common biochemical mechanism at the cellular level in regulating the activity of a number of metabolic processes in the cytoplasm. In vivo Ca^{2+}, as a secondary messenger, may be released from an intracellular store analogous to the sarcoplasmic reticulum or mitochondria of muscle, although direct support for an effect of cyclic-AMP on the level of Ca^{2+} in mitochondria from animal sources has not been reported (RASMUSSEN et al., 1972; BORLE, 1973). With mung bean mitochondria it has been found that cyclic-AMP reduces the uptake of Ca^{2+} with succinate as substrate (WILSON, 1975).

The plant literature is virtually devoid of studies concerned with establishing a role for mitochondria in the transport and regulation of Ca^{2+}. The lack both of a high affinity binding site for Ca^{2+} and of a respiratory stimulation associated with the accumulation of Ca^{2+} in plant mitochondria indicates different properties and possibly different function for the transport system from plant sources. The obligatory requirement for Pi in Ca^{2+} transport into plant mitochondria supports a role of salt deposition in the mitochondrial matrix. However, a study of the subcellular distribution of Ca^{2+} in developing roots and stem callus tissue cultures of bean show that concentrations of Ca^{2+}, in both stem callus and root mitochondria, represent only about 5% of the total cellular Ca^{2+} (RATHORE et al., 1972). The cell wall, cytoplasm and cell nucleus all contain higher concentrations of Ca^{2+}. Since the total amount of Ca^{2+} in the mitochondria typically is only a small portion of the cellular Ca^{2+}, it may be reasoned that mitochondria function by temporarily sequestering Ca^{2+}-Pi salt. This salt is later released to be permanently deposited in specific cellular compartments such as the central vacuole, or the outer surface of the sporangia in the case of slime mold (GUSTAFSON and THURSTON, 1974). Indirect support for this hypothesis is found in a number of external factors which can induce the abrupt release of accumulated Ca^{2+} back into the external medium including fatty acids, ATP synthesis and possibly Phospho-enol-pyruvate (JOHNSON and WILSON, 1973; GRAESSER, 1975).

Thus, mitochondria are seen as sites for amassment of dense Ca^{2+}-Pi aggregates, which later may be permanently deposited in other cellular compartments. In this connection it is interesting that the geotropic stimulation of oat coleoptile reported by SHEN-MILLER and MILLER (1972) showed a greater number of mitochondria on the bottom than on the top of geotropically stimulated oat coleoptiles. Since massive loading of mitochondria with Ca^{2+} plus Pi increases the mitochondrial density, it is interesting to speculate that the propensity for mitochondria to sediment in the bottom half of the coleoptile may be induced by the accumulation of dense salt deposits in the mitochondrial matrix, later to be released when the geotropic stimulation terminates.

References

ARSENIS, C.: Role of mitochondria in calcification. Mitochondrial activity distribution in the epiphyseal plate and accumulation of calcium and phosphate ions by chondrocite mitochondria. Biochem. Biophys. Res. Commun. **46**, 1928–1935 (1972)

ATKINSON, D.E.: Biological feedback control at the molecular level. Science **150**, 851–857 (1965)

AZZONE, G.F., CARAFOLI, E., LEHNINGER, A.L., QUAGLIARIELLO, E., SILIPRANDI, N.: Biochemistry and biophysics of mitochondrial membranes, pp. 714. New York: Academic Press 1972

BATRA, S.: Effect of some estrogens and progesterone on calcium uptake and calcium release by myometrial mitochondria. Biochem. Pharmacol. **22**, 803–809 (1973a)

BATRA, S.: The role of mitochondrial calcium uptake in contraction and relaxation of the human myometrium. Biochim. Biophys. Acta **305**, 428–432 (1973b)

BENTZEL, C.J., SOLOMON, A.K.: Osmotic properties of mitochondria. J. Gen. Physiol. **50**, 1547–1563 (1967)

BERT, L.M., BARTLEY, W.: The distribution and metabolism of intra- and extra-mitochondrial pyridine nucleotides in suspensions of liver mitochondria. Biochem. J. **75**, 303–315 (1960)

BONNER, W.C : Personal communication (1972)

BORLE, A.B.: Calcium metabolism at the cellular level. Fed. Proc. **32**, 1944–1950 (1973)

BRIERLEY, G.P., MURER, E., BACHMANN, E.: Studies on ion transport. III. The accumulation of calcium and inorganic phosphate by heart mitochondria. Arch. Biochem. Biophys. **105**, 89–102 (1964)

BYGRAVE, F.L.: The ionic environment and metabolic control. Nature **214**, 667–671 (1967)

CAMERON, D.A., PASCHALL, H.A., ROBINSON, R.A.: Changes in the fine structure of bone cells after the administration of parathyroid extract. J. Cell Biol. **33**, 1–14 (1967)

CARAFOLI, E., LEHNINGER, A.L.: A survey of the interaction of calcium ions with mitochondria from different tissues and species. Biochem. J. **122**, 681–690 (1971)

CARAFOLI, E., PATRIARCA, P., ROSSI, C.S.: A comparative study of the role of mitochondria and the sarcoplasmic reticulum in the uptake and release of Ca^{++} by the rat diaphragm. J. Cell. Physiol. **74**, 17–29 (1969)

CHANCE, B.: The energy-linked reaction of calcium with mitochondria. J. Biol. Chem. **240**, 2729–2748 (1965)

CHAPPELL, J B.: Systems used for the transport of substrates into mitochondria. Brit. Med. Bull. **24**, 150–157 (1968)

CHAPPELL, J.B., CROFTS, A.R.: Ion transport and reversible volume changes of isolated mitochondria. In: Regulation of Metabolic Processes of Mitochondria (TAGER, J.M., PAPA, S., QUAGLIARIELLO, E., SLATER, E.C., eds.), pp. 293–316. Amsterdam: Elsevier Publ. Co. 1966

CHAPPELL, J.B., HAARHOFF, K.N.: The penetration of the mitochondria membranes by anions and cations. In: Biochemistry of mitochondria (SLATER, E.C., KANIUGA, Z., WOJTCZAK, L., eds.), pp. 75–91 New York: Academic Press 1967

CHEN, C., LEHNINGER, A.L.: Ca^{++} transport activity in mitochondria from some plant tissues. Arch. Biochem. Biophys. **157**, 183–196 (1973)

CHUDAPONGSE, P., HAUGAARD, N.: The effect of phosphoenolpyruvate on calcium transport by mitochondria. Biochim. Biophys. Acta **307**, 599–606 (1973)

COCKRELL, R.S., HARRIS, E.J., PRESSMAN, B.C.: Synthesis of ATP driven by a potassium gradient in mitochondria. Nature **215**, 1487–1488 (1967)

DESHPANDE, P.D., HICKMAN, D.D., VON KORFF, R.W.: Morphology of isolated rabbit heart muscle mitochondria and the oxidation of extramitochondrial DPNH. J. Biophys. Biochem. Cytol. **11**, 77–93 (1961)

DOUCE, R., BONNER, W.D., JR.: Oxalacetate control of Krebs cycle oxidations in purified plant mitochondria. Biochem. Biophys. Res. Commun. **47**, 619–624 (1972)

DOUCE, R., CHRISTENSEN, E.L., BONNER, W.D., JR.: Preparation of intact plant mitochondria. Biochim. Biophys. Acta **275**, 148–160 (1972)

DOUCE, R., MANNELLA, C., BONNER, W.D., JR.: The external NADH dehydrogenase of intact plant mitochondria. Biochim. Biophys. Acta **292**, 105–116 (1973)

DRAHOTA, Z., CARAFOLI, E., ROSSI, C.S., GAMBLE, R.L., LEHNINGER, A.L.: The steady-state maintenance of accumulated calcium in rat liver mitochondria. J. Biol. Chem. **240**, 2712–2720 (1965)

EARNSHAW, M.J., HANSON, J.B.: Inhibition of postoxidative calcium release in corn mitochondria by inorganic phosphate. Plant Physiol. **52**, 403–406 (1973)

EARNSHAW, M.J., MADDEN, D.M., HANSON, J.B.: Calcium accumulation by corn mitochondria. J. Exptl. Botany **24**, 824–840 (1973)

EARNSHAW, M.J., TRUELOVE, B.: Swelling and contraction of Phaseolus hypocotyl mitochondria. Plant Physiol. **43**, 121–129 (1968)

EVANS, H.J., SORGER, G.J.: Role of mineral elements with emphasis on the univalent cations. Ann. Rev. Plant Physiol. **17**, 14–76 (1966)

GEVERS, W., KREBS, H.A.: The effect of adenine nucleotides on carbohydrate metabolism in pigeon-liver homogenates. Biochem. J. **98**, 720–734 (1966)

GONZALES, F., KARMOVSKY, M.J.: Electron microscopy of osteoclasts in healing fractures of rat bone. J. Biophys. Biochem. Cytol. **9**, 299–316 (1961)

GRAESSER, R.J.: Intermediary metabolite regulation and control of calcium uptake by isolated mung bean mitochondria. Master's Thesis. Austin: Univ. Texas 1975

GUSTAFSON, R.A., THURSTON, E.L.: Calcium deposition of the mycomycete *Didymium Squamulosum.* Mycologia **66**, 397–412 (1974)

HANSON, J.B., BERTAGNOLLI, B.L., SHEPHERD, W.D.: Phosphate induced stimulation of acceptorless respiration in corn mitochondria. Plant Physiol. **50**, 347–354 (1972)

HANSON, J.B., HODGES, T.K.: Energy-linked reactions of plant mitochondria. Curr. Top. Bioenerg. **2**, 65–98 (1967)

HANSON, J.B., MALHOTRA, S.S., STONER, C.D.: Action of calcium on corn mitochondria. Plant Physiol **40**, 1033–1040 (1965)

HARRIS, E.J., HAFER, M.D., PRESSMAN, B.C.: Stimulation of mitochondria respiration and phosphorylation by transport-inducing antibiotics. Biochemistry **6**, 1348–1360 (1967)

HENDERSON, P.J.F.: Ion transport by energy-conserving biological membranes. Ann. Rev. Biochem. **40**, 393–428 (1971)

HENSLEY, J.R.: Personal communication (1975a)

HENSLEY, J.R., HANSON, J.B.: The action of valinomycin and uncoupling corn (*Zea mays* L.) mitochondria. Plant Physiol. **56**, 13–18 (1975b)

HODGES, T.K., HANSON, J.B.: Calcium accumulation by maize mitochondria. Plant Physiol. **40**, 101–109 (1965)

HOHMAN, W., SCHRAER, H.: The intracellular distribution of calcium in the mucosa of the avian shell gland. J. Cell Biol. **30**, 317–331 (1966)

IKUMA, H., BONNER, W.D., JR.: Properties of higher plant mitochondria. I. Isolation and some characteristics of tightly coupled mitochondria from dark-grown mung bean hypocotyls. Plant Physiol. **42**, 67–75 (1967)

JOHNSON, H.M., WILSON, R.H.: The accumulation and the release of divalent cations across mitochondrial membranes. Am. J. Botany **60**, 858–862 (1973)

JOHNSON, H.M., WILSON, R.H.: Sr^{++} uptake by bean (*Phaseolus vulgaris*) mitochondria. Biochim. Biophys. Acta **267**, 398–408 (1972)

JUNG, D.W., HANSON, J.B.: Atractyloside inhibition of adenine nucleotide transport in mitochondria from plants Biochim. Biophys. Acta **325**, 189–192 (1973)

KENEFICK, D.G., HANSON, J.B.: Contracted state as an energy source for Ca binding and Ca^{+} inorganic phosphate accumulation by corn mitochondria. Plant Physiol. **41**, 1601–1609 (1966)
KIRK, B.I., HANSON, J.B.: The stoichiometry of respiration-driven potassium transport in corn mitochondria. Plant Physiol. **51**, 357–362 (1973)
KLINGENBERG, M.: Mitochondria metabolite transport. FEBS Lett. **6**, 145–154 (1970)
LARDY, H.A., FERGUSON, S.M.: Oxidative phosphorylation in mitochondria. Ann. Rev. Biochem. **38**, 991–1034 (1969)
LEE, D.C., WILSON, R.H.: Swelling in bean shoot mitochondria induced by a series of potassium salts or organic anions. Physiol. Plantarum **27**, 195–201 (1972)
LEHNINGER, A.L.: The coupling of Ca^{2+} transport to election transport in mitochondria. In: Molecular Basis of Election Transport (SCHULTZ, J., CAMERON, B.F., eds.), Vol. IV, pp. 131–151. New York: Academic Press 1972
LEHNINGER, A.L.: Mitochondria and calcium ion transport. Biochem. J. **119**, 129–138 (1970)
LEHNINGER, A.L., CARAFOLI, E.: The interaction of La^{3+} with mitochondria in relation to respiration-coupled Ca^{2+} transport. Arch. Biochem. Biophys. **143**, 506–515 (1971)
LEHNINGER, A.L., CARAFOLI, E., ROSSI, C.S.: Energy-linked ion movements in mitochondria systems. Advan. Enzymol. **29**, 259–319 (1967)
LE'JOHN, H.B.: On the involvement of Ca^{2+} and Mn^{2+} in the regulation of mitochondrial glutamic dehydrogenase from *Blastocladiella*. Biochem. Biophys. Res. Commun. **32**, 278–283 (1968)
LORIMER, G.H., MILLER, R.J.: The osmotic behavior of corn mitochondria. Plant Physiol. **44**, 839–844 (1969)
MARTIN, J.H., MATTHEWS, J.L.: Mitochondrial granules in chondrocytes. Cal. Tiss. Res. **3**, 184–189 (1969)
MARTIN, J.H., MATTHEWS, J.L.: Mitochondrial granules in chondrocytes, osteoblasts, and osteocytes. Clin. Orthop. Related Res. **68**, 273–278 (1970)
MELA, L.: Inhibition and activation of calcium transport in mitochondria. Effect of lanthanides and local anesthetic drugs. Biochemistry **8**, 2481–2486 (1969)
MILLARD, D.L., WISKICH, J.T., ROBERTSON, R.N.: Ion uptake by plant mitochondria. Proc. Natl. Acad. Sci. US **52**, 996–1004 (1964)
MILLARD, D.L., WISKICH, J.T., ROBERTSON, R.N.: Ion uptake and phosphorylation in mitochondria: effect of monovalent ions. Plant Physiol. **40**, 1129–1135 (1965)
MILLER, R.J., DUMFORD, W.S., KOEPPE, D.E: Effects of gramicidin on corn mitochondria. Plant Physiol. **46**, 471–474 (1970)
MITCHELL, P.: Chemiosmotic Coupling and Energy Transduction. Bodmin: Glynn Research Ltd. 1968
MITCHELL, P., MOYLE, J.: Estimation of membrane potential and pH difference across the cristae membrane of rat-liver mitochondria. Europ. J. Biochem. **7**, 471–484 (1969)
MOORE, C.L.: Specific inhibition of mitochondrial Ca^{++} transport by ruthenium red. Biochem. Biophys. Res. Commun. **42**, 298–305 (1971)
MURAKAMI, S., PACKER, L.: Light-induced changes in the conformation and configuration of the thylokoid membrane of Ulva and Parphyra chloroplasts in vivo. Plant Physiol. **45**, 289–299 (1970)
PACKER, L., MURAKAMI, S., MEHARD, C.W.: Ion transport in chlorplasts and plant mitochondria. Ann. Rev. Plant Physiol. **21**, 271–304 (1970)
PAPA, S., LOFRUMENTO, N E., SECCHI, A.G., QUAGLIARIELLO, E.: Utilization of respiratory substrates in calf-retina mitochondria. Biochim. Biophys. Acta **131**, 288–294 (1967)
PENG, C.F., PRICE, D.W., BHUVANESWARAN, C., WADKINS, C.L.: Factors that influence phosphoenolpyruvate-induced calcium efflux from rat liver mitochondria. Biochem. Biophys. Res. Commun. **56**, 134–141 (1974)
PHILLIPS, M.L., WILLIAMS, G.R.: Effects of 2-butylmalonate, 2-phenylsuccinate, benzylmalonate and P-iodobenzlmalonate on the oxidation of substrate by mung bean mitochondria. Plant Physiol. **51**, 225–228 (1973)
PRESSMAN, B.C.: Induced active transport of ions in mitochondria. Proc. Natl. Acad. Sci. US **53**, 1076–1083 (1965)
PRESSMAN, B.C., HARRIS, E.J., JAGGER, W.S., JOHNSON, J.H.: Antibiotic-mediated transport of alkali ions across lipid barriers Proc. Natl. Acad. Sci. US **58**, 1949–1956 (1967)
PRESTIPINO, G., CECCARELLI, D., CONTI, F., CARAFOLI, E.: Interactions of a mitochondrial Ca^{2+}-binding glycoprotein with lipid bilayer membranes. FEBS Lett. **45**, 99–103 (1974)

PULLMAN, M.E., SCHATZ, G.: Mitochondrial oxidations and energy coupling. Ann. Rev Biochem. **36**, 539–610 (1967)

RAMIREZ-MITCHELL, R., JOHNSON, H.M., WILSON, R.H.: Strontium deposits in isolated plant mitochondria. Exptl. Cell Res. **76**, 449–451 (1973)

RASMUSSEN, H.D.: Mitochondrial ion transport: mechanism and physiological significance. Fed. Proc. **25**, 903–911 (1966)

RASMUSSEN, H.D.: Cell communication, calcium ion, and cyclic adenosine monophosphate. Science **170**, 404–412 (1970)

RASMUSSEN, H.D., GOODMAN, D.B.P., TENENHOUSE, A.: The role of cyclic AMP and calcium in cell activation. C.R.C. Crit. Rev. Biochem. **1**, 95–148 (1972)

RATHORE, V.S., BAJAJ, Y.P.S., WITTWER, S.H.: Subcellular localization of zinc and calcium in bean (*Phaseolus vulgaris* L.) tissue. Plant Physiol. **49**, 207–211 (1972)

REYNAFARJE, B., LEHNINGER, A.L.: High affinity and low affinity binding of Ca^{++} by rat liver mitochondria. J. Biol. Chem. **244**, 584–593 (1969)

REYNAFARJE, B., LEHNINGER, A.L.: Ca^{++} transport by mitochondria from L1210 mouse ascites tumor cells. Proc. Natl. Acad. Sci. US **70**, 1744–1748 (1973)

REYNOLDS, E.S.: Liver parenchymal cell injury. III. The nature of calcium associated electron-opaque masses in rat liver mitochondria following poisoning with carbon tetrachloride. J. Cell Biol. **25**, 53–75 (1965)

ROBINSON, B.H., WILLIAMS, G.R.: The effect on mitochondrial oxidations of inhibitors of the dicarboxylate anion transport system. FEBS Lett. **5**, 301–304 (1969)

ROSSI, C.S., LEHNINGER, A.L.: Stoichiometry of respiratory stimulation, accumulation of Ca^{++} and phosphate, and oxidative phosphorylation in rat liver mitochondria. J. Biol. Chem. **239**, 3971–3980 (1964)

SCHRAER, R., ELDER, J.A., SCHRAER, H.: Aspects of mitochondrial function in calcium movement and calcification. Fed. Proc. **32**, 1938–1943 (1973)

SCHUSTER, S.M., OLSON, M.S.: The regulation of pyruvate dehydrogenase in isolated beef heart mitochondria. The role of calcium, magnesium and permeant anions. J. Biol. Chem. **249**, 7159–7165 (1974)

SHEN-MILLER, J., MILLER, C.: Intracellular distribution of mitochondria after geotropic stimulation of the oat coleoptile. Plant Physiol. **50**, 51–54 (1972)

SLATER, E.C.: The energy coupling between energy-yielding and energy-utilizing reactions in mitochondria. Quant. Rev. Biophys. **4**, 35–71 (1971)

STONER, C.D., HANSON, J.B.: Swelling and contraction of corn mitochondria. Plant Physiol. **41**, 255–266 (1966)

THORNE, R.F.W., BYGRAVE, F.L.: The role of mitochondria in modifying the cellular ionic environment of calcium-induced respiratory activities in mitochondria isolated from various tumor cells. Biochem. J. **144**, 551–558 (1974)

TRUELOVE, B., HANSON, J.B.: Calcium-activated phosphate uptake in contracting corn mitochondria. Plant Physiol. **41**, 1004–1013 (1966)

VAN DAM, K., MEYER, A.J.: Oxidation and energy conservation by mitochondria. Ann. Rev. Biochem. **40**, 115–160 (1971)

VASINGTON, F.D., GASSOTTI, P., TIOZZO, R., CARAFOLI, E.: The effect of ruthenium red on energy metabolism in mitochondria. In: Biochemistry and Biophysics of Mitochondrial Membranes (AZZONE, G.F., CARAFOLI, E., LEHNINGER, A.L., QUAGLIARIELLO, E., eds.), pp. 215–228. New York: Academic Press 1972

WILSON, R.H.: Unpublished data (1975)

WILSON, R.H., DEVER, J., HARPER, W., FRY, R.: The effect of valinomycin on respiration and volume changes in plant mitochondria. Plant and Cell Physiol. **13**, 274–282 (1972)

WILSON, R.H., HANSON, J.B.: The effect of respiratory inhibition on NADH, succinate and malate oxidation in corn mitochondria. Plant Physiol. **44**, 1335–1341 (1969)

WILSON, R.H., HANSON, J.B., MOLLENHAUER, H.H.: Active swelling and acetate uptake in corn mitochondria. Biochemistry **8**, 1203–1213 (1969)

WILSON, R.H., MINTON, G.A.: The comparative uptake of Ba^{2+} and other alkaline earth metals by plant mitochondria. Biochem. Biophys. Acta **333**, 22–27 (1974)

WYN-JONES, R.L., LUNT, O.R.: The function of calcium in plants. Botan. Rev. **33**, 407–426 (1967)

3. Energy Transfer between Cell Compartments

H. STROTMANN and S. MURAKAMI

1. Introduction

In cells of eukaryotic green plants two organelles exhibit salient significance in energy production, the mitochondria and the chloroplasts. At least 90% of the ATP generated by the cell comes from these two organelles.

Mitochondria and chloroplasts display striking similarities in morphological and physiological respects. They are both surrounded by two membranes. In the case of mitochondria, the inner membrane is invaginated and contains electron carriers and the coupling system which is responsible for the formation of ATP. On the other hand, the extensive thylakoid system of chloroplasts, by which light energy is converted into chemical energy, likewise originates from the inner envelope membrane and is situated in the chloroplast interior. In mitochondria, oxygen serves to accept electrons, which may be donated by reduced nicotinamide adenine dinucleotide (NADH). In chloroplasts, oxidized nicotinamide adenine dinucleotide phosphate ($NADP^+$) is the physiological electron acceptor and electrons are donated by water. Formation of ATP from ADP and phosphate is coupled to and driven by electron flow. Coupling is closely related to the osmotic properties of the energy-conserving membranes.

While ATP is produced in mitochondria and chloroplasts, numerous energy-requiring reactions proceed outside of these organelles, e.g. the syntheses of proteins, fatty acids, polysaccharides, and the active transport of substrates and ions across the cell membrane. This points to the necessity of energy transfer from the organelles to the cytoplasm. Since the mitochondrion exhibits low biosynthetic activity, nearly all the ATP formed by oxidative phosphorylation has to be exported.

In contrast, the major function of the chloroplast is the formation of carbohydrates from CO_2, which consumes ATP and reduced nicotinamide adenine dinucleotide phosphate (NADPH) generated by the light-dependent reactions of photosynthesis. That the chloroplast can do this work independently of the cytoplasmic environment is shown by the high CO_2 fixation capacity of intact isolated chloroplasts in artificial media. Accordingly, ATP export seems to be a marginal function of the chloroplast. The possible extent of ATP release may be related to the stoichiometry of ATP production (and consumption) and NADP reduction (and re-oxidation) in photosynthesis. Operation of the Calvin cycle requires 3 mol ATP and 2 mol NADPH per each mol of CO_2 fixed. The ATP/e_2^- ratio of non-cyclic systems exceeds the value of 1.5 in broken chloroplasts (IZAWA and GOOD, 1968; REEVES and HALL, 1973). However, the stoichiometry is lower in intact chloroplasts and it seems to be variable within a certain range (HEBER, 1973; HEBER and KIRK, 1975). In vivo additional ATP may be provided by cyclic and pseudocyclic photophosphorylation, regulated by the internal energy requirement (SIMONIS and URBACH, 1973). This may give rise to further doubt about an efficient energy flow from the chloroplasts to the cytoplasm.

However, several active processes taking place outside the chloroplasts, e.g. the uptake and metabolization of sugars and ions were demonstrated to depend on light energy absorbed by the photosynthetic apparatus. Furthermore, in the dark the chloroplast requires external energy in order to maintain its metabolic functions. Based on reasonable assumptions, rates of energy exchange between chloroplasts and cytoplasm were calculated by HEBER (1975). Assuming that energy consumption of cell compartments follows the protein distribution, HEBER estimated an ATP requirement for the chloroplasts of spinach leaf cells in the dark close to 20 μmol/mg chlorophyll/h, which has to be imported from the cytoplasm. Energy requirement of the cytoplasm was believed to exceed 10 μmol ATP/mg chlorophyll/h in the dark. Taking into account that several additional energy-dependent processes occur in the cytoplasm in the light, and that normal respiration is inhibited in photosynthesizing cells, HEBER concluded that more than 10 μmol ATP/mg chlorophyll/h have to be transferred from the chloroplasts to the cytoplasm in the light. This may amount to 3% of the total ATP produced by photophosphorylation.

There are several possibilities by which energy may traverse the envelopes of the organelles. In this respect the outer and the inner membranes seem to possess completely different properties. The outer mitochondrial membrane and the outer envelope membrane of chloroplasts are freely permeable to adenine nucleotides (HELDT and RAPLEY, 1970a). In contrast, the inner membrane of mitochondria and the inner envelope membrane of chloroplasts represent strong diffusion barriers toward adenylates. This is in accordance with our understanding of the general properties of biological membranes: the larger and more polar molecules are, the more limited should be their diffusion through membranes. In a physiological environment ATP and ADP carry 4 and 3 negative charges, respectively. Compared to the sucrose molecule which acts as an osmotically active substance in mitochondria and chloroplasts, adenine nucleotides should be much less diffusible molecules (HELDT and SAUER, 1971).

A possibility of ATP transport across the impermeable inner membranes is that it is a carrier-mediated translocation. Another feasible mechanism is an indirect transfer via transport metabolites. As will be shown in the following chapters, both carrier-mediated and indirect energy transport are relevant to intracellular energy exchange.

Our knowledge about mitochondrial energy exchange seems to be well established, owing mainly to the work of KLINGENBERG'S group during the past decade. A detailed presentation of this subject is contained in the chapter by HELDT of this volume. Energy exchange across the chloroplast envelope is at present less clearly understood. Since 1964 many experimental results and ideas have been contributed particularly by HEBER and his coworkers. Energetic communication between chloroplasts and cytoplasm seems to be a complex process in which several transporting systems may be involved. Its comprehension requires a more detailed description in this article.

2. Exchange of Energy between Mitochondria and Cytoplasm

Most of the investigations on energy transport across mitochondrial membranes were performed with isolated mitochondria from animal sources. However, plant mitochondria seem to behave quite similarly in this respect (ONISHI et al., 1967; BERTAGNOLLI and HANSON, 1973; JUNG and HANSON, 1973). Although significant differences were also observed (PASSHAM et al., 1973), there is the question as to whether these are real or due to artificial damage of the organelles during isolation (JUNG and HANSON, 1973).

The mitochondrial inner membrane is penetrated by ADP and ATP via a specific carrier, called adenine nucleotide translocase (HELDT et al., 1965). It operates in counter exchange, i.e. for one molecule leaving the mitochondrion an exogenous adenine nucleotide molecule is taken up. The nucleotides cross the membrane directly as shown by use of ^{14}C-labeled compounds (PFAFF et al., 1965). Adenine nucleotide translocase is specific for ATP and ADP; AMP as well as guanine, uracil, cytidine, and pyridine nucleotides are not transported (PFAFF and KLINGENBERG, 1968). The translocase is inhibited by atractyloside, a glycoside isolated from the roots of *Atractylis gummifera* (HELDT, 1969a) and by the antibiotic bongkrekic acid, which is produced in the fermentation of cocoa products (HENDERSON and LARDY, 1970; KLINGENBERG and BUCHHOLZ, 1973, see also Chap. II, 6).

In respiring mitochondria the specifity of adenine nucleotide translocase is quite different for exogenous ADP and ATP. If added simultaneously ADP and ATP compete for exchange, with ADP being ten-fold more active than ATP, but endogenous ADP and ATP are released in proportion to their intramitochondrial content (PFAFF and KLINGENBERG, 1968). The *Km* values reported for exogenous ADP and ATP are 1.3 μM and 2.5 μM, respectively; 4 μM and 12 μM were found in the presence of Mg^{2+} ions. Since Mg^{2+} increased the *Km* values and changed the ratio between them, it was concluded that the free nucleotides are transported rather than the Mg-adenylate complexes (PFAFF et al., 1969).

By uncoupling and anaerobiosis the specifity for exogenous ADP is completely abolished and ATP is transported at the same rate as ADP. Thus ADP specifity—i.e. inhibition of exchange of external ATP—requires energy from the respiratory chain (PFAFF and KLINGENBERG, 1968).

The differential affinity of the translocator system for exogenous ADP and ATP leads to an asymmetric distribution of the two nucleotides on the two sides of the membrane in respiring mitochondria. This was demonstrated using isolated mitochondria. In the equilibrium state the external ATP/ADP ratio was found to be about ten times higher than the internal ratio, and the phosphorylation potential ATP/ADP × phosphate was about 50 times higher in the external than internal system (KLINGENBERG et al., 1969). Similar results were obtained by studying the equilibrium situation of the intact cell. Using freeze stop and non-aqueous fractionation of mitochondria and extramitochondrial components, ELBERS et al. (1974) found a 5–10 times higher ATP/ADP ratio in the cytoplasm than in the mitochondria of rat liver cells. The adenine nucleotide translocating system of mitochondria may be visualized as a one-way energy-pumping system, translocating ATP against an energy gradient from the mitochondrion to the cytoplasm. Obviously pumping requires energy from respiration. This may be the reason why actual P/O ratios of oxidative phosphorylation are below the theoretical expectation (KLINGENBERG et al., 1969).

3. Exchange of Energy between Chloroplasts and Cytoplasm

3.1 Evidence for Energy Exchange between Chloroplasts and Cytoplasm

3.1.1 Active Uptake of Sugars and Ions

Several energy-requiring reactions of green plant cells although located outside the chloroplasts, have been shown to be supported by energy from the chloroplasts. Among these the assimilation of exogenously added sugars and the uptake of ions have been extensively studied (Vol. 2, Part A: Chaps. 6 and 9; Part B: Chap. 4.2).

Glucose fed to algal cells is taken up and converted to polysaccharides (KANDLER, 1954). The overall process requires 3 mol of ATP per mol of glucose (TANNER and KANDLER, 1969). ATP can be supplied by respiration under aerobic conditions in the dark, or by photophosphorylation under anaerobiosis in the light (KANDLER, 1954). Since far-red light is active in the latter case, it was concluded that cyclic photophosphorylation serves as the energy donor (TANNER et al., 1968).

Energy seems to be needed for at least two consecutive steps, namely the active uptake of the glucose molecule into the cell and its incorporation into the polysaccharide fraction. Active uptake may be studied independently by using 3-0-methylglucose or 6-deoxyglucose, which are taken up and accumulated but not metabolized by *Chlorella* cells (TANNER, 1969). In the dark, uptake of these sugar analogs leads to stimulation of respiration which allows the calculation of ATP requirement for active uptake. The stoichiometry is about 1 mol ATP per 1 mol sugar (DECKER and TANNER, 1972). In the absence of oxygen light is able to support sugar uptake (TANNER, 1969; cf. also Vol. 2, Part B: Chap. 5.3.1.1.2).

The initial step in intracellular glucose metabolization may be ATP-dependent phosphorylation catalyzed by hexokinase. Since the chloroplast contains only low activities of hexokinase (HEBER et al., 1967a), this reaction at least in part seems to take place in the cytoplasm. Accordingly under light conditions for active uptake as well as for metabolization of glucose, ATP generated from photophosphorylation has to be exported from the chloroplast into the cytoplasm. The rate of ATP transport from the chloroplast related to this process may exceed 10 μmol/mg chlorophyll/h (HEBER, 1974).

In a similar manner active uptake of several cations and anions seems to depend on ATP produced by the chloroplasts. For such investigations those ions which are not further metabolized in the cell, are of particular interest, e.g. potassium and chloride.

Uptake of K^+ is stimulated by light, 3-(3,4-dichlorophenyl)-1,1-dimethylurea (DCMU) does not affect the light component and far-red light is active, indicating again that cyclic photophosphorylation may be involved (MACROBBIE, 1965, 1966; RAVEN, 1969, 1971).

Controversial results were obtained when the relation between chloride uptake and photosynthesis was investigated. In *Nitella, Tolypella* and *Hydrodictyon* cells light-dependent Cl^- uptake was found to be inhibited by DCMU, but uncoupling of photophosphorylation by m-chlorocarbonyl cyanide phenylhydrazone (CCCP) which effectively decreased the rates of CO_2 fixation as well as K^+ uptake, left Cl^- uptake almost unchanged (MACROBBIE, 1966; RAVEN, 1967; SMITH, 1968). The energy transfer inhibitor Dio-9 sometimes even stimulated Cl^- uptake (RAVEN et al., 1969). From this result a direct dependence of Cl^- uptake on a high-energy state generated by non-cyclic electron flow without ATP requirement was suggested.

However such a link can hardly be understood, because uncoupling would not only inhibit ATP production but also abolish the creation of any kind of high-energy state in chloroplasts.

In several other organisms Cl^- uptake was shown to exhibit a dependence on ATP from cyclic photophosphorylation comparable to K^+ uptake (WEIGL, 1967; BARBER, 1968; JESCHKE and SIMONIS, 1969; LILLEY and HOPE, 1971; HOPE et al., 1974). Thus at least in certain plants active uptake of Cl^- in the light requires export of ATP from the chloroplasts, too. However, ATP consumption due to chloride uptake seems to be low. From the calculation performed by HOPE et al. (1974) one may conclude that Cl^- uptake does not require more than 1 μmol ATP/mg chlorophyll/h (cf. also Vol. 2, Part A: Chap. 9.4).

3.1.2 Transient Changes of Intracellular Adenine Nucleotide Levels

ATP concentrations of algal and leaf cells are known to be increased when the cells are transferred from dark to light conditions. Concomitantly ADP and sometimes AMP concentrations are found to be decreased. After turning off the light, the reverse behavior of the adenylates is observed (KANDLER and HABERER-LIESENKÖTTER, 1963; PEDERSEN et al., 1966; BOMSEL and PRADET, 1967; STROTMANN and HELDT, 1969). The increase of the ATP concentration at the expense of ADP in the light appears to be caused by photophosphorylation in the chloroplasts.

In order to obtain information on the communication between chloroplastic and cytoplasmic adenine nucleotides under such transient conditions, HEBER and his coworkers designed experiments which allow the independent analysis of chloroplasts and cytoplasm in vivo. After rapid-freeze stop and lyophilization, cell components were fractionated in non-polar solvents; this avoids secondary and artificial mixing of compartmented substrate pools (THALACKER and BEHRENS, 1959).

At this point it should not be ignored that some error may be inherent in this separation technique, due to imperfect purity of the fractions. Indeed BIRD et al. (1973) reported that their purest chloroplast fraction contained as much as 11% of cytoplasmic contamination. Another uncertainty concerns the calculation basis of the cytoplasmic space, because the so-called cytoplasmic fraction contains several different cell constituents including mitochondria and vacuoles. In spite of these disadventages which obviously had been recognized by the investigators, non-aqueous fractionation is a useful tool, since it is the only practicable method as yet which provides insight into the intracellular substrate distribution in situ.

First attempts along this line were carried out by SANTARIUS et al. (1964), HEBER et al. (1964) and ULLRICH et al. (1965). These authors studied the incorporation of radioactivity into adenylate and phosphate pools of both cytoplasm and chloroplasts during ^{32}Pi incubation of *Elodea* shoots. With short-time incubation (30 to 180 s) specific labeling of inorganic phosphate was found to be several times higher in the cytoplasm than in the chloroplasts. On the other hand, in the chloroplasts, specific activity of ATP was much higher than of inorganic phosphate. From this rather surprising result the authors concluded a fast exchange of adenine nucleotides between cytoplasm and chloroplasts and a relatively slow translocation of inorganic phosphate (SANTARIUS et al., 1964; HEBER et al., 1964).

This view was further confirmed by measurements of adenine nucleotide pool changes of chloroplasts and cytoplasm (SANTARIUS and HEBER, 1965; HEBER, 1970; HEBER and SANTARIUS, 1970). With transition from dark to light a rapid increase of ATP levels, parallel in both compartments, was observed; ADP exhibited the

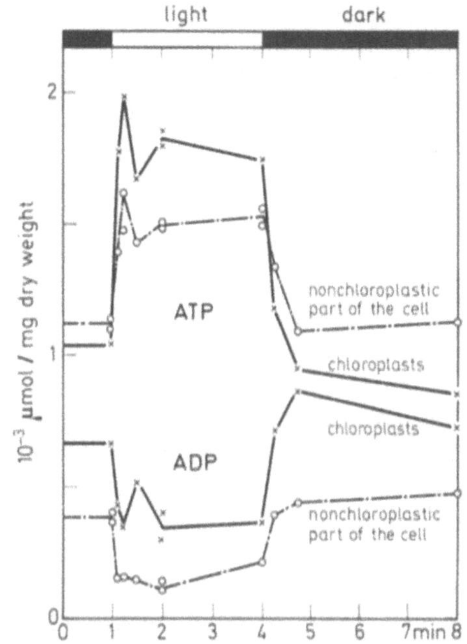

Fig. 1. Changes in the levels of ATP and ADP in chloroplasts and cytoplasm of leaf cells of *Elodea densa* caused by illumination and darkening. (From HEBER and SANTARIUS, 1970)

opposite changes in the chloroplasts and in the cytoplasm as well. With light to dark transition the reverse changes were observed: ATP concentrations decreased and ADP concentrations increased simultaneously in both cell compartments (Fig. 1).

Although these results clearly indicated a fast translocation of adenine nucleotides across the chloroplast envelope, this view had to be restricted in one point. Performing similar experiments with tobacco leaves, KEYS (1968) confirmed the conversion of ADP to ATP in chloroplasts and the non-chloroplast part of the cells upon illumination. However, at light to dark transition ATP was converted to AMP in the chloroplasts but not in the cytoplasm and AMP remained in the chloroplasts for 2 min in the dark. During a new illumination period AMP was converted to ATP in the chloroplasts. Accordingly AMP seemed not to be rapidly exchanged across the chloroplast envelope. If there is a discrimination between the three adenine nucleotides in respect to their translocation properties, free diffusion through the chloroplast envelope can be excluded and specific translocating systems for ADP and ATP have to be considered.

3.2 Direct Translocation of Adenine Nucleotides across the Chloroplast Envelope via Adenine Nucleotide Translocator

For more than a decade of biochemical work on chloroplasts the problem of isolating really intact chloroplasts able to carry out complete photosynthesis, including CO_2 fixation at appreciable rates, remained unsolved. A main difficulty was to keep the envelope membranes physiologically intact during isolation of the chloroplasts. When WALKER and his colleagues (WALKER, 1964, 1965; WALKER and HILL, 1967; COCKBURN et al., 1968) and JENSEN and BASSHAM (1966) succeeded in preparing chloroplasts which were able to reduce CO_2 at rates comparable to those observed in vivo, the way was open for re-investigation of energy transport across the envelope membranes of isolated chloroplasts. Using such intact isolated chloroplasts from spinach BASSHAM et al. (1968) studied the distribution of photosynthetic intermediates between chloroplasts and incubation medium by following ^{14}C and ^{32}P labeling of various metabolites in the pellet fraction and in the supernatant after centrifugation. In the absence of CO_2 in the light they found high ^{32}P activities of supernatant ATP and ADP. Upon the addition of bicarbonate ^{32}P activity of both ATP and ADP rapidly decreased in the supernatant. This result led the authors to conclude that ADP and ATP can freely diffuse from the chloroplasts and re-enter them when required for CO_2 reduction. Evidently this interpretation was highly improbable, since diffusion should largely deplete the chloroplasts of nucleotides during isolation and incubation. How then should such chloroplasts be able to reduce CO_2 at high rates? On the other hand, how should bicarbonate induce active uptake of ADP and ATP?

Isolated intact chloroplasts from *Acetabularia* retain about 0.1 μmol adenine nucleotides/mg chlorophyll, which are not removed by isotonic washing (STROTMANN and HELDT, 1969). HEBER and SANTARIUS (1970) showed that carefully isolated aqueous chloroplasts from spinach retain about 50% of their in situ adenine nucleotides.

A new approach to the problem of adenine nucleotide transport in isolated chloroplasts was untertaken by using ^{14}C-labeled nucleotides. First experiments in this context were carried out with chloroplasts from the siphonal green alga *Acetabularia* (STROTMANN and HELDT, 1969; STROTMANN and BERGER, 1969), which could be isolated with a high degree of purity and intactness by a very gentle method. Nucleotide translocation measurements were performed using the filtering centrifugation technique through silicon oil layers, which had been successfully applied in similar investigations on mitochondria (PFAFF, 1965).

Upon incubation with labeled adenylates, exogenous nucleotides were found to be taken up and exchanged against endogenous nucleotides; thus the total adenine nucleotide concentrations both in the chloroplasts and in the medium were kept on a constant level. ^{14}C-ATP pre-loaded chloroplasts released radioactivity when incubated with unlabeled nucleotides. The rate of ^{14}C release depended on the species and concentration of exogenous adenine nucleotides. The highest exchange rate was found with exogenous ATP and the lowest with AMP. These results were fully confirmed with isolated spinach chloroplasts (HELDT, 1969b).

These experiments showed that an adenine nucleotide translocator system similar to the mitochondrial adenine nucleotide translocase is operating in the chloroplast envelope membrane. However, the two systems differ in crucial respects. The chloroplast adenine nucleotide translocator is insensitive to atractyloside (HELDT, 1969b). In contrast to the mitochondrial carrier, the chloroplast system exhibits specifity for exogenous ATP. This led HELDT (1969b) to speculate that the major function of the translocator may be to supply ATP to the chloroplasts in the dark rather than to export ATP from the chloroplasts in the light.

Another problem in explaining the energy flow in vivo by operation of the adenine nucleotide translocator is the low capacity of the carrier. From their transient experiments HEBER and SANTARIUS (1970) calculated a rate of ATP export

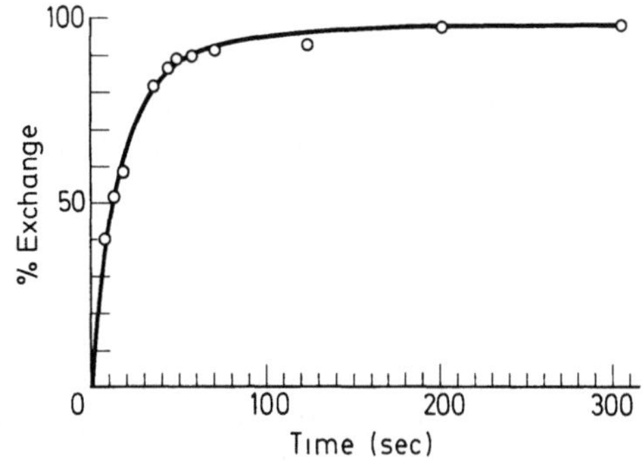

Fig. 2. Exchange of endogenous adenine nucleotides in spinach chloroplasts with ATP. First the adenine nucleotide pool of isolated chloroplasts was labeled by incubation for 60 min at 0° with ^{14}C-ATP (100 μM). After washing twice with isotonic medium, the labeled chloroplasts were incubated in the presence of unlabeled ATP for varying times at 20°. The rate of translocation was calculated in the following way: it is assumed that there is a simple exchange of a pool, which may be called the active adenine nucleotide pool, following a first order reaction. This pool may be smaller than the total amount of the ^{14}C-labeled endogenous adenine nucleotides. From the halftime of the exchange of this pool ($t_{1/2}=8$ s), the first order reaction constant (k) is obtained ($K=\ln 2/t_{1/2}$). Multiplication of k with the size of the active adenine nucleotide pool (A) yields the activity of translocation (v_t):

$$v_t = k \cdot A.$$

The value for A results from multiplication of the amount of adenine nucleotides assayed enzymatically (35.7 nmol/mg chlorophyll) by correcting factors for the extent of exchange during pre-labeling ($f_1=0.70$), for unspecific leakage of adenine nucleotides ($f_2=0.59$) and for the size of the rapidly exchangeable pool ($f_3=0.98$). Thus the active pool in this experiment amounts to $A=35.7\times0.70\times0.59\times0.98=14.4$ nmol/mg chlorophyll. From these data a translocation activity $v_t=4.5$ μmol/mg chlorophyll/h is calculated. (From HELDT, 1969b)

from the chloroplasts in the light between 7 and 9 μmol/mg chlorophyll/h. The maximum transport rate of spinach adenine nucleotide translocator is about half of this rate. HELDT (1969b) calculated a rate of 4.5 μmol/mg chlorophyll/h for exogenous ATP (see Fig. 2). In *Acetabularia* it was found to be one order of magnitude lower (STROTMANN and BERGER, 1969), probably due to the generally lower metabolic activity of this organism.

Corresponding to these results, addition of adenylates to intact chloroplasts had only a slight stimulating effect on the rate of CO_2 reduction (JENSEN and BASSHAM, 1968). Furthermore STOKES and WALKER (1971) reported that intact chloroplasts, which were poisoned by an uncoupler or energy transfer inhibitor, were no longer able to reduce exogenously added PGA in the light. This inhibition could not be released by ATP, whereas in a reconstituted envelope-free chloroplast system ATP was able to restore photoreduction of PGA.

3.3 Indirect Transport of Energy across the Chloroplast Envelope via Shuttle Systems

Looking for further possibilities of energy transfer between chloroplasts and cytoplasm, one has to take indirect mechanisms into consideration. An a priori improbable transmembrane phosphorylation mechanism could be ruled out by recent experiments performed in our laboratory (MATERN, 1975). Comparing translocation rates of ^{14}C- and ^{32}P-labeled ATP in spinach chloroplasts, MATERN could find no difference, although a considerably higher amount of ^{32}P than ^{14}C labeling was detected in the chloroplasts. However, it was clearly demonstrated that the excess ^{32}P in the chloroplasts was inorganic phosphate, probably formed by ATP hydrolysis outside and independently transported into the chloroplasts via the phosphate translocator described by HELDT and RAPLEY (1970b). By a quite similar effect stimulation of O_2 evolution of chloroplasts by ATP and ADP was explained (COCKBURN et al., 1967). These authors found a dependency of O_2 evolution on exogenous inorganic phosphate under certain conditions. Phosphate could be replaced to some extent by pyrophosphate, ATP and ADP, due to external hydrolysis of these compounds, which might be catalyzed by contamination with unspecific phosphatases.

As a further reasonable possibility, an energy transfer via transport metabolites has to be considered. Two main suppositions have to be fulfilled for such a mechanism to function: (a) an ATP-linked reaction with an appropriate equilibrium has to take place both in the chloroplast and in the cytoplasm, and (b) a substrate and a product of this reaction or a substrate/product couple of a preceding or consecutive reaction have to be able to traverse the chloroplast envelope.

A central ATP-linked reaction which is well known to be involved in photosynthesis and cytoplasmic glycolysis, is the phosphoglycerate kinase reaction. In this reaction ATP is consumed for phosphorylation of 3-phosphoglycerate (PGA), leading to the formation of 1,3-diphosphoglycerate (1,3-di-PGA) and ADP, and vice versa.

However, the substrate 1,3-di-PGA seems not to be a suitable transport metabolite, due to the limited permeability of the chloroplast envelope for this molecule and its low intracellular concentration. Accordingly HEBER (1975) could not find satisfactory recovery of PGA photoreduction in uncoupled intact chloroplasts by the addition of ATP and phosphoglycerate kinase to the external system.

From in vivo studies of dark-light transients in intact leaf cells, it is well known that levels of PGA decrease in the light and increase in the dark, both in the chloroplasts and in the cytoplasm (URBACH et al., 1965). Dihydroxyacetone phosphate (DHAP) exhibits parallel changes in the opposite direction (HEBER et al., 1967b). These results led to the conclusion that these two metabolites can rapidly exchange between chloroplasts and cytoplasm.

This view was confirmed by two independent lines of evidence derived from experiments on isolated chloroplasts. Intact isolated chloroplasts are capable of photoreducing exogenously added PGA (BAMBERGER and GIBBS, 1965; WALKER and HILL, 1967; STOCKING et al., 1969; HEBER and SANTARIUS, 1970) with rates up to 450 μmol/mg chlorophyll/h (HEBER, 1974). The secondary product of this reaction, DHAP, rapidly appears in the medium (BASSHAM et al., 1968).

A better understanding of the mechanism of PGA and DHAP transport through the chloroplast envelope was made possible by the discovery of the so-called phosphate translocator. HELDT and RAPLEY (1970b) found a rapid uptake of labeled phosphate and PGA by intact spinach chloroplasts. Uptake of each was competitively inhibited by the other and by DHAP; to a lesser extent also by glyceraldehydephosphate (GAP), α-glycerophosphate and phosphoenolpyruvate. Chloroplasts pre-loaded with ^{32}P-inorganic phosphate or ^{14}C-PGA released radioactivity upon the addition of these substrates in unlabeled form. Obviously phosphate, PGA and DHAP are transported in counter exchange by the same carrier located in the chloroplast envelope (see also Chap. II, 3).

If photosynthetically produced DHAP can exchange with the cytoplasmic PGA pool (URBACH et al., 1965), indirect ATP transport may be brought about by the following reaction complex:

$$\text{PGA} + \text{ATP} \rightleftharpoons \text{1,3-di-PGA} + \text{ADP} \quad (1)$$

$$\text{1,3-di-PGA} + \text{NAD(P)H} + \text{H}^+ \rightleftharpoons \text{GAP} + \text{NAD(P)}^+ + \text{Pi} \quad (2)$$

$$\text{GAP} \rightleftharpoons \text{DHAP} \quad (3)$$

Since this reaction complex can take place in the chloroplasts and in the cytoplasm, the exchange of cytoplasmic PGA with chloroplastic DHAP would lead to ATP consumption in the chloroplasts (in the forward direction) and ATP release in the cytoplasm (in the reverse direction). This results in net transport of energy from the chloroplasts to the cytoplasm. STOCKING and LARSON (1969), HEBER and SANTARIUS (1970) and KRAUSE (1971) presented evidence for the operation of this shuttle mechanism in vitro (see also Chap. II, 2.18).

However, this simple model which seems convincing at first sight includes considerable complications. One concerns the intracellular distribution of adenine nucleotides in vivo. In leaf cells the cytoplasmic ATP/ADP ratio is considerably higher than the ATP/ADP ratio in the chloroplasts, even in the light (KEYS and WHITTINGHAM, 1969; HEBER and SANTARIUS, 1970). Thus, ATP produced by photophosphorylation has to be transported against a gradient, when exported into the cytoplasm.

Following the interpretation given by HEBER (1975) such an uphill transport of energy may nevertheless be possible. Comparing the intracellular concentrations of ATP and ADP on the one hand and the levels of DHAP and PGA on the other hand, HEBER found that the gradient of DHAP/PGA ratios between cytoplasm and chloroplasts is still larger than the gradient of ATP/ADP ratios under equilibrium conditions in the light. This substrate imbalance is thought to facilitate net export of energy.

Another problem arises from the hydrogen balance of the DHAP/PGA shuttle. As can be seen from the reaction equations, translocation of ATP is accompanied by a stoichiometric transfer of reduced pyridine nucleotides in the same direction. Actually in a system containing intact isolated chloroplasts, PGA, ADP, phosphate, NAD^+, and the enzymes glyceraldehydephosphate dehydrogenase and phosphoglycerate kinase, light-dependent reduction of external NAD^+ was observed by STOCKING and LARSON (1969).

KRAUSE (1971) and HEBER and KRAUSE (1971) clearly demonstrated that indirect ATP export from the chloroplasts via DHAP/PGA shuttle is under the control of the external NAD^+/NADH ratio. A high NAD^+/NADH ratio promotes export, while a ratio of 1 stops it completely (Table 1).

Thus, for operation of the DHAP/PGA shuttle in vivo a mechanism is required by which cytoplasmic NADH is effectively re-oxidized. In intact cells the rapid

Table 1. Control of indirect ATP transport across the chloroplast envelope by the external NAD/NADH potential and stoichiometry between light dependent phosphate uptake (phosphorylation of ADP) and NAD reduction related to ATP transport in isolated intact spinach chloroplasts.

The incubation medium contained 3.5×10^{-3}M NAD+NADH at different ratios, 1.3×10^{-2}M ADP, 2×10^{-3}M PGA. (From KRAUSE, 1971)

NAD/NADH	NAD reduction (μmol/mg chl/h)	NAD reduction (μmol/mg chl/h)	Phosphate uptake (μmol/mg chl/h)
50	52	52	44
10	18	25	32
1	0	14	17

light-dependent transient changes in NADPH and NADH levels of the chloroplasts are not communicated to the cytoplasm (HEBER and SANTARIUS, 1965), indicating a strict compartmentation of the pyridine nucleotide pools. This was further confirmed by investigations on the permeability properties of the chloroplast envelope toward pyridine nucleotides (HEBER et al., 1967a; COCKBURN et al., 1967; ROBINSON and STOCKING, 1968). Consequently, a direct re-entrance of NADH into the chloroplasts can be excluded.

On the other hand cytoplasmic NADH is directly accessible to the mitochondrial respiratory chain, because plant mitochondria in contrast to mammalian mitochondria possess a NADH oxidase at the outer surface of the inner membrane, which enables them to oxidize exogenous NADH (HACKETT, 1961; BAKER and LIEBERMAN, 1962; CUNNINGHAM, 1964; IKUMA and BONNER, 1967; WILSON and HANSON, 1969; STOREY, 1970). However, HEBER (1975) believes that re-oxidation of cytoplasmic NADH by the respiratory chain is improbable under conditions of continuous photosynthesis, because respiration is inhibited by light (HOCH et al., 1963; OWENS and HOCH, 1963; RIED, 1968, 1969, 1972). The extent of inhibition is unclear. It is thought usually not to exceed 15% (RIED, pers. commun.), but complete inhibition has also been observed (CHEVALLIER and DOUCE, 1976).

A mechanism by which cytoplasmic NADH may be transferred back into the chloroplasts via the transport metabolites malate and oxaloacetate (OAA), was proposed by HEBER and KRAUSE (1971, 1972) and HEBER (1975). The chloroplast envelope contains a third translocator system, by which dicarboxylic acids are transported between the chloroplasts and their surroundings (HELDT and RAPLEY, 1970b). The dicarboxylate translocator facilitates the transfer of malate, succinate, α-ketoglutarate, fumarate, aspartate, glutamate, and oxaloacetate via counter exchange. Malate dehydrogenase which catalyzes the oxidation of malate to OAA with NAD^+ as the oxidant, was reported to be present in the cytoplasm of leaf cells as well as in the chloroplasts (HEBER and KRAUSE, 1971, 1972). Accordingly isolated chloroplasts are capable of photoreducing exogenously added OAA (HEBER and KRAUSE, 1971). In contrast to PGA photoreduction which is inhibited, OAA reduction is stimulated by uncoupling of photophosphorylation (HEBER and KRAUSE, 1972). This is easily understood, since OAA reduction does not require ATP.

Evidence for operation of the malate/OAA shuttle in vitro leading to hydrogen transfer across the chloroplast envelope was presented by HEBER and KRAUSE

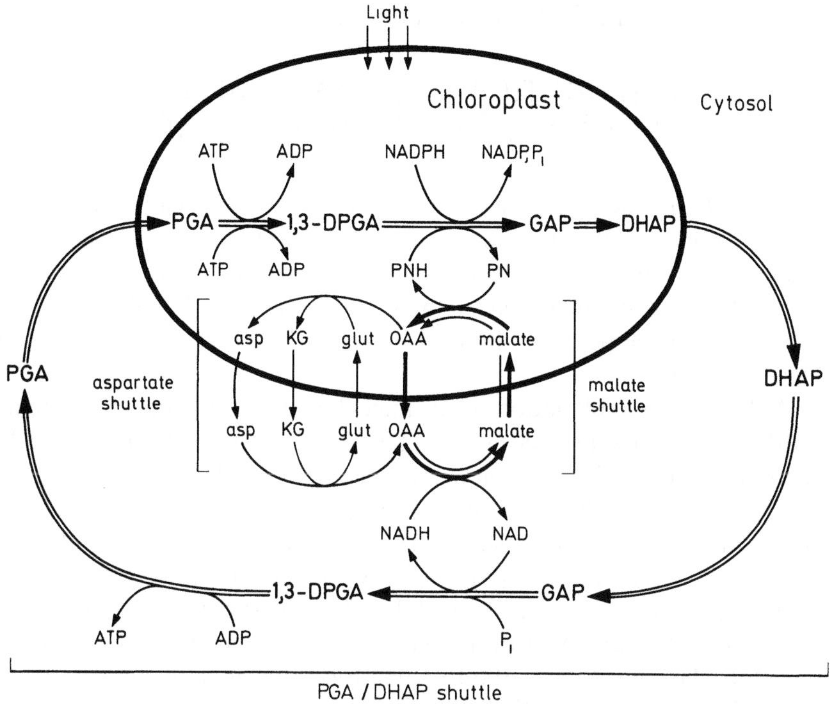

Fig. 3. Shuttle mechanisms leading to the transport of energy and hydrogen across the envelope of chloroplasts. (From HEBER, 1975)

(1971). In a chloroplast system containing malate, NAD^+ and malate dehydrogenase in the external medium, they could find light-dependent reduction of exogenous NAD^+. According to HEBER and KRAUSE's interpretation, external malate dehydrogenase adjusts the equilibrium between malate/OAA and NAD^+/NADH in the dark. OAA is transported into the chloroplasts via the dicarboxylate translocator and reduced to malate by photosynthetically generated NADH. The malate which again appears in the medium in exchange against OAA, is believed to shift the external malate dehydrogenase equilibrium toward the formation of NADH.

When PGA which acts as a hydrogen sink in the chloroplasts was also supplied to this system, an oxidation of NADH was observed in the medium instead of NADH formation. From these results HEBER and KRAUSE concluded that hydrogen may be transported via malate/OAA shuttle in both directions, depending on the physiological situation.

However, in explaining hydrogen re-cycling into the chloroplasts in vivo another thermodynamic difficulty has to be overcome. From studies in intact leaf cells it is known that pyridine nucleotides are much more reduced in the chloroplasts than in the cytoplasm, at least in the light (HEBER and SANTARIUS, 1965).

Accordingly a combined DHAP/PGA and malate/OAA shuttle system transporting ATP from the chloroplasts into the cytoplasm would require an uphill transfer of hydrogen into the chloroplasts. HEBER (1975) suggested that this may be forced by a proton gradient established between the chloroplast stroma and the cytoplasm in the light. The equilibrium expression of the malate/OAA shuttle is

$$\left(\frac{[\mathrm{OAA}]\cdot[\mathrm{PNH}]\cdot[\mathrm{H}^+]}{[\mathrm{malate}]\cdot[\mathrm{PN}^+]}\right)_{\mathrm{chl}} = \left(\frac{[\mathrm{OAA}]\cdot[\mathrm{NADH}]\cdot[\mathrm{H}^+]}{[\mathrm{malate}]\cdot[\mathrm{NAD}^+]}\right)_{\mathrm{cyt}} = \mathrm{const.}$$

(HEBER, 1975). The H^+ concentration of the chloroplast stroma is decreased in the light due to light dependent H^+ uptake into the thylakoid space, and this change in proton concentration is not communicated to the cytoplasm (HELDT et al., 1973). From these facts it was suggested that the pH gradient across the chloroplast envelope may support the uphill transfer of reducing equivalents (HEBER, 1975).

Although this explanation seems logical, a critical comment on the malate/OAA shuttle is necessary at this point. The equilibrium of the malate dehydrogenase reaction is far on the side of malate and NAD^+ formation (HOLLDORF, 1964). Accordingly intracellular OAA concentration is very low compared to malate concentration. In spinach leaves HEBER and KRAUSE (1972) determined 1 to 5×10^{-5} M OAA and 0.5 to 1×10^{-2} M malate. Following the assumption that the two substrates are transported by the dicarboxylate carrier in competition (HELDT and RAPLEY, 1970b), in vivo the probability of OAA translocation across the chloroplast envelope is 100 to 1,000 times lower than the transport of malate. The probability may be still lower, because OAA seems to exhibit a smaller affinity for the dicarboxylate carrier than malate (HELDT and RAPLEY, 1970b). Thus assuming a rather high translocation activity of the dicarboxylate carrier of 100 µmol/mg chlorophyll/h, the combined DHAP/PGA and malate/OAA shuttle system may transfer less than 5 µmol ATP/mg chlorophyll/h.

As another possibility of hydrogen transport, an aspartate/malate shuttle has been discussed by HEBER (1975). Since glutamate oxaloacetate transaminase is contained in both cytoplasm and chloroplasts (SANTARIUS and STOCKING, 1969), OAA formed in the chloroplasts may be converted into aspartate and α-ketoglutarate.

All the substrates involved in this process are exchangeable across the chloroplast envelope via the dicarboxylate translocator; furthermore the overall equilibrium of this system may be more favourable to effective translocation. Actually HEBER (1975) presented evidence that a malate/aspartate shuttle can bring about hydrogen transfer into the chloroplasts under certain conditions. The significance of this mechanism with respect to in vivo operation has still to be examined.

A shuttle mechanism discussed by KELLY and GIBBS (1973) leads to a one-way net transfer of reducing equivalents from the chloroplasts to the cytoplasm, but not the reverse. The transported substrate couple is DHAP and PGA. In contrast to the DHAP/PGA shuttle described above, in this system DHAP oxidation in the cytoplasm by $NADP^+$ is catalyzed by the non-phosphorylating glyceraldehydephosphate dehydrogenase, which occurs in this compartment.

4. Conclusion

The most surprising result in intracellular energy transport is the maintenance of different phosphorylation potentials in the cell compartments. It is lower in the mitochondria than in the cytoplasm; and a similar gradient exists between chloroplasts and the cytoplasm, although net energy flow occurs from the mitochondria and chloroplasts uphill to the cytoplasm. Probably this is one of the most important physiological meanings of cell compartmentation. The low phosphorylation potentials in mitochondria and chloroplasts enable the ATP-generating systems to operate at high capacities. On the other hand, the high phosphorylation potential in the cytoplasm provides favourable conditions for energy requiring reactions.

The gradient in phosphorylation potential across the mitochondrial membrane results from the differential specifity for cytoplasmic ADP and ATP of the mitochondrial adenine nucleotide translocase and is supported by respiratory energy. It is suggested that it is established—at least in part—by the charge imbalance between the two sides of the mitochondrial membrane, which is created by respiratory electron flow (KLINGENBERG et al., 1969). The resulting membrane potential, which is positive outside, inhibits the exchange of external ATP^{4-} and favors the exchange of exogenous ADP^{3-} against the internal ATP^{4-}. Since uncoupling dissipates charge separation, the abolition of ADP specifity by uncouplers is well understood in the scope of this interpretation.

The mechanism by which ATP is transported from the chloroplasts to the cytoplasm against an energy gradient is less clearly understood. If the driving force is a DHAP/PGA gradient as suggested by HEBER (1975) (see above), the problem is shifted to the question of how this gradient comes about. A critical judgment of the present situation leads to the conclusion, that at least two independent mechanisms might be involved in energy translocation from chloroplasts to cytoplasm: direct transport via adenylate translocator and indirect transport mediated by the DHAP/PGA shuttle. A not well-resolved point in operation of the shuttle is the removal of cytoplasmic reducing equivalents which are concomitantly transferred. Although refused by HEBER (1975), a regulation of the cytoplasmic redox level of pyridine nucleotides by the respiratory chain should be reconsidered, because the effectiveness of light inhibition of respiration may be overestimated.

Points against efficient involvement of the adenine nucleotide translocator system in energy transfer are its unfavorable specifity, i.e. its high affinity to external ATP compared to ADP and its low exchange capacity. However, the few experimental results available on the adenylate translocator do not permit a final judgment about its physiological significance. Recent experiments from our laboratory (MATERN, 1975) showed a drastic inhibition of ^{14}C-ATP uptake by spinach chloroplasts in the light, which might indicate a change in the properties of the carrier. In this context an interesting observation by STOCKING et al. (1969) should be mentioned. These authors found that exogenous ADP was able to stimulate PGA photoreduction by intact chloroplasts when added in a preceding dark period, but failed to do so when added in the light.

To have more precise understanding of energy transport across the chloroplast envelope, knowledge of the biochemical properties of the envelope membrane itself is required (see Chap. I). A promising way toward studying this problem was opened when isolation of envelope membranes was successfully achieved by MACKENDER and LEECH (1970, 1972), DOUCE et al. (1973), POINCELOT (1973) and HASHIMOTO and MURAKAMI (1975). It was established that the envelope membrane contains Mg^{2+} dependent ATPase activity (DOUCE et al., 1973; POINCELOT, 1973). Recently adenylate kinase activity was observed in the envelope membranes isolated from spinach chloroplasts (MURAKAMI and STROTMANN, unpublished). Thus, biochemical study of isolated chloroplast membranes, especially in respect to energy transport and metabolic regulation of chloroplast-cytoplasmic interaction, is coming to be an important project which should be pursued in the future.

References

BAKER, J.E., LIEBERMAN, M.: Cytochrome components and electron transfer in sweet potato mitochondria. Plant Physiol. **37**, 90–97 (1962)

BAMBERGER, E.S., GIBBS, M.: Effect of phosphorylated compounds and inhibitors on CO_2 fixation by intact spinach chloroplasts. Plant Physiol. **40**, 919–926 (1965)

BARBER, J.: Light-induced uptake of potassium and chloride by *Chlorella pyrenoidosa*. Nature **217**, 876–878 (1968)

BASSHAM, J.A., KIRK, M., JENSEN, R.G.: Photosynthesis by isolated chloroplasts. I. Diffusion of labelled photosynthetic intermediates between isolated chloroplasts and suspending medium. Biochim Biophys. Acta **153**, 211–218 (1968)

BERTAGNOLLI, B.L., HANSON, J.B.: Functioning of the adenine nucleotide transporter in the arsenate uncoupling of corn mitochondria. Plant Physiol. **52**, 431–435 (1973)

BIRD, I.F., CORNELIUS, M.J., DYER, T.A., KEYS, A.J.: The purity of chloroplasts isolated in non-aqueous media. J. Exptl. Botany. **24**, 211–215 (1973)

BOMSEL, J.-L., PRADET, A.: Étude des adénosine-5′-mono-, di- et tri-phosphates dans les tissus végétaux. II. Évolution in vivo de l'ATP, l'ADP et l'AMP dans les feuilles de blé en fonction de différentes conditions de milieu. Physiol. Vég. **5**, 223–236 (1967)

CHEVALLIER, D., DOUCE, R.: Interactions between mitochondria and chloroplasts in cells. I. Action of cyanide and of 3-(3,4-dichlorophenyl)-1,1-dimethylurea on the spore of *Funaria hygrometrica*. Plant Physiol., **57**, 400–402 (1976)

COCKBURN, W., BALDRY, C.W., WALKER, D.A.: Some effects of inorganic phosphate on O_2 evolution by isolated chloroplasts. Biochim. Biophys. Acta **143**, 614–624 (1967)

COCKBURN, W., WALKER, D.A., BALDRY, C.W.: The isolation of spinach chloroplasts in pyrophosphate medium. Plant Physiol. **43**, 1415–1418 (1968)

CUNNINGHAM, W.P.: Oxidation of externally added NADH by isolated corn root mitochondria. Plant Physiol. **39**, 699–703 (1964)

DECKER, M., TANNER, W.: Respiratory increase and active hexose uptake of *Chlorella vulgaris*. Biochim. Biophys. Acta **266**, 661–669 (1972)

DOUCE, R., HOLTZ, R.B., BENSON, A.A.: Isolation and properties of the envelope of spinach chloroplasts. J. Biol. Chem. **248**, 7215–7222 (1973)

ELBERS, R., HELDT, H.W., SCHMUCKER, P., SOBOLL, S., WIESE, H.: Measurement of the ATP/ADP ratio in mitochondria and in the extramitochondrial compartment by fractionation of freeze-stopped liver tissue in nonaqueous media. Hoppe-Seylers Z. Physiol. Chem. **355**, 378–393 (1974)

HACKETT, D.P.: Effects of salts on DPNH oxidase activity and structure of sweet potato mitochondria. Plant Physiol. **36**, 445–452 (1961)

HASHIMOTO, H., MURAKAMI, S.: Dual character of lipid composition of the envelope membrane of spinach chloroplasts. Plant and Cell Phys. **16**, 895–902 (1975)

HEBER, U.: Flow of metabolites and compartmentation phenomena in chloroplasts. In: Transport and Distribution of Matter in Cells of Higher Plants (K. MOTHES, E. MÜLLER, A. NELLES, D. NEUMANN, eds.), pp. 151–180. Berlin: Akademie-Verlag 1970

HEBER, U.: Stoichiometry of reduction and phosphorylation during illumination of intact chloroplasts. Biochim. Biophys. Acta **305**, 140–152 (1973)

HEBER, U.: Metabolite exchange between chloroplasts and cytoplasm. Ann. Rev. Plant Physiol. **25**, 393–421 (1974)

HEBER, U.: Energy transfer within leaf cells. In: Proceedings of the 3rd Intern. Congr. Photosynthesis Res. (M. AVRON, ed.) Vol. II, pp. 1335–1348. Amsterdam: Elsevier 1975

HEBER, U., HALLIER, U.W., HUDSON, M.A.: Untersuchungen zur intrazellulären Verteilung von Enzymen und Substraten in der Blattzelle. II. Lokalisation von Enzymen des reduktiven und des oxydativen Pentosephosphat-Zyklus in den Chloroplasten und Permeabilität der Chloroplasten-Membran gegenüber Metaboliten. Z. Naturforsch. **22b**, 1200–1215 (1967a)

HEBER, U., KIRK, M.R.: Flexibility of coupling and stoichiometry of ATP formation in intact chloroplasts. Biochim. Biophys. Acta **376**, 136–150 (1975)

HEBER, U., KRAUSE, G.H.: Transfer of carbon, phosphate energy, and reducing equivalents across the chloroplast envelope. In: Photosynthesis and Photorespiration (M.D. HATCH, C.B. OSMOND, R.O. SLATYER, eds.), pp. 218–225. New York-London-Sydney-Toronto: Wiley Interscience 1971

HEBER, U., KRAUSE, G.H.: Hydrogen and proton transfer across the chloroplast envelope. In: Proc. the 2nd Intern. Congr. Photosynthesis Res. (G. FORTI, M. AVRON, A. MELANDRI, eds.), pp. 1023–1033. The Hague: Junk 1972

HEBER, U., SANTARIUS, K.A.: Compartmentation and reduction of pyridine nucleotides in relation to photosynthesis. Biochim. Biophys. Acta **109**, 390–408 (1965)

HEBER, U., SANTARIUS, K.A.: Direct and indirect transfer of ATP and ADP across the chloroplast envelope. Z. Naturforsch. **25b**, 718–728 (1970)

HEBER, U., SANTARIUS, K.A., HUDSON, M.A., HALLIER, U.W.: Untersuchungen zur intrazellulären Verteilung von Enzymen und Substraten in der Blattzelle. I. Intrazellulärer Transport von Zwischenprodukten der Photosynthese im Photosynthese-Gleichgewicht und im Dunkel-Licht-Dunkel Wechsel. Z. Naturforsch. **22b**, 1189–1199 (1967b)

HEBER, U., SANTARIUS, K.A., URBACH, W., ULLRICH, W.: Photosynthese und Phosphathaushalt. Intrazellulärer Transport von ^{14}C- und ^{32}P-markierten Intermediärprodukten zwischen den Chloroplasten und dem Cytoplasma und seine Folgen für die Regulation des Stoffwechsels. Z. Naturforsch. **19b**, 576–587 (1964)

HELDT, H.W.: The inhibition of adenine nucleotide translocation by atractyloside. In: Inhibitors, Tools in Cell Research. 20. Mosbach Coll. (TH. BÜCHER, H. SIES, eds.), pp. 301–317. Berlin-Heidelberg-New York: Springer 1969a

HELDT, H.W.: Adenine nucleotide translocation in spinach chloroplasts. FEBS Letters **5**, 11–14 (1969b)

HELDT, H.W. JACOBS, H., KLINGENBERG, M.: Endogenous ADP of mitochondria, an early phosphate acceptor of oxidative phosphorylation as disclosed by kinetic studies with ^{14}C-labelled ADP and ATP and with atractyloside. Biochem. Biophys. Res. Commun. **18**, 174–179 (1965)

HELDT, H.W., RAPLEY, L.: Unspecific permeation and specific uptake of substances in spinach chloroplasts. FEBS Letters **7**, 139–142 (1970a)

HELDT, H.W., RAPLEY, L.: Specific transport of inorganic phosphate, 3-phosphoglycerate and dihydroxyacetone-phosphate, and of dicarboxylates across the inner membrane of spinach chloroplasts. FEBS Letters **10**, 143–148 (1970b)

HELDT, H.W., SAUER, F.: The inner membrane of the chloroplast envelope as the site of specific metabolite transport. Biochim. Biophys. Acta **234**, 83–91 (1971)

HELDT, H.W., WERDAN, K., MILOVANCEV, M., GELLER, G.: Alkalization of the chloroplast stroma caused by light-dependent proton flux into the thylakoid space. Biochim. Biophys. Acta **314**, 224–241 (1973)

HENDERSON, P.J.F., LARDY, H A.· Bongkrekic acid, an inhibitor of the adenine nucleotide translocase of mitochondria. J. Biol. Chem. **245**, 1319–1326 (1970)

HOCH, G., OWENS, O.v.H., KOK, B.: Photosynthesis and respiration. Arch. Biochem. Biophys. **101**, 171–180 (1963)

HOLLDORF, A.W.: Gleichgewichtskonstanten, Freie Energien und Redoxpotentiale biologisch wichtiger Reaktionen. In: Biochemisches Taschenbuch (H.M. RAUEN, ed.), pp. 121–150. Berlin-Göttingen-Heidelberg-New York: Springer 1964

HOPE, A B., LÜTTGE, U., BALL, E.: Chloride uptake in strains of *Scenedesmus obliquus*. Z. Pflanzenphysiol. **72**, 1–10 (1974)

IKUMA, H., BONNER, W.D., JR.: Properties of higher plant mitochondria. I. Isolation of some characteristics of tightly-coupled mitochondria from dark-grown mung bean hypocotyls. Plant Physiol. **42**, 67–75 (1967)

IZAWA, S., GOOD, N.E.: The stoichiometric relation of phosphorylation to electron transport in isolated chloroplasts. Biochim. Biophys. Acta **162**, 380–391 (1968)

JENSEN, R.G., BASSHAM, J.A.: Photosynthesis by isolated chloroplasts. Proc. Natl. Acad. Sci. US **56**, 1095–1101 (1966)

JENSEN, R.G., BASSHAM, J.A.: Photosynthesis by isolated chloroplasts. II. Effect of addition of cofactors and intermediate compounds. Biochim. Biophys. Acta **153**, 219–226 (1968)

JESCHKE, W.D., SIMONIS, W.: Über die Wirkung von CO_2 auf die lichtabhängige Cl^--Aufnahme bei *Elodea densa*: Regulation zwischen nichtcyclischer und cyclischer Photophosphorylierung. Planta **88**, 157–171 (1969)

JUNG, D.W., HANSON, J.B.: Atractyloside inhibition of adenine nucleotide transport in mitochondria from plants. Biochim. Biophys. Acta **325**, 189–192 (1973)

KANDLER, O.: Über die Beziehungen zwischen Phosphathaushalt und Photosynthese. II. Gesteigerter Glucoseeinbau im Licht als Indikator einer lichtabhängigen Phosphorylierung. Z. Naturforsch. **9b**, 625–644 (1954)

KANDLER, O., HABERER-LIESENKÖTTER, I.: Über den Zusammenhang zwischen Phosphathaushalt und Photosynthese. V. Regulation der Glykolyse durch die Lichtphosphorylierung bei Chlorella. Z. Naturforsch. **18b**, 718–730 (1963)

KELLY, G.J., GIBBS, M.: A mechanism for the indirect transfer of photosynthetically reduced nicotinamide adenine dinucleotide phosphate from chloroplasts to the cytoplasm. Plant Physiol. **52**, 674–676 (1973)

KEYS, A.J.: The intracellular distribution of free nucleotides in the tobacco leaf. Formation of adenosine 5′-phosphate from adenosine 5′-triphosphate in the chloroplasts. Biochem. J. **108**, 1–8 (1968)

KEYS, A.J., WHITTINGHAM, C.P.: Nucleotide metabolism in chloroplast and non-chloroplast components of tobacco leaves. In: Progress in Photosynthesis Reseaarch (H. METZNER, ed.), Vol. I, pp. 352–358. Tübingen: Laupp 1969

KLINGENBERG, M., BUCHHOLZ, M.: On the mechanism of bongkrekate effect on the mitochondrial adenine nucleotide carrier as studied through the binding of ADP. Europ. J. Biochem. **38**, 346–358 (1973)

KLINGENBERG, M., HELDT, H.W., PFAFF, E.: The role of adenine nucleotide translocation in the generation of phosphorylation energy. In: The Energy Level and Metabolic Control in Mitochondria (S. PAPA, J.M. TAGER, E. QUAGLIARIELLO, E.C. SLATER, eds.), pp. 237–253. Bari: Adriatica Editrice 1969

KRAUSE, G.H.: Indirekter ATP-Transport zwischen Chloroplasten und Zytoplasma während der Photosynthese. Z. Pflanzenphysiol. **65**, 13–23 (1971)

LILLEY, R.MCC., HOPE, A.B.: Chloride transport and photosynthesis in cells of *Griffithsia*. Biochim. Biophys. Acta **226**, 161–171 (1971)

MACKENDER, R.O., LEECH, R.M.: Isolation of chloroplast envelope membranes. Nature **228**, 1347–1349 (1970)

MACKENDER, R.O., LEECH, R.M.: The isolation and characterization of plastid envelope membranes. In: Proc. of the 2nd Intern. Congr. Photosynthesis Res. (G. FORTI, M. AVRON, A. MELANDRI, eds.), pp. 1431–1440. The Hague: Junk 1972

MACROBBIE, E.A.C.: The nature of the coupling between light energy and active ion transport in *Nitella translucens*. Biochim. Biophys. Acta **94**, 64–73 (1965)

MACROBBIE, E.A.C.: Metabolic effects on ion fluxes in *Nitella translucens*. I. Active influxes. Aust. J. Biol. Sci. **19**, 363–370 (1966)

MATERN, A.: Adeninnukleotid-Transport über die Chloroplasten-Hüllmembran Staatsexamensarbeit Hannover 1975

ONISHI, T., KRÖGER, A., HELDT, H.W., PFAFF, E., KLINGENBERG, M.: The response of the respiratory chain and adenine nucleotide system to oxidative phosphorylation in yeast mitochondria. Europ. J. Biochem. **1**, 301–311 (1967)

OWENS, O.V.H., HOCH, G.: Enhancement and de-enhancement effect in *Anacystis nidulans*. Biochim. Biophys. Acta **75**, 183–186 (1963)

PASSHAM, H.C., SOUVERIJN, J.H.M., KEMP, A., JR.: Adenine nucleotide translocation in Jerusalem-artichoke mitochondria. Biochim. Biophys. Acta **305**, 88–94 (1973)

PEDERSEN, T.A., KIRK, M., BASSHAM, J.A.: Light-dark transients in levels of intermediate compounds during photosynthesis in air adapted Chlorella. Physiol. Plantarum **19**, 219–231 (1966)
PFAFF, E.: Unspezifische Permeabilität und spezifischer Austausch der Adeninnukleotide als Beispiel mitochondrialer Compartmentierung. Diss. Marburg 1965
PFAFF, E., HELDT, H.W., KLINGENBERG, M.: Adenine nucleotide translocation of mitochondria. Kinetics of adenine nucleotide exchange. Europ. J. Biochem. **10**, 484–493 (1969)
PFAFF, E., KLINGENBERG, M.: Adenine nucleotide translocation of mitochondria. 1. Specifity and control. Europ. J. Biochem. **6**, 66–79 (1968)
PFAFF, E., KLINGENBERG, M., HELDT, H.W.: Unspecific permeation and specific exchange of adenine nucleotides in liver mitochondria. Biochim. Biophys. Acta **104**, 312–315 (1965)
POINCELOT, R.P.: Isolation and lipid composition of spinach chloroplast envelope membranes. Arch. Biochem. Biophys. **159**, 134–142 (1973)
RAVEN, J.A.: Light-stimulation of active ion transport in *Hydrodictyon africanum*. J. Gen. Physiol. **50**, 1627–1640 (1967)
RAVEN, J.A.: Action spectra for photosynthesis and light stimulated ion transport in *Hydrodictyon africanum*. New Phytol. **68**, 45–62 (1969)
RAVEN, J.A.: Cyclic and non-cyclic photophosphorylation as energy sources for active K influx in *Hydrodictyon africanum*. J. Exptl. Botany **22**, 420–433 (1971)
RAVEN, J.A., MACROBBIE, E.A.C., NEUMANN, J.: The effect of Dio-9 on photosynthesis and ion transport in *Nitella, Tolypella* and *Hydrodictyon*. J. Exptl. Botany **20**, 221–235 (1969)
REEVES, S.G., HALL, D.O.: The stoichiometry ($ATP/2e^-$ ratio) of non-cyclic photophosphorylation in isolated spinach chloroplasts. Biochim. Biophys. Acta **314**, 66–78 (1973)
RIED, A.: Interactions between photosynthesis and respiration in *Chlorella*. I. Types of transients of oxygen exchange after short light exposures. Biochim. Biophys. Acta **153**, 653–663 (1968)
RIED, A.: Studies on light-dark transients in *Chlorella*. In: Progress in Photosynthesis Research (H. METZNER, ed.), Vol. I, pp. 521–530. Tübingen: Laupp 1969
RIED, A.: Improved action spectra of light reaction I and II. In: Proc. 2nd Intern. Congr. Photosynthesis Res. (G. FORTI, M. AVRON, A. MELANDRI, eds.), pp. 763–772. The Hague: Junk 1972
ROBINSON, J.M., STOCKING, C.R.: Oxygen evolution and the permeability of the outer envelope of isolated whole chloroplasts. Plant Physiol. **43**, 1597–1604 (1968)
SANTARIUS, K.A., HEBER, U.: Changes in intracellular levels of ATP, ADP, AMP and Pi and regulatory function of the adenylate system in leaf cells during photosynthesis. Biochim. Biophys. Acta **102**, 39–54 (1965)
SANTARIUS, K.A., HEBER, U., ULLRICH, W., URBACH, W.: Intracellular translocation of ATP, ADP and inorganic phosphate in leaf cells of *Elodea densa* in relation to photosynthesis. Biochem. Biophys. Res. Commun. **15**, 139–146 (1964)
SANTARIUS, K.A., STOCKING, C.R.: Intracellular localization of enzymes in leaves and chloroplast membrane permeability to compounds involved in amino acid syntheses. Z. Naturforsch. **24b**, 1170–1179 (1969)
SIMONIS, W., URBACH, W.: Photophosphorylation in vivo. Ann. Rev. Plant Physiol. **24**, 89–114 (1973)
SMITH, F.A.: Metabolic effects on ion fluxes in *Tolypella intricata*. J. Exptl. Botany **19**, 442–451 (1968)
STOCKING, C.R., LARSON, S.: A chloroplast cytoplasmic shuttle and the reduction of extraplastid NAD. Biochem. Biophys. Res. Commun. **37**, 278–282 (1969)
STOCKING, C.R., ROBINSON, J.M., WEIER, T.E.: Interrelationships between the chloroplast and its cellular environment. In: Progress in Photosynthesis Research (H. METZNER, ed.), Vol. I, pp. 258–266. Tübingen: Laupp 1969
STOKES, D.M., WALKER, D.A.: Relative impermeability of the intact chloroplast envelope to ATP. In: Photosynthesis and Photorespiration (M.D. HATCH, C.B. OSMOND, R.O. SLATYER, eds.), pp. 226–231. New York-London-Sydney-Toronto: Wiley-Interscience 1971
STOREY, B.T.: The respiratory chain of plant mitochondria. VIII. Reduction kinetics of the respiratory chain carriers of mung bean mitochondria with reduced nicotinamide adenine dinucleotide. Plant Physiol. **46**, 625–630 (1970)
STROTMANN, H., BERGER, S.: Adenine nucleotide translocation across the membrane of isolated *Acetabularia* chloroplasts. Biochem. Biophys. Res. Commun. **35**, 20–26 (1969)

STROTMANN, H., HELDT, H.W.: Phosphate containing metabolites participating in photosynthetic reactions of *Chlorella pyrenoidosa*. In: Progress in Photosynthesis Research (H. METZNER, ed.), Vol. III, pp. 1131–1140. Tübingen: Laupp 1969

TANNER, W.: Light-driven active uptake of 3-O-methylglucose via an inducible hexose uptake system of *Chlorella*. Biochem. Biophys. Res. Commun. **36**, 278–283 (1969)

TANNER, W., KANDLER, O.: The lack of relationship between cyclic photophosphorylation and photosynthetic CO_2 fixation. In: Progress in Photosynthesis Research (H. METZNER, ed.), Vol. III, pp. 1217–1223. Tübingen: Laupp 1969

TANNER, W., LOOS, E., KLOB, W., KANDLER, O.: The quantum requirement for light dependent anaerobic glucose assimilation by *Chlorella vulgaris*. Z. Pflanzenphysiol. **59**, 301–303 (1968)

THALACKER, R., BEHRENS, M.: Über den Reinheitsgrad der in einem nichtwäßrigen spezifischen Gewichtsgradienten gewonnenen Chloroplasten. Z. Naturforsch. **14b**, 443–446 (1959)

ULLRICH, W., URBACH, W., SANTARIUS, K.A., HEBER, U.: Die Verteilung des Orthophosphates auf Plastiden, Cytoplasma und Vakuole in der Blattzelle und ihre Veränderung im Licht-Dunkel-Wechsel. Z. Naturforsch. **20b**, 905–910 (1965)

URBACH, W., HUDSON, M.A., ULLRICH, W., SANTARIUS, K.A., HEBER, U.: Verteilung und Wanderung von Phosphoglycerat zwischen den Chloroplasten und dem Cytoplasma während der Photosynthese. Z. Naturforsch. **20b**, 890–898 (1965)

WALKER, D.A.: Improved rates of carbon dioxide fixation by illuminated chloroplasts. Biochem. J. **92**, 22c–23c (1964)

WALKER, D.A.: Correlation between photosynthetic activity and membrane integrity in isolated pea chloroplasts. Plant Physiol. **40**, 1157–1161 (1965)

WALKER, D.A., HILL, R.: The relation of oxygen evolution to carbon assimilation with isolated chloroplasts. Biochim. Biophys. Acta **131**, 330–338 (1967)

WEIGL, J.: Beweis für die Beteiligung von beweglichen Transportstrukturen (Trägern) beim Ionentransport durch pflanzliche Membranen und die Kinetik des Anionentransports bei *Elodea* im Licht und Dunkeln. Planta **75**, 327–342 (1967)

WILSON, R.H., HANSON, J.B.: The effect of respiratory inhibitors on NADH, succinate and malate oxidation in corn mitochondria. Plant Physiol. **44**, 1335–1341 (1969)

IV. Theory of Membrane Transport

Mass Transport across Membranes

D. WOERMANN

It is the aim of this article to give an introduction to some of the basic concepts used in the theoretic treatment of transport processes across membranes. Section 1 is devoted to the thermodynamic treatment of membrane transport close to equilibrium in which the membrane is treated as a "black box". The information about processes taking place within the membrane in its stationary state is derived from measurements of state parameters of its bulk phases as function of time. In order to gain insights into the physical basis of observed transport properties it is necessary to construct a suitable model of the membrane which is amenable to theoretical analysis. In Section 2 a brief review of the major membrane models is given and the model of the membrane with narrow pores is discussed in detail.

The article is not intended as literature survey. Stress is placed on the understanding of the theoretical concepts on which the treatment of transport processes across membranes is based. The author hopes that the reader will not be unduly discouraged by the perhaps "difficult" appearance of this paper. The material is presented in such a way that the article should be suitable for self-study.

1. Nonequilibrium Thermodynamics of Membrane Transport

1.1 Introduction

In Section 1 of this article a brief presentation is given of nonequilibrium thermodynamics of discontinuous systems applied to transport processes across membranes in the steady state. The considerations are restricted to systems in which only small deviations from equilibrium occur. The presentation is based on the treatment of membrane transport developed by KEDEM and KATCHALSKY (1958, 1961, 1963a, b, c) and is similar to the approach used by SAUER (1973). For the general theory of nonequilibrium thermodynamics the reader is referred to a number of monographs covering that subject (e.g. DE GROOT and MAZUR, 1962, 1974; KATCHALSKY and CURRAN, 1965; HAASE, 1969).

Brief introduction to the principles of irreversible thermodynamics are given e.g. by DAINTY (1963), HOPE (1971) and NOBEL (1974), see also DAINTY (1976) and WALKER (1976).

1.2 Transport of Nonelectrolytes

1.2.1 Description of the Membrane System

Let us consider a closed isothermal system in which a membrane separates two homogeneous liquid phases of different composition. They are called bulk phases of the membrane and are marked by the symbols (′) and (″). The bulk phases are formed by uncharged species labeled by the index $i=0, 1, 2, \ldots, n$. The composition of the bulk phases is specified by the values of the mole fraction x_i of the $n+1$ species.

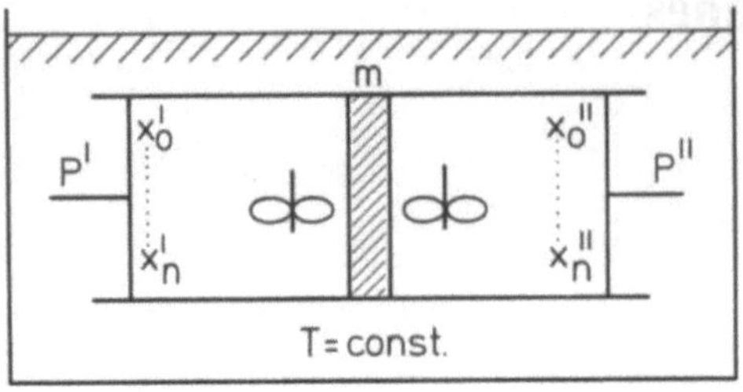

Fig. 1. Membrane system schematically. m=membrane; P=pressure; x_i=mole fraction of species i; $i=0, 1, \ldots, n$; T=thermodynamic temperature. Quantities refering to the left bulk phase are characterized by the superscript (′) and that refering to the right bulk phase by the superscript (″). ⏀=mechanical stirrer; ⊣⊢=movable pistons to maintain a pressure difference between the bulk phases

All species can permeate the membrane and are present in both bulk phases. There can exist a pressure difference between the bulk phases. Chemical reactions are excluded.

The whole system is immersed in a constant temperature bath to maintain isothermal conditions. The homogeneity of the bulk phases is maintained by mechanical stirring. The unstirred layers at the two interfaces membrane/bulk phase are considered part of the membrane. The membrane is maintained in a stationary state in which the amount of each permeating species entering the membrane per unit time across one of the interphases membrane/bulk phase is equal to the amount of the same species leaving it across the other interphase. This is accomplished by choosing the volume of the bulk phases much bigger than that of the membrane. Figure 1 shows the membrane system schematically.

1.2.2 Entropy Balance

Each bulk phase of the membrane is an open phase because it can exchange heat, work and matter with its surroundings. The exchange of matter takes place across the membrane. The entropy of each bulk phase depends on all the variables describing the internal state of the phase, such as internal energy, volume and amount of substance of each species present. The generalized GIBBS equation Eq. (1) can be used to describe the change of entropy of each phase with time if it is assumed that each phase is in internal equilibrium although the whole system is not in equilibrium. (A phase is in internal equilibrium if no changes of state of the phase take place as long as it does not come into contact with other phases.)

$$\frac{dS^\alpha}{dt} = \frac{1}{T}\frac{dU^\alpha}{dt} + \frac{P^\alpha}{T}\frac{dV^\alpha}{dt} - \sum_{i=0}^{n}\frac{\mu_i^\alpha}{T}\frac{dn_i^\alpha}{dt}. \tag{1}$$

S=entropy; U=internal energy; V=volume; μ_i=chemical potential of species i; T=thermodynamic temperature; n_i=amount of substance of species i (mol); t=time; α=phase index; α=(′), (″). For a review of the laws of equilibrium thermodynamics on which nonequilibrium thermodynamics is based the first three chapters of KATCHALSKY and CURRAN (1965) may be consulted.

The stationary state of the membrane is characterized by the vanishing of the time derivatives of all extensive quantities of the membrane:

$$\frac{dS^{\mathrm{m}}}{dt}=0, \quad \frac{dU^{\mathrm{m}}}{dt}=0, \quad \frac{dV^{\mathrm{m}}}{dt}=0, \quad \frac{dn_i^{\mathrm{m}}}{dt}=0. \tag{2}$$

The change of entropy with time of the whole membrane system in the stationary state of the membrane is given by:

$$\frac{dS}{dt}=\frac{dS'}{dt}+\frac{dS''}{dt}. \tag{3}$$

The terms dS'/dt and dS''/dt are described by Eq. (1).

1.2.3 Energy Balance

According to the first law of thermodynamics the change of internal energy with time of a closed phase which can exchange heat and work with its surroundings is given by:

$$\frac{dU}{dt}=\frac{dQ}{dt}-P\frac{dV}{dt}. \tag{4}$$

The term dQ/dt describes the exchange of heat between the closed phase and its surroundings. The term $-P\,dV/dt$ takes into account the exchange of quasistatic work of compression or expansion. Since the bulk phases of the membrane are open phases Eq. (4) cannot be applied to them without modifications. The change of internal energy with time of the open phases (′) and (″) can be expressed in the form:

$$\frac{dU^{\alpha}}{dt}=\frac{dQ^{\alpha}}{dt}-P^{\alpha}\frac{dV^{\alpha}}{dt}+J_U^{\alpha} \quad \text{with} \quad \alpha=('),('').\tag{5}$$

J_U^{α} represents the energy flow into phase α caused by the exchange of matter between the bulk phases across the membrane. Eq. (5) has to be looked upon as the equation of definition of the energy flow J_U^{α} (SAUER, 1973). In the stationary state of the membrane the energy flows J_U' and J_U'' are related to one another by the energy balance equation. Since the whole membrane system is a closed system the energy balance is given by:

$$\frac{dU'}{dt}+\frac{dU''}{dt}=\frac{dQ'}{dt}+\frac{dQ''}{dt}-P'\frac{dV'}{dt}-P''\frac{dV''}{dt}. \tag{6}$$

Eq. (6) is valid if it is assumed that the membrane is in its stationary state and that changes of kinetic energy of the bulk phases can be neglected. Changes of kinetic energy of the bulk phases have to be taken into account if the exchange of matter between the bulk phases causes high volume flows across the membrane.

In the stationary state of the membrane the energy flows J_U' and J_U'' are related by Eq. (7) which follows from Eqs. (5) and (6).

$$J_U'+J_U''=0. \tag{7}$$

1.2.4 Mass Balance

In the nonequilibrium situation of the membrane system the composition of the bulk phases will change with time due to flow of matter across the membrane. The flow of species i into phase (′) and phase (″) respectively is defined by

$$J_i' \equiv \frac{dn_i'}{dt} \qquad J_i'' \equiv \frac{dn_i''}{dt} \qquad i=0,1,2,\ldots,n. \tag{8}$$

In the stationary state of the membrane the flows J_i' and J_i'' are not independent of each other but are related by:

$$J_i' + J_i'' = 0. \tag{9}$$

The flows defined by Eq. (8) refer to the fixed membrane. In Section 1.2.7 a new set of flows will be introduced referring to another frame of reference.

1.2.5 Entropy Production

The expression Eq. (3) describing the change of entropy with time of the whole system in the stationary state of the membrane can be rewritten combining the Eqs. (1), (3), (5) and the Eqs. (7)–(9):

$$\frac{dS}{dt} - \frac{1}{T}\frac{dQ'}{dt} - \frac{1}{T}\frac{dQ''}{dt} = \sum_{i=0}^{n} J_i'' \frac{\Delta\mu_i}{T}. \tag{10}$$

The difference of the chemical potential between phase (′) and phase (″) is symbolized by Δ ($\Delta = ' - ''$).

Looking at Eq. (10) as a balance equation the right hand side of this equation can be interpreted as the entropy production $d_i S/dt$ caused by irreversible processes taking place inside the membrane. The terms $1/T \cdot dQ^\alpha/dt$ on the left hand side describe entropy changes due to reversible heat exchanges between the bulk phases and the temperature bath. Entropy production is expressed by a sum of products of flows and forces. The flows J_i'' characterize the rate of processes whereas the forces $\Delta\mu_i/T$ describe deviations from the equilibrium state of the membrane system. The flow and the force forming one term of the sum in the expression of the entropy production are called conjugate.

The second law of thermodynamics demands that entropy production be positive in the nonequilibrium situation of the membrane system and become zero when the system reaches its equilibrium state.

$$T\frac{d_i S}{dt} \equiv \sum_{i=0}^{n} J_i'' X_i \geq 0 \quad \text{with} \quad X_i \equiv \Delta\mu_i. \tag{11}$$

$d_i S/dt$ = entropy production; X_i = generalized force.

1.2.6 Linear Laws

The dependence of the flows on the thermodynamic state of the bulk phases of the membrane will now be discussed. It can be shown (SAUER, 1973) that in general the flows J_i'' depend not only on all the forces X_j but also on an "arbitrary" reference state which is assumed to be a possible equilibrium state of the system.

$$J_i'' = J_i''(r, X_0, X_1, \ldots, X_j, \ldots, X_n) \quad (i=0,1,2,\ldots,n)$$
$$\text{(nonequilibrium state).} \tag{12}$$

r symbolizes the set of independent intensive variables characterizing the thermodynamic state of the reference state. In the equilibrium state of the membrane system the flows vanish.

$$J_i'' = J_i''(r, X_0=0, \ldots, X_j=0, \ldots, X_n=0)=0 \quad (i=0,1,2,\ldots,n)$$
$$\text{(equilibrium state).} \tag{13}$$

In order to describe the dependence of the flows on the forces for a given reference state the function given by Eq. (12) is expanded into a power series (TAYLOR series of a function of many variables) around the equilibrium (eqm). For sufficiently small values of X_j the series can be truncated after the first order terms.

$$J_i'' = \sum_{j=0}^{n} L_{ij}(r)\, X_j \quad (i=0,1,2,\ldots,n)$$
$$\text{(linear laws)} \tag{14}$$

with

$$L_{ij} = \left(\frac{\partial J_i''}{\partial X_j}\right)_{r,\,\text{eqm.}}$$

The zero order terms do not appear in the TAYLOR series because the flows vanish in the equilibrium state.

The Eqs. (14) represent the linear laws of nonequilibrium thermodynamics. The range of the forces X_j for which the linear laws are a good approximation has to be determined experimentally for each force separately. The L-coefficients introduced by these equations are called phenomenological coefficients. They can be written

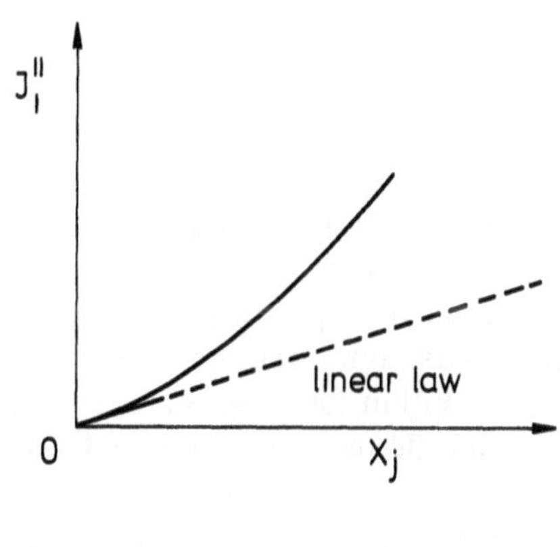

Fig. 2. Flow J_i'' as function of the generalized force X_j schematically

in the form of a matrix and characterize completely the transport properties of the membrane system close to equilibrium, for a given reference state.

$$\begin{pmatrix} L_{00} & L_{01} & L_{02} & \cdots & L_{0n} \\ L_{10} & L_{11} & L_{12} & \cdots & L_{1n} \\ \vdots & \vdots & \vdots & & \vdots \\ L_{n0} & L_{n1} & L_{n2} & \cdots & L_{nn} \end{pmatrix} \tag{15}$$

Using the theory of fluctuations ONSAGER was able to show that the matrix of the phenomenological coefficients is symmetric:

$$L_{ij}(r) = L_{ji}(r)$$

(ONSAGER reciprocity relation). (16)

As a consequence of the reciprocity law of ONSAGER the number of independent

If Eq. (14) is substituted into Eq. (11) we obtain the entropy production in a form which is quadratic in the forces:

$$T\frac{d_i S}{dt} = \sum_{i=0}^{n} \sum_{j=0}^{n} L_{ij}(r) X_i X_j \geq 0.$$

The requirement that this expression be positive in the nonequilibrium state has mathematical consequences for the L-matrix. We mention only:

1. The diagonal coefficients L_{ii} must be positive

$$L_{ii} > 0 \quad (i = 0, 1, 2, \ldots, n).$$

2. The off-diagonal coefficients (cross-coefficients) may be either positive or negative. For the cross-coefficients a number of inequalities can be derived, for example

$$L_{ii} L_{jj} - L_{ij}^2 > 0 \quad (i, j = 0, 1, 2, \ldots, n).$$

For systems close to equilibrium (small values of X_j) it is convenient to choose as reference state that equilibrium state of the system in which the composition of both bulk phases is equal to the arithmetic mean value of the mole fraction of each species.

$$r = (T, P, \bar{x}_0, \bar{x}_1, \ldots, \bar{x}_{n-1})$$

with

$$\bar{x}_i = 1/2(x_i' + x_i''); \quad \sum_{i=0}^{n} \bar{x}_i = 1.$$

For several reasons it is advantageous to start the thermodynamic treatment of transport processes across membranes by considering the entropy production:

1. Entropy production gives the flows and forces that have to be used in the thermodynamic treatment of transport processes across membranes.

2. Within the range of validity of the linear laws derived from the entropy production the reciprocity law of ONSAGER holds.

3. The entropy production is a convenient starting point for the development of the general phenomenological theory which also treats systems far from equilibrium (SAUER, 1973).

From an experimental point of view the reciprocity law of ONSAGER can be useful. It reduces the number of independent experiments necessary to establish the matrix of the phenomenological coefficients. On the other hand, it can be used as a check on the consistency of experimental results provided all experiments have been carried out to establish the complete matrix of the phenomenological coefficients.

1.2.7 Linear Laws for Dilute Solutions

In experiments with biological membranes one component of the bulk phases of the membrane is usually present in large surplus. This species is called solvent and the other species are called solutes. The solvent is labelled by the index 0. In such systems it is easier to measure the volume flow J_V and the flows of the solutes relative to the volume flow J_i^* rather than the flows J_i'' relative to the membrane. The new flows J_V and J_i^* which are introduced for purely experimental reasons are linear combinations of the flows J_i'' and defined by the Eqs. (17) and (18). Thereby the number of independent flows is not changed.

$$J_V \equiv \sum_{i=0}^{n} J_i'' V_i, \tag{17}$$

$$J_i^* \equiv J_i'' - \bar{c}_i J_V \quad (i=1, 2, 3, \ldots, n). \tag{18}$$

V_i is the partial molar volume of species i. It is assumed that V_i is not a function of concentration and pressure. $\bar{c}_i$ is the molarity of the solutes in the reference state which is assumed to be a possible equilibrium state of the membrane system.

Eq. (17) is obtained by applying the thermodynamic relation

$$V = \sum_{i=0}^{n} n_i V_i \quad \text{with } T, P = \text{const.}$$

to phase (″) and differentiating the relation with respect to time.

Eq. (18) can be rationalized by considering the following identity:

$$J_i^* \simeq V'' \frac{dc_i''}{dt} = V'' \frac{d}{dt}\left(\frac{n_i''}{V''}\right) = J_i'' - c_i'' J_V. \tag{19}$$

For systems near equilibrium the concentrations in the reference state $\bar{c}_i$ can be identified with the arithmetic mean of the bulk concentrations ($\bar{c}_i = 1/2(c_i' + c_i'')$; $i = 0, 1, 2, \ldots, n$). In this state the thermodynamic homogeneity relation holds.

$$\bar{c}_0 V_0 + \sum_{i=1}^{n} \bar{c}_i V_i = 1 \quad (T, P = \text{const.}). \tag{20}$$

We introduce the new flows J_V and J_i^* into the entropy production Eq. (11) by eliminating the flows J_0'' and J_i'' and taking into account Eq. (20). In this way the entropy production can be written as:

$$T\frac{d_i S}{dt} = J_V X_V + \sum_{i=1}^{n} J_i^* X_i^* \tag{21}$$

with

$$X_V = \sum_{i=1}^{n} \bar{c}_i \Delta \mu_i + \left(1 - \sum_{i=1}^{n} \bar{c}_i V_i\right) \frac{\Delta \mu_0}{V_0} \tag{22}$$

and

$$X_i^* = \Delta \mu_i - \frac{V_i}{V_0} \Delta \mu_0 . \tag{23}$$

From an experimental point of view it is desirable to have the forces X_V and X_i^* in a form in which the pressure dependence and the concentration dependence of the chemical potential are separated. The pressure difference between the bulk phases is a parameter which can be changed independently from concentration differences. The pressure and concentration dependence of the chemical potential can be separated by taking into account the thermodynamic relation:

$$\Delta \mu_i = V_i \Delta P + \Delta_c \mu_i . \tag{24}$$

$\Delta_c \mu_i$ describes the part of $\Delta \mu_i$ due to the difference of chemical composition of the bulk phases at constant P and T. Using Eq. (24) the expressions of the forces X_V and X_i^* can be written in the form:

$$X_V = \Delta P + \sum_{i=1}^{n} \bar{c}_i \left(\Delta_c \mu_i - \frac{V_i}{V_0} \Delta_c \mu_0\right) + \frac{\Delta_c \mu_0}{V_0}, \tag{25}$$

$$X_i^* = \Delta_c \mu_i - \frac{V_i}{V_0} \Delta_c \mu_0 . \tag{26}$$

From the transformed expression of the entropy production in Eq. (21) the linear laws can be derived using the method outlined in Section 1.2.6.

$$J_V = L_{VV}(r) X_V + \sum_{i=1}^{n} L_{Vi}(r) X_i^*, \tag{27a}$$

$$J_i^* = L_{iV}(r) X_V + \sum_{j=1}^{n} L_{ij}^*(r) X_j^* \quad (i = 1, 2, \ldots, n) \tag{27b}$$

(linear laws)

with

$$L_{VV} = \left(\frac{\partial J_V}{\partial X_V}\right)_{r,\,\text{eqm.}} \qquad L_{Vi} = \left(\frac{\partial J_V}{\partial X_i^*}\right)_{r,\,\text{eqm.}}$$

$$L_{iV} = \left(\frac{\partial J_i}{\partial X_V}\right)_{r,\,\text{eqm}} \qquad L_{ij}^* = \left(\frac{\partial J_i}{\partial X_j^*}\right)_{r,\,\text{eqm.}} .$$

The reciprocity law of ONSAGER is invariant to linear transformations of flows and forces (MEIXNER and REIK, 1959; HAASE, 1969). Since the new flows J_V and J_i^* and the new forces X_V and X_i^* are connected with the flows J_i'' and the forces X_j by linear transformations it can be concluded that:

$$L_{Vi}(r) = L_{iV}(r) \quad \text{and} \quad L_{ij}^*(r) = L_{ji}^*(r). \tag{28}$$

The force X_V conjugate with the volume flow J_V has a complicated form. SAUER (1973) has shown that in the linear laws derived from the entropy production Eq. (21), X_V can be replaced by the pressure difference $\Delta P(=P'-P'')$ without violating the reciprocity law of ONSAGER.

$$X_V \to \Delta P. \tag{29}$$

The expression of the force X_i^* can be simplified if the concentration difference between the bulk phases is small and the solutions are very dilute $\left(\sum_{i=1}^{n} \bar{c}_i V_i \ll 1;\right.$ consequently $\bar{c}_i V_i \ll 1$; $\Delta\mu_i = RT \cdot \Delta c_i/\bar{c}_i$; R = gas constant$\left.\right)$:

$$\bar{c}_i X_i^* = \bar{c}_i \left(\Delta_c \mu_i - \frac{V_i}{V_0} \Delta_c \mu_0\right) \to RT \Delta c_i = \pi_i. \tag{30}$$

π_i is called osmotic difference of species i.

For the characterization of the osmotic behavior of membranes (see Sect. 1.4.7) it is useful to introduce a new phenomenological coefficient for each species i of the bulk phases called reflection coefficient $\sigma_i \cdot \sigma_i$ is defined by:

$$\sigma_i \bar{c}_i \equiv -\frac{L_{Vi}}{L_{VV}}. \tag{31}$$

This definition is a generalization of the reflection coefficient introduced by STAVERMAN (1951, 1952a, b) for multicomponent systems (SAUER, 1973). If Eq. (31) is introduced into the linear laws [Eq. (27)] and X_V is replaced by ΔP we obtain the linear laws in the following form:

$$J_V = L_{VV}\left(\Delta P - \sum_{i=1}^{n} \sigma_i \bar{c}_i X_i^*\right), \tag{32}$$

$$J_i^* = -\sigma_i \bar{c}_i J_V + \sum_{j=1}^{n} L_{ij}^{**} X_j^* \tag{33}$$

or

$$J_i'' = (1-\sigma_i)\bar{c}_i J_V + \sum_{j=1}^{n} L_{ij}^{**} X_j^*, \tag{34}$$

with

$$L_{ij}^{**} = L_{ij}^* - \frac{L_{iV} L_{jV}}{L_{VV}}. \tag{35}$$

For very dilute solutions and small concentration differences between the bulk phases the coefficients L_{ii}^{**} ($i=1,2,\ldots,n$) are related to the permeability p_i of component i by

$$L_{ii}^{**} \equiv \bar{c}_i p_i . \tag{36}$$

The permeabilities p_i are introduced with reference to FICK's first law of diffusion ($J_i = \tilde{p}_i \Delta c_i$; $\tilde{p}_i = p_i RT$).

1.2.8 Influence of Nonpermeable Components

Our considerations will now be generalized by taking into account membrane systems in which the bulk phases are composed of permeable and nonpermeable uncharged species. The permeable species are labeled by the index $i, j = 0, 1, 2, \ldots, n$ and the nonpermeable species by the index $t = (n+1), (n+2), \ldots, w$.

The presence of the nonpermeable species does not change the form of the entropy production given by Eq. (11) and its transformed form Eq. (21) because the flows of the nonpermeable components are zero.

$$J_t' = J_t'' = 0 \qquad (t = (n+1), (n+2), \ldots, w).$$

Consequently the linear laws of Eqs. (14) and (27) derived from Eqs. (11) and (21) respectively can formally be applied to membrane systems with permeable and nonpermeable species in the bulk phases.

The choice of the reference state on which the phenomenological coefficients of the linear laws depend has to be changed. It is no longer possible to choose as reference state an equilibrium state common to both bulk phases (SAUER, 1973). In a membrane system with impermeable species present in its bulk phases there can exist differences in composition and pressure between the bulk phases in the equilibrium state (osmotic equilibrium). If the composition of phase (′) in the equilibrium state of the membrane is chosen as reference state and the system is close to equilibrium the reference state can be characterized by the variables:

$$r = (T, P', x_0', \ldots, x_{w-1}', \bar{x}_{n+1}, \ldots, \bar{x}_w)$$

with

$$\bar{x}_t = \tfrac{1}{2}(x_t' + x_t'') \qquad (t = (n+1), (n+2), \ldots, w).$$

For systems in which it is useful to distinguish in the bulk phases between solvent and solutes this references state is characterized by

$$r = (T, P', c_0', \ldots, c_w', \bar{c}_{n+1}, \bar{c}_{n+2}, \ldots, \bar{c}_w)$$

with

$$\bar{c}_t = \tfrac{1}{2}(c_t' + c_t'').$$

In dilute solutions the force X_V conjugate to the volume flow J_V can be approximated by Eq. (37) (SAUER, 1973).

$$X_V = \Delta P - \sum_{t=n+1}^{w} \bar{c}_t \left(\Delta_c \mu_t - \frac{V_t}{V_0} \Delta_c \mu_0 \right). \tag{37}$$

If nonpermeable components are present in high dilution $\left(\sum_{t=n+1}^{w} \bar{c}_t V_t \ll 1\right)$ Eq. (37) takes on the form

$$X_V = \Delta P - \sum_{t=n+1}^{w} \pi_t \tag{38}$$

with

$$\pi_t = RT\Delta c_t .$$

1.3 Transport of Nonelectrolytes with Chemical Reaction

1.3.1 Description of the Membrane System

In this section a membrane system is considered which is similar to that described in Section 1.2.1. The bulk phases are formed by uncharged species ($i=0, 1, \ldots, n$) which can permeate the membrane. Only one single chemical reaction is permitted to occur between several of the species of the bulk phases. This chemical reaction takes place only inside the membrane and is inhibited in the bulk phases.

1.3.2 Mass Balance

The stoichiometric equation of the chemical reaction can be represented by the symbolic Eq. (39).

$$\sum_{i=0}^{n} \nu_i B_i = 0. \tag{39}$$

B_i is the chemical symbol of species i. ν_i is the stoichiometric number of species i. The sign of convention is such that ν_i is positive if the species i appears on the right-hand side of the chemical reaction (products) and is negative if the species i appears on the left-hand side (reactants). If a species i is not involved in the chemical reaction its stoichiometric number is zero.

In the stationary state of the membrane ($dn_i^m/dt=0$) the flow of species i into phase (′) and phase (″) is linked with the rate of its chemical production.

$$J_i' + J_i'' = J_{c,i} \qquad (i=0, 1, 2, \ldots, n) \tag{40}$$

with

$$J_{c,i} \equiv \nu_i \frac{d\xi}{dt} .$$

$J_{c,i}$ is the rate of chemical production of species i. The quantity ξ is called the extent of reaction. It has the dimension mol; $d\xi/dt$ is the turn over velocity.

If we introduce the reaction flow defined by Eq. (41) the mass balance equation is given by Eq. (42).

$$J_c \equiv d\xi/dt , \tag{41}$$

$$J_i' + J_i'' = \nu_i J_c \qquad (i=0, 1, 2, \ldots, n) . \tag{42}$$

If species i is not involved in the chemical reaction ($\nu_i=0$) Eq. (42) is reduced to Eq. (9).

1.3.3 Entropy Production

The entropy production caused by the irreversible processes taking place inside the membrane is obtained by combining the Eqs. (1), (3), (5), (7), (8) and (42):

$$T\frac{d_i S}{dt} = \sum_{i=0}^{n} J_i'' \Delta\mu_i + J_c A' \geq 0 \tag{43}$$

with

$$A' \equiv -\sum_{i=0}^{n} \nu_i \mu_i'. \tag{44}$$

A' represents the affinity of the chemical reaction in phase (′).

1.3.4 Linear Laws

The flows J_i'' and J_c depend on the forces $X_j(\equiv \Delta\mu_j)$ $(j=0,1,2,\ldots,n)$ and A' and on a reference state which is assumed to be a possible equilibrium state of the membrane system.

$$\begin{aligned} J_i'' &= J_i''(r, X_0, X_1, \ldots, X_n, A') \quad (i=0,1,2,\ldots,n),\\ J_c &= J_c(r, X_0, X_1, \ldots, X_n, A')\\ &\text{(nonequilibrium state).}\end{aligned} \tag{45}$$

In the equilibrium state the flows J_i'' and J_c vanish:

$$\begin{aligned} J_i'' &= J_i''(r, X_0=0, X_1=0, \ldots, X_n=0, A'=0)=0 \quad (i=0,1,2,\ldots,n),\\ J_c &= J_c(r, X_0=0, X_1=0, \ldots, X_n=0, A'=0)=0\\ &\text{(equilibrium state).}\end{aligned} \tag{46}$$

For a given reference state the functions given by Eq. (45) are expanded into a power series around the equilibrium state. For small enough values of X_j and A' the series can be truncated after the first order term and the linear laws are obtained.

$$\begin{aligned} J_i'' &= \sum_{j=0}^{n} L_{ij}(r) X_j + L_{ic}(r) A' \quad (i=0,1,2,\ldots,n),\\ J_c &= \sum_{j=0}^{n} L_{cj}(r) X_j + L_{cc}(r) A'\\ &\text{(linear laws)}\end{aligned} \tag{47}$$

with

$$L_{ij} \equiv \left(\frac{\partial J_i}{\partial X_j}\right)_{r,\,\text{eqm}} \qquad L_{ic} \equiv \left(\frac{\partial J_i}{\partial A'}\right)_{r,\,\text{eqm.}}$$

$$L_{cj} \equiv \left(\frac{\partial J_c}{\partial X_j}\right)_{r,\,\text{eqm.}} \qquad L_{cc} \equiv \left(\frac{\partial J_c}{\partial A'}\right)_{r,\,\text{eqm}}.$$

For membrane systems close to equilibrium (small values of X_j and A') the equilibrium state of the system in which the composition of both bulk phases is equal to its arithmetic mean value can be chosen as the reference state.

The reciprocal law of ONSAGER is valid for the coefficient L_{ij} and L_{cj}:

$$\begin{aligned} L_{ij}(r) &= L_{ji}(r), \\ L_{cj}(r) &= L_{jc}(r). \end{aligned} \tag{48}$$

For systems in which it is practical to distinguish in the bulk phases between solvent and solutes linear laws can be derived from the transformed expression of entropy production, Eq. (49).

$$T\frac{d_i S}{dt} = J_V X_V + \sum_{i=1}^{n} J_i^* X_i^* + J_c A'. \tag{49}$$

Eq. (49) is obtained from Eq. (43) by linear transformation using the Eqs. (17) and (18) and (20) (see Sect. 1.2.7). The driving forces X_V and X_i^* are given by the Eqs. (25) and (26). The Eq. (50) represent the linear laws derived from Eq. (49).

$$\begin{aligned} J_V &= L_{VV}(r) X_V + \sum_{i=1}^{n} L_{Vi}(r)\, X_i^* + L_{Vc}(r) A', \\ J_i^* &= L_{iV}(r) X_V + \sum_{j=1}^{n} L_{ij}^*(r) X_j^* + L_{ic}(r) A' \quad (i = 1, 2, \ldots, n), \\ J_c &= L_{cV}(r) X_V + \sum_{i=1}^{n} L_{ci}(r) X_i^* + L_{cc}(r) A' \end{aligned} \tag{50}$$

(linear laws).

The reciprocity laws of ONSAGER are again valid:

$$L_{Vi} = L_{iV}, \quad L_{ij}^* = L_{ji}^*, \quad L_{ic} = L_{ci}, \quad L_{cV} = L_{Vc}. \tag{51}$$

1.3.5 Active Transport

Let us return to the linear laws, Eq. (47), to discuss their physical significance. The occurrence of a chemical reaction inside the membrane makes possible a coupling between the flow of species i and the chemical reaction. This type of coupling is expressed by the coefficient L_{ic}. It is obvious that the coefficient L_{ic} will be different from zero for components involved in the chemical reaction. Reactants are transformed to products inside the membrane causing a flow of reactants into the membrane and of products out of the membrane. But it is also possible that such a coupling occurs for components not involved in the chemical reaction. This is not selfevident.

This type of coupling is closely related to the phenomenon of active transport (KEDEM, 1961; SAUER, 1973). We follow the considerations of SAUER to define the terms active and passive transport. Within the range of the validity of the linear laws active transport of a certain species – say species n – relative to the membrane can be defined by

$$\begin{aligned} &J_n^{ac} \equiv L_{nc}(r) A' \quad \text{(linear system)} \\ &\text{for } v_n = 0; \; X_j = 0 \; (j = 0, 1, 2, \ldots, n); \; A' \neq 0 \end{aligned} \tag{52}$$

and passive transport of this species by

$$J_n^{\text{pas}} = \sum_{j=0}^{n} L_{nj}(r)\, X_j \quad \text{(linear system)}$$
$$\text{with } A' = 0 \text{ (inhibited chemical reaction)}; \; X_j \neq 0. \tag{53}$$

This means species n is transported actively if a flow of this species is generated by the affinity of a chemical reaction taking place inside the membrane phase in which species n is not involved provided all other driving forces X_j are zero. In this case the transport coefficient $L_{nc}(r)$ must be different from zero.

For linear systems the principle of superposition is applicable and the total flow of component n with $X_j \neq 0$ and $A' \neq 0$ is given by

$$J_n'' = J_n^{\text{ac}} + J_n^{\text{pas}} \tag{54}$$
$$\text{(linear system)}.$$

Since generally the flows depend on the frame of reference the active and passive flow of a species will be different in different frames of reference.

The Eqs. (52)–(54) contain instructions how to decide experimentally whether a species n is transported actively by a given membrane: the membrane is placed between two identical bulk phases ($X_j = 0$, $A' \neq 0$) and it is observed whether a flow of species n occurs (J_n^{ac}). Eq. (54) can then be used to split the flow J_n'' observed under the condition $X_j \neq 0$ and $A' \neq 0$ into an active and a passive component.

SAUER has generalized the definition of active and passive transport by Eq. (55):

$$J_n^{\text{ac}} \equiv J_n''(r, X_0 = 0, X_1 = 0, \ldots, X_n = 0, A')$$
$$J_n^{\text{pas}} \equiv J_n''(r, X_0, X_1, \ldots, X_n, A' = 0). \tag{55}$$
$$\text{(nonlinear system)}$$

For nonlinear systems the principle of superposition is no longer applicable and consequently Eq. (54) no longer holds.

The theoretical analysis of SAUER shows further that in discontinuous systems active transport cannot occur with symmetric membranes and that an asymmetry of the membrane is a necessary requirement (MEIER, 1973; MEIER et al., 1974).

1.4 Transport of Electrolytes

1.4.1 Description of the Membrane System

The membrane system considered in this section is shown schematically in Figure 3. The bulk phases of the membrane are composed of a solvent which is present in large excess and several electrolytes. The electrolytes are completely dissociated. All species are present in both bulk phases and can permeate the membrane. They are labeled by the index $i = 0, 1, 2, \ldots, e, \ldots, n$ (solvent: index 0). Chemical reactions are excluded. An electric current can be passed across the membrane. This is accomplished by placing into the bulk phases two identical electrodes which react selectively and reversibly with ionic species n (or e).

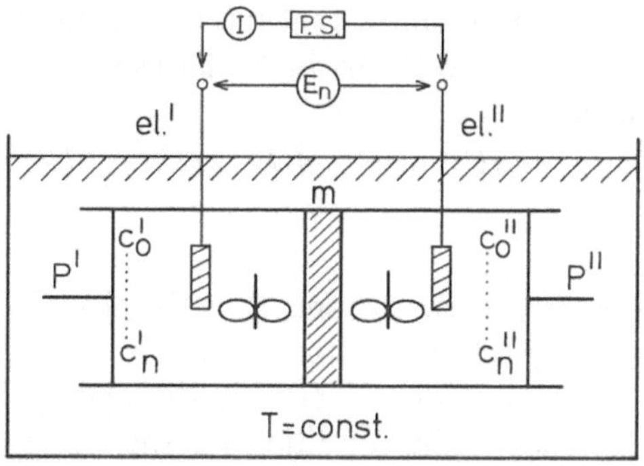

Fig. 3. Membrane system with electric current flow schematically; *el.* = electrode reacting specifically and reversibly with ionic species n; P = pressure; c_i = concentration of species i; $i = 0, 1, \ldots, n$. Quantities refering to the left bulk phase are characterized by the superscript (′) and that refering to the right bulk phase by the superscript (″). *P.S.* = power source; I = amperemeter; E_n = voltmeter

The use of this type of electrode is necessary to prevent the generation of additional ionic species in the bulk phases by electrode reactions during the passage of electric current. The electrodes have to be identical to allow passage of electric current in both directions across the membrane without complications. Examples of such electrodes are selective solid-state electrodes for halide ions (e.g. Ag/AgCl electrodes).

1.4.2 Entropy Balance

Each bulk phase is electrically neutral. This is expressed by the requirements of electroneutrality.

$$\sum_{i=0}^{n} z_i n_i = 0. \tag{56}$$

z_i is the valence of species i including the sign. z_i is positive for cations, negative for anions and zero for nonelectrolytes. n_i is the amount of substance of species i (mol).

The condition of electroneutrality reduces the number of independent variables in the generalized GIBBS equation Eq. (1). It is useful to eliminate from that equation the ionic species n which reacts specifically and reversibly with the current-carrying electrodes (SAUER, 1973).

$$\frac{dS^\alpha}{dt} = \frac{1}{T}\frac{dU^\alpha}{dt} + \frac{P^\alpha}{T}\frac{dV^\alpha}{dt} - \sum_{i=0}^{n-1} \frac{1}{T}\left(\mu_i^\alpha - \frac{z_i}{z_n}\mu_n^\alpha\right)\frac{dn_i^\alpha}{dt}, \quad \alpha = ('), (''). \tag{57}$$

1.4.3 Energy Balance

The passage of electric current across the membrane makes it necessary to add another term to the first law of thermodynamics which takes into account the electric work done on the whole membrane system.

If it is assumed that the membrane is in its stationary state and changes of the kinetic energy of the bulk phases can be neglected, the energy balance takes on the form:

$$\frac{dU'}{dt}+\frac{dU''}{dt}=\frac{dQ'}{dt}+\frac{dQ''}{dt}-P'\frac{dV'}{dt}-P''\frac{dV''}{dt}+I\,E_n. \tag{58}$$

I is the electric current; E_n is the electric potential difference between the electrodes reacting selectively and reversibly with ionic species n.

The additional term in the energy balance, Eq. (58), causes a change in the stationary state condition for the energy flow:

$$J_u'+J_u''=I\,E_n. \tag{59}$$

1.4.4 Mass Balance

The definition of the flow of the ionic species not involved in the electrode reaction relative to the membrane is given by:

$$J_i'\equiv\frac{dn_i'}{dt}\qquad J_i''\equiv\frac{dn_i''}{dt}\qquad (i=0,1,2,\ldots,n-1). \tag{60}$$

For the ionic species n the situation is different. The amount of substance of this species in the bulk phases does not only change due to transport processes across the membrane but also due to the electrode reactions caused by the flow electric current. The flow of species n relative to the membrane is given by:

$$J_n''=\frac{dn_n''}{dt}+\frac{I}{z_n F}. \tag{61}$$

F is the Faraday number.

Eq. (61) follows from the expression for the electric current of Eq. (62) and Eq. (63) which is a consequence of the requirement of electroneutrality (56).

$$I=\sum_{i=0}^{n-1} z_i F J_i''+z_n F J_n'', \tag{62}$$

$$\frac{dn_n''}{dt}=-\sum_{i=0}^{n-1}\frac{z_i}{z_n}J_i''. \tag{63}$$

1.4.5 Entropy Production

The entropy production of the system is given by Eq. (64) which is obtained by combining the modified form of the generalized GIBBS equation Eq. (57) with the first law of thermodynamics for open phases, Eq. (5), the steady-state condition for the energy flow, Eq. (59), and the equation of definition of particle flow, Eq. (60).

$$T\frac{d_i S}{dt}=\sum_{i=0}^{n-1} J_i''\left(\Delta\left(\mu_i-\frac{z_i}{z_n}\mu_n\right)\right)+I\,E_n. \tag{64}$$

SAUER (1973) has pointed out that the forces $\Delta(\mu_i - z_i/z_n \cdot \mu_n)$ conjugate to the flows J_i'' are measurable quantities. This will now be shown:

Introducing Eq. (62) into the expression of entropy production of Eq. (64) one obtains:

$$T\frac{d_i S}{dt} = \sum_{i=0}^{n} J_i'' \left(\Delta \left(\mu_i - \frac{z_i}{z_n} \mu_n \right) + z_i F E_n \right). \tag{65}$$

If the electrodes which react specifically and reversibly with the ionic species n are exchanged for another pair of electrodes which react specifically and reversibly with the ionic species e the entropy production is given by:

$$T\frac{d_i S}{dt} = \sum_{i=0}^{n} J_i'' \left(\Delta \left(\mu_i - \frac{z_i}{z_e} \mu_e \right) + z_i F E_e \right). \tag{66}$$

Since the entropy production is unchanged by the replacement of the electrodes it can be concluded from Eqs. (65) and (66) that:

$$\Delta \left(\mu_e - \frac{z_e}{z_n} \mu_n \right) = z_e F (E_e - E_n). \tag{67}$$

This argument demonstrates that in principle it is possible to measure the forces conjugate to the flows J_i'' with the aid of a set of suitably chosen pairs of electrodes which react specifically and reversibly with the ionic species present in the bulk phases.

Taking into account Eq. (67) the expression of entropy production Eq. (64) takes on the form:

$$T\frac{d_i S}{dt} = \sum_{i=0}^{n} J_i'' z_i F E_i. \tag{68}$$

The electric potential difference E_i measured between the bulk phases by two identical electrodes which react specifically and reversibly with ionic species i is proportional to the difference of the electrochemical potential of this ionic species.

$$z_i F E_i = z_i F \Delta \varphi^{\text{el.}} = \Delta \eta_i \quad \text{with} \quad \Delta = {}' - {}''. \tag{69}$$

η_i is the electrochemical potential defined by $\eta_i \equiv \mu_i + z_i F \varphi$. φ is the electric potential. The index el. refers to the electrode phase.

Substituting Eq. (14) into the entropy production Eq. (68) we obtain

$$T\frac{d_i S}{dt} = \sum_{i=0}^{n} J_i'' \Delta \eta_i. \tag{70}$$

Eq. (69) can be understood on the basis of the following development (SCHLÖGEL, 1964b): Since the electrodes react specifically and reversibly with the ionic species i this ionic species can permeate the phase boundary electrode/solution without hindrance. Therefore the electrochemical equilibrium for the ionic species i is established there. That means: $\eta_i^{\text{el.}\prime} = \eta_i'$ and $\eta_i^{\text{el.}\prime\prime} = \eta_i''$

$$\eta_i^{\text{el}\prime} - \eta_i^{\text{el}\prime\prime} = \eta_i' - \eta_i''. \tag{71}$$

$\eta_i^{el.'}$ and $\eta_i^{el.''}$ refer to the electrochemical potential of ionic species i in the electrode phase immersed into the left and right bulk phases respectively. In the term $(\eta_i^{el\,'} - \eta_i^{el\,''})$ the contribution of the chemical potential cancels under isobaric conditions because the electrodes are identical. If there exists a pressure difference between the bulk phases we neglect the influence of pressure on the chemical potential of the ionic species i in the electrode phase. Consequently Eq. (69) follows from Eq. (71).

1.4.6 Generalization

The expression of the entropy production, Eq. (72) [see Eq. (43)], is the starting point for the thermodynamic treatment of stationary transport processes in an isothermal membrane system in which the bulk phases are composed of several permeable electrolytes and nonelectrolytes (species $i=0, 1, 2, \ldots, n$) and a single chemical reaction takes place between several of these species inside the membrane.

$$T\frac{d_i S}{dt} = \sum_{i=0}^{n} J_i'' \Delta\mu_i + J_c A'. \tag{72}$$

For each ionic species we replace the difference of the chemical potentials between the bulk phases $\Delta\mu_i$ by the corresponding difference of the electrochemical potential $\Delta\eta_i$. For nonelectrolytes $\Delta\mu_i$ is identical with $\Delta\eta_i$ because for nonelectrolytes the valence z_i is equal to zero. The flows J_i'' are defined by Eq. (60). If electric current is passed across the membrane using electrodes which react specifically and reversibly with ion species n it has to be borne in mind that the flow of this ionic species relative to the membrane is given by Eq. (61). In the linear laws derived from Eq. (72) the same formal replacement of $\Delta\mu_i$ by $\Delta\eta_i$ has to be made for each ionic species.

The electric current does not appear in Eq. (72). If an electric current is passed across the membrane it can be expressed by the independent ion flows J_i''. For systems near equilibrium the electric current is given by:

$$I = \sum_{i=0}^{n} z_i F J_i'' = \sum_{i=0}^{n} \sum_{j=0}^{n} z_i F L_{ij} X_j + \sum_{i=0}^{n} z_i F L_{ic} A' \quad \text{with} \quad X_j = \Delta\eta_j. \tag{73}$$

If the thermodynamic states of the bulk phases are identical—that means $\Delta\mu_j = 0$ and $\Delta P = 0$—the forces X_j are proportional to the electric potential difference between the current carrying electrodes:

$$X_j = \Delta\eta_j = z_j F E_n \quad \text{with} \quad E_n = \varphi' - \varphi'' = \varphi^{el.'} - \varphi^{el.''}.$$

The potential drop in the bulk phases due to ohmic resistance of the electrolyte solution is neglected.

Under these experimental conditions the electric current is given by:

$$I = \sum_{i=0}^{n} \sum_{j=0}^{n} z_i z_j L_{ij} E_n + \sum_{i=0}^{n} z_i F L_{ic} A'. \tag{74}$$

The factor $\sum_{i=0}^{n} \sum_{j=0}^{n} z_i z_j L_{ij}$ describes the electric conductivity of the membrane. The term $\sum_{i=0}^{n} z_i F L_{ic} A'$ is the short circuit current (SAUER, 1973).

1.4.7 Linear Laws for Dilute Solutions; Special Case

To illustrate the application of the general equations we consider a simple isothermal membrane system in which the bulk phases are formed by aqueous solutions of a single electrolyte dissociating into two ion species (binary electrolyte: e.g. KCl, $CaCl_2$). Each electrolyte molecule is completely dissociated into ν_+ cations and ν_- anions. Two identical electrodes which react specifically and reversibly with the anion species of the electrolyte (for example Ag/AgCl-Electrode) are immersed into the bulk phases to allow passage of electric current across the membrane. The same type of electrodes is used to measure the electric potential difference between the bulk phases under the condition of zero electric current flow. According to Eq. (64) the entropy production of this system in the stationary state is given by Eq. (75).

$$T\frac{d_i S}{dt}=J_0'' \Delta\mu_0+J_+'' \Delta\left(\mu_+-\frac{z_+}{z_-}\mu_-\right)+IE_- . \tag{75}$$

The index 0 refers to the solvent.

The requirement of electroneutrality of the electrolyte molecule demands:

$$\nu_+ z_+ +\nu_- z_- =0. \tag{76}$$

The chemical potential of the electrolyte (index s) is given by:

$$\mu_s=\nu_+\mu_+ +\nu_-\mu_- . \tag{77}$$

The flow of the cations relative to the membrane is a direct measure of the flow of the electrolyte since the cations are not involved in the electrode reactions:

$$J_s''=\frac{J_+''}{\nu_+}. \tag{78}$$

Substituting the Eqs. (76–78) into the expression of entropy productions Eq. (75) one obtains:

$$T\frac{d_i S}{dt}=J_0'' \Delta\mu_0+J_s'' \Delta\mu_s+IE_- . \tag{79}$$

The linear transformation of the entropy production (79) using the Eqs. (80) and (81) together with the homogeneity relation Eq. (20) and Eq. (24) yields Eq. (82).

$$J_V=V_0 J_0''+V_s J_s'', \tag{80}$$

$$J_s^*=J_s''-\bar{c}_s J_V, \tag{81}$$

$$T\frac{d_i S}{dt}=J_V X_V+J_s^* X_s^*+IE_- \tag{82}$$

with

$$X_V=\Delta P+(\bar{c}_s\Delta_c\mu_s+\bar{c}_0\Delta_c\mu_0),$$

$$X_s^*=\left(\Delta_c\mu_s-\frac{V_s}{V_0}\Delta_c\mu_0\right).$$

For systems near equilibrium $\bar{c}_s$ can be identified with the arithmetic mean of the bulk concentrations.

In deriving linear laws from the entropy production [Eq. (82)] one makes use of the fact that the pressure difference ΔP can be introduced into the linear laws instead of the complicated force X_V without violating the reciprocity law of ONSAGER.

$$J_V = L_{VV}\Delta P + L_{Vs}X_s^* + L_{VE}E_-, \tag{83}$$

$$J_s^* = L_{Vs}\Delta P + L_{ss}^* X_s^* + L_{sE}E_-, \tag{84}$$

$$I = L_{VE}\Delta P + L_{sE}X_s^* + L_{EE}E_- \tag{85}$$

with

$$L_{ij} = L_{ji}.$$

For highly dilute solutions the expression of force X_s^* can be simplified:

$$\bar{c}_s X_s^* \to (\nu_+ + \nu_-) RT\Delta c_s.$$

The transport properties of the membrane for a given reference state are completely described if the six transport coefficients L_{VV}, L_{Vs}, L_{VE}, L_{ss}^*, L_{sE} and L_{EE} are known. From the Eqs. (83)–(85) it can be seen how the experiments have to be designed to obtain the complete set of transport coefficients.

For example, the coefficient L_{VV} can be determined by measuring the volume flow J_V at a given pressure difference between the bulk phases of the membrane under isotonic conditions ($X_s^* = 0$) and by short circuiting the electric potential difference E_- ($E_- = 0$):

$$L_{VV} = \left(\frac{J_V}{\Delta P}\right)_{X_s^*, E_-}.$$

The subscripts indicate the parameters which are kept zero during the experiment.

Since measurements under galvanostatic conditions (I=const.) are easier to carry out than those under potentiostatic conditions (E_- =const.) it is experimentally convenient to choose the electric current as an independent variable. For this experimental condition the Eqs. (83)–(85) have to be rewritten by solving Eq. (85) for E_- and substituting this expression into the Eqs. (83) and (84).

$$J_V = \tilde{L}_{VV}\Delta P + \tilde{L}_{Vs}X_s^* + \tilde{L}_{VI}I, \tag{86}$$

$$J_s^* = \tilde{L}_{Vs}\Delta P + \tilde{L}_{ss}X_s^* + \tilde{L}_{sI}I, \tag{87}$$

$$E_- = -\tilde{L}_{VI}\Delta P - \tilde{L}_{sI}X_s^* + \tilde{L}_{II}I \tag{88}$$

with

$$\tilde{L}_{VV} = L_{VV} - \frac{L_{VE}^2}{L_{EE}}; \quad \tilde{L}_{Vs} = L_{Vs} - \frac{L_{sE}L_{VE}}{L_{EE}}; \quad \tilde{L}_{VI} = \frac{L_{VE}}{L_{EE}};$$

$$\tilde{L}_{ss} = L_{ss}^* - \frac{L_{sE}^2}{L_{EE}}; \quad \tilde{L}_{sI} = \frac{L_{sE}}{L_{EE}}; \quad \tilde{L}_{II} = \frac{1}{L_{EE}}.$$

In the literature one often finds the Eqs. (86) and (87) written in a different form introducing the reflection coefficient σ_s of the solute defined by Eq. (91) and the permeability p_s of the solute defined by Eq. (92).

$$J_V = \tilde{L}_{VV}(\Delta P - \sigma_s \bar{c}_s X_s^*) + \tilde{L}_{VI}I, \tag{89}$$

$$J_s^* = -\sigma_s \bar{c}_s J_V + p_s \bar{c}_s X_s^* + \left(\tilde{L}_{sI} - \frac{\tilde{L}_{Vs}\tilde{L}_{VI}}{\tilde{L}_{VV}}\right) I \tag{90}$$

with

$$\sigma_s = -\frac{\tilde{L}_{Vs}}{\bar{c}_s \tilde{L}_{VV}}, \tag{91}$$

$$p_s = \frac{1}{\bar{c}_s}\left(\tilde{L}_{ss} - \frac{\tilde{L}_{Vs}^2}{\tilde{L}_{VV}}\right). \tag{92}$$

The six phenomenological coefficients which appear in the Eqs. (86)–(88) and characterize the transport properties of the membrane near equilibrium in a given reference state can be obtained from a set of experiments characterized by the Eqs. (93)–(98) (WIEDNER and WOERMANN, 1975).

$$\tilde{L}_{VV} = \left(\frac{J_V}{\Delta P}\right)_{I, X_s^*}, \tag{93}$$

$$\tilde{L}_{ss} = \left(\frac{J_s^*}{X_s^*}\right)_{\Delta P, I}, \tag{94}$$

$$\tilde{L}_{II} = \left(\frac{E_-}{I}\right)_{\Delta P, X_s^*}, \tag{95}$$

$$\tilde{L}_{VI} = \left(\frac{J_V}{I}\right)_{\Delta P, X_s^*} = -\left(\frac{E_-}{\Delta P}\right)_{X_s^*, I}, \tag{96}$$

$$\tilde{L}_{sI} = \left(\frac{E_-}{X_s^*}\right)_{\Delta P, I} = \left(\frac{J_s^*}{I}\right)_{\Delta P, X_s^*}, \tag{97}$$

$$\tilde{L}_{Vs} = \left(\frac{J_V}{X_s^*}\right)_{\Delta P, I} = \left(\frac{J_s^*}{\Delta P}\right)_{X_s^*, I}. \tag{98}$$

The equality of the two ratios on the right side of the Eqs. (96)–(98) is a consequence of the reciprocity law of ONSAGER.

In experiments with zero electric current flow the volume flow J_V is equal to volume changes of the bulk phases (and equal to the volume flow across the membrane). If an electric current is passed across the membrane is no longer true. The observed changes of the bulk phases include the volume changes of the electrodes caused by the electrode reaction. For example: with Ag/AgCl-electrodes $dV''/dt = J_V + (V_{Ag} - V_{AgCl})^{el.''}\, I/F$; V_{Ag}, V_{AgCl} = molar volume of solid Ag and AgCl respectively.

Near equilibrium there are other measurable quantities of the membrane which are functions of the coefficients given by Eqs. (93)–(98) for example:

$$\left(\frac{\Delta P}{X_s^*}\right)_{J_V, I} = -\frac{\tilde{L}_{Vs}}{\tilde{L}_{VV}}. \tag{99}$$

The ratio $(\Delta P/X_s^*)_{J_v, I}$ characterizes the osmotic properties of the membrane and is closely related to the reflection coefficient of the electrolyte defined by Eq. (91).

In order to see what kind of information about the transport properties of a membrane is contained in the reflection coefficient let us consider an experiment in which a volume flow and a flow of electrolyte across the membrane is generated by a pressure difference between its bulk phases under the condition $X_s^* = 0$ and $I = 0$. Under these experimental conditions Eq. (100) holds which follows from Eq. (90).

$$\sigma_s = -\frac{1}{\bar{c}_s}\left(\frac{J_s^*}{J_V}\right)_{X_s^*, I} = 1 - \frac{1}{\bar{c}_s}\left(\frac{J_s''}{J_V}\right)_{X_s^*, I}. \tag{100}$$

If it is observed that the flow of the electrolyte relative to the volume flow is zero ($J_s^* = 0$) the reflection coefficient has the value zero. That means electrolyte and water pass the membrane in the same ratio in which they are present in the bulk phases. The membrane does not act as a filter.

If on the other hand the flow of electrolyte relative to the membrane disappears ($J_s'' = 0$) the reflection coefficient has the value 1. The membrane is impermeable to the electrolyte.

The statement that the reflection coefficient of a substance is equal to 1 does not necessarily mean that the membrane is impermeable to this substance. This is only true if the coefficients $\tilde{L}_{ss}$, $\tilde{L}_{Vs}$ and $\tilde{L}_{sI}$ are zero. On the other hand, if a substance cannot penetrate the membrane its reflection coefficient is equal to 1. There are no limitations of the value the reflection coefficient can have (SCHLÖGL, 1964c). This can be seen from Eq. (100). The reflection coefficient can be larger than 1 because there is no necessity that J_V and J_s'' must have the same direction under the condition $\Delta P \neq 0$, $X_s^* = 0$ and $I = 0$. But the reflection coefficient can also be negative if J_V and J_s'' have the same direction.

The sign of the reflection coefficient determines the direction of the osmotic volume flow observed under isobaric conditions at zero electric current flow.

$$J_V = -\tilde{L}_{VV} \sigma_s \bar{c}_s X_s^* \quad \text{with} \quad \Delta P = 0, \; I = 0.$$

In a system with a positive reflection coefficient of the solute the volume flow has the same direction as the gradient of the chemical potential of the solvent between the bulk phases (positive osmosis). In systems with negative reflection coefficient the situation is reversed (negative osmosis). The sign of the reflection coefficient also determines the sign of the filtration effect in a hyperfiltration experiment (SCHLÖGL, 1964g; PUSCH and WOERMANN, 1970).

Furthermore the reflection coefficient characterizes the transport of solute induced by a volume flow under the condition $X_s^* = 0$, $I = 0$:

$$J_s = \bar{c}_s (1 - \sigma_s) J_V.$$

This phenomenon is called "solvent drag" (ANDERSEN and USSING, 1957).

1.5 Summary

The main contributions of nonequilibrium thermodynamics to membrane transport can be summarized as follows (WEI, 1966; KEDEM, 1972):

1. Nonequilibrium thermodynamics provides a frame in which all model theories about transport mechanisms have to be embedded in order to be consistent with thermodynamics. Since nonequilibrium thermodynamics is a phenomenological theory – the membrane is treated as a black box – it cannot give any insight into transport mechanisms.

2. From the linear laws of nonequilibrium thermodynamics the membrane biologist can develop the type of experiments necessary to describe completely the transport properties of a given membrane system near equilibrium. With biological membrane systems it will often not be possible to determine the complete matrix of the phenomenological coefficients for experimental reasons. However the thermodynamic treatment of transport processes will provoke thoughts about the interactions possible between flows and driving forces in the system under consideration.

2. Transport Processes across Membranes with Narrow Pores

2.1 Introduction

In the thermodynamic treatment of transport processes across membranes very little is postulated concerning the properties of the membrane. The membrane should have certain permeabilities and the ability to maintain a stationary state. In order to gain insight into the physical basis of observed transport properties of a membrane it is necessary to construct a suitable model which is amenable to theoretical analysis. The membrane is replaced by a fictitious model reflecting only the characteristic properties of the real membrane. It will be one of the aims of the theoretical treatment of the membrane models to express the important phenomenological transport coefficients in terms of the parameters characteristic of the model. In the literature one finds several different membrane models. Special attention will be focused on the model of the membrane with narrow pores because this model has been worked out in detail (SCHMID, 1950, 1951a, b, 1952; TEORELL, 1951, 1953; SCHLÖGL, 1964a, 1966) and has been used widely to interpret experimental results obtained with biological membranes (HOPE, 1971; WALKER, 1976). The theoretical treatment of this model uses concepts which are also employed in a modified form in the treatment of other membrane models. In the model of the membrane with narrow pores the membrane is treated as a separate macroscopic phase. It is assumed that the matrix of the membrane forming the boundaries of the pores can be a lattice-like structure or can consist of a statistically crosslinked high polymer network. It can carry fixed electric charges: ionic groups or absorbed ions. The pores which are filled with a fluid medium are so narrow that the different components of the fluid are homogeneously distributed over the cross section of the pore by the thermal motion of the molecules. This requires pore diameters in the range from 10 to 10^3 nm (magnitude of radius of the ion cloud in the DEBYE-HÜCKEL theory of strong electrolytes). The topology of the pores can be complicated. The pores may branch out statistically and their cross-sections can have any form. It is assumed that the resistance of the membrane to transport processes is evenly distributed over the membrane phase and not located at the phase boundary membrane/solution.

The thickness of the zone of transition between the equilibrium solution and the membrane phase should be small compared with that of the membrane. Further, the distribution equilibrium at the phase boundary/solution should practically not be disturbed by the transport processes across the membrane. The transport of matter is described by a modified NERNST-PLANCK equation. Membranes with ion exchange properties formed by crosslinked high polymer networks with ionic groups attached by adsorption or by covalent bonds belong to this type of membrane.

The model of the membrane with wide pores (diameter larger than several 10^3 nm) is applicable to membranes in which the different components of the fluid within the pores cannot be distributed homogeneously over the cross-section of the pores by the thermal motion of the molecules. The charges of the ionic groups attached to the wall of the pores are electrically compensated within a narrow zone of fluid adjacent to the wall. The fluid in the center of the pores remains electrically neutral. The theoretical treatment of this model is complicated (MANEGOLD, 1955; KUHN et al., 1963; DRESNER, 1963; KOBATAKE and FUJITA, 1964; LÄUGER, 1964; JACAZIO et al., 1972; GROSS and OSTERLE, 1968).

For thick non-aqueous liquid membranes with and without dissolved oleophilic components the model of the "oil membrane" has been developed. In this model the selective permeabilities of different species is explained on the basis of their different solubilities and mobilities in the non-aqueous phase. The theoretical treatment of oil membranes has been reviewed recently by SANDBLOM and ORME (1973). If the liquid membrane is a liquid ion-exchange membrane the fixed charge model (TEORELL, 1951, 1953) – a special case of the narrow pores model – can be used. However, it has to be taken into account that the distribution of the ion exchange sites within the non-aqueous phase depends on the local electric field.

The theoretical treatments of the membrane models mentioned so far are based on a continuum concept. For very thin membranes (lipid bilayer membranes) which are only a few layers of molecules thick another approach can be used. This is due to the fact that the translocation of ions across thin membranes can be imagined to take place in a relatively small number of kinetic steps. DAVSON and DANIELLI (1952) outlined a "relaxation" theory which pictures the membrane as a series of potential barriers for the transport of particles. This approach has been developed by EYRING and coworkers (PARLIN and EYRING, 1954; JOHNSON et al., 1954) into the absolute rate theory of membrane permeation. It has been applied to an analysis of electric properties of unmodified and modified (by carrier molecules) lipid bilayer membranes (LÄUGER and NEUMCKE, 1973; CIANI et al., 1973). In lipid bilayers the rate theory can be considered as an alternative to the NERNST-PLANCK approach. Both treatments lead to comparable results (CIANI, 1965; SCHLÖGL, 1969).

Due to the simplifying assumptions on which model calculations are based, only qualitative agreement between these calculations and experimental results can be expected. A chosen model can be considered adequate for a given membrane only when the predictions of the theoretical treatment of the model are in reasonably good agreement with experimental results obtained under a wide range of conditions.

2.2 Transport Equations

The model of the membrane with narrow pores is further characterized by the following simplifying assumptions:

1. The solution contained in the pores is so dilute that it can be treated as an ideal one.

2. The concentrations of all components of the medium within the pores are only a function of the coordinate x perpendicular to the phase boundary membrane/solution.

3. If the matrix of the membrane carries electric charges of one kind the solution within the pores contains a surplus of mobile ions of sign opposite to that of the fixed charges. The density of space charge ρ_{el} formed by these mocile ions is only a function of the coordinate x.

4. Coupling between the motion of dissolved particle species is neglected. Not neglected are coupling effects between the dissolved particle species and the solvent, the dissolved particle species and the membrane matrix, and between the solvent and the membrane matrix.

The density of the space charge within the pores is given by:

$$\rho_{el} = -F\omega X. \tag{101}$$

X is the equivalent concentration of the fixed ion groups per unit volume of the solution within the pores (mol · cm^{-3}); F is the Faraday constant and ω the sign of the charges of the fixed ionic groups.

For the whole membrane the condition of electroneutrality has to be fulfilled. That means

$$\sum_{i=1}^{n} z_i C_i + \omega X = 0. \tag{102}$$

The solvent molecules are characterized by the index 0 and the species of the solutes by the index i ($i = 1, 2, 3, \ldots, n$). C_i is the concentration of species i per unit volume of the solution within the pores (mol · cm^{-3}); z_i is its valence including the sign (cations positive, anions negative, non-electrolytes zero). The summation is extended over all mobile ion species ($i = 1, 2, \ldots, n$). The ionic species whose electric charge has the opposite sign to that of the fixed ion groups are called counterions, the ionic species with the same sign are called coions.

For the description of the flow of the dissolved species i across the membranes with narrow pores SCHLÖGL has introduced the following expression:

$$\Phi_i = C_i v - D_i(dC_i/dx + z_i C_i d\bar{\psi}/dx) \quad (i=1,2,\ldots,n). \tag{103}$$

Φ_i represents the molar flow density ($\mathrm{mol \cdot cm^{-2} \cdot s^{-1}}$) and v the volume flow density ($\mathrm{cm \cdot s^{-1}}$) across the membrane. D_i stands for effective diffusion coefficient inside the membrane ($\mathrm{cm^2 \cdot s^{-1}}$). electric potential inside the membrane measured in units of $\frac{RT}{F}$ $\left(\bar{\psi} = \frac{F}{RT}\bar{\varphi};\ \bar{\varphi}\right.$ is the electric potential inside the membrane; $RT/F = 0{,}025$ Volt at $\left.20°\,\mathrm{C}\right)$.

The first term ($C_i v$) on the right hand side of Eq. (103) describes particle transport by convection caused by a streaming medium within the pores in the direction of the x axis. The second term ($-D_i\,dC_i/dx$) describes a diffusional flow caused by a concentration gradient inside the membrane phase. The third term takes into account the transport of charged particles by electrical transfer in the presence of an electric potential gradient ($-d\bar{\varphi}/dx$) inside the membrane. It is assumed that the EINSTEIN relation between the effective diffusion coefficient D_i and the ion mobility u_i ($\mathrm{cm^2 \cdot V^{-1} \cdot s^{-1}}$) holds:

$$D_i = \frac{u_i}{F} RT.$$

The potential gradient can be caused by a diffusion potential or by ohmic resistance if an electric current is passed across the membrane.

According to SCHLÖGL the volume flow density (streaming velocity) is given by:

$$v = d_h(-d\bar{P}/dx + RT\omega X\,d\bar{\psi}/dx). \tag{104}$$

The streaming velocity v of the solution across the membrane is proportional to the forces acting on a unit volume of the medium within the pores. The proportionality factor d_h is called the specific mechanical permeability. The forces are a gradient of hydrostatic pressure inside the membrane ($d\bar{P}/dx$) and an electrical force ($-\rho_{el} \cdot d\bar{\varphi}/dx$) acting on the surplus charge of the solution within the pores. The surplus charge density is given by Eq. (101) and the electric potential $\bar{\psi}$ is again expressed in units of RT/F.

The Eqs. (101)–(104) are called the extended NERNST-PLANCK equations. Their applicability is not limited to membrane systems near equilibrium since the forces are written in differential form. This allows us to include in our considerations systems which are far from equilibrium (e.g. systems in which the concentration of one species in one bulk phase of the membrane is finite and approaching zero in the other). The Eqs. (103) and (104) can be established on the basis of the thermodynamics of irreversible processes in continuous systems (SCHLÖGL, 1964c). Eq. (103) differs from the original NERNST-PLANCK equation (NERNST, 1888, 1889; PLANCK, 1890a, b, 1930) by the convection term $C_i v$.

In order to treat the ion transport across unmodified bilayer membranes in terms of the NERNST-PLANCK equation, an additional term describing the image forces interaction of an ion with the membrane has to be introduced into Eq. (103). In this way the microscopic nature of the bilayer film is taken into account explicitly (NEUMCKE and LÄUGER, 1969; LÄUGER and NEUMCKE, 1973).

Within the frame of the narrow-pore membrane model the properties of the membrane are characterized by the parameters d_h, C_i, D_i and ωX. In general they are functions of the electrolyte concentrations in the bulk phases of the membrane. The determination of the characteristic parameters under physiological conditions is discussed by SCHLÖGL (1966). In the following it will be assumed that the values of these parameters are known for the membrane system under discussion.

In principle the equilibrium concentrations C_i and the fixed ion concentration X can be obtained by determining analytically the composition of the medium within the pores and the water content of the membrane. The effective diffusion coefficient D_i can be determined for example from measurements of the flow of a radioactive isotope across the membrane if the bulk phases are in the same thermodynamic state. The specific mechanical permeability d_h can be obtained from measurements of volume flow across the membrane generated by a pressure difference.

2.3 Equilibrium at the Phase Boundary Membrane/Solution

Let us consider a membrane with ion-exchange properties in equilibrium with a dilute aqueous solution containing several electrolytes and nonelectrolytes. All species of the equilibrium solution can permeate the phase boundary membrane/solution. The equilibrium concentrations C_i in the membrane phase are related to the concentrations c_i in the bulk phases by the requirement of electrochemical equilibrium for each species i:

$$\bar{\eta}_i = \eta_i \quad (i=0, 1, 2, \ldots, n) \tag{105}$$

with

$$\eta_i \equiv \mu_i + z_i F \varphi \tag{106}$$

and

$$\mu_i = \mu_i^0(T, P) + RT \ln a_i \,.$$

$\bar{\eta}_i$ and η_i is the electrochemical potential of species i in the membrane phase and in the bulk phase respectively; μ_i is the chemical potential; the standard value μ_i^0 refers to an unspecified reference state and is a function of P and T only. φ is the electric potential; $a_i \,(=c_i y_i/c^+)$ is the activity of the species i. y_i is the activity coefficient. It characterizes the deviations of the system from ideality and is in general a function of P, T and composition. $c^+ = 1\ \text{mol/cm}^3$.

To obtain information about the equilibrium at the phase boundary membrane/solution, Eq. (105) is applied to every species of the equilibrium solution assuming ideal solutions and neglecting the small pressure dependence of the electrochemical potential. With these assumptions a simplified form of the DONNAN relation is obtained:

$$C_i = c_i \exp(-z_i \delta\psi) \quad (i=1, 2, \ldots, n) \quad \text{with} \quad \delta\psi = \bar{\psi} - \psi \,. \tag{107}$$

$\bar{\psi}$ and ψ is the electric potential in the membrane phase and in the bulk phase respectively, measured in units of RT/F. The jump of the electric potential at the phase boundary membrane/solution, $\delta\psi$ is called DONNAN potential.

According to Eq. (107) the concentrations C_i of uncharged species ($z_i = 0$) is equal to c_i. Individual distribution coefficients of uncharged (as well as charged) species are not taken into account in this approximation. Since $\delta\psi$ has the same sign as the electric charge of the fixed ion groups (ω) (see below and SCHLÖGL, 1964d) the following conclusions can be drawn from Eq. (107):

1. The concentration of counterions in the membrane phase is higher and that of coions smaller than the corresponding concentration in the equilibrium solution.
2. Higher valent counterions are accumulated in the membrane phase to a greater degree than lower valent counterions.
3. Higher valent coions are more effectively excluded from the membrane phase than lower valent coions.

If the equilibrium solution contains only a single binary z, $-z$ valent electrolyte (e.g. KCl, $MgSO_4$) it follows from Eq. (107) that:

$$C_+ C_- = c_s^2 \quad \text{with} \quad c_s = c_+ = c_- . \tag{108}$$

The index s refers to the salt.

Combining Eq. (108) with the requirement of electroneutrality of the membrane phase Eq. (102), we obtain

$$\begin{aligned} C_+ &= \sqrt{\frac{X^2}{4z^2} + c_s^2} - \frac{\omega X}{2z}, \\ C_- &= \sqrt{\frac{X^2}{4z^2} + c_s^2} + \frac{\omega X}{2z}. \end{aligned} \tag{109}$$

Eq. (109) can be used to calculate the concentration of the mobile ions in the membrane phase for a given concentration of a binary z, $-z$ valent electrolyte in the bulk phase if the fixed ion concentration is known. These equations show that the exclusion of coions from the membrane phase is the more effective the greater the fixed ion concentration of the membrane. This statement is generally true and not limited to binary z, $-z$ valent electrolytes.

The jump of the electric potential at the phase boundary membrane/solution (DONNAN potential) is caused by a diffuse double layer. At equilibrium it prevents more counterions passing from the counterion-rich membrane phase into the bulk phase than are moved in the opposite direction by the thermal motion of the ions. In the electric field generated by noncompensated fixed charges at the interface of the membrane the counterions return to the membrane phase sooner or later depending on their kinetic energy.

Therefore the potential of the membrane phase with respect to the equilibrium solution has the same sign as the electric charges of the fixed ion groups. The extension of the diffuse double layer into the bulk phase is of the same order of magnitude as the radius of the ion cloud in the DEBYE-HÜCKEL theory of strong electrolytes at corresponding electrolyte concentrations.

At the phase boundary membrane/solution there exists also a hydrostatic pressure jump. It can be interpreted as an osmotic pressure caused by the concentration difference of the osmotically active particle species between the membrane and the bulk phase (SCHLÖGL, 1955). In this picture the phase boundary membrane/solution is considered a semipermeable membrane, impermeable to the fixed ion groups but permeable for all species of the equilibrium solution. Across such a

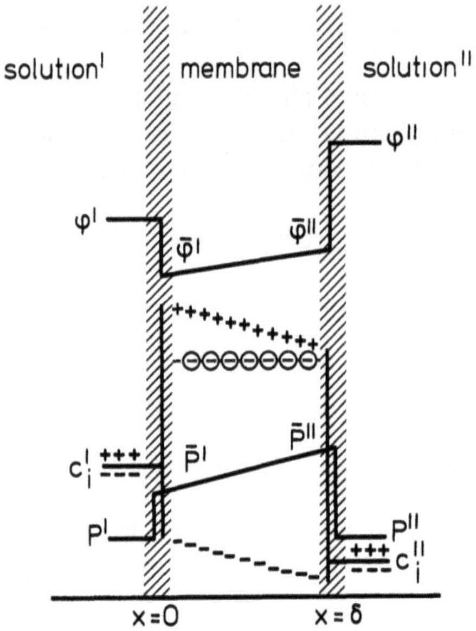

Fig. 4. Concentration, electric potential and pressure profile inside a cation exchange membrane-schematically. The membrane separates two different concentrated solutions of a single binary electrolyte under isobaric conditions. ⊖ = fixed ion groups ($\omega=-1$); + = counterions; − = coions; φ = electric potentials; c_i = concentration of species i; P = pressure; x = space coordinate; δ = thickness of membrane. Quantities refering to the left bulk phase are characterized by the superscript (′) and that to the right bulk phase by the superscript (″). The corresponding quantities refering to the membrane phase are indicated by bar (e.g. $\bar{P}$)

hypothetical membrane a DONNAN equilibrium is established. Assuming that the membrane matrix including the fixed ion groups are osmotically not active and that the solutions are very dilute the pressure jump is given by:

$$\delta P = RT \sum_{i=1}^{n} (C_i - c_i) \quad \text{with} \quad \delta P = \bar{P} - P. \tag{110}$$

The medium within the pores is always under higher hydrostatic pressure irrespective of the sign of the electric charges of the fixed ion groups. The matrix of membranes consisting of a crosslinked high polymer network will also contribute to the pressure jump δP because the matrix is osmotically active due to its numerous degrees of freedom. The osmotic contributions of the high polymer network cancel approximately if the difference of pressure jumps occurring at the phase boundaries of a membrane separating two electrolyte solutions of different composition are considered.

Figure 4 shows schematically the concentration, electric potential, and pressure profiles of a membrane system in which a cation-exchange membrane separates two different concentrated solutions of a single binary electrolyte.

2.4 Integration of the Transport Equations

The exact integration of the extended NERNST-PLANCK equations has been carried out only for the special case in which two mobile ionic species (two counterion species, or one counterion and one coion species) are present in the membrane phase. In addition, the membrane has to be in the stationary state in which the flows Φ_i and the concentrations C_i do not depend on time. In this state no matter is accumulated in the membrane phase and therefore the flows Φ_i are also independent of the space coordinate x. The stationary state of the membrane can be maintained provided the changes of the composition of the bulk phases are sufficiently slow so that

$$\frac{c_i^\alpha}{dc_i^\alpha/dt} \gg \frac{\delta^2}{D_i} \qquad \alpha = ('), ('')$$

δ = thickness of membrane.

The exact integral of the original NERNST-PLANCK equation (SCHLÖGL, 1954, 1964f) as well as the exact integral of the extended NERNST-PLANCK equation for the special case just mentioned (SCHLÖGL, 1966) are obtained in parametric form. This makes quantitative evaluation tedious. The integration of the extended NERNST-PLANCK equation yields results which are easier to handle if it is assumed that the profile of the electric potential inside the membrane is linear. With this "constant field assumption" systems can be treated in which the membrane contains more than two ionic species GOLDMAN has used the constant field assumption to obtain an approximate integral of the original NERNST-PLANCK equation for 1, −1 valent electrolytes (GOLDMAN, 1943; TEORELL, 1951, 1953).

Since that time this approximation has been used extensively in the theory of membrane transport. If the bulk phases have identical compositions and an electric current is passed across the membrane or a volume flow is generated by a hydrostatic pressure difference $d\bar{\psi}/dx$ is constant. But if the bulk phases have different compositions $\bar{\psi}(x)$ shows some curvature in general so that the constant field assumption is an approximation. The implication of this approximation is discussed in detail by SCHLÖGL (1966).

2.4.1 Bulk Phases of Equal Composition

Let us consider an isothermal system in which a membrane with ion exchange properties and a finite mechanical permeability ($d_h \neq 0$) separates two homogeneous dilute aqueous electrolyte solutions of the same composition. The bulk phases are well stirred to keep their composition homogeneous during the experiment. The effect of unstirred layers at the interface membrane/solution is neglected. Electric current can be passed across the membrane by placing two identical electrodes which react reversibly and selectively with one of the ionic species present in the solution (e.g. ionic species n) into the bulk phases and by connecting them to an external current source. For simplicity it is assumed that the current flow causes no potential drop in the bulk phases. A hydrostatic pressure difference across the membrane is allowed.

Due to the symmetry of the system the difference of the pressure jumps ($\delta P' - \delta P''$) and the electric potential jumps ($\delta\varphi' - \delta\varphi''$) at the phase boundaries membrane/solution are zero. The same is true for the difference of the electric potential jumps at the phase boundary electrode/solution.

This means:

$$\Delta P = P' - P'' = \bar{P}' - \bar{P}''$$

and

$$E_n = \varphi^{\text{el.}'} - \varphi^{\text{el.}''} = \varphi' - \varphi'' = \frac{RT}{F}(\bar{\psi}' - \bar{\psi}'').$$

With these boundary conditions the equations of motion, Eqs. (103) and (104) can be integrated without difficulty since $d\bar{\psi}/dx$ is constant ($d\bar{\psi}/dx = -\Delta\bar{\psi}/\delta = -FE_n/RT\delta$; δ = thickness of the membrane).

$$\Phi_\iota = C_\iota v + z_\iota D_\iota C_i \frac{F}{RT\delta} E_n, \tag{111}$$

$$v = \frac{d_h}{\delta}(\Delta P - F\omega X E_n). \tag{112}$$

Depending on the magnitude of the fixed ion concentration X the electric potential difference E_n can be a very effective driving force for the volume flow. A combination of Eqs. (111) and (112) yields:

$$\Phi_\iota = \frac{FC_\iota}{RT\delta}(z_\iota D_i - d_h RT\omega X) E_n + \frac{d_h C_i}{\delta}\Delta P. \tag{113}$$

The electric current density i across the membrane is given by Eq. (114). Eq. (114) follows from Eq. (113) if Eq. (102) is taken into account.

$$i = F\sum_{i=1}^{n} z_\iota \Phi_\iota = \frac{F^2}{\delta}\left(\frac{1}{RT}\sum_{\iota=1}^{n} z_\iota^2 D_\iota C_\iota + X^2 d_h\right) E_n - \frac{F}{\delta}\omega X d_h \Delta P. \tag{114}$$

Starting from the Eqs. (112)–(114) expressions will be derived relating several phenomenological coefficients of a membrane (see Sect. 1.4.7) with the model parameters d_h, C_ι, D_ι and ωX.

2.4.1.1 Electroosmotic Volume Flow

When a direct electric current is passed across the membrane under isobaric conditions a volume flow (electroosmotic volume flow) across the membrane is observed generally. The electric field generated by the external current source exerts a force on the medium within the pores if it carries a surplus charge due to the fixed ion groups of the membrane matrix. The sign of the surplus charge is opposite to that of the fixed ion groups. The streaming velocity (volume flow density) caused by the electric force is determined by the hydrodynamic resistance of the membrane. Near the equilibrium state of the membrane system the electroosmotic volume flow is characterized by the ratio $(J_V/I)_{\Delta P}$ ($J_V = av$; $I = ia$; $a =$ membrane area), which is given by:

$$\left(\frac{J_V}{I}\right)_{\Delta P} = \frac{L_{VE}}{L_{EE}} = -\frac{\omega D_0}{F\left(\sum_{i=1}^{n} z_i^2 D_i C_i + D_0 X\right)} \tag{115}$$

with

$$D_0 \equiv d_h RTX. \tag{116}$$

The sign of $(J_V/I)_{\Delta P}$ is determined by the sign of the fixed ion groups (ω). The parameter D_0 has the dimension of a diffusion coefficient ($cm^2 \cdot s^{-1}$).

2.4.1.2 *Electric Conductivity*

The specific conductivity of the membrane κ^m is given by:

$$\kappa^m = L_{EE}\frac{\delta}{A} = \frac{F^2}{RT}\left(\sum_{i=1}^{n} z_i^2 D_i C_i + D_0 X\right). \tag{117}$$

The first term on the right hand side of Eq. (117) takes into account the specific ionic conductivity of the membrane. The second term describes the specific conductivity caused by the translation of the medium within the pores relative to the membrane matrix (convection conductivity) (SCHMID, 1951b). A volume flow can contribute to the conductivity of the membrane if the streaming medium carries a surplus charge.

2.4.1.3 *Electric Transference*

The transport number $\bar{t}_i$ of the mobile ionic species i in the membrane phase is defined as the fraction of the total electric current carried across the membrane by this ionic species under isobaric conditions.

$$\bar{t}_i \equiv \frac{I_i}{I} = \frac{z_i \Phi_i}{\sum_{i=1}^{n} z_i \Phi_i}. \tag{118}$$

Taking into account the Eqs. (113), (114), (116) and (119) together with the requirement $\Delta P = 0$ the transport number $\bar{t}_i$ is given by:

$$\bar{t}_i = \frac{C_i(z_i^2 D_i - \omega z_i D_0)}{\sum_{i=1}^{n} z_i^2 D_i C_i + D_0 X}. \tag{119}$$

In the limit of infinite dilution of the solutions forming the bulk phases the membrane contains no coions and the electric current is carried across the membrane only by counterions ($\bar{t}_{count} = 1$; $\bar{t}_{co} = 0$).

2.4.1.4 Streaming Potential

If a hydrostatic pressure difference is applied across the membrane separating two bulk phases of equal composition an electric potential difference between the bulk phases—called streaming potential—is observed. The volume flow across the membrane generated by the pressure difference displaces the space charge of the medium within the pores and thus produces the electric potential difference between the bulk phases.

From Eq. (114) the ratio $\left(\frac{E_n}{\Delta P}\right)_I$ can be calculated.

$$\left(\frac{E_n}{\Delta P}\right)_I = \frac{\omega D_0}{F\left(\sum_{i=1}^{n} z_i^2 D_i C_i + D_0 X\right)}. \tag{120}$$

The electric potential of the high pressure phase has the same sign as the charge of the fixed ion groups (ω). The Eqs. (115) and (120) are in agreement with the ONSAGER reciprocity relation [see Eq. (96)].

The volume flow causing a streaming potential can also be generated by a concentration difference of a nonelectrolyte between the bulk phases (HINGSON and DIAMOND, 1972).

Due to the presence of the fixed ion groups within the membrane the concentration of the coions in the membrane is smaller than that in the bulk phases (DONNAN distribution). Thus the flow of salt across the membrane associated with the volume flow remains small. This results in a filtration of the electrolyte solution: the salt concentration in the unstirred layer at the interface membrane/high pressure phase increases during the experiment. The concentration difference between the unstirred layers of the membrane generates an additional electric potential difference (concentration potential) which is superimposed on the streaming potential. Both potential differences have the same sign. Therefore Eq. (120) is only valid for the earliest phase following the application of the hydrostatic pressure difference (SCHMID and SCHWARZ, 1952; WOERMANN, 1974).

2.4.2 Bulk Phases of Different Composition

2.4.2.1 Approximate Integration of the Transport Equations (Constant Field Assumption)

An approximate integration of the extended NERNST-PLANCK equations for systems with different composition of the bulk phases can be carried out assuming a linear profile of the electric potential inside the membrane (constant field assumption). For the integration it is further assumed that the effective diffusion coefficient D_i, the specific mechanical permeability d_h as well as the fixed ion concentration X are constant.

The following presentation follows a treatment given by SCHLÖGL (1966).

For the integration of Eq. (103) it is convenient to introduce a function $\varepsilon_i(x)$ defined by:

$$\varepsilon_i(x) \equiv \exp\left\{z_i \bar{\psi}(x) - \frac{v}{D_i} x\right\}. \tag{121}$$

If the function $\varepsilon_i(x)$ is introduced into Eq. (103) one obtains:

$$\frac{\Phi_i}{D_i} = -\frac{dC_i}{dx} - C_i \frac{d}{dx} \ln \varepsilon_i(x)$$

$$\frac{\Phi_i}{D_i} \varepsilon_i(x)\, dx = -\varepsilon_i(x)\, dC_i - C_i\, d\varepsilon_i(x) = dC_i\, \varepsilon_i(x) \qquad (i=1,2,\ldots,n). \tag{122}$$

In the stationary state of the membrane the flows Φ_i are constant. Integrating both sides of Eq. (122) across the membrane and setting the electric potential inside the membrane at the left membrane interphase equal to zero ($\bar{\psi}'=0$) Eq. (123) is obtained.

$$\Phi_i = D_i \frac{C_i' - C_i''\, \varepsilon_i(\delta)}{\int\limits_0^\delta \varepsilon_i(x)\, dx} \tag{123}$$

with

$$\varepsilon_i(\delta) = \exp\left\{-z_i\, \Delta\bar{\psi} - \frac{v}{D_i}\delta\right\}.$$

So far no approximations have been made. To carry out the integration in the denominator of Eq. (123) the constant field assumption is used:

$$\bar{\psi}(x) \simeq \frac{\bar{\psi}'' - \bar{\psi}'}{\delta} x \quad \text{with} \quad \bar{\psi}'=0. \tag{124}$$

The integration yields:

$$\int\limits_0^\delta \varepsilon_i(x)\, dx = \frac{\varepsilon_i(\delta)-1}{\ln \varepsilon_i(\delta)} \delta. \tag{125}$$

The Eqs. (126)–(128) summarize the results of the approximate integration.

$$\Phi_i = \frac{D_i}{\delta} \frac{C_i' - C_i''\, \varepsilon_i(\delta)}{L(\varepsilon_i(\delta))} \qquad i=1,2,\ldots,n \tag{126}$$

$$L(\varepsilon_i(\delta)) \equiv \frac{\varepsilon_i(\delta)-1}{\ln \varepsilon_i(\delta)}, \tag{127}$$

$$\varepsilon_i(\delta) = \exp\left\{-z_i\, \Delta\bar{\psi} - \frac{v}{D_i}\delta\right\}. \tag{128}$$

The flow Φ_i can be calculated from the Eqs. (126)–(128) if the concentrations C_i' and C_i'' of the medium within the pores at the left and right phase boundary membrane/solution as well as the electric potential difference $\Delta\bar{\psi}$ ($=\bar{\psi}' - \bar{\psi}''$) inside the membrane and the volume flow density v across the membrane are known. Since $L(\varepsilon_i(\delta))$ is a positive quantity the sign (direction) of the flows is determined by the sign of the differences $(C_i' - C_i'\varepsilon_i(\delta))$.

Integration of the transport equation of the volume flow Eq. (104) can be carried out directly:

$$v = \frac{d_h}{\delta}(\Delta\bar{P} - RT\omega X \Delta\bar{\psi})$$

or

$$v = \frac{D_0}{\delta}\left(\frac{\Delta\bar{P}}{RTX} - \omega\Delta\bar{\psi}\right) \quad \text{with} \quad D_0 \equiv d_h RTX. \tag{129}$$

$\Delta\bar{P}\,(=\bar{P}' - \bar{P}'')$ is the pressure difference inside the membrane between the left and right interface.

The pressure difference $\Delta\bar{P}$ can be introduced into the Eqs. (126)–(128) instead of the volume flow density v by substituting Eq. (129) into Eq. (128):

$$\varepsilon_i(\delta) = \exp\left\{-\alpha_i\,\Delta\bar{\psi} - \beta_i\,\frac{\Delta\bar{P}}{RTX}\right\} \tag{130}$$

with

$$\alpha_i = z_i - \omega\frac{D_0}{D_i} \quad \text{and} \quad \beta_i = \frac{D_0}{D_i}. \tag{131}$$

$\Delta\bar{\psi}$ and $\Delta\bar{P}$ cannot be measured directly but can be calculated from the known characteristic parameters of the model of the membrane with narrow pores. If it is assumed that the distribution equilibrium at the phase boundary membrane/solution is maintained although transport processes are taking place across the membranes the relationship between the electric potential difference $\Delta\bar{\psi}$ and the corresponding difference $\Delta\psi\,(=\psi' - \psi'')$ between the bulk phases is given by:

$$\Delta\psi = -\delta\psi' + \Delta\bar{\psi} + \delta\psi''. \tag{132}$$

For ideal dilute solutions the jumps of the electric potential $\delta\psi\,(=\bar{\psi} - \psi)$ at the phase boundary membrane/solution can be calculated from Eq. (107). $\Delta\psi$ may be identified with the electric potential difference between the bulk phases – expressed in units of RT/F – measured with two identical calomel electrodes with KCl-bridges. Consequently $\Delta\bar{\psi}$ is given by:

$$\Delta\bar{\psi} = \frac{F}{RT}\Delta\varphi + \frac{1}{z_i}\ln\frac{c_i'\,C_i''}{c_i''\,C_i'}. \tag{133}$$

However, if two identical electrodes reacting reversibly and specifically with ionic species n are used for the electric potential measurement the Eqs. (134) and (135) hold.

$$E_n = \Delta\varphi^{\mathrm{el}} = \frac{RT}{Fz_n}\ln\frac{c_n'}{c_n''} + \frac{RT}{F}\Delta\psi, \tag{134}$$

$$\Delta\bar{\psi} = \frac{F}{RT}E_n + \frac{1}{z_n}\ln\frac{C_n''}{C_n'}. \tag{135}$$

An expression analogous to Eq. (132) holds for the relationship between the pressure difference $\Delta\bar{P}$ inside the membrane and the pressure difference $\Delta P\,(=P'-P'')$ between the bulk phases:

$$\Delta P = -\delta P' + \Delta\bar{P} + \delta P''. \tag{136}$$

The difference of the hydrostatic pressure jumps $(\delta P''-\delta P')$ at the phase boundaries membrane/solution for ideal dilute solutions is given by [see Eq. (110)]:

$$\delta P'' - \delta P' = RT\sum_{i=1}^{n}(c_i' - c_i'') - RT\sum_{i=1}^{n}(C_i' - C_i''). \tag{137}$$

If Eq. (136) is combined with Eq. (137) one obtains an expression relating $\Delta\bar{P}$ with experimentally determinable quantities:

$$\Delta\bar{P} = \Delta P - RT\sum_{i=1}^{n}(c_i' - c_i'') + RT\sum_{i=1}^{n}(C_i' - C_i''). \tag{138}$$

2.4.2.2 Unidirectional Flows and Flow Ratio

Let us consider an isothermal membrane system in which a net flow of species i takes place across the membrane.

The species i in the two bulk phases are labeled by adding two different radioactive isotopes (tracer species i^* and i^{**}) and the stationary flows of the isotopes across the membrane are measured separately. The experiments are carried out under condition such that the concentration of each of the tracer species is practically zero in one of the bulk phases and constant in the other one during the time of the experiment. To be specific let the species i in the bulk phase (′) be labeled by the radioactive isotope i^* and that in the bulk phase (″) by the isotope i^{**} ($c_i^{*\prime}=$const.; $c_i^{*\prime\prime}\simeq 0$; $c_i^{**\prime}\simeq 0$; $c_i^{**\prime\prime}=$const.). The flow density of tracer species i^* from bulk phase (′) to (″) is indicated by $\vec{\Phi}_i^*$ and that of the tracer species i^{**} in the opposite direction by $\overleftarrow{\Phi}_i^{**}$. If we apply Eq. (123) to the tracer flows taking into account the boundary conditions $c_i^{**\prime}=0$, $c_i^{*\prime\prime}=0$ ($C_i^{**\prime}=0$; $C_i^{*\prime\prime}=0$) we obtain the following expressions for the flows $\vec{\Phi}_i^*$ and $\overleftarrow{\Phi}_i^{**}$ and the flow ratio $\vec{\Phi}_i^*/\overleftarrow{\Phi}_i^{**}$

$$\vec{\Phi}_i^* = D_i\frac{C_i^{*\prime}}{\int_0^\delta \varepsilon_i(x)\,dx}, \tag{139}$$

$$\overleftarrow{\Phi}_i^{**} = -D_i\frac{C_i^{**\prime\prime}\,\varepsilon_i(\delta)}{\int_0^\delta \varepsilon_i(x)\,dx}, \tag{140}$$

$$-\frac{\vec{\Phi}_i^*}{\overleftarrow{\Phi}_i^{**}} = \frac{C_i^{*\prime}}{C_i^{**\prime\prime}}\exp\left\{z_i\,\Delta\bar{\psi} + \frac{v}{D_i}\,\delta\right\}. \tag{141}$$

The Eqs. (140)–(141) only hold if the additional assumption is made that tracer coupling can be neglected (KEDEM and ESSIG, 1965; SAUER, 1973; LI et al., 1974).

If the distribution equilibrium at the phase boundary membrane/solution is maintained we can use the DONNAN relation to replace the concentrations $C_i^{*\prime}$ and $C_i^{**\prime\prime}$ by the corresponding concentrations in the bulk phase.

$$-\frac{\vec{\Phi}_i^*}{\overleftarrow{\Phi}_i^{**}}=\frac{c_i^{*\prime}}{c_i^{**\prime\prime}}\exp\left\{z_i\,\Delta\psi+\frac{v}{D_i}\,\delta\right\}. \tag{142}$$

$\Delta\psi\ (=\psi'-\psi'')$ is the electric potential difference between the bulk phases measured in units of $\frac{RT}{F}$.

Eq. (142) is known as the USSING relation (USSING, 1952, 1965; MEARES and USSING, 1959a, b). For the evaluation of this equation it is not necessary to know the concentration of the tracers in the bulk phases. Since only the ratios $\vec{\Phi}_i^*/c_i^{*\prime}$ and $\overleftarrow{\Phi}_i^{**}/c_i^{**\prime\prime}$ appear in Eq. (142) the amount of tracers per unit volume and the change of these quantities with time can be expressed in any appropriate units, for example in counts per unit volume per unit time.

The double tracer technique makes it possible to decompose the net flow density of particle species i across the membrane Φ_i into two unidirectional flow densities $\vec{\Phi}_i$ and $\overleftarrow{\Phi}_i$.

$$\Phi_i=\vec{\Phi}_i+\overleftarrow{\Phi}_i. \tag{143}$$

$\vec{\Phi}_i$ is the flow density from bulk phase (′) into bulk phase (″), $\overleftarrow{\Phi}_i$ is the flow in the opposite direction. These flows are related to the tracer flow densities $\vec{\Phi}_i^*$ and $\overleftarrow{\Phi}_i^{**}$ by:

$$\vec{\Phi}_i=\frac{c_i'}{c_i^{*\prime}}\,\vec{\Phi}_i^*\quad\text{and}\quad\overleftarrow{\Phi}_i=\frac{c_i''}{c_i^{**\prime\prime}}\,\overleftarrow{\Phi}_i^{**}. \tag{144}$$

If the unidirectional flows are introduced into Eq. (4.32) we obtain:

$$-\frac{\vec{\Phi}_i}{\overleftarrow{\Phi}_i}=\frac{c_i'}{c_i''}\exp\left\{z_i\,\Delta\psi+\frac{v}{D_i}\,\delta\right\}. \tag{145}$$

For membranes with vanishingly small mechanical permeabilities ($d_h\to 0$, $v\to 0$) Eq. (145) can be simplified:

$$-\frac{\vec{\Phi}_i}{\overleftarrow{\Phi}_i}=\frac{c_i'}{c_i''}\exp\left\{\frac{z_i F\Delta\varphi}{RT}\right\}. \tag{146}$$

It follows from Eq. (146) that the electric potential difference between the bulk phases at which the net flow of ion species i across the membrane vanishes ($\vec{\Phi}_i=-\overleftarrow{\Phi}_i$) is given by:

$$\Delta\varphi_i=\frac{RT}{z_i F}\ln\frac{c_i''}{c_i'}. \tag{147}$$

This potential differences is called equilibrium potential of ionic species i. An alternative form of Eq. (146) is obtained by combining Eqs. (145) and (146):

$$-\frac{\vec{\Phi}_i}{\overleftarrow{\Phi}_i}=\exp\left\{\frac{z_i F(\Delta\varphi-\Delta\varphi_i)}{RT}\right\}. \tag{148}$$

Eq. (148) can be used to calculate a relation between the unidirectional flow of ionic species i and the membrane conductivity for this ionic species under the condition $\vec{\Phi}_i=-\overleftarrow{\Phi}_i$ (HODGKIN, 1951). Let us assume that the membrane system is in a state in which $\Delta\varphi=\Delta\varphi_i$ and consequently $\vec{\Phi}=-\overleftarrow{\Phi}_i=\Phi_i^{\mathrm{uni}}$: Now this state is disturbed by applying an infinitesimally small electric current dI causing an infinitesimally small change $d\Delta\varphi$ of the membrane potential. The partial specific conductivity κ_i^{m} of the membrane for ionic species i is related to the partial derivative $(\partial i_i/\partial\Delta\varphi)$ by

$$\kappa_i^{\mathrm{m}}=\delta\left(\frac{\partial i_i}{\partial\Delta\varphi}\right).$$

i_i is the partial electric current density of ionic species i in the membrane. For the total specific conductivity of the membrane. Eq. (149), holds.

$$\kappa^{\mathrm{m}}=\sum_{i=1}^{n}\kappa_i^{\mathrm{m}}. \tag{149}$$

The summation is extended over all ionic species present. i_i is given by:

$$i_i=z_i F(\vec{\Phi}_i+\overleftarrow{\Phi}_i)=z_i F\overleftarrow{\Phi}_i\left(\frac{\vec{\Phi}_i}{\overleftarrow{\Phi}_i}+1\right).$$

From this the derivative $(\partial i_i/\partial\Delta\varphi)$ can be calculated:

$$\frac{\partial i_i}{\partial\Delta\varphi}=z_i F\overleftarrow{\Phi}_i\frac{\partial}{\partial\Delta\varphi}\left(\frac{\vec{\Phi}_i}{\overleftarrow{\Phi}_i}\right)_{\Delta\varphi=\Delta\varphi_i}+z_i F\left(\frac{\vec{\Phi}_i}{\overleftarrow{\Phi}_i}+1\right)\left(\frac{\partial\overleftarrow{\Phi}_i}{\partial\Delta\varphi}\right)_{\Delta\varphi=\Delta\varphi_i}. \tag{150}$$

The first term on the right hand side of Eq. (150) can be calculated from Eq. (148). The second term vanishes for $\Delta\varphi=\Delta\varphi_i$ because $\vec{\Phi}_i=-\overleftarrow{\Phi}_i$.

Finally the following expressions for κ_i^{m} and κ^{m} are obtained:

$$\kappa_i^{\mathrm{m}}=\delta\frac{z_i^2F^2}{RT}\Phi_i^{\mathrm{uni}}$$
$$\kappa^{\mathrm{m}}=\frac{\delta F^2}{RT}\sum_{i=1}^{n}z_i^2\Phi_i^{\mathrm{uni}}. \tag{151}$$

2.4.2.3 *Membrane Potential*

In this section there will be discussed the electric potential difference between the bulk phases of a membrane separating two aqueous electrolyte solutions of different composition under isothermal and isobaric conditions. No electric current is passed across the membrane. This potential difference – called membrane potential – may be identified with the electric potential difference between the bulk phases measured by two identical calomel electrodes with KCl-bridges. The following considerations are limited to membranes with vanishingly small mechanical permeabilities.

The calculation of the membrane potential of membranes with vanishingly small mechanical permeabilities ($d_h \rightarrow 0$) based on the constant field assumption is relatively simple because the influence of the volume flow caused by the osmotic differences between the bulk phases on the membrane potential can be neglected ($v \rightarrow 0$). (For the calculation of the membrane potential of membranes with finite mechanical permeability see SCHLÖGL (1966).)

Under these conditions $\varepsilon_i(\delta)$ is simplified to:

$$\varepsilon_i(\delta) = \exp\{-z_i \Delta\bar{\psi}\}$$

and the flow of ion species i across the membrane [see Eq. (126)] is given by:

$$\Phi_i = \frac{D_i}{\delta} z_i \Delta\bar{\psi} \frac{C_i' - C_i'' \exp\{-z_i \Delta\bar{\psi}\}}{1 - \exp\{-z_i \Delta\bar{\psi}\}}. \tag{152}$$

Since no electric current is passed across the membrane, Eq. (153), holds:

$$\sum_{i=0}^{n} z_i \Phi_i = 0. \tag{153}$$

Combining the Eqs. (152) and (153) the electric potential difference $\Delta\bar{\psi}$ can be calculated. The case in which the bulk phases contain only binary $z, -z$ valent electrolytes ($z_+ = -z_- = z$) is of practical importance. Under this condition the electric potential difference is given by:

$$\Delta\bar{\psi} = \frac{1}{z} \ln \frac{\sum\limits_{\text{cations}} \tilde{p}_+ C_+'' + \sum\limits_{\text{anions}} \tilde{p}_- C_-'}{\sum\limits_{\text{cations}} \tilde{p}_+ C_+' + \sum\limits_{\text{anions}} \tilde{p}_- C_-''}. \tag{154}$$

$\tilde{p}_i$ is the permeability of species i defined by $\tilde{p}_i \equiv a/\delta D_i$.

To obtain the potential difference between the bulk phases the potential jumps at the phase boundary membrane/solution have to be added [see Eq. (132)].

The development of Eq. (154) is based on the constant field assumption. But if the bulk phases of the membrane contain only a *single* binary $z, -z$ valent electrolyte a simple expression for the membrane potentials can be derived without making use of the constant field assumption. This treatment which is again limited to membranes with vanishing small mechanical permeabilities has been used by SCHLÖGL (1964e) to develop a graphical method for the approximate determination of the parameters ωX and D_+/D_- from measurements of the membrane potential (FRÖMMTER, 1970). This method is a modification of a procedure proposed by MEYER and SIEVERS (1936).

2.4.2.4 Osmotic Properties

Let us consider an isothermal-isobaric system in which a membrane with ion exchange properties and a finite mechanical permeability ($d_h \neq 0$) separates two aqueous solutions of electrolytes and nonelectrolytes which can permeate the membrane. The solutions have different compositions and no electric current is passed across the membrane.

In general the particle flows in such a system will be associated with a volume flow across the membrane which can be measured by observing the volume changes of the bulk phases. If a pressure difference is applied across the membrane adjusted in such a way that the osmotic volume flow is suppressed, the system reaches a state which is commonly called "apparent osmotic equilibrium" ($v=0$, $I=0$). The corresponding pressure difference ΔP_0 is called "apparent osmotic pressure". In this state of the membrane system the molar flow of species i is given by:

$$\Phi_i = \frac{z_i D_i}{\delta} \frac{C'_i - C''_i \lambda_0^{\omega z_i}}{\lambda_0^{\omega z_i} - 1} \omega \ln \lambda_0 \tag{155}$$

with

$$\lambda_0 \equiv \exp\left\{-\frac{\Delta \bar{P}_0}{RTX}\right\}. \tag{156}$$

The index 0 refers to the apparent osmotic equilibrium state. λ_0 has to be calculated from the transcendental Eq. (157) by numerical approximation (SCHLÖGL, 1966).

Eq. (155) follows from the Eqs. (126)–(129) together with $\omega \bar{\Delta}\psi_0 = \Delta \bar{P}_0 / RTX$. With the aid of Eqs. (153) and (155) an expression is obtained from which λ_0 and finally the apparent osmotic pressure can be calculated (SCHLÖGL, 1966).

$$\sum_{i=1}^{n} z_i^2 D_i^2 \frac{C'_i - C''_i \lambda_0^{\omega z_i}}{\lambda_0^{\omega z_i} - 1} = 0 \qquad (i=0, v=0), \tag{157}$$

$$\Delta \bar{P}_0 = -RTX \ln \lambda_0, \tag{158}$$

$$\Delta P_0 = \Delta \bar{P}_0 + RT \sum_{i=1}^{n} (c'_i - c''_i) - RT \sum_{i=1}^{n} (C'_i - C''_i). \tag{159}$$

If the bulk phases contain only binary z, $-z$ valent electrolytes Eq. (157) takes on a simple form which can be solved for λ_0.

$$\lambda_0^{\omega z} = \frac{\sum\limits_{\text{cations}} D_+ C'_+ + \sum\limits_{\text{anions}} D_- C''_-}{\sum\limits_{\text{cations}} D_+ C''_+ + \sum\limits_{\text{anions}} D_- C'_-}.$$

For systems near equilibrium the osmotic properties of the membrane system can be characterized by the reflection coefficients of the solutes (see Sect. 1.2.7 and 1.4.7). If the bulk phases contain only a single electrolyte (index s) its reflection coefficient σ_s is given by:

$$\sigma_s = \left(\frac{\Delta P}{\pi_s}\right)_{v,I}. \tag{160}$$

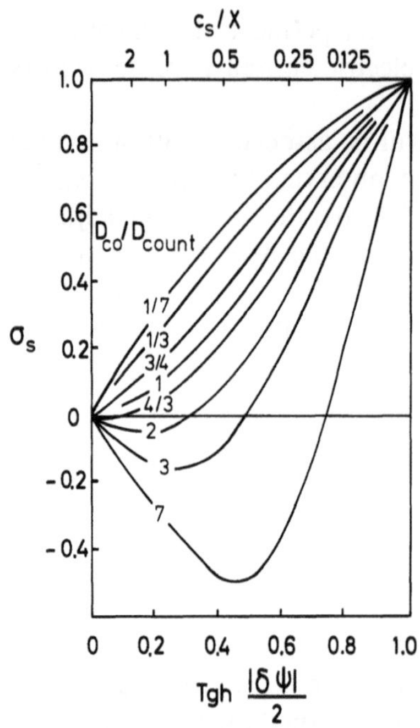

Fig. 5. Reflection coefficient σ_s of a single 1, −1 valent electrolyte as function of the ratio c_s/X. Parameter D_{co}/D_{count}. The curves are calculated theoretically (adapted from SCHLÖGL, 1964). X = fixed ion concentration; c_s = electrolyte concentration; D_{co}, D_{count} = effective diffusion coefficient of coion and counterion species respectively; Tgh $|\delta\psi|/2$ = hyperbolic tangent of $|\delta\psi|/2$; $|\delta\psi|$ = absolute value of $\delta\psi(=\bar{\psi}-\psi)$; $\bar{\psi}$, ψ = normalized electric potental of the membrane phase and the bulk phase respectively

ΔP is the pressure difference which has to be applied between the bulk phases of the membrane to suppress the osmotic volume flow (apparent osmotic pressure ($\Delta P = \Delta P_0$)). π_s is the osmotic difference between the bulk phases. For ideal dilute solution π_s is given by $\pi_s = (\nu_+ + \nu_-) RT \Delta c_s$.

For membrane systems in which the bulk phases contain only a single binary z_+, z_- valent electrolyte the reflection coefficient can be calculated in terms of the parameter D_{co}, D_{count}, X and $\bar{c}_s$ without making use of the constant field assumption (SCHLÖGL, 1955, 1964f). D_{co} and D_{count} are the effective diffusion coefficient of the coions and the counterions within the membrane. $\bar{c}_s$ is the mean electrolyte concentration of the bulk phases ($\bar{c}_s = 1/2 \cdot (c_s' + c_s'')$). We will summarize the results of the theoretical treatment and limit our considerations to the special case of a 1, −1 valent electrolyte.

It is one of the results of the theoretical analysis that the reflection coefficient is a function of the parameter D_{co}/D_{count} and $\bar{c}_s/X$. In Figure 5 plots of theoretically calculated σ_s values as function of $\bar{c}_s/X$ are shown. The ratio D_{co}/D_{count} is used as parameter. For small values of $\bar{c}_s/X$ ($\bar{c}_s/X \ll 1$) the reflection coefficient approaches the value 1. The apparent osmotic pressure becomes equal to the osmotic pressure and the membrane is impermeable to the solute. In the limit of infinite dilution the membrane contains only counterions and no coions (DONNAN equilibrium). Therefore the electrolyte flow across the membrane must vanish. With increasing mean electrolyte concentration of the bulk phases the coion concentration in the membrane increases and consequently also the electrolyte flow across the membrane for a given osmotic difference. The reflection coefficient decreases. For large values of $\bar{c}_s/X$ ($\bar{c}_s/X \gg 1$) the electrolyte content of the medium within the pores of the membrane practically does not differ from that of the bulk phases and the membrane has almost completely lost its ability to exclude the electrolyte from the membrane. The reflection coefficient approaches the value zero.

Further it can be seen from Figure 5 that the reflection coefficient is positive as long as the counterions are more mobile than the coions ($D_{co}/D_{count} > 1$). This means that under isothermal conditions volume flow will be observed which is directed from the diluted to the more concentrated bulk phase (positive osmosis). If on the other hand the coions are more mobile than the counterions ($D_{co.}/D_{count} < 1$) the reflection coefficients will be negative for sufficiently large values of $\bar{c}_s/X$. That means that under isobaric conditions a volume flow will be observed which is directed from the concentrated bulk phase to the more dilute one (negative osmosis). The solvent is transported against its gradient of chemical potential between the bulk phases (SCHÖNBORN and WOERMANN, 1967).

2.5 Summary

After a brief review of the major membrane models the model membrane with narrow pores is discussed in detail. To make a theoretical analysis of this model possible, several assumptions involving drastic simplifications of the real membrane are introduced. From the model calculations insights into the physical basis of observed transport phenomena are gained.

Acknowledgements. The approach used to develop the treatment of transport phenomena across membranes has been much influenced by R. SCHLÖGL's book "Stofftransport durch Membranen" (1964a) and by F. SAUER's article on nonequilibrium thermodynamics of membrane transport (1973). There are basically two reasons for this emphasis. From a didactic point of view it seemed a very appealing approach for an introduction into this subject. But probably more to the point, the author was a student of SCHLÖGL's shortly after the publication of his book and grew up in the field of membrane transport with his approach. With SAUER the author collaborated for some years and learnt a good deal from his thinking about thermodynamics of irreversible processes.

The author is indebted to H. GRUEN for critical reading of the manuscript.

List of Symbols

Latin Letters

a	= membrane area (cm^2)
a_i	= activity of species i (1)
A	= affinity ($J \cdot mol^{-1}$)
B_i	= chemical symbol of species i (1)
c_i	= concentration of species i ($mol \cdot cm^{-3}$)
c^+	= standard value of concentration ($1\ mol \cdot cm^{-3}$)
$\bar{c}_i$	= arithmetic mean value of the concentration of species i ($mol \cdot cm^{-3}$)
C_i	= concentration of species i in the medium within the pores ($mol \cdot cm^{-3}$)
d_h	= specific mechanical permeability ($cm \cdot J^{-1} \cdot s^{-1}$)
D_i	= effective diffusion coefficient of species i in the membrane ($cm^2 \cdot s^{-1}$)
D_0	= measure of mechanical permeability of the membrane ($cm^2 \cdot s^{-1}$)
e (index)	= permeable ionic species reacting specifically and reversibly with a given electrode

E_i	= electric potential difference between two identical electrodes reacting specifically and reversibly with ionic species i (V)
el (index)	= electrode phase
F	= Faraday number 96487 ($A \cdot s \cdot mol^{-1}$)
i (index)	= permeable species i
i^*, i^{**} (index)	= different radioactive isotopes of species i
i	= electric current density ($A \cdot cm^{-2}$)
i_i	= partial electric current density of ion species i in the membrane phase ($A \cdot cm^{-2}$)
I	= electric current (A)
j (index)	= permeable species j
J_c	= turn over velocity ($mol \cdot s^{-1}$)
$J_{c,i}$	= rate of chemical production of species i ($mol \cdot s^{-1}$)
J_i	= flow of species i relative to the membrane ($mol \cdot s^{-1}$)
J_i^*	= flow of species i relative to the volume flow ($mol \cdot s^{-1}$)
J_u	= energy flow ($J \cdot s^{-1}$)
J_V	= volume flow ($cm^3 \cdot s^{-1}$)
L_{ij}, $\tilde{L}_{ij}$, L_{ij}^*, L_{ij}^{**}	= phenomenological coefficient
$L(\varepsilon_i(\delta))$	= quantity defined by Eq. (127) (1)
m (index)	= membrane
n (index)	= permeable species n; ionic species reacting specifically and reversibly with a given electrode
n_i	= amount of substance of species i (mol)
0 (index)	= solvent
p_i	= permeability of species i ($mol \cdot cm^3 \cdot J^{-1} s^{-1}$)
$\tilde{p}_i$	= permeability of species i ($cm^3 \cdot s^{-1}$)
P, $\bar{P}$	= pressure in the bulk phase and the membrane phase respectively ($J \cdot cm^{-3}$)
Q	= heat (J)
R	= gas constant 8.314 ($J \cdot K^{-1} mol^{-1}$)
r	= reference state
s (index)	= salt
S	= entropy ($J \cdot K^{-1}$)
t (index)	= nonpermeable species
t	= time (s)
$\bar{t}_i$	= transference number of ion species i in the membrane (1)
T	= thermodynamic temperature (K)
U	= internal energy (J)
v	= volume flow density ($cm \cdot s^{-1}$)
V	= volume (cm^3)
V_i	= partial molar volume of species i ($cm^3 \cdot mol^{-1}$)
w (index)	= nonpermeable species
x	= space coordinate running perpendicular to phase boundary membrane/solution (cm)
x_i	= mole fraction of species i (1)
$\bar{x}_i$	= arithmetic mean value of mole fraction of species i (1)

X	= equivalent concentration of fixed ionic groups in the medium within the pores (mol · cm^{-3})
X_i^*, X_i	= generalized force conjugate to the flow J_i^* and J_i respectively (J · mol^{-1})
X_V	= driving force conjugate to the flow J_v (J · cm^{-3})
y_i	= activity coefficient (1)
z, (z_i)	= valency of species i (1)
+, − (index)	= cationic species and anionic species respectively

Greek Letters

α (index)	= phase index
α_i	= quantity defined by Eq. (131) (1)
β_i	= quantity defined by Eq. (131) (1)
δ	= thickness of membrane (cm)
Δ	= difference between phase (′) and phase (″) quantities (Δ = ′ − ″)
$\varepsilon_i(x)$, $\varepsilon_i(\delta)$	= quantity defined by Eq. (121) and Eq. (123) respectively (1)
η_i, $\bar{\eta}_i$	= electrochemical potential of species i in the bulk phase and in the membrane phase respectively (J/mol)
κ^{m}	= specific conductivity of the membrane material (Ω^{-1} · cm^{-1})
κ_i^{m}	= partial specific conductivity of ion species i in the membrane (Ω^{-1} · cm^{-1})
λ	= quantity defined by Eq. (156) (1); λ_0 refers to the condition $J_V = 0$, $I = 0$
μ_i	= chemical potential of species i (J · mol^{-1})
μ_i^0	= unspecified standard state of chemical potential of species i (J · mol^{-1})
ν_i	= stoichiometric number of species i (1)
ξ	= extent of reaction (mol)
π_i	= osmotic difference of species i (J · cm^{-3})
$\rho_{el.}$	= density of space charge (A · s · cm^{-3})
σ_i	= reflection coefficient of species i (1)
φ, $\bar{\varphi}$	= electric potential of bulk phase and membrane phase respectively (V)
ϕ_i	= molar flow density of species i across the membrane (mol · cm^{-2} · s^{-1})
$\vec{\phi}_i$, $\overleftarrow{\phi}_i$	= unidirectional flow density of species i across the membrane (mol · cm^{-2} · s^{-1}); → from the left of the right bulk phase; ← opposite direction
ψ, $\bar{\psi}$	= normalized electric potential of bulk phase and membrane phase respectively (1)
ω	= sign of the fixed charges (1)
(′), (″)	= phase index of the left and right bulk phase respectively
1 atm	= 101 325 kg · m^{-1} s^{-2} = 0.101 325 J · cm^{-3}
1 J	= 0.2389 cal

References

ANDERSEN, B., USSING, H.H.: Solvent drag on non-electrolytes during osmotic flow through isolated toad skin and its response to antidiuretic hormone. Acta Physiol. Scand. **39**, 228–239 (1957)

CIANI, S.M.: A rate theory analysis of steady diffusion in a fixed charge membrane. Biophysik **2**, 368–378 (1965)

CIANI, S.M., EISENMANN, G., LAPRODE, R., SZABO, S.: Theoretical analysis of carrier-mediated electrical properties of bilayer membranes. In: Membranes (G. EISENMANN, ed.), Vol. 2. New York: Marcel Dekker 1973

DAINTY, J.: Water relations in plant cells. Adv. Bot. Res. **1**, 279–326 (1963)

DAINTY, J.: Water Relations of Plant Cells. In Encyclopedia of Plant Physiology, New Series, Vol. 2A (U. LÜTTGE, M.G. PITTMAN, eds.), pp. 12–35. Berlin-Heidelberg-New York: Springer 1976

DAVSON, H., DANIELLI, J.F.: The permeability of natural membranes, pp. 324–335. Cambridge: University Press 1952

DRESNER, L.: Electrokinetic phenomena in charged microcapillaries. J. Phys. Chem. **67**, 1635–1641 (1963)

FRÖMMTER, E.: Elektrophysiologische Untersuchungen am proximalen Tubulus der Rattenniere. Habilitationsschrift Frankfurt 1970

GOLDMAN, D.E.: Potential impedance and rectification in membranes. J. Gen. Physiol. **27**, 37–60 (1943)

GROOT, S.R. DE, MAZUR, P.: Non-equilibrium thermodynamics. Amsterdam: North Holland 1962

GROOT, S.R. DE, MAZUR, P.: Anwendung der Thermodynamik irreversibler Prozesse. Mannheim-Wien-Zürich: Bibliographisches Institut 1974

GROSS, R.J., OSTERLE, J.F.: Membrane transport characteristics of ultrafine capillaries. J. Chem. Phys. **49**, 228–234 (1968)

HAASE, R.: Thermodynamics of Irreversible Processes. Reading, Mass.-Menlo Park, Calif.-London-Don Mills, Ontario: Addison-Wesley 1969

HINGSON, D.J., DIAMOND, J.M.: Comparison of nonelectrolyte permeability. Patterns in several epithelia. J. Membrane Biol. **10**, 93–135 (1972)

HODGKIN, A.L.: The ionic basis of electric activity in nerve and muscle Biol. Rev. **26**, 339–409 (1951)

HOPE, A.B.: Ion Transport and Membranes. London: Butterworths 1971

JACAZIO, G., PROBSTEIN, R.F., SONIN, A.A., YUNG, D.: Porous materials for reverse osmosis membranes. Fluid Mechanics Laboratory, Dept. Mech. Eng., Massachusetts Inst. Tech. 1972

JOHNSON, F.H., EYRING, H., POLISSAR, J.: The Kinetic Basis of Molecular Biology. New York: Wiley 1954

KATCHALSKY, A., CURRAN, P.F.: Nonequilibrium Thermodynamics in Biophysics. Cambridge, Mass.: Harvard Univ. Press 1965

KEDEM, O.: Criteria of Active Transport in Membrane Transport and Metabolism (A. KLEINZELLER, A. KOTYK, eds.), pp. 87–90. New York: Academic Press 1961

KEDEM, O.: From irreversible thermodynamics to network thermodynamics. J. Membrane Biol. **10**, 213–219 (1972)

KEDEM, O., ESSIG, A.: Isotope flow and flux ratios in biological membranes. J. Gen. Physiol. **48**, 1047–1070 (1965)

KEDEM, O., KATCHALSKY, A.: Thermodynamic analysis of the permeability of biological membranes to non-electrolytes. Biochem. Biophys. Acta **27**, 229–246 (1958)

KEDEM, O., KATCHALSKY, A.: A physical interpretation of the phenomenological coefficients of membrane permeability. J. Gen. Physiol. **45**, 143–179 (1961)

KEDEM, O., KATCHALSKY, A.: Permeability of composite membranes. I. Electric current, volume flow and flow of solute through membranes. Trans. Faraday Soc. **59**, 1918–1930 (1963a)

KEDEM, O., KATCHALSKY, A.: Permeability of composite membranes. II. Parallel elements. Trans. Faraday Soc. **59**, 1931–1940 (1963b)

KEDEM, O., KATCHALSKY, A.: Permeability of composite membranes. III. Series array of elements. Trans. Faraday Soc. **59**, 1941–1953 (1963c)

KOBATAKE, Y., FUJITA, H.: Flows through charged membranes J. Chem. Phys. **40**, 2212–2222 (1964)

KUHN, W., LÄUGER, P., VOELLMY, H., BLOCK, R., MAJER, H.: Volumen- und Stofftransport durch weitporige Membranen. Ber. Bunsenges. Physik. Chem. **67**, 364–372 (1963)

LÄUGER, P.: Stoff- und Volumentransport durch Membranen mit elektrisch geladenem Gerüst. Ber. Bunsenges. Physik. Chem. **68**, 352–361 (1964)

LÄUGER, P., NEUMCKE, B.: Theoretical analysis of ion conductance in lipid bilayer membranes. In: Membranes (G. EISENMANN, ed.), Vol. 2, pp. 1–59. New York: Marcel Dekker 1973

LI, J.H., DESOUSA, R.C., ESSIG, A.: Kinetics of tracer flows and isotape interaction in an ion exchange membrane. J. Membrane Biol. **19**, 93–104 (1974)

MANEGOLD, E.: Kapillarsysteme, Bd. 1 und Bd. 2, Straßenbau. Heidelberg: Chemie und Technik Verlagsgesellschaft 1955

MEARES, P., USSING, H.H.: The fluxes of sodium and chloride ions across a cation exchange resin membrane. Part 1–Effect of a concentration gradient. Trans. Faraday Soc. **55**, 142–155 (1959a)

MEARES, P., USSING, H.H.: The fluxes of sodium and chloride ions across a cation exchange resin membrane. Part 2–Diffusion with electric current. Trans. Faraday Soc. **55**, 244–254 (1959b)

MEIER, J.: Kopplung von Stofftransport und chemischer Reaktion an einer zusammengesetzten asymmetrischen Membran. Ein Beispiel für aktiven Transport. Doktorarbeit Frankfurt (M.) 1973

MEIER, J., SAUER, F., WOERMANN, D.: Coupling of mass transfer and chemical reaction across an asymmetric sandwich membrane. In: Membrane Transport in Plants (U. ZIMMERMANN, J. DAINTY, eds.), pp. 28–35. Berlin-Heidelberg-New York: Springer 1974

MEIXNER, J., REIK, H.G.: Thermodynamik der irreversiblen Prozesse. In: Handbuch der Physik, Vol. III/2, pp. 413–523. Berlin-Göttingen-Heidelberg: Springer 1959

MEYER, K.H., SIEVERS, J.F.: La permeabilité des membranes. II. Helv. Chim. Acta **19**, 665–677 (1936)

NERNST, W.: Zur Kinetik der in Lösung befindlichen Körper. Z. Physik. Chem. **2**, 613–636 (1888)

NERNST, W.: Die elektromotorische Wirksamkeit der Ionen. Z. Physik. Chem. **4**, 129–181 (1889)

NEUMCKE, B., LÄUGER, P.: Nonlinear electrical effects in lipid bilayer membranes. II. Integration of the generalized Nernst-Planck-equation. Biophys. J. **9**, 1160 (1969)

NOBEL, P.S.: Introduction to Biophysical Plant Physiology, pp. 138–156. San Francisco: W.H. Freeman and Company 1974

PARLIN, B., EYRING, H.: Membrane permeability and electrical potential. In: Ion Transport across Membranes (H.T. CLARKE, ed.), pp. 103–119. New York: Academic Press 1954

PLANCK, M.: Über die Erregung von Elektrizität und Wärme in Elektrolyten. Ann. Physik u. Chem. N.F. **39**, 161–186 (1890a)

PLANCK, M.: Über die Potentialdifferenz zwischen zwei verdünnten Lösungen binärer Elektrolyte. Ann. Physik u. Chem. N.F. **40**, 561 (1890b)

PLANCK, M.: Über die Grenzschicht verdünnter Elektrolyte. Sitzungsber. Preuß. Akad. d. Wiss. Phys. Math. Kl. Berlin XX, pp. 367–373 (1930)

PUSCH, W., WOERMANN, D.: Study of the relation between reflection coefficient and solute rejection efficiency using a strong basic anion exchange membrane. Ber. Bunsenges. Physik. Chem. **74**, 444–449 (1970)

SANDBLOM, J., ORME, F.: Liquid membranes as electrodes and biological models. In: Membranes (G. EISENMANN, ed.), Vol. 2, pp. 125–177. New York: Marcel Dekker 1973

SAUER, F.: Nonequilibrium thermodynamics of kidney tubule transport. Handbook of Physiology, Section Renal Physiology. Amer. Physiol. Soc. pp. 399–414 (1973)

SCHLÖGL, R.: Elektrodiffusion in freier Lösung und geladenen Membranen. Z. Physik. Chem. (Frankfurt) **1**, 305–339 (1954)

SCHLÖGL, R.: Zur Theorie der anomalen Osmose. Z. Physik. Chem. (Frankfurt) **3**, 73–102 (1955)

SCHLÖGL, R.: Stofftransport durch Membranen. Darmstadt: Steinkopff 1964a; pp. 42–43 1964b; pp. 58–63 1964c; p. 70 1964d; pp. 76–79 1964e; pp 79–88 1964f; p. 100 1964g

SCHLÖGL, R.: Membrane permeation in systems far from equilibrium. Ber. Bunsenges. Physik. Chem. **70**, 400–414 (1966)

SCHLÖGL, R.: Non-linear transport behaviour in very thin membranes. Quart. Rev. **2**, 305–313 (1969)

SCHMID, G.: Zur Elektrochemie feinporiger Kapillarsysteme. I. Übersicht. Z. Elektrochem. **54**, 424–430 (1950)

SCHMID, G.: Zur Elektrochemie feinporiger Kapillarsysteme. II. Elektroosmose. Z. Elektrochem. **55**, 229–237 (1951a)

SCHMID, G.: Zur Elektrochemie feinporiger Kapillarsysteme. III. Elektrische Leitfähigkeit. Z. Elektrochem. **55**, 295–307 (1951b)

SCHMID, G.: Zur Elektrochemie feinporiger Kapillarsysteme. VI. Konvektionsleitfähigkeit. Z. Elektrochem. **56**, 181–193 (1952)

SCHMID, G., SCHWARZ, H.: Zur Elektrochemie feinporiger Kapillarsysteme. V. Strömungspotentiale, Donnan-Behinderung des Elektrolytdurchgangs bei Strömung. Z. Elektrochem. **56**, 35–44 (1952)

SCHÖNBORN, M., WOERMANN, D.: Über das osmotische Verhalten von Ionenaustauschermembranen. Ber. Bunsenges. Physik. Chem. **71**, 843–855 (1967)

STAVERMAN, A.J.: The theory of measurement of osmotic pressure. Rec. Trav. Chim. Pays-Bas **70**, 344–352 (1951)

STAVERMAN, A.J.: Apparent osmotic pressure of solutions of hetero disperse polymers. Rec. Trav. Chim. Pays-Bas **71**, 623–633 (1952)

STAVERMAN, A.J.: Non-equilibrium thermodynamics of membrane processes. Trans. Faraday Soc. **48**, 176–185 (1952)

TEORELL, T.: Zur quantitativen Behandlung der Membranpermabilität. Z. Elektrochem. **55**, 460–469 (1951)

TEORELL, T.: Transport processes and electrical phenomena in ionic membranes. Progr. Biophys. **3**, 305–369 (1953)

USSING, H.H.: Some aspects of the application of tracers in permeability studies. Adv. Enzymol. **13**, 21–65 (1952)

USSING, H.H.: Transport of Electrolytes and Water across Epithelia. The Harvey Lectures. Series 59, pp. 1–30. New York: Academic Press 1963

WALKER, N.A: Membrane Transport: Theoretical Background. In Encyclopedia of Plant Physiology, New Series, Vol. 2A (U. LÜTTGE, M.G. PITTMAN, eds.), pp. 36–52. Berlin-Heidelberg-New York: Springer 1976

WEI, J.: Irreversible thermodynamics in engineering. Ind. Eng. Chem. **58**, 55–60 (1966)

WIEDNER, G., WOERMANN, D.: Transport coefficients of an ion exchange membrane and their concentration dependance. Ber. Bunsenges. Physik. Chem. **79**, 868–878 (1975)

WOERMANN, D.: Transport processes across membranes with narrow pores. In: Membrane Transport in Plants (U. ZIMMERMANN, J. DAINTY, eds.). Berlin-Heidelberg-New York: Springer 1974

Author Index

Aach, H.G., Heber, U. 110, *127*, 260, *281*
Aaronson, A.I., Wilt, F.H 62, *79*
Aaronson, S see Patni, N.J. 16, *46*
Aas, M. 246, *250*
Aas, M., Bremer, J. 246, *250*
Abelson, H T., Penman, S. 59, *79*
Adams, E C., Hertig, A.T. 294, *326*
Adelberg, E A. see Davis, E M 62, *80*
Agutter, P.S. 19, *40*
Aimi, J 57, *79*
Akazawa, T. see Sugiyama, T. 108, *134*
Albanese, I. see Giudice, G 74, *81*
Albersheim, P., Anderson-Prouty, A.J. 323, *326*
Albersheim, P see Keegstra, K 307, *332*
Albersheim, P. see Villemez, C L. 6, 16, *49*, 319, *343*
Alberti, R.A. 241, *250*
Alexeev, A.B., Betina, M.I., Swolinsky, S L , Yazikov, A , Zubarev, T.N. 65, *79*
Alibert, G., Ranjeva, R , Boudet, A. 268, *281*
Alivisatos, S G.A see Kaufman, S. 238, *252*
Allaway, W G. see Osmond, C B. 146, *182*
Allen, M B., Arnon, D I , Capindale, J B , Whatley, F.R , Durham, L.J. 116, *127*
Allen, M.B., Whatley, F R , Arnon, D I. 116, *128*
Allen, M.B. see Arnon, D.I 111, 116, 124, *128*, 347, *370*
Allen, M.B. see Whatley, F.R. 86, 118, *136*
Allison, A.C., Davies, P., de Petris, S. 275, *281*
Allison, V F. see Kellems, R E 25, 28, *43*
Amelunxen, F., Arbeiter, H 314, *326*
Amelunxen, F., Wahlig, T , Arbeiter, H. 314, *326*
Amrhein, N., Zenk, M.H. 268, *281*
Andersen, B., Ussing, H H. 440, *462*
Andersen, K S., Bain, J M , Bishop, D.G., Smillie, R.M 171, *177*
Andersen, K.S. see Bishop, D.G 171, *177*
Andersen, K.S see Smillie, R.M. 171, 172, *183*
Anderson, B.J. see Giaquinta, R.T 350, *372*
Anderson, C.M.A. see Bradbeer, J.W. 194, 195, *224*
Anderson, J M , Boardman, N.K., Spencer, D. 171, *177*
Anderson, J.M , Woo, K.C., Boardman, N K 171, 172, *177*
Anderson, J.M see Woo, K.C 171, *184*
Anderson, L. see Elder, J.H. 309, *329*
Anderson, M.M , McCarty, R.E., Zimmer, E.A. 354, *370*
Anderson, N G., Harris, W.W., Barber, A.A , Rankin, C.T., Jr , Candler, E L. 10, *40*
Anderson, W P see Higinbotham, N. 265, *283*
Anderson-Prouty, A J see Albersheim, P 323, *326*
Andrews, T J., Hatch, M.D 155, 156, *177*
Andrews, T.J , Johnson, H.S , Slack, C R , Hatch, M.D. 151, 153, 155, *177*, 210, *223*
Andrews, T J , Lorimer, G H , Tolbert, N.E 109, *128*, 176, *177*, 185, 194, 195, 221, *223*, *224*
Andrews, T.J. see Badger, M R 174, 175, 176, *177*
Andrews, T J. see Lorimer, G.H. 109, *132*, 176, *182*, 185, 195, 197, 222, *229*
Apel, K., Schweiger, H.G 67, 68, *79*
Appelmans, F., Wattiaux, R., De Duve, C. 6, *40*
Aquila, H see Klingenberg, M. 239, *252*
Arbeiter, H see Amelunxen, F. 314, *326*
Armond, P see Arntzen, C.J. 350, *370*
Arnason, B. see Darzynkiewicz, Z 63, *80*
Arnon, D I. 7, *40*, 86, 116, *128*, 193, *224*, 347, *370*
Arnon, D.I , Allen, M.B., Whatley, F.R. 111, 116, 124, *128*, 347, *370*
Arnon, D.I., Whatley, F.R., Allen, M B. 116, *128*
Arnon, D.I. see Allen, M B. 116, *127*, *128*
Arnon, D.I. see Buchanan, B.B. 96, *129*
Arnon, D.I. see Kalberer, P.P. 86, *132*
Arnon, D.I see Schürmann, P. 118, *134*
Arnon, D.I see Tagawa, H. 355, *375*
Arnon, D.I. see Whatley, F R. 86, 118, *136*
Arnott, H.J., Walne, P.L. 310, *326*
Arntzen, C J., Briantais, J -M. 171, *177*
Arntzen, C.J , Dilley, R A , Crane, F.L. 34, 35, *40*
Arntzen, C J., Dilley, R.A., Neumann, J 171, *177*
Arntzen, C.J., Vernotte, C , Briantais, J -M , Armond, P 350, *370*
Aronoff, S. 190, *224*
Arsenis, C. 391, *394*
Asaki, T. see Hirai, M. 270, *283*

Asaki, T. see Nakano, M. 270, 271, *285*
Ash, G.F see Wessels, N.K 58, *84*
Ash, J.F. see Wessels, N.K 292, 294, *343*
Ashworth, J.M. see Wiener, E. 270, *287*
Assel, S. van, Brachet, J. 64, *79*
Assimacopoulos-Jeannet, F see Orci, L. 291, *337*
Atkins, C.A 211, 215, *224*
Atkins, C.A., Canvin, D.T. 188, 217, *224*
Atkins, C.A. see Graham, D. 165, *179*
Atkinson, D.E. 392, *394*
Ausländer, W., Heathcote, P., Junge, W. 353, *371*
Ausländer, W., Junge, W. 353, 354, *371*
Ausländer, W see Junge, W 350, 352, 353, *373*
d'Auzac, J. see Ribaillier, D. 258, 260, 261, 263, *285*
Avrameas, S 17, *40*
Avrameas, S. see Leduc, E.H. 17, *44*
Avrameas, S. see Leduc, E N 17, *44*
Avron, M. 348, *371*
Avron, M., Chance, B. 353, 358, *371*
Avron, M., Gibbs, M. 116, 118, *128*
Avron, M., Krogmann, D W , Jagendorf, A T. 363, *371*
Avron, M. see Bamberger, E.S. 358, 368, *371*
Avron, M. see Forti, G 348, *372*
Avron, M. see Jagendorf, A.T. 116, *131*
Avron, M see Karlish, S.J.D. 351, 364, *373*
Avron, M. see Pick, U 365, *374*
Avron, M. see Rottenberg, H. 357, 360, 362, 363, *374*
Avron, M. see Schuldiner, S. 362, 363, *375*
Axelos, M see Péaud-Lenöel, C. 320, *338*
Axline, S G , Reaven, E.P. 294, *326*
Azzi, A , Chappell, J B., Robinson, B.H. 245, *250*
Azzone, G F , Carafoli, E , Lehninger, A L , Quagliariello, E , Siliprandi, N. 377, *394*

Bachmann, E see Brierley, G.P. 389, *394*
Badger, M.R , Andrews, T.J. 174, 175, 176, *177*
Badger, M.R., Andrews, T J., Osmond, C.B 175, 176, *177*
Baglioni, C see Bishop, J.O 75, *79*
Baglioni, C see Gross, K.W 73, *81*
Bahr, J T , Jensen, R.G. 108, *128*, 174, 175, 176, *177*, 195, *224*
Bain, J.M. see Andersen, K S. 171, *177*
Bajaj, Y.P.S. see Rathore, V.S. 393, *397*
Baker, D A., Hall, J.L 265, *281*
Baker, J.E., Lieberman, M 408, *412*
Bakker, E.P. see Fiolet, J.T.W. 362, 363, *372*
Bal, A.K , Payne, J.F. 315, *326*
Baldry, C W , Bucke, C., Coombs, J. 152, *177*
Baldry, C.W., Bucke, C., Walker, D A 111, *128*
Baldry, C.W , Cockburn, W., Walker, D.A. 112, *128*
Baldry, C W., Walker, D A , Bucke, C. 103, 104, 106, *128*
Baldry, C.W. see Bucke, C. 103, 104, *129*
Baldry, C.W. see Cockburn, W. 89, 90, 102, 103, 104, 105, 106, 112, 113, 118, 121, 124, 125, *129*, 404, 406, 408, *412*
Baldry, C W see Coombs, J. 146, 152, 174, 175, *178*
Baldry, C.W. see Walker, D.A. 89, 104, *135*
Baldwin, R.L. see Jimenez, E 197, 201, *228*
Ball, E see Hope, A B 402, *413*
Baltus, E., Quertier, J., Ficq, A., Brachet, J 73, *79*
Bamberger, E S., Ehrlich, B.A., Gibbs, M. 115, *128*, 140, *142*
Bamberger, E.S., Gibbs, M. 103, *128*, 406, *412*
Bamberger, E S., Rottenberg, H., Avron, M 358, 368, *371*
Bamberger, E.S. see Gibbs, M. 103, *130*
Bandlow, W. 27, *40*
Bannister, T.T. see Selman, B.R. 350, *375*
Bannwarth, H., Schweiger, H.G. 67, *79*
Barber, A.A. see Anderson, N.G 10, *40*
Barber, J. 402, *412*
Barber, J , Mills, J., Nicolson, J 126, *128*, 357, *371*
Barber, J., Telfer, A , Nicolson, J 126, *128*
Barber, J. see Telfer, A. 352, 364, *375*
Barker, R see Ryan, F.J. 195, *231*
Barlow, S.A. see Moore, T C. 292, *336*
Barnett, J.R., Preston, R.D 321, *326*
Bartley, W. see Bert, L M. 387, *394*
Bartley, W see Werkheiser, W C 27, *50*, 89, *135*, 360, *376*
Bartnicki-Garcia, S. 288, 308, *326*
Bartnicki-Garcia, S. see Ruiz-Herrera, J. 320, 322, *340*
Bassham, J.A. 96, 123, *128*
Bassham, J.A., Benson, A A., Calvin, M 194, *224*
Bassham, J.A , Calvin, M 86, 89, *128*, 144, *177*
Bassham, J.A , El-Badry, A M , Kirk, M.R., Ottenheym, H C.J., Springer-Lederer, H. 116, *128*
Bassham, J.A., Jensen, R.G 103, 104, 105, 106, *128*
Bassham, J A , Kirk, M R 193, 195, *224*
Bassham, J A., Kirk, M R , Jensen, R G. 103, 104, 105, 106, 109, *128*, 140, *142*, 209, *224*, 404, 406, *412*
Bassham, J.A , Sharp, P , Morris, I 108, *128*, *129*
Bassham, J A see Benson, A.A. 86, *129*
Bassham, J A. see Chan, H W.-S 194, 209, *225*

Bassham, J.A see Chu, D K. 105, *129*, 195, *225*
Bassham, J.A. see Jensen, R G. 86, 105, 111, 115, 116, 117, *132*, 191, 194, 209, *228*, 404, 405, *414*
Bassham, J.A. see Levine, G 116, *132*
Bassham, J.A. see Pedersen, T.A 96, *133*, 402, *415*
Bassham, J.A. see Schou, L. 194, *232*
Basu, S. see Keenan, T.W. 321, 323, *333*
Batenburg, J.J. see VanGolde, L.M G 301, 302, *343*
Bathelt, H. see Schötz, F. 259, *286*
Batra, S 389, 392, *394*
Baudhuin, P. 5, *40*
Baudhuin, P , Evarard, P., Berthet, J 5, *40*
Bauer, H., Sigarlakie, E. 261, *281*
Bauer, W.D. see Keegstra, K. 307, *332*
Baumgartner, B. 269, *281*
Baumgartner, B , Kende, H., Matile, P. 272, *281*
Baur, P.S., Walkinshaw, C H 261, 268, *281*
Bazzaz, M.B., Govindjee, I. 171, *177*
Bech-Hansen, C.W. see Fowke, L.C. 311, *329*
Beck, C., Meyenburg, H.K. von 280, *281*
Beck, E. see Eickenbusch, J.D. 196, *226*
Becker, J.D. see Fock, H. 188, *226*
Becker, J.E. see Bushnell, D.E 59, *80*
Becker, W M see Graves, L.B. 206, *227*
Becker, W.M. see Newcomb, E H 6, *46*
Becker, W.M. see Trelease, R.N 29, *49*
Beer, M see Colvin, J.R. 290, 322, *328*
Beeson, M. see Goldstein, L. 72, *81*
Beevers, H., Stiller, M.L., Butt, V.S. 258, 260, *281*
Beevers, H. see Breidenbach, R W. 6, *41*, 199, *225*
Beevers, H. see Cooper, T.G. 199, *225*, 248, *251*
Beevers, H. see Huang, A H C 176, *180*, 203, *228*
Beevers, H see Kagawa, T. 302, *332*
Beevers, H. see Lord, J.M. 290, 301, *334*
Beevers, H. see MacLennan, D.H. 260, *284*, 319, *334*
Beevers, H. see Miflin, B.J. 88, *132*
Beevers, H. see Moore, T.S. 290, 301, *336*
Beevers, H. see Stabenau, H. 206, *232*
Behrens, M. see Thalacker, R. 402, *416*
Behrens, N.H., Carminatti, H., Staneloni, R J , Leloir, L.F., Cantarella, A I 321, *326*
Beisenherz, G. 193, *224*
Ben-Arie, R., Ordin, L., Kindinger, J.I 321, *326*
Benedetti, E.L. see Emmelot, P 6, 16, *42*
Benedict, C.R. see Cooper, T.G. 107, *130*
Benedict, C R. see Whelan, T. 165, *184*
Bennett, G. see Leblond, C.P. 323, *333*
Benson, A.A. 11, 28, *40*
Benson, A.A , Bassham, J.A., Calvin, M., Goodale, T.C., Haas, V.A., Stepka, W. 86, *129*
Benson, A.A., Bassham, J.A., Calvin, M , Hall, A G., Hirsch, H.E., Kawaguchi, S., Lynch, V., Tolbert, N.E. 86, *129*
Benson, A A., Calvin, M 86, 89, *129*, 185, 193, *224*
Benson, A.A., Strickland, E H. 28, *40*
Benson, A.A. see Bassham, J.A. 194, *224*
Benson, A.A. see Delawan, L.A. 195, *226*
Benson, A.A. see Douce, R 6, 7, 23, 24, *41*, 412, *412*
Benson, A.A. see McCarty, R E 24, *45*
Benson, A.A. see Schou, L. 194, *232*
Benson, A.A. see Weier, T.E. 33, *50*
Ben-Tal, Y., Varner, J.E. 316, *326*
Bentzel, C J , Solomon, A.K 379, *394*
Beránek, R. see Smith, D.S 291, *341*
Berezney, R., Funk, L.K., Crane, F.L 18, *40*
Berezney, R , Macaulay, L.K., Crane, F.L. 19, *40*
Berger, C. see Gerhardt, B. 205, *227*
Berger, S., Schweiger, H.G. 69, 70, *79*
Berger, S. see Strotmann, H 25, *49*, 404, 405, *415*
Berger, W. see Franke, W.W 65, *81*
Bergeron, J.J M., Ehrenreich, J.H., Siekevitz, P., Palade, G E. 301, *326*
Bergeron, J.J M , Evans, W.H., Geschwind, I.I. 323, *326*
Berjak, P 268, *281*, 317, *326*
Berjak, P., Villiers, T.A. 278, *281*, 317, *326*
Berlin, J.D., Ramsey, J.C. 308, *326*
Berlin, J.D. see Watson, M.W. 308, *343*
Berman-Kurtz, M. see Mascarenhas, J.P. 18, *45*
Berridge, M V., Ralph, R.K., Letham, D.S. 324, *326*
Berry, J. see Björkman, O. 148, 176, *178*
Berry, J.A., Downton, W.J.S., Tregunna, E.B. 151, 155, 162, 170, *177*
Berry, J.A. see Downton, W.J S. 170, *179*
Berry, W.L. see Thomson, W W 315, *342*
Bert, L.M , Bartley, W. 387, *394*
Bertagnolli, B.L., Hanson, J.B. 241, *250*, 400, *412*
Bertagnolli, B.L see Hanson, J.B. 379, 382, 386, *395*
Berthet, J. see Baudhuin, P. 5, *40*
Berthillier, G., Got, R. 290, *326*
Berthillier, G see Frot-Coutaz, J 290, *330*
Bertini, F., Brandes, D., Buetow, D.E. 280, *281*
Bertini, F. see Brandes, D. 280, *282*
Bertsch, W see Cohen, W.S. 356, *371*

Bertsch, W see Cohn, D.E 350, *371*
Berzborn, R.J 349, 354, *371*
Berzborn, R.J see Hauska, G.A 349, 350, 354, *372*
Berzborn, R J see Racker, E 349, *374*
Betina, M I see Alexeev, A.B. 65, *79*
Betina, N I see Chirkova, L.I 65, *80*
Betz, H. see Matern, H 269, 272, *284*
Bhuvaneswaran, C see Peng, C F. 390, 393, *396*
Bibby, B T., Dodge, J.D 205, *224*
Bidwell, R.G.S., Levin, W.B , Shephard, D.C 212, *224*
Bidwell, R.G S. see Hellebust, J A 111, *131*
Bidwell, R.G.S. see Mangat, B.S. 212, 215, *230*
Bidwell, R G.S see Oaks, A 258, 260, 261, *285*, 290, *337*
Bidwell, R.G S. see Shephard, D.C 66, *83*
Bidwell, R.G.S. see Steward, F.C. 258, 272, *286*
Bidwell, R.G.S see Yemm, E W. *234*
Biegelmayer, C., Ruis, H 6, *40*
Bieleski, R.L. 259, 263, *281*
Bieri, V see Wallach, D.F H. 40, *49*
Biggins, J see Park, R B 21, 22, 33, *46*
Biggs, M.L see Preiss, J 96, *133*
Bikle, D , Tilney, L G., Porter, K R. 291, *326*
Billmire, J E. see Patni, N J. 16, *46*
Bird, I F., Cornelius, M.J., Dyer, T.A., Keys, A.J. 402, *412*
Bird, I F , Cornelius, M.J., Keys, A.J., Kumarasinghe, S , Whittingham, C P. 220, *224*
Bird, I F., Cornelius, M.J , Keys, A.J , Whittingham, C P 94, *129*, 201, 208, *224*
Bird, I F , Porter, H K , Stocking, C R 94, *129*
Birmingham, M see Whittingham, C P. 194, *233*
Birnstiel, M , Speirs, J., Purdom, I., Jones, K , Loening, U.E 55, *79*
Birnstiel, M.L see John, A A. 55, *82*
Bisalputra, T., Shields, C.M., Markham, J W 205, *224*
Bischoff, R see Ishikawa, H. 294, *332*
Bishop, D G , Andersen, K S , Smillie, R.M. 171, *177*
Bishop, D.G see Andersen, K S. 171, *177*
Bishop, D G see Smillie, R M. 171, 172, *183*
Bishop, J A. see Varmus, H E. 77, *84*
Bishop, J.O., Pemberton, R , Baglioni, C 75, *79*
Bittiger, H. see Herth, W. 308, 312, *331*
Bjorklund, R F see Cloules, G.L. 37, *41*
Bjorkman, O 165, 174, *177*
Björkman, O , Gauhl, E 147, 148, 151, 155, 175, *177*
Björkman, O , Nobs, M , Pearcy, R , Boynton, J., Berry, J 148, 176, *178*
Bjorkman, O., Troughton, J H , Nobs, M. 147, 176, *178*
Bjorkman, O see Osmond, C.B. 176, *182*, 186, 222, *230*
Black, C C 144, 152, *178*, 186, 202, 212, 214, 222, *224*
Black, C.C , Chen, T M., Brown, R H 176, *178*
Black, C.C , Mollenhauer, H H. 203, *224*
Black, C.C see Chen, T.M. 149, 150, 152, 153, 155, 157, 164, *178*
Black, C C. see Edwards, G E. 146, 148, 149, 150, 151, 153, 155, 156, *179*, 203, *226*
Black, C C see Kanai, R 148, 151, 152, 155, *181*
Black, C.C see Liu, A Y 176, *182*, 203, *229*
Black, C.C. see Mayne, B C 171, *182*
Black, C C see Salin, M L 149, 150, *183*
Blair, P V , Oda, T , Green, D E , Fernández-Morán, H 27, *40*
Blair, P V see Fernández-Morán, H. 27, *42*
Blanchear, M.C. see Pande, S V. 240, *253*
Blasie, J K. see Cain, J 12, *41*
Blinks, L R., Skow, R.K. 106, *129*
Blobel, G , Potter, V R. 18, *40*
Blobel, G see Monneron, A 19, *45*
Block, R. see Kuhn, W. 441, *463*
Blondin, G.A 38, *40*
Blum, J J., Buetow, D.E. 280, *281*
Boardman, N.K see Anderson, J.M 171, 172, *177*
Boardman, N K. see Woo, K.C 171, *184*
Boehm, C. see Fuhrmann, G F. 15, 16, *42*
Boehm, J. 95, *129*
Bogorad, L. 20, *40*
Bohme, H., Reimer, S., Trebst, A. 354, *371*
Boller, T., Dürr, M., Wiemken, A 265, *281*
Boloukhere, M. 65, *79*
Bolund, L. see Ringertz, N.R 55, 62, *83*
Bombik, B M. see Burger, M M. 63, *80*
Bomsel, J.-L., Pradet, A. 402, *412*
Bonner, J 288, *326*
Bonner, J see Olson, A C. 323, 325, *337*
Bonner, R D , Jr. see Douce, R 378, *395*
Bonner, W C 378, *394*
Bonner, W D., Voss, D.O 238, *251*
Bonner, W D. see Douce, R 238, *251*
Bonner, W D., Jr. see Douce, R 384, 387, 391, *395*
Bonner, W.D., Jr. see Ikuma, H 387, *395*, 408, *414*
Bonner, W M. 60, *79*
Bonnett, H.T., Newcomb, E H. 295, 308, *326*
Bonneville, M A. see Staehelin, L A 292, *342*
Bonotto, S , Puiseux-Dao, S , Kirchmann, M , Brachet, J 64, 65, 66, *79*
Bonotto, S see Brachet, J 64, *79*
Boos, W 13, 39, *40*

Borle, A B. 393, *394*
Bos, C.J. see Emmelot, P 6, 16, *42*
Bouck, G B. 17, *40*, 304, 305, 306, *326*, *327*
Bouck, G B. see Cronshaw, J. 292, 314, *328*
Bouck, G R. 205, *224*
Boudet, A see Alıbert, G. 268, *281*
Boulter, D., Ellıs, R J., Yarwood, A 25, *40*
Boulter, D. see Payne, P.I 273, *285*
Boulton, A.A. 14, 15, *41*
Bouteılle, M. see Leduc, E N 17, *44*
Bowen, C.C. see Stoermer, E.F. 310, *342*
Bowes, B.G. 273, *282*, 317, *327*
Bowes, G., Ogren, W.L 109, *129*, 175, 176, *178*, 195, *224*
Bowes, G., Ogren, W.L., Hageman, R.H. 109, *129*, 175, *178*, 194, 195, *224*
Bowes, G. see Ogren, W L. 109, *133*, 185, 195, *230*
Bowles, D.J , Northcote, D.H. 314, 319, *327*
Boyer, P.D , Cross, R.L., Momsen, W. 370, *371*
Boyer, P.D. see Cross, R.L. 370, *371*
Boynton, J. see Bjorkman, O. 148, 176, *178*
Brachet, J. 56, 64, 66, 67, 69, 70, 71, 73, 77, *79*
Brachet, J., Bonotto, S. 64, *79*
Brachet, J., Denıs, H , Vıtry, F de 60, 66, *80*
Brachet, J., Fıcq, A., Tencer, R. 73, *80*
Brachet, J., Lang, A. 56, *80*
Brachet, J. see Assel, S van 64, *79*
Brachet, J. see Baltus, E 73, *79*
Brachet, J. see Bonotto, S. 64, 65, 66, *79*
Brachet, J. see Goffeau, A. 67, *81*
Brachet, J. see Hulın, N 55, *82*
Bracker, C.E. 294, 299, 317, *327*
Bracker, C.E , Grove, S N. 256, *282*, 298, *327*
Bracker, C E , Grove, S.N., Heintz, C E., Morré, D.J 298, *327*
Bracker, C E., Heıntz, C E , Grove, S.N. 306, *327*
Bracker, C.E. see Grove, S.N. 288, 290, 298, 300, 308, *330*
Bracker, C E. see Hodges, T.K 6, 8, 15, 16, *43*
Bracker, C.E see Morré, D.J. 29, 30, *45*, 294, 295, 297, 298, 299, 303, 324, *336*, *337*
Bracker, C.E see Van Der Woude, W.J. 298, 301, 307, 308, 312, 319, 320, *343*
Bradbeer, J W , Anderson, C M.A 194, 195, *224*
Bradbeer, J W , Racker, E. 109, *129*
Bradley, M.O. see Wessels, N K. 292, 294, *343*
Brandes, D , Bertını, F. 280, *282*
Brandes, D see Bertını, F. 280, *281*
Brandle, E., Zetsche, K. 65, 68, *80*
Branton, D 15, 34, *41*
Branton, D., Moor, D. 14, *41*
Branton, D., Park, R.B. 34, 35, *41*
Branton, D. see Deamer, D.W. 21, *41*
Branton, D. see Pınto Da Sılva, P 301, 321, 323, *338*
Bravdo, B A. 188, *224*
Bravdo, B A., Canvın, D T. 212, *224*
Breckenrıdge, B.M.L. see Burger, M M 63, *80*
Breıdenbach, R W , Beevers, H. 199, *225*
Breıdenbach, R.W , Kahn, A., Beevers, H 6, *41*, 199, *225*
Bremer, J., Norum, K.R , Farstad, M 246, *251*
Bremer, J. see Aas, M. 246, *250*
Bresch, H. 74, *80*
Bretscher, M S. 10, 36, *41*
Brıantaıs, J.-M see Arntzen, C.J 171, *177*, 350, *370*
Brıerley, G.P , Murer, E., Bachmann, E 389, *394*
Brody, M , Whıte, J.E 205, *225*
Brouwer, A., Smıts, G G., Tas, J., Meıjer, A J , Tager, J M 246, *251*
Brouwer, A. see La Noue, K.F 245, *252*
Brouwer, A. see Meıjer, A.J. 245, *253*
Brown, C.L. see Wodzıckı, T.J. 279, *287*
Brown, D.D. see Gurdon, J B. 59, *81*
Brown, D.L see Weıer, T.E. 20, 21, *50*
Brown, E.B see Tregunna, D.L. 204, *233*
Brown, H.T., Morrıs, G H 95, *129*
Brown, J E see Coombs, J 146, 152, *178*
Brown, J.M. see Walker, D A. 175, *184*
Brown, R.H. see Black, C.C. 176, *178*
Brown, R.H see Chen, T.M. 153, 164, *178*
Brown, R.M., Jr. 309, *327*
Brown, R.M., Jr., Franke, W W., Kleınıg, H., Falk, H , Sıtte, P. 308, 309, 312, *327*
Brown, R.M , Jr , Herth, W , Franke, W , Romanovicz, D. 308, *327*
Brown, R M., Jr. see Herth, W. 308, 312, *331*
Brown, W.V see Smıth, B.N. 147, *183*
Brummond, D.O. see Larsen, G.L. 290, *333*
Bruner, W.E. see Kean, E.L. 72, *82*
Bruın, W J., Nelson, E.B., Tolbert, N.E 204, 205, *225*
Bruın, W J. see Tolbert, N E 204, *232*
Brzozowska, J , Hanower, P , Chezeau, R. 260, *282*
Buchanan, B B , Kalberer, P.P , Arnon, D.I 96, *129*
Buchanan, B.B., Schürmann, P. 105, *129*
Buchanan, B.B., Schürmann, P., Kalberer, P.P 96, *129*
Buchanan, B.B. see Kalberer, P.P. 86, *132*
Buchanan, B.B see Schurmann, P 118, *134*
Bücher, Th., Klıngenberg, M 248, *251*
Bucher, Th. see Zebe, E. 248, *254*

Buchholz, M. see Klingenberg, M. 238, 239, 240, *252*, 400, *414*
Bucke, C., Long, S.P. 152, *178*
Bucke, C., Oliver, I.R. 157, *178*
Bucke, C., Walker, D.A., Baldry, C.W. 103, 104, *129*
Bucke, C. see Baldry, C.W. 103, 104, 106, 111, *128*, 152, *177*
Bucke, C. see Coombs, J. 175, *178*
Buetow, D.E. see Bertini, F. 280, *281*
Buetow, D.E see Blum, J.J. 280, *281*
Buetow, D.E. see Malkoff, D B 280, *284*
Bull, T.A. see Hatch, M.D 151, 155, *180*
Bunning, G see Ritenour, G.L. 208, *231*
Burger, M.M., Bombik, B M., Breckenridge, B.M.L., Sheppard, J.R. 63, *80*
Burger, M M. see Oseroff, A.R. 37, *46*
Burger, M M see Schnaebli, H.P. 63, *83*
Burgess, J. 322, *327*
Burgess, J., Fleming, E N. 311, *327*
Burke, J.J see Trelease, R.N 29, *49*
Burny, A. see Chantrenne, H 75, *80*
Burny, A see Marbaix, G 59, 75, *82*
Burr, B., Burr, F.A 25, *41*
Burr, F.A. see Burr, B. 25, *41*
Burr, G.O see Kortschak, H.P. 144, *181*
Burris, R H. see Clagett, C.O. 199, *225*
Burris, R.H. see Tolbert, N.E. 199, 208, *232*
Burris, R.H. see Umbreit, W.W 175, *183*
Burris, R H see Wang, D. 197, 201, *233*
Bushnell, D.E., Becker, J E, Potter, V R. 59, *80*
Butler, W L see Yamashita, T. 356, *376*
Butow, R.A see Kellems, R.E. 25, 28, *43*
Butt, V S. see Beevers, H. 258, 260, *281*
Butt, V.S. see Leek, A.E. 208, *229*
Buvat, R. 256, 275, 277, *282*, 317, *327*
Buvat, R see Coulomb, C 273, *282*
Bygrave, F.L. 392, *394*
Bygrave, F.L. see Thorne, R.F.W 392, *397*

Cain, J., Santillan, G., Blasie, J.K. 12, *41*
Calvin, M. see Bassham, J.A. 86, 89, *128*, 144, 177, 194, *224*
Calvin, M. see Benson, A.A. 86, 89, *129*, 185, 193, *224*
Calvin, M see Pon, N G. 108, *133*
Calvin, M. see Schou, L 194, *232*
Calvin, M. see Wilson, A.T. 193, 195, *234*
Camarata, M see Giudice, G 74, *81*
Cameron, D.A., Paschall, H.A., Robinson, R A 390, *394*
Campbell, N., Thomson, W.W., Platt, K. 315, *327*
Campbell, W.H. see Chen, T.M. 149, 150, 152, 153, 155, 157, *178*
Campbell, W H see Salin, M.L. 149, 150, *183*
Candler, E.L see Anderson, N.G. 10, *40*
Canny, M.J see Kuo, J. 162, *181*
Cantarella, A I. see Behrens, N.H 321, *326*
Cantatore, P, Nicotra, A., Loria, P., Saccone, C. 74, *80*
Canvin, D T, Fock, H. 188, 211, 215, *225*
Canvin, D.T., Fock, H, Przybylla, K, Lloyd, N.D.H. 215, 216, *225*
Canvin, D.T. see Atkins, C A. 188, 217, *224*
Canvin, D.T. see Bravdo, B A 212, *224*
Canvin, D.T. see D'Aoust, A.L. 188, 211, 212, 213, 214, 215, *226*
Canvin, D.T. see Fock, H. 186, 188, 204, 207, 219, *226*, *227*
Canvin, D T see Hew, C.-S. 188, 212, 213, 214, *228*
Canvin, D.T. see Ludwig, L.J. 188, 211, 212, 213, 214, 215, *230*
Canvin, D.T. see Mahon, J D. 176, *182*, 191, 200, 211, 216, 217, 218, 220, 221, *230*
Capindale, J.B. see Allen, M.B. 116, *127*
Capindale, J.B see Whatley, F.R. 86, 118, *136*
Carafoli, E, Lehninger, A L. 387, 388, 389, *394*
Carafoli, E., Patriarca, P., Rossi, C S 392, *394*
Carafoli, E. see Azzone, G.F 377, *394*
Carafoli, E. see Drahota, Z. 389, *395*
Carafoli, E see Lehninger, A L 377, 387, 388, 389, *396*
Carafoli, E see Prestipino, G 388, *397*
Carafoli, E. see Vasington, F.D. 388, *397*
Carasso, N., Favard, P, Mentré, P., Poux, N. 298, *327*
Carasso, N., Ovtracht, L., Favard, P 298, 319, *327*
Carasso, N. see Favard, P. 298, *329*
Carmeli, C. 367, *371*
Carminatti, H see Behrens, N.H 321, *326*
Carothers, Z.B. 298, *327*
Carr, D J. see O'Brien, T.P 161, *182*
Carroll, F.E., Carroll, G.C 277, 279, *282*
Carroll, G.C see Carroll, F.E. 277, 279, *282*
Carvalho, A. see Rhoades, M.M. 147, *183*
Caspersson, T. 53, *80*
Cassady, W E., Wagner, R.P. 27, *41*
Catanzarite, V. see Rechsteiner, M. 77, *83*
Catesson, A.-M., Czaninski, Y 268, *282*
Catsky, J see Sestak, Z 187, *232*
Ceccarelli, D see Prestipino, G. 388, *397*
Ceron, G., Johnson, E.M. 66, *80*

Cetorelli, J.J. see Roberts, R.M. 301, *339*
Chafe, S.C., Durzan, D.J. 259, 261, 268, *282*
Chafe, S.C., Wardrop, A.B. 321, *327*
Chamberlin, A.H.L. 308, *327*
Chamberlain, J T., Metz, C.B 74, *80*
Champigny, M.L., Miginiac-Maslow, M. 113, *129*
Chan, H.W -S., Bassham, J.A. 194, 209, *225*
Chance, B 387, 388, 389, 392, *394*
Chance, B. see Avron, M. 353, 358, *371*
Chance, B. see Parsons, D.F. 6, 24, 27, 28, *46*, 236, *253*
Chance, B. see Racker, E. 27, *47*
Chance, B. see Strichartz, G.R 364, *375*
Chang, H. see Chen, Y. 7, *41*
Chang, H.C. see Marayama, H 175, *182*
Chantrenne, H., Burny, A., Marbaix, G. 75, *80*
Chantrenne, H. see Marbaix, G. 59, 75, *82*
Chapman, E.A., Graham, D. 213, 215, *225*
Chapman, E.A., Osmond, C.B. 169, *178*
Chappell, J.B. 27, 28, *41*, 210, *225*, 242, *251*, 383, 384, *394*
Chappell, J.B., Crofts, A.R. 378, 387, *394*
Chappell, J B , Haaroff, K N. 169, *178*, 383, 387, *394*
Chappell, J B. see Azzi, A 245, *250*
Chappell, J B see McGivan, J.D 245, *253*
Chardard, R 298, *328*
Charvat, I see Cronshaw, J. 268, 273, *282*
Chase, J.F.A. see West, D.W. 246, *254*
Chatterjee, S. see Rao, M.V. 72, *83*
Cheetham, R.D. see Franke, W.W. 302, 304, *330*
Chemilka, E. see Darzynkiewicz, Z. 63, *80*
Chen, C., Lehninger, A.L. 380, 387, 388, 389, *394*
Chen, J.C W 308, *328*
Chen, T M., Brown, R H., Black, C.C 153, 164, *178*
Chen, T.M., Campbell, W.H., Dittrich, P., Black, C.C. 149, 152, 155, *178*
Chen, T M., Dittrich, P., Campbell, W.H., Black, C C 149, 150, 153, 157, *178*
Chen, T.M. see Black, C.C. 176, *178*
Chen, Y., Lin, C , Chang, H., Guilfoyle, T.H., Key, J.L. 7, *41*
Cheng, G.P., Rosenblum, I.Y., Sallach, H.J. 199, *225*
Cheng, K G., Miller, A.G. Colman, B. 206, *225*
Chemiae, G.M., Martin, I.F. 356, *371*
Chenoweth, A.R F. see Stevenninck, R.F.M. 259, *286*
Chen-She, S.H., Lewis, D H., Walker, D.A 95, 96, 97, 98, *129*
Cherry, J.H. see Clark, J.E 301, 324, *328*
Cherry, J.H. see Hardin, J.W 324, *331*
Cheung, Y N.S. see Nobel, P.S. 25, *46*, 110, *133*
Chevallier, D., Douce, R. 408, *412*
Chezeau, R. see Brzozowska, J. 260, *282*
Chirkova, L.I., Pikalov, A.V., Kisseleva, E.V., Betina, N.I., Khristolubova, N.B., Nikoro, Z S., Sandakhiev, L S 65, *80*
Chlapowski, F J. see Staehelin, L.A. 292, *342*
Chollet, R. 202, *225*
Chollet, R., Ogren, W L. 149, 150, 176, *178*, 202, *225*
Chrispeels, M.J. 303, *328*
Chrispeels, M J. see Gardiner, M. 308, *330*
Chrispeels, M.J. see Sadava, D 307, *340*
Christensen, A.K., Fawcett, D W 29, *41*
Christensen, E L. see Douce, R. 378, *395*
Chu, D.K., Bassham, J A. 105, *129*, 195, *225*
Chudapongse, P., Haugaard, N 390, 393, *395*
Ciani, S.M. 442, *462*
Ciani, S.M , Eisenman, G , Laprode, R., Szabo, S. 442, *462*
Clagett, C.O., Tolbert, N.E., Burris, R.H. 199, *225*
Clagett, C.O. see Tolbert, N.E. 199, 208, *232*
Clandinin, M.T., Cossins, E.A. 201, 207, *225*
Clark, A F see Villemez, C.L. 290, 321, *343*
Clark, J.E., Morré, D.J , Cherry, J.H. 301, 324, *328*
Claude, A. 58, *80*
Cleland, R. 307, 323, *328*
Cleland, R., Karlsnes, A.M. 307, *328*
Clement, A.C., Tyler, A 74, *80*
Cleuter, Y. see Marbaix, G. 59, 75, *82*
Cloules, G.L., Björklund, R F. 37, *41*
Clowes, F.A L., Juniper, B E. 322, *328*
Coartney, J.S. 325, *328*
Cockburn, B.J., Wellburn, A.R. 25, *41*
Cockburn, W., Baldry, C.W., Walker, D.A. 89, 90, 102, 103, 104, 111, 112, 113, 118, 121, 124, *129*, 406, 408, *412*
Cockburn, W., Walker, D.A., Baldry, C W. 89, 90, 103, 104, 105, 106, 112, 124, 125, *129*, 404, *412*
Cockburn, W. see Baldry, C.W. 112, *128*
Cockburn, W. see Everson, R.G. 94, *130*
Cockburn, W. see Gibbs, M. 94, 96, 108, *130*
Cockburn, W. see Walker, D.A. 89, 104, *135*
Cocking, E C. 269, *282*
Cocking, E C. see Willison, J.H M. 319, 321, *344*
Cockrell, R.S., Harris, E.J., Pressman, B.C. 383, *395*
Codd, G.A., Lord, J M., Merrett, M.J. 204, 205, *225*
Codd, G.A., Merrett, M.J. 204, 205, *225*
Codd, G.A , Schmid, G.H., Kowallik, W. 206, *225*
Codd, G.A., Stewart, W.D.P. 206, *225*
Codd, G.A. see Lord, J.M. 204, *229*

Cohan, M S. see Tolbert, N E 197, *232*
Cohen, W.S , Cohn, D E., Bertsch, W. 356, *371*
Cohen, W.S see Cohn, D E. 350, *371*
Cohn, D E., Cohen, W S , Bertsch, W. 350, *371*
Cohn, D E see Cohen, W S. 356, *371*
Cohn, N.S. 29, *41*
Colbeau, A., Nachbaur, J., Vignais, P M 24, *41*
Cole, K see McBridge, D.L. 317, *335*
Coleman, J.O D., Palmer, J M. 238, 246, *251*
Coleman, R., Finean, J B. 16, *41*
Collins, N , Merrett, M J 206, *225*
Colman, B., Miller, A.G., Grodzinski, B. 204, *225*
Colman, B. see Cheng, K G 206, *225*
Colot, H V. see Raff, R.A. 73, *83*
Colvin, J.R. 321, 322, *328*
Colvin, J.R., Beer, M. 290, 322, *328*
Colvin, J R. see Leppard, G G. 312, *334*
Conard, V. see Legros, F. 66, *82*
Connor, A.B. see Roberts, R M. 301, *339*
Connover, T E. see Racker, E. 27, *47*
Conolly, T N. see Good, N.E 115, *130*
Conti, F. see Prestipino, G 388, *397*
Cook, G M.W., Stoddart, R W. 301, *328*
Coolbaugh, R.C. see Moore, T.C. 292, *336*
Coombs, J , Baldry, C.W, 152, *178*
Coombs, J., Baldry, C.W , Brown, J.E 146, 152, *178*
Coombs, J., Baldry, C.W , Bucke, C 175, *178*
Coombs, J., Maw, S.L., Baldry, C.W 174, *178*
Coombs, J., Whittingham, C.P 193, 194, 195, 196, 197, *225*
Coombs, J see Baldry, C.W 152, *177*
Coombs, J. see Lauritis, J A 310, *333*
Coombs, J see Whittingham, C.P. 194, *233*
Cooper, T G , Beevers, H. 199, *225*, 248, *251*
Cooper, T G , Filmer, D., Wishnick, M., Lane, M.D 107, *130*
Cooper, T G., Tchen, T.T., Wood, H.G , Benedict, C.R. 107, *130*
Cooper, T.G , Wood, H G. 174, *178*
Cooper, T G see Filmer, D 107, *130*
Cornelius, M J see Bird, I.F. 94, *129*, 201, 208, 220, *224*, 402, *412*
Cortat, M., Matile, P., Wiemken, A. 315, *328*
Cossins, E.A. see Clandinin, M.T. 201, 207, *225*
Cossins, E.A see Shah, S P.J. 201, *232*
Cost, K., Frenkel, A.W. 352, *371*
Coty, W.A., Pedersen, P.L 242, *251*
Coulomb, C 273, 277, *282*
Coulomb, C., Buvat, R. 273, *282*
Coulomb, C. see Coulomb, P 275, *282*, 317, *328*
Coulomb, P. 268, 270, 271, *282*
Coulomb, P , Coulomb, C 275, *282*
Coulomb, P , Coulomb, C , Coulon, J. 317, *328*
Coulomb, P , Coulon, J 312, *328*
Coulomb, P. see Meunier, D 27, *45*
Coulomb, P. see Truchet, G. 277, *287*
Coulon, J see Coulomb, P. 312, 317, *328*
Cova, D see Meldolesi, J. 301, *335*
Craig, I.W , Gibor, A 66, *80*
Craig, L C 37, *41*
Craig, S. see Hatch, M.D 126, *130*, 149, 153, 155, 156, 158, 160, 170, 171, *180*
Cran, D.G , Dyer, A F. 298, *328*
Crane, F.L. see Arntzen, C J 34, 35, *40*
Crane, F.L see Berezney, R 18, 19, *40*
Crane, F L see Stiles, J.W 27, *48*
Cresti, M , Pacini, E , Sarfatti, G 275, 277, *282*
Crofts, A R 357, 359, *371*
Crofts, A.R. see Chappell, J B. 378, 387, *394*
Crofts, A.R. see Deamer, D W. 362, *371*
Crofts, A R see Jackson, J B. 364, *373*
Crofts, A R see Packer, L 110, *133*
Crofts, A R. see Walker, D A 89, 104, 108, 113, 122, 124, *135*, 348, *376*
Cronshaw, J , Bouck, G B 292, 314, *328*
Cronshaw, J , Charvat, I. 268, 273, *282*
Cronshaw, J., Wardrop, A B 313, 314, *328*
Cronshaw, J see Esau, K. 322, *329*
Cronshaw, J. see Sexton, R. 268, *286*
Cross, R.L , Boyer, P.D. 370, *371*
Cross, R.L see Boyer, P D. 370, *371*
Crotty, W.J., Ledbetter, M.C. 298, *328*
Cuatrecasas, P 323, *328*
Cunningham, W.B 238, *251*
Cunningham, W.P. 408, *412*
Curran, P F see Katchalsky, A. 419, 420, *462*
Currier, H B. 320, *328*
Cynkin, M.A. see Gibbs, M. 111, *130*
Czaninski, Y. see Catesson, A.-M 268, *282*
Czaninski, Y. see Giraud, G 205, *227*

Dahlstrom, A. 291, *328*
Dainko, J.L see Schlenk, F. 258, 261, *286*
Dainko, J.L. see Svihla, G. 259, 261, *286*
Dainty, J 319, *328*, 419, *462*
Dainty, J see Gutknecht, J. 265, *283*
D'Alessio, G , Trim, A R 18, *41*
Daly, O see Siegel, S.M. 258, *286*

Damadian, R 265, *282*
Danielli, J.F. 37, *41*
Danielli, J.F , Davson, H. 32, *41*, 441, *462*
Danielli, J.F see Jeon, K.W. 71, *82*
Dankert, M see Garcia, R C. 321, *330*
D'Aoust, A L. 214, *226*
D'Aoust, A.L., Canvin, D T. 188, 211, 212, 213, 214, 215, *226*
Darzynkiewicz, Z., Chemilka, E., Arnason, B. 63, *80*
Darzynkiewicz, Z. see Ringertz, N R 55, 62, *83*
Dashek, W.V 308, *328*
Dashek, W V. see Rosen, W.G. 308, *340*
Dauwalder, M., Whaley, W.G. 301, 308, 314, *328*
Dauwalder, M., Whaley, W G., Kephart, J.E. 7, *41*, 315, 328, *329*
Dauwalder, M. see Whaley, W.G. 30, *50*, 288, 304, 308, *343*
Davidson, R.L. 62, *80*
Davie, C.A.M. see Hall, J.L. 268, *283*
Davies, D D., Davies, S. 205, *226*
Davies, P. see Allison, A.C. 275, *281*
Davies, S. see Davies, D.D. 205, *226*
Davis, D D. see Wilkinson, A P 201, *233*
Davis, E M , Adelberg, E A 62, *80*
Davis, M.E see Stewart, K.D 205, *232*
Davis, R.H 258, 265, *282*
Davson, H., Danielli, J.F. 441, *462*
Davson, H see Danielli, J F. 32, *41*
Day, D.A , Wiskich, I.T 238, 246, *251*
Day, P.R see Poincelot, R P 23, 24, 25, *47*
Day, P.R. see Zelitch, I. 205, *234*
De, D N see Maitra, S.C 292, *334*
Deak, I.I , Defendi, V. 62, *80*
Deamer, D.W , Branton, D 21, *41*
Deamer, D W , Packer, L 357, *371*
Deamer, D.W., Prince, R.C., Crofts, A.R. 362, *371*
Dean, B.B. see Kolattukudy, P E. 161, *181*
Decker, J P 188, 213, *226*
Decker, M , Tanner, W. 401, *412*
De Duve, C. see Appelmans, F. 6, *40*
Dee, A.M see Mayne, B C 171, *182*
Defendi, V. see Deak, I.I 62, *80*
De Guzman, N.T. see Pressman, B.C. 38, *47*
De Haan, E.J. see Gimpel, J.A. 242, *251*
De Haan, E.J , Tager, J.M 243, *251*
Deichgräber, G see Schnepf, E 306, *341*
Deitsch, A.D see Miranda, A F. 292, *335*
Delawan, L.A., Benson, A.A. 195, *226*
Delbruck, A. see Zebe, E. 248, *254*
Delektorskaya, L.N see Zbarsky, I.B. 19, *50*
Delektorsky, V.V see Zbarsky, I B. 19, *50*
Denis, H see Brachet, J. 60, 66, *80*
Dennick, R.G. 290, 292, *329*
Denny, P., Tyler, A 73, *80*
Denton, R M see Halestrap, A P 246, *251*
Depetris, S see Taylor, R B 11, *49*
Deshpande, P D., Hickman, D D., von Korff, R W. 387, *395*
Desousa, R C. see Li, J.H. 453, *463*
Deumling, B see Franke, W.W 18, 19, *42*, 302, 304, *330*
Deumling, B. see Morré, D J. 299, *336*
Devault, D see Izawa, S 355, *373*
Devaux, P see McConnell, H.M. 12, *45*
Dever, J. see Wilson, R.H. 382, *397*
Diamond, J M see Hingson, D J 450, *462*
Dickenson, P B 266, *282*
Diers, L., Schotz, F., Meyer, B 259, 261, *282*
Diers, L. see Schötz, F. 259, *286*
Dilley, R.A. 348, 351, *371*
Dilley, R.A., Shavit, N. 365, *371*
Dilley, R A., Vernon, L.P. 349, 351, 357, *371*
Dilley, R.A. see Arntzen, C.J. 34, 35, *40*, 171, *177*
Dilley, R.A see Giaquinta, R.T. 350, *372*
Dilley, R.A see Schwelitz, F D. 310, *341*
Dilley, R.A see Selman, B.R. 350, *375*
Dirksen, E.R see Straprans, I 294, *342*
Dittrich, P see Chen, T.M. 149, 150, 152, 153, 157, *178*
Diwan, J.J see King, M.J. 245, *252*
Dobberstein, B., Kiermayer, O 303, 304, 310, *329*
Dobberstein, B see Kiermayer, O. 304, 310, 321, *333*
Dodge, A.D. see Harris, N. 278, *283*
Dodge, J D 308, 309, 310, 314, *329*
Dodge, J.D see Bibby, B.T. 205, *224*
Dodge, J D. see Oakley, B.R. 205, *230*
Dohler, G , Koch, R. 206, *226*
Doi, E., Ohtsuru, C., Matoba, T 270, *282*
Donaldson, R.P., Tolbert, N E., Schnarrenberger, C 6, *41*, 193, *226*
Dornhoff, G M. see Samish, Y.B 212, *231*
Douce, R., Benson, A.A 24, *41*
Douce, R , Bonner, W.D., Jr 384, 391, *395*
Douce, R., Christensen, E.L., Bonner, W.D., Jr. 378, *395*
Douce, R , Holtz, R.B., Benson, A A 6, 7, 23, 24, *41*, 412, *412*
Douce, R., Mannella, C.A., Bonner, W.D , Jr 238, *251*, 387, *395*
Douce, R see Chevallier, D 408, *412*
Douce, R see McCarty, R.E. 24, *45*

Douglas, S.D see Pinto Da Silva, P 301, 321, 323, *338*
Downes, R.W. 175, 176, *178*
Downton, W.J.S. 153, 157, 160, 164, 170, *178*
Downton, W.J.S., Berry, J.A., Tregunna, B. 170, *179*
Downton, W.J.S., Hawker, J.S. 157, *179*
Downton, W J.S., Tregunna, E.B. 147, 164, *179*
Downton, W J.S. see Berry, J.A. 151, 155, 162, 170, *177*
Downton, W J.S. see Woo, K.C. 171, *184*
Draber, W. see Hauska, G.A. 354, 355, *372*
Draber, W see Trebst, A. 352, *375*
Drahota, Z., Carafoli, E., Rossi, C S., Gamble, R.L , Lehninger, A.L. 389, *395*
Dresner, L. 441, *462*
Droop, M.R., McGill, S. 204, *226*
Drum, R.W., Pankratz, H.S. 310, *329*
Duckett, J.G. see Peel, M.C 298, *338*
Duffus, W.P H see Taylor, R.B. 11, *49*
Dumford, W.S. see Miller, R.J. 383, *396*
Dupont, J see Moreau, F 24, 27, 28, *45*
Dupraw, E.J 17, *41*
Durand, R. see Klingenberg, M 242, *252*
Durham, L.J see Allen, M.B 116, *127*
Dürr, M. see Boller, T. 265, *281*
Durr, M. see Wiemken, A. 258, 260, 265, *287*
Durzan, D.J see Chafe, S.C. 259, 261, 268, *282*
Duszynski, J. see Morel, F. 240, *253*
Duve, C. de 256, 268, *282*
Dyer, A.F see Cran, D G. 298, *328*
Dyer, T.A. see Bird, I F 402, *412*

Earnshaw, M J , Hanson, J B. 384, 385, *395*
Earnshaw, M.J., Madden, D.M , Hanson, J B. 384, 387, *395*
Earnshaw, M.J., Truelove, B 378, *395*
Easterday, R.L. see Marayama, H 175, *182*
Ecker, R.E. see Smith, L.D. 74, *83*
Eckert, W.A. see Franke, W.W 294, 295, 297, 298, 299, 303, *330*
Edelman, J., Jefford, T.G 261, 266, *282*
Edelman, J. see Shibko, S. 96, *134*
Edge, H. see Hall, D.O. 354, *372*
Edidin, M. 11, 12, 37, *41*
Edidin, M. see Frye, L.D. 11, *42*
Edwards, G.E , Black, C.C 146, 148, 149, 150, 151, 153, 155, *179*, 203, *226*
Edwards, G.E., Gutierrez, M 150, *179*
Edwards, G.E., Kanai, R., Black, C C. 151, 155, 156, *179*
Edwards, G.E. see Gutierrez, M. 126, *130*, 149, 150, 153, 155, 156, 157, 160, 171, 175, *179*
Edwards, G.E see Huber, S C. 126, *131*, 149, 150, 167, *180*
Edwards, G E see Kanai, R 149, 153, *181*
Edwards, G.E see Ku, S.B. 152, 153, 171, *181*
Edwards, G.E. see Mayne, B.C. 171, *182*
Ege, T., Hamberg, H , Krondahl, U., Ericsson, J , Ringertz, N R. 76, 77, *80*
Ege, T., Krondahl, U , Ringertz, N.R. 58, 76, 77, *80*
Egle, K., Fock, H. 188, *226*
Egle, K., Schenk, W 204, 214, *226*
Egle, K see Fock, H. 186, 187, 188, 204, 207, 220, *226*, *227*
Egle, K. see Mahon, J D 191, *230*
Egle, K see Schaub, H 204, *231*
Ehrenreich, J H see Bergeron, J J.M. 301, *326*
Ehrlich, B.A see Bamberger, E.S 115, *128*, 140, *142*
Eickenbusch, J.D , Beck, E. 196, *226*
Eickenbusch, J.D., Scheibe, R., Beck, E 196, *226*
Eisenman, G. see Ciani, S M. 442, *462*
Eisenman, G. see Szabo, G. 12, *49*
Eisinger, W., Ray, P.M. 319, *329*
El-Badry, A M see Bassham, J.A 116, *128*
Elbein, A.D , Forsee, W T. 321, *329*
Elbein, A.D. see Forsee, W.T. 321, *329*
Elbers, R., Heldt, H.W., Schmucker, P , Soboll, S , Wiese, H. 241, *251*, 400, *412*
Elder, J.A see Schraer, R. 391, 392, *397*
Elder, J.H , Lembi, C.A., Anderson, L., Morré, D.J. 309, *329*
Elder, J.H., Morré, D.J. 301, *329*
Elder, J.H. see Morré, D J. 290, 291, 299, 301, 302, 304, 312, *336*
Eley, J H., Jr see Schacter, B. 103, 104, 105, *134*
El Kheir, Y. see Fairbairn, J.W. 261, *283*
Ellis, R.J see Boulter, D. 25, *40*
Ellyard, P.W., Gibbs, M 194, 195, *226*
Ellyard, P W see Gibbs, M. 103, *130*
El-Sharkawy, M , Hesketh, J. 108, *130*
El-Sharkawy, M A., Loomis, R.S , Williams, W.A 187, *226*
Elson, Ch see Lerner, E. 240, *252*
Emmelot, P., Bos, C.J., Benedetti, E.L., Rümke, P.H. 6, 16, *42*
Engelman, D.M. see Stoeckenius, W. 31, 36, *49*
Engels, F.M. 308, *329*
Entsch, B. see Smillie, R.M. 171, *183*
Epstein, E. 39, *42*, 264, *283*, 319, *329*
Ericsson, J see Ege, T. 76, 77, *80*
Ermen, B see Franke, W W. 18, 19, *42*

Ernster, L , Kuylenstierna, B 28, *42*
Esau, K., Cronshaw, J., Hoefert, L.L 322, *329*
Essig, A see Kedem, O. 453, *462*
Essig, A see Li, J.H. 453, *463*
Essner, E. see Golfischer, S. 7, *42*
Estabrook, R.W. see Racker, E 27, *47*
Estensen, B.D , Plagemann, P.G.W. 292, *329*
Estensen, R.D. see Ladda, R. 57, 76, *82*
Ettl, H see Manton, I. 306, *334*
Evans, H.J , Sorger, G J. 392, *395*
Evans, M C.W. see Telfer, A. 351, 363, 369, *375*
Evans, W. see Mollenhauer, H.H. 310, *335*
Evans, W.H. see Bergeron, J.J.M. 323, *326*
Evans, W.R. see Schwelitz, F.D. 310, *341*
Evarard, P. see Baudhuin, P. 5, *40*
Everson, R.G. 108, *130*
Everson, R.G., Cockburn, W., Gibbs, M. 94, *130*
Everson, R.G , Slack, C.R 165, *179*
Everson, R.G. see Gibbs, M 94, 96, 103, 108, *130*
Evert, R.F. see Fahn, A. 314, *329*
Evins, W.H., Varner, J.E. 316, *329*
Eyring, H see Johnson, F.H. 442, *462*
Eyring, H see Parlin, B. 442, *463*
Eytan, G , Ohad, I. 193, *226*

Fagan, J. see McCarty, R.E. 369, *373*
Fahn, A., Evert, R.F. 314, *329*
Fahn, A see Rachmilevitz, T. 315, *338*
Fairbairn, J.W., Hakim, F., El Kheir, Y. 261, *283*
Fairbanks, G., Steck, T.L., Wallach, D.F.H. 15, *42*
Fairbanks, G. see Steck, T.L. 15, *48*
Falk, H 303, *329*
Falk, H. see Brown, R.M , Jr. 308, 309, 312, *327*
Falk, H see Franke, W.W 60, 65, *81*, 295, *330*
Falk, H. see Franke, W W. 295, *330*
Falk, R.H. see Paolillo, D.J., Jr 21, *46*
Fan, D.F., MacLachlan, G.A. 315, *329*
Fan, W J.W see Varmus, H.E. 77, *84*
Farineau, J. 146, 147, *179*
Farquhar, M.G. see Friend, D.S. 295, *330*
Farstad, M. see Bremer, J 246, *251*
Favard, P., Carasso, N. 298, *329*
Favard, P see Carasso, N 298, 319, *327*
Fawcett, D.W. 29, *42*
Fawcett, D.W. see Christensen, A.K 29, *41*
Fekete, M.A.R. de, Vieweg, G.H. 98, *130*, 157, *179*
Fekete, M.A R. de see Vieweg, G.H. 157, *184*
Feldherr, C.M. 17, *42*, 60, *80*
Ferguson, S.M. see Lardy, H.A. 6, *44*, 377, *396*
Fernández-Morán, H. 27, *42*
Fernández-Morán, H see Blair, P.V. 27, *40*
Fernández-Morán, H., Oda, T , Blair, P.V., Green, D E. 27, *42*
Fernbach, B.R. see Wallach, D.F.H. 37, *49*
Ferrari, T.E., Yoder, O.C., Filner, P 259, *283*
Feulgen, R., Rossenbeck, H.S. 55, *80*
Ficq, A. 54, *80*
Ficq, A see Baltus, E 73, *79*
Ficq, A see Brachet, J 73, *80*
Figier, J. 268, *283*, 317, *329*
Filmer, D., Cooper, T.G. 107, *130*
Filmer, D. see Cooper, T G. 107, *130*
Filner, P see Ferrari, T.E. 259, *283*
Filner, P. see Kelker, H C. 273, *283*
Filner, P see Zielke, H R 273, *287*
Finean, J.B. 13, 15, *42*
Finean, J.B. see Coleman, R. 16, *41*
Fineran, B.A. 273, 275, *283*, 298, 300, *329*
Fiolet, J.T.W., Bakker, E.P., Van Dam, K 362, 363, *372*
Firn, R.D., Kende, H. 316, *329*
Fitzgerald, M.P. see Lilley, R.McC. 88, 91, 92, 93, *132*
Fiume, L., Wieland, T 59, *81*
Fleischer, B , Fleischer, S. 301, *329*
Fleischer, B see VanGolde, L.M G 301, 302, *343*
Fleischer, S , Packer, L 3, *42*
Fleischer, S see Fleischer, B. 301, *329*
Fleischer, S. see VanGolde, L.M.G. 301, 302, *343*
Fleming, E.N. see Burgess, J. 311, *327*
Flickinger, C J. 71, *81*
Fliege, R. see Heldt, H.W. 139, 141, 142, *142*, 167, 168, *180*
Floyd, G.L. see Stewart, K.D. 205, *232*
Fock, H 186, *226*
Fock, H., Becker, J.D., Egle, K. 188, *226*
Fock, H , Canvin, D T., Grant, B.R. 186, 188, 204, 207, *226*
Fock, H , Egle, K 187, 204, 220, *227*
Fock, H., Hilgenberg, W., Egle, K. 207, *227*
Fock, H , Krotkov, B. 221, *227*
Fock, H , Mahon, J.D., Canvin, D.T., Grant, B.R. 219, *227*
Fock, H., Schaub, H., Hilgenberg, W , Egle, K. 186, 207, *227*
Fock, H. see Canvin, D.T. 188, 211, 215, 216, *225*
Fock, H. see Egle, K. 188, *226*
Fock, H. see Lawlor, D.W 188, 213, 214, 215, *229*
Fock, H. see Mahon, J.D. 176, *182*, 191, 200, 211, 216, 217, 218, 220, 221, *230*

Follett, E.A C. 77, *81*
Fonyo, A. 242, *251*
Forrester, M.L , Krotkov, G., Nelson, C D. 186, 188, 213, 214, *227*
Forsee, W.T., Elbein, A.D. 321, *329*
Forsee, W.T. see Elbein, A D. 321, *329*
Forti, G., Avron, M , Melandri, A 348, *372*
Foskett, D E. see Hepler, P.K 292, *331*
Foster, P. see Thomson, W W. 20, *49*
Fowke, L C., Bech-Hansen, C.W , Gambourg, O L 311, *329*
Fowke, L.C., Pickett-Heaps, J D 312, *329*
Fowler, C.F., Kok, B. 350, 352, 353, *372*
Fox, C.F see Steck, T L. 10, 12, *48*
Fraenkel, G., Friedman, S. 247, *251*
Francavilla, A see Papa, S. 246, *253*
Franke, W W 61, *81*, 298, 300, *329*
Franke, W W , Berger, W , Falk, H., Spring, H , Scheer, U , Herth, W., Trendelenburg, M , Schweiger, H 65, *81*
Franke, W W , Deumling, B., Ermen, B , Jarasch, E.-D , Kleinig, H.J 18, 19, *42*
Franke, W.W , Eckert, W A. 298, 299, 303, *330*
Franke, W W , Eckert, W A , Krien, S. 294, 295, 297, 298, 299, 303, *330*
Franke, W.W , Herth, W 308, *330*
Franke, W.W , Herth, W., VanDerWoude, W.J., Morré, D J 288, 290, 292, 315, *330*
Franke, W W , Kartenbeck, J 298, *330*
Franke, W.W., Kartenbeck, J., Krien, S., VanDerWoude, W J , Scheer, U , Morré, D J 294, 301, *330*
Franke, W W , Kartenbeck, J , Zentgraf, H , Scheer, U , Falk, H 60, *81*, 295, *330*
Franke, W W , Morré, D J., Deumling, B , Cheetham, R.D , Kartenbeck, J , Jarasch, E D., Zentgraf, H W. 302, 304, *330*
Franke, W.W , Scheer, U 61, *81*
Franke, W.W., Scheer, V 17, *42*
Franke, W.W., Scheer, U , Herth, W 312, *330*
Franke, W.W see Brown, R.M., Jr 308, 309, 312, *327*
Franke, W W. see Herth, W 292, 308, 312, *331*
Franke, W W see Kartenbeck, J 294, *332*
Franke, W.W. see Morré, D.J. 299, *336*
Franke, W.W. see Philipp, E.-I 18, *47*
Franke, W W. see Spring, H 64, 69, *83*, *84*
Franke, W.W. see Trendelenburg, M.F 69, *84*
Frantz, C , Roland, J -C., Williamson, F A , Morré, D J. 303, *330*
Frederick, S E., Gruber, P.J , Newcomb, E H. 198, *227*
Frederick, S.E , Gruber, P J , Tolbert, N.E 205, *227*
Frederick, S E , Newcomb, E H 199, 203, 207, *227*
Frederick, S E. see Gruber, P J. 205, *227*
Frederick, S.E see Newcomb, E.H 93, *133*, 170, *182*
Freed, J J , Lebowitz, M M. 292, *330*
Frenkel, A W see Cost, K 352, *371*
Frey-Wyssling, A., Grieshaber, E , Mühlethaler, K 31, *42*, 316, *330*
Frey-Wyssling, A., López-Sáez, J F , Mühlethaler, K 288, 308, *330*
Frey-Wyssling, A , Steinmann, E. 33, *42*
Friedman, S. see Fraenkel, G. 247, *251*
Friend, D S , Farquhar, M G 295, *330*
Fritz, I B 246, *251*
Frömmter, E 456, *462*
Frot-Coutaz, J., Berthillier, G , Got, R 290, *330*
Fry, R see Wilson, R H. 382, *397*
Frye, L.D., Edidin, M. 11, *42*
Fuhrmann, G.F., Wehrli, E , Boehm, C. 15, 16, *42*
Fuhrman, J S. see McCarty, R E. 363, 369, *373*
Fujita, H see Kobatake, Y. 441, *463*
Fuller, R C. see Smillie, R M 88, 105, *134*
Fulton, C. 294, *330*
Funk, L.K. see Berezney, R. 18, *40*

Gaastra, P. 189, *227*
Gaensslen, R E 360, *372*
Gaensslen, R E , McCarty, R.E 357, 359, 360, 361, *372*
Gahan, P B. 268, *283*
Gall, J.G see Pardue, M.L 55, *83*
Gallant, D. see Rougier, M 301, 307, *340*
Galliard, T. 11, *42*
Galmiche, J.M 146, *179*, 218, *227*
Galpin, M. see Pitt, D 270, 271, *285*
Galston, A.W. see Sabnis, D D. 24, *47*
Gamble, R.L. see Drahota, Z 389, *395*
Gambourg, O L see Fowke, L.C 311, *329*
Garcia, R C., Recondo, E , Dankert, M 321, *330*
Gardiner, M., Chrispeels, M J 308, *330*
Gassotti, P see Vasington, F D 388, *397*
Gauhl, E see Bjorkman, O 147, 148, 151, 155, 175, *177*
Gautheron, D C., Julliard, J.H , Godinot, C 245, *251*
Gavey, K L. see Scallen, T.J 290, 292, *340*
Gawlik, S R see Rosen, W.G. 308, *340*
Geller, G see Heldt, H W. 107, *131*, 410, *413*
Geller, G. see Werdan, K 107, 108, *135*, 137, *143*
Georgiev, G P 61, *81*
Gergis, M S. 205, *227*
Gerhardt, B , Berger, C. 205, *227*
Geschwind, I I see Bergeron, J J M. 323, *326*

Geuskens, M 73, *81*
Geuskens, M , Jonghe D'Ardoye, V de 74, *81*
Gevers, W., Krebs, H A. 393, *395*
Gezelius, K 273, *283*
Ghosh, H P see Preiss, J. 95, 98, *133*
Giaquinta, R T., Dilley, R.A., Anderson, B.J 350, *372*
Gibbs, M 89, 104, 109, *130*
Gibbs, M., Bamberger, E.S , Ellyard, P.W., Everson, R.G 103, *130*
Gibbs, M., Cynkin, M A. 111, *130*
Gibbs, M., Latzko, E., Everson, R.G , Cockburn, W 94, 96, 108, *130*
Gibbs, M. see Avron, M. 116, 118, *128*
Gibbs, M. see Bamberger, E.S 103, 115, *128,* 140, *142,* 406, *412*
Gibbs, M see Ellyard, P.W 194, 195, *226*
Gibbs, M. see Everson, R.G 94, *130*
Gibbs, M. see Kelly, G.J. 122, *132,* 410, *414*
Gibbs, M. see Laber, L J. 146, 147, *181*
Gibbs, M. see Latzko, E. 147, *182*
Gibbs, M see Morgenthaler, J J. 88, *133*
Gibbs, M. see O'Neal, D. 157, 158, 168, *182*
Gibbs, M see Plaut, Z 109, *133,* 194, 195, 196, *231*
Gibbs, M see Robinson, J M 191, 194, *231*
Gibbs, M see Schacter, B 103, 104, 105, *134*
Gibbs, M. see Shain, Y 109, *134,* 194, 195, 196, *232*
Gibor, A. see Craig, I W 66, *80*
Gifford, E M., Stewart, K D. 275, *283*
Gifford, R.M 172, *179*
Gilmour, R.S., Paul, J 76, *81*
Gimenéz-Martin, G see Risueno, M.C. 312, 317, *339*
Gimmler, H., Schafer, G., Kraminer, A , Heber, U. 25, *42,* 92, 106, 110, 111, *130,* 199, *227*
Gimmler, H see Heber, U. 106, *130,* 194, 201, 207, 222, *228*
Gimpel, J.A., De Haan, E.J., Tager, J.M 242, *251*
Giovanelli, J. see Murray, D.R. 204, 205, *230*
Giraud, G , Czaninski, Y. 205, *227*
Girbardt, M. 299, 308, *330*
Giudice, G , Sconzo, G., Albanese, I., Ortolani, G , Camarata, M. 74, *81*
Glaser, C see Kauss, H. 324, *332*
Godinot, C. see Gautheron, D C. 245, *251*
Godman, G C see Miranda, A F 292, 294, *335*
Goffeau, A , Brachet, J. 67, *81*
Goldman, D.E. 447, *462*
Goldman, R D , Pollack, R., Hopkins, N.H 76, *81*
Goldstein, L 72, *81*
Goldstein, L , Trescott, D.H. 72, *81*
Goldstein, L., Wise, G E., Beeson, M 72, *81*
Goldstein, L. see Jelinek, W. 72, *82*
Goldstein, L. see Prescott, D. 72, *83*
Goldsworthy, A 109, *130,* 186, 188, 214, 215, *227*
Goldsworthy, A , Street, H.E 98, *130*
Golfischer, S , Essner, E., Novikoff, A.B. 7, *42*
Gonzales, F , Karmovsky, M.J 391, 392, *395*
Good, N E , Izawa, S., Hind, G 368, *372*
Good, N E., Winget, G D., Winter, W , Conolly, T N., Izawa, S , Singh, R.M M. 115, *130*
Good, N.E see Izawa, S. 367, *373,* 398, *414*
Good, N.E see Saha, S. 354, *375*
Good, N E see Winget, G D. 363, *376*
Goodale, T.C see Benson, A A 86, *129*
Goodchild, D.J. see Osmond, C B. 146, 170, *183*
Goodchild, D J see Sane, P V 22, *47*
Goodchild, D.J. see Slack, C R 147, 148, 151, 153, 155, 156, 157, 170, *183,* 210, *232*
Goodman, D B P. see Rasmussen, H D. 393, *397*
Goodman, R N see Preston, R.D. 321, *338*
Gordon, M see Sabnis, D.D. 24, *47*
Gorter, E., Grendel, F 32, *42*
Got, R see Berthillier, G. 290, *326*
Got, R. see Frot-Coutaz, J. 290, *330*
Gotto, A.M see Zelitch, I 196, 200, 209, *234*
Gould, J M , Izawa, S 352, 353, 356, 365, *372*
Goulding, K.H. see Merrett, M J 196, *230*
Govindjee, I. 348, *372*
Govindjee, I see Bazzaz, M.B. 171, *177*
Gracen, V E see Gutierrez, M 153, 155, 160, 171, *179*
Gracen, V E. see Hilliard, J H 199, 203, *228*
Graesser, R.J 390, 391, 393, *395*
Graham, D , Atkins, C.A , Reed, M.L., Patterson, B D , Smillie, R.M 165, *179*
Graham, D , Hatch, M.D , Slack, C.R , Smillie, R.M 151, *179*
Graham, D , Reed, M.L. 108, *130*
Graham, D see Chapman, E A. 213, 215, *225*
Graham, J M. see Wallach, D F H. 37, *49*
Grant, B R see Fock, H. 186, 188, 204, 207, 219, *226, 227*
Graves, L.B., Hanzely, L , Trelease, R.N. 205, 206, *227*
Graves, L.B., Trelease, R.N , Grill, A., Becker, W M. 206, *227*
Grebe, K. see Klingenberg, M. 239, 240, *252*
Green, B.R. 69, *81*
Green, D E 3, 31, *42*
Green, D E , Perdue, J.F 28, *42*
Green, D E see Blair, P.V 27, *40*

Green, D.E. see Fernández-Morán, H. 27, *42*
Green, D.G. 38, 39, *42*
Greenawalt, J.W see Schnaitman, C.A. 28, *48*, 236, *254*
Greenberg, E. see Preiss, J. 96, *133*
Greenhouse, G. see Raff, R.A. 73, *83*
Greenhouse, G.E. see Selvig, S A. 74, *83*
Greenwood, A.D. see Heath, I.B. 306, *331*
Greenwood, A D. see Peel, M.C. 298, *338*
Greider, M.H. see Malaisse-Lagae, F. 291, *334*
Gremel, D. see Randall, D.D. 195, 201, 221, *231*
Grendel, F. see Gorter, E. 32, *42*
Greville, G.D. 348, *372*
Grieshaber, E. see Frey-Wyssling, A. 31, *42*, 316, *330*
Griffin, W J. see Pritchard, G.G. 204, *231*
Griffiths, D.A. 273, *283*
Griffiths, H.B. see Heath, I.B. 306, *331*
Grill, A. see Graves, L.B. 206, *227*
Grodzinski, B see Colman, B. 204, *225*
Groot, G S P see Meijer, A J 242, 243, *253*
Groot, S.R. de, Mazur, P. 419, *462*
Gross, K.W., Jacobs-Lorena, M., Baglioni, C., Gross, P.R. 73, *81*
Gross, K.W. see Raff, R A. 73, *83*
Gross, P.R. see Gross, K.W 73, *81*
Gross, P.R. see Raff, R.A 73, *83*
Gross, P.R. see Selvig, S.A. 74, *83*
Gross, P.R see Skoultchi, A. 73, *83*
Gross, R.J , Osterle, J.F. 441, *462*
Grove, S N., Bracker, C.E. 308, *330*
Grove, S.N., Bracker, C.E., Morré, D J 288, 290, 298, 300, 308, *330*
Grove, S.N see Bracker, C.E 256, *282*, 298, 306, *327*
Gruber, P.J., Frederick, S.E., Tolbert, N.E. 205, *227*
Gruber, P.J. see Frederick, S E. 198, 205, *227*
Grünhagen, H.H , Witt, H.T. 353, *372*
Grunwald, T. see Rottenberg, H. 357, 360, 361, 362, 363, 368, *374*
Guerin, B. see Klingenberg, M. 242, *252*
Guidotti, G see Rosenberg, S.A. 15, 16, *47*
Guilfoyle, T.H see Chen, Y 7, *41*
Guilliermond, A. 259, 263, *283*
Guilliermond, A., Mangenot, G., Plantefol, L. 255, *283*
Guiot, H.F.L. see Kemp, J.A 239, *252*
Gunning, B.E S 20, *43*
Gunning, B.E.S , Pate, J.S. 315, *330*
Gunning, B.E.S. see Pate, J.S 170, *183*
Guntaka, R V. see Varmus, H.E. 77, *84*
Gurdon, J.B. 62, *81*
Gurdon, J B., Brown, D.D. 59, *81*
Gurdon, J.B., Lane, C.D , Woodland, H.R., Marbaix, G 59, 75, *81*
Gurdon, J B., Woodland, H.R. 60, *81*
Gurney-Smith, M. see Hills, G.J 307, *331*
Gurney-Smith, M. see Roberts, K. 307, *339*
Gustafson, R.A., Thurston, E L. 392, 393, *395*
Gutierrez, M., Gracen, V.E., Edwards, G.E. 153, 155, 160, 171, *179*
Gutierrez, M., Huber, S.C , Ku, S B., Kanai, R , Edwards, G.E. 126, *130*, 150, 153, 155, 157, 175, *179*
Gutierrez, M , Kanai, R., Huber, S C., Ku, S B., Edwards, G.E. 149, 150, 155, 156, *179*
Gutierrez, M. see Edwards, G.E. 150, *179*
Gutierrez, M see Ku, S B 152, 153, 171, *181*
Gutknecht, J , Dainty, J 265, *283*
Haaroff, K N see Chappell, J B. 169, *178*, 383, 387, *394*
Haas, V A. see Benson, A.A. 86, *129*
Haase, R. 419, 427, *462*
Haberer-Liesenkötter, I. see Kandler, O. 402, *414*
Haberlandt, G. 147, 148, *179*
Hackett, D.P. 408, *412*
Hafer, M.D. see Harris, E.J 383, *395*
Hageman, R.H. see Bowes, G. 109, *129*, 175, *178*, 194, 195, *224*
Hageman, R.H see Laing, W.A. 176, *181*, 186, 193, 195, 213, 221, 222, *229*
Hageman, R.H. see Ritenour, G.L. 208, *231*
Hageman, R.H. see Tolbert, N.E 6, *49*, 109, 134, 193, 197, 199, 209, *233*
Hagihara, B. see Onishi, T. 238, *253*
Hakim, F. see Fairbairn, J.W. 261, *283*
Halestrap, A P., Denton, R.M 246, *251*
Hall, A G see Benson, A.A. 86, *129*
Hall, D O. 90, 117, 121, *130*
Hall, D.O., Edge, H., Kalina, M 354, *372*
Hall, D.O. see Palmer, J M 169, *183*
Hall, D.O. see Reeves, S G. 367, *374*, 398, *415*
Hall, J.L., Davie, C.A.M. 268, *283*
Hall, J L. see Baker, D.A. 265, *281*
Hall, J L see Sexton, R. 268, *286*
Hall, M A. see Ordin, L. 291, 320, 321, *337*
Hallier, U.W. see Heber, R. 90, 96, 105, 110, 121, *130*, *131*, 401, 406, 408, *413*
Halliwell, B. 208, *227*
Halliwell, B. see Leek, A E. 208, *229*
Halperin, M.L. see Robinson, B H 243, *254*
Halperin, W 268, 273, *283*
Halvorson, H.O. 272, *283*
Hamberg, H. see Ege, T 76, 77, *80*
Hammerling, J 56, 64, *81*

Hanower, P see Brzozowska, J 260, *282*
Hansbury, E see Scallen, T.J. 290, 292, *340*
Hanson, J.B., Bertagnollı, B.L., Shepherd, W.D. 379, 382, 386, *395*
Hanson, J.B., Hodges, T.K. 377, 384, 387, *395*
Hanson, J.B., Leonard, R.T., Mollenhauer, H.H. 319, *330*
Hanson, J.B., Malhotra, S.S., Stoner, C.D. 384, *395*
Hanson, J.B. see Bertagnollı, B.L. 241, *250*, 400, *412*
Hanson, J.B. see Earnshaw, M.J. 384, 385, 387, *395*
Hanson, J.B. see Hensley, J.R. 382, *395*
Hanson, J.B. see Hodges, T.K. 377, 384, 387, 390, 391, *395*
Hanson, J.B. see Jung, D.W. 384, *395*, 400, *414*
Hanson, J.B. see Kenefıck, D.G. 385, 387, *396*
Hanson, J.B. see Kırk, B.I. 378, 380, 385, *396*
Hanson, J.B. see Stoner, C.D 378, *397*
Hanson, J.B see Truelove, B. 384, *397*
Hanson, J B. see Wılson, R.H. 236, *254*, 378, 379, 381, 387, *397*, 408, *416*
Hanzely, L. see Graves, L.B. 205, 206, *227*
Hanzon, V. see Sjostrand, F.S. 294, *341*
Hardın, J.W , Cherry, J.H , Morré, D.J., Lembı, C.A. 324, *331*
Harley, J.L. see MacLennan, D.H. 260, *284*, 319, *334*
Harper, W see Wılson, R H 382, *397*
Harrıs, B. see Osmond, C.B. 151, 176, *182*, 202, 203, *230*
Harrıs, E.J., Hafer, M.D., Pressman, B D 383, *395*
Harris, E.J. see Cockrell, R.S. 383, *395*
Harrıs, E J see Pressman, B.C 381, *396*
Harrıs, H. 62, *81*
Harrıs, N., Dodge, A.D. 278, *283*
Harrıs, P J., Northcote, D H. 314, 319, *331*
Harrıs, R.D see Urry, D.W. 38, *49*
Harrıs, W.M., Spurr, A.R. 21, *43*
Harrıs, W M. see Spurr, A.R. 21, *48*
Harrıs, W.W. see Anderson, N.G. 10, *40*
Hart, C.E. see Kortschak, H.P. 144, *181*
Harth, E., Oettmeıer, W., Trebst, A. 356, *372*
Harth, E., Reımer, S., Trebst, A. 350, *372*
Harth, E see Trebst, A. 352, *375*
Harvey, E.B. 57, *81*
Hashımoto, H., Murakamı, S. 23, 24, *43*, 412, *412*
Hasılık, A., Müller, H., Holzer, H 269, 272, *283*
Hass, B.S see Mollenhauer, H H 294, *335*
Hassıd, W.Z. 290, *331*
Hassıd, W.Z. see Tsai, C M 290, 320, *342*
Hatch, M.D. 144, 145, 146, 151, 152, 153, 155, 162, 164, 165, 172, 174, 175, *179*, *180*, 210, *228*
Hatch, M.D., Kagawa, T. 151, 152, 153, 155, 156, 157, 158, 160, 172, *180*
Hatch, M.D., Kagawa, T., Craıg, S. 126, *130*, 149, 153, 155, 156, 158, 160, 170, 171, *180*
Hatch, M.D., Mau, S. 151, 152, 153, 155, 157, *180*, 210, *228*
Hatch, M.D , Mau, S., Kagawa, T 155, *180*
Hatch, M D., Osmond, C B., Slatyer, R.O. 144, *180*
Hatch, M D , Slack, C.R. 144, 155, 156, 167, 175, *180*, 201, 202, 210, *228*
Hatch, M.D., Slack, C.R., Bull, T.A. 151, 155, *180*
Hatch, M.D see Andrews, T J 151, 153, 155, 156, *177*, 210, *223*
Hatch, M.D. see Graham, D. 151, *179*
Hatch, M.D see Hawker, J S. 261, *283*
Hatch, M.D see Johnson, H.S. 123, *132*, 146, 151, 155, *181*, 210, *228*
Hatch, M.D see Kagawa, T. 148, 149, 150, 151, 153, 155, 157, 158, 167, 168, 169, *181*
Hatch, M.D. see Slack, C R 147, 148, 151, 153, 155, 156, 157, 170, *183*, 210, *232*
Haugaard, N. see Chudapongse, P. 390, 393, *395*
Hauska, G.A. 354, *372*
Hauska, G.A., McCarty, R.E., Berzborn, R.J., Racker, E. 349, 350, 354, *372*
Hauska, G.A , Reımer, S , Trebst, A. 355, *372*
Hauska, G.A., Trebst, A., Draber, W 354, 355, *372*
Hauska, G.A. see Racker, E. 349, *374*
Hawker, J.S., Hatch, M D 261, *283*
Hawker, J S. see Downton, W.J S 157, *179*
Hawkins, E.K. 312, *331*
Healy, P.L. see Unzelman, J.M. 314, *342*
Heasley, S see Varmus, H.E. 77, *84*
Heath, I B , Greenwood, A.D., Grıffıths, H.B 306, *331*
Heath, O.V.S. 107, *130*
Heath, O.V.S., Orchard, B. 188, *228*
Heathcote, P. see Ausländer, W 353, *371*
Heber, M. see Heber, U. 7, *43*, 193, *228*
Heber, U. 85, 86, 87, 88, 89, 102, 105, 110, 122, 123, 124, 126, *130*, 140, 142, *142*, 149, 168, *180*, 210, *228*, 261, *283*, 398, 399, 401, 402, 406, 407, 408, 409, 410, 411, *412*, *413*
Heber, U., Hallıer, U.W , Hudson, M.A 90, 96, 105, 110, 121, *130*, 401, 408, *413*
Heber, U., Kırk, M R. 398, *413*
Heber, U , Kırk, M R , Gımmler, H., Schäfer, G. 106, *130*, 194, 201, 207, 222, *228*
Heber, U , Krause, G.H 106, 107, 109, 123, *131*, 167, *180*, 208, 209, 210, 211, *228*, 407, 408, 409, 410, *413*
Heber, U., Pon, N.G., Heber, M. 7, *43*, 193, *228*
Heber, U., Santarıus, K.A 90, 93, 103, 117, 119, 121, 122, *131*, 140, *142*, 402, 403, 404, 406, 407, 408, 409, *413*

Heber, U., Santarius, K A., Hudson, M.A., Hallier, U W. 105, *131*, 406, *413*
Heber, U., Santarius, K.A , Urbach, W , Ullrich, W 102, *131*, 399, 402, *413*
Heber, U , Willenbrink, J. 86, 94, 102, 105, *131*
Heber, U. see Aach, H.G 110, *127*, 260, *281*
Heber, U see Gimmler, H 25, *42*, 92, 106, 110, 111, *130*, 199, *227*
Heber, U. see Santarius, K.A 402, *415*
Heber, U. see Ullrich, W 259, *287*, 402, *416*
Heber, U. see Urbach, W. 102, *135*, 406, 407, *416*
Heftmann, E 270, *283*
Heilporn, V., Limbosch, S 69, *81*
Heininger, B. see Kluge, M. 260, *284*
Heinrich, G 314, *331*
Heintz, C.E see Bracker, C.E 298, 306, *327*
Heldt, H.W. 25, *43*, 124, *131*, 139, 142, *142*, 235, 237, 239, *251*, 400, 404, 405, *413*
Heldt, H.W., Fliege, R., Lehner, K., Milovancev, M., Werdan, K. 139, 141, 142, *142*, 167, 168, *180*
Heldt, H W., Jacobs, H , Klingenberg, M. 239, *251*, 400, *413*
Heldt, H W , Klingenberg, M. 238, *251*
Heldt, H.W , Klingenberg, M., Milovancev, M 241, *251*
Heldt, H W , Rapley, L 25, *43*, 86, 89, 103, 104, 105, 106, 109, 113, 124, *131*, 137, *143*, 363, *372*, 399, 406, 408, 410, *413*
Heldt, H.W., Sauer, F. 86, 92, 94, 109, *131*, 137, *143*, 399, *413*
Heldt, H W , Sauer, F , Rapley, L. 25, 39, *43*, 89, 94, 104, 109, 116, 124, 125, *131*
Heldt, H W , Schwalbach, K 238, *252*
Heldt. H.W., Werdan, K 107, *131*
Heldt. H W., Werdan, K., Milovancev, M , Geller, G 107, *131*, 410, *413*
Heldt, H.W. see Elbers, R. 241, *251*, 400, *412*
Heldt, H.W. see Klingenberg, M. 239, *252*, 400, 411, *414*
Heldt, H.W. see Onishi, T. 238, 241, *253*, 400, *414*
Heldt, H.W. see Pfaff, E. 240, *253*, 400, *415*
Heldt, H.W. see Strotmann, H. 103, 104, *134*, 142, *143*, 402, 404, *416*
Heldt, H.W. see Werdan, K. 25, *50*, 104, 107, 108, 124, 126, *135*, 137, 139, 140, *143*
Helender, H.F 29, *43*
Hellebust, J A., Bidwell, R G.S 111, *131*
Henderson, P.J F. 377, 380, 382, *395*
Henderson, P.J.F , Lardy, H A. 239, *252*, 400, *413*
Hendler, R W 31, *43*, 290, *331*
Hengeveld, E. see Winter, H. 307, *344*
Hennig, W. 55, *82*
Hensley, J.R. 387, *395*
Hensley, J R., Hanson, J B. 382, *395*
Hepler, P.K., Foskett, D.E. 292, *331*
Hepler, P K , Jackson, W T 291, 308, *331*
Hepler, P K , Newcomb, E H 292, *331*
Hepler, P.K see McIntosh, J R 291, *335*
Heppel, L A 37, *43*
Hershko, R., Mamont, P., Shields, R., Tomkins, G.M 63, *82*
Hertel, H see Lembi, C A 5, 15, 16, *44*
Hertel, R., Thomson, K -St , Russo, V E A 323, 324, *331*
Hertel, R. see Lembi, C.A. 291, 323, *334*
Herth, W., Franke, W.W., Stadler, J., Bittiger, H., Keilich, G., Brown, R M., Jr 308, 312, *331*
Herth, W , Franke, W W , VanDerWoude, W J 292, *331*
Herth, W see Brown, R M , Jr 308, *327*
Herth, W. see Franke, W W. 65, *81*, 288, 290, 292, 308, 312, 315, *330*
Herth, W see Spring, H. 64, *84*
Hertig, A T. see Adams, E.C. 294, *326*
Hesketh, J.D 108, *131*
Hesketh. J D see El-Sharkawy, M 108, *130*
Hesketh, J D. see Hofstra, G 213, 214, *228*
Heslop-Harrison, Y., Knox, R.B 315, 316, *331*
Hess, J.L , Tolbert, N E. 191, 194, 204, 205, *228*
Hess, W.M see Strobel, G.A 323, *342*
Hettiaratchi, D R.P., O'Callaghan, J.R 319, *331*
Hew, C -S , Krotkov, G 187, *228*
Hew, C.-S , Krotkov, G , Canvin, D T 188, 212, 213, 214, *228*
Hew, C -S. see O'Neal, D. 157, 158, 168, *182*
Heyn, A.N.J 288, 311, 319, *331*
Heywood, P. 306, *331*
Hickman, D.D. see Deshpande, P.D. 387, *395*
Hiedeman-van Wyk, D., Kannangara, C G 349, *372*
Higinbotham, N., Anderson, W P 265, *283*
Hilgenberg, W. see Fock, H 186, 207, *227*
Hill, F G. 304, 305, *331*
Hill, F G , Outka, D 304, 305, *331*
Hill, R 86, *131*
Hill, R., Walker, D.A. 116, *131*
Hill, R. see Walker, D.A 102, 119, *135*, 403, 404, 406, *416*
Hiller, R.G. see Whittingham, C.P. 194, *233*
Hilliard, J H., Gracen, V E., West, S.H 199, 203, *228*
Hills, G.J 307, *331*
Hills, G.J., Gurney-Smith, M , Roberts, K 307, *331*
Hills, G J see Roberts, K 307, *339*
Hind, G , Jagendorf, A T. 347, *373*
Hind, G , McCarty, R.E 348, 359, 364, *373*

Hind, G., Nakatani, H.Y., Izawa, S. 126, *131*, 357, *373*
Hind, G., Whittingham, C.P. 359, *373*
Hind, G. see Good, N.E. 368, *372*
Hind, G. see Izawa, S. 351, *373*
Hind, G. see Jagendorf, A.T. 347, *373*
Hingson, D.J., Diamond, J.M. 450, *462*
Hinkle, P.C. 353, *373*
Hirabayashi, S. see Kisaki, T. 201, *229*
Hirai, M., Asaki, T. 270, *283*
Hirano, H., Parkhouse, B., Nicholson, G.L., Lennox, E.S., Singer, S.J. 301, *331*
Hirsch, H.E. see Benson, A.A. 86, *129*
Hittelman, W.N. see Rao, P.N. 62, *83*
Hoch, G., Owens, O.H., Kok, B. 186, 188, 212, 213, *228*, 408, *413*
Hoch, G. see Owens, O.v.H. 408, *414*
Hochachka, P.W., Somero, G.N. 165, *180*
Hodge, A.J., McLean, J.D., Mercer, F.V. 170, *180*
Hodges, T.K., Hanson, J.B. 377, 384, 387, 390, 391, *395*
Hodges, T.K., Leonard, R.T. 15, *43*
Hodges, T.K., Leonard, R.T., Bracker, C.E., Keenan, T.W. 6, 8, 15, 16, *43*
Hodges, T.K. see Hanson, J.B. 377, 384, 387, *395*
Hodgkin, A.L. 455, *462*
Hoefert, L.L. see Esau, K. 322, *329*
Hoek, J.B. see Meijer, A.J. 245, *253*
Hofstra, G., Hesketh, J.D. 213, 214, *228*
Hohler, T. see Mahon, J.D. 176, *182*, 191, 220, *230*
Hohman, W., Schraer, H. 391, *395*
Holborow, K. see Lilley, R.McC. 105, 108, 121, 126, *132*
Holldorf, A.W. 410, *413*
Holmgren, P., Jarvis, P. 186, 187, 188, 212, 213, *228*
Holtz, R.B., Stewart, P.S., Patton, S., Schisler, L.C. 16, *43*
Holtz, R.B. see Douce, R. 6, 7, 23, 24, *41*, 412, *412*
Holzer, H., Weintraub, H., Mayne, R., Mochan, B. 63, *82*
Holzer, H. see Hasilik, A. 269, 272, *283*
Holzer, H. see Matern, H. 269, 272, *284*
Holzer, H. see Ishikawa, H. 294, *332*
Holzwarth, G. 12, *43*
Homan, P.H., Schmid, G.H. 170, *180*
Hooymans, J.J.M. 319, *332*
Hope, A.B. 419, 441, *462*
Hope, A.B., Luttge, U., Ball, E. 402, *413*
Hope, A.B. see Lilley, R.McC. 402, *414*
Hopkins, N.H. see Goldman, R.D. 76, *81*
Horner, H.T., Lersten, N.R. 314, *332*
Horwitz, A.F. 12, *43*
Hruban, Z., Rechcigl, M. 198, *228*
Hsie, A.W. see Schroder, H. 76, *83*
Huang, A.H.C. 6, 7, *43*
Huang, A.H.C., Beevers, H. 176, *180*, 203, *228*
Huang, C.M. see Keenan, T.W. 301, 323, *332*, *333*
Huang, C.M. see Morré, D.J. 297, 299, 300, 301, 304, *336*
Hubbell, W.L., McConnell, H.M. 11, *43*
Huber, S.C., Edwards, G.E. 126, *131*, 149, 150, 167, *180*
Huber, S.C., Kanai, R., Edwards, G.E. 149, 150, *180*
Huber, S.C. see Gutierrez, M. 126, *130*, 149, 150, 153, 155, 156, 157, 175, *179*
Hubert, E. see Huez, G. 59, 74, 75, *82*
Hubert, E. see Marbaix, G. 59, 75, *82*
Hudgin, R.L. see Schachter, H. 301, *340*
Hudson, M.A. see Heber, U. 90, 96, 105, 110, 121, *130*, *131*, 401, 406, 408, *413*
Hudson, M.A. see Urbach, W. 102, *135*, 406, 407, *416*
Huez, G., Marbaix, G., Hubert, E., Leclercq, M., Nudel, V., Soreq, H., Salomon, R., Lebleu, G., Revel, M., Littauer, U.J. 59, 74, 75, *82*
Huez, G. see Marbaix, G. 59, 75, *82*
Huffaker, R.C., Peterson, L.W. 272, *283*
Hulin, N., Brachet, J. 55, *82*
Hume, S.P., Lamb, J.F., Weingart, R. 77, *82*
Humphreys, T.E. 261, *283*, 319, *332*
Hyde, B.B. 314, *332*

Ikuma, H. 3, *43*
Ikuma, H., Bonner, W.D., Jr 387, *395*, 408, *414*
Imai, A. see Kisaki, T. 109, *132*, 201, 222, *229*
Indge, K.J. 258, 260, 263, 269, *283*
Isaac, P.K. see Thomas, P.L. 273, *286*
Ishikawa, H., Bischoff, R., Holzer, H. 294, *332*
Iten, W., Matile, P. 270, 273, 279, *283*
Ito, S. see Loewenstein, W.R. 19, *45*
Izawa, S. 359, *373*
Izawa, S., Good, N.E. 367, *373*, 398, *414*
Izawa, S., Hind, G. 351, *373*
Izawa, S., Kraayenhof, R., Ruuge, E.H., Devault, D. 355, *373*
Izawa, S., Ort, D.R. 356, *373*
Izawa, S. see Good, N.E. 115, *130*, 368, *372*
Izawa, S. see Gould, J.M. 352, 353, 356, 365, *372*
Izawa, S. see Hind, G. 126, *131*, 357, *373*
Izawa, S. see Ort, D.R. 356, *374*
Izawa, S. see Saha, S. 354, *375*
Izawa, S. see Winget, G.D. 363, *376*

Jabbal, I. see Schachter, H. 301, *340*
Jacazio, G., Probstein, R.F., Sonin, A.A., Yung, D. 441, *462*

Jackson, J.B., Crofts, A R. 364, *373*
Jackson, J B., Crofts, A R., von Stedıngk, L.-V. 364, *373*
Jackson, W.A., Volk, R.J. 109, *131*, 175, 176, *181*, 185, 186, 188, 202, 211, 212, 214, 215, *228*
Jackson, W.A see Mulchı, C.L 186, 188, *230*
Jackson, W.A. see Ozbun, J.L. 188, 202, *231*
Jackson, W.A see Volk, R J. 214, *233*
Jackson, W.T see Hepler, P.K. 291, 308, *331*
Jacob, J L. see Rıbaıllıer, D. 258, 260, 261, 263, *285*
Jacobs, H. see Heldt, H W. 239, *251*, 400, *413*
Jacobs, M. see Urry, D.W. 38, *49*
Jacobs, W.P see Sabnıs, D.D. 291, 292, *340*
Jacobs-Lorena, M. see Gross, K.W. 73, *81*
Jaffe, M.J. see Toriyama, H. 261, *286*
Jaffe, M.J. see Wıllıamson, F.A. 5, 6, *50*
Jagendorf, A T. 347, *373*
Jagendorf, A.T., Avron, M. 116, *131*
Jagendorf, A.T , Hınd, G 347, *373*
Jagendorf, A T , Margulıes, M 354, *373*
Jagendorf, A.T , Urıbe, E.G 347, 365, *373*
Jagendorf, A.T see Avron, M. 363, *371*
Jagendorf, A.T see Hınd, G. 347, *373*
Jagendorf, A.T. see Ryrıe, I J 369, *375*
Jagendorf, A T see Uribe, E G. 347, *375*
Jagger, W.S. see Pressman, B.C. 381, *396*
Jagow, G. v. 241, *252*
Jagow, G.v., Klıngenberg, M 238, *252*
James, W.O., Leech, R.M 86, *131*
Jamıeson, J D., Palade, G E 316, *332*
Jamıeson, J.D. see Meldolesı, J. 301, *335*
Jans, B 263, 266, 267, *283*
Jans, B. see Matile, P. 258, 263, 266, 267, 270, *284*
Jarasch, E.-D. see Franke, W.W 18, 19, *42*, 302, 304, *330*
Jarasch, E.-D. see Phılıpp, E.-I. 18, *47*
Jarett, L see McKeel, D W 8, *45*
Jarlfors, U. see Smith, D.S. 291, *341*
Jarvis, P G see Holmgren, P. 186, 187, 188, 212, 213, *228*
Jarvis, P.G. see Ludlow, M M. 188, *229*
Jarvıs, P.G. see Sestak, Z 187, *232*
Jeanrenaud, B see Orcı, L. 291, *337*
Jefford, T.G see Edelman, J 261, 266, *282*
Jelinek, W , Goldsteın, L. 72, *82*
Jelsema, C L , Morré, D.J. 302, *332*
Jelsema, C.L., Morré, D J., Ruddat, M. 316, *332*
Jelsema, C.L., Ruddat, M., Morré, D.J., Wıllıamson, F.A. 324, *332*
Jelsema, C.L see Morré, D.J 290, 291, 299, 301, 302, 304, 312, *336*
Jensen, R.G. 108, *132*
Jensen, R.G., Bassham, J A. 86, 105, 111, 115, 116, 117, *132*, 191, 194, 209, *228*, 404, 405, *414*
Jensen, R G see Bahr, J.T 108, *128*, 174, 175, 176, *177*, 195, *224*
Jensen, R G see Bassham, J A. 103, 104, 105, 106, 109, *128*, 140, *142*, 209, *224*, 404, 406, *412*
Jeon, K.W., Lorch, I.J , Danıellı, J F 71, *82*
Jeschke, W D , Sımonis, W 402, *414*
Jımenez, E., Baldwın, R L., Tolbert, N.E., Wood, W.A 197, 201, *228*
Jımenez, E. see Orth, G.M. 194, *230*
John, A.A , Bırnstıel, M.L , Jones, K W. 55, *82*
Johnson, E M. see Ceron, G 66, *80*
Johnson, F.H., Eyrıng, H., Polissar, J. 442, *462*
Johnson, H.M., Wılson, R.H. 380, 384, 385, 386, 387, 389, 390, 391, 393, *395*
Johnson, H M see Ramırez-Mitchell, R 385, *397*
Johnson, H S. 210, *228*
Johnson, H.S., Hatch, M.D. 123, *132*, 146, 151, 155, *181*, 210, *228*
Johnson, H S see Andrews, T J 151, 153, 155, *177*, 210, *223*
Johnson, J.H see Pressman, B C. 381, *396*
Johnson, K.D., Kende, H. 290, 316, *332*
Jolliffe, P.A., Tregunna, E.B. 213, *229*
Jones, D D , Morré, D.J. 314, *332*, *336*
Jones, K.W. see Bırnstıel, M 55, *79*
Jones, K W see John, A A 55, *82*
Jones, R.L 315, 316, *332*
Jones, R.L , Prıce, J.M. 315, 316, *332*
Jonghe d'Ardoye, V de see Geuskens, M 74, *81*
Jordan, E.G. 315, *332*
Joy, K W. see Ritenour, G.L. 208, *231*
Jullıard, J H see Gautheron, D C 245, *251*
Jung, D W., Hanson, J B 384, *395*, 400, *414*
Jung, P , Tanner, W. 321, *332*
Junge, W , Ausländer, W 350, 352, 353, *373*
Junge, W see Ausländer, W 353, 354, *371*
Junge, W see Schlıephake, W. 352, 364, *375*
Junıper, B.E., Roberts, R.M. 314, *332*
Juniper, B E see Clowes, F A L. 322, *328*

Kadenbach, B. 28, *43*
Kagawa, T , Hatch, M.D 148, 149, 150, 151, 153, 155, 157, 158, 167, 168, 169, *181*
Kagawa, T , Lord, J M , Beevers, H 302, *332*
Kagawa, T. see Hatch, M.D 126, *130*, 149, 151, 152, 153,

155, 156, 157, 158, 160, 170, 171, 172, *180*
Kagawa, T. see Moore, T.S. 290, 301, *336*
Kagawa, T. see Lord, J.M. 290, 301, *335*
Kagawa, Y., Racker, E. 27, *43*
Kahane, I. see Marchesı, V.T. 13, 15, *45*
Kahn, A see Breıdenbach, R.W. 6, *41*, 199, *225*
Kalberer, P.P., Buchanan, B.B., Arnon, D.I 86, *132*
Kalberer, P P. see Buchanan, B.B. 96, *129*
Kalına, M see Hall, D.O. 354, *372*
Kalsova, A. see Schnepf, E 314, *341*
Kamıya, N. 291, 292, *332*
Kanaı, R., Black, C.C. 148, 151, 152, 155, *181*
Kanaı, R., Edwards, G E. 149, 153, *181*
Kanai, R. see Edwards, G.E. 151, 155, 156, *179*
Kanaı, R. see Gutıerrez, M. 126, *130*, 149, 150, 153, 155, 156, 157, 175, *179*
Kanaı, R. see Huber, S.C. 149, 150, *180*
Kanaı, R. see Ku, S.B. 171, *181*
Kandler, O 401, *414*
Kandler, O., Haberer-Lıesen-kötter, I 402, *414*
Kandler, O. see Tanner, W. 401, *416*
Kanduc, D see Papa, S. 242, *253*
Kannangara, C.G. see Hıede-man-van Wyk, D. 349, *372*
Kanno, Y see Loewensteın, W.R. 19, *45*
Kapıl, R.N , Pugh, T D , New-comb, E.H. 203, *229*
Karlısh, S.J D., Avron, M. 351, 352, 364, *373*
Karlsnes, A M. see Cleland, R. 307, *328*
Karmovsky, M.J. see Gonzales, F. 391, 392, *395*
Karnovsky, M L. see Pıerre, J.W. 4, 6, 9, *47*
Karpılov, Y.S. 144, 148, 170, 171, *181*
Kartenbeck, J., Franke, W.W. 294, *332*
Kartenbeck, J see Franke, W.W. 60, *81*, 294, 295, 298, 301, 302, 304, *330*
Kasamatsu, H. see Matsumoto, L. 74, *82*
Kashnıng, D.M., Kasper, C.B 18, *43*
Kasper, C.B. 6, 18, 19, *43*
Kasper, C.B. see Kashnıng, D.M. 18, *43*
Katchalsky, A., Curran, P.F. 419, 420, *462*
Katchalsky, A. see Kedem, O. 419, *462*, *463*
Kato, K.H., Sugıyama, M. 64, 73, *82*
Kaufman, S., Alıvısatos, S.G.A. 238, *252*
Kauss, H 321, *332*
Kauss, H., Glaser, C 324, *332*
Kawaguchi, K see Onıshi, T. 238, *253*
Kawaguchı, S. see Benson, A.A 86, *129*
Kean, E.L., Bruner, W.E 72, *82*
Kearney, P.C , Tolbert, N.E. 109, *132*, 194, 209, *229*
Kearney, P.C. see Rabson, R. 197, 201, *231*
Keck, K. 56, *82*
Kedem, O. 431, 440, *462*
Kedem, O , Essıg, A. 453, *462*
Kedem, O., Katchalsky, A. 419, *462*, *463*
Keegstra, K , Talmadge, K.W., Bauer, W.D., Albersheım, P 307, *332*
Keenan, T.W., Huang, C.M., Morré, D J 323, *332*
Keenan, T.W , Morré, D J. 299, 300, 301, 302, 321, *333*
Keenan, T W., Morré, D.J., Basu, S 321, 323, *333*
Keenan, T W., Morré, D.J., Huang, C.M. 301, 323, *333*
Keenan, T.W. see Hodges, T.K. 6, 8, 15, 16, *43*
Keenan, T W see Morré, D J 7, 30, *45*, 290, 291, 297, 299, 300, 301, 302, 304, 312, 323, *336*, *337*
Keenan, T W. see Phılıpp, E - I 18, *47*
Keenan, T W see Yunghans, W N 300, *344*
Keılıch, G. see Herth, W. 308, 312, *331*
Kelker, H.C , Filner, P 273, *283*
Kellems, R.E., Allison, V.F., Butow, R.A. 25, 28, *43*
Kellems, R.E., Butow, R A. 28, *43*
Kelly, G.J., Gibbs, M. 122, *132*, 410, *414*
Kelly, G.J., Latzko, E. 96, *132*
Kemp, A , Jr see Passham, H C 241, *253*, 400, *414*
Kemp, A , Jr see Vaartȷes, W J. 240, *254*
Kemp, J.A., Out, T.A., Guıot, H.F.L., Souverıjn, J.H.M 239, *252*
Kende, H see Baumgartner, B. 272, *281*
Kende, H see Fırn, R.D. 316, *329*
Kende, H. see Johnson, K.D. 290, 316, *332*
Kenefıck, D.G., Hanson, J.B. 385, 387, *396*
Kennedy, R A., Laetsch, W M 146, *181*
Kephart, J.E. see Dauwalder, M. 7, *41*, 315, 328, *329*
Kephart, J.E see Whaley, W.G. 30, *50*, 288, 304, 308, 318, *343*
Key, J.L. see Chen, Y. 7, *41*
Keys, A J 403, *414*
Keys, A.J., Whıttıngham, C.P. 407, *414*
Keys, A J see Bırd, I.F. 94, *129*, 201, 208, 220, *224*, 402, *412*
Keys, A.J. see Roberts, G R. 110, 111, *134*
Khanna, R , Sınha, S.K 146, *181*
Khrıstolubova, N.B. see Chır-kova, L.I. 65, *80*
Kidby, D K. see Schıberı, A. 15, *48*
Kıermayer, O , Dobbersteın, B. 304, 310, 321, *333*
Kıermayer, O see Dobberstein, B 303, 304, 310, *329*
Kıermayer, O see Staehelin, L.A. 300, *342*
Kıeth, A D , Mehlhorn, R.J. 12, *44*
Kıeth, A.D see Mehlhorn, R J 12, *45*

Kindinger, J.I. see Ben-Arie, R. 321, *326*
Kindinger, J I. see VanDer-Woude, W.J. 291, 312, 320, 321, *343*
King, M.J , Diwan, J J. 245, *252*
Kinzel, H. see Stadelmann, E J 260, *286*
Kirby, E.G., Roberts, R M 314, *333*
Kirchmann, M see Bonotto, S. 64, 65, 66, *79*
Kirk, B I , Hanson, J B 378, 380, 385, *396*
Kirk, J T O., Tilney-Bassett, R.A.E. 11, *44*
Kirk, M R. see Bassham, J A 103, 104, 105, 106, 109, 116, *128*, 140, *142*, 193, 195, 209, *224*, 404, 406, *412*
Kirk, M R. see Heber, U 106, *130*, 194, 201, 207, 222, *228*, 398, *413*
Kirk, M R see Pedersen, T.A 96, *133*, 402, *415*
Kirk, P.R., Leech, R.M. 109, *132*
Kisaki, T., Imai, A., Tolbert, N.E 109, *132*, 201, 222, *229*
Kisaki, T., Tolbert, N.E. 109, *132*, 199, 200, 207, *229*
Kisaki, T , Yano, N , Hirabayashi, S 201, *229*
Kisaki, T , Yoshida, N , Imai, A 201, *229*
Kisaki, T see Tolbert, N.E 6, *49*, 109, *134*, 193, 197, 199, 209, *233*
Kisseleva, E.V see Chirkova, L.I 65, *80*
Kit, S 290, *333*
Kleinig, H 19, *44*
Kleinig, H see Brown, R.M , Jr. 308, 309, 312, *327*
Kleinig, H.J. see Franke, W W 18, 19, *42*
Kletzien, R.F., Perdue, J F 292, *333*
Klingenberg, M. 235, 236, 239, 240, 241, 245, *252*, 383, *396*
Klingenberg, M., Buchholz, M. 238, 239, 240, *252*, 400, *414*
Klingenberg, M., Durand, R., Guerin, B 242, *252*
Klingenberg, M , Grebe, K , Heldt, H W 239, *252*
Klingenberg, M , Heldt, H W , Pfaff, E. 400, 411, *414*
Klingenberg, M , Pfaff, E. 89, *132*, 239, *252*
Klingenberg, M , Riccio, P , Aquila, H., Schmiedt, B., Grebe, K., Topitsch, P. 239, *252*
Klingenberg, M , Scherer, B., Stengel-Rutkowski, L., Buchholz, M., Grebe, K 239, 240, *252*
Klingenberg, M. see Bucher, Th. 248, *251*
Klingenberg, M. see Heldt, H.W. 238, 239, 241, *251*, 400, *413*
Klingenberg, M see Jagow, G.v. 238, *252*
Klingenberg, M. see McGivan, J.D. 244, *253*
Klingenberg, M. see Onishi, T. 238, 241, *253*, 400, *414*
Klingenberg, M. see Palmieri, F. 242, 243, 244, *253*
Klingenberg, M. see Pfaff, E. 27, *46*, 236, 240, *253*, 360, *374*, 400, *415*
Klingenberg, M. see Quagliariello, E. 244, *254*
Klingenberg, M see Scherer, B 240, *254*
Klob, W see Tanner, W. 401, *416*
Kloppstech, K , Schweiger, H.G 68, *82*
Kluge, M , Heininger, B. 260, *284*
Knights, B A see Longley, R P 14, 15, *45*
Knox, R.B. see Heslop-Harrison, Y 315, 316, *331*
Kobatake, Y., Fujita, H 441, *463*
Koch, R. see Döhler, G 206, *226*
Koch, W see Schnepf, E. 316, *341*
Koehler, D.E., Varner, J E. 316, *333*
Koeppe, D E , Miller, R J 238, *252*
Koeppe, D.E. see Malone, C 235, *253*
Koeppe, D.E see Miller, R.J 383, *396*
Kohn, L.D., Warren, W A. 200, 209, *229*
Kok, B see Fowler, C.F. 350, 352, 353, *372*
Kok, B see Hoch, G. 186, 188, 212, 213, *228*, 408, *413*
Kolattukudy, P.E., Dean, B B. 161, *181*
Koller, D. see Samish, Y. 189, 212, *231*
Koller, D see Whiteman, P.C. 213, *233*
Komnick, H., Stockem, W., Wohlfarth-Bottermann, K.E. 292, *330*
Korff, R.W. von see Deshpande, P.D. 387, *395*
Korn, E.D. 31, 35, *44*
Korn, E D see Pollard, T.D. 292, *338*
Kortschak, H P., Hart, C E , Burr, G O 144, *181*
Kosuge, T see Preiss, J 95, 96, 126, *133*
Kowallik, W. see Codd, G A. 206, *225*
Kraayenhof, R. 362, 365, 366, *373*
Kraayenhof, R. see Izawa, S. 355, *373*
Kram, R. see Moens, W. 64, *82*
Kraminer, A see Gimmler, H 25, *42*, 92, 106, 110, 111, *130*, 199, *227*
Krasne, S see Szabo, G. 12, *49*
Krause, G H 103, 122, *132*, 407, 408, *414*
Krause, G H see Heber, U 106, 107, 109, 123, *131*, 167, *180*, 208, 209, 210, 211, *228*, 407, 408, 409, 410, *413*
Krebs, H A., Veech, R.L 210, *229*
Krebs, H A. see Gevers, W. 393, *395*
Krebs, H.A. see Williamson, D.H 248, *254*
Kreutz, W 33, 34, *44*
Kreutz, W., Menke, W. 33, *44*
Kreutzberg, G W 291, *333*
Krien, S. see Franke, W W. 294, 295, 297, 298, 299, 301, 303, *330*
Krippahl, G. see Warburg, O 193, *233*
Kroger, A see Onishi, T. 238, 241, *253*, 400, *414*

Krogmann, D W see Avron, M. 363, *371*
Kroh, M. 315, *333*
Krondahl, U. see Ege, T 58, 76, 77, *80*
Krotkov, G , Renuckles, V C., Thımann, K.V 188, *229*
Krotkov, G. see Fock, H 221, *227*
Krotkov, G see Forrester, M.L. 186, 188, 213, 214, *227*
Krotkov, G. see Hew, C -S. 187, 188, 212, 213, 214, *228*
Krotkov, G. see Lister, G.R 188, *229*
Krotkov, G. see Tregunna, E.B. 186, 188, 213, *233*
Ku, S.B., Gutıerrez, M , Edwards, G.E. 152, 153, *181*
Ku, S B., Gutıerrez, M., Kanaı, R., Edwards, G.E. 171, *181*
Ku, S B. see Gutıerrez, M. 126, *130*, 149, 150, 153, 155, 156, 157, 175, *179*
Kuhn, W., Läuger, P , Voellmy, H , Block, R., Majer, H. 441, *463*
Kulıkowskı, R R see Mascarenhas, J.P. 18, *45*
Kumarasınghe, S see Bird, I.F 220, *224*
Kuo, J., O'Brien, T.P., Canny, M.J. 162, *181*
Kuster, E. 255, 259, *284*
Kuylenstıerna, B. see Ernster, L 28, *42*

Labavıtch, J M , Ray, P M. 307, *333*
Laber, L.J., Latzko, E., Gıbbs, M. 146, 147, *181*
Laber, L J see Latzko, E 147, *182*
Lacy, P.E see Malaısse-Lagae, F. 291, *334*
Ladda, R., Estensen, R.D. 57, 76, *82*
Laetsch, W.M. 21, *44*, 147, 157, 161, 170, *181*
Laetsch, W.M see Kennedy, R A 146, *181*
Laing, W A., Ogren, W.L , Hageman, R.H. 176, *181*, 186, 193, 195, 213, 221, 222, *229*
Lake, J.V 188, *229*
Lamant, A , Roland, J C 324, *333*
Lamb, J.F see Hume, S.P 77, *82*
Lampen, J O. 25, *44*
Lamport, D.T.A. 307, *333*
Lance, C. see Moreau, F. 24, 27, 28, *45*, 236, *253*
Landre, P 312, *333*
Lane, C.D. see Gurdon, J.B. 59, 75, *81*
Lane, M.D. see Cooper, T.G 107, *130*
Lane, M D. see Marayama, H. 175, *182*
Lang, A see Brachet, J. 56, *80*
La Noue, K.F., Meıjer, A J., Brouwer, A. 245, *252*
La Noue, K F., Walajtys, E I., Wıllıamson, J.R 245, *252*
Lansıng, A.I. see Ray, T.K 304, *339*
Laprode, R. see Cıanı, S.M. 442, *462*
Lardy, H A , Ferguson, S.M 6, *44*, 377, *396*
Lardy, H.A see Henderson, P.J.F 239, *252*, 400, *413*
Larsen, G.L., Brummond, D.O 290, *333*
Larson, S see Stockıng, C.R 103, 122, *134*, 140, *143*, 407, *415*
Latıes, G.G. 264, *284*
Latzko, E , Laber, L.J , Gıbbs, M. 147, *182*
Latzko, E see Gıbbs, M. 94, 96, 108, *130*
Latzko, E see Kelly, G J 96, *132*
Latzko, E. see Laber, L J 146, 147, *181*
Latzko, E see O'Neal, D 157, 158, 168, *182*
Läuchlı, A. 259, *284*
Lauchlı, A. see Schroter, K. 259, *286*
Làuger, P 441, *463*
Lauger, P , Neumcke, B 442, 443, *463*
Lauger, P. see Kuhn, W 441, *463*
Lauger, P. see Neumcke, B. 443, *463*
Lauquın, G see Morel, F 240, *253*
Lauquın, G. see Vıgnaıs, P.V 239, *254*
Laurıtıs, J.A., Coombs, J , Volcanı, B E. 310, *333*
Law, J.H., Snyder, W.R. 11, *44*
Lawlor, D.W 188, *229*
Lawlor, D.W , Fock, H. 188, 213, 214, 215, *229*
Leadbeater, B.S C 306, *333*
Leadbeater, B.S C see Leedale, G.F 306, 310, *334*
Le Beaux, Y.J 294, *326*
Lebleu, B. see Huez, G 59, 74, 75, *82*
Leblond, C.P., Bennett, G. 323, *333*
Lebowıtz, M.M see Freed, J.J. 292, *330*
Leclercq, M see Huez, G. 59, 74, 75, *82*
Leclercq, M see Marbaıx, G. 59, 75, *82*
Ledbetter, M C , Porter, K.R 14, 17, *44*, 294, *334*
Ledbetter, M.C. see Crotty, W J 298, *328*
Ledbetter, M.C see Voeller, B.R 317, *343*
Leduc, E.H., Scott, G B., Avrameas, S. 17, *44*
Leduc, E.N., Avrameas, S., Bouteılle, M 17, *44*
Lee, D.C , Wılson, R H 379, 380, 383, *396*
Leech, J.H see Mollenhauer, H.H 314, *336*
Leech, R.M. 86, *132*
Leech, R M see James, W O 86, *131*
Leech, R.M. see Kırk, P.R 109, *132*
Leech, R.M see Mackender, R.O 22, 23, 24, *45*, 412, *414*
Leech, R.M see Rıdley, S M. 93, *133*
Leech, R M. see Thomson, W.W 20, *49*
Leedale, G.F. 310, *334*
Leedale, G.F., Leadbeater, B.S.C., Massalskı, A 306, 310, *334*
Leedale, G F. see Manton, I 308, 309, *334*
Leek, A E., Hallıwell, B., Butt, V S 208, *229*
Lefebre, P see Rebhun, L. 64, *83*

Legros, F., Saines, M, Renard, M, Conard, V 66, *82*
Lehner, K. see Heldt, H W. 139, 141, 142, *142*, 167, 168, *180*
Lehninger, A L. 377, 387, 388, 391, 392, *396*
Lehninger, A.L., Carafoli, E. 388, *396*
Lehninger, A.L., Carafoli, E., Rossi, C.S. 377, 387, 389, *396*
Lehninger, A.L. see Azzone, G F. 377, *394*
Lehninger, A.L. see Carafoli, E. 387, 388, 389, *394*
Lehninger, A.L see Chen, C 380, 387, 388, 389, *394*
Lehninger, A.L. see Drahota, Z. 389, *395*
Lehninger, A L. see Reynafarje, B. 386, 388, *397*
Lehninger, A.L. see Rossi, C.S. 387, 389, *397*
Leigh, R.A., Williamson, F A., Wyn Jones, R.G. 16, *44*
Lein, S. see Racker, E. 349, *374*
Le'John, H.B. 392, *396*
Leloir, L F. 290, 321, *334*
Leloir, L.F. see Behrens, N H. 321, *326*
LeMarchand, Y. see Orci, L. 291, *337*
Lembi, C A., Morré, D J., Thomson, K S., Hertel, R. 5, 15, 16, *44*, 291, 323, *334*
Lembi, C.A. see Elder, J.H. 309, *329*
Lembi, C A see Hardin, J.W. 324, *331*
Lembi, C A. see Morré, D.J. 298, *336*
Lembi, C.A see Roland, J.-C 6, 16, *47*, 291, 299, 300, 303, 324, *339*
Lembi, C.A. see St. Thomson, K. 5, 15, 16, *44*
Lembi, C A. see VanDerWoude, W.J. 15, *49*, 291, 312, 320, 321, 324, *342*, *343*
Lenard, J., Singer, S.J. 11, 13, 35, 36, *44*
Lennarz, W J, Scher, M G 321, *334*
Lennarz, W.J. see Waechter, C J. 321, *343*
Lenney, J.F, Matile, P, Wiemken, A, Schellenberg, M., Meyer, J. 271, 272, 280, *284*
Lennox, E S. see Hirano, H 301, *331*
Lentz, T L. 291, *334*
Leonard, R T, VanDerWoude, W.J. 6, 7, 15, 16, *44*
Leonard, R.T. see Hanson, J B. 319, *330*
Leonard, R.T. see Hodges, T.K. 6, 8, 15, 16, *43*
Leopold, A.C see Thomson, K.-S. 323, *342*
Leppard, G.G, Colvin, J.R. 312, *334*
Lerman, J.C. 165, *182*
Lerner, E, Shug, A.L, Elson, Ch., Shrago, E. 240, *252*
Lersten, N R. see Horner, H.T. 314, *332*
Letham, D S. see Berridge, M.V 324, *326*
Levin, W.B. see Bidwell, R.G.S. 212, *224*
Levin, W.B. see Mangat, B.S. 212, 215, *230*
Levine, G., Bassham, J A. 116, *132*
Levine, R.P see Neumann, J. 106, *133*
Levine, S, Pictet, R., Rutter, W. 63, *82*
Lewis, D.H. see Chen-She, S H 95, 96, 97, 98, *129*
Lewis, D R see Northcote, D.H 14, *46*
Leznoff, C C see Robinson, B.H. 243, *254*
Li, J.H, Desousa, R.C, Essig, A. 453, *463*
Liebecq, C. see Sluse, F.E 243, *254*
Lieberman, I see Ray, T.K 304, *339*
Lieberman, M see Baker, J.E 408, *412*
Lilley, R McC., Fitzgerald, M.P., Rienits, K.G., Walker, D.A. 88, 91, 92, 93, *132*
Lilley, R McC, Holborow, K., Walker, D A. 105, 108, 121, 126, *132*
Lilley, R McC, Hope, A.B 402, *414*
Lilley, R McC, Schwenn, J D, Walker, D A. 91, 93, 111, 115, 116, 125, *132*
Lilley, R.McC., Walker, D.A. 107, 108, 111, 121, *132*
Lilley, R.McC. see Schwenn, J.D. 91, 93, 111, 113, 114, *134*
Lilley, R.McC see Walker, D A. 108, *135*
Limbosch, S see Heilporn, V. 69, *81*
Lin, C. see Chen, Y. 7, *41*
Lin, C.H see Stocking, C.R. 25, *49*
Lin, D.C., Nobel, P S. 108, *132*
Ling, G.N. 265, *284*
Lis, H. see Sharon, N. 324, *341*
Lister, G.R., Krotkov, G., Nelson, C.D. 188, *229*
Littauer, U J. see Huez, G. 59, 75, *82*
Littauer, U.Z. see Marbaix, G 59, 75, *82*
Liu, A.Y., Black, C C. 176, *182*, 203, *229*
Liu, L.L see Thomson, W W 315, *342*
Lloyd, N.D.H. see Canvin, D.T. 215, 216, *225*
Loeb, J. 53, *82*
Loening, U E. see Birnstiel, M. 55, *79*
Loescher, W.H., Nevins, D J. 319, *334*
Loewenstein, W.R., Kanno, Y 19, *45*
Loewenstein, W.R., Kanno, Y., Ito, S. 19, *45*
Lofrumento, N E see Papa, S. 242, *253*, 387, *396*
Logliscі, M see Papa, S. 242, *253*
Long, M.M see Urry, D.W. 38, *49*
Long, S.P see Bucke, C. 152, *178*
Longley, R.P., Rose, A H., Knights, B.A. 14, 15, *45*
Loomis, R S see El-Sharkawy, M.A 187, *226*
Loomis, W D. 3, *45*
Loos, E see Tanner, W 401, *416*
López-Sáez, J.F. see Frey-Wyssling, A. 288, 308, *330*
López-Sáez, J.F. see Risueno, M.C 312, *339*
Lorch, I J see Jeon, K W 71, *82*
Lord, J M, Codd, G.A., Merrett, M.J 204, *229*

Lord, J.M , Kagawa, T , Moore, T.S , Beevers, H 290, 301, *334*
Lord, J.M., Merrett, M.J. 204, 205, *229*
Lord, J M. see Codd, G.A. 204, 205, *225*
Lord, J M. see Kagawa, T 302, *332*
Lord, J.M. see Merrett, M.J. 204, *230*
Lord, J.M. see Moore, T.S. 290, 301, *336*
Loria, P. see Cantatore, P 74, *80*
Lorimer, G.H., Andrews, T.J. 109, *132*, 176, *182*, 222, *229*
Lorimer, G H , Andrews, T.J., Tolbert, N.E. 185, 195, 197, *229*
Lorimer, G.H., Miller, R.J. 379, *396*
Lorimer, G.H. see Andrews, T.J 109, *128*, 176, *177*, 185, 194, 195, 221, *223*, *224*
Loud, A V. 5, *45*
Loughman, B C. 98, *132*
Louisot, P see Richard, M. 303, *339*
Lowenstein, J.M 250, *253*
Lucas, I.A M. see Peel, M C. 298, *338*
Lucas, J.J. see Waechter, C J 321, *343*
Ludlow, M.M., Jarvis, P.G. 188, *229*
Ludlow, W.M. 175, *182*
Ludwig, L.J. 187, 188, 214, *230*
Ludwig, L.J., Canvin, D.T. 188, 211, 212, 213, 214, 215, *230*
Ludwig, L.J. see Whitehouse, D G. 117, *136*
Luduena, M A. see Wessels, N.K. 292, 294, *343*
Lunardi, J see Morel, F. 240, *253*
Lund, P. see Williamson, D H. 248, *254*
Lunt, O.R. see Wyn-Jones, R.L. 393, *397*
Luscher, A., Matile, P. 269, 270, *284*
Luscher, K., Matile, P. 65, *82*
Luttge, U. 170, *182*, 259, 263, 264, *284*
Lüttge, U. see Hope, A.B. 402, *413*
Lutzac, M. see Thomas, D des S. 291, 315, *342*
Lynch, V. see Benson, A.A. 86, *129*

Macaulay, L.K. see Berezney, R. 19, *40*
MacBradley, M.A. see Wessels, N.K 58, *84*
Mache, R. see Waygood, E.R 175, *184*
Mackay, N.C. see Paolillo, D.J., Jr. 21, *46*
Mackender, R.O , Leech, R.M. 22, 23, 24, *45*, 412, *414*
MacLachlan, G A., Porter, H.K. 95, 96, 116, *132*
MacLachlan, G.A. see Fan, D.F. 315, *329*
MacLachlan, G A. see Shore, G 312, 324, *341*
MacLachlan, G.A. see Spencer, F.S. 324, *342*
MacLennan, D.H., Beevers, H., Harley, J.L 260, *284*, 319, *334*
MacRobbie, E A.C. 264, 265, *284*, 319, *334*, 401, *414*
MacRobbie, E A C. see Raven, J.A. 401, *415*
Madden, D.M. see Earnshaw, M.J. 384, 387, *395*
Magnusson, R P., McCarty, R E. 369, *373*
Magnusson, R.P. see Portis, A R., Jr 358, 368, 369, *374*
Mahlberg, P G 317, *334*
Mahlberg, P G , Turner, F R , Walkinshaw, C H , Venketeswaran, S 277, *284*, 317, *334*
Mahon, J D , Egle, K , Fock, H 191, *230*
Mahon, J.D., Fock, H., Canvin, D.T. 191, 200, 211, 216, 217, 218, 221, *230*
Mahon, J.D., Fock, H., Höhler, T., Canvin, D T. 176, *182*, 191, 220, *230*
Mahon, J.D. see Fock, H. 219, *227*
Maitra, S.C., De, D.N. 292, *334*
Majer, H. see Kuhn, W. 441, *463*
Malacinski, G M. 74, *82*
Malaisse, W.J. see Malaisse-Lagae, F 291, *334*
Malaisse-Lagae, F., Greider, M.H., Malaisse, W.J., Lacy, P.E. 291, *334*
Malhotra, S.S. see Hanson, J B. 384, *395*
Malkoff, D B., Buetow, D.E. 280, *284*
Malone, C., Koeppe, D E., Miller, R.I. 235, *253*
Mamont, P. see Hershko, R. 63, *82*
Manegold, E. 441, *463*
Mangat, B.S., Levin, W.B , Bidwell, R.G.S. 212, 215, *230*
Mangenot, G. see Guilliermond, A. 255, *283*
Manavathu, E see Thomas, D des S. 291, 315, *342*
Manley, E.R., Skrdlant, H.B. 292, *334*
Mannella, C.A. see Douce, R. 238, *251*, 387, *395*
Manton, I. 308, 309, *334*
Manton, I., Leedale, G F. 308, 309, *334*
Manton, I., Rayns, D G , Ettl, H , Parke, M. 306, *334*
Marayama, H., Easterday, R.L., Chang, H.C., Lane, M.D. 175, *182*
Marbaix, G., Huez, G., Burny, A., Cleuter, Y., Hubert, E., Leclercq, M., Chantrenne, H., Soreq, H., Nudel, U., Littauer, U.Z. 59, 75, *82*
Marbaix, G. see Chantrenne, H. 75, *80*
Marbaix, G see Gurdon, J.B. 59, 75, *81*
Marbaix, G. see Huez, G. 59, 74, 75, *82*
Marchant, R., Moore, R.T. 311, *334*
Marchant, R., Robards, A.W. 311, *334*
Marchant, R. see Smith, D.G. 261, *286*
Marchesi, V T , Segrest, J P., Kahane, I 13, 15, *45*
Marchesi, V T see Tillack, T.W. 321, *342*
Margulies, M. see Jagendorf, A.T 354, *373*
Marinos, N.G. 317, *334*
Marker, A.F.H. see Whittingham, C.P. 194, *233*

Markham, J.W. see Bısalputra, T 205, *224*
Marshall-Graves, J.A. 62, *82*
Martın, F.W., Telek, L. 315, *334*
Martın, H G., Thorne, K J.I 321, *335*
Martın, I F. see Chenıae, G M 356, *371*
Martin, J.H., Matthews, J.L. 391, 392, *396*
Martın, P 259, *284*
Marty, F 266, 275, 279, *284*, 298, 317, 318, 319, *335*
Marx-Fıgını, M 323, *335*
Mascarenhas, J.P , Berman-Kurtz, M., Kulıkowskı, R.R. 18, *45*
Mason, T G see Phıllıs, E. 95, *133*
Mason, T L. see Schatz, G. 28, *48*
Massalskı, A see Leedale, G.F 306, 310, *334*
Master, R.W P. see Schweıger, H.G. 67, *83*
Matern, A. 406, 411, *414*
Matern, H., Betz, H , Holzer, H 269, 272, *284*
Mathıeu, Y 90, 121, *132*
Matile, P. 13, 14, 15, 16, *45*, 261, 268, 269, 270, 271, 278, *284*, 317, *335*
Matıle, P , Jans, B , Rıckenbacher, R. 258, 263, 266, 267, 270, *284*
Matıle, P , Moor, H 274, *284*, 317, 319, *335*
Matıle, P , Wıemken, A 269, 271, *284*, *285*
Matıle, P., Wınkenbach, F 272, 273, 278, 279, 280, *285*
Matıle, P. see Baumgartner, B. 272, *281*
Matıle, P. see Cortat, M 315, *328*
Matıle, P. see Iten, W 270, 273, 279, *283*
Matıle, P see Lenney, J.F 271, 272, 280, *284*
Matıle, P see Lüscher, A. 269, 270, *284*
Matıle, P see Luscher, K 65, *82*
Matıle, P. see Meyer, J 271, *285*
Matıle, P. see Wilden, W. van der 271, 272, *287*
Matoba, T see Doı, E 270, *282*
Matsumoto, L , Kasamatsu, H , Pıko, L., Vınograd, J 74, *82*
Matthews, J.L see Martın, J.H. 391, 392, *396*
Mattox, K.R see Stewart, K.D. 205, *232*
Mau, S see Hatch, M.D. 151, 152, 153, 155, 157, *180*, 210, *228*
Maw, S.L. see Coombs, J 174, *178*
Mayne, B.C , Dee, A M., Edwards, G E. 171, *182*
Mayne, B.C., Edwards, G E , Black, C C 171, *182*
Mayne, R see Holzer, H 63, *82*
Mazıa, D see Wılt, F.H 74, *84*
Mazliak, P. 301, *335*
Mazliak, P. see Meunıer, P. 24, *45*
Mazur, P see Groot, S R. de 419, *462*
McBrıde, D.L , Cole, K 317, *335*
McCarty, R.E. 352, 364, *373*
McCarty, R.E , Douce, R , Benson, A A 24, *45*
McCarty, R E , Fagan, J 369, *373*
McCarty, R.E , Fuhrman, J S., Tsuchıya, Y 363, 369, *373*
McCarty, R E., Pıttman, P.R , Tsuchıya, Y 369, *374*
McCarty, R E , Racker, E 349, 369, *374*
McCarty, R.E see Anderson, M.M 354, *370*
McCarty, R E see Gaensslen, R E. 357, 359, 360, 361, *372*
McCarty, R E. see Hauska, G.A 349, 350, 354, *372*
McCarty, R.E see Hınd, G. 348, 359, 364, *373*
McCarty, R E see Magnusson, R P 369, *373*
McCarty, R.E see Portıs, A.R , Jr 358, 359, 360, 361, 363, 365, 367, 368, 369, *374*
McClure, W K see Robınson, P.M. 317, *339*
McConnell, H M., Devaux, P , Scandella, C 12, *45*
McConnell, H M see Hubbell, W L. 11, *43*
McCormıck, A.V. see Stokes, D M 108, *134*
McCormıck, A V see Walker, D A 121, *135*
McFadden, B.A. 194, *230*
McFadden, B.A. see Tabıta, F R. 105, *134*
McGıll, S see Droop, M R. 204, *226*
McGinnıs, G.D. see Shafızadeh, F 307, 321, *341*
McGıvan, J D , Chappell, J.B 245, *253*
McGıvan, J.D., Klıngenberg, M 244, *253*
McGuıre, E.J. see Schachter, H 301, *340*
McIntosh, J R , Hepler, P.K., Van Wıe, D.G. 291, *335*
McKeel, D.W., Jarett, L 8, *45*
McLaughlın, S.G A see Szabo, G. 12, *49*
McLean, J.D. see Hodge, A.J 170, *180*
McNab, J M see Vıllemez, C L 6, 16, *49*, 319, *343*
Meares, P., Ussing, H.H. 454, *463*
Medurı, B see Papa, S 246, *253*
Mehard, C.W. see Packer, L. 235, *253*, 348, *374*, 380, *396*
Mehler, A H 196, *230*
Mehlhorn, R J , Kıeth, A.D 12, *45*
Mehlhorn, R J see Kıeth, A D 12, *44*
Meıdner, H 214, *230*
Meıer, H see Reıd, J S.G. 273, *285*, 315, 321, *339*
Meıer, J. 432, *463*
Meıer, J , Sauer, F , Woermann, D 432, *463*
Meıjer, A J , Brouwer, A , Reıjngoud, D.J., Hoek, J B., Tager, J.M. 245, *253*
Meijer, A.J., Van Dam, K 235, *253*
Meıjer, A.J., Groot, G S.P., Tager, J M 242, 243, *253*
Meijer, A.J., Vıgnaıs, P M 245, *253*
Meıjer, A J see Brouwer, A 246, *251*
Meıjer, A J see La Noue, K F 245, *252*
Meıxner, J , Reık, H G. 427, *463*

Mela, L. 388, *396*
Melandri, A. see Forti, G 348, *372*
Meldolesi, J, Cova, D. 301, *335*
Meldolesi, J., Jamieson, J.D., Palade, G.E 301, *335*
Menke, W. see Kreutz, W. 33, *44*
Mense, R.M. see Varner, J E. 315, *343*
Mentré, P. see Carasso, N 298, *327*
Mercer, F V. see Hodge, A J. 170, *180*
Merlin, L M see Morré, D.J 301, *336*
Merrett, M.J., Goulding, K H. 196, *230*
Merrett, M.J., Lord, J.M 204, *230*
Merrett, M.J. see Codd, G.A. 204, 205, *225*
Merrett, M.J. see Collins, N. 206, *225*
Merrett, M.J. see Lord, J.M. 204, 205, *229*
Merritt, W D., Morré, D.J. 312, *335*
Merritt, W D see Morré, D J. 290, 291, 298, 299, 301, 302, 312, *336*
Mesquita, J.F. 273, 275, 277, *285*, 317, *335*
Metz, C.B. see Chamberlain, J.T. 74, *80*
Meunier, D., Pianeta, C., Coulomb, P 27, *45*
Meunier, D., Mazliak, P. 24, *45*
Meyenburg, H.K. von see Beck, C. 280, *281*
Meyer, A J see Van Dam, K 210, *226*, 377, *397*
Meyer, B see Diers, L. 259, *282*
Meyer, J, Matile, P 271, *285*
Meyer, J. see Lenney, J.F. 271, 272, 280, *284*
Meyer, J see Wilden, W. van der 271, 272, *287*
Meyer, K.H., Sievers, J.F. 456, *463*
Meyer, L. see Winter, H. 307, *344*
Miflin, B.J., Beevers, H 88, *132*
Migniac-Maslow, M. see Champigny, M.L. 113, *129*
Millard, D L, Wiskich, J.T, Robertson, R.N. 385, *396*
Miller, A.G see Cheng, K.G 206, *225*
Miller, A G see Colman, B 204, *225*
Miller, C. see Shen-Miller, J. 323, *341*, 394, *397*
Miller, R A, Ruddle, F H 76, *82*
Miller, R.I. see Malone, C. 235, *253*
Miller, R.J, Dumford, W.S, Koeppe, D E 383, *396*
Miller, R J see Koeppe, D.E 238, *252*
Miller, R J see Lorimer, G H 379, *396*
Mills, J see Barber, J 126, *128*, 357, *371*
Milovancev, M see Heldt, H W. 107, *131*, 139, 141, 142, *142*, 167, 168, *180*, 241, *251*, 410, *413*
Milovancev, M see Werdan, K. 126, *135*, 140, *143*
Minton, G.A. see Wilson, R.H 385, 386, *397*
Miranda, A.F, Godman, G.C., Deitsch, A D, Tanenbaum, S.W 292, *335*
Miranda, A F, Godman, G C, Tanenbaum, S W 294, *335*
Mitchell, P 347, 349, 366, 370, *374*, 380, 389, *396*
Mitchell, P., Moyle, J. 389, *396*
Mochan, B see Holzer, H 63, *82*
Moens, W, Vokaer, A, Kram, R. 64, *82*
Moestrup, O, Thomsen, H A 309, *335*
Molisch, H 99, 100, *132*
Mollenhauer, H.H 30, *45*, 289, 291, 307, 311, 313, 314, *335*
Mollenhauer, H H., Evans, W. 310, *335*
Mollenhauer, H.H., Morré, D J. 291, 292, 293, 294, 295, 301, 304, 312, 314, 319, *336*
Mollenhauer, H H., Hass, B.S., Morré, D.J 294, *335*
Mollenhauer, H.H., Whaley, W G. 314, *336*
Mollenhauer, H.H., Whaley, W G., Leech, J.H. 314, *336*
Mollenhauer, H.H see Black, C C. 203, *224*
Mollenhauer, H.H see Hanson, J B. 319, *330*
Mollenhauer, H.H. see Morré, D.J. 7, 29, 30, *45*, 290, 294, 295, 296, 297, 298, 299, 301, 303, 314, 317, 318, *336*, *337*
Mollenhauer, H H see Schweltiz, F D 310, *341*
Mollenhauer, H.H see Whaley, W.G 288, 308, 318, *343*
Mollenhauer, H H see Wilson, R.H. 236, *254*, 378, 379, 381, *397*
Momsen, W. see Boyer, P D. 370, *371*
Monneron, A., Blobel, G., Palade, G.E 19, *45*
Moor, D see Branton, D 14, *41*
Moor, H., Mühlethaler, K. 13, *45*, 321, *336*
Moor, H. see Matile, P. 274, *284*, 317, 319, *335*
Moore, C L 388, *396*
Moore, R.T. see Marchant, R 311, *334*
Moore, T.C, Barlow, S A., Coolbaugh, R C. 292, *336*
Moore, T S, Lord, J M, Kagawa, T., Beevers, H. 290, 301, *336*
Moore, T S see Lord, J.M. 290, 301, *334*
Moreau, F., Dupont, J., Lance, C 24, 27, 28, *45*
Moreau, F., Lance, C. 27, *45*, 236, *253*
Morel, F., Lauquin, G., Lunardi, J, Duszynski, J, Vignais, P V 240, *253*
Morel, F see Vignais, P V 239, *254*
Morgenthaler, J.J, Price, C A, Robinson, M, Gibbs, M 88, *133*
Morré, D.J. 10, *45*, 290, 297, 298, 301, 302, 303, 315, 323, *336*
Morré, D J., Bracker, C E. 294, 324, *336*
Morré, D.J, Elder, J H, Jelsema, C.L., Keenan, T.W., Merritt, W D. 290, 291, 299, 301, 302, 304, 312, *336*
Morré, D.J, Franke, W W., Deumling, B., Nyquist, S.E., Ovtracht, L. 299, *336*

Morré, D J., Jones, D D , Mollenhauer, H.H. 314, *336*
Morré, D J., Keenan, T W , Huang, C.M. 297, 299, 300, 301, 304, *336*
Morré, D.J., Keenan, T.W., Mollenhauer, H H. 7, 30, *45*, 299, 301, *336*
Morré, D.J , Merlın, L.M., Keenan, T.W. 301, *336*
Morré, D.J , Merrıtt, W D., Lembı, C.A. 298, *336*
Morré, D.J , Mollenhauer, H.H. 290, 295, 296, 297, 298, 299, 303, 317, 318, *336*
Morré, D.J., Mollenhauer, H.H., Bracker, C.E. 29, 30, *45*, 294, 295, 297, 298, 299, 303, *337*
Morré, D.J., Nyquıst, S., Rivera, E. 290, *337*
Morré, D J., VanDerWoude, W.J. 294, 302, 304, 306, 307, 312, 319, 320, *337*
Morré, D.J., Yunghans, W.N., Vıgıl, E L., Keenan, T W 323, *337*
Morré, D.J see Bracker, C.E. 298, *327*
Morré, D J. see Clark, J E. 301, 324, *328*
Morré, D.J. see Elder, J H. 301, 309, *329*
Morré, D J. see Franke, W W. 288, 290, 292, 294, 301, 302, 304, 315, *330*
Morré, D.J. see Frantz, C. 303, *330*
Morré, D.J see Grove, S N 288, 290, 298, 300, 308, *330*
Morré, D.J. see Hardın, J.W. 324, *331*
Morré, D.J see Jelsema, C.L 302, 316, 324, *332*
Morré, D.J see Jones, D.D 314, *332*
Morré, D J. see Keenan, T.W 299, 300, 301, 302, 321, 323, *332*, *333*
Morré, D.J. see Lembi, C A. 5, 15, 16, *44*, 291, 323, *324*
Morré, D.J. see Merrıtt, W.D. 312, *335*
Morré, D.J. see Mollenhauer, H H. 291, 292, 293, 294, 295, 301, 304, 312, 314, 319, *335*, *336*
Morré, D H. see Olson, A.C. 323, 325, *337*
Morré, D J. see Roland, J.-C. 6, 16, *47*, 291, 299, 300, 303, 324, *339*
Morré, D.J. see VanDerWoude, W.J. 15, *49*, 291, 298, 301, 307, 308, 312, 319, 320, 321, 324, *342*, *343*
Morré, D.J see Wıllıamson, F A. 5, 6, *50*
Morré, D.J. see Yunghans, W N 300, *344*
Morris, G H. see Brown, H T 95, *129*
Morrıs, I. see Bassham, J.A. 108, *128*, *129*
Moss, D N 187, 188, 214, *230*
Mothes, K see Schulze, C 266, *286*
Mott, R.L. see Steward, F C. 264, *286*
Moyle, J. see Mıtchell, P 389, *396*
Mudd, J.B. see Panter, R A. 247, *253*
Muhle, H. see Schröder, H 365, 366, *375*
Mühlethaler, K. 34, 35, *45*, 307, 312, 319, 321, *337*
Mühlethaler, K. see Frey-Wysslıng, A. 31, *42*, 288, 308, 316, *330*
Mühlethaler, K. see Moor, H. 13, *45*, 321, *336*
Muhlethaler, K see Parthasarathy, M V. 292, *338*
Mulchı, C.L., Volk, R J., Jackson, W A. 186, 188, *230*
Müller, E., Nelles, A., Neumann, D 263, *285*
Müller, E. see Neumann, D 259, 261, 263, 267, *285*
Muller, H see Hasılık, A. 269, 272, *283*
Munn, E.A. 235, 237, *253*
Murakamı, S., Packer, L. 379, *396*
Murakami, S. see Hashımoto, H 23, 24, *43*, 412, *412*
Murakami, S. see Packer, L. 235, *253*, 348, *374*, 380, *396*
Murer, E. see Brıerley, G.P. 389, *394*
Murray, D.R., Giovanellı, J., Smıllıe, R M. 204, 205, *230*

Nachbaur, J see Colbeau, A. 24, *41*
Nagaı, R , Rebhun, L.I. 292, *337*
Naım, V. see Nelson, N. 359, *374*
Nakano, M., Asakı, T. 270, 271, *285*
Nakao, M., Packer, L 3, *45*
Nakatanı, H.Y. see Hind, G. 126, *131*, 357, *373*
Nakayama, N see Sugıyama, T. 108, *134*
Nelles, A see Müller, E. 263, *285*
Nelson, C.D. see Forrester, M.L. 186, 188, 213, 214, *227*
Nelson, C.D. see Lıster, G R 188, *229*
Nelson, C.D. see Tregunna, E B 186, 188, 213, *233*
Nelson, E B., Tolbert, N.E. 204, 205, *230*
Nelson, E.B. see Bruın, W J. 204, 205, *225*
Nelson, E.B see Tolbert, N E. 204, *232*
Nelson, H. see Nelson, N. 359, *374*
Nelson, N , Nelson, H., Naım, V , Neumann, J. 359, *374*
Nelson, N. see Racker, E. 349, *374*
Nernst, W. 443, *463*
Neumann, D , Müller, E 259, 261, 263, 267, *285*
Neumann, D. see Muller, E. 263, *285*
Neumann, J , Levıne, R.P 106, *133*
Neumann, J. see Arntzen, C.J 171, *177*
Neumann, J see Nelson, N. 359, *374*
Neumann, J. see Raven, J.A. 401, *415*
Neumcke, B , Läuger, P. 443, *463*
Neumcke, B. see Lauger, P. 442, 443, *463*
Nevıns, D.J. see Loescher, W H 319, *334*
Newcomb, E H. 20, *46*, 291, 294, 295, 316, *337*
Newcomb, E H , Becker, W M 6, *46*
Newcomb, E H., Frederıck, S E. 93, *133*, 170, *182*
Newcomb, E.H. see Bonnett, H.T 295, 308, *326*

Newcomb, E H. see Frederick, S E 198, 199, 203, 207, *227*
Newcomb, E.H. see Hepler, P K. 292, *331*
Newcomb, E.H see Kapil, R.N. 203, *229*
Neyfakh, A A see Yudin, A.L. 72, *84*
Nichiporovich, A.A. see Zak, E.G 205, *234*
Nicolson, G.L. 12, *46*
Nicolson, G.L. see Hirano, H. 301, *331*
Nicolson, G.L. see Singer, S J. 11, 12, 38, *48*
Nicolson, J. see Barber, J. 126, *128*, 357, *371*
Nicotra, A. see Cantatore, P. 74, *80*
Nikoro, Z.S. see Chirkova, L.I. 65, *80*
Nilshammar-Holmvall, M. see Sundberg, I. 263, *286*
Nir, I., Pease, D.C. 354, *374*
Nissen, P 259, 264, *285*
Nobel, P.S. 88, *133*, 163, 165, *182*, 419, *463*
Nobel, P.S., Cheung, Y -N.S 25, *46*, 110, *133*
Nobel, P S., Wang, C.-T. 25, *46*, 92, 106, 110, *133*, *135*
Nobel, P.S. see Lin, D.C. 108, *132*
Nobel, P.S. see Packer, L. 110, *133*
Nobs, M. see Björkman, O. 147, 148, 176, *178*
Northcote, D.H. 290, 292, 295, 297, 312, 315, 319, *337*
Northcote, D.H., Lewis, D R 14, *46*
Northcote, D H , Pickett-Heaps, J D 314, *337*
Northcote, D H. see Bowles, D J 314, 319, *327*
Northcote, D.H see Harris, P.J 314, 319, *331*
Northcote, D.H. see Pickett-Heaps, J.D 292, 314, *338*
Northcote, D.H. see Wooding, F.B.P. 292, 313, *344*
Northcote, D.H. see Wright, K. 314, *344*
Norum, K.R. see Bremer, J 246, *251*
Novikoff, A.B. see Golfischer, S. 7, *42*
Nudel, U see Marbaix, G. 59, 75, *82*
Nudel, V. see Huez, G. 59, 74, 75, *82*
Nurse, P see Wiemken, A. 258, 260, 269, 281, *287*
Nyquist, S E see Morré, D.J. 290, 299, *336*, *337*

Oakley, B.R., Dodge, J.D. 205, *230*
Oaks, A , Bidwell, R.G.S. 258, 260, 261, *285*, 290, *337*
O'Brien, T P 292, 308, 311, 312, 313, 314, *337*
O'Brien, T.P., Carr, D.J. 161, *182*
O'Brien, T.P., Thimann, K.V. 292, *337*
O'Brien, T.P. see Kuo, J. 162, *181*
O'Callaghan, J.R. see Hettiaratchi, D R.P 319, *331*
Ochoa, S. see Vishniac, W. 116, *135*
Ochoa, S. see Zelitch, I. 199, 208, *234*
Oda, T. see Blair, P V. 27, *40*
Oda, T see Fernández-Morán, H. 27, *42*
O'Day, D.H. 270, *285*
Oeser, A. see Tolbert, N.E. 196, 200, 209, *233*
Oeser, A. see Schnarrenberger, C 191, 210, *232*
Oeser, A. see Tolbert, N E. 6, *49*, 109, *134*, 193, 197, 199, 209, *233*
Oesterhelt, D., Stoeckenius, W. 370, *374*
Oettmeier, W. see Harth, E 356, *372*
Ogren, W L , Bowes, G. 109, *133*, 185, 195, *230*
Ogren, W L. see Bowes, G. 109, *129*, 175, 176, *178*, 194, 195, *224*
Ogren, W L see Chollet, R. 149, 150, 176, *178*, 202, *225*
Ogren, W.L. see Laing, W.A. 176, *181*, 186, 193, 195, 213, 221, 222, *229*
Ohad, I. see Eytan, G. 193, *226*
Ohtsuru, C. see Doi, E. 270, *282*
Olesen, P. 163, *182*
Oliver, I.R. see Bucke, C. 157, *178*
Olson, A C , Bonner, J , Morré, D.H. 323, 325, *337*
Olson, M.S. see Schuster, S.M. 393, *397*
O'Neal, D , Hew, C.S., Latzko, E., Gibbs, M. 157, 158, 168, *182*
Ongun, A., Stocking, C R 110, *133*
Onishi, T., Kawaguchi, K., Hagihara, B. 238, *253*
Onishi, T., Kroger, A., Heldt, H.W., Pfaff, E., Klingenberg, M. 238, 241, *253*, 400, *414*
Öpik, H. 316, *337*
Orchard, B. see Heath, O.V S. 188, *228*
Orci, L., LeMarchand, Y , Singh, A., Assimacopoulos-Jeannet, F., Rouiller, C., Jeanrenaud, B. 291, *337*
Ord, M.J. 72, *83*
Ordin, L., Hall, M.A. 291, 320, 321, *337*
Ordin, L. see Ben-Arie, R. 321, *326*
Ordin, L see Pinsky, A. 291, *338*
Ordin, L see VanDerWoude, W J. 291, 312, 320, *343*
Orme, F. see Sandblom, J. 441, *463*
Ort, D.R., Izawa, S. 356, *374*
Ort, D.R see Izawa, S. 356, *373*
Orth, G.M., Tolbert, N.E., Jimenez, E. 194, *230*
Orth, G.M see Yu, Y.L 195, *234*
Ortolani, G see Giudice, G. 74, *81*
Osborne, D.J. see Ridge, I. 307, *339*
Oseroff, A R., Robbins, P W., Burger, M M. 37, *46*
Osmond, C.B. 162, 163, 171, 172, 173, 176, *182*
Osmond, C.B , Allaway, W G. 146, *182*
Osmond, C.B., Björkman, O. 176, 186, 222, *230*
Osmond, C.B , Harris, B. 151, 176, *182*, 202, 203, *230*
Osmond, C.B., Smith, F.A. 163, *182*
Osmond, C B., Troughton, J.H., Goodchild, D.J. 146, 170, *183*

Osmond, C B. see Badger, M R. 175, 176, *177*
Osmond, C.B see Chapman, E A. 169, *178*
Osmond, C.B see Hatch, M D 144, *180*
Osmond, C.B see Polya, G.M 171, *183*
Osmond, C.B. see Ting, I.P 155, 175, *183*
Osmond, C B see Woo, K C. 171, *184*
Osterhout, W.J.V. 38, *46*
Osterle, J.F. see Gross, R.J 441, *462*
Ottenheym, H.C.J. see Bassham, J A 116, *128*
Ouitrakul, R see Saha, S. 354, *375*
Out, T A. see Kemp, J.A. 239, *252*
Outka, D E. see Hill, F.G. 304, 305, *331*
Outka, D.E , Williams, D C. 309, *337*
Ovtracht, L., Thiéry, J.P 301, *337*
Ovtracht, L. see Carasso, N 298, 319, *327*
Ovtracht, L. see Morré, D.J 299, *336*
Owens, O.H., Hoch, G. 408, *414*
Owens, O.H see Hoch, G 186, 188, 212, 213, *228*, 408, *413*, *414*
Oxender, D L 37, *46*
Oxender, D L , Quay, S 37, *46*
Ozbun, J L , Volk, R J., Jackson, W A 188, 202, *231*

Pacini, E. see Cresti, M 275, 277, *282*
Packer, L , Crofts, A R. 110, *133*
Packer, L , Murakami, S., Mehard, C.W. 235, *253*, 348, *374*, 380, *396*
Packer, L , Siegenthaler, P.A., Nobel, P.S 110, *133*
Packer, L see Deamer, D W. 357, *371*
Packer, L. see Fleischer, S 3, *42*
Packer, L. see Murakami, S 379, *396*
Packer, L. see Nakao, M. 3, *45*
Palade, G.E 25, 26, *46*
Palade, G.E see Bergeron, J.J.M 301, *326*
Palade, G.E. see Jamieson, J.D 316, *332*
Palade, G.E. see Meldolesi, J 301, *335*
Palade, G E. see Monneron, A 19, *45*
Pallaghy, C.K. 259, *285*
Pallas, J.E see Samish, Y B. 212, *231*
Palisano, J.R., Walne, P L 315, *337*
Pall, M.T. 323, *337*
Palmer, J M., Hall, D O 169, *183*
Palmer, J.M. see Coleman, J O D. 238, 246, *251*
Palmieri, F , Prezioso, G., Quagliariello, E , Klingenberg, M. 242, 244, *253*
Palmieri, F., Quagliariello, E , Klingenberg, M. 242, 243, 244, *253*
Palmieri, F., Stipani, I , Quagliariello, E , Klingenberg, M 242, 243, 244, *253*
Palmieri, F. see Quagliariello, E. 243, 244, *254*
Pande, S.V , Blanchaer, M.C. 240, *253*
Pankratz, H.S. see Drum, R W 310, *329*
Pankratz, H.S see Stoermer, E F 310, *342*
Panter, R A , Mudd, J B 247, *253*
Paolillo, D J., Jr. 21, *46*
Paolillo, D J., Jr., Falk, R H 21, *46*
Paolillo, D J., Jr , Mackay, N C., Reighard, J.A. 21, *46*
Paolillo, D.J , Jr., Reighard, J A. 21, 23, *46*
Papa, S., Francavilla, A., Paradies, G., Meduri, B. 246, *253*
Papa, S , Lofrumento, N.E., Kanduc, D., Paradies, G , Quagliariello, E 242, *253*
Papa, S , Lofrumento, N.E , Loglisci, M , Quagliariello, E 242, *253*
Papa, S., Lofrumento, N E., Secchi, A.G , Quagliariello, E 387, *396*
Paradies, G see Papa, S. 242, 246, *253*
Pardee, A.B 37, *46*
Pardue, M.L., Gall, J.G 55, *83*
Parish, R W 271, *285*
Parish, R.W., Rickenbacher, R 191, *231*
Park, D see Robinson, P.M. 317, *339*
Park, R B. 12, 33, *46*
Park, R B , Biggins, J. 21, 22, 33, *46*
Park, R.B , Pfeifhofer, A O. 21, 23, *46*
Park, R.B., Pon, N.G. 33, *46*
Park, R.B., Sane, P.V. 3, *46*
Park, R B see Branton, D. 34, 35, *41*
Park, R B see Sane, P V. 22, *47*
Parke, M see Manton, I 306, *334*
Parkhouse, B see Hirano, H. 301, *331*
Parkin, J 95, *133*
Parlin, B , Eyring, H. 442, *463*
Parsons, D F., Williams, G.R., Chance, B 6, 27, 28, *46*
Parsons, D.F., Williams, G.R., Thompson, W., Wilson, D., Chance, B 24, 27, *46*, 236, *253*
Parsons, D F., Yano, Y 236, *253*
Parsons, D F. see Racker, E 27, *47*
Parthasarathy, M V., Mühlethaler, K 292, *338*
Paschall, H A see Cameron, D A. 390, *394*
Passham, H C , Souverijn, J H M., Kemp, A , Jr 241, *253*, 400, *414*
Pate, J S., Gunning, B.E.S 170, *183*
Pate, J S see Gunning, B.E S. 315, *330*
Patni, N J , Billmire, J E , Aaronson, S 16, *46*
Patriarca, P. see Carafoli, E 392, *394*
Patterson, B D see Graham, D 165, *179*
Patton, S see Holtz, R B 16, *43*
Paul, J see Gilmour, R S 76, *81*

Payne, J F see Bal, A K. 315, *326*
Payne, P I, Boulter, D 273, *285*
Pearcy, R. see Bjorkman, O. 148, 176, *178*
Pease, D.C 4, *46*
Pease, D.C. see Nir, I. 354, *374*
Péaud-Lenöel, C, Axelos, M. 320, *338*
Pedersen, P L. see Coty, W.A 242, *251*
Pedersen, T.A., Kirk, M., Bassham, J A 96, *133*, 402, *415*
Peel, M.C., Lucas, I A.M., Duckett, J.G., Grennwood, A D. 298, *338*
Pemberton, R. see Bishop, J O. 75, *79*
Peng, C.F., Price, D.W., Bhuvaneswaran, C., Wadkins, C.L 390, 393, *396*
Penman, S. see Abelson, H.T 59, *79*
Perdue, J.F see Green, D.E. 28, *42*
Perdue, J.F see Kletzien, R F 292, *333*
Perevoshchikova, K.A see Zbarsky, I.B. 19, *50*
Perry, R. 58, *83*
Peterkofsky, A., Racker, E. 108, *133*
Peterson, L W. see Huffaker, R C. 272, *283*
Petris, S de see Allison, A C 275, *281*
Pfaff, E 404, *415*
Pfaff, E., Heldt, H.W, Klingenberg, M 240, *253*, 400, *415*
Pfaff, E., Klingenberg, M 400, *415*
Pfaff, E., Klingenberg, M, Heldt, H W. 400, *415*
Pfaff, E, Klingenberg, M., Ritt, E., Vogell, W. 27, *46*, 236, *254*, 360, *374*
Pfaff, E see Klingenberg, M. 89, *132*, 239, *252*, 400, 411, *414*
Pfaff, E. see Onishi, T 238, 241, *253*, 400, *414*
Pfeiffer, H.H. 261, *285*
Pfeifhofer, A.O. see Park, R B 21, 23, *46*
Philipp, E.-I., Franke, W W, Keenan, T W, Stadler, J., Jarasch, E-D 18, *47*
Phillips, M.L., Williams, G R 246, *254*, 384, *396*
Phillis, E., Mason, T.G. 95, *133*
Pianeta, C. see Meunier, D 27, *45*
Pick, U., Rottenberg, H., Avron, M. 365, *374*
Pickett-Heaps, J.D 291, 292, 308, 312, 314, 315, *338*
Pickett-Heaps, J D, Northcote, D.H. 292, 314, *338*
Pickett-Heaps, J.D. see Fowke, L.C. 312, *329*
Pickett-Heaps, J.D. see Northcote, D.H 314, *337*
Pictet, R. see Levine, S 63, *82*
Pierre, J.W. De, Karnovsky, M.L. 4, 6, 9, *47*
Pikalov, A V. see Chirkova, L.I. 65, *80*
Pikalov, A.V. see Sandakhiev, L.S. 65, *83*
Piko, L see Matsumoto, L. 74, *82*
Pilet, P-E see Roland, J-C. 307, *340*
Pinsky, A., Ordin, L 291, 321, *338*
Pinteric, L. see Schachter, H. 301, *340*
Pinto Da Silva, P, Douglas, S D, Branton, D. 301, 321, 323, *338*
Pitman, M G 162, *183*
Pitt, D. 268, *285*
Pitt, D, Galpin, M. 270, 271, *285*
Pittman, P R see McCarty, R E 369, *374*
Plagemann, P G W see Estensen, B.D. 292, *329*
Planck, M. 443, *463*
Plantefol, L. see Guilliermond, A. 255, *283*
Plaut, Z, Gibbs, M. 109, *133*, 194, 195, 196, *231*
Platt, K. see Campbell, N. 315, *327*
Poincelot, R.P. 23, 24, 25, *47*, 107, 108, *133*, 152, 165, *183*, 209, *231*, 412, *415*
Poincelot, R.P., Day, P.R. 23, 24, 25, *47*
Poincelot, R.P. see Racusen, D 23, 24, *47*
Polissar, J see Johnson, F H 442, *462*
Pollack, R. see Goldman, R.D. 76, *81*
Pollard, T D., Korn, E.D. 292, *338*
Pollard, T D, Weihing, R R 292, 294, *338*
Pollard, W.E. see Roberts, R.M. 301, *339*
Polya, G.M, Osmond, C B. 171, *183*
Pon, N.G., Rabin, B.R., Calvin, M 108, *133*
Pon, N.G. see Heber, U 7, *43*, 193, *228*
Pon, N G see Park, R B. 33, *46*
Poole, R.J. 319, *338*
Porter, H.K. see Bird, I.F. 94, *129*
Porter, H K see MacLachlan, G.A 95, 96, 116, *132*
Porter, K R. 314, *338*
Porter, K R see Bikle, D. 291, *326*
Porter, K.R. see Ledbetter, M.C 14, 17, *44*, 294, *334*
Porter, K.R. see Shay, J W. 76, 77, *83*
Porter, K R. see Veomatt, G. 76, *84*
Porter, K R. see Voeller, B.R. 317, *343*
Portis, A.R, Jr., Magnusson, R.P., McCarty, R E. 358, 368, 369, *374*
Portis, A.R, Jr., McCarty, R E 359, 360, 361, 363, 365, 367, 368, *374*
Poste, G 77, *83*
Poste, G, Reeve, P 57, 76, *83*
Potter, V.R. see Blobel, G 18, *40*
Potter, V R see Bushnell, D E 59, *80*
Poux, N 259, 268, 273, *285*, 296, 298, 315, 317, *338*
Poux, N. see Carasso, N 298, *327*
Pradet, A. see Bomsel, J.-L. 402, *412*
Prat, R 321, *338*
Prat, R, Roland, J-C 311, 321, *338*
Prat, R. see Roland, J.-C. 311, *340*
Preiss, J., Biggs, M.L, Greenberg, E. 96, *133*

Preiss, J., Ghosh, H.P., Wittkop, J. 95, 98, *133*
Preiss, J., Kosuge, T. 95, 96, 126, *133*
Prescott, D.M., Goldstein, L 72, *83*
Prescott, D M see Shay, J.W 76, 77, *83*
Prescott, D.M. see Veomatt, G. 76, *84*
Prescott, D.M see Wise, G.E. 76, *84*
Pressman, B.C. 28, 38, *47*, 381, 382, *396*
Pressman, B C., De Guzman, N.T. 38, *47*
Pressman, B.C., Harris, E.J., Jagger, W.S., Johnson, J.H. 381, *396*
Pressman, B.C. see Cockrell, R.S. 383, *395*
Pressman, B C. see Harris, E.J. 383, *395*
Prestipiño, G., Ceccarelli, D., Conti, F., Carafoli, E 388, *397*
Preston, R D 307, 321, 322, *338*
Preston, R.D., Goodman, R.N. 321, *338*
Preston, R.D. see Barnett, J.R. 321, *326*
Preston, R.D. see Robinson, D.G 319, 320, 321, *339*
Prezioso, G. see Palmieri, F. 242, 244, *253*
Prezioso, G. see Quagliariello, E. 244, *254*
Price, C.A. 3, 7, 9, 10, *47*
Price, C A see Morgenthaler, J.J 88, *133*
Price, D.W see Peng, C F. 390, 393, *396*
Price, J M see Jones, R.L. 315, 316, *332*
Prince, R.C. see Deamer, D.W 362, *371*
Pritchard, G.G., Griffin, W.J., Whittingham, C.P. 204, *231*
Pritchard, G.G. see Whittingham, C.P. 194, 204, *233*
Probstein, R.F. see Jacazio, G 441, *462*
Przybylla, K. see Canvin, D.T. 215, 216, *225*
Puchkova, L.I. see Sandakhiev, L S. 65, *83*
Pugh, T.D. see Kapil, R.N. 203, *229*
Puiseux-Dao, S 64, *83*
Puiseux-Dao, S. see Bonotto, S. 64, 65, 66, *79*
Pujarniscle, S. 269, 270, *285*
Pullman, M.E , Schatz, G. 377, *397*
Purdom, I. see Birnstiel, M. 55, *79*
Pusch, W., Woermann, D 440, *463*

Quagliariello, E , Palmieri, F. 243, *254*
Quagliariello, E., Palmieri, F., Prezioso, G., Klingenberg, M 244, *254*
Quagliariello, E. see Azzone, G.F. 377, *394*
Quagliariello, E. see Palmieri, F. 242, 243, 244, *253*
Quagliariello, E. see Papa, S. 242, *253*, 387, *396*
Quay, S. see Oxender, D L. 37, *46*
Queiroz, O. 260, *285*
Quertier, J. see Baltus, E 73, *79*

Raben, J see VanGolde, L.M.G. 301, 302, *343*
Rabin, B.R. see Pon, N.G. 108, *133*
Rabinowitch, E I 85, 86, 87, 89, 95, 106, *133*
Rabson, R., Tolbert, N.E , Kearney, P.C 197, 201, *231*
Rachmilevitz, T , Fahn, A 315, *338*
Racker, E 3, 27, *47*, 370, *374*
Racker, E., Chance, B., Parsons, D.F. 27, *47*
Racker, E., Hauska, G.A., Lein, S., Berzborn, R.J., Nelson, N. 349, *374*
Racker, E., Stoeckenius, W 370, *374*
Racker, E., Tyler, D.D , Estabrook, R.W , Connover, T.E , Parsons, D.F., Chance, B. 27, *47*
Racker, E. see Bradbeer, J.W 109, *129*
Racker, E. see Hauska, G A. 349, 350, 354, *372*
Racker, E. see Kagawa, Y 27, *43*
Racker, E see McCarty, R.E. 349, 369, *374*
Racker, E. see Peterkofsky, A. 108, *133*
Racusen, D., Poincelot, R.P. 23, 24, *47*
Radunz, A 34, *47*
Raff, M.C. see Taylor, R.B. 11, *49*
Raff, R A., Colot, H.V., Selvig, S.E., Gross, P.R. 73, *83*
Raff, R.A., Greenhouse, G., Gross, K.W., Gross, P.R. 73, *83*
Ralph, R.K. see Berridge, M.V. 324, *326*
Rambourg, A. 303, *339*
Ramirez-Mitchell, R., Johnson, H.M., Wilson, R.H 385, *397*
Ramsey, J.C see Berlin, J.D 308, *326*
Ramsey, R.R., Tubbs, P.K. 246, *254*
Randall, D.D , Tolbert, N.E 201, *231*
Randall, D.D., Tolbert, N.E , Gremel, D 195, 201, 221, *231*
Randall, D.D. see Rehfeld, D.W 203, *231*
Ranjeva, R. see Alibert, G 268, *281*
Rankin, C T , Jr see Anderson, N.G. 10, *40*
Ranson, S.L see Thomas, M 85, 86, *134*
Rao, M V., Chatterjee, S. 72, *83*
Rao, P.N , Hittelman, W.N , Wilson, B A. 62, *83*
Rapley, L see Heldt, H.W 25, 39, *43*, 86, 89, 94, 103, 104, 105, 106, 109, 113, 124, 125, *131*, 137, *143*, 363, *372*, 399, 406, 408, 410, *413*
Rasmussen, H D. 392, *397*
Rasmussen, H D., Goodman, D.B.P., Tenenhouse, A. 393, *397*
Rathore, V.S , Bajaj, Y.P.S., Wittwer, S.H. 393, *397*
Rattray, J B.M. see Schiberi, A. 15, *48*
Raven, J A 212, 213, 214, *231*, 401, *415*
Raven, J.A., MacRobbie, E.A C., Neumann, J. 401, *415*

Raven, J.A , Smith, F.A 265, 266, *285*
Rawlence, D.J., Taylor, A R.A 308, *339*
Ray, M M. see Ray, P.M. 6, 7, 16, *47*, 308, 312, *339*
Ray, P.M. 319, 324, *339*
Ray, P M , Shininger, T.L , Ray, M.M. 6, 7, 16, *47*, 308, 312, *339*
Ray, P.M. see Eisinger, W. 319, *329*
Ray, P.M. see Labavitch, J.M. 307, *333*
Ray, P.M. see Robinson, D.G. 319, *339*
Ray, T.K., Lieberman, I., Lansing, A.I. 304, *339*
Rayns, D.G. see Manton, I. 306, *334*
Reaven, E.P. see Axline, S.G. 294, *326*
Rebhun, L., Rosenbaum, J , Lefebre, P., Smith, G. 64, *83*
Rebhun, L.I. see Nagai, R. 292, *337*
Rechcigl, M see Hruban, Z. 198, *228*
Rechsteiner, M., Catanzarite, V. 77, *83*
Recondo, E. see Garcia, R.C. 321, *330*
Reed, M.L. see Graham, D. 108, *130*, 165, *179*
Reeve, P. see Poste, G. 57, 76, *83*
Reeves, S.G., Hall, D.O. 367, *374*, 398, *415*
Rehfeld, D W., Randall, D D , Tolbert, N.E. 203, *231*
Rehfeld, D.W , Tolbert, N.E 199, 210, *231*
Reid, J.S G , Meier, H. 273, *285*, 315, 321, *339*
Reighard, J A see Paolillo, D.J., Jr. 21, 23, *46*
Reijngoud, D.J see Meijer, A.J 245, *253*
Reik, H.G. see Meixner, J. 427, *463*
Reimann, B.E.F. 310, *339*
Reimer, S. see Bohme, H 354, *371*
Reimer, S. see Hauska, G.A. 355, *372*
Reimer, S. see Harth, E. 350, *372*
Reimer, S. see Trebst, A. 356, *375*
Reiner, J.M 190, 191, *231*
Reinwald, E. see Rumberg, B 351, 368, *374*
Reiss, J. 268, *285*
Renard, M. see Legros, F. 66, *82*
Renuckles, V.C. see Krotkov, G. 188, *229*
Revel, M see Huez, G. 59, 74, 75, *82*
Reynafarje, B., Lehninger, A L. 386, 388, *397*
Reynolds, E.S. 392, *397*
R-Garciá, M.I. see Risueno, M.C. 317, *339*
Rhoades, M.M , Carvalho, A. 147, *183*
Ribaillier, D., Jacob, J.L., d'Auzac, J. 258, 260, 261, 263, *285*
Riccio, P. see Klingenberg, M. 239, *252*
Richard, M., Louisot, P 303, *339*
Richardson, J.A see Thomas, M 85, 86, *134*
Richardson, K.E , Tolbert, N.E. 195, *231*
Rickenbacher, R. see Matile, P. 258, 263, 266, 267, 270, *284*
Rickenbacher, R. see Parish, R.W. 191, *231*
Ridge, I., Osborne, D.J. 307, *339*
Ridley, S.M , Leech, R.M. 93, *133*
Ried, A. 408, *415*
Rienits, K G see Lilley, R McC 88, 91, 92, 93, *132*
Ringertz, N R , Bolund, L 55, *83*
Ringertz, N R., Darzynkiewicz, Z., Bolund, L. 55, 62, *83*
Ringertz, N.R. see Ege, T. 58, 76, 77, *80*
Risueno, M.C., Gimenéz-Martín, G., López-Sáez, J.F. 312, *339*
Risueno, M C , Sogo, J M , Gimenéz-Martín, G., R-Garciá, M I 317, *339*
Ritenour, G L , Joy, K.W., Bunning, G., Hageman, R H. 208, *231*
Ritt, E see Pfaff, E. 27, *46*, 236, *254*, 360, *374*
Rivera, E see Morré, D.J. 290, *337*
Robards, A W. 161, *183*, 292, 314, 322, *339*
Robards, A.W. see Marchant, R 311, *334*
Robbins, P.W. see Oseroff, A.R. 37, *46*
Roberts, G.R., Keys, A.J , Whittingham, C.P 110, 111, *134*
Roberts, K., Gurney-Smith, M , Hills, G J. 307, *339*
Roberts, K see Hills, G.J. 307, *331*
Roberts, R.M., Connor, A.B , Cetorelli, J.J. 301, *339*
Roberts, R.M., Pollard, W.E. 301, *339*
Roberts, R.M. see Juniper, B.E 314, *332*
Roberts, R.M. see Kirby, E G. 314, *333*
Robertson, J D. 32, 36, *47*
Robertson, R.N. see Millard, D.L. 385, *396*
Robinson, B.H., Williams, G R. 384, *397*
Robinson, B.H., Williams, G R., Halperin, M L., Leznoff, C.C 243, *254*
Robinson, B.H. see Azzi, A. 245, *250*
Robinson, D.G., Preston, R.D. 319, 320, 321, *339*
Robinson, D.G., Ray, P.M. 319, *339*
Robinson, J.M., Gibbs, M 191, 194, *231*
Robinson, J.M , Stocking, C R 408, *415*
Robinson, J M. see Stocking, C.R. 406, 411, *415*
Robinson, M see Morgenthaler, J J 88, *133*
Robinson, P.M., Park, D., McClure, W K 317, *339*
Robinson, R.A see Cameron, D.A. 390, *394*
Rocha, V., Ting, I.P. 191, 210, *231*
Rocha, V. see Ting, I.P. 210, *232*
Roelofsen, P.A. 288, 307, 308, *339*
Roland, J.-C. 298, 300, 301, 303, 307, 311, 312, *339*
Roland, J -C., Lembi, C.A., Morré, D.J. 6, 16, *47*, 291, 299, 300, 303, 324, *339*

Roland, J.-C., Pilet, P.-E. 307, *340*
Roland, J.-C., Prat, R 311, *340*
Roland, J.-C., Sandoz, D. 312, *340*
Roland, J.-C., Vian, B 301, 303, *340*
Roland, J.-C. see Frantz, C 303, *330*
Roland, J.-C. see Lamant, A. 324, *333*
Roland, J.-C. see Prat, R. 311, 321, *338*
Roland, J.-C. see Rougier, M. 301, 307, *340*
Roland, J.-C see Vian, B 298, 300, 303, 312, 314, *343*
Romanovicz, D see Brown, R.M., Jr 308, *327*
Roncari, D A.K. 292, *340*
Rosado-Alberio, J., Weier, T E., Stocking, C R 21, 26, *47*
Rose, A.H. see Longley, R.P 14, 15, *45*
Roseman, S. 290, 301, 323, *340*
Roseman, S. see Schachter, H. 301, *340*
Rosen, W.G. 288, 308, *340*
Rosen, W.G., Gawlik, S R., Dashek, W.V., Siegesmund, K.A. 308, *340*
Rosen, W G., Thomas, H R. 315, *340*
Rosenbaum, J see Rebhun, L 64, *83*
Rosenberg, L.L see Whatley, F R 86, 118, *136*
Rosenberg, S A., Guidotti, G. 15, 16, *47*
Rosenblum, I.Y. see Cheng, G P. 199, *225*
Rosing, J., Slater, E C. 366, *374*
Rossenbeck, H.S. see Feulgen, R. 55, *80*
Rossi, C.S., Lehninger, A L 387, 389, *397*
Rossi, C.S see Carafoli, E. 392, *394*
Rossi, C S see Drahota, Z. 389, *395*
Rossi, C S see Lehninger, A L 377, 387, 389, *396*
Rothe, G.M. 205, *231*
Rothfield, L.I 31, *47*
Rottenberg, H, Grunwald, T 361, 363, 368, *374*
Rottenberg, H, Grunwald, T, Avron, M 357, 360, 362, 363, *374*
Rottenberg, H. see Bamberger, E.S. 358, 368, *371*
Rottenberg, H. see Pick, U 365, *374*
Rottenberg, H. see Schuldiner, S 362, 363, *375*
Rougier, M. 314, *340*
Rougier, M., Vian, B., Gallant, D., Roland, J.C. 301, 307, *340*
Rougier, M. see Vian, B. 322, *343*
Rouiller, C. see Orci, L. 291, *337*
Roush, A.H. 261, *286*
Ruddat, M. see Jelsema, C.L. 316, 324, *332*
Ruddat, M. see Vigil, E.L 315, 316, *343*
Ruddle, F.H see Miller, R A 76, *82*
Ruesink, A.W. 16, *47*, 258, *286*
Ruis, H see Bieglmeyer, C. 6, *40*
Ruiz-Herrera, J., Bartnicki-Garcia, S. 320, 322, *340*
Rumberg, B, Reinwald, E., Schröder, H, Siggel, U. 351, 368, *374*
Rumberg, B, Schroder, H 367, *375*
Rumberg, B, Siggel, V 358, 368, *375*
Rumberg, B. see Schröder, H 365, 366, *375*
Rümke, P H. see Emmelot, P 6, 16, *42*
Russo, V E A see Hertel, R. 323, 324, *331*
Rutter, W. see Levine, S. 63, *82*
Ruuge, E H. see Izawa, S. 355, *373*
Ryan, C A, Shumway, L.K. 261, *286*
Ryan, F J, Barker, R, Tolbert, N E 195, *231*
Ryan, F J., Tolbert, N E 195, *231*
Ryan, F.J. see Tolbert, N E. 196, 207, 208, 209, 222, *233*
Ryan, T.E. see Shamoo, A.E. 38, *48*
Ryrie, I.J, Jagendorf, A T 369, *375*
Sabnis, D.D., Gordon, M., Galston, A.W. 24, *47*
Sabnis, D.D., Jacobs, W P 291, 292, *340*
Saccone, C see Cantatore, P. 74, *80*
Sachs, J. 95, 99, *134*
Sackett, W.M. see Whelan, T 165, *184*
Sadava, D, Chrispeels, M.J. 307, *340*
Sadava, D., Walker, F., Chrispeels, M.J 307, *340*
Saha, S, Ouitrakul, R, Izawa, S, Good, N.E. 354, *375*
Saines, M. see Legros, F 66, *82*
Salin, M L, Campbell, W H, Black, C.C. 149, 150, *183*
Sallach, H J see Cheng, G P. 199, *225*
Sallach, H J see Willis, J.E 199, *233*
Salomon, R. see Huez, G. 59, 74, 75, *82*
Samish, Y., Koller, D 189, 212, *231*
Samish, Y.B., Pallas, J E., Dornhoff, G M, Shibles, R M 212, *231*
Sandakhiev, L.S., Puchkova, L.I., Pikalov, A V. 65, *83*
Sandakhiev, L.S. see Chirkova, L I 65, *80*
Sandblom, J, Orme, F 441, *463*
Sandoz, D. see Roland, J.-C 312, *340*
Sane, P.V, Goodchild, D.J, Park, R B. 22, *47*
Sane, P V. see Park, R.B 3, *46*
Sanger, L. see Stein, O. 291, *342*
Santarius, K A, Heber, U. 402, *415*
Santarius, K.A., Heber, U., Ullrich, W, Urbach, W 402, *415*
Santarius, K A., Stocking, C.R 410, *415*
Santarius, K.A. see Heber, U 90, 93, 102, 103, 105, 117, 119, 121, 122, *131*, 140, *142*, 399, 402, 403, 404, 406, 407, 408, 409, *413*
Santarius, K A see Ullrich, W 259, 263, *287*, 402, *416*

Santarius, K A see Urbach, W 102, *135*, 406, 407, *416*
Santillan, G see Cain, J. 12, *41*
Sarfatti, G see Cresti, M. 275, 277, *282*
Sassen, M.M A 308, *340*
Satô, S see Toriyama, H 261, *286*
Sauer, F. 419, 421, 423, 425, 427, 428, 431, 433, 435, 436, 453, 459, *463*
Sauer, F. see Heldt, H W. 25, 39, *43*, 86, 89, 92, 94, 104, 109, 116, 124, 125, *131*, 137, *143*, 399, *413*
Sauer, F. see Meier, J 432, *463*
Sawa, T. see Silverberg, B.A 205, *232*
Scala, J. see Schwab, D.W. 315, *341*
Scallen, T.J., Srikantaiah, M.V , Seetharam, B., Hansbury, E , Gavey, K.L 290, 292, *340*
Scandella, C. see McConnell, H.M. 12, *45*
Schachter, H. 301, 323, *340*
Schachter, H., Jabbal, I., Hudgin, R.L., Pinteric, L , McGuire, E.J., Roseman, S. 301, *340*
Schacter, B , Eley, J H , Jr., Gibbs, M 103, 104, 105, *134*
Schafer, G. see Gimmler, H. 25, *42*, 92, 106, 110, 111, *130*, 199, *227*
Schafer, G. see Heber, U. 106, *130*, 194, 201, 207, 222, *228*
Schatz, G , Mason, T.L. 28, *48*
Schatz, G see Pullman, M E 377, *397*
Schaub, H , Egle, K. 204, *231*
Schnaub, H. see Fock, H. 186, 207, *227*
Scheer, U see Franke, W W. 60, 61, 65, *81*, 294, 295, 301, 312, *330*
Scheer, U see Spring, H. 64, 69, *83*, *84*
Scheer, U. see Trendelenburg, M.F. 69, *84*
Scheer, V. see Franke, W.W. 17, *42*
Scheibe, R see Eickenbusch, J.D 196, *226*
Schellenberg, M see Lenney, J F. 271, 272, 280, *284*
Schellenberg, M. see Wilden, W. van der 271, 272, *287*
Schenk, W. see Egle, K 204, 214, *226*
Scher, M G see Lennarz, W J 321, *334*
Scherer, B., Klingenberg, M. 240, *254*
Scherer, B. see Klingenberg, M. 239, 240, *252*
Schiberi, A., Rattray, J.B M., Kidby, K.K 15, *48*
Schiefer, H.G. see Stoffel, W. 236, *254*
Schisler, L C see Holtz, R.B. 16, *43*
Schlenk, F , Dainko, J.L., Svihla, G. 258, 261, *286*
Schlenk, F. see Svihla, G 259, 261, *286*
Schliephake, W , Junge, W., Witt, H.T 352, 364, *375*
Schlögl, R. 435, 439, 440, 441, 442, 443, 444, 445, 446, 450, 456, 457, 458, 459, *463, 464*
Schmid, G 441, 449, *464*
Schmid, G., Schwarz, H. 450, *464*
Schmid, G H. see Codd, G.A. 206, *225*
Schmid, G.H. see Homan, P.H. 170, *180*
Schmidt-Ullrich, R. see Wallach, D F H. 40, *49*
Schmiedt, B see Klingenberg, M. 239, *252*
Schmitt, F O 291, 292, *340*
Schmucker, P see Elbers, R 241, *251*, 400, *412*
Schnaebli, H.P , Burger, M M 63, *83*
Schnaitman, C.A , Greenawalt, J W 28, *48*, 236, *254*
Schnarrenberger, C., Oeser, A., Tolbert, N.E 191, 210, *232*
Schnarrenberger, C. see Donaldson, R.P. 6, *41*, 193, *226*
Schnepf, E. 256, 266, *286*, 298, 314, 315, *340*
Schnepf, E., Deichgräber, G 306, *341*
Schnepf, E., Kalsova, A. 314, *341*
Schnepf, E., Koch, W 316, *341*
Schnepf, E see Schulze, C. 266, *286*
Schnepf, E. see Wollenweber, E 314, *344*
Schnitka, T.K , Seligman, A M 5, *48*
Schonborn, M., Woermann, D. 459, *464*
Schötz, F., Diers, L., Bathelt, H. 259, *286*
Schötz, F see Diers, L. 259, 261, *282*
Schou, L , Benson, A A , Bassham, J.A., Calvin, M 194, *232*
Schraer, H. see Hohman, W 391, *395*
Schraer, H. see Schraer, R. 391, 392, *397*
Schraer, R , Elder, J A., Schraer, H. 391, 392, *397*
Schröder, H , Hsie, A.W. 76, *83*
Schroder, H , Muhle, H , Rumberg, B 365, 366, *375*
Schröder, H. see Rumberg, B 351, 367, 368, *374*, *375*
Schroeder, T E. 294, *341*
Schroter, K., Läuchli, A., Sievers, A. 259, *286*
Schroter, K., Sievers, A 308, *341*
Schuldiner, S., Rottenberg, H., Avron, M 362, 363, *375*
Schulze, C., Schnepf, E , Mothes, K. 266, *286*
Schürmann, P., Buchanan, B.B., Arnon, D.I. 118, *134*
Schurmann, P. see Buchanan, B B. 96, 105, *129*
Schuster, S M., Olson, M S 393, *397*
Schwab, D W , Simmons, E , Scala, J 315, *341*
Schwalbach, K see Heldt, H.W. 238, *252*
Schwartz, M. 348, 365, 366, 367, *375*
Schwarz, H see Schmid, G 450, *464*
Schwarzenbach, A.M. 31, *48*
Schweiger, E., Wallraft, H.G., Schweiger, H.G. 66, *83*
Schweiger, H G , Master, R.W P , Werz, G 67, *83*
Schweiger, H.G. see Apel, K. 67, 68, *79*
Schweiger, H.G. see Bannwart, H. 67, *79*

Schweiger, H G. see Berger, S 69, 70, *79*
Schweiger, H G see Franke, W.W 65, *81*
Schweiger, H.G see Klopp-stech, K. 68, *82*
Schweiger, H.G. see Schweiger, E. 66, *83*
Schwelitz, F.D., Evans, W R., Mollenhauer, H.H., Dilley, R.A. 310, *341*
Schwenn, J.D , Lilley, R.McC., Walker, D.A. 91, 93, 111, 113, 114, *134*
Schwenn, J.D. see Lilley, R.McC. 91, 93, 111, 115, 116, 125, *132*
Sconzo, G see Giudice, G. 74, *81*
Scott, G.B see Leduc, E H 17, *44*
Scott, R.E see Tillack, T.W. 321, *342*
Secchi, A.G. see Papa, S. 387, *396*
Seetharam, B. see Scallen, T.J 290, 292, *340*
Segrest, J.P. see Marchesi, V.T. 13, 15, *45*
Seligman, A M. see Schnitka, T K. 5, *48*
Selman, B.R., Bannister, T.T. 350, *375*
Selman, B R , Bannister, T T , Dilley, R.A 350, *375*
Selvig, S.A., Greenhouse, G E., Gross, P R. 74, *83*
Selvig, S.E. see Raff, R.A. 73, *83*
Semenenko, V.E. 188, *232*
Sestak, Z., Catsky, J., Jarvis, P.G. 187, *232*
Sexton, R , Cronshaw, J , Hall, J.L 268, *286*
Shafizadeh, F., McGinnis, G.D. 307, 321, *341*
Shah, S P.J , Cossins, E.A 201, *232*
Shain, Y., Gibbs, M. 109, *134*, 194, 195, 196, *232*
Shamoo, A.E., Ryan, T.E. 38, *48*
Shamoo, A.F. 38, *48*
Sharon, N., Lis, H. 324, *341*
Sharp, P. see Bassham, J.A 108, *128, 129*
Shavit, N. see Dilley, R.A. 365, *371*
Shay, J W., Porter, K.R., Pres-cott, D.M. 76, 77, *83*
Shay, J see Veomatt, G 76, *84*
Shen, G.M. see Shen, Y K. 347, *375*
Shen, Y K , Shen, G.M. 347, *375*
Shen-Miller, J., Miller, C. 323, *341*, 394, *397*
Shephard, D C., Bidwell, R.G.S. 66, *83*
Shephard, D.C see Bidwell, R.G.S. 212, *224*
Shepherd, W D. see Hanson, J.B. 379, 382, 386, *395*
Sheppard, J.R. see Burger, M.M. 63, *80*
Shibko, S., Edelman, J 96, *134*
Shibles, R.M see Samish, Y.B. 212, *231*
Shields, C.M. see Bisalputra, T. 205, *224*
Shields, R. see Hershko, R 63, *82*
Shininger, T L. see Ray, P.M. 6, 7, 16, *47*, 308, 312, *339*
Shomer-Ilan, A., Waisel, Y. 146, *183*
Shore, G., MacLachlan, G.A. 312, 324, *341*
Shrago, E. see Lerner, E. 240, *252*
Shug, A.L. see Lerner, E. 240, *252*
Shumway, L.K., Weier, T.E. 21, *48*
Shumway, L.K. see Ryan, C.A 261, *286*
Siegel, S.M , Daly, O 258, *286*
Siegel, M.R., Sisler, H.D. 28, *48*
Siegenthaler, P A see Packer, L. 110, *133*
Siegesmund, K.A. see Rosen, W.G. 308, *340*
Siekevitz, P. see Bergeron, J J M 301, *326*
Sievers, A. 273, *286*, 308, *341*
Sievers, A. see Schröter, K 259, *286*, 308, *341*
Sievers, J.F. see Meyer, K.H. 456, *463*
Sigarlakie, E see Bauer, H. 261, *281*
Siggel, U. see Rumberg, B 351, 368, *374*
Siggel, V see Rumberg, B 358, 368, *375*
Siliprandi, N. see Azzone, G F 377, *394*
Silverberg, B.A 205, 206, *232*
Silverberg, B.A., Sawa, T 205, *232*
Simmons, E see Schwab, D.W 315, *341*
Simoni, R.D 39, *48*
Simonis, W., Urbach, W. 399, *415*
Simonis, W see Jeschke, W.D. 402, *414*
Singer, S.J 11, 12, 35, 36, 37, 38, 39, *48*
Singer, S.J., Nicolson, G. 11, 12, 38, *48*
Singer, S.J. see Hirano, H 301, *331*
Singer, S.J see Lenard, J. 11, 13, 35, 36, *44*
Singh, A 294, *341*
Singh, A. see Orci, L 291, *337*
Singh, R.M.M. see Good, N E. 115, *130*
Sinha, S.K. see Khanna, R. 146, *181*
Sisler, H.D see Siegel, M R. 28, *48*
Sitte, P see Brown, R.M., Jr. 308, 309, 312, *327*
Sjöstrand, F.S 4, 26, 33, *48*
Sjostrand, F.S., Hanzon, V. 294, *341*
Skoultchi, A., Gross, P.R. 73, *83*
Skow, R K see Blinks, L.R 106, *129*
Skrdlant, H.B. see Manley, E R 292, *334*
Slack, C.R. 148, *183*
Slack, C.R , Hatch, M D. 155, *183*
Slack, C R., Hatch, M.D., Goodchild, D.J. 147, 148, 151, 153, 155, 156, 157, 170, *183*, 210, *232*
Slack, C.R. see Andrews, T J 151, 153, 155, *177*, 210, *223*
Slack, C.R. see Everson, R.G 165, *179*
Slack, C.R. see Graham, D 151, *179*
Slack, C.R see Hatch, M D. 144, 151, 155, 156, 167, 175, *180*, 201, 202, 210, *228*
Slater, D.W. see Slater, I 74, *83*

Slater, E.C. 377, *397*
Slater, E C see Rosing, J. 366, *374*
Slater, I., Slater, D.W. 74, *83*
Slatyer, R.O. 175, *183*
Slatyer, R.O. see Hatch, M.D. 144, *180*
Slautterback, D.B. 295, *341*
Sluse, F.E., Liebecq, C. 243, *254*
Smaoui, A. 199, *232*
Smillie, R.M., Andersen, K.S., Tobin, N.F., Entsch, B., Bishop, D.G. 171, *183*
Smillie, R.M., Bishop, D.G., Andersen, K.S. 172, *183*
Smillie, R.M., Fuller, R.C. 88, 105, *134*
Smillie, R.M. see Andersen, K.S. 171, *177*
Smillie, R.M. see Bishop, D.G. 171, *177*
Smillie, R.M. see Graham, D. 151, 165, *179*
Smillie, R.M. see Murray, D.R 204, 205, *230*
Smith, B.N., Brown, W.V. 147, *183*
Smith, D.G., Marchant, R. 261, *286*
Smith, D.S 291, *341*
Smith, D.S., Järlfors, U., Beránek, R. 291, *341*
Smith, F A 165, *183*, 401, *415*
Smith, F.A see Osmond, C.B. 163, *182*
Smith, F.A. see Raven, J.A. 265, 266, *285*
Smith, G. see Rebhun, L. 64, *83*
Smith, L.D., Ecker, R.E. 74, *83*
Smith, M M., Stone, B.A. 319, *341*
Smits, G.G. see Brouwer, A 246, *251*
Snyder, W.R. see Law, J.H 11, *44*
Soboll, S. see Elbers, R. 241, *251*, 400, *412*
Sogo, J.M. see Risueno, M.C. 317, *339*
Solomon, A.K. see Bentzel, C.J. 379, *394*
Somero, G.N. see Hochachka, P.W. 165, *180*
Sonin, A.A. see Jacazio, G. 441, *462*
Soreq, H. see Huez, G 59, 74, 75, *82*
Soreq, H. see Marbaix, G. 59, 75, *82*
Sorger, G.J see Evans, H.J. 392, *395*
Sorokin, S P 294, *342*
Souverijn, J H M. see Kemp, J.A. 239, *252*
Souverijn, J.H M. see Passham, H C 241, *253*, 400, *414*
Souverijn, J.H.M. see Vaartjes, W.J. 240, *254*
Spanswick, R.M. 161, *183*
Speirs, J. see Birnstiel, M. 55, *79*
Spencer, D.L., Unt, H. 117, *134*
Spencer, D see Anderson, J.M 171, *177*
Spencer, F.S., Ziola, B , MacLachlan, G.A 324, *342*
Spirin, A.S. 61, *83*
Spooner, B.S. 294, *342*
Spooner, B.S. see Wessels, N.K. 292, 294, *343*
Spooner, B.S. see Yamada, K.M. 292, 308, *344*
Spooner, D F. see Wessels, N.K. 58, *84*
Spring, H., Scheer, U., Franke, W W , Trendelenburg, M.F. 69, *83*
Spring, H., Trendelenburg, M.F , Scheer, U., Franke, W W , Herth, W. 64, *84*
Spring, H see Franke, W.W 65, *81*
Spring, H see Trendelenburg, M.F. 69, *84*
Springer-Lederer, H see Bassham, J.A. 116, *128*
Spurr, A.R., Harris, W M. 21, *48*
Spurr, A.R. see Harris, W M 21, *43*
Srikantaiah, M V see Scallen, T.J. 290, 292, *340*
Stabenau, H. 206, *232*
Stabenau, H , Beevers, H 206, *232*
Stadelmann, E.J. 15, *48*
Stadelmann, E.J., Kinzel, H. 260, *286*
Stadler, J see Herth, W. 308, 312, *331*
Stadler, J. see Philipp, E.-I. 18, *47*
Staehelin, L A. 299, *342*
Staehelin, L A., Chlapowski, F.J., Bonneville, M.A. 292, *342*
Staehelin, L.A., Kiermayer, O. 300, *342*
Staneloni, R.J see Behrens, N.H 321, *326*
Stauffer, J F see Umbreit, W.W. 175, *183*
Staverman, A J. 427, *464*
Steck, T.L. 7, 8, 10, 11, 12, 13, 15, 17, 18, *48*
Steck, T.L., Fairbanks, G., Wallach, D.F.H. 15, *48*
Steck, T.L., Fox, C.F. 10, 12, *48*
Steck, T.L., Wallach, D.F.H. 15, *48*
Steck, T.L. see Fairbanks, G. 15, *42*
Stedingk, L.-V. von see Jackson, J.B 364, *373*
Stein, O., Sanger, L., Stein, Y. 291, *342*
Stein, Y. see Stein, O. 291, *342*
Steinert, G., Vangansen, P. 55, *84*
Steinert, G. see Steinert, M. 60, *84*
Steinert, M., Steinert, G. 60, *84*
Steinmann, E. see Frey-Wyssling, A. 33, *42*
Stengel-Rutkowski, L. see Klingenberg, M. 239, 240, *252*
Stepka, W. see Benson, A.A. 86, *129*
Stern, H. 18, *48*
Stern, H. see Stieglitz, H. 315, *342*
Steveninck, R.F.M., Chenoweth, A.R.F. 259, *286*
Steward, F.C., Bidwell, R G S 258, 272, *286*
Steward, F C., Mott, R.L. 264, *286*
Steward, F.C., Sutcliffe, J F. 263, *286*
Stewart, K.D., Floyd, G.L., Mattox, K.R., Davis, M.E. 205, *232*
Stewart, K.D. see Gifford, E.M. 275, *283*
Stewart, P.S. see Holtz, R.B. 16, *43*
Stewart, W.D.P. see Codd, G.A. 206, *225*

Stich, H. 70, *84*
Stieglitz, H., Stern, H. 315, *342*
Stiehl, H.H , Witt, M T 353, *375*
Stiles, J.W , Wilson, J.T., Crane. F.L. 27, *48*
Stiller, M.L. see Beevers. H. 258, 260, *281*
Stipani, I. see Palmieri, F. 242, 243, 244, *253*
Stockem, W see Komnick, H. 292, *330*
Stocking, C.R. 88, *134*, 191, *232*
Stocking, C.R., Larson, S 103, 122, *134*, 140, *143*, 407, *415*
Stocking, C.R , Lin, C H. 25, *49*
Stocking, C R., Robinson, J.M., Weier, T E 406, 411, *415*
Stocking, C.R see Bird, I.F 94, *129*
Stocking, C.R see Ongun, A. 110, *133*
Stocking, C R. see Robinson, J M 408, *415*
Stocking, C R. see Rosado-Alberio, J 21, 26, *47*
Stocking, C R see Santarius, K A. 410, *415*
Stoddart, R.W. see Cook, G M W 301, *328*
Stoeckenius, W , Engelman, D.M. 31, 36, *49*
Stoeckenius, W see Oesterhelt, D. 370, *374*
Stoeckenius, W see Racker, E. 370, *374*
Stoermer, E.F , Pankratz, H S., Bowen, C C. 310, *342*
Stoffel. W., Schiefer, H.G. 236, *254*
Stokes. D.M , Walker, D.A 106, 119, *134*, 405, *415*
Stokes. D.M., Walker, D A , McCormick, A.V 108, *134*
Stokes, D.M. see Walker, D A 121, *135*
Stone, B A see Smith, M.M. 319, *341*
Stoner, C D., Hanson, J B. 378, *397*
Stoner, C D. see Hanson, J.B. 384, *395*
Storey. B.T 408, *415*
Straprans, I , Dirksen, E R. 294, *342*
Street, H E see Goldsworthy, A. 98, *130*
Strichartz, G.R , Chance, B. 364, *375*
Strickland. E H. see Benson, A.A 28, *40*
Strobel, G.A , Hess, W M. 323, *342*
Strotmann, H., Berger, S. 25, *49*, 404, 405, *415*
Strotmann, H , Heldt, H W 103, 104, *134*, 142, *143*, 402, 404, *416*
St Thomson, K see Lembi, C.A 5, 15, 16, *44*
Sugiyama, M. see Kato, K.H 64, 73, *82*
Sugiyama, T. 155, *183*
Sugiyama, T., Nakayama, N , Akazawa, T 108, *134*
Sundberg, I., Nilshammar-Holmvall, M. 263, *286*
Sutcliffe, J F. see Steward, F.C 263, *286*
Svihla, G., Dainko, J.L., Schlenk, F 259, 261, *286*
Svihla, G see Schlenk, F. 258, 261, *286*
Swolinsky, S.L see Alexeev, A B. 65, *79*
Szabo, G., Eisenman, G., McLaughlin, S G.A., Krasne, S. 12, *49*
Szabo, S see Ciani, S M 442, *462*

Tabita, F.R., McFadden, B.A 105, *134*
Tagawa, H , Tsujimoto, H.Y , Arnon, D I 355, *375*
Tager, J M see Brouwer, A 246, *251*
Tager, J.M. see Gimpel, J A 242, *251*
Tager. J M see Haan. E J de 243, *251*
Tager, J M see Meijer, A.J. 242, 243, 245, *253*
Talmadge, K W see Keegstra, K. 307, *332*
Tan, C K. see Waygood, E.R 175, *184*
Tanenbaum, S W see Miranda, A F. 292, 294, *335*
Tanner, W 401, *416*
Tanner, W , Kandler, O 401, *416*
Tanner, W , Loos, E , Klob, W , Kandler, O 401, *416*
Tanner, W see Decker, M 401, *412*
Tanner, W see Jung, P 321, *332*
Tas, J see Brouwer, A. 246, *251*
Tautvydas, K.J. 18, *49*
Taylor, A R.A. see Rawlence, D J 308, *339*
Taylor. E L see Wessels, N K 58, *84*, 292, 294, *343*
Taylor, R B., Duffus, W.P.H , Raff, M C., Depetris, S. 11, *49*
Tchen, T.T see Cooper, T G 107, *130*
Telfer, A , Barber, J. 352, 364, *375*
Telfer, A., Evans, M.C W 351, 363, 369, *375*
Telfer, A see Barber, J. 126, *128*
Telek, L see Martin, F W 315, *334*
Tencer, R. see Brachet, J. 73, *80*
Tenenhouse, A. see Rasmussen, H D 393, *397*
Teorell, T 441, 447, *464*
Terra, N. de 72, *84*
Thalacker, R , Behrens, M 402, *416*
Thiéry, J -P 301, *342*
Thiéry, J -P see Ovtracht, L. 301, *337*
Thimann, K.V 63, *84*
Thimann, K V. see Krotkov, G 188, *229*
Thimann, K V. see O'Brien, T P 292, *337*
Thomas, D des S , Lutzac, M , Manavathu, E. 291, 315, *342*
Thomas, H R see Rosen, W.G 315, *340*
Thomas, M , Ranson, S L , Richardson, J A 85, 86, *134*
Thomas, P L , Isaac, P K 273, *286*
Tomkins, G.M. see Hershko, R 63, *82*
Thompson, C.M., Whittingham, C.P 194, 195, 196, 200, 209, *232*
Thompson, D see Parson, D.F 236, *253*
Thompson, W see Parsons, D F 24, 27, *46*
Thomsen, H A see Moestrup, O 309, *335*

Thomson, K S , Leopold, A.C 323, *342*
Thomson, K S see Hertel, R. 323, 324, *331*
Thomson, K S. see Lembi, C.A. 291, 323, *334*
Thomson, W W., Berry, W L , Liu, L.L. 315, *342*
Thomson, W.W , Liu, L.L. 315, *342*
Thomson, W W., Foster, P., Leech, R M. 20, *49*
Thomson, W.W see Campbell, N. 315, *327*
Thornber, J P 25, *49*
Thorne, K.J.I. see Martin, H G. 321, *335*
Thorne, R.F.W., Bygrave, F L. 392, *397*
Thorne, S W. see Woo, K.C. 171, *184*
Thornton, R.M 273, 279, *286*
Thurston, E.L see Gustafson, R A. 392, 393, *395*
Tillack, T W., Scott, R.E., Marchesi, V T. 321, *342*
Tilney, L.G. 292, *342*
Tilney, L.G. see Bikle, D 291, *326*
Tilney-Bassett, R.A E see Kirk, J T O. 11, *44*
Ting, I P , Osmond, C B. 155, 175, *183*
Ting, I P , Rocha, V 210, *232*
Ting, I.P see Rocha, V 191, 210, *231*
Tiozzo, R. see Vasington, F D 388, *397*
Tobin, N.F. see Smillie, R M. 171, *183*
Tolbert, N.E 6, 7, 8, 29, *49*, 109, *134*, 169, *183*, 185, 186, 191, 193, 194, 197, 198, 199, 201, 204, 209, 211, 222, *232*
Tolbert, N E., Clagett, C O., Burris, R H. 199, 208, *232*
Tolbert, N E., Cohan, M S. 197, *232*
Tolbert, N.E., Nelson, E.B , Bruin, W.J. 204, *232*
Tolbert, N E , Oeser, A , Kisaki, T., Hageman, R H , Yamazaki, R.K 6, *49*, 193, 197, 199, 209, *233*
Tolbert, N E., Oeser, A , Yamazaki, R.K , Hageman, R.H., Kisaki, T 109, *134*, 197, 199, *233*
Tolbert, N E , Ryan, F J 196, 207, 208, 209, 222, *233*
Tolbert, N.E , Yamazaki, R K 197, 199, 221, 222, *233*
Tolbert, N.E., Yamazaki, R.K , Oeser, A. 196, 200, 209, *233*
Tolbert, N E., Zill, I P 204, *233*
Tolbert, N.E. see Andrews, T.J. 109, *128*, 176, *177*, 185, 194, 195, 221, *223*, *224*
Tolbert, N E see Benson, A.A 86, *129*
Tolbert, N.E see Bruin, W.J 204, 205, *225*
Tolbert, N E see Clagett, C.O. 199, *225*
Tolbert, N.E see Donaldson, R P. 6, *41*, 193, *226*
Tolbert, N.E. see Frederick, S.E. 205, *227*
Tolbert, N.E see Gruber, P J. 205, *227*
Tolbert, N.E. see Hess, J L. 191, 194, 204, 205, *228*
Tolbert, N.E. see Jimenez, E. 197, 201, *228*
Tolbert, N.E see Kearney, P C 109, *132*, 194, 209, *229*
Tolbert, N.E see Kisaki, T. 109, *132*, 199, 200, 201, 207, 222, *229*
Tolbert, N.E see Lorimer, G H 185, 195, 197, *229*
Tolbert, N.E. see Nelson, E B 204, 205, *230*
Tolbert, N.E. see Orth, G.M 194, *230*
Tolbert, N E. see Rabson, R 197, 201, *231*
Tolbert, N E see Randall, D.D 195, 201, 221, *231*
Tolbert, N E see Rehfeld, D W. 199, 203, 210, *231*
Tolbert, N.E see Richardson, K.E 195, *231*
Tolbert, N.E see Ryan, F.J. 195, *231*
Tolbert, N.E. see Schnarrenberger, C 191, 210, *232*
Tolbert, N.E see Yamazaki, R.K 199, 210, *234*
Tolbert, N E see Yu, Y L. 195, *234*
Tollenaar, D 95, *134*
Topitsch, P. see Klingenberg, M 239, *252*
Toriyama, H., Jaffe, M J 261, *286*
Toriyama, H., Satô, S. 261, *286*
Tourte, M. 205, *233*
Trebst, A. 3, 8, *49*, 348, 350, 352, 354, 356, *375*
Trebst, A., Harth, E , Draber, W. 352, *375*
Trebst, A., Reimer, S 356, *375*
Trebst, A. see Bohme, H 354, *371*
Trebst, A. see Harth, E 350, 356, *372*
Trebst, A see Hauska, G A. 354, 355, *372*
Tregunna, D.L., Brown, E.B 204, *233*
Tregunna, E B., Krotkov, G., Nelson, C D 186, 188, 213, *233*
Tregunna, E.B. see Berry, J.A. 151, 155, 162, 170, *177*
Tregunna, E.B. see Downton, W.J.S 147, 164, 170, *179*
Tregunna, E B. see Jolliffe, P A. 213, *229*
Trelease, R N., Becker, W.M., Burke, J.J. 29, *49*
Trelease, R N. see Graves, L.B 205, 206, *227*
Trendelenburg, M F., Spring, H., Scheer, U., Franke, W W. 69, *84*
Trendelenburg, M.F. see Franke, W.W 65, *81*
Trendelenburg, M.F. see Spring, H 64, 69, *83*, *84*
Trescott, D H. see Goldstein, L 72, *81*
Trewavas, A 272, *286*, *287*
Trim, A.R see D'Alessio, G. 18, *41*
Troughton, J H 164, *183*
Troughton, J H. see Bjorkman, O. 147, 176, *178*
Troughton, J H see Osmond, C.B. 146, 170, *183*
Truchet, G., Coulomb, P. 277, *287*
Truelove, B., Hanson, J.B. 384, *397*
Truelove, B. see Earnshaw, M J. 378, *395*
Tsai, C M., Hassid, W Z. 290, 320, *342*
Tsuchiya, Y see McCarty, R.E. 363, 369, *373*, *374*

Tsujimoto, H.Y. see Tagawa, H 355, *375*
Tubbs, P.K. see Ramsey, R.R. 246, *254*
Tubbs, P.K see West, D.W. 246, *254*
Turner, F.R. see Mahlberg, P.G. 277, *284*, 317, *334*
Tyler, A see Clement, A C. 74, *80*
Tyler, A. see Denny, P. 73, *80*
Tyler, D.D. 242, *254*
Tyler, D.D. see Racker, E. 27, *47*
Tyree, M.T. 161, 162, 163, *183*

Ueda, K. 317, *342*
Ullrey, D. see Wallach, D.F.H. 8, *49*
Ullrich, W., Urbach, W., Santarius, K.A., Heber, U. 259, 263, *287*, 402, *416*
Ullrich, W see Heber, U 102, *131*, 399, 402, *413*
Ullrich, W. see Santarius, K.A 402, *415*
Ullrich, W. see Urbach, W 102, *135*, 406, 407, *416*
Umbreit, W.W , Burris, R.H., Stauffer, J.F. 175, *183*
Unt, H. see Spencer, D.L. 117, *134*
Unzelman, J M , Healy, P L 314, *342*
Uphof, J.C.Th. 308, *342*
Urbach, W., Hudson, M.A., Ullrich, W., Santarius, K A., Heber, U. 102, *135*, 406, 407, *416*
Urbach, W see Heber, U. 102, *131*, 399, 402, *413*
Urbach, W see Santarius, K A. 402, *415*
Urbach, W. see Simonis, W. 399, *415*
Urbach, W. see Ullrich, W. 259, 263, *287*, 402, *416*
Uribe, E G., Jagendorf, A T. 347, *375*
Uribe, E.G. see Jagendorf, A.T. 347, 365, *373*
Urry, D W 12, *49*
Urry, D W , Long, M M., Jacobs, M., Harris, R.D. 38, *49*
Ussing, H H. 454, *464*
Ussing, H.H. see Andersen, B 440, *462*
Ussing, H.H. see Meares, P. 454, *463*

Vaartjes, W.J., Kemp, A., Jr., Souverijn, J.H M., van den Bergh, S.G. 240, *254*
Van Dam, K , Meyer, A J. 210, 226, 377, *397*
Van Dam, K. see Fiolet, J.T.W. 362, 363, *372*
Van Dam, K. see Meijer, A.J. 235, *253*
Van den Bergh, S.G. see Vaartjes, W J. 240, *254*
Vanden Driessche, T 66, *84*
Vanderkooi, G. 11, 12, *49*
VanDerWoude, W.J. 6, 16, *49*
VanDerWoude, W.J., Lembi, C.A., Morré, D.J. 15, *49*, 320, 324, *342*
VanDerWoude, W J., Lembi, C.A., Morré, D.J., Kindinger, J.I., Ordin, L. 291, 312, 320, 321, *343*
VanDerWoude, W J., Morré, D.J., Bracker, C E 298, 301, 307, 308, 312, 319, 320, *343*
VanDerWoude, W J. see Franke, W.W 288, 290, 292, 294, 301, 315, *330*
VanDerWoude, W.J. see Herth, W. 292, *331*
VanDerWoude, W J see Leonard, R T. 6, 7, 15, 16, *44*
VanDerWoude, W.J. see Morré, D J 294, 302, 304, 306, 307, 312, 319, 320, *337*
Vangansen, P. see Steinert, G 55, *84*
VanGolde, L.M.G., Fleischer, B , Fleischer, S 301, *343*
VanGolde, L.M.G., Raben, J , Batenburg, J.J , Fleischer, B., Zambrano, F., Fleischer, S 301, 302, *343*
Van Heyningen, W E 323, *331*
Van Wie, D.G. see McIntosh, J.R. 291, *335*
Varmus, H.E., Guntaka, R.V., Fan, W J.W , Heasley, S., Bishop, J.A. 77, *84*
Varner, J E., Mense, R.M. 315, *343*
Varner, J.E. see Ben-Tal, Y 316, *326*
Varner, J.E. see Evins, W.H 316, *329*
Varner, J E. see Koehler, D E. 316, *333*
Vasington, F.D., Gassotti, P., Tiozzo, R., Carafoli, E. 388, *397*
Veech, R L. see Krebs, H.A 210, *229*
Venketeswaran, S. see Mahlberg, P.G. 277, *284*, 317, *334*
Veomatt, G , Prescott, D.M., Shay, J., Porter, K R. 76, *84*
Verma, S.P. see Wallach, D.F H. 40, *49*
Vernon, L.P. see Dilley, R.A. 349, 351, 357, *371*
Vernotte, C. see Arntzen, C.J. 350, *370*
Vian, B. 300, *343*
Vian, B., Roland, J.-C 298, 300, 303, 312, 314, *343*
Vian, B., Rougier, M. 322, *343*
Vian, B see Roland, J -C. 301, 303, *340*
Vian, B. see Rougier, M. 301, 307, *340*
Vieweg, G.H., Fekete, M.A.R. de 157, *184*
Vieweg, G.H see Fekete, M.A.R de 98, *130*, 157, *179*
Vigil, E L. 198, 199, *233*
Vigil, E L , Ruddat, M 315, 316, *343*
Vigil, E.L. see Morré, D.J 323, *337*
Vignais, P.M see Colbeau, A 24, *41*
Vignais, P.M see Meijer, A J. 245, *253*
Vignais, P M. see Vignais, P.V. 239, *254*
Vignais, P.V., Vignais, P.M., Lauquin, G., Morel, F. 239, *254*
Vignais, P V see Morel, F 240, *253*
Villemez, C L. 319, 321, *343*
Villemez, C.L., Clark, A.F. 290, 321, *343*
Villemez, C L., McNab, J.M., Albersheim, P. 6, 16, *49*, 319, *343*
Villiers, T.A 275, *287*
Villiers, T A. see Berjak, P 278, *281*, 317, *326*
Vinograd, J see Matsumoto, L 74, *82*

Vıntéjoux, C 268, *287*
Vıshnıac, W , Ochoa, S 116, *135*
Vıtry, F. de see Brachet, J. 60, 66, *80*
Voeller, B.R., Ledbetter, M.C., Porter, K.R. 317, *343*
Voellmy, H. see Kuhn, W 441, *463*
Vogell, W. see Pfaff, E. 27, *46*, 236, *254*, 360, *374*
Vokaer, A. see Moens, W. 64, *82*
Volcanı, B.E. see Laurıtis, J.A. 310, *333*
Volk, R.J., Jackson, W.A. 214, *233*
Volk, R.J. see Jackson, W.A. 109, *131*, 175, 176, *181*, 185, 186, 188, 202, 211, 212, 214, 215, *228*
Volk, R.J. see Mulchı, C L. 186, 188, *230*
Volk, R.J see Ozbun, J.L. 188, 202, *231*
Voss, D.O. see Bonner, W.D. 238, *251*

Wadkins, C.L. see Peng, C.F. 390, 393, *396*
Waechter, C.J , Lucas, J.J., Lennarz, W.J. 321, *343*
Waggoner, P.E. 188, 189, *233*
Wagner, R.P. see Cassady, W E 27, *41*
Wahlıg, T. see Amelunxen, F. 314, *326*
Waısel, Y see Shomer-Ilan, A. 146, *183*
Walajtys, E I. see La Noue, K.F. 245, *252*
Walker, D A 85, 86, 87, 88, 89, 90, 93, 94, 95, 96, 103, 104, 105, 108, 111, 115, 119, 122, 123, 124, 126, *135*, 194, *233*, 403, *416*
Walker, D.A , Brown, J.M. 175, *184*
Walker, D.A., Cockburn, W., Baldry, C.W. 89, 104, *135*
Walker, D.A., Crofts, A.R. 89, 104, 108, 113, 122, 124, *135*, 348, *376*
Walker, D.A., Hıll, R. 102, 119, *135*, 403, 404, 406, *416*
Walker, D A., Lılley, R McC 108, *135*
Walker, D.A , McCormıck, A.V., Stokes, D.M. 121, *135*
Walker, D A see Baldry, C W. 103, 104, 106, 111, 112, *128*
Walker, D A. see Bucke, C. 103, 104, *129*
Walker, D.A. see Chen-She, S.H 95, 96, 97, 98, *129*
Walker, D A. see Cockburn, W 89, 90, 102, 103, 104, 105, 106, 111, 112, 113, 118, 121, 124, 125, *129*, 404, 406, 408, *412*
Walker, D.A. see Hıll, R. 116, *131*
Walker, D.A. see Lılley, R.McC. 88, 91, 92, 93, 105, 107, 108, 111, 115, 116, 121, 125, 126, *132*
Walker, D A. see Schwenn, J.D 91, 93, 111, 113, 114, *134*
Walker, D.A. see Stokes, D.M. 106, 108, 119, *134*, 405, *415*
Walker, D.A. see Whıtehouse, D.G. 117, *136*
Walker, F. see Sadava, D. 307, *340*
Walker, N A. 359, *376*, 419, 441, *464*
Walkınshaw, C.H. see Baur, P.S. 261, 268, *281*
Walkınshaw, C H see Mahlberg, P G. 277, *284*, 317, *334*
Wallach, D.F.H. 15, 16, 31, *49*
Wallach, D.F.H , Bıeri, V., Verma, S P., Schmıdt-Ullrich, R. 40, *49*
Wallach, D.F.H., Graham, J.M., Fernbach, B.R. 37, *49*
Wallach, D.F.H , Zahler, P.H. 11, 13, 35, *49*
Wallach, D F.H., Ullrey, D. 8, *49*
Wallach, D F.H. see Faırbanks, G. 15, *42*
Wallach, D.F.H. see Steck, T.L. 15, *48*
Wallraft, H.G. see Schweıger, E. 66, *83*
Walne, P.L. see Arnott, H.J. 310, *326*
Walne, P.L. see Palısano, J.R. 315, *337*
Wang, C -T., Nobel, P S. 92, 106, 110, *135*
Wang, C -T see Nobel, P.S 25, *46*, 92, 110, *133*
Wang, D., Burrıs, R.H 197, 201, *233*
Wang, D., Waygood, E.R 197, 201, *233*
Warburg, O 204, *233*
Warburg, O., Krippahl, G. 193, *233*
Ward, E. see Ward, R.T. 294, *343*
Ward, R.T , Ward, E. 294, *343*
Wardrop, A.B. 273, *287*
Wardrop, A.B. see Chafe, S.C. 321, *327*
Wardrop, A.B. see Cronshaw, J 313, 314, *328*
Warren, W.A. see Kohn, L.D 200, 209, *229*
Watson, M.W., Berlın, J.D. 308, *343*
Wattendorfer, J. 259, *287*
Wattıaux, R. see Appelmans, F. 6, *40*
Waygood, E R., Mache, R., Tan, C.K. 175, *184*
Waygood, E.R. see Wang, D. 197, 201, *233*
Weast, R.C. 162, 163, *184*
Wehrlı, E. see Fuhrmann, G.F. 15, 16, *42*
Wehrmeyer, W. 20, *49*
Weı, J 440, *464*
Weier, T.E. 21, *50*
Weıer, T E , Benson, A A. 33, *50*
Weier, T.E., Brown, D.L. 20, 21, *50*
Weıer, T.E. see Rosado-Alberio, J. 21, 26, *47*
Weıer, T.E. see Shumway, L.K 21, *48*
Weıer, T.E. see Stockıng, C R. 406, 411, *415*
Weıgl, J 402, *416*
Weıhıng, R.R see Pollard, T D. 292, 294, *338*
Weıngart, R see Hume, S.P. 77, *82*
Weinsteın, S see Wıgler, M. 76, *84*
Weıntraub, H. see Holzer, H. 63, *82*
Weiss, R L 260, *287*
Wellburn, A.R. see Cockburn, B.J. 25, *41*
Wells, K. 17, *50*
Werdan, K. 140, *143*
Werdan, K., Heldt, H.W. 25, *50*, 104, 107, 124, 126, *135*, 139, 140, *143*

Werdan, K., Heldt, H W., Geller, G 107, 108, *135*, 137, *143*
Werdan, K., Heldt, H W , Milovancev, M 126, *135*, 140, *143*
Werdan, K. see Heldt, H W. 107, *131*, 139, 141, 142, *142*, 167, 168, *180*, 410, *413*
Werkheiser, W.C., Bartley, W 27, *50*, 89, *135*, 360, *376*
Werz, G 56, 64, 65, *84*
Werz, G see Schweiger, H G 67, *83*
Wessels, N.K. 294, *343*
Wessels, N.K., Spooner, B S., Ash, J.F., Bradley, M.O., Luduena, M.A , Taylor, E L., Wrenn, J T., Yamada, K.M. 292, 294, *343*
Wessels, N K , Spooner, D F , Ash, G.F., MacBradley, M.A., Taylor, E.L., Wreng, J., Yamada, K.M. 58, *84*
Wessels, N K see Yamada, K.M 292, 308, *344*
West, D.W., Chase, J F.A , Tubbs, P.K. 246, *254*
West, K.R , Wiskich, J.T. 117, *136*
West, S.H see Hilliard, J H 199, 203, *228*
Wettstein, D. von 20, *50*
Whaley, W G , Dauwalder, M , Kephart, J E 30, *50*, 288, 304, 308, *343*
Whaley, W G., Kephart, J E., Mollenhauer, H.H. 318, *343*
Whaley, W G., Mollenhauer, H.H. 288, 308, *343*
Whaley, W G. see Dauwalder, M. 7, *41*, 301, 308, 314, 315, *328*, *329*
Whaley, W.G see Mollenhauer, H.H. 314, *336*
Whatley, F.R., Allen, M.B., Rosenberg, L L., Capindale, J B , Arnon, D.I. 86, 118, *136*
Whatley, F.R. see Allen, M B 116, *127*, *128*
Whatley, F.R. see Arnon, D I. 111, 116, 124, *128*, 347, *370*
Whelan, T., Sackett, W M , Benedict, C R. 165, *184*
White, J E. see Brody, M. 205, *225*
Whitehouse, D G , Ludwig, L.J , Walker, D A 117, *136*
Whiteman, P.C , Koller, D. 213, *233*
Whittingham, C.P., Birmingham, M., Hiller, R G. 194, *233*
Whittingham, C P., Coombs, J., Marker, A.F.H. 194, *233*
Whittingham, C P., Pritchard, G.G. 194, 204, *233*
Whittingham, C P. see Bird, I.F 94, *129*, 201, 208, 220, *224*
Whittingham, C P. see Coombs, J. 193, 194, 195, 196, 197, *225*
Whittingham, C.P. see Hind, G. 359, *373*
Whittingham, C.P. see Keys, A.J. 407, *414*
Whittingham, C.P see Pritchard, G G 204, *231*
Whittingham, C P. see Roberts, G.R 110, 111, *134*
Whittingham, C.P see Thompson, C M. 194, 195, 196, 200, 209, *232*
Wiedner, G , Woermann, D. 439, *464*
Wieland, T. see Fiume, L. 59, *81*
Wiemken, A 271, 280, *287*
Wiemken, A , Durr, M. 258, 260, 265, *287*
Wiemken, A., Nurse, P 258, 260, 269, 281, *287*
Wiemken, A see Boller, T 265, *281*
Wiemken, A. see Cortat, M. 315, *328*
Wiemken, A see Lenney, J F 271, 272, 280, *284*
Wiemken, A. see Matile, P 269, 271, *284*, *285*
Wiemken, A see Wiemken, V 272, *287*
Wiemken, A see Wilden, W van der 271, 272, *287*
Wiemken, V. 273, *287*
Wiemken, V , Wiemken, A. 272, *287*
Wiener, E , Ashworth, J M. 270, *287*
Wiersema, P.K see Winter, H. 307, *344*
Wiese, H. see Elbers, R 241, *251*, 400, *412*
Wiggins, P M 265, *287*
Wigler, M , Weinstein, S 76, *84*
Wilden, W. van der, Matile, P , Schellenberg, M., Meyer, J., Wiemken, A. 271, 272, *287*
Wilkinson, A P., Davis, D.D 201, *233*
Willenbrink, J see Heber, U. 86, 94, 102, 105, *131*
Williams, D.C 310, *343*
Williams, D.C. see Outka, D E 309, *337*
Williams, G R. see Parsons, D.F. 6, 24, 27, 28, *46*, 236, *253*
Williams, G R see Phillips, M.L 246, *254*, 384, *396*
Williams, G.R see Robinson, B.H 243, *254*, 384, *397*
Williams, W A. see El-Sharkawy, M.A. 187, *226*
Williamson, D H., Lund, P., Krebs, H A 248, *254*
Williamson, F A., Morré, D.J , Jaffe, M J 5, 6, *50*
Williamson, F A. see Frantz, C. 303, *330*
Williamson, F.A see Jelsema, C L 324, *332*
Williamson, F A. see Leigh, R A. 16, *44*
Williamson, J.R. 235, *254*
Williamson, J.R. see La Noue, K F 245, *252*
Willis, J E , Sallach, H J. 199, *233*
Willison, J.H.M , Cocking, E C 319, 321, *344*
Wilson, A T , Calvin, M 193, 195, *234*
Wilson, B.A see Rao, P N 62, *83*
Wilson, D. see Parsons, D F. 24, 27, *46*, 236, *253*
Wilson, E.B. 53, *84*
Wilson, J T see Stiles, J.W. 27, *48*
Wilson, K. 307, 322, 323, *344*
Wilson, R H. 378, 380, 393, *397*
Wilson, R.H., Dever, J , Harper, W , Fry, R 382, *397*
Wilson, R.H., Hanson, J.B 387, *397*, 408, *416*
Wilson, R.H., Hanson, J.B., Mollenhauer, H H. 236, *254*, 378, 379, 381, *397*

Wilson, R H , Minton, G A 385, 386, *397*
Wilson, R H see Johnson, H M. 380, 384, 385, 386, 387, 389, 390, 391, 393, *395*
Wilson, R.H see Lee, D.C. 379, 380, 383, *396*
Wilson, R.H. see Ramirez-Mitchell, R. 385, *397*
Wilt, F.H. 74, *84*
Wilt, F.H., Mazia, D. 74, *84*
Wilt, F.H see Aaronson, A.I. 62, *79*
Winget, G D , Izawa, S , Good, N.E. 363, *376*
Winget, G.D. see Good, N.E. 115, *130*
Winkenbach, F. 280, *287*
Winkenbach, F. see Matile, P. 272, 273, 278, 279, 280, *285*
Winter, H , Meyer, L , Hengeveld, E., Wiersema, P.K 307, *344*
Winter, W. see Good, N.E. 115, *130*
Wintermans, J F G M 11, *50*
Winzler, R.J. 15, *50*
Wise, G.E 76, *84*
Wise, G E , Prescott, D M 76, *84*
Wise, G E see Goldstein, L 72, *81*
Wishnick, M see Cooper, T.G. 107, *130*
Wiskich, I T see Day, D A. 238, 246, *251*
Wiskich, J.T. 169, *184*
Wiskich, J.T see Millard, D.L 385, *396*
Wiskich, J.T. see West, K R. 117, *136*
Witt, H.T 348, 350, 364, *376*
Witt, H.T. see Grunhagen, H.H 353, *372*
Witt, H.T. see Schliephake, W 352, 364, *375*
Witt, M T. see Stiehl, H.H. 353, *375*
Wittkop, J see Preiss, J 95, 98, *133*
Wittwer, S.H. see Rathore, V S 393, *397*
Wodzicki, T J , Brown, C L 279, *287*
Woermann, D 450, *464*
Woermann, D. see Meier, J. 432, *463*
Woermann, D. see Pusch, W. 440, *463*
Woermann, D. see Schonborn, M. 459, *464*
Woermann, D. see Wiedner, G 439, *464*
Wohlfarth-Bottermann, K.E. see Komnick, H. 292, *330*
Wolf, F T 186, *234*
Wollenweber, E., Schnepf, E. 314, *344*
Woo, K C., Anderson, J M., Boardman, N K., Downton, W J.S., Osmond, C.B., Thorne, S.W. 171, *184*
Woo, K.C see Anderson, J M. 171, 172, *177*
Wood, H.G see Cooper, T G. 107, *130*, 174, *178*
Wood, W A see Jimenez, E 197, *228*
Wooding, F B P 313, 319, *344*
Wooding, F.B.P., Northcote, D H. 292, 313, *344*
Woodland, H.R. see Gurdon, J.B. 59, 60, 75, *81*
Woolley, D E. 292, *344*
Wreng, J see Wessels, N K. 58, *84*
Wrenn, J T. see Wessels, N.K 292, 294, *343*
Wright, K., Northcote, D.H. 314, *344*
Wyn Jones, R.G. see Leigh, R.A. 16, *44*
Wyn-Jones, R.L., Lunt, O.R. 393, *397*

Yamazaki, R K see Tolbert, N E 6, *49*, 109, *134*, 193, 196, 197, 199, 200, 209, 221, 222, *233*
Yamada, K M., Spooner, B.S., Wessels, N K 292, 308, *344*
Yamada, K M see Wessels, N K 58, *84*, 292, 294, *343*
Yamamoto, M 294, *344*
Yamashita, T., Butler, W.L. 356, *376*
Yamazaki, R K , Tolbert, N E 199, 210, *234*
Yano, N see Kisaki, T 201, *229*
Yano, Y. see Parson, D F 236, *253*
Yarwood, A. see Boulter, D. 25, *40*
Yazikov, A see Alexeev, A B. 65, *79*
Yemm, E W , Bidwell, R.G.S. *234*
Yoder, O.C. see Ferrari, T.E. 259, *283*
Yoshida, N. see Kisaki, T. 201, *229*
Yu, Y.L , Tolbert, N E., Orth, G M. 195, *234*
Yudin, A.L , Neyfakh, A.A. 72, *84*
Yung, D. see Jacazio, G. 441, *462*
Yunghans, W N., Keenan, T W , Morré, D J 300, *344*
Yunghans, W N see Morré, D.J. 323, *337*

Zahler, P H. see Wallach, D.F H. 11, 13, 35, *49*
Zak, E.G., Nichiporovich, A.A. 205, *234*
Zambrano, F see VanGolde, L.M.G. 301, 302, *343*
Zandonella, P. 273, *287*
Zbarsky, I B , Perevoshchikova, K.A., Delektorskaya, L.N , Delektorsky, V.V. 19, *50*
Zebe, E , Delbruck, A., Bücher, Th 248, *254*
Zelitch, I. 109, *136*, 176, *184*, 186, 187, 188, 189, 191, 193, 194, 196, 197, 200, 202, 208, 209, 211, 212, 213, 220, 221, 222, *234*
Zelitch, I , Day, P.R. 205, *234*
Zelitch, I , Gotto, A M 196, 200, 209, *234*
Zelitch, I., Ochoa, S. 199, 208, *234*
Zenk, M.H. see Amrhein, N 268, *281*
Zentgraf, H.W. see Franke, W W 60, *81*, 295, 302, 304, *330*
Zetsche, K. see Brandle, E 68, *80*
Zielke, H R , Filner, P. 273, *287*
Zill, I.P. see Tolbert, N.E. 204, *233*
Zimmer, E A see Anderson, M M 354, *370*
Ziola, B see Spencer, F S 324, *342*
Zirkle, C 255, *287*
Zubarev, T N see Alexeev, A.B 65, *79*

Subject Index

accumulation of amines, in illuminated thylakoids 359
accumulation of ions, mechanisms of 264
Acetabularia, morphogenesis of 64
acid photophatase, localization of 271
acids, localization in vacuoles 260
acridine orange, binding of 55
acridines, as indicators of ΔpH 362
actin-like proteins, of microfilaments 294
actinomycin, binding of 55
  inhibition of transcription by 58, 65, 66
active swelling, of mitochondria 379
active uptake of sugars 401
acyl coenzyme A carnitine transferases 246
adaptive significance, of C_4 pathway 176
adenine nucleotide carrier 116, 124, 239–241, 383, 400, 404
adenine nucleotide transport, in isolated chloroplasts 116–121, 404
  in mitochondria 238, 384
adenine nucleotides, pool changes in chloroplasts and cytoplasm 402
adenylate cyclase, secretion of 323
adenylate kinase, absence in the mitochondrial matrix 237
  localization of 154
adenylate translocator 116, 124, 239–241, 383, 400, 404
adenylates, as inhibitors of electron flow 363
ADP, as stimulator of proton uptake 363
ADP-glucose pyrophosphorylase 95, 98
alanine aminotransferase, localization of 154
alanine, role in C_4 photosynthesis 156
alanine transport, in C_4 photosynthesis 159
aleurone bodies, role in protein storage 261
alkaloids,deposition in cell walls and plastids 263
  deposition in vacuoles 261, 263
  detoxification of 267
  localization by autoradiography 259
α-amanitin, suppression of transcription by 56, 58, 59, 65
amine uptake by thylakoids, mechanism of 359
amino acid carriers 25, 110, 111, 243, 383
amino acids, storage in vacuoles 260
  transport of 109, 155, 159, 164, 243, 383
amino group balance, in C_4 photosynthesis 166
9-aminoacridine, as indicator of ΔpH 362
aminotransferases, in the glycolate pathway 199
  role in C_4 metabolism 153, 155, 156
ammonium uptake, by chloroplasts 361
amplification, of ribosomal genes 69
amyloplasts 21
anion exchange 125, 139, 142
anisotropy, of membranes 13, 23, 349, 350, 432
anthocyanins, localization of 257
antibodies, effect on components of the electron transport chain 354
antimycin A, effect on photosynthesis 113
  effect on respiratory electron transport 113, 237, 238
anucleate cell fragments, capacity for development 75
  properties of 53, 56
apparent osmotic equilibrium, definition of 457
apparent photosynthesis (APS), definition of 187
araban, in primary cell walls 307
arginine, accumulation in vacuoles 265
aspartate aminotransferase, localization of 154, 210
aspartate/malate shuttle 142, 209, 248, 410
aspartate, role in C_4 photosynthesis 153, 154
  role in hydrogen transfer 410, 142
aspartate transport, in C_4 photosynthesis 159, 166
assimilation rates, of cells from C_4 plants 149, 150
asymmetry, of thylakoid membranes 23, 350
ATP, as inhibitor of electron flow 363
  as stimulator of proton uptake 363
ATP-ase, role in ion transfer 25, 265
ATP/ADP ratio, in chloroplasts 407
  in mitochondria 241, 400
ATP, direct transport of 116–121, 142, 400, 403–405
ATP, indirect transfer of 123, 405, 407–410
ATP-synthesis, energized by ion gradients 347, 365, 379, 380
  relation to proton transport 365–367, 380
ATP/2e ratio 367, 398
atractyloside, as inhibitor of the mitochondrial adenylate carrier 239, 389, 400
  role in Ca^{++}transport 391
autocatalysis, as a mechanism to accumulate intermediates 85, 124
autolysis 278, 279
autophagy, extent of 280
autophagy, mechanism of 273, 274, 278

autoradiography, role in localization studies 54, 58, 259
auxin, binding of 63, 323, 324

back exchange experiments
in translocator studies 140
bacteriorhodopsin, relation to phosphorylation 370
role as proton pump 370
barium, transport of 386
Benson-Calvin cycle = Calvin cycle = PCR cycle = C_3 pathway 96, 104, 106, 109, 144, 153, 159
benzoquinone, as carrier of protons 353
benzylviologen, as electron acceptor 352
bicarbonate, uptake by chloroplasts 107
bongkrekic acid 239, 400
bridging proteins, in membrane transport 38
bromodeoxyuridine, effect on cell differentiation 63
bundle sheath cells, CO_2 fixation 149
in relation to C_4 photosynthesis 125, 147–149, 154, 155, 159, 161, 168
buoyant densities, for membranes 6
for organelles 193

C_3 pathway = Benson-Calvin cycle = PCR cycle = Calvin cycle 96, 104, 106, 109, 144, 153, 159
C_3 pathway, in C_4 plants 144, 149, 153, 168
C_3 plants, carboxylation capacity of 108
CO_2 compensation point of 109
metabolite transport in 126
C_4 acids, chloroplast fluxes 168
C_4 photosynthesis, definition of 144
metabolite transport in 125, 144, 152, 159, 160, 167
mitochondrial involvement 158
physiological function 174
role of photosynthetic carbon cycle in 146
C_4 plants, leaf anatomy 147
photorespiration of 175
subdivision of 153
Ca^{++} desposition 391–393
Ca^{++}, role at cell surface 324
subcellular distribution of 393
Ca^{++} transport 377, 387, 388, 390, 391
in mitochondria 384
regulation of flux 389
calcification, of bone, cartilage, avian shells 391
Calvin cycle = Benson-Calvin cycle, PCR cycle = C_3 pathway 96, 104, 106, 109, 144, 153, 159
CAM plants, accumulation of malate in 260
glycolate metabolism in 203
camptothecin, as inhibitor of RNA processing 59
carbohydrates, as membrane constituents 10, 14, 15, 301
carbon flux, for light-respiratory CO_2 evolution 221
from the glycolate pathway into 3-PGA 219
through the glycolate pathway 190, 218
carbon isotope discrimination, in C_4 plants 147, 164
carbonic anhydrase 106, 108
carbonylcyanide p-trifluoromethoxyphenylhydrazone, effect on proton permeability of membranes 352
carboxylic acids, transport of 109, 141, 161, 167, 168, 198, 206, 242, 246, 409
carrier, of adenine nucleotides 116, 124, 239–241, 384, 400, 404
carriers 25, 38, 89, 110, 123, 137, 141, 168, 242, 245, 246, 383, 403, 406
carrier transport, of amino acids 25, 110, 111, 141, 243, 265, 383
of protons 349
caryoplasts, properties of 75, 77
catabolism, of cellular constituents 273
cation exchange, in chloroplasts 349
in the vacuole 266
cation transport 126, 383, 387, 389, 391
CDP-choline-1,2-diglyceride transferase, localization of 290
cell fractionation, aqueous 86, 87, 110, 148
nonaqueous 87, 88
cell fusion, effect on cell differentiation 62
cell plate formation 308
cell sap, composition of 256, 257
definition of 255
cell separation 149
cell surface, role of 319
cell wall, assembly of microfibrillar phase 307
glycoprotein components of 307, 308
synthesis of 307, 308, 319
cell wall formation, in desmids 310
in diatoms 309
cellulase, secretion of 291
cellulose, in primary cell walls 307
synthesis of 319–321
cellulose synthetase, membrane association of 290
centrifugal filtration 89, 113, 137
centrifugation, density gradient 9
isopycnic 10
charge balance, in ion transport 166, 167
chelating agents 115, 118
in chloroplast isolation 115
chemical hypothesis, of coupling 386, 387
chemiosmotic theory 347–349, 365, 370, 379, 381, 382, 389
chitin synthetase, localization of 320, 322

chloramphenicol, as inhibitor of chloroplast and mitochondrial protein synthesis 60
chloride uptake, during H^+ transport 357
chloroplast DNA 21
chloroplast envelope 21, 22
chemical composition 22
impermeability 105, 106, 113, 121
permeability 86, 87, 92, 93, 95, 99, 101, 102, 105–109
rupture 90
chloroplast isolation, from C_3 plants 86–88
from C_4 plants 148, 157
chloroplast protein synthesis, control by the nucleus 67
chloroplasts, ATP/ADP ratio 407
classes of 93, 126
intactness 93, 126
chloroplast translocators 110, 124, 137–142, 403
cholin kinase, localization of 290
chromoplasts 21
chymotrypsin inhibitor protein, localization in vacuoles 261
circadian rhythms 66
cisternae, dictyosomal 30
citric acid, in laticifers 260
CO_2 affinity, of RuBP carboxylase 174
CO_2 compensation point 164, 188
CO_2 concentration mechanism 165, 174
CO_2 evolution, in the light 186
into CO_2 free air 187
CO_2 pools, in C_3 plants 174
in C_4 plants 165
CO_2 release, by C_4 plants 164
CO_2, transport of 107
$^{14}CO_2$, fixation kinetics 145
Coccoliths, formation of 309, 310
coions, 442, 445
communication, between chloroplastic and cytoplasmic adenine nucleotides 402
compartmentation, of PCR cycle enzymes 153, 154, 193
of carboxylases 152
of metabolites 101, 105, 106, 258, 121, 260, 236
of photorespiratory enzymes 193, 198, 200, 201
of pyridine nucleotides 121, 408
concanavalin A, effect on RNA and DNA synthesis 63
concentration gradients, of C_4 metabolites 162, 164
conditions favoring glycolate formation 193
conformational changes, energy storage by 370
constant field assumption 450
convection 443
conductivity 449
cooperative photosynthesis 144, 173
cordycepin, as inhibitor of the polyadenylation of m-RNS 59
counter exchange 139, 142
counter ion fluxes, during H^+ transport 358
counterions 442, 445
coupling factor 1, conformational change of 369
role in H^+ efflux from thylakoids 369
cresol red, as indicator of a transmembrane proton gradient 352
cristae, elaboration of 170
cyanide, reaction with plastocyanine 355
cyclic AMP, effect on cell differentiation 63, 76
role as secondary messenger 393
cyclic electron flow 355
cyclic GMP, effect on cell proliferation 63
cyclic photophosphorylation 355, 398, 401
cycloheximide, as inhibitor of protein synthesis 60
cytochemical localization, of phosphatases 268
cytochalasin 292, 293, 304
role in enucleation 57, 76
cytokinesis 322
cytokinins, binding of 324
cytolysomes 275, 277
cytonucleoprotein 72
cytoplasmic ATP/ADP ratio 407
cytoplasmic streaming, mechanism of 292
cytoplast, properties of 75, 77
cytosol, organization of 292
role in biosynthesis 290

dark respiration, inhibition by light 186, 408
DBMIB, inhibition of electron transport by 352
decarboxylation, of malate 151, 168
density gradient 9
6-deoxyglucose, uptake of 401
4-deoxyphlorizin, as energy transfer inhibitor 363
desmotubules 322
detoxification, by the vacuome 257
DHAP/PGA ratio 407
shuttle 122, 407
DHAP, transport of 102, 103, 137, 406, 407
diazonium-benzene sulfonate, as inhibitor of oxygen evolution 350
dibromomethylisopropyl-p-benzo-quinone, inhibition of electron transport by 352
dicarboxylate carrier, in chloroplasts 108, 109, 123, 124, 141, 408
in mitochondria 242, 383, 384
dichlorophenyl-1,1-dimethylurea, as inhibitor of electron transport 354
dictosomes, distribution of 323
staining properties of 303
differential grinding, separation of mesophyll and bundle sheath cell contents by 148
differentiation, of membranes 298, 303, 317
of meristem cells 279
diffusion, as means of C_4 transport 163
coefficients 161, 163
of CO_2 to bundle sheath chloroplasts 164

digestive processes, in lysosomes 269
in vacuoles 268, 272, 273, 277, 280
digitonin, effect on proton uptake by thylakoids 354
dihydrocoptisine, cation exchange by 266
dihydroxyacetone phosphate, transport of 102, 103, 137, 406, 407
dihydroxyacetone phosphate/phosphoglycerate shuttle 122, 407
dimorphic chloroplasts, in C_4 plants 170
2,4-dinitrophenol, inhibition of mitrochondrial swelling by 383
distribution, of enzymes of C_4 photosynthesis 151, 157
of glycolate pathway enzymes between the mesophyll cell and the bundle sheath cells 203
of glycolate pathway enzymes between peroxisomes and other cell organelles 197
of hydrolases 269
divalent cation transport, across thylakoids 126, 357
into mitochondria 384
DNA-synthesis 60
control of 62
DONNAN equilibrium 383, 446
DONNAN potential 444, 445
double layer, electrical 445
double tracer technique 454

electric conductivity, of membranes 449
electric current 432, 434, 439, 448
electric transference 449
electrochemical gradient, role in energy conservation 348, 358, 381
electrochemical potential 380, 381
electrochromic shift at 518 nm, as indicator of a primary photochemical reaction 350
electrode measurements, of cation uptake 361
electrogenic pumps 265
electron carriers, localization of 351
electron donation, by photosystem I 354
electron flow, in relation to proton fluxes 368
electron microscopy, of vacuolar depositions 259
electron probe analysis 259
electron transport, control of, by ΔpH 368
electroneutrality 433, 437, 442
electroosmosis 448
enchylema, definition of 255
endocytosis 277
endomembranes, composition of 300
continuity of 298
formation of racudes from 275
ontogenetic realtionships between 298, 302
transition states 298, 299, 317
endomembrane system, organization of 297, 298
endoplasmic reticulum, biosynthetic role of 301
function of 301, 302
synthesis of 302
energy balance 421, 433
energy conservation, in chloroplasts 347
in mitochondria 377
energy flow 421, 398
energy storage, in ΔpH 365, 367, 379, 380
by conformational changes 370
energy transfer, via adenylate transport 116, 142, 238
via transport metabolites 122, 406
entropy balance 420
entropy production 422, 424, 426, 430, 431, 434, 435, 436, 437
enucleation 57, 71, 75
envelope, mitochondrial 235, 236, 398
nuclear 16, 66, 67
of erythrozytes 14
plastid 20, 21, 86, 137, 398
enzyme distribution 54, 94, 147, 151, 154, 157, 197, 203, 237, 269, 290, 291
enzyme markers 5, 192
enzymes of C_4 acid decarboxylation, localization of 147
enzymes of C_4 acid synthesis, localization of 147, 153
enzymes, activation of 126
localization in mitochondria 236, 237, 238
of photosynthetic carbon cycle, localization of 154
of phospholipid synthesis 290
of sucrose synthesis 94
of the glycolate pathway 197, 203
secretion of 315, 316
vacuolar localization of 269
equilibrium density gradient centrifugation 10
erythrozyte envelope 14
ethidium bromide, as inhibitor of DNA synthesis 60, 69
N-ethylmaleimide, as inhibitor of phosphorylation 369
etioplasts 20
excretion, of glycolate 204, 206
extraction, stepwise 258
extrinsic proteins, of membranes 12

fatty acid carrier 383
FCCP, effect on proton permeability 352, 365
ferredoxin, involvement in cyclic electron flow 354, 355
fibrils, formation within secretory vesicles 312
field indicating absorption change, calibration of 364
flow across membranes, mass balance 422
flow ratio 453
fluid mosaic membrane model 38
fluorescence in ΔpH determinations 362
flux calculations, for C_4 metabolites 162
flux, of carbon into serine 217

free energy change, of ATP synthesis 366
freeze-fracture-etching 299
fructosans, localization in vacuoles 261
fructose bisphosphatase 96, 116
function, of photorespiration 222
fusion, of vesicles 301

$\Delta G'$, of ATP synthesis 366
galactans, in primary cell walls 307
galactolipids, in chloroplast envelopes 22
  role in thylakoid function 354
GIBBS equation 420
glands, secretory 314
glucan synthetase, localization of 291
glucans, synthesis of 320
glutamic dehydrogenase (NAD), regulation of, by divalent cations 392
glutamate carrier 141, 243, 245
glyceraldehyde-3-phosphate dehydrogenase, irreservible 122
  reversible 407
glycerate kinase 201
glycerophosphate shuttle 248
glycine, transport of 110
glycolate, as photosynthetic product 194
glycolate dehydrogenase 205
glycolate, formation from acetate 196
  formation from glyoxylate 196
  formation in bundle sheath cells of C_4 plants 202
  formation in the transketolase reaction 195
  synthesis by isolated chloroplasts 194
glycolate-glyoxylate shuttle 208
glycolate metabolism, in C_4 plants 202
  in CAM plants 203
glycolate oxidase 199
glycolate pathway 186, 197
  in C_4 plants 176
glycolate pathway enzymes in algae 205
glycolipids 301, 323
  biosynthesis of 303
  localization of synthesis 321
  role in polysaccharide synthesis 321
glycophorin 13
glycoproteins 15, 301, 303, 323
glycosyl transferase, in GOLGI vesicles 312
glyoxylate reductase, NADPH dependent, in chloroplasts 209
  NADH dependent, in peroxisomes 209
GOLGI apparatus, enzymic equipment of 301
  function of 308, 309, 312, 313, 314, 319, 323, 298
  role in membrane synthesis 301
  transitional nature of 300
GOMORI technique, in phosphatase localization 268
grafting experiments, in *Acetabularia* 67
gramicidin, role in ion transfer 383

H^+/ATP ratio 366, 367
H^+/e^- ratio 349, 351
H^+ transport 126, 348, 380
H^+ uptake, by thylakoid membranes 348, 349, 351, 353, 357, 365
hemicelluloses, synthesis of 312, 319
hemoglobin, synthesis of 75
heptose monophosphates, transport of 105
hexoseamine 24
hexose monophosphates, transport of 105
HILL reaction 90, 102, 119, 121, 123
homeostasis 280
hormones, inhibition of Ca^{++} transport by 389
hybridisation experiments 55, 75
hydolases, intracellular localization 268
  localization in cell walls 271
  localization in vacuoles 270, 271
hydroxylamine, as inhibitor of water oxidation 356
hydroxymethyltransferase, intracellular localization of 201
hydroxyproline, in glycoproteins 307
hydroxypyruvate reductase (NAD), function in photorespiration 200
α-hydroxysulfonates, as inhibitors of glycolate oxidase 197

image force interaction 443
immune response 301, 323
immunological methods, application in membrane studies 349
impermeability of the chloroplast envelope, to coenzymes and sugar bisphosphates 104
induction of photosynthesis 89, 103, 106, 118, 124
informosomes 61
inhibition of Ca^{++} transport, by lanthanides 388
  by ruthenium red 388
inhibition of DNA synthesis, by ethidium bromide 60, 69
inhibition of energy transfer, by 4-deoxyphlorizin 363
inhibition of glycolate oxidation, by α-hydroxysulfonates 197
inhibition of mitochondrial swelling, by 2,4-dinitrophenol 383
inhibition of NADH oxydation, by rotenone 238
inhibition of oxygen evolution, by DCMU 354
  by diazonium-benzene sulfonate 350
  by hydroxylamine 356
inhibition of phosphorylation, by N-ethylmaleimide 369
inhibition of photosynthesis, by DBMIB 352
  by glyceraldehyde 106
  by Mg^{++} 116
  by orthophosphate 89, 90, 103–105, 112, 124

inhibition of photosynthesis, by iodoacetate 104
by ribose-5-phosphate 112
by SO_4^{2-} 112
reversal of 103, 104, 112, 113, 125
inhibition of protein synthesis, by cycloheximide 60
by puromycin 60
inhibition of ribulose bisphosphate carboxylase, by oxygen 165, 176
inhibition of RNA polyadenylation, by cordycepin 59
inhibition of RuBP oxygenase, by CO_2 176
inhibition of the glycine/serine interconversion, by isonicotyl hydrazide 197
inhibition of transcription, by rifampicin 60
inhibition of true photosynthesis, by oxygen 213
inhibition of water oxidation, by Tris 356
inhibitor proteins, of proteases 272
inner membrane, of chloroplasts 20, 21, 137, 102
of mitochondria 236
of nuclei 18
inorganic compounds, localization in vacuoles 263
intact isolated chloroplasts 404
intact isolated mitochondria 378
intactness, of cell organelles 93, 126, 378
intercellular compartmentation, in C_4 plants 147, 148, 149, 151, 152, 153
intercellular transport, in C_4 photosynthesis 160, 163, 167, 168
interconversion, of glycine and serine 200
intermembrane space 236
internal membranes, function in intracellular transport 295
intracellular localization 236, 291
of enzymes of the C_4 pathway 157
of glycolate pathway enzymes 197
of hydrolases 269
of reactions of the C_4 pathway 152
intracellular transport, methods for study 88
intrinsic proteins, of membranes 12
ion carriers 138, 242, 400
high affinity system 264
low affinity system 264
ion exchange, as transport mechanism 266
ionophores 38, 381, 382
effects on H^+ uptake and ΔpH 364
effect on membrane potential 364
ion transport, in energy conservation 347, 377
into the vacuoles 264
through plasmodesmata 161
isolation, of cell organelles 191
of chloroplast envelopes 22, 23, 412
of membranes 3, 9
of mitochondria 378
of peroxisomes 191
of plastids 87
isonicotyl hydrazide, as inhibitor of the glycine/serine interconversion 197
isopycnic centrifugation 10, 191
isopycnic densities, for membranes 6
for organelles 193
isotope analysis, of compartmentation 258, 260

α-ketoglutarate carrier 141, 243
kinetic analysis, of compartmentation 259
of metabolism 145
kinetic complexity, of chloroplast DNA 69
kinetics, of $^{14}CO_2$ fixation 145
KRANZ anatomy, in C_4 plants 147, 161

lag phase of photosynthesis 89, 103, 124
lampbrush chromosomes 69
lanthanides, as inhibitors of Ca^{++} transport 388
latency, of enzymes 270, 271
laticifers 260, 263, 269
L-coefficients, definition of 423
diagonal 424
off-diagonal 424
leaf anatomy, in C_4 plants 147
lectins 301, 324
binding of 63
light dependent Cl^- uptake 401
light dependent reduction, of exogeneous NAD^+ 409
light respiration 185
CO_2 evolution in 207
substrate of 216
lignin 315
linear law 423–427, 430, 431, 437
lipid bilayers 11, 32, 37, 442
lipid constituents, incorporation into membranes 302
lipid-globular protein mosaic model 35
lipids, amphipathic 11
lipogenesis 250
lysosomes 268, 272

malate accumulation, in CAM plants 260
malate/aspartate shuttle 142, 209, 248, 410
malate, decarboxylation of 151, 168
malate dehydrogenases, localization of 123, 210
malate/oxaloacetate shuttle 123, 408, 409
malic dehydrogenase (NAD), localization of 210, 123
malic dehydrogenase (NADP), localization of 154, 123, 210
malic enzyme (NAD) 154, 156
malic enzyme (NADP), localization of 153, 154
α-mannosidase, membrane association of 272
marker enzymes, for membranes 6
for organelles 193
mass balance 422, 429

mastigonemes, assembly of 304
migration of 305, 306
matrix space 235, 236, 255, 271
mechanical permeability 443, 454, 456
membrane, of spherosomes 31
of the endoplasmatic reticulum 29, 255, 295, 297
of thylakoids 33, 349, 351, 355
structure of 10–40
vacuolar 31, 257
membrane asymmetry 13, 15, 23, 349, 432
membrane biogenesis 295, 303
membrane carbohydrates 10
membrane channels 38
membrane composition 10, 14, 17, 22, 27
membrane differentation 256, 298–300, 302, 303, 317
membrane flow 304, 322
membrane isolation 3, 9
membrane lipids 11, 15, 19, 24, 28
membrane models 31
membrane potential 380, 388
effect on apparent H^+/e^- ratio 351
in illuminated thylakoids 364
in mitochondria 240
of the tonoplast 265
membrane probes 11
membrane proteins, synthesis of 301, 302
membrane symmetry 432
membrane transport, influence of nonpermeable components 428
membranes, dimensions of 16, 20, 21, 27, 34, 35, 299, 300
of the GOLGI apparatus 29, 30, 298
partial specific conductivity of 455
total specific conductivity of 455
with fine pores 440
with wide pores 441
mesophyll cells, CO_2 fixation of 149
in C_4 plants 125, 147
metabolite carriers 25, 38, 89, 98, 103, 111, 113, 115, 124, 137, 167, 236, 242, 246, 384, 400, 406
metabolite compartmentation 105, 106, 110, 111, 113, 118, 121, 144, 236, 257, 258
metabolites, transport of 101–107, 125, 137, 144, 206, 235, 236, 398
metabolite transport, through plasmodesmata 161
in relation to anatomy 161
3-O-methylglucose 401
Mg^{++} release, during light-dependent H^+ uptake 126, 357
Mg^{++} transport in mitochondria 385
microbodies, in algae 205
microbody membrane 29
microcentrifugation technique, for determination of ΔpH 360
microfibrils, assembly of 309, 322
role of 291
microfilaments 58
binding of meromyosin by 294
role of 292, 294
microtubules, as guide elements for transport 291
mitochondria 28, 235
ATP/ADP ratio in 241
isolation of 191, 378
permeability of 236, 378
mitochondrial enzymes 236
localization of 237
mitochondrial involvement, in C_4 photosynthesis 169
mitochondrial membrane 26, 28
chemical composition 27
morphology 26
permeability of 236, 378
mitochondrial metabolism 235, 236
mitochondrial translocators 239–250, 383
mitosis, control of 72
monophosphates, heptose 105
hexose 105
pentose 104
morphogenetic substances, in Acetabularia 65, 66
m-RNA, binding to proteins 73
stability of 75
mucilages, secretion of 314
multinet growth 322

NAD synthesis, localization of 77
NADP-ME-type of C plants, definition of 153, 159
nectar secretions 315
NERNST-PLANCK-equations 443, 446, 447
neutral red, accumulation in vacuoles 259, 266
NH_4^+ uptake, by thylakoids 358
nitrate reductase, stability of 273
nonaqueous cell fractionation, separation of mesophyll and bundle sheath chloroplasts 148
nonaqueous chloroplast isolation 87, 88, 148
noncyclic photophosphorylation 126
nonpermeable components 105, 106, 113, 121, 137, 236, 428
nuclear activity, control by cytoplasma 62
nuclear envelope, chemical composition 17
function of 301
morphology 16
permeability 19
synthesis of 302
nuclear membrane, role of 60, 301
nuclear proteins, cytoplasmic origin of 60
nuclear transplantation 72
nuclei, role in energy production 53
role in nucleic acid synthesis 54
role in protein synthesis 53
nucleic acids, synthesis of 69

nucleolar organizers 58, 59
nucleus, isolation of 18

oil-membrane 441
oils, essential 314
sites of accumulation 314
oligomycin, as energy transfer inhibitor 384, 385, 389
restoration of mitochondrial swelling by 383
ONSAGER reciprocity relation 424, 425, 427, 431, 439
organelle identification 4
organelle ultrastructure, in C_4 plants 170
orthophosphate, sequestration of 97, 98, 113
transport of 103, 111, 138, 139, 242, 243
osmometric response, of mitochondria 379
osmosis, negative 440, 459
positive 440, 459
osmotic difference 427
osmotic shock, of chloroplasts 90
osmotic swelling, of mitochondria 378, 379
osmotic volume changes 88, 92, 378, 379, 382, 385
osmotica, for organelle isolation 86
outer membrane, of chloroplasts 21, 137
of mitochondria 26, 235
oxidation of cytoplasmic NADH, by the chloroplast 409, 410
by the respiratory chain 239, 248, 408
oxidation of glyoxylate 208
oxidative phosphorylation 237, 240, 241
oxygen evolution 102, 106, 119–121
oxygen inhibition of photosynthesis 165, 176, 185, 202

P/2e ratio 356, 357
paper chromatography 89
parallel carboxylation 146
passive penetration, of cations into mitochondria 378
passive swelling, of mitochondria 378
PCK-type of C_4 plants, definition of 153, 156, 159
PCR-cycle, in C_4 plants 144, 149, 153, 168
pectic substances, synthesis of 319
pellicle formation, in euglenoids 310
penetration of mitochondria, by adenylates 238–241
by anionic substrates 242–250
by cations 378
penetration of the chloroplast envelope, by anionic substrates 101–104, 109, 111, 137–142
by CO_2 106
pentose monophosphates, transport of 104
pentose phosphate pathway, oxidative 96, 123, 290
reductive see PCR cycle
PEP carboxykinase, localization of 154, 169
role in C_4 metabolism
PEP carboxylase, localization of 151, 154
peripheral reticulum, of chloroplasts 21, 170
permeability, and membrane potential 456
definition of 438
of the chloroplast envelope 25, 28, 102, 109, 111, 137, 149, 167
of the mitochondrial envelope 200, 236, 377, 400
of the nuclear envelope 60
permeation of peptides and proteins, through membranes 25, 28, 60
peroxisomes, in C_4 plants 203
role in glycolate metabolism 186, 189, 197
structure of 29
PGA transport 102, 122, 137, 140, 172, 409
PGA/DHAP shuttle 122, 406, 407, 409
phagocytosis 277
phenazonium methosulfate, role as artificial proton carrier 354, 355
phenolic glycosides, localization in vacuoles 261, 266
phenomenological coefficients see L-coefficients
phenylene diamines, as uncouplers of phosphorylation 356, 367
pH gradient, across the chloroplast envelope 140, 409, 410
across the inner mitochondrial membrane 242, 379
across the thylakoid membrane 358–363
across the tonoplast 266
phlorizin, as energy transfer inhibitor 363
phosphatases, localization in vacuoles 268
phosphate carrier 89, 98, 103, 111, 113, 124, 137, 167, 242, 406
phosphate, storage in vacuoles 263
phosphate translocator 25, 89, 98, 103, 111, 113, 115, 124, 137, 167, 242, 406
phosphate transport see phosphate translocator
phosphocholine cytidyltransferases, localization of 290
phosphoenolpyruvate, as inhibitor of phosphate transport 139
effect on Ca^{++} transport 309
transport of 103, 159, 167
phosphofructokinase, localization in chloroplasts 96
phosphoglucomutase, distribution of 291
3-phosphoglycerate, transport of 102, 103, 122, 137, 140, 172, 406, 409
phosphoglycerate phosphatase, role in glycerate formation 201
phosphoglycolate phosphatase, localization of 95
phospholipids, as membrane constituents 16, 19, 23, 24, 28
phosphorylation, oxidative 380
photosynthetic 116, 118, 171, 172, 347, 365
relation to proton transport 365
phosphorylation potential 241, 411
phosphorylation rate, as related to ΔpH 367

photochemical activities, of C_4 chloroplasts 170
photophosphorylation 116, 118, 171, 172, 347, 365
effect on ΔpH 365
photorespiration 109, 175, 185
magnitude of 188
photorespiratory CO_2 formation 187,199, 108
photorespiratory oxygen consumption 186, 195, 199
photosynthesis, compartmentation of 93, 95, 101, 125, 144, 198
efficiency of 108
gas exchange in 187
photosynthetic control 117
photosystem I, reduction by 354
photosystem II activity, deficiency of 171, 172
sensitivity of 349
phytohemagglutinins, effect on RNA and DNA-synthesis 63
plasmalemma, chemical composition 13, 14, 288
morphology 13
sensitivity of 258
sterol/phospholipid ratio 16
plasma membranes, specific staining of 303
plasmodesmata 322
role in intercellular transport 161, 164
plastid isolation, aqueous 87, 191
nonaqueous 87, 157
plastid membranes 19–26, 137, 170
plastocyanin, inactivation by cyanide 355
localization of 354
plastoquinone, as carrier of protons 349, 353, 358, 368
PMS, role of as artificial proton carrier 354, 355
polyadenylation, of m-RNA 74
polyanions, as vacuolar constituents 266
polyphosphates, localization in vacuoles 263
polysaccharide synthesizing enzymes, distribution of 291
P/O ratio 378
pool size, of CO_2 in C_4 plants 165
pore complex, of the nuclear envelope 17, 61
pores, hydrophilic 37
postillumination CO_2 burst, in leaves 188
potassium transfer, across the mitochondrial membrane 378, 381
as mediated by valinomycin 381
primary cell walls, composition of 307
formation of 311
processing, of RNA species 59
proplastids 20
protein, extrinsic 12
half-life of 272
in primary cell walls 307
integral 12
intrinsic 12
peripheral 12, 37
periplasmic 37
protein degradation, in senescence 272
protein synthesis, in the absence of the nucleus 67
stimulation of 73
protein turnover 272
proteins, transport of 53
proton carriers 349, 533–355
proton gradient 242, 266, 347, 360, 379, 409, 410
magnitude of 363, 369
proton motive force 380, 381, 382
definition of 366
proton transport 126, 348–357, 380
relation to phosphorylation 365, 369
protoplast preparation, by enzymic wall digestion 149
provacuolar bodies 317
pulse chase experiments 58, 146, 173
purines, in vacuoles 261
puromycin, as inhibitor of protein synthesis 60
pyocyanine, as mediator of cyclic electron flow 360
pyrophosphatase 91, 93, 111, 113
localization of 154
pyrophosphate, hydrolysis of 113
impermeability of 113
pyruvate, chloroplast fluxes 167
pyruvate carboxylase, inhibition by Ca^{++} 393
pyruvate, P_i dikinase, localization of 154
role in C_4 metabolism 126, 156

Q-value for CO_2 evolution in the light 213
quantasomes, as membrane subunits 33

radiotracer studies, in C_4 metabolism 146
in glycolate metabolism 197
in transport work 88
rate of photorespiration 211, 212
rate of photosynthetic CO_2 fixation 107, 187, 220
rates of energy exchange between chloroplasts and cytoplasm 399
ratio H^+/ATP 366, 367
ratio H^+/e^- 349, 351
ratio of O_2 uptake to CO_2 evolution 207
ratio $P/2e^-$ 356, 357
reassimilation, of photorespiratory CO_2 211, 213
receptor molecules, in cell surface 323
reconstitution, of a chloroplast system 119
reflection coefficient, definition of 427
for dilute solutions 438
of monovalent electrolytes 458
relative specific activity (RSA) of $^{14}CO_2$ evolved in the light 214
respiration, in the light 185, 408
mitochondrial 186, 237, 408
respiratory control 378
respiratory states 390

reversible electrodes 435, 447
rhamnogalactan, in primary cell walls 307
ribonucleic acids, transport of 54, 58, 60, 61, 72
ribosomal genes, redundancy of 68, 69
ribulose bisphosphate carboxylase, activity of 108
in C_4 plants 151, 153
oxygenase function of 109, 194
ribulose-1,5-bisphosphate, retention of by chloroplasts 105
rifampicin, as inhibitor of transcription 60
RNA, degradition of 273
transport of 54, 58, 60, 61, 72
RNA polymerases, localization of 54, 59
RNase, localization in vacuoles 269
RNA-synthesis, control of 62
rRNA, stability of 68
rRNA synthesis, control of 59
rotenone, as inhibitor of NADH oxidation 238
rough endoplasmic reticulum 298
rubber globules, localization of 266
ruthenium red, as inhibitor of Ca^{++} transport 388

S-adenosyl-methionine, vacuolar deposition of 261
salt flux in mitochondria, mechanism of 381
sanguinarine, uptake of 266, 267
scale formation 309
secondary walls, formation of 312
secretion 297, 306, 315, 316
of cellulase 291
of cellulose synthetase 312
of insulin 291
of metabolic products 206–308
of proteins 291
secretory vesicles 303, 304, 306, 308, 311, 312, 321
sedoheptulose-1,7-bisphosphate, retention of by chloroplasts 105
senescence 278, 279, 280
permeability changes in 278
sequestration, of organelles 275
serine, transport of 110
series carboxylation 146
shrinkage, of chloroplasts 92
shuttle transfer of metabolites 96, 103, 121, 122, 123, 140, 142, 152, 155, 159, 198, 209, 248, 408
sidedness of membranes 8, 13, 349, 350, 351
silicone fluid technique, for ΔpH determination 363
silicone layer filtering centrifugation 89, 137, 363
slimes, secretion of 314
smooth endoplasmic reticulum 298
solvent drag 440
specific activity, of the $^{14}CO_2$ evolved in the light 214
specific transport 124, 137, 238, 242, 306
spheroplasts, rupture of 258
spherosomes, origin of 316
starch prints 99, 100
starch synthesis 85, 95, 96, 97
effect of exogenous sugars on 96
in C_4 plants 157
stationary state 421, 422, 429, 434, 447
statoliths, composition of 259
storage pools, in vacuoles 258–263
stratification, of organelles 269
streaming potential 450
strontium, transport of 386
suberin lamellae, between mesophyll and bundle sheath cells 161, 164
subunit membrane model 33
sucrose synthesis, localization of 86, 93, 94
in C_4 plants 157
sugar bisphosphates, immobility of 105
sugars, in vacuoles 261
swelling, of chloroplasts 92
of mitochondria 379
symmetry relation of ONSAGER 450
symplast transport 161

tannins, localization in vacuoles 261, 266
synthesis of 268
thermodynamic flows, dependence on reference state 423
thylakoid membranes, asymmetry of 23, 350
role in proton transport 349, 358
thylakoid swelling, in the presence of amines 359
tip growth 308
tonoplast, breakdown of 278
electropotential 265
invaginations 275
origin of 317
pumps in 265
role in autophagy 273
role of 275
tracer coupling 453
translocation activity, of the dicarboxylate carrier 410
translocators, of adenine nucleotides 116, 123, 124, 239, 240, 241, 381, 400, 404
of amino acids 25, 109, 110, 243, 245, 383
of dicarboxylates 25, 109, 123, 124, 141, 243
of fatty acids 246
of phosphate 25, 89, 98, 103, 111, 113, 115, 124, 137, 167, 242, 246, 406
of pyruvate 246
of tricarboxylates 25, 109, 123, 124, 141, 143
transmembrane H^+ gradient 242, 266, 347, 360, 358, 369, 379, 409, 410
determination of 360
transmembrane phosphorylation 406
transpiratory loss of water, by C_4 plants 176

transport, passive 246, 432
transport capacity, relation to ultrastructure 170
transport equations, for membranes with pores 442
integration of 446
transport of adenine nucleotides 124, 142, 238, 400, 403, 405
transport of amino acids 25, 110, 109, 198, 199, 243, 245
transport of electrolytes 102, 137, 235, 348, 377, 432
entropy balance 433
transport of glycolate pathway intermediates 206
transport of inorganic phosphate (see phosphate translocator)
transport of intermediates, in C_4 plants 144, 159, 180, 167, 168
of the tricarboxylic acid cycle 242
transport of nonelectrolytes 37, 106, 419
transport of pyruvate 168, 246
tricarboxylate carrier 243, 383
triosephosphate dehydrogenase, role in indirect translocation of NADPH 122
triosephosphate isomerase, effect on indirect ATP translocation 103
triosephosphate, transport of 97, 98, 99, 102, 103, 122, 125, 137, 406, 409
tris buffer, as inhibitor of water oxidation 356
trypsin digestion, of membranes 349
turn over velocity 429

UDP glucose, localization of synthesis 290
UDPG, synthesis of 290
UDPG pyrophosporylose, distribution of 291
uncoupling, by phenylendiamines 357
of phosphorylation 119, 359, 362, 381, 383, 389
unidirectional flow 453
unit membrane model 32
uphill transfer, of hydrogen into the chloroplasts 409

vacuolar accumulation, of secondary metabolites 266
vacuolar enzymes 270, 271
vacuolar membranes, resistance of 258
vacuolar transport, mechanism of 264
vacuoles, as ion traps 266
as storage organ 317
composition of 259, 260, 317, 319
function of 257
isolation of 258, 265, 269
lysomal nature of 268, 269, 270
role in waste deposition 317
vacuome, definition of 255
valinomycin 380, 382, 383
effect on H^+ diffusion potential 352, 367
effects on membrane potential 364
effects on proton transport 380
role in potassium transfer 381
secondary effects of 352
vesicles, endocytotic and exocytotic 291
function in intracellular transport 293, 295
volume determination, by tritiated water 361
volume flow 425, 440, 443, 448

water extrusion, through GOLGI vesicles 316

xyloglucan, in primary cell walls 307

zones of exclusion 292, 294

Encyclopedia of Plant Physiology, New Series

Editors: A. PIRSON, M.H. ZIMMERMANN

Volume 1
Transport in Plants I
Phloem Transport

Edited by M.H. ZIMMERMANN, J.A. MILBURN
With contributions by numerous experts
93 figures. XIX, 535 pages. 1975

Contents: Structural Considerations in Phloem Transport. – Nature of Substances in Phloem. – Phloem Transport: Assessment of Evidence. – Possible Mechanisms of Phloem Transport. – Phloem Loading: Storage and Circulation. – Author Index. – Subject Index.

Volume 2
Transport in Plants II (in 2 parts)

Edited by U. LÜTTGE, M.G. PITMAN
With contributions by numerous experts

Part A: Cells
With a foreword by R.N. ROBERTSON
97 figures. XVI, 419 pages. 1976

Contents: Theoretical and Biophysical Approaches. – Particular Cell Systems. – Regulation, Metabolism and Transport.

Part B: Tissues and Organs
129 figures. Approx. 50 tables. XII, 475 pages. 1976

Contents: Pathways of Transport in Tissues. – Particular Tissue Systems. – Control and Regulation of Transport in Tissues and Integration in Whole Plants.

Volume 4
Physiological Plant Pathology
Edited by R. HEITEFUSS, P.H. WILLIAMS
With contributions by numerous experts
91 figures. Approx. 840 pages. 1976

Contents: General. – Spore Germination and Its Regulation. – Cytology and Physiology of Penetration and Establishment. – Forces by Which the Pathogen Attacks the Host Plant. – Physiology of Host Response to Infection. – Modification of the Host Response. – Biotrophic Parasites in Culture. – Genetics of Host-parasite Interactions. – Subject Index. – Author Index.

Distribution rights for India: The Universal Book Stall, New Delhi

Springer-Verlag
Berlin Heidelberg New York

MIX
Papier aus verantwortungsvollen Quellen
Paper from responsible sources
FSC® C105338

If you have any concerns about our products,
you can contact us on
ProductSafety@springernature.com

In case Publisher is established outside the EU,
the EU authorized representative is:
Springer Nature Customer Service Center GmbH
Europaplatz 3, 69115 Heidelberg, Germany

Printed by Libri Plureos GmbH
in Hamburg, Germany